JAGUAR
E TYPE MODELS
3·8 & 4·2 LITRE
SERIES 1 & 2
SERVICE MANUAL

Jaguar Cars Limited reserve the right to make changes in design, or to make additions to or improvements upon their products without incurring any obligation to install the same on vehicles previously built.

ISSUED BY

JAGUAR CARS LIMITED

Publication No. E/123/8 - E/123B/3 - E156/1

CONTENTS

	PAGE
JAGUAR 3·8 "E" TYPE GRAND TOUR MODELS SERVICE MANUAL	5
SUPPLEMENTARY INFORMATION FOR 4·2 LITRE "E" TYPE AND 2+2 CARS - SERIES 1	381
SUPPLEMENTARY INFORMATION FOR 4·2 LITRE "E" TYPE AND 2+2 CARS - SERIES 2	551

INDEX TO SECTIONS

SECTION TITLE	SECTION REFERENCE
GENERAL INFORMATION	A
ENGINE	B
CARBURETTERS AND FUEL SYSTEM	C
COOLING SYSTEM	D
CLUTCH	E
GEARBOX	F
PROPELLER SHAFT	G
REAR AXLE	H
STEERING	I
FRONT SUSPENSION	J
REAR SUSPENSION	K
BRAKES	L
WHEELS AND TYRES	M
BODY AND EXHAUST SYSTEM	N
HEATING AND WINDSCREEN WASHING EQUIPMENT	O
ELECTRICAL AND INSTRUMENTS	P

JAGUAR

3·8 "E" TYPE
GRAND TOURING MODELS
SERVICE MANUAL

Jaguar Cars Limited reserve the right to make changes in design, or to make additions to or improvements upon their products without incurring any obligation to install the same on vehicles previously built.

> Note: All references in this Manual to "right-hand side" and "left-hand side" are made assuming the person to be looking from the rear of the car or unit.

ISSUED BY

JAGUAR CARS LIMITED, COVENTRY, ENGLAND

Telephone
ALLESLEY 2121 (P.B.X.)

Code
BENTLEY'S SECOND

Telegraphic Address
"JAGUAR," COVENTRY. Telex. 31622

Publication No. E/123/8

SECTION A

ENGINE
3·8 "E" TYPE
GRAND TOURING MODELS

INDEX

	Page
Car Identification	A.3
General Data:	
Dimensions and weights	A.4
Capacities	A.4
Performance Data	A.5
Operating Instructions:	
Instruments	A.8
Controls and Accessories	A.9
Wheel Changing	A.14
Starting and driving	A.16
Summary of Maintenance	A.17
Recommended Lubricants	A.18
Multi-grade Engine Oils	A.18
Recommended Hydraulic Fluids	A.18
Special Service Tools	A.19
Service Departments	A.19
Conversion Tables	A.20

Page A.1/2

GENERAL INFORMATION

CAR IDENTIFICATION

It is imperative that the Car and Engine numbers, together with any prefix or suffix letters, are quoted in any correspondence concerning this vehicle. If the unit in question is the Gearbox the Gearbox number and any prefix or suffix letters must also be quoted. This also applies when ordering spare parts.

Car Number .. ..
Stamped on the right-hand frame/cross member above hydraulic damper mounting.

Engine Number .. ..
Stamped on the right-hand side of the cylinder block above the oil filter and at the front of the cylinder head casting.

/8 or /9 following the engine number denotes the compression ratio.

Fig. 1   The identification numbers are also stamped on a plate situated in the engine compartment

Gearbox Number .. ..
Stamped on a shoulder at the left-hand rear corner of the gearbox casing and on the top cover.

Body Number .. ..
Stamped on a plate attached to the right-hand side of the scuttle.

Key Numbers .. ..
The keys provided operate the ignition switch and door locks.

Page A.3

GENERAL INFORMATION

GENERAL DATA

DIMENSIONS AND WEIGHTS

Wheel base	8′ 0″ (2·44 m.)
Track, Front	4′ 2″ (1·27 m.)
Track, Rear	4′ 2″ (1·27 m.)
Overall length	14′ 7 5/16″ (4·45 m.)
Overall width	5′ 5¼″ (1·66 m.)
Overall height (Fixed head coupé)	4′ 0⅛″ (1·22 m.)
(Open 2-seater)	3′ 10½″ (1·18 m.)
Weight (dry) approximate (Fixed head coupé)	22½ cwts. (1123 kg.)
(Open 2-seater)	22 cwts. (1098 kg.)
Turning circle	37′ 0″ (11·27 m.)
Ground clearance	5½″ (140 mm.)

CAPACITIES

	Imperial	U.S.	Litres
Engine (refill)	15 pints	18 pints	8·5
Gearbox	2½ ,,	3 ,,	1·42
Rear axle	2¾ ,,	3¼ ,,	1·54
Cooling system (including heater)	32 ,,	38½ ,,	18·18
Petrol tank	14 galls.	16¾ galls.	63·64

Page A.4

GENERAL INFORMATION

PERFORMANCE DATA

The following table gives the relationship between engine revolutions per minute and road speed in miles and kilometres per hour.

The safe maximum engine speed is 5,500 revolutions per minute.

Engines must not, under **Any Circumstances** be allowed to exceed this figure.

It is recommended that engine revolutions in excess of 5,000 per minute should not be exceeded for long periods. Therefore, if travelling at sustained high speed on motorways, the accelerator should be released occasionally to allow the car to overrun for a few seconds.

ROAD SPEED		ENGINE REVOLUTIONS PER MINUTE	ENGINE REVOLUTIONS PER MINUTE	ENGINE REVOLUTIONS PER MINUTE
Kilometres per hour	Miles per hour	Top Gear 3.31:1	Top Gear 3.07:1	Top Gear 3.54:1
16	10	436	405	466
32	20	873	810	932
48	30	1309	1215	1398
64	40	1745	1614	1862
80	50	2182	2008	2319
96	60	2618	2398	2775
112	70	3054	2780	3221
128	80	3490	3156	3667
144	90	*3800	3521	4110
160	100	*4200	3877	4542
176	110	*4600	4227	4963
192	120	*5000	4562	5416
208	130	*5410	4887	—
225	140	—	5200	—
240	150	—	5506	—

*The figures marked thus make allowance for changes in tyre radius due to the effect of centrifugal force.

Page A.5

GENERAL INFORMATION

OPERATING INSTRUCTIONS

Fig. 2. Instruments and Controls—Right-hand drive

1. Ammeter.
2. Fuel contents gauge.
3. Lighting switch.
4. Oil pressure gauge.
5. Water temperature gauge.
6. Mixture control and warning light.
7. Revolution counter.
8. Flashing direction indicator warning lights.
9. Speedometer.
10. Brake fluid warning light.
11. Headlamp dipper switch.
12. Heater—Air Control.
13. Heater—Temperature control.
14. Interior light switch.
15. Panel light switch.
16. Heater fan switch.
17. Ignition switch.
18. Cigar lighter.
19. Starter switch.
20. Map light switch.
21. Windscreen wiper switch.
22. Windscreen washer switch.
23. Clock adjuster.
24. Horn button.
25. Speedometer trip control.
26. Flashing direction indicator and headlamp flashing switch.

Page A.6

GENERAL INFORMATION

Fig. 3. Instruments and Controls—Left-hand drive

1. Headlamp dipper switch.
2. Brake fluid warning light.
3. Speedometer.
4. Flashing direction indicator warning lights.
5. Revolution counter.
6. Water temperature gauge.
7. Oil pressure gauge.
8. Lighting switch.
9. Fuel contents gauge.
10. Ammeter.
11. Mixture control and warning light.
12. Flashing direction indicator and headlamp flashing switch.
13. Speedometer trip control.
14. Horn button.
15. Clock adjuster.
16. Heater—air control.
17. Heater—temperature control.
18. Windscreen washer switch.
19. Windscreen wiper switch.
20. Map light switch.
21. Starter switch.
22. Cigar lighter.
23. Ignition switch.
24. Heater fan switch.
25. Panel light switch.
26. Interior light switch.

Page A.7

GENERAL INFORMATION

INSTRUMENTS

Ammeter

Records the flow of current into or out of the battery. Since compensated voltage control is incorporated, the flow of current is adjusted to the state of charge of the battery; thus when the battery is fully charged the dynamo provides only a small output and little charge is registered on the ammeter, whereas when the battery is low a continuous high charge is shown.

Oil Pressure Gauge

The electrically operated pressure gauge records the oil pressure being delivered by the oil pump to the engine; it does not record the quantity of oil in the sump. The minimum pressure at 3000 r.p.m. when hot should not be less than 40 lbs. per square inch.

Note: After switching on, a period of approximately 20 seconds will elapse before the correct reading is obtained.

Water Temperature Gauge

The electrically operated water temperature gauge records the temperature of the coolant by means of a bulb screwed into the inlet manifold water jacket.

Fuel Level Gauge

Records the quantity of fuel in the supply tank. Readings will only be obtained when the ignition is switched "on". An amber warning light situated in the speedometer lights up intermittently when the petrol level in the tank becomes low. When the petrol is almost exhausted the warning light operates continuously.

Note: After switching on, a period of approximately 20 seconds will elapse before the correct reading is obtained.

Electric Clock

The clock is built in the revolution counter instrument and is powered by the battery. The clock hands may be adjusted by pushing up the winder and rotating. Starting is accomplished in the same manner.

Revolution Counter

Records the speed of the engine in revolutions per minute.

Speedometer

Records the vehicle speed in miles per hour, total mileage and trip mileage (kilometres on certain export models). The trip figures can be set to zero by pushing the winder upwards and rotating clockwise.

Headlamp Warning Light

A red warning light marked "Headlamps" situated in the speedometer, lights up when the headlamps are in full beam position and is automatically extinguished when the lamps are in the dipped beam position.

Ignition Warning Light

A red warning light (marked "Ignition") situated in the speedometer lights up when the ignition is switched "on" and the engine is not running, or when the engine is running at a speed insufficient to charge the battery. The latter condition is not harmful, but always switch "off" when the engine is not running.

Fuel Level Warning Light

An amber warning light (marked "Fuel") situated in the speedometer lights up intermittently when the fuel level in the tank becomes low. When the fuel is almost exhausted the warning light operates continuously.

Flashing Indicator Warning Lights

The warning lights are in the form of green arrows located on the facia panel situated behind the steering wheel.

When the flasher indicators are in operation one of the arrows lights up on the side selected.

Mixture Control Warning Light

A red warning light situated above the mixture control on the facia panel behind the steering wheel serves to indicate if the mixture is in operation. This warning light is illuminated immediately control lever is moved from "off" position.

To change the bulb, accessible behind the facia panel, pull bulb holder away from "clip in" attachment, and unscrew bulb by turning anti-clockwise. For full instructions on the use of the mixture control see "Starting and Driving," page A16.

Page A.8

GENERAL INFORMATION

Brake Fluid Level and Handbrake Warning Light

A warning light (marked "Brake Fluid—Handbrake") situated on the facia behind the steering wheel, serves to indicate if the level in either of the two brake fluid reservoirs has become low, provided the ignition is "on". As the warning light is also illuminated when the handbrake is applied, the handbrake must be fully released before it is assumed that fluid level is low. If with the ignition "on" and the handbrake fully released the warning light is illuminated the brake fluid must be "topped up" and the reason for the loss investigated and corrected immediately. IT IS ESSENTIAL that the correct specification of brake fluid be used when topping up.

As the warning light is illuminated when the handbrake is applied and the ignition is "on" a two-fold purpose is served. Firstly, to avoid the possibility of driving away with the handbrake applied. Secondly, as a check that the warning light bulb has not "blown"; if on first starting up the car with the handbrake fully applied, the warning light does not become illuminated the bulb should be changed immediately.

CONTROLS AND ACCESSORIES

Accelerator Pedal

Controls the speed of the engine.

Brake Pedal

Operates the vacuum servo assisted disc brakes on all four wheels.

Clutch Pedal

Connects and disconnects the engine and the transmission. Never drive with the foot resting on the pedal and do not keep the pedal depressed for long periods in traffic. Never coast the car with a gear engaged and clutch depressed.

Headlamp Dipper

Situated on the facia panel behind the steering wheel. The switch is of the "flick-over" type, and if the headlamps are on main beam, moving the lever will switch the dipped beam on, and main beam off. They will remain so until the switch lever is reversed.

Gear Lever

Centrally situated and with the gear positions indicated on the control knob. To engage reverse gear first press the gear lever against the spring pressure before pushing the lever forward. Always engage neutral and release the clutch when the car is at rest.

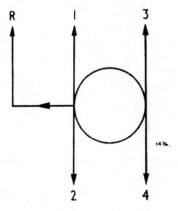

Fig. 4. The gear positions

Handbrake Lever

Positioned centrally between seats. The handbrake operates mechanically on the rear wheels only and is provided for parking, driving away on a hill and when at a standstill in traffic. To apply the brake, pull the lever upward and the trigger will automatically engage with the ratchet. The handbrake is released by pressing in the knob, and pushing the lever downward.

Seat Adjustment

Both front seats are adjustable for reach. Push the lock bar, situated beside the inside runner, towards the inside of the car and slide into the required position. Release the lock bar and slide until the mechanism engages with a click.

Page A.9

GENERAL INFORMATION

Steering Wheel Adjustment

Rotate the knurled ring at the base of the steering wheel hub in an anti-clockwise direction when the steering wheel may be slid into the desired position. Turn the knurled ring clockwise to lock the steering wheel.

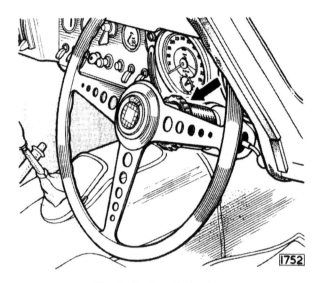

Fig. 5. Steering wheel adjustment

Door Locks

The doors may be opened from the outside by pressing the button incorporated in the door handle. The doors are opened from the inside by pulling the interior handles rearward.

Both doors can be locked from the inside by pushing the interior handles forward and allowing them to return to their original position; this feature only applies if the doors are fully closed before operating the interior handles. Both doors can be locked from the outside by means of the ignition key, the locks are incorporated in the push buttons of the door handles.

To lock the right-hand door insert the key in the lock, rotate anti-clockwise as far as possible and allow the lock to return to its original position—the door is now locked. To unlock the right-hand door turn key clockwise as far as possible and allow the lock to return to its original position.

To lock the left-hand door rotate key clockwise; to unlock, rotate key anti-clockwise.

KEYLESS LOCKING is obtainable by first pushing the interior door handle fully forward and allowing it to return to its original position. If the door is now closed from the outside with the push button of the handle **fully depressed** the door will become locked.

Warning.—If the doors are to be locked by this method the ignition key should be removed beforehand (or the spare key kept on the driver's person) as the only means of unlocking the doors is with this key.

Horn

Depress the circular button in the centre of the steering wheel to operate the horns.

Note.—The horns will not operate if ignition is off.

Ignition Switch

Inserting the key provided in the switch and turning clockwise will switch on the ignition.

Never leave the ignition on when the engine has stopped, a reminder of such circumstances is provided by the ignition warning light situated in the speedometer.

Interior Light Switch

Lift the switch lever (marked "Interior") to illuminate the car interior.

Lighting Switch

From "Off" can be rotated into two positions, giving in the first location side and tail, and in the second location head, side and tail.

Panel Light Switch

Lift the switch lever (marked "Panel") to enable the instruments to be read at night and to provide illumination of the switch markings. The switch has two positions "Dim" and "Bright" to suit the driver's requirements. The panel lights will only operate when the side lights are switched on.

Starter Switch

Press the button (marked "Starter") with the ignition switched on, to start the engine. Release the switch immediately the engine fires and never operate the starter when the engine is running.

Page A.10

GENERAL INFORMATION

Flashing Direction Indicator

The "flashers" are operated by a lever behind the steering wheel. To operate the flashing direction indicators on the right-hand side of the car, move the lever clockwise; to operate the left-hand side indicators, move the lever anti-clockwise. While the flashing indicators are in operation, one of the warning lights on the facia panel behind the steering wheel will flash on the side selected.

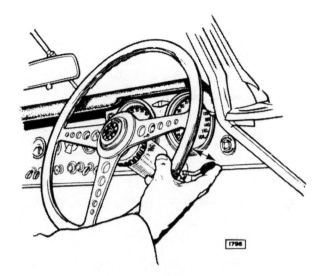

Fig. 7. Method of "flashing" the headlamps

Fig. 6. The flashing direction indicator control

Map Light

Lift the switch lever (marked "Map") to illuminate the lamp situated above the instrument panel. To provide ease of entry into the car at night the map light is switched on when either one of the doors is opened, and is extinguished when the door is closed.

Headlamp Flasher

To "flash" the headlamps as a warning signal, lift and release the flashing indicator lever in quick succession. The headlamps can be "flashed" when the lights are "off" or when they are in the dipped beam position; they will not "flash" in the main beam position.

Braking Lights

Twin combined tail and brake lights automatically function when the footbrake is applied.

Luggage Compartment Illumination

The luggage compartment is illuminated by the interior light when this lamp is switched on.

Cigar Lighter

To operate, press holder (marked "Cigar") into the socket and remove the hand. On reaching the required temperature, the holder will return to the extended position. Do not hold the lighter in the "pressed-in" position.

Windscreen Wipers

The wipers are controlled by a three position switch (marked "Wiper"). Lift the switch to the second position (Slow) which is recommended for all normal adverse weather conditions and snow.

For conditions of very heavy rain and for fast driving in rain lift the switch to the third position (Fast). This position should not be used in heavy snow or with a drying windscreen, that is, when the load on motor is in excess of normal; the motor incorporates a protective cut-out switch which under conditions of excessive load cuts off the current supply until normal conditions are restored.

When the switch is placed in the "Off" position the wipers will automatically return to a position along the lower edge of the screen.

Page A.11

GENERAL INFORMATION

Windscreen Washer

The windscreen washer is electrically operated and comprises a glass water container mounted in the engine compartment, which is connected to jets at the base of the windscreen. Water is delivered to the jets by an electrically driven pump incorporated in the water container.

Operation

The windscreen washer should be used in conjunction with the windscreen wipers to remove foreign matter that settles on the windscreen.

Lift the switch lever (marked "Washer") and release immediately, when the washer should operate at once and continue to function for approximately seven seconds. Allow a lapse of time before operating the switch a second time.

For full instructions on the use of the Windscreen Washing Equipment see Section "O".

Heating and Ventilating Equipment

The car heating and ventilating equipment consists of a heating element and an electrically driven fan mounted on the engine side of the bulkhead. Air from the heater unit is conducted:

(a) To a built in duct fitted with two doors situated behind the instrument panel.
(b) To vents at the bottom of the windscreen to provide demisting and defrosting.

The amount of fresh air can be controlled at the will of driver and is introduced into the system by operating the "Air" control lever and switching on the fan.

For full instruction on the use of the Heating and Ventilating Equipment see Section "O".

Steering Column—Adjustment for Rake

The steering column can be adjusted for rake. To adjust, release nut and bolt at the top of the column located behind the instrument panel, and adjust to suit requirements. Re-tighten nut fully after adjustment.

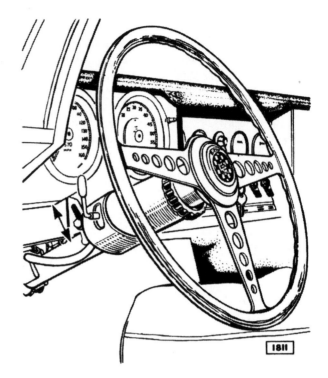

Fig. 8. Steering column adjustment for rake

Bonnet Lock
(Early Cars)

The bonnet is locked by means of the two locks situated at the sides of the bonnet.

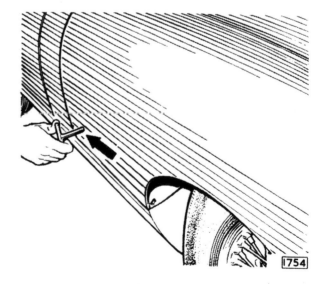

Fig. 9. Unlocking the bonnet (early cars)

Page A.12

GENERAL INFORMATION

To open the bonnet insert the "T" handle key provided in the lock and on the right-hand side turn key clockwise, and on the left-hand side turn key anti-clockwise.

This will release the bonnet which will now be retained by the safety catch. Insert the fingers under the rear edge of the bonnet and press in the safety catch.

To close the bonnet push down to the safety catch position. Hold the bonnet depressed and insert the "T" handle in the lock. On the right-hand side turn the key anti-clockwise and on the left-hand side turn the key clockwise.

(Later Cars)

The bonnet is locked by means of two locks situated at the sides of the bonnet. To open the bonnet, turn the two small levers located on the right and left-hand door hinge posts anti-clockwise and pull to full extent.

This will release the bonnet, which will now be retained by the safety catch.

Insert the fingers under the rear edge of the bonnet and press in the safety catch.

To close the bonnet, push down to the safety catch position, push in the two levers and turn clockwise.

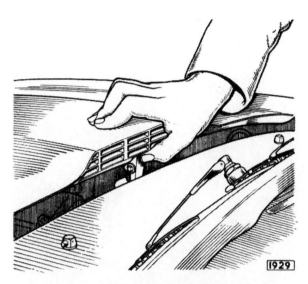

Fig. 11. Releasing the bonnet safety catch

Radiator Fan

The radiator fan is electrically driven, the cutting in speed being controlled automatically by means of a thermostatic switch incorporated in the engine cooling system. The fan will not operate with the ignition switched off.

When the coolant reaches a temperature of approximately 80°C., the thermostatic switch closes and starts the fan motor. The fan motor will continue to run until the temperature has fallen below approximately 72°C.

For full information on the Radiator Fan see "Cooling System (Section D)".

Interior Driving Mirror (Open 2-seater)

This is of the dipping type. Move lever, situated under mirror, forward for night driving, to avoid being dazzled by the lights of a following car.

Interior Driving Mirror (Fixed Head Coupe)

This is of the dipping type. Move lever, situated under the mirror, to the left for night driving, to avoid being dazzled by the lights of a following car.

Fuel Tank Filler

The fuel tank filler is situated in a recess in the left-hand rear wing and is provided with a hinged cover.

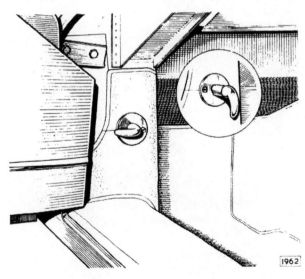

Fig. 10. Unlocking the bonnet (later cars)

Page A.13

GENERAL INFORMATION

Luggage Compartment (Open 2-seater)

The luggage compartment is unlocked by pulling the black knob situated inside the car on seat back panel right-hand side.

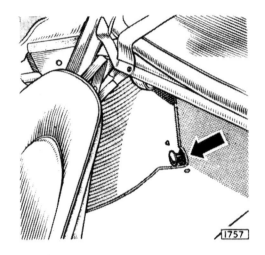

Fig. 12. Luggage compartment lock control (Open 2-seater)

Luggage Compartment (Fixed Head Coupe)

The luggage compartment is unlocked by lifting the recessed chromium plated lever situated in body trim panel beside right-hand seat. To operate, insert finger in recess and lift out lever to full extent. Retain the lid in the open position by means of the prop.

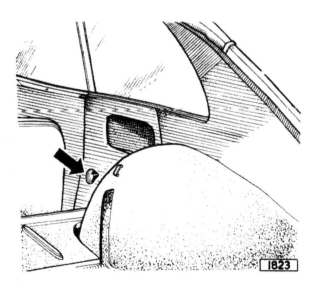

Fig. 13. Luggage compartment lock control (Fixed Head Coupe)

Seat Back Panel

The back panel behind the seat normally serves as a partition between the driving and luggage compartment. The panel can be lowered to give an increased boot floor area if required for extra storage.

To lower panel, release the two side fixing bolts and lower panel to check strap limits. Return panel to vertical position when extra boot space is not required.

Spare Wheel and Jacking Equipment

The spare wheel is housed in a well under the luggage compartment, and is accessible after removal of the square lid.

The copper hammer and jack are retained in clips in the luggage compartment. The jack handle is retained in clips under the spare wheel.

Tools

The tools are contained in a tool roll placed in the spare wheel compartment.

WHEEL CHANGING

Whenever possible the wheel changing should be carried out with the car standing on level ground, and in all cases with the handbrake fully applied.

Unlock the luggage compartment by pulling the black knob situated inside car at right-hand side of seat back panel.

The spare wheel is housed in a compartment underneath the luggage boot floor; the wheel changing equipment is retained in clips.

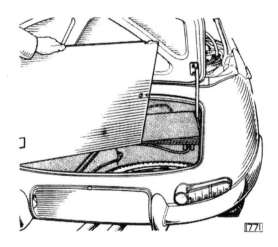

Fig. 14. Spare Wheel Housing (Open 2-seater)

Page A.14

GENERAL INFORMATION

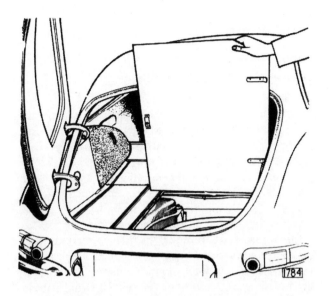

Fig. 15. Spare wheel housing (Fixed Head Coupe)

Fig. 17. Hub cap—left hand side

Remove the copper and hide mallet from the tool kit. Using the mallet, slacken but do not remove the hub caps; the hub caps are marked Right (off) side or Left (near) side, and the direction of rotation to remove, that is, clockwise for the right-hand side and anti-clockwise for the left-hand side.

The jacking sockets will be found centrally located on either side of the car. Place jack under car with pad located in the socket and raise car until wheels are clear of ground. Remove hub cap and withdraw wheel. Mount the spare wheel on the splined hub. Refit the hub cap and tighten as much as possible by rotating cap in the required direction, that is, anti-clockwise for the right-hand side and clockwise for the left-hand side.

Lower the jack and finally tighten the hub fully with the copper and hide mallet.

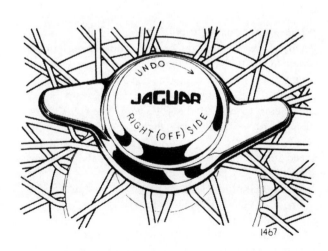

Fig. 16. Hub cap—right-hand side

Fig. 18. The jack in position for raising the left-hand side of the car. The position of the lever shown in the inset controls the operation of the jack screw

Page A.15

GENERAL INFORMATION

STARTING AND DRIVING

Prior to Starting

Ensure that the coolant level in the radiator and the oil level in the sump are correct. Check for sufficient petrol in the tank.

Starting from Cold

A manual mixture control is provided located in facia panel behind steering wheel. This control has six positions; the "fully rich" position being at the top of the slide marked "COLD". Moving the lever progressively downwards weakens the mixture strength. The two positions from "HOT" give a fast idle condition; the last position "RUN" being off.

A red warning light is incorporated in the control which lights up immediately the lever is moved from "RUN" position.

When starting from cold the mixture control should be moved to the fully rich "COLD" position. Switch on the ignition and press the starter button, but do not touch the accelerator. Release the starter button as soon as the engine fires—this is important. If for any reason the engine does not start do not operate the starter button again until both the engine and the starter motor have come to rest.

As soon as the engine speed increases slide the control progressively to the intermediate "HOT" position.

Drive off at a moderate speed progressively sliding the mixture control to the "RUN" position until the knob is at the bottom of the slide and the red warning light is extinguished.

Always return the control to "RUN" position as soon as possible. Unnecessary use of the mixture control will result in reduced engine life.

Starting in Moderate Temperature

In warm weather or if the engine is not absolutely cold, it is usually possible to start the engine with the mixture control in one of the intermediate "HOT" positions. Do not touch the accelerator pedal.

Starting When Hot

Do not use the mixture control. If the engine does not start immediately slightly depress the accelerator pedal when making the next attempt.

Warming Up

Do not operate the engine at a fast speed when first started but allow time for the engine to warm up and the oil to circulate. A thermostat is incorporated in the cooling system to assist rapid warming up. In very cold weather run the engine at 1,500 r.p.m. with the car stationary until a rise in temperature is indicated on the temperature gauge.

Driving

(a) Careful adherence to the "Running-in" Instructions given will be amply repaid by obtaining the best performance and utmost satisfaction from the car.

(b) The habit should be formed of reading the oil pressure gauge, water temperature gauge and ammeter occasionally as a check on the correct functioning of the car. Should an abnormal reading be obtained an investigation should be made immediately.

(c) Always start from rest in first or second gear; on a hill always use first gear. To start in a higher gear will cause excessive clutch slip and premature wear. Never drive with a foot resting on the clutch pedal and do not keep the clutch depressed for long periods in traffic.

(d) The synchromesh gearbox provides a synchronized change in second, third and top. When changing gear the movement should be slow and deliberate.

When changing down a smoother gear change will be obtained if the accelerator is left depressed to provide the higher engine speed suitable to the lower gear. Always fully depress the clutch pedal when changing gear.

(e) Gear changing may be slightly stiff on a new car but this will disappear as the gearbox becomes "run-in".

(f) Always apply the footbrake progressively; fierce and sudden application is bad for the car and tyres. The handbrake is for use when parking the car, when driving away on a hill and when at a standstill in traffic.

"Running-in" Instructions

Only if the following important recommendations are observed will the high performance and continued good running of which the Jaguar is capable be obtained.

During the "running-in" period do not allow the engine to exceed the following speeds and particularly do not allow the engine to labour on hills; it is preferable to select a lower gear and use a higher speed rather than allow the engine to labour at low speed:—

First 1,000 miles (1,600 km.) .. .. 2,500 r.p.m.
From 1,000—2,000 miles (1,600—
    3,200 km.) .. .. 3,000 r.p.m.

Have the engine sump drained and refilled and the oil filter attended to as recommended at the free service, that is, after the first 500 miles (800 km.).

Page A.16

GENERAL INFORMATION

SUMMARY OF MAINTENANCE

Daily

Check radiator coolant level.

Check engine oil level.

Weekly

Check tyre pressures.

Check fluid level in brake and clutch master cylinder reservoirs.

Monthly

Check battery electrolyte level and connections.

Change over road wheels.

Every 2,500 miles (4,000 km.)

Drain engine sump and refill.

Renew oil filter element.

Check gearbox oil level and top-up if necessary.

Check rear axle oil level and top-up if necessary.

Lubricate steering housing.

Lubricate steering tie-rod ball joints.

Lubricate wheel swivels.

Lubricate propeller shaft universal joints (early cars).

Lubricate propeller shaft splines (early cars).

Lubricate carburetter hydraulic piston dampers.

Lubricate rear half shaft universal joints.

Lubricate distributor and check contact points.

Clean, adjust and test sparking plugs.

Check clutch free travel and adjust if necessary.

Check handbrake adjustment (early cars only).

Check carburetter slow running.

Every 5,000 miles (8,000 km.)

Carry out 2,500 miles service

Tune carburetters.

Clean fuel line filter.

Lubricate door hinges.

Check dynamo belt and adjust if necessary.

Renew oil filter element.

Examine brake friction pads for wear.

Clear drain holes in bottoms of doors.

Adjust top timing chain (if required).

Check front wheel alignment.

Lubricate rear suspension wishbone pivot bearings.

Carry out oil can lubrication of (a) seat runners and adjusting mechanism, (b) handbrake lever ratchet, (c) door locks, (d) boot hinges and lock, (e) bonnet hinges and catches, (f) windscreen wiper arms, (g) accelerator linkage, (h) fuel filler cover hinge, (i) handbrake cable compensator, (j) brake pedal bearing, (k) carburetter linkage.

Lubricate generator end bush (later cars only).

Every 10,000 miles (16,000 km.)

Carry out the 2,500 miles and 5,000 miles service.

Drain and refill gearbox.

Drain and refill rear axle.

Lubricate wheel hub bearings.

Check and tighten all chassis and body nuts, screws and bolts.

Renew air cleaner element.

Check wheel bearing end float and adjust if necessary.

Renew sparking plugs.

Clean fuel tank filter.

Page A.17

GENERAL INFORMATION

RECOMMENDED LUBRICANTS

Component	Mobil	Castrol	Shell	Esso	B.P.	Duckham	Regent Caltex/Texaco
Engine	Mobiloil Special*	Castrolite* or Castrol XL	Shell Super Oil	Esso Extra Motor Oil 5W/20*† Esso Extra Motor Oil 10W/30*† Esso Extra Motor Oil 20W/40*†	Viscostatic	Q20–50 or Q5500*	Havoline 20W/40 or 10W/30*
Upper cylinder lubrication	Mobile upper lube	Castrollo	Shell UCL or Donax U	Esso UCL	UCL	Adcoid Liquid	Regent UCL
Gearbox Distrubutor oil can points Oil can lubrication	Mobiloil A	Castrol XL	X-100 30	Esso Motor Oil 20W/30	Energol SAE 30	NOL 30	Havoline 30
Rear Axle	Mobilube GX90	Castrol Hypoy	Spirax 90 E.P.	Esso Gear Oil G.P.90/140	Gear Oil SAE 90EP	Hypoid 90	Multigear Lubricant EP 90
Propeller shafts Rear axle half shafts Front wheel bearings Rear wheel bearings Distributor cam	Mobilgrease MP	Castrolease LM	Retinax A	Esso Multi-purpose Grease H	Energrease L.2	LB 10	Marfak All purpose
Steering housing Steering tie-rods Wheel swivels Door hinges Rear wishbone pivots	Mobilgrease MP	Castrolease LM	Retinax A	Esso Multi-purpose Grease H	Energrease L.2	LB 10	Marfak All purpose

*These oils should not be used in worn engines requiring overhaul. If an SAE 30 or 40 oil has previously been used in the engine, a slight increase in oil consumption may be noticed but this will be compensated by the advantages gained.

†According to availability in country of operation.

RECOMMENDED HYDRAULIC FLUIDS
Braking System and Clutch Operation

Preferred Fluid
Castrol/Girling Crimson Clutch/Brake Fluid. (SAE J1703A).

Alternative Fluids
Recognised brands of brake fluid conforming to specification SAE J1703A.

Page A.18

GENERAL INFORMATION

SPECIAL SERVICE TOOLS

Specialised Service Tools illustrated in this Service Manual which bear a Churchill Tool number, can be obtained direct from Messrs. V. L. Churchill & Co. Ltd., at the address given below.

V. L. Churchill & Co. Ltd.,
London Road,
Daventry,
Northants.
P.O. Box No. 3.

SERVICE DEPARTMENTS

Factory:

The Service Division,
Jaguar Cars Limited,
Coventry, England.
Telephone No. Allesley 2121 (P.B.X.)

London:

Messrs. Henlys Ltd.,
The Hyde,
Hendon,
London, N.W.9
Telephone No. Colindale 6565

U.S.A.:

The Technical Service Department,
600 Willow Tree Road, Leonia,
New Jersey 07605, U.S.A.

Canada:

The Technical Service Department,
British Motor Holdings (Canada) Limited
4445 Fairview Street, P.O. Box 5033,
Burlington, Ontario.

Page A.19

GENERAL INFORMATION

CONVERSION TABLES

METRIC INTO ENGLISH MEASURE

1 millimetre is approximately $^1/_{25}''$, and is exactly ·03937″.
1 centimetre is approximately $\frac{3}{8}''$, and is exactly ·3937″.
1 metre is approximately $39\frac{3}{8}''$, and is exactly 39·37″ or 1·0936 yards.
1 kilometre is approximately $\frac{5}{8}$ mile, and is exactly ·6213 miles.
1 kilogramme is approximately $2\frac{1}{4}$ lbs., and is exactly 2·21 lbs.
1 litre is approximately $1\frac{3}{4}$ pints, and is exactly 1·76 pints.
To convert metres to yards, multiply by 70 and divide by 64.
To convert kilometres to miles, multiply by 5 and divide by 8 (approx.)
To convert litres to pints, multiply by 88 and divide by 50.
To convert grammes to ounces, multiply by 20 and divide by 567.
To find the cubical contents of a motor cylinder, square the diameter (or bore), multiply by 0·7854 and multiply the result by the stroke.
1 M.P.G.—0·3546 kilometres per litre or 2.84 litres per kilometre.

MILES INTO KILOMETRES

Kilo.	Miles	Kilo.	Miles	Kilo.	Miles	Kilo.	Miles	Kilo.	Miles
1	$\frac{5}{8}$	16	10	31	$19\frac{1}{4}$	46	$28\frac{5}{8}$	60	$37\frac{1}{4}$
2	$1\frac{1}{4}$	17	$10\frac{5}{8}$	32	$19\frac{7}{8}$	47	$29\frac{1}{4}$	70	$43\frac{1}{2}$
3	$1\frac{7}{8}$	18	$11\frac{1}{4}$	33	$20\frac{1}{2}$	48	$29\frac{7}{8}$	80	$49\frac{3}{4}$
4	$2\frac{1}{2}$	19	$11\frac{3}{4}$	34	$21\frac{1}{8}$	49	$30\frac{1}{2}$	90	$55\frac{7}{8}$
5	$3\frac{1}{8}$	20	$12\frac{3}{8}$	35	$21\frac{3}{4}$	50	$31\frac{1}{8}$	100	$62\frac{1}{4}$
6	$3\frac{3}{4}$	21	13	36	$22\frac{3}{8}$	51	$31\frac{3}{4}$	200	$124\frac{1}{4}$
7	$4\frac{3}{8}$	22	$13\frac{5}{8}$	37	23	52	$32\frac{1}{4}$	300	$186\frac{3}{8}$
8	5	23	$14\frac{1}{4}$	38	$23\frac{5}{8}$	53	$32\frac{7}{8}$	400	$248\frac{1}{2}$
9	$5\frac{5}{8}$	24	$14\frac{7}{8}$	39	$24\frac{1}{4}$	54	$33\frac{1}{2}$	500	$310\frac{1}{2}$
10	$6\frac{1}{4}$	25	$15\frac{1}{2}$	40	$24\frac{7}{8}$	55	$34\frac{1}{8}$	600	$372\frac{1}{2}$
11	$6\frac{7}{8}$	26	$16\frac{1}{8}$	41	$25\frac{1}{2}$	56	$34\frac{3}{4}$	700	435
12	$7\frac{1}{2}$	27	$16\frac{3}{4}$	42	$26\frac{1}{8}$	57	$35\frac{3}{8}$	800	$497\frac{1}{2}$
13	$8\frac{1}{8}$	28	$17\frac{3}{8}$	43	$26\frac{3}{4}$	58	36	900	$559\frac{1}{2}$
14	$8\frac{3}{4}$	29	18	44	$27\frac{3}{8}$	59	$36\frac{5}{8}$	1000	$621\frac{3}{8}$
15	$9\frac{3}{8}$	30	$18\frac{5}{8}$	45	28				

PINTS AND GALLONS TO LITRES

Pints	Gallons	Litres Approx.	Litres Exact	Pints	Gallons	Litres Approx.	Litres Exact
1	$\frac{1}{8}$	$\frac{1}{2}$	·57	40	5	23	22·75
2	$\frac{1}{4}$	1	1·14	48	6	27	27·30
3	$\frac{3}{8}$	$1\frac{3}{4}$	1·71	56	7	32	31·85
4	$\frac{1}{2}$	$2\frac{1}{4}$	2·27	64	8	$36\frac{1}{2}$	36·40
8	1	$4\frac{1}{2}$	4·54	72	9	41	40·95
16	2	9	9·10	80	10	$45\frac{1}{2}$	45·50
24	3	$13\frac{1}{2}$	13·65	88	11	50	50·05
32	4	18	18·20	96	12	$54\frac{1}{2}$	54·60

Page A.20

GENERAL INFORMATION

RELATIVE VALUE OF MILLIMETRES AND INCHES

mm.	Inches	mm.	Inches	mm.	Inches	mm.	Inches
1	0·0394	26	1·0236	51	2·0079	76	2·9922
2	0·0787	27	1·0630	52	2·0473	77	3·0315
3	0·1181	28	1·1024	53	2·0866	78	3·0709
4	0·1575	29	1·1417	54	2·1260	79	3·1103
5	0·1968	30	1·1811	55	2·1654	80	3·1496
6	0·2362	31	1·2205	56	2·2047	81	3·1890
7	0·2756	32	1·2598	57	2·2441	82	3·2284
8	0·3150	33	1·2992	58	2·2835	83	3·2677
9	0·3543	34	1·3386	59	2·3228	84	3·3071
10	0·3937	35	1·3780	60	2·3622	85	3·3465
11	0·4331	36	1·4173	61	2·4016	86	3·3859
12	0·4724	37	1·4567	62	2·4410	87	3·4252
13	0·5118	38	1·4961	63	2·4803	88	3·4646
14	0·5512	39	1·5354	64	2·5197	89	3·5040
15	0·5906	40	1·5748	65	2·5591	90	3·5433
16	0·6299	41	1·6142	66	2·5984	91	3·5827
17	0·6693	42	1·6536	67	2·6378	92	3·6221
18	0·7087	43	1·6929	68	2·6772	93	3·6614
19	0·7480	44	1·7323	69	2·7166	94	3·7008
20	0·7874	45	1·7717	70	2·7559	95	3·7402
21	0·8268	46	1·8110	71	2·7953	96	3·7796
22	0·8661	47	1·8504	72	2·8347	97	3·8189
23	0·9055	48	1·8898	73	2·8740	98	3·8583
24	0·9449	49	1·9291	74	2·9134	99	3·8977
25	0·9843	50	1·9685	75	2·9528	100	3·9370

RELATIVE VALUE OF INCHES AND MILLIMETRES

Inches	0	$\frac{1}{16}$	$\frac{1}{8}$	$\frac{3}{16}$	$\frac{1}{4}$	$\frac{5}{16}$	$\frac{3}{8}$	$\frac{7}{16}$
0	0·0	1·6	3·2	4·8	6·4	7·9	9·5	11·1
1	25·4	27·0	28·6	30·2	31·7	33·3	34·9	36·5
2	50·8	52·4	54·0	55·6	57·1	58·7	60·3	61·9
3	76·2	77·8	79·4	81·0	82·5	84·1	85·7	87·3
4	101·6	103·2	104·8	106·4	108·0	109·5	111·1	112·7
5	127·0	128·6	130·2	131·8	133·4	134·9	136·5	138·1
6	152·4	154·0	155·6	157·2	158·8	160·3	161·9	163·5

Inches	$\frac{1}{2}$	$\frac{9}{16}$	$\frac{5}{8}$	$\frac{11}{16}$	$\frac{3}{4}$	$\frac{13}{16}$	$\frac{7}{8}$	$\frac{15}{16}$
0	12·7	14·3	15·9	17·5	19·1	20·6	22·2	23·8
1	38·1	39·7	31·4	42·9	44·4	46·0	47·6	49·2
2	63·5	65·1	66·7	68·3	69·8	71·4	73·0	74·6
3	88·9	90·5	92·1	93·7	95·2	96·8	98·4	100·0
4	114·3	115·9	117·5	119·1	120·7	122·2	123·8	125·4
5	139·7	141·3	142·9	144·5	146·1	147·6	149·2	150·8
6	165·1	166·7	168·3	169·9	171·5	173·0	174·6	176·2

Page A.21

SECTION B

ENGINE
3·8 "E" TYPE
GRAND TOURING MODELS

INDEX

	Page
Air Cleaner	B.65
Bottom Chain Tensioner:	
Removal	B.59
Refitting	B.59
Camshafts:	
Removal	B.33
Refitting	B.33
Overhaul	B.33
Compression Pressures	B.31
Connecting Rod and Bearings:	
Removal	B.31
Overhaul	B.31
Refitting	B.31
Big-end bearing replacement	B.32
Crankshaft:	
Removal	B.34
Overhaul	B.34
Refitting	B.34
Crankshaft Damper and Pulley:	
Removal	B.35
Overhaul	B.35
Refitting	B.35
Cylinder Block:	
Overhaul	B.40

Page B.1/2

INDEX (continued)

	Page
Cylinder Head:	
Removal	B.41
Overhaul	B.42
Refitting	B.42
Data	B.5
Decarbonising and Grinding Valves	B.29
Engine—Removal and Refitting	B.19
Engine—To dismantle	B.21
Engine—To assemble	B.23
Engine Mountings	B.65
Engine Stabilizer	B.66
Exhaust Manifolds:	
Removal	B.43
Refitting	B.43
Flywheel:	
Removal	B.43
Overhaul	B.44
Refitting	B.44
Ignition Timing	B.44
Inlet Manifold:	
Removal	B.45
Refitting	B.45
Oil Filter:	
Removal	B.46
Refitting	B.47
Element replacement	B.47
Oil Pump:	
Removal	B.48
Dismantling	B.48
Overhaul	B.48
Re-assembling	B.50
Refitting	B.50

INDEX (continued)

	Page
Oil Pump:	
Removal	B.48
Dismantling	B.48
Overhaul	B.48
Re-assembling	B.50
Refitting	B.50
Oil Sump:	
Removal	B.50
Refitting	B.50
Pistons and Gudgeon Pins:	
Removal	B.51
Overhaul	B.51
Refitting	B.53
Routine Maintenance	B.14
Sparking Plugs:	
Service procedure	B.53
Analysing service conditions	B.53
Standard gap setting	B.55
Tappets, Tappet Guides and Adjusting Pads:	
Removal of tappets and adjusting pads	B.55
Overhaul	B.55
Timing Gear:	
Removal	B.57
Dismantling	B.57
Overhaul	B.57
Assembling	B.57
Refitting	B.58
Torsion Bar Reaction Tie Plate:	
Removal	B.66
Refitting	B.67
Valves and Springs:	
Removal	B.60
Overhaul	B.60
Valve clearance adjustment	B.61
Refitting	B.61
Valve Guides:	
Replacement	B.62
Valve Seat Inserts:	
Replacement	B.62
Valve Timing	B.63

ENGINE

All "E" Type models have the twin overhead camshaft XK type engine, fitted with the "S" type cylinder head with straight ports and 3/8" lift camshafts.

Compression Ratio	Engine Number Prefix	Colour of Cylinder Head
8 : 1 or 9 : 1	R	Gold

Compression ratios of 8 to 1 and 9 to 1 are specified for the "E" Type engine, the differences in compression ratio being obtained by varying the crown design of the piston.

The compression ratio of an engine is indicated by /8, /9 following the engine number.

DATA

Camshaft

Number of journals	Four per shaft
Journal diameter	1·00″ —·0005″ —·001″ (25·4 mm.—·013 mm.) —·025
Thrust taken	Front end
Number of bearings	Four per shaft (eight half bearings)
Type of bearing	White metal steel backed shell
Diameter clearance	·0005″ to ·002″ (·013 to ·05 mm.)
Permissible end float	·0045″ to ·008″ (·11 to ·20 mm.)
Tightening torque—Bearing cap nuts	15 lbs. ft. (175 lbs. ins.) (2·0 kg/m.)

Connecting Rod

Length centre to centre	7¾″ (19·68 cm.)
Big end—Bearing type	Lead bronze, steel backed shell

Page B.5

ENGINE

Bore for big end bearing	2·233″ to 2·2335″ (56·72 to 56·73 mm.)
Big end—Width	$1\frac{3}{16}″ - ·006″$ $- ·008″$ (30·16 mm. — ·15 mm.) — ·20
Big end—Diameter clearance	·0015″ to ·0033″ (·04 mm. to ·08 mm.)
Big end—side clearance	·0058″ to ·0087″ (·15 mm. to ·22 mm.)
Bore for small end bush	$1·00″ \pm ·0005″$ ($25·4$ mm. $\pm ·013$ mm.)
Small end bush—Type	Phosphor bronze—steel backed
Small end—Width	$1\frac{5}{64}″$ (27·4 mm.)
Small end bush—Bore diameter	·875″ + ·0002″ — ·0000″ (22·22 mm. + ·005 mm.) — ·000
Tightening torque—Con rod bolts	37 lbs. ft. (450 lbs. ins.) (5·1 kg./m.)

Crankshaft

Number of main bearings	Seven
Main bearing—Type	Lead bronze, steel backed shell
Journal diameter	Front, centre, rear 2·750″ to 2·7505″ (69·85 to 69·86 mm.) Intermediate 2·7495″ to 2·750″ (69·84 to 69·85 mm.)
Journal length Front	$1\frac{11}{16}″ \pm ·005″$ (42·86 mm. $\pm ·13$ mm.)
Centre	$1\frac{3}{4}″ + ·0005″$ + ·001″ (44·45 mm. + ·013 mm.) + ·025

Page B.6

ENGINE

Rear	$1\frac{7}{8}''$
	(47·63 mm.)
Intermediate	$1\frac{7}{32}'' \pm \cdot 002''$
	(30·96 mm. ± ·05 mm.)
Thrust taken	Centre bearing thrust washers
Thrust washer—Thickness	$\cdot 092'' \pm \cdot 001''$ and $\cdot 096'' \pm \cdot 001''$
	(2·33 mm. ± ·025 mm. and 2·43 mm. ± ·025 mm.)
End clearance	·004″ to ·006″
	(·10 to ·15 mm.)
Main bearing—Length	
Front ⎫	
Centre ⎬	$1\frac{1}{2}'' \pm \cdot 005''$
Rear ⎭	(38·1 mm. ± ·13 mm.)
Intermediate	$1'' \pm \cdot 005''$
	(25·4 mm. ± ·13 mm.)
Diameter clearance	·0025″—·0042″
	(·063 to ·106 mm.)
Crankpin—Diameter	$2\cdot086'' {+\cdot 0006'' \atop -\cdot 000''}$
	$(52\cdot98\text{ mm.} {+\cdot 015\text{ mm.} \atop -\cdot 000})$
Length	$1\frac{3}{16}'' {+\cdot 0007'' \atop -\cdot 0002''}$
	$(30\cdot16\text{ mm.} {+\cdot 018\text{ mm.} \atop -\cdot 006})$
Regrind undersize	·010″, ·020″, ·030″ and ·040″
	(·25, ·51, ·76 and 1·02 mm.)
Minimum diameter for regrind	—·040″
	(1·02 mm.)
Tightening torque—main bearing bolts	83 lbs. ft. (1,000 lbs. ins.)
	(11·5 kg./m.)

Cylinder Block

Material	"Brivadium" dry liners

Page B.7

ENGINE

Cylinder bores—Nominal	87 mm. +·0127 mm. (3·4252 +·0005″)
	−·0064 mm. −·00025″
Maximum rebore size	+·030″
	(·76 mm.)
Bore size for fitting liners	3·561″ to 3·562″
	(90·45 mm. to 90·47)
Outside diameter of liner	3·563″ to 3·566″
	(90·50 to 90·58 mm.)
Interference fit	·001″ to ·005″
	(·025 to ·125 mm.)
Overall length of liner	6 31/32″ (17·7 cm.)
Outside diameter of lead-in	3·558″ to 3·560″
	(90·37 to 90·42 mm.)
Size of bore honed after assembly—in cylinder block—Nominal	87 mm. (3·4252″)
Main line bore for main bearings	2·9165″ +·0005″
	−·0000″
	(74·08 +·013 mm.)
	−·000 mm.)

Cylinder Head

Type	Straight Port (Gold Top)
Material	Aluminium Alloy
Valve seat angle—Inlet	45°
—Exhaust	45°
Valve throat diameter—Inlet	1½″
	(38·1 mm.)
—Exhaust	1 3/8″
	(34·9 mm.)
Tightening torque—Cylinder head nuts	54 lbs. ft. (650 lbs. ins.)
	(7·5 kg./m.)
Firing order	1, 5, 3, 6, 2, 4

No. 1 cylinder being at the rear of the engine unit.

Page B.8

ENGINE

Gudgeon Pin

Type	Fully floating
Length	2·840″ to 2·845″ (72·14 to 72·26 mm.)
Inside diameter	$\frac{5}{8}''$ (15·87 mm.)
Outside diameter	·8750″ to ·8752″ (22·22 to 22·23 mm.)

Lubricating System

Oil pressure (hot)	40 lbs. per sq. in. at 3,000 r.p.m.
Oil pump—Type	Eccentric rotor
—Clearance at end of lobes	·006″ maximum (·15 mm.)
—End clearance	·0025″ maximum (·06 mm.)
—Clearance between outer rotor and body	·010″ maximum (·25 mm.)

Piston and Piston Rings

Make	Brico
Type	Semi-split skirt
Piston	
Skirt clearance *(measured at bottom of skirt at 90° to gudgeon pin axis)*	·0011″ to ·0017″ (·028 to ·043 mm.)
Gudgeon pin bore	·8749″ to ·8751″ (2·223 to 2·227 mm.)

Compression height

8 : 1 compression ratio	2·069″ to 2·064″ (52·42 to 52·55 mm.)
9 : 1 compression ratio	2·247″ to 2·242″ (56·94 to 57·07 mm.)

Page B.9

ENGINE

Piston rings—Number
- Compression 2
- Oil control 1

Piston rings—Width
- Compression .. .. 0·777" to ·0787"
  (1·97 to 2·00 mm.)
- Oil Control .. .. ·155" to ·156"
  (3·94 to 3·96 mm.)

Piston rings—Thickness
- Compression .. .. ·124" to ·130"
  (3·15 to 3·30 mm.)
- Oil control .. .. ·119" to ·127"
  (3·02 to 3·23 mm.)

Piston rings—Side clearance in groove
- Compression .. .. ·001" to ·003"
  (·02 to ·07 mm.)
- Oil Control .. .. ·001" to ·003"
  (·02 to ·07 mm.)

Piston rings—Gap when fitted to cylinder bore
- Compression .. .. ·015" to ·020"
  (·38 to ·51 mm.)
- Oil control .. .. ·011" to ·016"
  (·28 to ·41 mm.)

Sparking Plugs
Make .. .. Champion

Type
- 8 : 1 compression ratio UN12Y*
- 9 : 1 compression ratio UN12Y*
- Gap .. .. ·025"
  (·64 mm.)

* N.3 for racing.

Tappets and Tappet Guides

Tappet—Material .. .. Cast iron (chilled)

—Outside diameter .. .. 1·3738" to 1·3742"
(34·89 to 34·90 mm.)

Page B.10

ENGINE

Diameter clearance	·0008" to ·0019" (·02 to ·048 mm.)
Tappet guide—Material	Austenitic iron
—Inside diameter (before reaming)	1·353" to 1·357" (34·37 to 34·48 mm.)
—Reaming size (when fitted to cylinder head)	1·375" + ·0007" — ·0000" (34·925 mm. + ·018 mm. — ·000)
—Interference (shrink) fit in head	·003" (·07 mm.)

Timing Chains and Sprockets

Type	Duplex
Pitch	⅜" (9·5 mm.)
Number of pitches—Top chain	100
—Bottom chain	82
Crankshaft sprocket—Teeth	21
Intermediate sprocket, outer—Teeth	28
Intermediate sprocket, inner—Teeth	20
Camshaft sprocket—Teeth	30
Idler Sprocket	21

Valve Timing

Inlet valve opens	15° B.T.D.C.
Inlet valve closes	57° A.B.D.C.
Exhaust valve opens	57° B.B.D.C.
Exhaust valve closes	15° A.T.D.C.
	(with valve clearances set at ·010" (·25 mm.))

Valves and Valve Springs

Valves—Material, Inlet	Silicon chrome steel
Exhaust	Austenitic steel

Page B.11

ENGINE

Valve head diameter, Inlet	$1\frac{3}{4}'' \pm \cdot 002''$	
	(44·45 mm. ± ·05 mm.)	
Exhaust	$1\frac{5}{8}'' \pm \cdot 002''$	
	(41·27 mm. ·05 mm.)	
Valve stem diameter, Inlet and Exhaust	$\frac{5}{16}'' - \cdot 0025''$	
	$- \cdot 0035''$	
	(7·95 mm. — ·06 mm.)	
	— ·09 mm.)	
Valve lift	$\frac{3}{8}''$	
	Touring	**Racing**
Valve clearance—Inlet	·004″	·006″
	(·10 mm.)	(·15 mm.)
—Exhaust	·006″	·010″
	(·15 mm.)	·25 mm.)
Valve seat angle—Inlet	45°	
—Exhaust	45°	
Valve spring—Free length. Inner	$1\frac{21}{32}''$	
	(·42 mm.)	
Outer	$1\frac{15}{16}''$	
	(49·2 mm.)	
Valve spring—Fitted length. Inner	$1\frac{7}{32}''$	
	(30·96 mm.)	
Outer	$1\frac{5}{16}''$	
	(33·34 mm.)	
Valve spring—Fitted load		
Inner	30·33 lbs.	
	(13·76 kg.)	
Outer	48·375 lbs.	
	(21·94 kg.)	
Valve spring—Solid length (max.) Inner	·810″	
	(20·57 mm.)	
Outer	·880″	
	(22·35 mm.)	
Number of free coils Inner	6	
Outer	5	
Diameter of wire Inner	12 SWG (·104″)	
	(2·64 mm.)	
Outer	10 SWG (·128″)	
	(3·25 mm.)	

Valve Guide and Valve Seat Insert

Valve guides—Material	Cast iron
Valve guide—Length, Inlet	$1\frac{13}{16}''$
	(46·04 mm.)
Exhaust	$1\frac{15}{16}''$
	(49·21 mm.)

Page B.12

ENGINE

Valve guide—Inside diameter—Inlet .. .. .. $\frac{5}{16}"-·0005"$
$-·0015"$
(7·94 mm.—·013 mm.)
$-·038$ mm.)

Exhaust .. .. .. $\frac{5}{16}\pm·0005"$
(7·94 mm.±·01 mm.)

Interference fit in head .. .. ·0005" to ·0022"
(·013 to ·055 mm.)

Valve seat inserts—Material .. .. Cast iron (centrifugally cast)

Inside diameter Inlet .. .. .. $1\frac{1}{2}"+·003"$
$-·001"$
(38·1+·076 mm.)
$-·025$ mm.)

Exhaust .. .. 1·379" to 1·383"
(35·03 to 35·13 mm.)

Interference (shrink) fit in head .. ·003"
(·076 mm.)

Capacities

	Imperial	U.S.	Litres
Engine (refill)	15 pints	18 pints	$8\frac{1}{2}$

Page B.13

ENGINE

ROUTINE MAINTENANCE

DAILY
Checking the Engine Oil Level

Check the oil level with the car standing on level ground otherwise a false reading will be obtained.

Remove the dipstick and wipe it dry. Replace and withdraw the dipstick; if the oil level is on the knurled patch, with the engine hot or cold, no additional oil is required. If the engine has been run immediately prior to making an oil level check, wait one minute after switching off before checking the oil level.

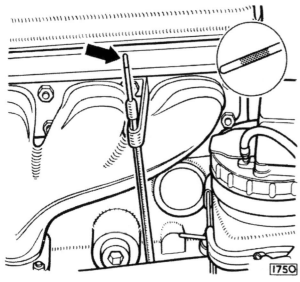

Fig. 1. Engine dipstick.

Note: Almost all modern engine oils contain special additives, and whilst it is permissible to mix the recommended brands it is undesirable. If it is desired to change from one brand to another this should be done when the sump is drained and the Oil Company's recommendation in regard to flushing procedure should be followed.

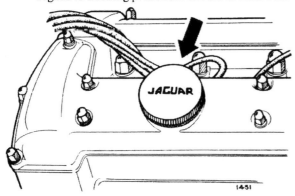

Fig. 2. Engine oil filler.

EVERY 2,500 MILES (4,000 KM.)
Changing the Engine Oil

Note: Under certain adverse operating conditions, conducive to oil dilution and sludge formation, more frequent oil changing than the normal 2,500 mile (4,000 km.) period is advised. Where the car is used mainly for low-speed city driving, stop-start driving particularly in cold weather or in dusty territory the oil should be changed at least every 1,000 miles (1,600 km.).

The draining of the sump should be carried out at the end of a run when the oil is hot and therefore will flow more freely. The drain plug is situated at the right-hand rear corner of the sump. When the engine oil is changed, the oil filter which is situated on the right-hand side of the engine, must also receive attention.

First drain the oil from the filter by removing the small hexagon-headed drain plug situated at the

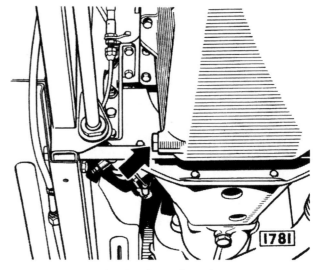

Fig. 3. Engine drain plug.

bottom of the filter head. Unscrew the central bolt and remove the canister and element. Thoroughly wash these parts in petrol and allow to dry out. When replacing the canister renew the circular rubber seal in the filter head. (Attention is drawn to the importance of renewing the filter element at 5,000 miles (8,000 km.) intervals.

These instructions apply only to felt filter elements. If a paper element is employed, this filter should be renewed at every oil change.

Page B.14

ENGINE

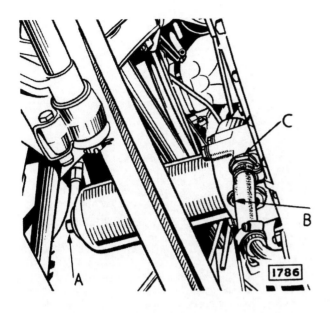

Fig. 4. Engine oil filter.
A, securing bolt; B, drain plug; C, oil pressure relief valve union

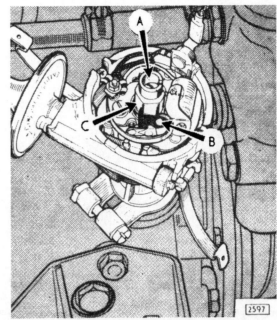

Fig. 5. Lift off the rotor arm and apply a few drops of oil around the screw 'A'. Apply one drop of oil to post 'B'. Lightly smear the cam 'C' with grease.

Distributor Lubrication

Take care to prevent oil or grease from getting on or near the contact breaker points.

Remove the moulded cap at the top of the distributor by springing back the two clips. Lift off the rotor arm and apply a few drops of engine oil around the screw (A Fig. 5) now exposed. It is not necessary to remove the screw as it has clearance to permit the passage of oil.

Apply **one** drop of oil to the post (B) on which the contact breaker pivots. Lightly smear the cam (C) with grease. Lubricate the centrifugal advance mechanism by injecting a few drops of engine oil through the aperture at the edge of the contact breaker base plate.

Distributor Contact Breaker Points

Check the gap between the contact points with feeler gauges when the points are fully opened by one of the cams on the distributor shaft. A combined screwdriver and feeler gauge is provided in the tool kit.

The correct gap ·014″—·016″ (·36—·41 mm.).

If the gap is incorrect, slacken the two screws (A Fig. 6) securing the fixed contact plate and turn the eccentric-headed adjustment screw (B) in its slot until the required gap is obtained. Tighten the securing screws and re-check the gap.

Examine the contact breaker points. If the contacts are burned or blackened, clean them with a fine carborundum stone or very fine emery cloth. After-

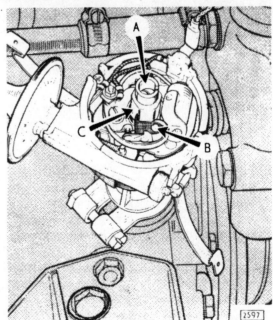

Fig. 6. Checking the gap between the distributor contact points. The two screws 'A' secure the fixed contact plate; the contact gap is adjusted by means of the eccentric headed screw 'B'.

Page B.15

ENGINE

wards wipe away any trace of grease or metal dust with a petrol moistened cloth.

Cleaning of the contacts is made easier if the contact breaker lever carrying the moving contact is removed. To do this, remove the nut, insulating piece and connections from the post to which the end of the contact breaker spring is anchored. The contact breaker lever can now be lifted off its pivot post.

Sparking Plugs

Every 2,500 miles (4,000 km.) or more often if operating conditions demand, withdraw, clean and reset the plugs.

The only efficient way to clean sparking plugs is to have them properly serviced on machines specially designed for this purpose. These machines operate with compressed air and utilise a dry abrasive material specially graded and selected to remove harmful deposits from the plug insulator without damaging the insulator surface. In addition the majority of the machines incorporate electrical testing apparatus enabling the plugs to be pressure tested to check their electrical efficiency and gas tightness.

The gap between the points should be ·025″ (·64 mm.). When adjusting the gap always move the side wire—never bend the centre wire.

The Champion Sparking Plug Co. supply a special combination gauge and setting tool, the use of which is recommended.

Every 10,000 miles (16,000 km.) a new set of plugs of the recommended type should be fitted. To save petrol and to ensure easy starting, the plugs should be cleaned and tested regularly.

EVERY 5,000 MILES (8,000 KM.)

Water Pump/Dynamo Belt Tension—(Early Cars)

When the belt is correctly tensioned it should be possible to depress the belt about half an inch (12 mm.) midway between the water pump and dynamo pulleys.

Adjustment is effected by slackening the three dynamo mounting bolts, moving the dynamo until the correct tension is obtained and tightening the bolts.

Do not overtighten the belt or this will cause undue wear of the belt and the water pump and dynamo bearings. Slackness of the belt may cause slippage with the possible result of a squealing noise from the belt and a reduced charging rate from the dynamo.

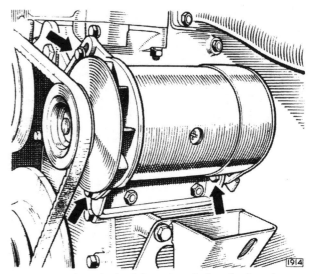

Fig. 7. To adjust the fan belt tension, slacken the three dynamo mounting bolts and move the dynamo to the desired position.

Later Cars
Dynamo and Water Pump Belt Replacement

The dynamo and water pump belt is kept at the correct tension by means of a spring loaded jockey pulley on the right hand side of the engine. If the belt has to be replaced carry out the following procedure:—

Slacken the two bolts securing the dynamo to the mounting bracket. Remove the nut and unscrew the bolt securing the top dynamo link to the dynamo. Slacken the bolt securing the dynamo link to the engine and press the dynamo as far as possible towards the engine. Place the new belt in position on the water pump, jockey and crankshaft pulleys and by pressing the jockey pulley towards the engine pass the belt over the dynamo pulley. Pass the dynamo securing top bolt through the link and screw into the lug of the dynamo. Pull the dynamo away from the engine as far as possible and tighten the top dynamo securing bolt and replace the lock nut. Tighten the bolt securing the dynamo link to the engine and also the two bottom dynamo mounting bolts.

Oil Filter Element (Felt type)

It is most important to renew the oil filter element every 5,000 miles (8,000 km.) as after this mileage it will have become choked with impurities.

To guard against the possibility of the filter being neglected to the extent where the element becomes completely choked, a balance valve is incorporated in the filter head which allows **unfiltered** oil to by-pass the element and reach bearings. This will be accompanied by a drop in the normal oil pressures of some 10 lb.

Page B.16

ENGINE

per sq. in. and if this occurs the filter element should be renewed as soon as possible.

The oil filter is situated on the right-hand side of the engine and before removing the canister it will be necessary to drain the filter by removing the small hexagon-headed drain plug situated at the bottom of the filter head.

To gain access to the element, unscrew the central bolt when the canister complete with the element can be removed. Thoroughly wash out the canister with petrol and allow to dry before inserting the new element.

When replacing the canister renew the circular rubber seal in the filter head.

Note: If a paper filter element is employed, it should be renewed at every oil change.

Top Timing Chain Tension

If the top timing chain is audible adjust the tension as follows:—

This operation requires the use of a special tool (Churchill Tool No. J.2) to enable the adjuster plate to be rotated. To gain access to the adjuster plate remove the breather housing attached to the front face of the cylinder head.

Slacken the locknut securing the serrated adjuster plate. Tension the chain by pressing the locking plunger inwards and rotating the adjuster plate in an anti-clockwise direction.

When correctly tensioned there should be slight flexibility on both outer sides of the chain below the camshaft sprockets, that is, the chain must not be dead tight. Release locking plunger, and securely tighten locknut. Refit the breather housing.

EVERY 10,000 MILES (16,000 KM.)

Air Cleaner

The air cleaner is of the paper element type and is situated in the engine compartment on the right-hand side adjacent to the carburetters.

No maintenance is necessary, but the element should be renewed every 10,000 miles (16,000 km.) or more

Fig. 8. The air cleaner.

frequently in dusty territories. To gain access to the element release the three spring clips retaining top cover to base. Remove two wing nuts attaching cleaner to air box and lift out element and cover. Remove serrated nut, and retainer plate from base of unit and withdraw element.

FUEL REQUIREMENTS FOR 9 TO 1 and 8 TO 1 COMPRESSION RATIO ENGINES

If the engine of your car is fitted with 9 to 1 compression ratio pistons (indicated by /9 after the engine number) use only Super grade fuel with a minimum octane rating of 98. (Research method). If a car is fitted with 8 to 1 compression ratio pistons (indicated by /8 after the engine number) use premium grade fuel with a minimum rating of 91. (Research method).

If, of necessity, the car has to be operated on lower octane fuel do not use full throttle otherwise detonation may occur with resultant piston trouble.

In the United Kingdom use "5 Star" (9 to 1) or "4 Star" (8 to 1) petrol.

Page B.17

ENGINE

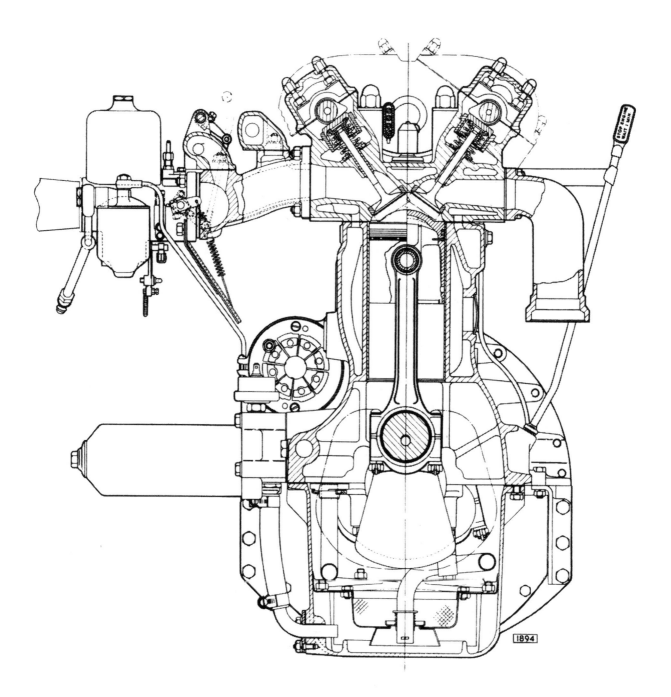

Fig. 9. Cross sectional view of the engine.

Page B.18

ENGINE

ENGINE REMOVAL

ENGINE REMOVAL

Remove the bonnet (for details see Section N.).
Disconnect the battery.
Drain the cooling system by turning the radiator drain tap and removing the filler cap. Conserve coolant if anti-freeze is in use.
Slacken the clip on the breather pipe, unscrew the two wing nuts and remove the top of the air cleaner.
Disconnect the petrol feed pipe below the centre carburetter.
Slacken the clips securing the water hoses from the cylinder head and radiator to the header tank.
Slacken the clips securing the heater hoses to the manifold.
Disconnect the brake vacuum pipe.
Disconnect the two electrical connections from the fan control thermostat in the header tank together with the anchoring clip.
Remove the two nuts and bolts securing the header tank mounting bracket to the front cross member.
Disconnect the radiator header tank overflow pipe and remove the header tank complete with mounting bracket.
Disconnect the throttle linkage at the rear carburetter.
Remove the green/blue cable from the water temperature transmitter.
Remove the white/black cable from the distributor to the C.B. coil terminal and the white cable from the S.W. coil terminal.
Disconnect the battery cable and the solenoid switch cable from the starter motor.
Remove the bolt from the oil filter canister and remove the canister, together with the filter, from below ensuring that the rubber sealing ring is renewed when refitting.
Remove the lower crankshaft pulley, complete with the crankshaft damper and drive belt. Remove the ignition timing pointer from the sump. Mark the pulley and the damper to facilitate refitting.
Remove the upper clip from the water pump hose.
Remove the white and brown cable from the oil pressure element.
Remove the revolution counter generator complete with cables.
Disconnect the brown/yellow cable from the "D" terminal on the dynamo and the brown/green cable from the "F" terminal.

Remove the four nuts and washers securing each downpipe to the exhaust manifold, unclip the downpipes at the silencer assembly and withdraw the pipes. Collect the sealing rings between the exhaust manifolds and the downpipes.
Remove the seats, radio (if fitted) and the ash tray.
Remove the three setscrews securing the propeller shaft tunnel cover to the body.
Apply the handbrake and remove the gear lever knob.
Slide the propeller shaft tunnel cover over the handbrake and gear levers. Withdraw the tunnel cover.
Turn back the carpet. Withdraw the plastic gearbox bellows having removed the six drive screws. Remove trim and screws securing the gearbox cover. Remove gearbox cover and the gear lever.
Remove the engine rear mounting plate.
Remove the four bolts and self-locking nuts securing the front propeller shaft universal joint to the gearbox flange.
Remove the two lower nuts securing the torsion bar reaction tie plate on each side and tap the bolts back flush with the face of the tie plate. With the aid of a helper, place a lever between the head of the bolt just released and the torsion bar. Exert pressure on the bolt head to relieve the tension on the upper bolt. Remove the nut and tap the upper bolt back flush with the face of the tie plate. Repeat for the other side and tap the tie plate off the four bolts.

Note: Failure to relieve the tension on the upper bolts when tapping them back against the face of the tie plate will result in stripping the threads. If this occurs, new bolts must be fitted and the torsion bars re-set.

Remove the two cables from the reverse light switch on the gearbox top cover. When refitting, these cables can be fitted to either terminal.
Remove the engine earth strap from the left hand side member.
Disconnect the clutch slave cylinder.
Support the engine by means of lifting tackle, utilizing the lifting straps (later cars) or the engine lifting plate (Churchill Tool No. J.8) in the case of early cars and by inserting the trolley jack under the gearbox from the front of the car.
Remove the self-locking nut and stepped washer from the engine stabiliser.

Page B.19

ENGINE

Remove the bolts from the front engine mountings.
Remove the speedometer cable.

Raise the engine on the lifting tackle and, keeping the engine level, move it towards the front of the car ensuring that the water pump pulley clears the sub-frame top cross member and that the bell housing clears the anchor brackets at the rear of the torsion bars.

Gradually lifting the front of the engine and lowering the rear, withdraw the engine from the front.

When withdrawing the engine ensure that the rears of the camshaft covers do not foul the bonnet drain channel and that the brake pipe is not damaged.

REFITTING

Refitting the engine is the reverse of the removal procedure.

Note: Care must be taken to ensure that the brake pipes are not damaged at the front sub-frame cross members and that the engine does not foul the torsion bar anchor brackets or displace the silver steel locating bars.

Replace the exhaust manifold sealing rings and if the cylinder head nuts have been removed they should be tightened to a torque of 54 lb.ft. (7·4 kgm.). Bleed the clutch hydraulic system, reset the manual mixture control and adjust the engine stabiliser.

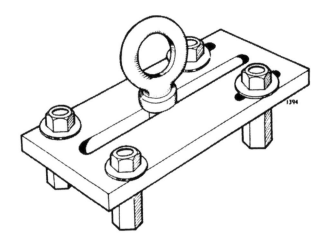

Fig. 11. The engine lifting plate (Churchill Tool No. J.8.). On later cars, engine lifting straps are fitted to the cylinder head.

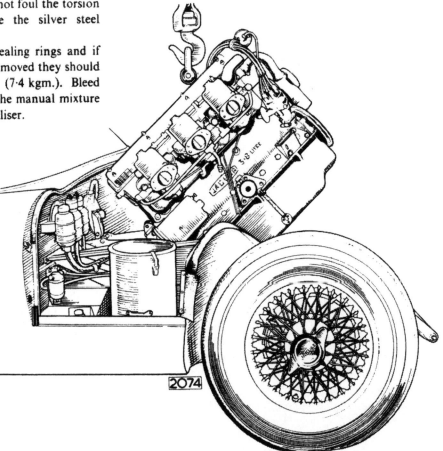

Fig. 10. Removing the engine from above.

Page B.20

ENGINE

ENGINE TO DISMANTLE

GENERAL

The following instructions apply when the engine components are removed in the following sequence with the engine unit out of the chassis. Dismantling of sub-assemblies and the removal of individual components when the engine is in the chassis frame are dealt with separately in this section.

All references made in this section to the top or bottom of the engine assume the engine to be in the normal upright position. References to the left- or right-hand side assume the engine to be upright and looking from the rear.

REMOVE STARTER

Unscrew the two nuts securing the starter to the clutch housing and withdraw the starter.

REMOVE GEARBOX

Unscrew the four setscrews and remove the cover plate from the front face of the clutch housing.

Remove the set bolts and nuts securing the clutch housing to the engine and withdraw the gearbox unit. The gearbox must be supported during this operation in order to avoid straining the clutch driven plate and constant pinion shaft.

REMOVE DISTRIBUTOR

Spring back clips and remove the cover complete with high tension leads. Disconnect the electrical cable from the distributor. Slacken the clamp plate bolt and withdraw distributor. Remove the setscrew and remove the clamp plate. Note the cork seal in recess at the top of the distributor drive hole.

REMOVE CYLINDER HEAD

Disconnect the distributor vacuum feed pipe from the front carburetter. Remove the high tension leads from the sparking plugs and lead carrier from the cylinder head studs. Remove the sparking plugs. Disconnect the camshaft oil feed pipe from the rear of the cylinder head. Remove the eleven dome nuts from each camshaft cover and lift off the covers.

Remove the four dome nuts securing the breather housing and withdraw housing. Release the tension on the camshaft chain by slackening the nut on the eccentric idler sprocket shaft, depressing the spring-loaded stop peg and rotating serrated adjuster plate clockwise. Anti-clockwise rotation of the serrated adjuster viewed from the front of the engine tightens the chain.

Break the locking wire on the two setscrews securing the camshaft sprockets to their respective camshafts.

Remove the setscrews and withdraw the sprockets from the camshafts with chain in position. Having

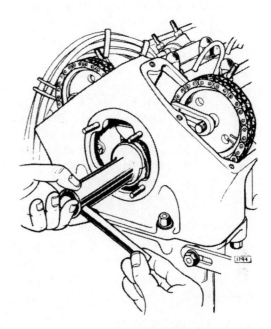

Fig. 12. Adjusting the top timing chain.

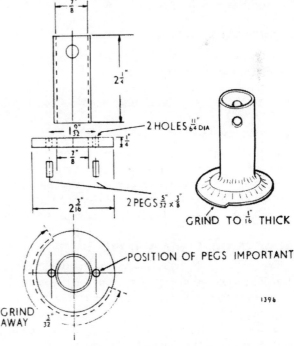

Fig. 13. The top timing chain adjusting tool.

Page B.21

ENGINE

once disconnected the camshaft sprockets do NOT rotate the engine or camshafts.

Slacken the fourteen cylinder head dome nuts and six nuts securing the front of the cylinder head a part of a turn at a time in the order shown in Fig. 18 until the nuts become free. Lift off the cylinder head complete with exhaust manifold and inlet manifolds. Remove and scrap the cylinder head gasket.

REMOVE CLUTCH AND FLYWHEEL

Unscrew the six setscrews securing the flange of the clutch cover to the flywheel and remove the clutch assembly. Note the balance marks 'B' stamped on the clutch cover and on the edge of the flywheel.

Knock back the tabs of locking plate securing the ten flywheel bolts. Unscrew the flywheel bolts and remove the locking plate. Remove flywheel from the crankshaft flange by gently tapping with a rawhide mallet.

REMOVE CRANKSHAFT DAMPER

Unscrew the large nut and remove the plain washer.

Insert two levers behind the damper and ease it off the split cone—a sharp tap on the end of the cone will assist removal.

REMOVE WATER PUMP

Unscrew the set bolts and three nuts, and remove the water pump from the timing cover. Note the gasket between the pump and timing cover.

REMOVE OIL FILTER

Detach the short length of flexible pipe between the oil filter head and the oil sump.

Unscrew the four set bolts securing the oil filter head to the cylinder block and remove filter head.

REMOVE SUMP

Drain the sump by removing the hexagon plug and washer from the right-hand side of the sump.

Remove the twenty-six setscrews securing the sump to the crankcase and the four nuts securing the sump to the timing cover. The sump can now be removed.

REMOVE OIL PUMP AND PIPES

Tap back the tab washers and unscrew the two set bolts securing the oil feed pipe from the oil pump to the bottom face of the crankcase. Withdraw the pipe from the pump.

Remove the nut and bolt securing the oil pump inlet pipe clip to the bracket on the main bearing cap.

Remove the nut and bolt securing the oil pump inlet pipe clip in the bracket on the oil pump.

Withdraw the pipe from the pump.

Tap back the tab washers and unscrew the three bolts securing the oil pump to the front main bearing cap. The oil pump can now be withdrawn.

Remove the coupling shaft from the squared end of the distributor and oil pump drive shaft.

REMOVE PISTONS AND CONNECTING RODS

As the pistons will not pass the crankshaft it will be necessary to withdraw the pistons and connecting rods from the top.

Remove the split pins from the connecting rod bolt nuts and unscrew nuts. Remove the connecting rod cap, noting that the corresponding cylinder numbers on the connecting rod and cap are together.

Withdraw the piston and connecting rod from top of cylinder block.

Note: Split skirt pistons MUST be fitted with the split opposite to the thrust side, that is, with the split on the left-hand or exhaust side of the engine. To facilitate correct fitting the pistons crowns are marked "Front".

REMOVE TIMING COVER

Remove the set bolts securing the timing cover to the front face of the cylinder block. Remove the timing cover, noting that the cover is located to the cylinder block by two dowels.

REMOVE TIMING GEAR ASSEMBLY

When removing the bottom timing chain tensioner from the engine, remove the hexagon head plug and tab washer from the end of the body. Insert an Allen key into the hole until it registers in the end of the restraint cylinder. Turn the Allen key clockwise until the restraint cylinder can be felt to be fully retracted within the body. The adjuster head will then be free of the chain.

Knock back the tab washers on the two set bolts securing the chain tensioner to the cylinder block.

Withdraw the bolts and remove the tensioner together with the conical gauze filter fitted in the tensioner oil feed hole in the cylinder block this should be cleaned in petrol.

Unscrew the four set bolts securing the front mounting bracket to the cylinder block. Release the tabs of the tab washers and remove the two screwdriver slotted setscrews from the rear mounting

Page B.22

ENGINE

bracket; these setscrews also secure the intermediate timing chain damper bracket.

The timing gear can now be removed.

REMOVE DISTRIBUTOR DRIVE GEAR

Tap back the tab washer securing the distributor drive gear nut and remove the nut and washer. Tap the squared end of the distributor drive shaft through the gear, noting that the gear is keyed to the shaft. Remove the gear and thrust washer and withdraw the drive shaft.

REMOVE CRANKSHAFT

Knock back the tab washers securing the fourteen main bearing cap bolts. Unscrew the bolts and remove the main bearing caps, noting the corresponding numbers stamped on the caps and bottom face of crankcase and also the thrust washers fitted to the recesses in the centre main bearing caps.

Detach the bottom half of the oil return thread cover from the top half by unscrewing the two Allen screws. Note that the two halves are located by hollow dowels.

The crankshaft can now be lifted out from the crankcase.

ENGINE—TO ASSEMBLE

GENERAL

All references in this section to the top or bottom of the engine assume the engine to be upright, irrespective of the position of the unit when the reference is made. References to the left- or right-hand side assume the engine to be upright and looking from the rear.

FIT DISTRIBUTOR DRIVE SHAFT BUSH

If a new bush is to be fitted, press the bush into the bore of the lug at front of cylinder block.

Ream the bush in position to a diameter of $\frac{3}{4}'' \begin{matrix} +\cdot0005'' \\ -\cdot00025'' \end{matrix}$ (19·05 mm. $\begin{matrix} +\cdot012 \text{ mm.} \\ -\cdot006 \text{ mm.} \end{matrix}$)

FIT CRANKSHAFT

Fit the main bearing shells to the top half of the main line bore in the cylinder block. Lay the crankshaft in the bearing shells. Fit the bottom half of the oil return thread cover to the top half which is bolted to the cylinder block behind the rear main bearing. The two halves are located by hollow dowels and secured with Allen screws. The clearance between the oil return thread cover and the oil return thread on the crankshaft should be ·0025" to ·0055" (·06 to ·14 mm.).

The two halves of the oil return thread cover are supplied only as an assembly together with the dowels and screws.

Fit the centre main bearing cap with a thrust washer, white metal side outward, to the recess in each side of cap. Tighten down the cap and check the crank-

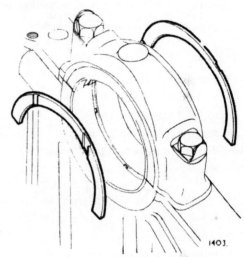

Fig. 14. The crankshaft thrust washers.

shaft end float, which should be ·004" to ·006" (·10 to ·15 mm.). The thrust washers are supplied in two thicknesses, standard and ·004" (·10 mm.) oversize and should be selected to bring the end float within permissible limits. The oversize thrust washers are stamped +·004" (·10 mm.) on the steel face.

Fit the main bearing caps with the numbers stamped on the caps with the corresponding numbers stamped on the bottom face of the crankcase.

Fit the main bearing cap bolts and tab washers and tighten to a torque of 83 lb.ft. (11·5 kgm.).

Test the crankshaft for free rotation.

The tab washers for the rear main bearing bolts are longer than the remainder and the plain ends should be tapped down around the bolt hole bosses.

Page B.23

ENGINE

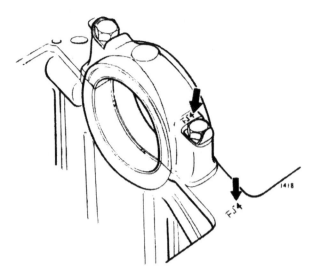

Fig. 15. Showing the corresponding numbers marked on the main bearing cap and the crankcase.

FIT PISTONS AND CONNECTING RODS

Turn the engine on its side. Remove the connecting rod caps and fit the pistons and connecting rods to their respective bores from the top of the cylinder block, using a suitable piston ring compressor. The cylinder number is stamped on the connecting rod and cap, No. 1 cylinder being at rear.

Note: Semi-split skirt pistons MUST be fitted with the split opposite the thrust side, that is, with the split on the left-hand or exhaust side of the engine. To facilitate correct fitting the piston crowns are marked "Front".

Fit the connecting rod caps to the connecting rods with the corresponding numbers together. Fit the castellated nuts and tighten to a torque of 37 lb.ft. (5·1 kgm.). Secure nut with split pins.

FIT CRANKSHAFT GEAR AND SPROCKET

Fit the Woodruff key to the inner slot and tap the distributor crankshaft gear into position with the widest part of the boss to the rear (see Fig. 25).

Fit the Woodruff key to the outer slot and tap the crankshaft timing gear sprocket into position. Fit the oil thrower and distance piece.

Turn the engine until Nos. 1 and 6 pistons are on T.D.C.

FIT DISTRIBUTOR AND OIL PUMP DRIVE GEAR

Ensure that the Woodruff key on the distributor drive shaft is in good condition and renew if necessary.

Place the drive shaft into position with the offset slot in the top of the shaft as shown in Fig. 16.

Withdraw the shaft slightly maintaining the same slot position and place the thrust washer and drive gear on the end of the shaft. Press the shaft into the drive gear ensuring that the key engages the keyway correctly.

Fit the pegged tab washer with the peg in the keyway of the drive gear.

Fully tighten the nut and secure with the tab washer. Check the end float of the shaft which should be ·004" to ·006" (·10 to ·15 mm.).

If no clearance exists fit a new oil pump/distributor drive gear which will restore the clearance. In an emergency if a new drive gear is not available, the thrust washer may be reduced in thickness by rubbing down on a piece of emery cloth placed on a surface plate.

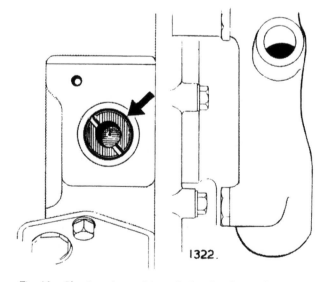

Fig. 16. Showing the position of the distributor drive shaft offset when No. 6 (front) piston is on Top Dead Centre.

Page B.24

ENGINE

FIT OIL PUMP AND PIPES

Fit the coupling shaft between the squared end of the distributor drive shaft and the driving gear of the oil pump. Secure the oil pump to the front main bearing cap by the three dowel bolts and tab washers. Check that there is appreciable end-float of the short coupling shaft. Fit the oil delivery pipe from the oil pump to the bottom face of the crankcase with a new 'O' ring and gasket. Fit the suction pipe with a new 'O' ring at the oil pump end.

TO ASSEMBLE TIMING GEAR

Fit the eccentric shaft to the hole in front mounting bracket. Insert the spring and locking plunger for the serrated plate to the hole in the front mounting bracket. Fit the serrated plate and secure with the shakeproof washer and nut. Fit the idler sprocket (21 teeth) to the eccentric shaft.

Fit the two intermediate sprockets (20 and 28 teeth) to their shaft with the larger sprocket forward and press the shaft through lower central hole in rear mounting bracket. Secure with the circlip at the rear of the bracket.

Fit the top timing chain (longer chain) to the small intermediate sprocket and the bottom timing chain (shorter chain) to the large intermediate sprocket.

Loop upper timing chain under the idler sprocket and offer up the front mounting bracket to the rear mounting bracket with the two chain dampers interposed between the brackets.

Fit the intermediate damper to the bottom of the rear mounting bracket with two screwdriver slotted setscrews and tab washers.

Pass the four securing bolts through the holes in the brackets, chain dampers and spacers noting that shakeproof washers are fitted under the bolt heads. Secure the two mounting brackets together with four stud nuts and shakeproof washers.

FIT TIMING GEAR

Turn the engine upside down. Fit the lower timing chain damper and bracket to the front face of the cylinder block with two set bolts and locking plate.

Turn the timing gear assembly upside down and offer it up to the cylinder block. Loop the bottom timing chain over the crankshaft sprocket and secure the mounting brackets to the front face of the cylinder block with the four long securing bolts and the two screwdriver slotted setscrews which also secure the intermediate timing chain damper bracket, but do not fully tighten these two setscrews until the four long securing bolts are tight.

TIMING CHAIN TENSIONER

Place the timing chain tensioner, backing plate and filter in position so that the spigot on the tensioner aligns with the hole in the cylinder block. Fit shims, as necessary, between the backing plate and cylinder block so that the timing chain runs centrally along the rubber slipper. Fit the tab washer and two securing bolts. Tighten the bolts and tap the tab washers against the bolt heads.

It is important that no attempt is made to release the locking mechanism until the adjuster has been finally mounted in the engine WITH THE TIMING CHAIN IN POSITION.

Remove the hexagon head plug and tab washer from the end of the body. Insert the Allen key into the hole until it registers in the end of the retraint cylinder. Turn the key clockwise until the tensioner head moves forward under spring pressure against the chain. Do not attempt to turn the key anti-clockwise, nor force the tensioner head into the chain by external pressure.

Refit the plug and secure with the tab washer.

FIT TIMING COVER

Fit the circular oil seal to the recess in the bottom face of timing cover, ensuring that seal is well bedded in its groove.

Fit the timing cover gasket with good quality jointing compound and secure the timing cover to the front face of the cylinder block with securing bolts.

FIT OIL SUMP

Fit a new sump gasket to the bottom face of the crankcase. Fit the cork seal to the recess in the rear main bearing cap.

Fit the sump to the crankcase and secure with the twenty-six set screws, four nuts and washers.

Note: The short setscrew must be fitted to the right-hand front corner of the sump (Fig. 39).

FIT FLYWHEEL AND CLUTCH

Turn the engine upright.

Check that the crankshaft flange and the holes for the flywheel bolts and dowels are free from burrs.

Turn the engine until Nos. 1 and 6 pistons are on T.D.C. and fit the flywheel to the crankshaft flange so that the 'B' stamped on the edge of the flywheel is at approximately the B.D.C. position. (This will ensure that the balance mark 'B' on the flywheel is

Page B.25

ENGINE

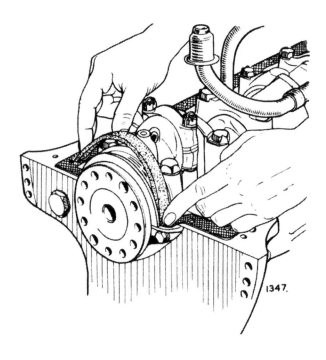

Fig. 17. Fitting the rear oil seal.

in line with the balance mark on the crankshaft which is a group of letters stamped on the crank throw just forward of the rear main journal).

Tap the two mushroom-headed dowels into position, fit the locking plate and flywheel securing set screws. Tighten the set screws to a torque of 67 lbs. ft. (9·2 kgm.) and secure with the locking plate tabs. Assemble the clutch driven plate to the flywheel, noting that one side of the plate is marked "Flywheel Side". Centralise the driven plate by means of a dummy shaft which fits the splined bore of the driven plate and the spigot bush in the crankshaft. (A constant pinion shaft may be used for this purpose). Fit clutch cover assembly so that the 'B' stamped adjacent to one of the dowel holes coincides with the 'B' stamped on the periphery of the flywheel. Secure the clutch assembly with the six set screws and spring washers, tightening the screws a turn at a time by diagonal selection. Remove the dummy shaft.

FIT CYLINDER HEAD

Before refitting the cylinder head it is important to observe that if the camshafts are out of phase with piston position fouling may take place between the valves and pistons. It is, therefore, essential to adhere

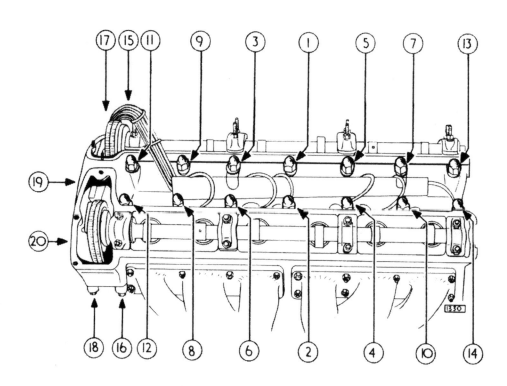

Fig. 18. Tightening sequence for the cylinder head nuts.

Page B.26

ENGINE

to the following procedure before fitting the cylinder head:—

Check that the grooves in the front flanges of the camshafts are vertical to the camshaft housing face and accurately position by engaging the valve timing gauge. If it is found necessary to rotate one of the camshafts the other camshaft must either be removed or the bearing cap nuts slackened to their fullest extent to allow the valves to be released.

Turn No. 6 (front) piston to the top dead centre position with the widest portion of the distributor drive shaft offset positioned as shown in Fig. 16.

Do NOT rotate the engine or camshafts until the camshaft sprockets have been connected to the camshafts. Fit the two camshaft sprockets complete with adjuster plates and circlips to the top timing chain and enter the guide pins in the slots in the front mounting bracket.

Fit the cylinder head gasket, taking care that the side marked "Top" is uppermost. Fit the cylinder head complete with manifolds to the cylinder block. Note that the second cylinder head stud from the front on the left-hand side is a dowel stud.

Fit the sparking plug lead carrier to the 3rd and 6th stud on the right-hand side. Fit plain washers to these and the two front stud positions and 'D' washers to the remaining studs. Tighten the fourteen large cylinder head dome nuts a part of a turn at a time to a torque of 54 lb.ft. (7·5 kgm.) in the order shown in Fig. 18. Also tighten the six nuts securing the front end of the cylinder head.

VALVE TIMING

Check that the No. 6 (front) piston is exactly in the T.D.C. position.

Through the breather aperture in the front of the cylinder head slacken the lock nut securing the serrated plate.

With the camshaft sprocket on the flanges of the camshafts, tension chain by pressing locking plunger inwards and rotating serrated plate by the two holes in an anti-clockwise direction.

When correctly tensioned there should be slight flexibility on both outer sides of the chain below the camshaft sprockets, that is the chain must not be dead tight. Release the locking plunger and securely tighten the locknut. Tap the camshaft sprockets off the flanges of the camshafts.

Accurately position the camshaft with the valve timing gauge, and check that the T.D.C. marks are in exact alignment.

Withdraw the circlips retaining the adjusting plates to the camshaft sprockets and pull the adjusting plates forward until the serrations disengage. Replace the sprockets on to the flanges of camshafts and align the two holes in the adjuster plate with the two tapped holes in each camshaft flange. Engage the serrations of the adjuster plates with the serrations in the sprockets.

Note: It is most important that the holes are in exact alignment, otherwise when the setscrews are fitted the camshafts will be moved out of position. If difficulty is experienced in aligning the holes exactly, the adjuster plates should be turned through 180°, which, due to the construction of the plate, will facilitate alignment.

Fit the circlips to the sprockets and one setscrew to the accessible hole in each adjuster plate. Turn the engine until the other two holes are accessible and fit the two remaining setscrews.

Finally, recheck the timing chain tension and timing in this order. Secure the four setscrews retaining the camshaft sprockets with new lock wire.

FIT CYLINDER HEAD OIL FEED PIPE AND OIL FILTER

Fit the cylinder head oil feed pipe from the tapped hole in the main oil gallery to the two tapped holes in the rear of the cylinder head. Secure the pipe with the three banjo bolts with a copper washer fitted to both sides of each banjo.

Fit the oil filter head to the cylinder block with the four setscrews and copper washers. New gasket(s) must always be fitted between the filter and cylinder block.

Fit the short length of flexible hose between the oil filter head and the oil sump and tighten two hose clips.

FIT CRANKSHAFT DAMPER AND PULLEY

Fit a Woodruff key to the crankshaft and the split cone. Fit the split cone to the crankshaft with the widest end towards the timing cover. Fit the damper to the cone and secure with the flat washer, chamfered side outwards, and large bolt.

FIT WATER PUMP

Fit the water pump to the timing cover with a new gasket and secure with six bolts, three nuts and spring washers.

Page B.27

ENGINE

FIT DYNAMO AND WATER PUMP BELT

Slacken the setscrew securing the dynamo adjusting link to the timing cover and swing link upwards.

Fit the dynamo belt to the crankshaft and water pump pulleys. Offer up dynamo and engage dynamo belt with pulley. Secure dynamo with the two mounting bolts and adjusting link at the water pump. Before finally tightening, adjust dynamo belt tension by pulling dynamo outwards until the belt can be flexed approximately ½" (12 mm.) either way in the middle of the vertical run. Tighten the adjusting setscrew, the two dynamo mounting bolts and the bolt securing the adjusting link to the water pump.

Note: Undue tension will create heavy wear of belt, pulleys, water pump and dynamo bearings.

FIT DISTRIBUTOR AND SPARKING PLUGS

Fit the cork seal to the recess at the top of the hole for the distributor. Secure the distributor clamping plate to the cylinder block with the setscrew. Slacken the clamping plate bolt.

Set the micrometer adjustment in the centre of the scale.

Enter the distributor into the cylinder block with the vacuum advance unit connection facing the cylinder block.

Rotate the rotor-arm until the driving dog engages with the distributor drive shaft.

Rotate the engine until the rotor-arm approaches the No. 6 (front) cylinder segment in the distributor cap. (Fig. 19).

Slowly rotate the engine until the ignition timing scale on the crankshaft damper is the appropriate number of degrees before the pointer on the sump.

Slowly rotate the distributor body until the points are just breaking.

Tighten the distributor plate pinch bolt.

A maximum of six clicks on the vernier adjustment from this setting, to either advance or retard, is allowed.

Fit the vacuum advance pipe from the distributor to the union on the front carburetter.

Fit the distributor cover and secure with the two spring clips. Fit the sparking plugs with new washers and attach high tension leads.

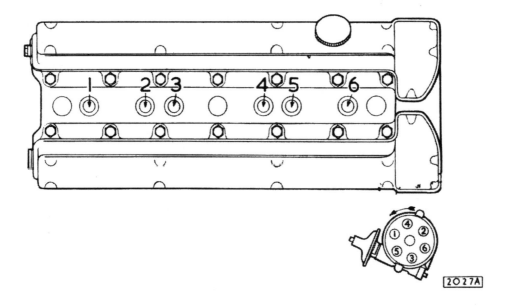

Fig. 19. View of the engine showing the firing order and cylinder numbers.

Page B.28

ENGINE

HIGH TENSION LEAD RENEWAL

If it is necessary to renew the high tension leads the following procedure should be followed:—

Remove the plug terminals and withdraw the leads from the conduit.

Remove the distributor cap terminals and the five spacing washers.

Cut the new high tension leads to suitable length.

Fit the leads to the conduit, No. 1 lead emerges from the rear of the conduit and the other leads from holes along the conduit.

Fit the plug terminals.

Fit the two thick fibre washers, arranging the leads in firing order (that is, 1, 5, 3, 6, 2, 4) in an anti-clockwise order, as the leads will enter the distributor cap.

Fit the three thin fibre spacers and place them equally along the leads.

Fit the distributor cap terminals.

FIT CAMSHAFT COVERS

Fit each camshaft cover to the cylinder head using a new gasket. Fit the eleven copper washers and dome nuts to the cover retaining studs but do not tighten fully.

Fit the revolution counter generator and flanged plug to the rear of right-hand and left-hand camshaft covers respectively with the rubber sealing rings seated in the recesses provided. Secure with the setscrews and copper washers. Tighten fully the dome nuts securing the camshaft covers.

FIT STARTER

Fit the starter motor to the clutch housing with the two bolts, nuts and spring washers.

FIT GEARBOX

Fit the gearbox and clutch housing to the rear of the crankcase with setscrews and shakeproof washers.

Fit the support brackets to each side, at the bottom face of the crankcase with two bolts, nuts and spring washers, and to the clutch housing with three bolts, nuts and shakeproof washers.

DECARBONISING AND GRINDING VALVES

REMOVE CYLINDER HEAD

Remove the cylinder head as described on page B.41.

REMOVE VALVES

With the cylinder head on the bench remove the inlet manifold, and the revolution counter generator.

Remove the four bearing caps from each camshaft and lift out the camshaft (note mating marks on each bearing cap).

Remove the twelve tappets and adjusting pads situated between tappets and valve stems. Lay out the tappets and pads in order, to ensure that they can be replaced in their original guides.

Obtain a block of wood the approximate size of the combustion chambers and place this under the valve heads in No. 1 cylinder combustion chamber. Press down the valve collars and extract the split cotters. Remove the collars, valve springs and spring seats. Repeat for the remaining five cylinders. Valves are numbered and must be replaced in their original locations, No. 1 cylinder being at the rear, that is the flywheel end.

DECARBONISE AND GRIND VALVES

Remove all traces of carbon and deposits from the combustion chambers from the induction and exhaust ports. The cylinder head is of aluminium alloy and great care should be exercised not to damage this with scrapers or sharp pointed tools. Use worn emery cloth and paraffin only. Thoroughly clean the water passages in the cylinder head. Clean the carbon deposits from the piston crowns and ensure that the top face of the cylinder block is quite clean particularly round the cylinder head studs. Remove any pitting in the valve seats, using valve seat grinding equipment. Reface the valves if necessary using valve grinding equipment; grind the valves to the seats, using a suction valve grinding tool.

Clean the sparking plugs and set gaps; if possible use approved plug cleaning and testing equipment. Clean and adjust distributor contact breaker points.

VALVE CLEARANCE ADJUSTMENT

Thoroughly clean all traces of valve grinding compound from the cylinder head and valve gear.

Page B.29

ENGINE

Assemble the valves to the cylinder head. **When checking the valve clearances the camshafts must be fitted one at a time as if one camshaft is rotated when the other camshaft is in position, fouling is likely to take place between the inlet and exhaust valves. Obtain and record all valve clearances by using a feeler gauge between the back of each cam and the appropriate valve tappet.**

Correct valve clearances are:—
Normal Touring Use
  Inlet .. .. .. ·004″ (·10 mm.).
  Exhaust .. .. .. ·006″ (·15 mm.).
Racing
  Inlet .. .. .. ·006″ (·15 mm.).
  Exhaust .. .. .. ·010″ (·25 mm.).

Adjusting pads are available rising in ·001″ (·03 mm.) sizes from ·085″ to ·110″ (2·16 to 2·79 mm.) and are etched on the surface with the letter 'A' to 'Z', each letter indicating an increase in size of ·001″ (·03 mm.). Should any valve clearance require correction, remove the camshaft, tappet and adjusting pad. Observe the letter etched on the existing adjusting pad if visible. If the letter is not visible measure the pad with a micrometer, and should the recorded clearance for this valve have shown say ·002″ (·05 mm.) excessive clearance, select a new adjusting pad bearing a letter two lower than the original pad.

As an example, assume that No. 1 inlet valve clearance is tested and recorded as ·007″ (·18 mm.). On removal of the adjusting pad, if this is etched with the letter 'D' then substitution with a pad bearing the letter 'G' will correct the clearance for No. 1 inlet valve.

When fitting the camshafts prior to fitting the cylinder head to the engine it is most important that the keyway in the front bearing flange of each camshaft is perpendicular (at 90°) to the adjacent camshaft cover face (using valve timing gauge) before tightening down the camshaft bearing cap nuts.

Tighten the camshaft bearing cap nuts to a torque of 15 lb.ft. (2·0 kgm.).

REFIT CYLINDER HEAD

Before attempting to refit the cylinder head refer to the instructions given on page B.42.

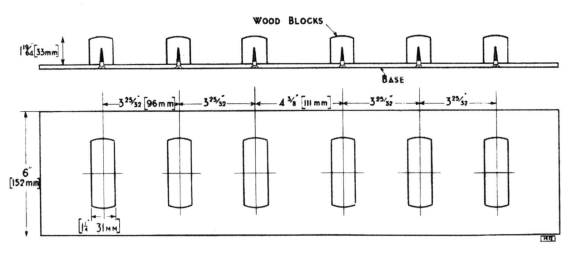

Fig. 20. Combustion chamber blocks for valve removal.

Page B.30

ENGINE

COMPRESSION PRESSURES

The compression pressures for all the six cylinders should be even and should approximate to the figures given below.

If one or more compressions are weak it will most probably be due to poor valve seatings when the cylinder head must be removed and the valves and valve seats refaced and reground.

COMPRESSION PRESSURES

8 to 1 compression ratio: 155 lbs per sq. in. (10·90 kg/cm²).

9 to 1 compression ratio: 180 lbs per sq. in. (12·65 kg/cm²).

Pressures must be taken with all the sparking plugs removed, carburetter throttles wide open and the engine at its normal operating temperature (70°C approximately).

Note: When taking compression pressures ensure that the ignition switch is 'off'; rotate the engine by operating the push button on the starter solenoid.

THE CONNECTING ROD AND BEARINGS

The connecting rods are steel stampings and are provided with precision shell big-end bearings and steel backed phosphor-bronze small end bushes. A longitudinal drilling through the connecting rod provides an oil feed from the big end to the small end bush.

REMOVAL

As the pistons will not pass the crankshaft it will be necessary to withdraw the pistons and connecting rods from the top.

Proceed as follows:—

Remove Cylinder Head

Remove the cylinder head as described on page B.41.

Remove Sump

Remove the sump as described on page B.50.

Remove Piston and Connecting Rod

Remove the split pins from the connecting rod bolt nuts and unscrew the nuts. Remove the connecting rod cap, noting that the corresponding cylinder numbers on the connecting rod and cap are on the same side. Remove the connecting rod bolts and withdraw the piston and connecting rod from the top of the cylinder block.

OVERHAUL

If connecting rods have been in use for a very high mileage, or if bearing failure has been experienced, it is desirable to renew the rod(s) owing to the possibility of fatigue.

The connecting rods fitted to an engine should not vary one with another by more than 2 drams (3·5 grammes). The alignment should be checked on an approved connecting rod alignment jig. Correct any misalignment as necessary. The big end bearings are of the precision shell type and under no circumstances should they be hand scraped or the bearing caps filed.

The small ends are fitted with steel-backed phosphor-bronze bushes which are a press fit in the connecting rod. After fitting, the bush should be reamed or honed to a diameter of ·875" to ·8752" (22·225 to 22·23 mm.). Always use new connecting bolts and nuts at overhauls.

When a new connecting rod is to be fitted, although the small end bush is reamed to the correct dimensions, it may be necessary to hone the bush to achieve the correct gudgeon pin fit.

REFITTING

Refitting is the reverse of the removal procedure. Pistons and connecting rods must be fitted to their respective cylinders (pistons and connecting rods are stamped with their cylinder number, No. 1 being at the rear) and the same way round in the bore.

The pistons must be fitted with split on the left-hand or exhaust side of the engine. To facilitate

Page B.31

ENGINE

correct fitting the piston crowns are marked "Front", see Fig. 42.

The cap must be fitted to the connecting rod so that the cylinder numbers stamped on each part are on the same side.

Tighten the connecting rod nuts to a torque of 37 lb.ft. (5·1 kgm.).

BIG-END BEARING REPLACEMENT

The big-end bearings can be replaced without removing the engine from the car but before fitting the new bearings the crankpin must be examined for damage or for the transfer of bearing metal. The oilway in the crankshaft must also be tested for blockage.

Remove the sump as described on page B.50.

Turn the engine until the big-end is approximately at the bottom dead centre position.

Remove the split pins from the connecting rod bolt nuts and unscrew the nuts. Remove the connecting rod cap, noting that the corresponding cylinder numbers on the connecting rod and cap are on the same side.

Lift the connecting rod off the crankpin and detach the bearing shell.

If all the big-end bearings are to be replaced they are most easily replaced in pairs, that is, in pairs of connecting rods having corresponding crank throws.

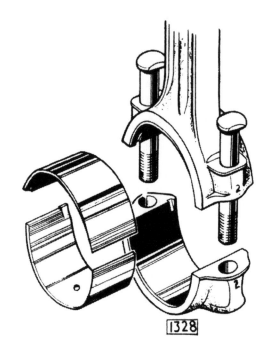

Fig. 21. The connecting rod and cap are stamped with the cylinder number.

THE CAMSHAFTS

The camshafts are manufactured of cast iron and each shaft is supported in four white metal steel backed bearings. End float is taken on the flanges formed at each side of the front bearing. Oil is fed from the main oil gallery to the camshaft rear bearing housings through an external pipe. Oil then passes through the rear bearing into a longitudinal drilling in the camshaft; cross drillings which break into this oilway feed the three remaining bearings.

Warning: Before carrying out any work on the camshafts the following points must be observed to avoid possible fouling between (a) the inlet and exhaust valves and (b) the valves and pistons.

(1) Do NOT rotate the engine or the camshafts with the camshafts sprockets disconnected. If, with the cylinder head removed from the engine, it is required to rotate a camshaft, the other camshaft must either be removed or the bearing cap nuts slackened to their fullest extent to allow the valves to be released.

(2) When fitting the camshafts to the cylinder head ensure that keyway in the front bearing flange of each camshaft is perpendicular (at 90°) to the adjacent camshaft cover face (use valve timing gauge) before tightening down the camshaft bearing cap nuts.

If this operation is being carried out with the cylinder head fitted to the engine, rotate the engine until No. 6 (front) piston is on Top Dead Centre in the firing position, that is with the distributor rotor opposite No. 6 cylinder segment, before fitting the camshafts.

Page B.32

ENGINE

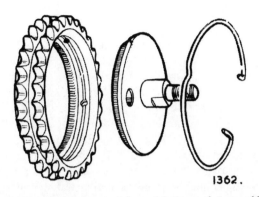

Fig 22. Exploded view of the camshaft sprocket assembly.

REMOVAL

Remove the eleven dome nuts and copper washers securing each camshaft cover and lift off the cover.

Unscrew the three Allen setscrews attaching the revolution counter generator to the right-hand side of the cylinder head and the sealing plug from the left-hand side (note the copper washers under the heads of the setscrew). Remove the circular rubber sealing rings.

Break the wire locking the camshaft adjuster plate setscrews.

Rotate the engine until No. 6 (front) piston is approximately on Top Dead Centre on compression stroke (firing position), that is, when the keyway in the front bearing flange of each camshaft is at 90° to the adjacent cover face (see Fig. 23).

Note the positions of the **inaccessible** adjuster plate setscrews and rotate the engine until they can be removed.

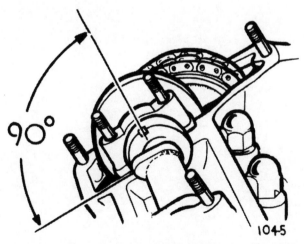

Fig. 23. When fitting a camshaft the keyway must be at 90° to the camshaft cover face.

Turn back the engine to the T.D.C. position with No. 6 firing and remove the two remaining setscrews.

Tap the sprockets off their respective camshaft flanges. Release the eight nuts securing the bearing caps a turn at a time. Remove the nuts, spring washers and 'D' washers from the bearing studs.

Remove the bearing caps, noting that the caps and cylinder head are marked with corresponding numbers. Also note that the bearing caps are located to the lower bearing housings with hollow dowels.

If the same bearing shells are to be replaced they should be fitted to their original positions.

The camshaft can now be lifted out from the cylinder head.

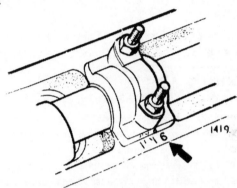

Fig. 24. Showing the corresponding numbers on the bearing cap and cylinder head.

REFITTING

Check that No. 6 (front) piston is exactly on T.D.C. on the compression stroke (firing position), that is, with the distributor rotor opposite No. 6 cylinder segment. (Fig. 19).

Replace the shell bearings—in their original positions if the same bearings are being refitted.

Replace each camshaft with the keyways in the front bearing flange at 90° to the adjacent cover face (using the valve timing gauge).

Refit the bearing caps to their respective positions and the 'D' washers, spring washers and nuts.

Tighten down the bearing caps evenly a turn at a time. Finally tighten the nuts to a torque of 15 lb.ft. (2·0 kgm.).

Set the valve timing as described on page B.63.

OVERHAUL

It is unlikely, except after very high mileages, to find wear in the camshafts and camshaft bearings. The camshaft bearings are of the precision shell type and under no circumstances should these be hand scraped or the bearing caps filed. Undersize bearings are not supplied.

Page B.33

ENGINE

THE CRANKSHAFT

The counterbalanced crankshaft is of manganese molybdenum steel and is supported in seven precision shell bearings. End thrust of the crankshaft is taken on two semi-circular white metal faced steel thrust washers fitted in recesses in the centre main bearing cap. A torsional vibration damper is fitted at the front end of the crankshaft.

Initially, the crankshaft is itself balanced both statically and dynamically and is then re-balanced as an assembly with the flywheel and clutch unit attached.

REMOVAL

Proceed as detailed under "Engine—To Dismantle" on page B.21.

OVERHAUL

Regrinding of the crankshaft journals is generally recommended when wear or ovality in excess of ·003" (·08 mm.) is found. Factory reconditioned crankshafts are available on an exchange basis, subject to the existing crankshaft being fit for satisfactory reconditioning, with undersize main and big end bearings —·010" (·25 mm.), —·020" (·51 mm.), —·030" (·76 mm.), and —·040" (1·02 mm.).

Grinding beyond the limits of ·040" (1·02 mm.) is not recommended and under such circumstances a new crankshaft should be obtained.

New crankshaft thrust washers should be fitted, these being in two halves located in recesses in the centre main bearing cap. Fit the main bearing cap with a thrust washer, white metal side outwards, to the recess in each side of the cap. Tighten down the cap and check the crankshaft end float, which should be ·004" to ·006" (·10 to ·15 mm.). The thrust washers are supplied in two thicknesses, standard and ·004' (·10 mm.) oversize and should be selected to bring the end float within the required limits. It is permissible to fit a standard size thrust washer to one side of the main bearing cap and an oversize washer to the other. Oversize thrust washers are stamped ·004" on the steel face.

Ensure that the oil passages in the crankshaft are clear and perfectly clean before re-assembling. If the original crankshaft is to be refitted remove the Allen headed plugs in the webs (which are secured by staking) and thoroughly clean out any accumulated sludge with a high pressure jet followed by blowing out with compressed air.

After refitting the plugs, secure by staking with a blunt chisel.

REFITTING

Proceed as detailed under "Engine—To Assemble" on page B.23.

ENGINE

CRANKSHAFT DAMPER AND PULLEY

A torsional vibration damper is fitted at the front end of the crankshaft.

The damper consists of a malleable iron ring bonded to a thick rubber disc. An inner member also bonded to the disc is attached to a hub which is keyed to a split cone on the front extension of the crankshaft.

The crankshaft damper and pulley are balanced as an assembly, mark each part before dismantling so that they can be refitted in their original positions.

REMOVAL

It will be necessary to remove the crankshaft damper from beneath the car.

Remove the dynamo and water pump belt by slackening the dynamo and moving it towards the engine.

Remove the locking washer securing the damper bolt by knocking back the tabs and unscrewing the two setscrews. Remove the other two setscrews securing the crankshaft pulley to the damper and remove the pulley.

Unscrew the large damper securing bolt and remove the flat washer.

Insert two levers behind the damper and ease it off the split cone—a sharp tap on the end of the cone will assist removal.

OVERHAUL

Examine the rubber portion of the damper for signs of deterioration and if necessary fit a new one. Also examine the crankshaft pulley for signs of wear and renew if necessary. The drive should be taken on the 'V' faces of the pulley; renew the pulley if a new fan belt bottoms in the 'V' groove.

REFITTING

Refitting is the reverse of the removal procedure.

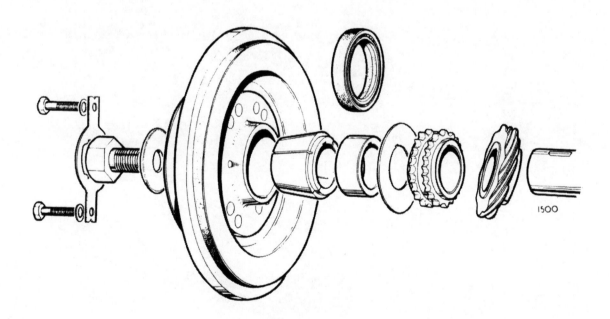

Fig. 25. The crankshaft damper and components.

Page B.35

ENGINE

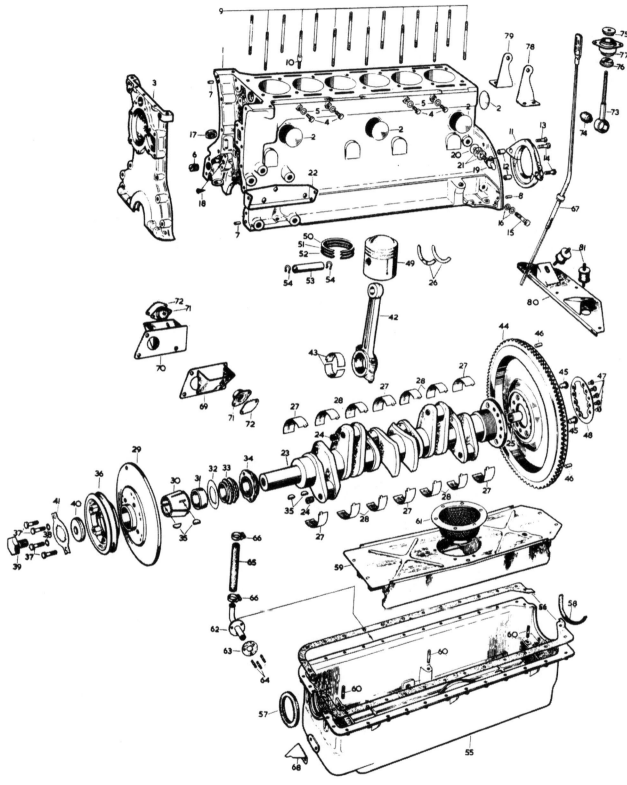

Fig. 26. Exploded view of the cylinder block assembly.

Page B.36

ENGINE

1. Cylinder block
2. Core plug
3. Timing cover
4. Setscrew
5. Copper washer
6. Plug
7. Dowel
8. Dowel
9. Stud
10. Dowel stud
11. Cover
12. Ring dowel
13. Setscrew
14. Bolt
15. Banjo bolt
16. Copper washer
17. Sealing ring
18. Gauze filter
19. Drain tap
20. Copper washer
21. Fibre washer
22. Mounting bracket
23. Crankshaft
24. Plug
25. Bush
26. Thrust washer
27. Main bearing
28. Main bearing
29. Crankshaft damper
30. Cone
31. Distance piece
32. Oil thrower
33. Sprocket
34. Gear
35. Key
36. Pulley
37. Bolt
38. Locking washer
39. Bolt
40. Washer
41. Tab washer
42. Connecting rod
43. Bearings
44. Flywheel
45. Dowel
46. Dowel
47. Setscrew
48. Locking plate
49. Piston
50. Compression ring
51. Compression ring
52. Scraper ring
53. Gudgeon pin
54. Circlip
55. Oil sump
56. Gasket
57. Seal
58. Cork seal
59. Baffle plate
60. Stud
61. Filter basket
62. Adaptor
63. Gasket
64. Stud
65. Hose
66. Clip
67. Dipstick
68. Pointer
69. Bracket
70. Bracket
71. Engine mounting
72. Plate
73. Link
74. Bush
75. Stepped washer
76. Stepped washer
77. Rubber mounting
78. Bracket
79. Bracket
80. Support bracket
81. Rubber mounting

Page B.37

ENGINE

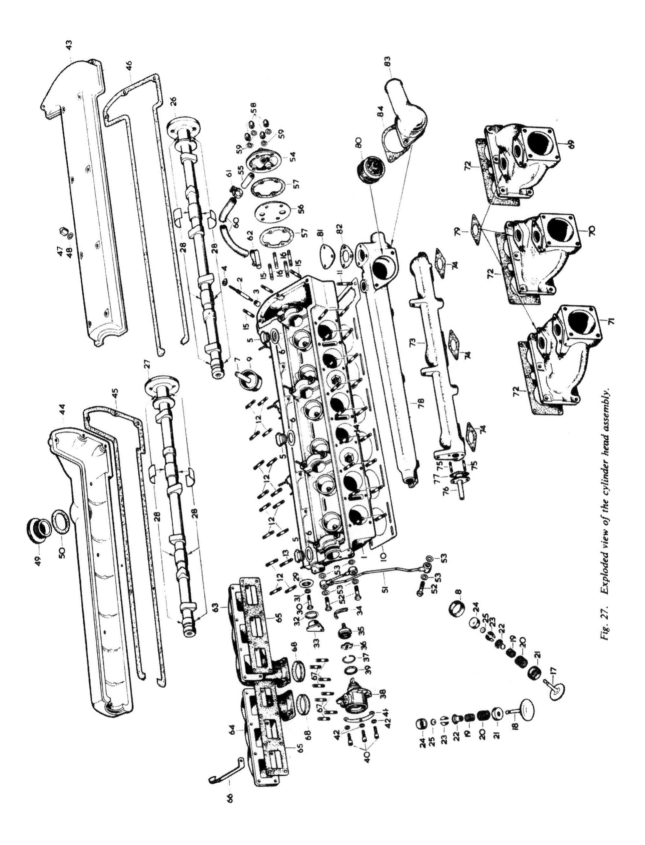

Fig. 27. Exploded view of the cylinder head assembly.

Page B.38

ENGINE

1. Cylinder head
2. Stud
3. Ring dowel
4. 'D' washer
5. Plug
6. Copper washer
7. Valve guide
8. Valve insert
9. Tappet guide
10. Gasket
11. Stud
12. Stud
13. Stud
14. Stud
15. Stud
16. Stud
17. Inlet valve
18. Exhaust valve
19. Valve spring
20. Valve spring
21. Seat
22. Collar
23. Cotter
24. Tappet
25. Adjusting pad
26. Inlet camshaft
27. Exhaust camshaft
28. Bearing
29. Oil thrower
30. Setscrew
31. Copper washer
32. Sealing ring
33. Sealing plug
34. Seal
35. Adaptor
36. Driving dog
37. Circlip
38. Generator
39. Sealing ring
40. Screw
41. Plate washer
42. Lock washer
43. Inlet camshaft cover
44. Exhaust camshaft cover
45. Gasket
46. Gasket
47. Dome nut
48. Copper washer
49. Filler cap
50. Fibre washer
51. Oil pipe
52. Banjo bolt
53. Copper washer
54. Breather housing
55. Pipe
56. Baffle
57. Gasket
58. Dome nut
59. Spring washer
60. Flexible pipe
61. Clip
62. Clip
63. Exhaust manifold
64. Exhaust manifold
65. Gasket
66. Clip
67. Stud
68. Sealing ring
69. Inlet manifold
70. Inlet manifold
71. Inlet manifold
72. Gasket
73. Air balance pipe
74. Gasket
75. Stud
76. Adaptor
77. Gasket
78. Water pipe
79. Gasket
80. Thermostat
81. Plate
82. Gasket
83. Elbow
84. Gasket

Page B.39

ENGINE

THE CYLINDER BLOCK

The cylinder block is of chromium iron and is integral with the crankcase. The main bearing housings are line bored and the caps are not interchangeable, corresponding numbers being stamped on the caps and the bottom face of the crankcase for identification purposes. Pressed in dry liners are fitted.

OVERHAUL

Check the top face of the cylinder block for truth. Check that the main bearing caps have not been filed and that the bores for the main bearings are in alignment. If the caps have been filed or if there is misalignment of the bearing housings the caps must be re-machined and the bearing housings line bored.

After removal of the cylinder head studs prior to reboring, check the area around the stud holes for flatness. When the edges of the stud holes are found to be raised they must be skimmed flush with the surrounding joint face, to ensure a dead flat surface on which to mount the boring equipment.

Reboring is normally recommended when the bore wear exceeds ·006″ (·15 mm.). Reboring beyond the limit of ·030″ (·76 mm.) is not recommended and when the bores will not clean out at ·030″ (·76 mm.), liners and standard size pistons should be fitted.

The worn liners must be pressed out from below utilizing the illustrated stepped block.

Before fitting the new liner, lightly smear the cylinder walls with jointing compound to a point half way down the bore and also smear the top outer surface of the liner.

Press the new liners in from the top and lightly skim the tops of the liners flush with the top face of the cylinder block.

Bore out and hone the liners to suit the grade (or grades) of pistons to be fitted. (See piston grades on page B.52).

The following oversize pistons are available: + ·010″ (·25 mm.), + ·020″ (·51 mm.) and + ·030″ (·76 mm.).

Following reboring the blanking plugs in the main oil gallery should be removed and the cylinder block oilways and the crankcase interior thoroughly cleaned. After cleaning, paint the crankcase interior with heat and oil resisting paint.

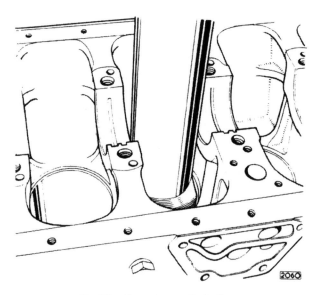

Fig. 28. Removing a cylinder liner.

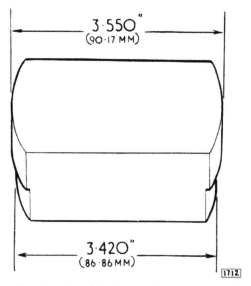

Fig. 29. Stepped block for cylinder liner removal.

ENGINE

THE CYLINDER HEAD

The cylinder head is manufactured of aluminium alloy and has machined hemispherical combustion chambers. Cast iron valve seat inserts, tappet guides and valve guides are shrunk into the cylinder head castings.

Warning: Before carrying out any work on the cylinder head the following points should be observed to avoid possible fouling between (a) the inlet and exhaust valves, and (b) the valves and pistons.

(1) Do NOT rotate the engine or the camshafts with the camshaft sprockets disconnected.

If, with the cylinder head removed from the engine, it is required to rotate a camshaft, the other camshaft must either be removed or the bearing cap nuts slackened to their fullest extent to allow the valves to be released.

(2) When fitting the camshafts to the cylinder head ensure that the keyway in the front bearing flange of each camshaft is perpendicular (at 90°) to the adjacent camshaft cover face before tightening down the camshaft bearing cap nuts. If this operation is being carried out with the cylinder head fitted to the engine, rotate the engine until No. 6 (front) piston is on Top Dead Centre in the firing position, that is with the distributor rotor opposite No. 6 cylinder segment, before fitting the camshafts.

Note: As the valves in the fully open position protrude below the cylinder head joint face, the cylinder head must not be placed joint face downwards directly on a flat surface; support the cylinder head on wooden blocks, one at each end.

REMOVAL

Remove the bonnet (as described in Section N "Body").

Drain the cooling system by turning the radiator drain tap, opening the cylinder block drain tap and removing the header tank filler cap.

Conserve the coolant if anti-freeze is in use.

Disconnect the battery.

Remove the wing nuts and remove the air cleaner elbow from the top of the air cleaner.

Disconnect the accelerator shaft from the ball joint on the throttle spindle.

Remove the setscrews from the backing plate and from the bush carrying plate.

Ensure that the push-in cage nuts do not fall out of the bulkhead.

Turn the throttle spindle to the fully open position and remove the short spindle from the ball joint socket.

Disconnect the distributor vacuum advance pipe from the front carburetter.

Disconnect the petrol feed pipe at the float chamber unions.

Remove the clip attaching the overflow pipes from the float chambers to the oil filter mounting bolt.

Disconnect the mixture control inner and outer cable.

Disconnect the cables from the revolution counter generator at the rear of the cylinder head.

Disconnect the top water hose and by-pass hose from the front of the inlet manifold water jacket.

Remove the high tension leads from the sparking plugs and the lead carrier from the cylinder head studs.

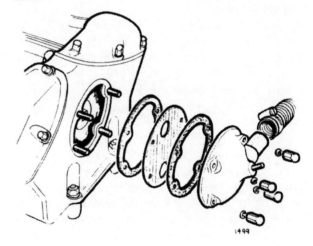

Fig. 30. Removal of the engine breather.

Disconnect the wires from the ignition coil and remove the coil.

Remove the sparking plugs.

Disconnect the exhaust manifolds from the engine.

Disconnect the two camshaft oil feed pipe unions from the rear of the cylinder head.

Disconnect the heater box from the rear of the inlet manifold water jacket.

Disconnect the heater pipe clips from the inlet manifold lower securing nuts.

Page B.41

ENGINE

Disconnect the cable from the water temperature gauge bulb in the inlet manifold water jacket.

Slacken the clip and remove the vacuum servo pipe from the connection at the rear of the inlet manifold.

Remove the eleven dome nuts from each camshaft cover and lift off the covers.

Remove the four nuts securing the breather housing to the front of the cylinder head and withdraw the housing and breather pipe observing the position of the baffle plate with the two holes vertical.

Release the tension on the top timing chain by slackening the nut on the eccentric idler sprocket shaft, depressing the spring-loaded stop peg and rotating the serrated adjuster plate clockwise.

Break the locking wire on the two setscrews securing the camshaft sprocket to the respective camshafts.

Remove one setscrew only from each of the camshaft sprockets; rotate the engine until the two remaining setscrews are accessible and remove these two screws.

Do NOT rotate the engine or the camshafts after having disconnected the sprockets.

The two camshaft sprockets may now be slid up the support brackets.

Slacken the fourteen cylinder head dome nuts a part of a turn at a time in the order shown (Fig. 18) until the nuts become free. Remove the six nuts securing the front of the cylinder head.

Lift off the cylinder head complete with inlet manifolds. Remove and scrap the cylinder head gasket.

OVERHAUL

As the cylinder head is of aluminium alloy, great care should be exercised when carrying out overhaul work, not to damage or score the machined surfaces. When removing carbon do not use scrapers or sharply pointed tools—use worn emery cloth and paraffin only.

Check the bottom face of the cylinder head for truth.

Remove all traces of carbon and deposits from the combustion chambers and the inlet and exhaust ports and regrind the valve and seats if necessary, as described under "Decarbonising and Grinding Valves" on page B.29.

If it is required to replace the valve guides, valve seat inserts or tappet guides, only the special replacement parts must be used. The replacement parts must be shrunk into the cylinder head in accordance with the instructions given under the appropriate headings in this section.

REFITTING

Fit Cylinder Head

Before refitting the cylinder head it is important to observe that if the camshafts are out of phase with piston position fouling may take place between the valves and pistons. It is, therefore, essential to adhere to the following procedure before fitting the cylinder head:—

Check that the keyways in the front flanges of the camshafts are vertical to the camshaft housing face and accurately position by engaging the valve timing gauge. If it is found necessary to rotate one of the camshafts the other camshaft must either be removed or the bearing cap nuts slackened to their fullest extent to allow the valves to be released.

Turn No. 6 (front) piston to the Top Dead Centre position with the distributor rotor arm opposite No. 6 cylinder segment. (Fig. 19).

Do NOT rotate the engine or camshafts until the camshaft sprockets have been connected to the camshafts.

Fit the cylinder head gasket, taking care that the side marked "Top" is uppermost. Fit the cylinder head complete with manifolds to the cylinder block. Note that the second cylinder head stud from the front on the left-hand side is a dowel stud.

Fit the sparking plug lead carrier to the 3rd and 6th stud from the front on the right-hand side. Fit plain washers to these and the two front stud positions. Fit 'D' washers to the remaining studs.

Tighten the fourteen large cylinder head dome nuts a part of a turn at a time to a torque of 54 lb.ft. (7·5 kgm.) in the order shown in Fig. 18. Also tighten the six nuts securing the front end of the cylinder head.

Valve Timing

Check that No. 6 (front) piston is exactly in the T.D.C. position.

Through the breather aperture in the front of the cylinder head slacken the locknut securing the serrated plate.

With the camshaft sprocket on the flanges of the camshafts, tension chain by pressing locking plunger inwards and rotating serrated plate by two holes in an anti-clockwise direction.

When correctly tensioned there should be slight flexibility on both outer sides of the chain below the camshaft sprockets, that is, the chain must not be

Page B.42

ENGINE

dead tight. Release the locking plunger and securely tighten the locknut. Tap the camshaft sprockets off the flanges of the camshafts.

Accurately position the camshafts with the valve timing gauge and check that the T.D.C. marks are in exact alignment.

Withdraw the circlips retaining the adjusting plates to the camshaft sprockets and pull the adjusting plates forward until the serrations disengage. Replace the sprockets on to the flanges of camshafts and align the two holes in the adjuster plate with the two tapped holes in each camshaft flange. Engage the serrations of the adjuster plates with the serrations in the sprockets.

Note: It is most important that the holes are in exact alignment, otherwise when the set-screws are fitted the camshafts will be moved out of position. If difficulty is experienced in aligning the holes exactly, the adjuster plates should be turned through 180°, which, due to the construction of the plate, will facilitate alignment.

Fit the circlips to the sprockets and one setscrew to the accessible hole in each adjuster plate. Turn the engine until the other two holes are accessible and fit the two remaining setscrews.

Finally, recheck the timing chain tension and valve timing in this order. Secure the four setscrews retaining the camshaft sprockets with new locking wire.

Fit Cylinder Head Oil feed Pipe

Fit the cylinder head oil feed pipe from the tapped hole in the main oil gallery to the two tapped holes in the rear of the cylinder head. Secure the pipe with the three banjo bolts with a new copper washer fitted to both sides of each banjo.

Fit Camshaft Covers

Fit each camshaft cover to the cylinder head using a new gasket. Fit the eleven copper washers and dome nuts to the cover retaining studs but do not tighten fully.

Fit the revolution counter generator and flanged plug to the rear of left-hand and right-hand camshaft covers respectively with the rubber sealing rings seated in the recesses provided and secure with the setscrews and copper washers. Tighten fully the dome nuts securing the camshaft covers.

Note on Refitting

When refitting the throttle linkage, note that the backing plate is offset and ensure that the backing plate assembly is aligned correctly before tightening up.

The remainder of the re-assembly is the reverse of the removal procedure.

THE EXHAUST MANIFOLDS

REMOVAL

Remove the eight brass nuts and spring washers securing the exhaust pipe flanges to the exhaust manifolds.

Remove the sixteen brass nuts and spring washers securing the exhaust manifolds to the cylinder head when the manifolds can be detached.

REFITTING

Refitting is the reverse of the removal procedure. Use new gaskets between the manifolds and the cylinder head and new sealing rings between the exhaust pipe and manifold flanges.

THE FLYWHEEL

The flywheel is a steel forging and has integral starter gear teeth. The flywheel is located to the crankshaft by two mushroom-headed dowels and is secured by ten setscrews retained by a circular locking plate

REMOVAL

Remove the engine as described on page B.19. Unscrew the four setscrews and remove the cover plate from the front face of the clutch housing.

Remove the bolts and nuts securing the clutch

Page B.43

ENGINE

housing to the engine and withdraw the gearbox unit.

Unscrew the six setscrews securing the flange of clutch cover to the flywheel and remove clutch assembly. Note the balance marks 'B' stamped on the clutch cover and on the periphery of the flywheel.

Knock back the tabs of locking plate securing the ten flywheel bolts. Unscrew the flywheel bolts and remove the locking plate. Remove flywheel from the crankshaft flange by gently tapping with a rawhide mallet.

OVERHAUL

If the starter gear is badly worn a new flywheel should be used, since the starter gear teeth are integral with the flywheel, and in this case it will be necessary to balance the flywheel and clutch as an assembly.

If a new flywheel is being fitted, check the flywheel and clutch balance as an assembly by mounting on a mandrel and setting up on parallel knife edges. Mark the relative position of clutch and flywheel. If necessary, remove the clutch and drill ⅜" (9·5 mm.) balance holes not more than ½" (12·7 mm.) deep at a distance of ⅜" (9·5 mm.) from the edge of the flywheel.

REFITTING

Turn the engine upright.

Check that the crankshaft flange and the holes for the flywheel bolts and dowels are free from burrs.

Turn the engine until Nos. 1 and 6 pistons are on T.D.C. and fit the flywheel to the crankshaft flange so that the 'B' stamped on the edge of the flywheel is at approximately the B.D.C. position. (This will ensure that the balance mark 'B' on the flywheel is in line with the balance mark on the crankshaft which is a group of letters stamped on the crank throw just forward of the rear main journal).

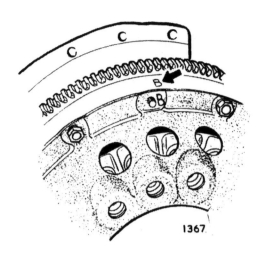

Fig. 31. Showing the balance marks 'B' on the clutch and flywheel.

Tap the two mushroom-headed dowels into position, fit the locking plate and flywheel securing setscrews. Tighten the setscrews to a torque of 67 lb.ft. (9·2 kgm.) and secure with the locking plate tabs. Assemble the clutch driven plate to the flywheel, noting that one side of the plate is marked "Flywheel Side". Centralise the driven plate by means of a dummy shaft which fits the splined bore of the driven plate and the spigot bush in the crankshaft. (A constant pinion shaft may be used for this purpose). Fit clutch cover assembly so that the 'B' stamped adjacent to one of the dowel holes coincides with the 'B' stamped on the periphery of the flywheel. Secure the clutch assembly with the six setscrews and spring washers, tightening the screws a turn at a time by diagonal selection. Remove the dummy shaft.

IGNITION TIMING

Set the micrometer adjustment in the centre of the scale.

Rotate the engine until the rotor-arm approaches the No. 6 (front) cylinder segment in the distributor cap. (Fig. 19).

Slowly rotate the engine until the ignition timing scale on the crankshaft damper is the appropriate number of degrees before the pointer on the sump.

Ignition Settings

Connect a 12 volt test lamp with one lead to the distributor terminal (or the CB terminal of the ignition coil) and the other to a good earth.

Page B.44

ENGINE

Slacken the distributor plate pinch bolt.

Switch on the ignition.

Slowly rotate the distributor body until the points are just breaking, that is, when the lamp lights up with the fibre heel leading the appropriate cam lobe in the normal direction of rotation.

Tighten the distributor plate pinch bolt.

A maximum of six clicks on the vernier adjustment from this setting, to either advance or retard, is allowed.

Static Ignition Timing

8 to 1 compression ratio	9° B.T.D.C.
9 to 1 compression ratio	10° B.T.D.C.

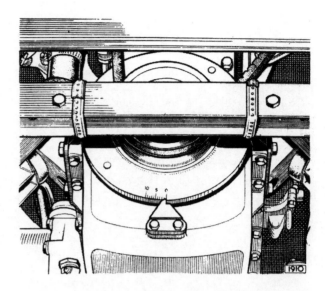

Fig. 32. Showing the timing scale marked on the crankshaft damper. The scale is marked in crankshaft degrees from 0° (top dead centre) to 10° advance (before top dead centre).

THE INLET MANIFOLD

The inlet manifold is in three separate aluminium castings each feeding two cylinders. They are water heated by the coolant from the cylinder head through cast in passages. A water outlet pipe attached to the inlet manifold houses the thermostat and has the top water hose and by-pass hose connected at the front end.

REMOVAL

Drain the radiator.

Remove the carburetters (as described in Section C "Carburetters and Fuel System").

Slacken the clips and disconnect the top water hose and by-pass hoses from the inlet manifold water outlet pipe.

Disconnect the cable to the temperature gauge indicator unit.

Disconnect the heater hose from the connection at the rear of the manifold.

Disconnect the servo pipe from the connection at the rear of the manifold.

Disconnect the accelerator shaft from the ball joint on the throttle spindle.

Remove the setscrews from the backing plate and bush carrying plate.

Ensure that push-in cage nuts do not fall out of the bulkhead. Turn the throttle spindle to the fully open position and remove the short spindle from the ball joint socket.

Remove the eighteen nuts and spring washers, detach the heater pipe clips from the lower studs when the inlet manifold can be withdrawn. Remove six nuts and spring washers and the water manifold. Remove six nuts and spring washers and the air balance pipe.

REFITTING

Refitting is the reverse of the removal procedure. It should be noted that when refitting the throttle linkage ensure that the backing plate is fitted with the cage nuts for the bush carrying plate offset towards the engine—ensure that the backing plate assembly is aligned correctly before tightening up.

Page B.45

ENGINE

THE OIL FILTER

The oil filter is of the full flow type and has a renewable element. The oil from the oil pressure relief valve is returned to the engine sump by an external rubber hose. The oil pressure relief valve is retained by the outlet adaptor to which the hose to the sump is secured.

A balance valve fitted in the filter head opens at a pressure differential of 10 to 15 lbs. per sq. in. (0·7 to 1·1 kg./cm.²) provides a safeguard against the possibility of the filter element becoming so choked that oil is prevented from reaching the bearings.

REMOVAL OF THE OIL FILTER

When removing the oil filter it is advisable to catch any escaping oil.

Remove the splash tray from below the brake vacuum reservoir. Remove the cable to the oil pressure transmitter unit in the oil filter head. Slacken the clip and

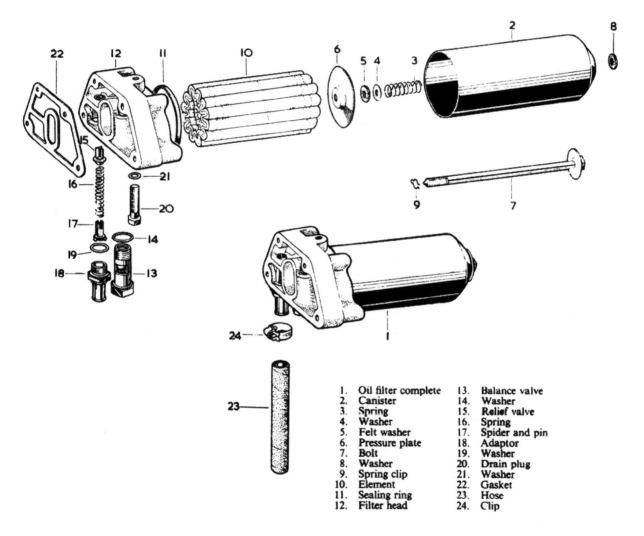

1. Oil filter complete
2. Canister
3. Spring
4. Washer
5. Felt washer
6. Pressure plate
7. Bolt
8. Washer
9. Spring clip
10. Element
11. Sealing ring
12. Filter head
13. Balance valve
14. Washer
15. Relief valve
16. Spring
17. Spider and pin
18. Adaptor
19. Washer
20. Drain plug
21. Washer
22. Gasket
23. Hose
24. Clip

Fig. 33. Exploded view of oil filter.

Page B.46

ENGINE

remove the rubber hose from below the filter head.

Detach the oil filter assembly from the side face of the cylinder block by removing the four bolts and withdraw the assembly from beneath the car. Collect the gasket fitted between the filter head and the cylinder block.

REFITTING THE OIL FILTER

Refitting is the reverse of the removal procedure but a new gasket must be fitted between the oil filter head and the cylinder block.

ELEMENT REPLACEMENT

It is most important to renew the oil filter element at the recommended periods as after this mileage it will have become choked with impurities.

To guard against the possibility of the filter being neglected to the extent where the element becomes completely choked, a balance valve is incorporated in the filter head which allows unfiltered oil to by-pass the element and reach the bearings. This will be accompanied by a drop in the normal oil pressure of some 10 lbs. per sq. in and if this occurs the filter element should be renewed as soon as possible.

The oil filter is situated at the right-hand side of the engine and it is advisable when removing the filter canister, to catch any escaping oil. Unscrew the centre bolt and remove the canister and element from beneath the car retaining the rubber sealing ring. Empty out the oil, thoroughly wash out the canister with petrol and allow to dry before inserting a new element.

When refitting the canister, renew the rubber sealing ring. Ensure that the ring is seating correctly in the groove between the canister and the filter head before tightening the centre bolt.

THE OIL PUMP

The oil pump is of the eccentric rotor type and consists of five main parts:— the body, the driving spindle with the inner rotor pinned to it, the outer rotor and the cover, which is secured to the main body by four bolts, finally being secured to the engine with additional dowel bolts. The inner rotor has one lobe less than the number of internal segments in the outer rotor. The spindle centre is eccentric to that of the bore in which the outer rotor is located, thus the inner rotor is able to rotate within the outer, and causes the outer rotor to revolve. The inlet connection is positioned in the pump cover, and the outlet connection in the body. These are both connected to the ports in the pump.

Consider the oil flow with the lobes of the inner rotor lying along the line of eccentricity. In this position oil is free to flow from the port into the space (dotted portion) between the rotors, and on the other side of the lobe (shaded portion) the oil is free to flow into the delivery port (see Fig. 34).

In the second position, the inner and outer rotors have rotated and caused the oil that was flowing from the inlet port into the space between them to be cut off from the port and transferred to the enclosed space

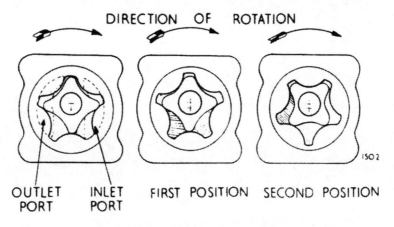

Fig. 34. Operation of eccentric rotor type oil pump.

Page B.47

ENGINE

between the ports. Similarly, the space which enclosed oil free to flow to the delivery port in the first position has decreased in size in the second position, and thus caused this oil to flow into the delivery port. The action of the pump is then a repetition of the above, oil flowing into the space between the rotors from the inlet port under atmospheric pressure and being discharged into the delivery port by reason of the space in which it is contained decreasing in size as it passes over the port.

REMOVAL

Remove the sump as described on page B.50.

Detach the suction and delivery pipe brackets and withdraw the pipes from the oil pump.

Tap back the tab washers and remove the three bolts which secure the oil pump to the front main bearing cap.

Withdraw the oil pump and collect the coupling sleeve at the top of the drive shaft.

DISMANTLING

Unscrew the four bolts and detach the bottom cover from the oil pump.

Withdraw the inner and outer rotors from the oil pump body. The inner rotor is pinned to the drive shaft and must not be dismantled.

OVERHAUL

Check the clearance between lobes of the inner and outer rotors which should be ·006″ (·15 mm.) maximum (see Fig. 35).

Check the clearance between the outer rotor and the pump body (see Fig. 36) which should not exceed ·010″ (·25 mm.).

Check the end-float of the rotors by placing a straight edge across the joint face of the body and measuring the clearance between the rotors and straight edge (see Fig. 37). This clearance should be ·0025″ (·06 mm.) and in an emergency can be restored by lapping the pump body and outer rotor on a surface plate to suit the inner rotor.

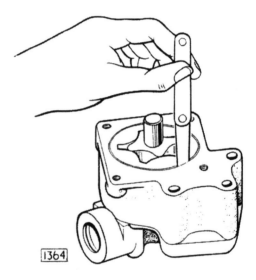

Fig. 36. Measuring the clearance between the outer rotor and the pump body.

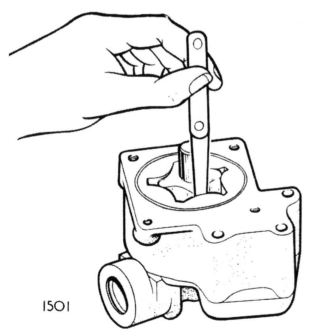

Fig. 35. Measuring the clearance between the inner and outer rotors.

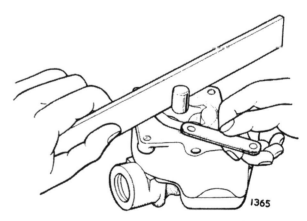

Fig. 37. Measuring the end float of the rotors.

Page B.48

ENGINE

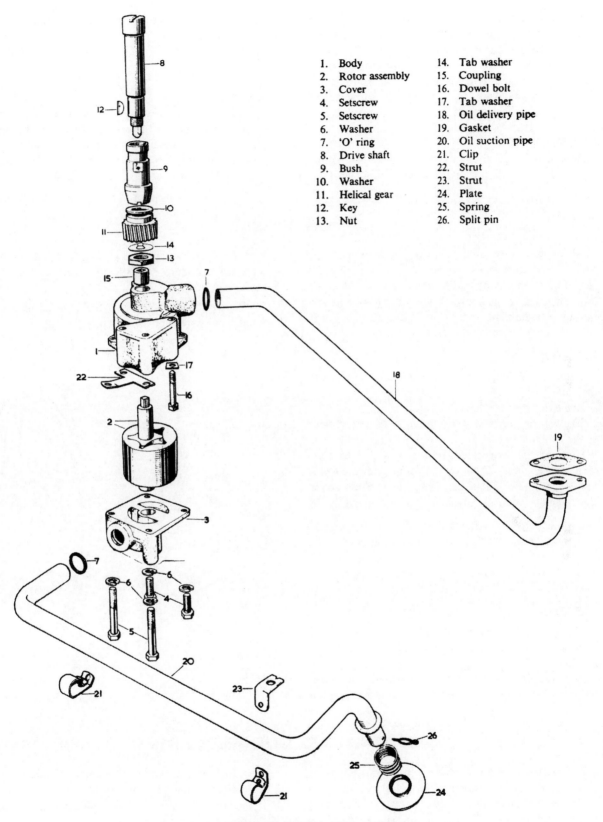

1. Body
2. Rotor assembly
3. Cover
4. Setscrew
5. Setscrew
6. Washer
7. 'O' ring
8. Drive shaft
9. Bush
10. Washer
11. Helical gear
12. Key
13. Nut
14. Tab washer
15. Coupling
16. Dowel bolt
17. Tab washer
18. Oil delivery pipe
19. Gasket
20. Oil suction pipe
21. Clip
22. Strut
23. Strut
24. Plate
25. Spring
26. Split pin

Fig. 38. Exploded view of the oil pump.

Page B.49

ENGINE

Examine the pump body and bottom cover for signs of scoring and the drive shaft bores for signs of wear; fit new parts as necessary.

Place the drive shaft in a vice fitted with soft jaws and check that the inner rotor is tight on the securing pin.

Note that the drive shaft, inner and outer rotors are supplied only as an assembly.

RE-ASSEMBLING

Re-assembly is the reverse of the dismantling procedure but it is important when fitting the outer rotor to the pump body to insert the chamfered end of the rotor foremost.

Always fit new "O" rings to the suction and delivery pipe bores.

REFITTING

Refitting is the reverse of the removal procedure.

Do not omit to fit the coupling sleeve to the squared end of the drive shaft before offering up the oil pump.

After fitting of the oil pump, check that there is appreciable end-float of the coupling sleeve.

OIL SUMP

All engine units are fitted with aluminium sumps which have an external connection for a rubber oil return hose the second end of which is attached to the oil filter head. A gauze bowl type filter is attached to the sump baffle plate.

REMOVAL

Remove the sump drain plug and drain the oil from the sump.

Remove the crankshaft damper.

Slacken the clip and disconnect the oil return hose at the oil filter head.

Unscrew the twenty-six setscrews and four nuts securing the sump. Remove the sump from the cylinder block noting that a short setscrew is fitted at the right-hand front corner of the sump as shown in Fig. 39.

Note: It may be necessary to slacken the engine stabiliser washers to allow the engine to be raised at the rear before the sump can be removed.

REFITTING

Scrape off all traces of old gaskets or sealing compound from the joint faces of the sump and crankcase.

Always fit new gaskets and rear oil seal when refitting the sump. If time permits, roll the rear oil seal into a coil and retain with string for a few hours. This will facilitate the fitting of the seal to its semi-circular recess.

Ensure that the short setscrew is fitted to the right hand front corner of the sump.

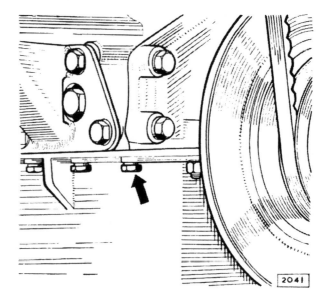

Fig. 39. Showing the location of the short setscrew.

PISTONS AND GUDGEON PINS

The pistons are made from low expansion aluminium alloy and are of the semi-split skirt type.

The pistons have three rings each, two compression and one oil control. The top compression ring only is chromium plated; both the top and second compression rings have a tapered periphery.

The fully floating gudgeon pin is retained in the piston by a circlip at each end.

Page B.50

ENGINE

REMOVAL

As the pistons will not pass the crankshaft it will be necessary to withdraw the pistons and connecting rods from the top. Proceed as follows:—

Remove Cylinder Head

Remove the cylinder head as described on page B.41.

Remove Sump

Remove the sump as described on page B.50.

Remove Piston and Connecting Rod

Remove the split pins from the connecting rod bolt nuts and unscrew nuts. Remove the connecting rod cap, noting the corresponding cylinder numbers on the connecting rod and cap. Remove the connecting rod bolts and withdraw the piston and connecting rod from the top of cylinder block.

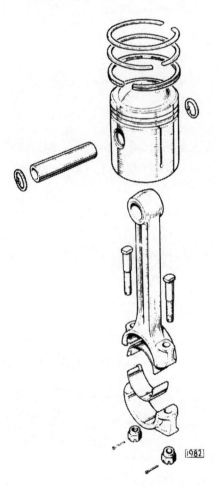

Fig. 40. Exploded view of the piston and connecting rod.

OVERHAUL

Pistons are supplied complete with gudgeon pins which have been selectively assembled and are, therefore, not interchangeable one with another.

The pistons fitted to an engine should not vary one with another by more than 2 drams (3·5 grammes).

Gudgeon Pin Fitting

Gudgeon pins are a finger push fit in the piston at normal room temperature 68°F (20°C).

When actually removing or refitting the gudgeon pin, the operation should be effected by immersing the piston, gudgeon pin and connecting rod little end in a bath of hot oil. When the piston and little end have reached a sufficient temperature (230°F. 110°C.) the gudgeon pin can be moved into position. Always use new circlips on assembly.

When assembling the engine, centralise the small end of the connecting rod between the gudgeon pin bosses in the piston and ensure that the connecting rod mates up with the crankshaft journal without any pressure being exerted on the rod.

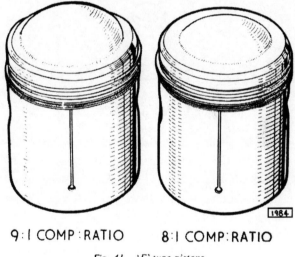

Fig. 41. 'E' type pistons.

Piston Grades

The following selective grades are available in standard size pistons only. When ordering standard size pistons the identification letter of the selective grade should be clearly stated. Pistons are stamped on the crown with the letter identification and the cylinder block is also stamped on the top face adjacent to the bores.

Page B.51

ENGINE

Grade Identification Letter	To suit cylinder bore size
F	3·4248″ to 3·4251″ (86·990 to 86·997 mm.)
G	3·4252″ to 3·4255″ (87·000 to 87·007 mm.)
H	3·4256″ to 3·4259″ (87·010 to 87·017 mm.)
J	3·4260″ to 3·4263″ (87·020 to 87·027 mm.)
K	3·4264″ to 3·4267″ (87·030 to 87·037 mm.)

Oversize Pistons

Fig. 42. Showing the marking on the piston crown.

Oversize pistons are available in the following sizes:—
+·010″ (·25 mm.) +·020″ (·51 mm.) +·030″ (·76 mm.).

There are no selective grades in oversize pistons as grading is necessarily purely for factory production methods.

Piston Rings

Check the piston ring gap with the ring as far down the cylinder bore as possible. Push the ring down the bore with a piston to ensure that it is square and measure the gap with a feeler gauge. The correct gaps are as follows:—

Compression rings ·015″ to ·020″ (·38 to ·51 mm.)
Oil control rings ·011″ to ·016″ (·28 to ·41 mm.)

With the rings fitted to the piston check the side clearance in the grooves which should be ·001″ to ·003″ (·025 to ·076 mm.).

One of the compression rings is hard chrome plated and this ring must be fitted to the top groove in the piston.

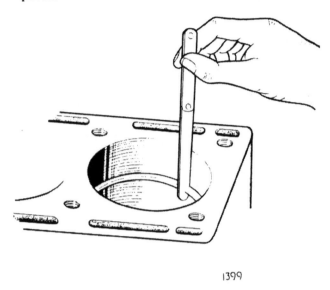

Fig. 43. Checking the piston ring gap.

Tapered Periphery Rings

All engine units are fitted with tapered periphery piston rings in at least one position and these must be fitted the correct way up.

Fig. 44. Showing the identification marks on the tapered periphery compression rings.

Page B.52

ENGINE

The narrowest part of the ring must be fitted uppermost; to assist in identifying the narrowest face a letter "T" or "Top" is marked on the side of the ring to be fitted uppermost.

The oil control ring is not tapered and can be fitted either way up.

REFITTING

Pistons and connecting rods must be fitted to their respective cylinders (piston and connecting rods are stamped with their cylinder number, No. 1 being at the rear) and the same way round in the bore.

The pistons must be fitted with split on the left-hand or exhaust side of the engine. To facilitate correct fitting the piston crowns are marked "Front", see Fig. 42.

Use a piston ring clamp when entering the rings into the cylinder bore.

The cap must be fitted to the connecting rod so that the cylinder numbers stamped on each part are on the same side.

Tighten the connecting rod nuts to a torque of 37 lb.ft. (5·1 kgm.).

SPARKING PLUGS

SERVICE PROCEDURE

To maintain peak sparking plug performance, plugs should be inspected, cleaned and re-gapped at regular intervals of 2,500 miles. Under certain fuel and operating conditions, particularly extended slow speed town driving, sparking plugs may have to be serviced at shorter intervals.

Disconnect the ignition cables from all sparking plugs.

Loosen the sparking plugs about two turns anti-clockwise using the proper sized deep-socket wrench.

Blow away the dirt from around the base of each plug.

Remove the sparking plugs and place them in a suitable holder, preferably in the order they were in the engine.

ANALYSING SERVICE CONDITIONS

Examine the gaskets to see if the sparking plugs were properly installed. If the gaskets were excessively compressed, installed on dirty seats or distorted, leakage has probably occurred during service which would tend to cause overheating of the sparking plugs. Gaskets properly installed will have flat clean surfaces. Gaskets which are approximately one-half their original thickness will be satisfactory but thinner ones should be renewed.

Examine the firing ends of the sparking plugs, noting the type of the deposits and the degree of electrode erosion. The typical conditions illustrated may indicate the use of a sparking plug with an incorrect heat range or faulty engine and ignition system operation. Remember that if sufficient voltage is not delivered to the sparking plug, no type of plug can fire the mixture in the cylinder properly.

Normal Condition

Look for powdery deposits ranging from brown to greyish tan. Electrodes may be worn slightly. These are signs of a sparking plug of the correct heat range used under **normal** conditions, that is mixed periods of high speed and low speed driving. Cleaning and re-gapping of the sparking plugs is all that is required.

Normal Condition

Watch for white to yellowish powdery deposits. This usually indicates long periods of constant speed driving or a lot of slow speed city driving. These deposits have no effect on performance if the sparking

Fig. 45. Normal condition.

Page B.53

ENGINE

plugs are cleaned **thoroughly** at approximately 2,500 miles intervals. Remember to "wobble" the plug during abrasive blasting in the Champion Service Unit. Then file the sparking surfaces vigorously to expose bright clean metal.

Oil Fouling

This is usually indicated by wet, sludgy deposits traceable to excessive oil entering the combustion chamber through worn cylinders, rings and pistons, excessive clearances between intake valve guides and stems, or worn and loose bearings, etc. Hotter sparking plugs may alleviate oil fouling temporarily, but in severe cases engine overhaul is called for.

Fig. 46. Oil fouling.

Petrol Fouling

This is usually indicated by dry, fluffy black deposits which result from incomplete combustion. Too rich an air-fuel mixture, excessive use of the mixture control or a faulty automatic choke can cause incomplete burning. In addition, a defective coil, contact breaker points, or ignition cable, can reduce the voltage supplied to the sparking plug and cause misfiring. If fouling is evident in only a few cylinders, sticking valves may be the cause. Excessive idling, slow speeds, or stop-and-go driving, can also keep the plug temperatures so low that normal combustion deposits are not burned off. In the latter case, hotter plugs may be installed.

Burned or Overheated Condition

This condition is usually identified by a white, burned or blistered insulator nose and badly eroded electrodes. Inefficient engine cooling and improper

Fig. 48. Badly burned sparking plug.

ignition timing can cause general overheating. Severe service, such as sustained high speed and heavy loads, can also produce abnormally high temperatures in the combustion chamber which necessitate the use of colder sparking plugs.

File the sparking surfaces of the electrodes by means of a point file. If necessary, open the gaps slightly and file vigorously enough to obtain bright, clean, parallel surfaces. For best results, hold the plug in a vice.

Reset the gaps using the bending fixture of the Champion Gap Tool. Do not apply pressure on the centre electrode as insulator fractures may result. Use the bending fixture to obtain parallel sparking surfaces for maximum gap life.

Visually inspect all sparking plugs for cracked or chipped insulators. Discard all plugs with insulator fractures.

Test the sparking ability of a used sparking plug on a comparator.

Fig. 47. Petrol fouling.

Page B.54

ENGINE

Clean the threads by means of wire hand or power-driven brush. If the latter type is used, wire size should not exceed ·005″ (·127 mm.) diameter. Do not wire brush the insulator nor the electrodes.

Clean gasket seats on the cylinder head before installing sparking plugs to ensure proper seating of the sparking plug gasket. Then, using a new gasket, screw in the plug by hand finger-tight.

Note: If the sparking plug cannot be seated on its gasket by hand, clean out the cylinder head threads with a clean-out tap or with another used sparking plug having three or four vertical flutes filed in its threads.

Grease the tap well to retain chippings which may fall into the combustion chamber. Tighten the sparking plugs to a torque of 27 lb.ft. (3·73 kgm.).

STANDARD GAP SETTING

The sparking plug gap settings recommended in this Service Manual have been found to give the best overall performance under all service conditions. They are based on extensive dynamometer testing and experience on the road, and are generally a compromise between the wide gaps necessary for best idling performance and the small gaps required for the best high speed performance.

All plugs should be reset to the specified gap by bending the side electrode only, using the special tool available from the Champion Sparking Plug Company.

SPARKING PLUG INSERTS (Fig. 50)

When it becomes necessary to fit a sparking plug insert in the event of a stripped thread proceed as detailed below.

Bore out the stripped thread to ·75″ (19·05 mm.) diameter and tap ½″ B.S.P.

Counterbore $\frac{57}{64}$″ (22·62 mm.) diameter to accommodate the larger diameter of the insert.

Fit the screwed insert ensuring that it sits firmly on the face at the bottom of the thread.

Drill and ream a $\frac{1}{8}$″ (3·17 mm.) diameter hole $\frac{3}{16}$″ (4·76 mm.) deep between the side of the insert and the cylinder head as shown. Drive in the locking pin and ensure that the pin is below the surface. To secure peen over the aluminium on the chamfered portion of the insert and also peen over the locking pin.

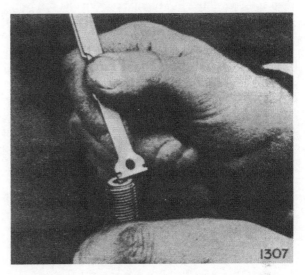

Fig. 49. Setting the gap with the special tool.

TAPPETS, TAPPET GUIDES AND ADJUSTING PADS

The chilled cast iron tappets are of cylindrical form and run in guides made of austenitic iron which are shrunk into the cylinder head. A steel pad for adjustment of the valve clearance is sandwiched between the underside of the tappet and top of the valve stem. The pads are available in a range of thicknesses, rising in ·001″ (·025 mm.) steps, from ·085″ to ·110″ (2·16 to 2·79 mm.) and are etched on the surface with the letter "A" to "Z", each letter indicating an increase in size of ·001″ (·025 mm.). (Page B.61).

REMOVAL OF TAPPETS AND ADJUSTING PADS

Remove the camshafts as described on page B.33. The tappets can now be withdrawn with a suction valve grinding tool.

Page B.55

ENGINE

Remove the adjusting pads. If valve clearance adjustment is not being carried out the adjusting pads must be refitted to their original positions.

OVERHAUL

Examine the tappets and tappet guides for signs of wear. The diametrical clearance between the tappet and tappet guide should be ·0008″ to ·0019″ (·02 to ·05 mm.).

Examine the adjusting pads for signs of indentation. Renew if necessary with the appropriate size when making valve clearance adjustment on re-assembly.

Tappet Guide Replacement

If it is found necessary to replace the tappet guides they must be fitted in accordance with the following instructions and only genuine factory replacement parts used.

(1) Remove the old tappet guide by boring out until the guide collapses. Take care not to damage the bore for the guide in the cylinder head.

(2) Carefully measure the diameter of the tappet guide bore in the cylinder head at room temperature—68°F (20°C).

(3) Grind down the 1·643″ (41·73 mm.) outside diameter of tappet guide to a diameter of ·003″ (·08 mm.) larger than the tappet guide bore dimension, that is to give an interference fit of ·003″ (·08 mm.).

(4) Also grind off the same amount from the "lead-in" at the bottom of tappet guide. The reduction in diameter from the adjacent diameter should be ·0032″ to ·0057″ (·08 to ·14 mm.).

(5) Heat the cylinder head in an oven for half an hour from cold at a temperature of 300°F (150°C).

(6) Fit the tappet guide, ensuring that the lip at top of guide beds evenly in the recess.

(7) After fitting, ream tappet guide bore to a diameter of $1\frac{3}{8}″ \; {+·0007″ \atop -·0000″}$ ($34·925 \; {+·018 \atop -·000}$ mm).

Note: It is essential that, when reamed, the tappet guide bore is concentric with the bore of the valve guide.

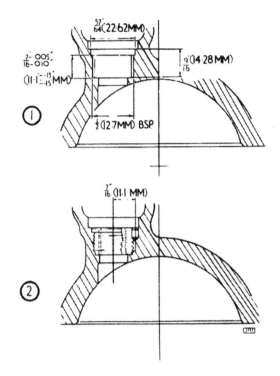

Fig. 50. Fitting dimensions for sparking plug inserts.

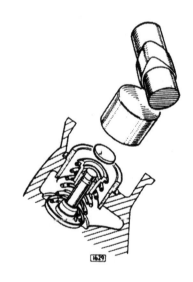

Fig. 51. Showing the tappet and adjustment pad.

Page B.56

ENGINE

THE TIMING GEAR

The camshafts are driven by Duplex endless roller chains in two stages.

The first stage or bottom timing chain drives the larger wheel of a double intermediate sprocket; the second stage or top timing chain passes round the smaller wheel of the intermediate sprocket, both camshaft sprockets, and is looped below an idler sprocket.

The idler sprocket has an eccentric shaft for top timing chain tension adjustment and the bottom chain is automatically tensioned by an hydraulic tensioner bolted to the cylinder block. Nylon or rubber vibration dampers are located at convenient points around the chains.

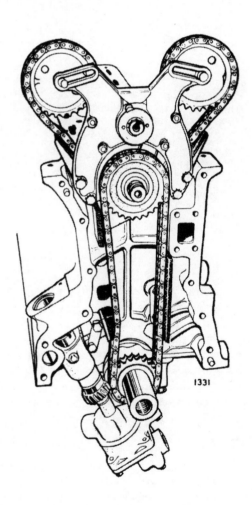

Fig. 52. The timing gear arrangement.

REMOVAL

Remove the cylinder head as described on page B.41.

Remove the radiator, cowl, header tank and cooling fan (as described in Section D "Cooling System").

Remove the damper as described on page B.35.

Withdraw the split cone.

Remove the sump as described on page B.50.

Unscrew the set bolts and nuts, and remove the water pump from the timing cover.

Note the gasket between the pump and the timing cover.

Remove the front cover as described on page B.22.

Remove the bottom timing chain tensioner as described on page B.59.

Unscrew the four setscrews securing the front mounting bracket to the cylinder block.

Remove the two screwdriver slotted setscrews securing the rear mounting bracket; these setscrews secure the intermediate damper bracket.

The timing gear assembly can now be removed.

DISMANTLING

Remove the nut and serrated washer from the front end of the idler shaft, and withdraw the plunger and spring.

Remove the four nuts securing the front mounting bracket to the rear bracket. Withdraw the front bracket from the studs.

Remove the bottom timing chain from the large intermediate sprocket.

To remove the intermediate sprockets, remove the circlip from the end of the shaft in the mounting bracket. Press the shaft out of the bracket, and withdraw the sprockets from the shaft.

To separate the two intermediate sprockets, press the boss of the small sprocket from the bore of the large sprocket, noting that they are keyed together. (On later models the intermediate sprocket is in one piece).

OVERHAUL

If the chain shows signs of stretching or wear new ones should be fitted. Replace any sprockets and dampers that show signs of wear.

ASSEMBLING

Fit the eccentric shaft to the hole in front mounting bracket. Insert the spring and locking plunger for the serrated plate to the hole in the front mounting bracket.

Page B.57

ENGINE

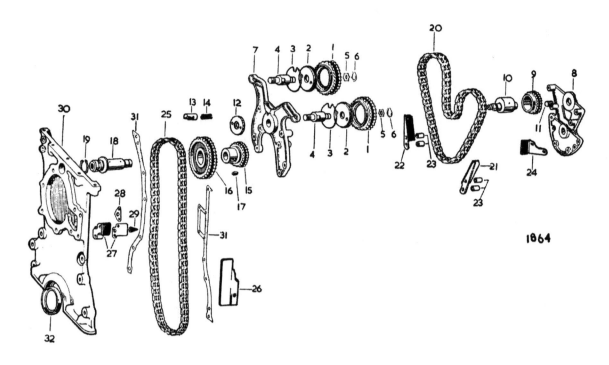

1. Camshaft sprocket
2. Adjusting plate
3. Circlip
4. Guide pin
5. Star washer
6. Circlip
7. Timing gear front mounting bracket
8. Timing gear rear mounting bracket
9. Idler sprocket
10. Eccentric shaft
11. Plug
12. Adjustment plate
13. Plunger pin
14. Spring
15. Intermediate sprocket of top timing chain
16. Intermediate sprocket of lower timing chain
17. Key
18. Shaft
19. Circlip
20. Top timing chain
21. Damper for top timing chain (left hand)
22. Damper for top timing chain (right hand)
23. Distance piece
24. Intermediate damper
25. Bottom timing chain
26. Vibration damper
27. Hydraulic chain tensioner
28. Shim
29. Filter gauze
30. Front timing cover
31. Gasket
32. Oil seal

Fig. 53. Exploded view of the timing gear.

Fit the serrated plate and secure with the shakeproof washer and nut. Fit the idler sprocket (21 teeth) to the eccentric shaft.

Fit the two intermediate sprockets (20 and 28 teeth) to their shaft with the larger sprocket forward and press the shaft through lower central hole in rear mounting bracket. Secure with the circlip at the rear of bracket.

Fit the top timing chain (longer chain) to the small intermediate sprocket and the bottom timing chain (shorter chain) to the large intermediate sprocket.

Loop the upper timing chain under the idler sprocket and offer up the front mounting bracket to the rear mounting bracket with the two chain dampers interposed between the brackets.

Fit the intermediate damper to the bottom of the rear mounting bracket with two screwdriver slotted setscrews and shakeproof washer.

Pass the four securing bolts through the holes in the brackets, chain dampers and spacers noting that shakeproof washers are fitted under the bolt heads. Secure the two mounting brackets together with four stud nuts and shakeproof washers.

REFITTING

Refitting the remainder of the assembly is the reverse of the removal procedure.

When refitting the timing chain tensioner refer to page B.59.

Page B.58

ENGINE

THE BOTTOM CHAIN TENSIONER

The bottom timing chain tensioner is of hydraulic type and consists of an oil resistant rubber slipper mounted on a plunger (A, Fig. 54) which bears on the outside of the chain. The light spring (C) cased by the restraint cylinder (B) and the plunger, in combination with oil pressure holds the slipper head against the chain keeping it in correct tension.

Return movement of the slipper head is prevented by the limit peg at the bottom end of the plunger bore engaging the nearest tooth in the helical slot of the restraint cylinder. The oil is introduced into the adjuster body (D) via a small drilling in the locating spigot and passing through a hole in the slipper head lubricates the chain. The backing plate (E) provides a suitable face along which the slipper head can work.

REMOVAL

Proceed as described under "Timing Gear—Removal" on page B.57 until the chain tensioner is accessible.

Remove the bottom plug which provides access to the hexagonal hole in the end of the restraint cylinder. Insert an Allen key (·125"A/F) into this and turn the key in a *clockwise* direction until the slipper head remains in the retracted position. Remove the securing bolts and detach the adjuster. A conical filter is fitted in the oil feed hole in the cylinder block and this should be removed and cleaned in petrol.

REFITTING

Fit the conical filter to the oil feed hole in the cylinder block.

Fit shims as necessary, between the backing plate and cylinder block so that the timing chain runs centrally along the rubber slipper.

Fit the tab washer and two securing bolts. Tighten the bolts and tap the tab washers against the bolt heads.

It is important that no attempt is made to release the locking mechanism until the adjuster has been finally mounted in the engine WITH THE TIMING CHAIN IN POSITION.

Remove the hexagon head plug and tab washer from the end of the body. Insert the Allen key into

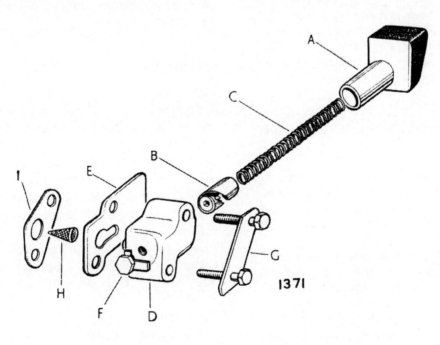

Fig. 54. Exploded view of the bottom timing chain tensioner.

Page B.59

ENGINE

the hole until it registers in the end of the restraint cylinder. Turn the key clockwise until the tensioner head moves forward under spring pressure against the chain. Do not attempt to turn the key anti-clockwise, nor force the tensioner head into the chain by external pressure.

Refit the plug and secure with the tab washer.

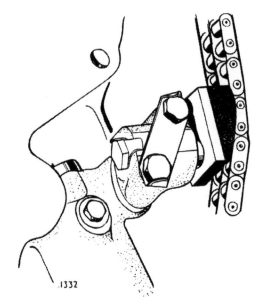

Fig. 55. Showing the bottom timing chain tensioner in position.

THE VALVES AND SPRINGS

The inlet valves are of silicon chrome steel and the exhaust valves are of austenitic steel. Double coil valve springs are fitted and are retained by a valve collar with split cotters.

Warning: As the valves in the fully open position protrude below the cylinder head joint face, the cylinder head must not be placed joint face downwards directly on a flat surface; support the cylinder head on wooden blocks, one at each end.

REMOVAL
Removal the cylinder head as described on page B.41

Remove Valves
With the cylinder head on the bench remove the inlet manifold, and the revolution counter generator.

Remove the four bearing caps from each camshaft and lift out the camshafts (note mating marks on each bearing cap).

Remove the twelve tappets and adjusting pads situated between tappets and valve stems. Lay out the tappets and pads in order, to ensure that they can be replaced in their original guides.

Obtain a block of wood the approximate size of the combustion chambers and place this under the valve heads in No. 1 cylinder combustion chamber. Press down the valve collars and extract the split cotters. Remove the collars, valve springs and spring seats. Repeat for the remaining five cylinders. Valves are numbered and must be replaced in the original locations, No. 1 cylinder being at the rear, that is, the flywheel end.

OVERHAUL
Valves
Examine the valves for pitting, burning or distortion and reface or renew the valves as necessary. Also reface the valve seats in the cylinder head and grind the valves to their seats using a suction valve tool. When refacing the valves or seat inserts do not remove more metal than is necessary to clean up the facings.

The valve seat angles are as follows:—inlet and exhaust, 45°.

Renew valves where the stem wear exceeds ·003" (·08 mm.). The clearance of the valve stem in the guide when new is ·001" to ·004" (·025 to ·10 mm.).

Valve Springs
Test the valve springs for pressure, either by comparison with the figures given in the "Valve Spring

Page B.60

ENGINE

Data" or by comparison with a new valve spring.

To test against a new valve spring, insert both valve springs to end between the jaws of a vice or under a press with a flat metal plate interposed between the two springs. Apply a load to compress the springs partly and measure their comparative lengths.

When fitting valve springs to the cylinder head compress the springs using Churchill tool No. J.6118.

VALVE CLEARANCE ADJUSTMENT

When checking the valve clearances, the camshafts must be fitted one at a time as if one camshaft is rotated when the other camshaft is in position, fouling is likely to take place between the inlet and exhaust valves. Obtain and record all valve clearances by using a feeler gauge between the back of each cam and the appropriate valve tappet.

Fig. 56. Fitting the valve springs utilizing the valve spring compressing tool Churchill tool No. J.6118.

Correct valve clearances are:—

Normal Touring Use
    Inlet   ..   ..   ·004" (·10 mm.)
    Exhaust   ..   ..   ·006" (·15 mm.)

Racing
    Inlet   ..   ..   ·006" (·15 mm.)
    Exhaust   ..   ..   ·010" (·25 mm)

Adjusting pads are available rising in ·001" (·03 mm.) sizes from ·085" to ·110" (2·16 to 2·79 mm.) and are etched on the surface with the letter 'A' to 'Z', each letter indicating an increase in size of ·001" (·03 mm.). Should any valve clearance require correction, remove the camshaft, tappet and adjusting pad. Observe the letter etched on the adjusting pad if visible. If the letter is not visible measure the pad with a micrometer and should the recorded clearance for this valve have shown say ·002" (·05 mm.) excessive clearance select a new adjusting pad bearing a letter two lower than the original pad.

As an example, assume that No. 1 inlet valve clearance is tested and recorded as ·007" (·18 mm.). On removal of the adjusting pad, if this is etched with the letter 'D' then substitution with a pad bearing the letter 'G' will correct the clearance for No. 1 inlet valve.

Valve Adjusting Pads

	ins.	mm.
A	·085	2·16
B	·086	2·18
C	·087	2·21
D	·088	2·23
E	·089	2·26
F	·090	2·29
G	·091	2·31
H	·092	2·34
I	·093	2·36
J	·094	2·39
K	·095	2·41
L	·096	2·44
M	·097	2·46
N	·098	2·49
O	·099	2·51
P	·100	2·54
Q	·101	2·56
R	·102	2·59
S	·103	2·62
T	·104	2·64
U	·105	2·67
V	·106	2·69
W	·107	2·72
X	·108	2·74
Y	·109	2·77
Z	·110	2·79

When fitting the camshafts prior to fitting the cylinder head to the engine it is most important that the keyway in the front bearing flange of each camshaft is perpendicular (at 90°) to the adjacent camshaft cover face before tightening down the camshaft bearing cap nuts. Tighten the camshaft bearing cap nuts to a torque of 15 lb.ft. (2·0 kgm.).

REFITTING

Before attempting to refit the cylinder head refer to the instructions given on page B.42.

Page B.61

ENGINE

THE VALVE GUIDES

The valve guides are of cast iron and are chamfered at the upper ends. The outside diameter of the guide is reduced at the upper end to provide a "lead-in" when fitting the guide to the cylinder head. The inlet and exhaust guides are of different lengths, the inlet being the shorter of the two.

REPLACEMENT

Examine the valve guides for evidence of wear in the bore. The clearance between the valve stem and the guide when new is ·001" to ·004" (·025 to ·10 mm.).

If it is found necessary to replace worn valve guides they must be fitted in accordance with the following instructions and only genuine factory replacement parts used.

(1) Press out, or drive out with a piloted drift, the old valve guide from the top of the cylinder head.

(2) Ream the valve guide bore in the cylinder head to a diameter of

·505" $^{+·0005''}_{-·0002''}$ (12·83 mm. $^{+·012 \text{ mm.}}_{-·005 \text{ mm.}}$)

(3) Heat the cylinder head by immersing in boiling water for 30 minutes.

(4) Coat the valve guide with graphite grease and press in, or drive in with a piloted drift, from the combustion chamber end. The correct fitted position for both inlet and exhaust guides is with the top of the guide (chamfered end) $\frac{5}{16}$" (8 mm.) above the spot facing for the valve spring seat. (See Fig. 57).

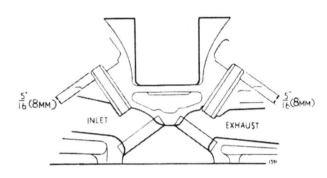

Fig. 57. Showing the fitted position of the valve guide.

THE VALVE SEAT INSERTS

The valve seat inserts are centrifugally cast iron and are shrunk into the cylinder head.

REPLACEMENT

If it is found necessary to replace the valve seat inserts they must be fitted in accordance with the following instructions and only genuine factory replacement parts used.

(1) Remove the old valve seat insert by boring out until the insert collapses. Take care not to damage the recess for insert in the cylinder head.

(2) Carefully measure diameter of insert recess in cylinder head at room temperature 68°F. (20°C.).

(3) Grind down outside of insert to a diameter of ·003" (·08 mm.) larger than recess dimension, that is, to give an interference fit of ·003" (·08 mm.).

(4) Heat the cylinder head in an oven for one hour from cold at a temperature of 300°F. (150° C.).

(5) Fit insert, ensuring that it beds evenly in its recess.

(6) After the valve seat insert has been fitted the following instructions should be carried out to ensure that the valve clearance can be obtained within the range of the adjusting pads, that is, ·085" to ·110" (2·16 to 2·79 mm.).

(a) Assemble the camshafts to the cylinder head. Fit the appropriate valve to the insert in question and, with the valve seat faces touching, check the distance between the top of the valve stem and the **back** of the cam. This should be ·320" (8·13 mm.) **plus** the appropriate valve clearance. (The figure of ·320" (8·13 mm.) includes an allowance for an adjusting pad thickness of ·095" (2·41 mm.) to ·097" (2·46 mm.) which will, if necessary, permit the fitting of thicker or thinner adjusting pads when making the final valve clearance adjustment).

(b) If the distance is greater than the figure of ·320" (8·13 mm.), plus the appropriate valve clearance, grind the valve seat of the insert with suitable valve grinding equipment until the correct distance is obtained.

Page B.62

ENGINE

Example: Assume that the valve insert in question is an exhaust and the distance between the top of the valve stem and the back of the cam is found to be ·344″ (8·74 mm.).

Adding the exhaust valve clearance of ·006″ (·15mm.) to ·320″ (8·13 mm.) equals ·326″ (8·28 mm.). In this case the valve seat of the insert will have to be ground down to reduce the distance between the top of valve stem and the back of the cam by ·018″ (·46 mm.) that is, ·344″ minus ·326″ (8·74 minus 8·28 mm.).

(c) After assembling the cylinder head, check and adjust the valve clearances in the normal manner.

VALVE TIMING

Turn the engine so that No. 6 (front) piston is exactly in the T.D.C. position on compression stroke (firing position) that is, with the distributor rotor arm opposite No. 6 cylinder segment. (See Fig. 19).

See Figs. 32 or 61 for location of T.D.C. marks.

It is important to tension the top timing chain before attempting to check or set the valve timing. Proceed as follows:—

Through the breather aperture in the front of the cylinder head slacken the locknut securing the serrated plate (Fig. 58).

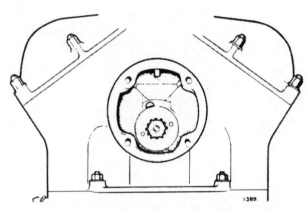

Fig. 58. Showing the serrated plate for adjustment of the top timing chain tension.

Tension the chain by pressing locking plunger inwards and rotating serrated plate by the two holes in an anti-clockwise direction. Turn the engine each way slightly and recheck the chain tension. When correctly tensioned there should be slight flexibility on both outer sides below the camshaft sprockets, that is, the chain must not be dead tight. Release the locking plunger and securely tighten the locknut.

Remove the locking wire from the setscrews securing the camshaft sprockets. Note the positions of the **inaccessible** setscrews and rotate the engine until they can be removed. Remove the setscrew from each sprocket and turn the engine back to the T.D.C. position with the No. 6 firing and remove the remaining screws. Tap the camshaft sprockets off the flanges of the camshafts.

Accurately position the camshafts with the valve timing gauge, and check that the T.D.C. marks are in exact alignment.

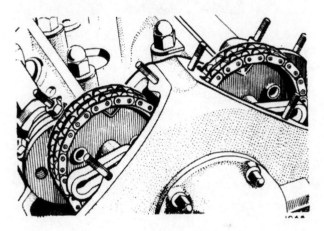

Fig. 59. Showing the camshaft sprockets disconnected from the camshafts.

Withdraw the circlips retaining the adjusting plates to the camshaft sprockets and press the adjusting plates forward until the serrations disengage. Replace the sprockets on the flanges of camshafts and align the two holes in the adjuster plate with the two tapped holes in each camshaft flange. Engage the serrations of the adjuster plates with the serrations in the sprockets.

Note: It is most important that the holes are in exact alignment, otherwise when the setscrews are fitted, the camshafts will be moved out of position. If difficulty is experienced in aligning the holes exactly the adjuster plates should be turned through 180°, which due to the construction of the plate will facilitate alignment.

Page B.63

ENGINE

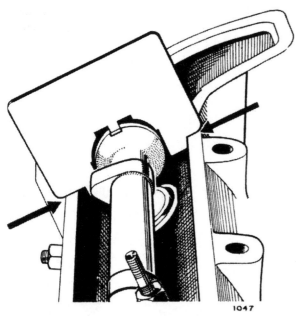

Fig. 60. The valve timing gauge in position. Ensure that the gauge is seated at the points indicated by the arrows.

Fit the circlips to the sprockets and one setscrew to the accessible holes in each adjuster plate. Turn the engine until the other two holes are accessible and fit the two remaining setscrews.

Finally, recheck the timing chain tension and valve timing in this order. Secure the four setscrews for camshaft sprockets with new locking wire.

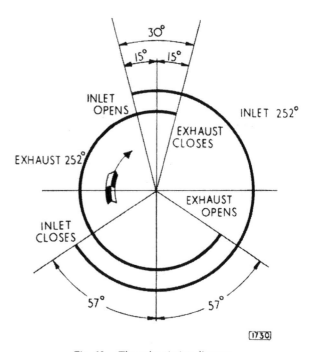

Fig. 62. The valve timing diagram.

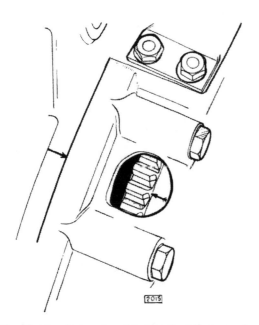

Fig. 61. Showing the location of the Top Dead Centre marks on the left hand side of the combined engine and transmission unit.

Page B.64

ENGINE

ENGINE MOUNTINGS

The engine is supported at the front on two rubber mountings which are attached to brackets on the front subframe. The rear is supported on two rubber mountings between the gearbox rear cover and a mounting plate attached to the body underframe.

FRONT ENGINE MOUNTINGS
Removal

Either place a sling around the front of the engine or attach a lifting plate to the cylinder head, as described in "Engine Removal" Page B.19. Unscrew the large set bolt and remove the spring washer, plain washer and bolt securing the front engine mounting bracket to the mounting rubber. Repeat for the other side.

Raise the engine so that the front mounting brackets are just clear of the mounting rubbers.

Remove the two bolts and self-locking nuts securing the front engine mounting to the support bracket on the front subframe. Repeat for the other side.

Refitting

Refitting is the reverse of the removal procedure.

REAR ENGINE MOUNTINGS
Removal

Remove the eight nuts and spring washers at the exhaust manifold flanges and the bolts from the five body mountings and withdraw the exhaust system from below.

Remove the small asbestos heat shield attached to the rear engine mounting plate.

Support the engine either by slinging or on a lifting plate as described in "Engine Removal" page B.19.

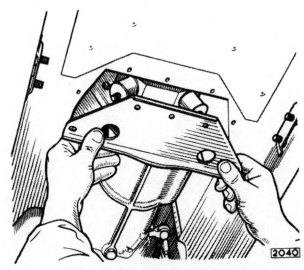

Fig. 63. The engine rear mounting plate.

Remove the self locking nuts securing the lower ends of the rear engine mountings to the mounting plate through the holes in the plate.

Remove the five bolts securing the mounting plate to the body and withdraw the plate.

Remove the propeller shaft tunnel cover and the gearbox cowl as described in "Engine Removal" page B.19.

Remove the self locking nuts securing the top ends of the engine mountings to the gearbox rear cover.

Withdraw the rear engine mountings from below.

Refitting

Refitting is the reverse of the removal procedure. New exhaust manifold sealing rings should be fitted.

AIR CLEANER

The air cleaner is of the paper element type mounted on the right hand side of the engine compartment and connected to the carburetters by means of an elbow trumpet plate. Servicing instructions are given in "Routine Maintenance" on page B.17.

REMOVAL

Unscrew the two butterfly nuts at the carburetter trumpet plate. Remove the air cleaner elbow. Unfasten the three spring clips and withdraw the air cleaner element assembly. Remove the serrated nut from the base of the assembly and withdraw the paper element.

Remove the nut and shakeproof washer at the base of the air cleaner canister. Remove the two setscrews at the side of the canister and withdraw the canister.

REFITTING

Refitting is the reverse of the removal procedure.

Page B.65

ENGINE

THE ENGINE STABILISER

The engine stabiliser is situated at the rear of the engine and consists of a rubber/steel mounting attached to the body which is connected to brackets on the clutch housing via a rubber bushed link pin. The link pin is threaded at its upper end and is connected to the rubber mounting by means of flanged washers and a self-locking nut.

ADJUSTMENT

It is MOST IMPORTANT that the stabiliser is assembled in the following manner, as failure to observe this procedure may cause engine vibration and/or fouling of the gearbox in its cowl, due to the engine having been pulled up on its mountings.

(a) Screw the lower flanged washer (D, Fig. 63) up the stabiliser pin until the flange contacts the bottom of the stabiliser rubber mounting (C). The washer is slotted on its upper face and can be screwed up the pin by engaging a thin bladed screwdriver in the slot through the centre hole of the rubber mounting.

(b) Fit the upper flanged washer (B) and tighten down with the self-locking nut (A).

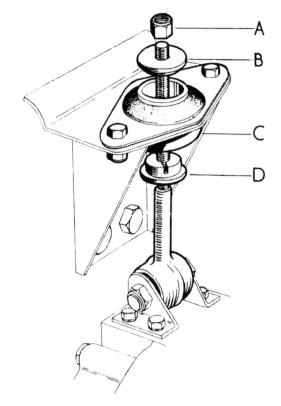

Fig. 64. The engine stabiliser.

TORSION BAR REACTION TIE PLATE

REMOVAL

The following instructions should be carried out either on a ramp or over a pit.

Remove the eight nuts and spring washers at the exhaust manifold flanges, remove the bolts from the five body mountings and withdraw the exhaust system from below discarding the manifold sealing rings.

Jack up the front of the car using a block of hard wood 16″ × 1⅛″ × 1″ (406·4 × 28·6 × 25·4 mm.) under the subframe lower cross tube as shown in (Section J "Front Suspension") until the front wheels are clear of the ground. Do NOT jack up the car unless the block of wood is in place.

Remove the lower bolt and self locking nut from the torsion bar reaction bracket and drive the locating bar (see Fig. 65) through the bracket from the front in

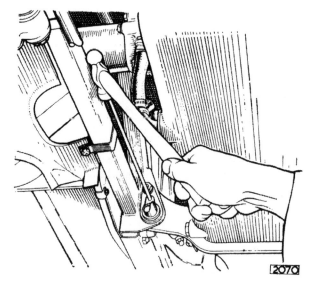

Fig. 65. Driving in the locating bars. (The torsion bar has been cut away for illustrative purposes).

Page B.66

ENGINE

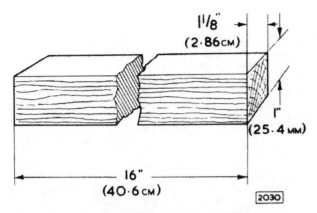

Fig. 66. The jacking block dimensions.

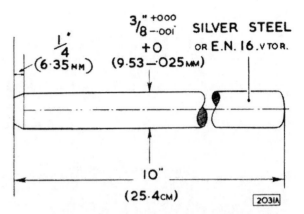

Fig. 67. The tie plate locating bar dimensions.

place of the bolt so that approximately ¼" (6·35 mm.) protrudes.

Repeat for the lower bolt on the other side and drive the second locating bar into its place.

Remove the self locking nuts from both top bolts and tap the bolts back until they are flush with the tie plate.

Remove the bolt and self locking nuts securing the tie plate to the body under frame on each side.

Withdraw the tie plate over the bolts and locating bars and remove from below ensuring that the locating bars do not lose their position.

REFITTING

Refitting is the reverse of the removal procedure, new exhaust manifold sealing rings should be fitted.

Note: If the locating bars are accidently displaced, and the torsion bar setting is lost, the torsion height will have to be reset as described in Section J "Front Suspension".

Page B.67

SECTION C

CARBURETTERS AND FUEL SYSTEM
INDEX

CARBURETTERS

	Page
Description	
Throttle spindle glands	C.3
Idling	C.3
Manual mixture control	C.4
Data	C.4
Routine Maintenance	
Lubricate carburetter piston damper	C.6
Lubrication of throttle linkage	C.6
Checking carburetter slow running	C.6
Cleaning carburetter filters	C.6
Fuel feed line filter	C.7
Fuel tank filter	C.7
Carburetters	
Removal	C.7
Refitting	C.8
Cleaning the suction chamber and piston	C.8
Carburetter tuning	C.8
Fast idle setting	C.9
Float chamber fuel level	C.10
Centring the jet	C.10
Setting the carburetter mixture control warning light switch	C.11

THE FUEL SYSTEM

	Page
The Fuel Pump	
Description	C.11
Removal	C.11
Refitting	C.12
Servicing instructions	C.13
Fuel Tank	
Removal	C.13
Refitting	C.14
Fuel Tank Gauge Unit	
Removal	C.15
Refitting	C.15

Page C.1/2

CARBURETTERS

DESCRIPTION

The "E" Type is fitted with triple S.U. HD.8 type carburetters. A manual mixture control is provided which operates on all three carburetters.

A reminder that the starting device is in operation is provided by a red warning light adjacent to the mixture control slide. When the control is returned to the "Run" position the warning light is extinguished.

The HD type carburetter differs from the earlier type in that the jet glands are replaced by a flexible diaphragm, and idling mixture is conducted along a passage way, in which is located a metering screw, instead of being controlled by a throttle disc.

The jet (18) (Fig. 1), which is fed through its lower end, is attached to a synthetic rubber diaphragm (10) by means of the jet cup (9) and jet return spring cup (13), the centre of the diaphragm being compressed between these two parts; at its outer edge it is held between the diaphragm casing (14) and the float chamber arm. The jet is controlled by the jet return spring (12) and the jet actuating lever (15), the latter having an external adjusting screw which limits the upward travel of the jet and thus controls the mixture adjustment; screwing it in (clockwise) enriches the mixture, and unscrewing it weakens the mixture.

Throttle Spindle Glands

Provision is made for the use of throttle spindle glands consisting of the cork gland itself (25) (Fig. 1), a dished retaining washer (28), a spring (27) and a shroud (26). This assembly should not require servicing and can only be removed by dismantling the throttle spindle and disc.

Idling

The carburetter idles on the main jet and the mixture is conducted along the passage way (8) (Fig. 1) connecting the choke space to the other side of the throttle disc.

The quantity of the mixture passing through the passage way and, therefore, the idling speed of the engine, is controlled by the "slow-run" valve (5), the quality or relative richness of the mixture being determined by the jet adjusting screw. It follows that when idling, the throttle remains completely closed against the bore of the carburetter.

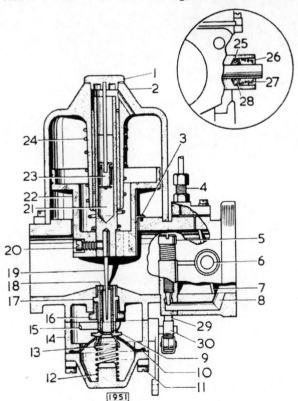

1. Damper cap
2. Suction chamber
3. Piston guide
4. Union for vacuum advance/retard
5. Slow running volume screw
6. Throttle spindle
7. Throttle butterfly
8. Slow run passage
9. Jet cup
10. Diaphragm
11. Float chamber securing screw
12. Jet return spring
13. Return spring cup
14. Diaphragm casing
15. Actuating lever
16. Nut—jet bearing
17. Jet bearing
18. Jet
19. Needle
20. Needle retaining screw
21. Oil reservoir
22. Piston
23. Damper
24. Piston return spring
25. Throttle spring gland
26. Shroud for spring
27. Spring
28. Washer
29. Push rod
30. Cam

Fig. 1. Sectioned view of H.D.8 carburetter

Page C.3

CARBURETTER

Manual Mixture Control

The manual mixture control mechanism is arranged to operate in two stages. The first stage provides a certain amount of throttle opening and the second stage richens the mixture by moving the jet away from the jet needle.

The first stage operates when the mixture control cable moves the jet lever (A, Fig. 2) towards the rear of the engine. This operates the cam (B) and pulls the brass fast idle push rod (C) down. The fast idle screw (D) then presses down on the fast idle lever (E) opening the throttle slightly.

The second stage brings the jet lever back further, opening the throttle further and bringing the lever (F) into contact with mixture adjusting screw arm (G), moving it upwards and thus moving the jet and diaphragm downwards away from the jet needle.

Should it be necessary to disconnect the mixture control rods (H) they should be reconnected when all three jet levers are against their stop, the rod fork ends being adjusted to line up with the clevis pin holes in the jet levers.

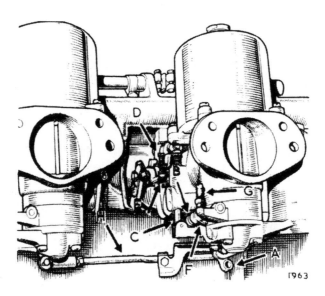

Fig. 2. Carburetter controls

DATA

Type	S.U. HD 8 (triple)
Size	2″ (5·08 cm.)
Jet needle type:	
8 to 1 compression ratio	UM (With Standard Air Cleaner)
9 to 1 compression ratio	UM (With Standard Air Cleaner)
Jet size	0·125″ (3·17 mm.)

Note: The jet needle type is stamped on the side or top face of the parallel portion of the needle.

Page C.4

CARBURETTER

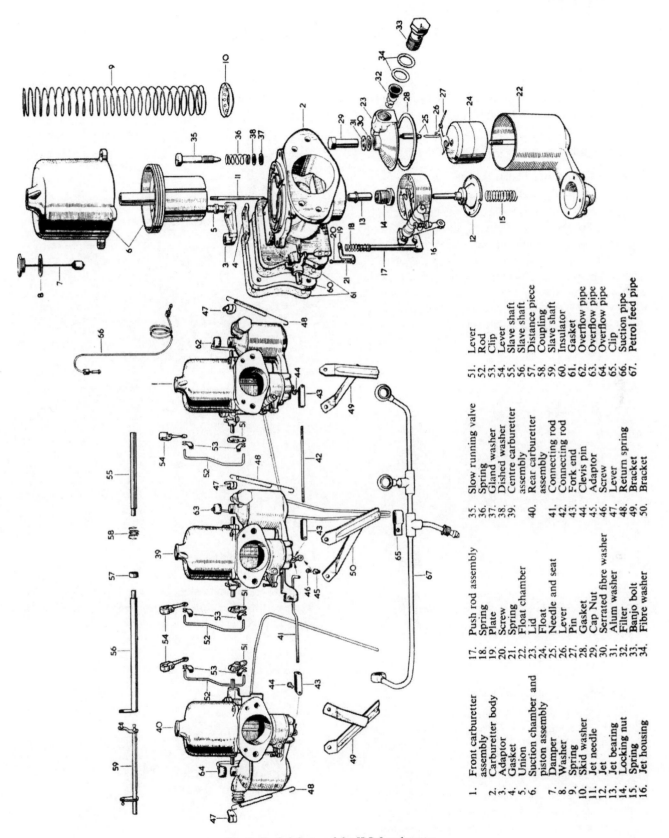

Fig. 3. Exploded view of the H.D.8 carburetter

1. Front carburetter assembly
2. Carburetter body
3. Adaptor
4. Gasket
5. Union
6. Suction chamber and piston assembly
7. Damper
8. Washer
9. Spring
10. Skid washer
11. Jet needle
12. Jet
13. Jet bearing
14. Locking nut
15. Spring
16. Jet housing
17. Push rod assembly
18. Spring
19. Plate
20. Screw
21. Spring
22. Float chamber
23. Lid
24. Float
25. Needle and seat
26. Lever
27. Pin
28. Gasket
29. Cap Nut
30. Serrated fibre washer
31. Alum washer
32. Filter
33. Banjo bolt
34. Fibre washer
35. Slow running valve
36. Spring
37. Gland washer
38. Dished washer
39. Centre carburetter assembly
40. Rear carburetter assembly
41. Connecting rod
42. Connecting rod
43. Fork end
44. Clevis pin
45. Adaptor
46. Screw
47. Lever
48. Return spring
49. Bracket
50. Bracket
51. Lever
52. Rod
53. Clip
54. Lever
55. Slave shaft
56. Slave shaft
57. Distance piece
58. Coupling
59. Slave shaft
60. Insulator
61. Gasket
62. Overflow pipe
63. Overflow pipe
64. Overflow pipe
65. Clip
66. Suction pipe
67. Petrol feed pipe

Page C.5

CARBURETTER

ROUTINE MAINTENANCE

Warning: If it is desired to clean out the float chamber, do not use compressed air as this may cause rupture of the rubber jet diaphragm.

EVERY 2,500 MILES (4,000 KM).

Lubricate Carburetter Piston Damper

Each carburetter is fitted with a hydraulic piston damper which unless periodically replenished with oil, will cause poor acceleration and spitting back through the carburetter on rapid opening of the throttle.

To replenish with oil, unscrew the cap on top of the suction chambers, and lift out the damper valve which is attached to the cap. Fill the hollow piston spindle, which can be seen down inside the bore of the suction chamber, with S.A.E.20 engine oil.

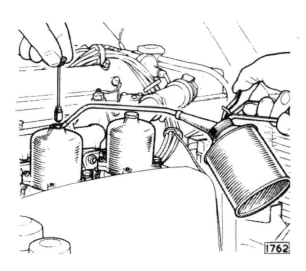

Fig. 4. Topping up a hydraulic piston damper

Lubrication of Throttle Linkage

All moving parts of the throttle linkage should be lubricated with engine oil, especially the brass fast idle push rod. (See Fig. 5).

Checking Carburetter Slow Running

The idling speed of the engine should be 500 r.p.m. when the engine is at its normal working temperature.

If adjustment is required turn the three slow running volume screws (see Fig. 9) **by exactly equal amounts** until the idling speed, observed on the revolution counter instrument, is correct.

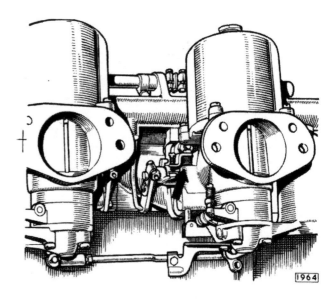

Fig. 5. The fast idle push rod

EVERY 5,000 MILES (8,000 KM.)

Cleaning Carburetter Filters

Removal of the bolt securing the petrol pipe banjo union to each float chamber will expose the filters. Remove the filter and clean in petrol; do not use a cloth as particles will stick to the gauze.

When refitting, insert the filter with the spring first and ensure that the fibre washers are replaced, one to each side of the banjo union.

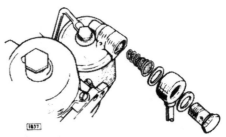

Fig. 6. The carburetter float chamber filter

Page C.6

CARBURETTER

Fuel Feed Line Filter

The filter is attached to the bulkhead (right-hand side) and is of the glass bowl type with a flat filter gauze.

At the recommended intervals, or more frequently if the glass bowl shows signs of becoming full of sediment, slacken the locking nut, swing the retaining clip to one side and remove the bowl, sealing washer and filter gauze.

Clean the filter gauze and bowl by washing in fuel. Examine the sealing washer and if necessary fit a new one.

EVERY 10,000 MILES (16,000 KM.)

Fuel Tank Filter

The filter is incorporated in the fuel tank drain plug on the underside of the fuel tank. (4, Fig. 17).

At the recommended intervals, drain the fuel tank by removing the drain plug and filter assembly. Wash the filter thoroughly in fuel but do NOT use compressed air and ensure that the filter is not damaged in any way.

Examine the cork washer on the drain plug and replace if necessary. Lubricate the fuel filter 'O' ring and feed the filter over the fuel inlet pipe, screw the drain plug into position.

The fuel drained from the tank should be filtered before refilling the tank to remove any sediment.

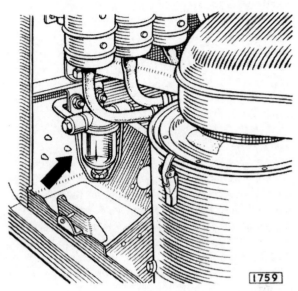

Fig. 7. The fuel feed line filter

———oo———

CARBURETTERS

Removal

Remove the two butterfly nuts at the carburetter trumpets and remove the air cleaner elbow. Remove the carburetter trumpet plate from the carburetters by removing the six nuts and spring washers and three gaskets. Remove the three banjo union bolts and six fibre washers from the float chambers. Ensure that the three float chamber filters are not mislaid. Remove the float chamber drain pipe clip from the oil filter. Remove the three butterfly return springs. Disconnect the three throttle links from the clips on the throttle spindle levers. Disconnect the mixture control outer and inner cables. Remove the vacuum advance pipe from the front carburetter. Remove four nuts and

Fig. 8. Refitting the mixture control rods with the jet levers against the stops

Page C.7

CARBURETTER

spring washers and return spring bracket from each carburetter and remove the three carburetters together. If necessary remove the mixture control linkage from each carburetter by discarding the split pins and withdrawing the clevis pins.

Refitting

Refitting is the reverse of the removal procedure, but new thin gaskets should be fitted to either side of the heat insulating gasket and also to the carburetter trumpet flanges.

CLEANING THE SUCTION CHAMBER AND PISTON

This should be done at approximate intervals of every twelve months or if the carburetter is dismantled for any reason. After detaching, clean the main inside bore of the suction chamber and the two outside diameters of the piston with a rag moistened in fuel or thinners and then reassemble in a dry and clean condition with a few spots of thin oil on the piston rod only. Do NOT use metal polish to clean the suction chamber and piston.

CARBURETTER TUNING

Before tuning the carburetters, the sparking plug gaps and contact breaker gaps should be checked and adjusted if necessary. The distributor centrifugal advance mechanism and vacuum advance operation should be checked and ignition timing set to the figure given in Section B "Engine", with the centrifugal advance mechanisms in the static position. For final road test, adjustment of not more than six clicks of the micrometer adjustment at the distributor to either advance or retard is permitted. The ignition setting is important since if retarded or advanced too far the setting of the carburetters will be affected. As the needle size is determined during engine development, tuning of the carburetters is confined to the correct idling setting.

If after tuning the carburetters, the idling setting and engine performance is not satisfactory, it will be necessary to check the cylinder compressions and the valve clearances.

Tuning

The air intake should be removed and the engine run until it has attained its normal operating temperature. Release the three pinch bolts securing the two piece throttle levers to the carburetter throttle spindles.

Taking one carburetter at a time close each throttle butterfly valve fully by rotating the throttle spindle in a clockwise direction looking from the

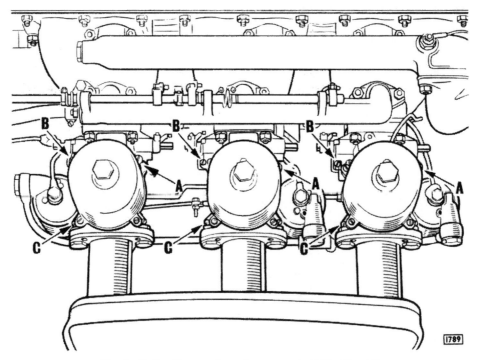

Fig. 9. "A"—Slow running volume screws. "B"—Fast idle screws. "C"—Mixture adjusting screws

Page C.8

CARBURETTER

front; with the throttle held closed tighten the pinch bolt keeping the two piece throttle lever in the midway position. Repeat for the other two carburetters, then operate the accelerator linkage and observe if all the throttles are opening simultaneously by noting the movement of the full throttle stops of the left-hand side of the throttle spindles.

Screw down the slow running volume screws (A, Fig. 9) on to their seatings and then unscrew two full turns. Remove the piston and suction chambers; disconnect the mixture control linkage by removing the clevis pins from the connecting rod fork ends underneath the front and rear carburetters. Unscrew the mixture adjusting screws (C) until each jet is flush with the bridge of its carburetter. Replace the pistons and suction chambers and check that each piston falls freely on to the bridge of its carburetter (by means of the piston lifting pin). Turn down the mixture adjusting screws 2½ turns.

Restart the engine and adjust to the desired idling speed of 500 r.p.m. by moving each slow running screw an equal amount. By listening to the hiss in the intakes, adjust the slow running screws until the intensity of the hiss is similar on all intakes. This will synchronise the mixture flow of the three carburetters.

When this is satisfactory the mixture should be adjusted by screwing all the mixture adjusting screws up (weaker) or down (richer) by the same amount until the fastest idling speed is obtained consistent with even firing.

As the mixture is adjusted, the engine will probably run faster and it may therefore be necessary to screw down the slow running volume screws in order to reduce the speed.

Now check the mixture strength by lifting the piston of the front carburetter by approximately $\frac{1}{32}''$ (·8 mm) when, if:
(a) the engine speed increases and **continues to run faster,** this indicates that the mixture is too rich.
(b) the engine speed immediately decreases, this indicates that the mixture is too weak.
(c) the engine speed **momentarily** increases very slightly, this indicates that the mixture is correct.

Repeat the operation at the remaining two carburetters and after adjustment recheck the front carburetter since the carburetters are interdependent.

When the mixture is correct, the exhaust note should be regular and even. If it is irregular, with a splashy type of misfire and colourless exhaust, the mixture is too weak. If there is a regular or rythmical type of misfire in the exhaust beat together with a blackish exhaust, then the mixture is too rich.

When reconnecting the mixture control cable allow $\frac{1}{16}''$ (1·5 mm.) free travel at the bottom of the facia control before the jet levers begin to move.

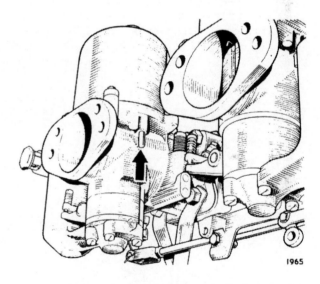

Fig. 11. *The carburetter piston lifting pin; the first part of the movement is spring loaded free travel*

Fast Idle Setting

Set the mixture control knob on the facia panel to the highest position in the slide immediately short of the position where the mixture adjusting screw levers (C Fig. 9) begin to move. This will be approaching the mid-travel position of the control knob and approximates to $\frac{5}{8}''$ (16 mm.) movement at the bottom of the

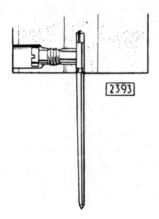

Fig. 10. *Positioning the jet needle with the lower edge of the groove flush with the base of the piston*

Page C.9

CARBURETTER

jet levers. Adjust the fast idle screws (B) on the throttle stops to give an engine speed of about 1,000 r.p.m. (when hot). This operation is best carried out by lightly nipping a ·002" (·051 mm.) feeler gauge under each screw when the mixture control knob is at the bottom of the slide.

Float Chamber Fuel Level

When the fuel level setting is correct a $\frac{7}{16}$" (11·1 mm.) test bar will just slide between the lid face and the inside curve of the float lever fork when the needle valve is in the "shut-off" position (see Fig. 12).

If the float lever fails to conform with this check figure, it must be carefully bent at the start of the fork section, in the necessary direction, for correction. Take care to keep both prongs of the fork level with each other and maintain the straight portion of the lever dead flat.

It is not advisable to alter the fuel level unless there is trouble with flooding; although too high a level can cause slow flooding, particularly when a car is left ticking over on a steep drive, it should be remembered that flooding can also be caused by grit in the fuel jamming open the needle valve, undue friction in the float gear, excessive engine vibration, or a porous float.

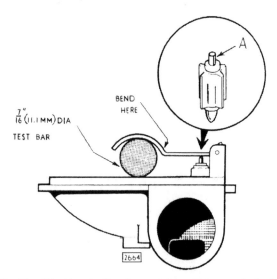

Fig. 12. Checking the float lever setting, which controls the fuel level in the float chamber. When setting the fuel level ensure that the spring loaded plunger (A) in the "Delrin" needle is not compressed

CENTRING THE JET

Warning: Take care not to bend the carburetter needle when carrying out this operation.

Remove the carburetter from the engine as described on page C.7.

Remove the four setscrews securing the float chamber to the carburetter body. Remove the float chamber, jet housing and jet. Remove the hydraulic damper.

With a ring spanner slacken the jet locking nut approximately half a turn. Replace the jet and diaphragm assembly.

The jet is correctly centred when the piston falls freely and hits the jet "bridge" with a metallic click. To centre the jet, push the jet and diaphragm assembly as high as possible with the hand and with a pencil or rod gently press the piston down on to the jet bridge; centralisation will be facilitated if the side of the carburetter body is tapped lightly. Tighten the jet locking nut.

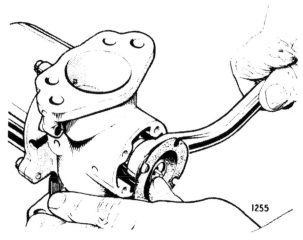

Fig. 13. Centring the jet

The actual centring must be carried out with the setscrew holes in the jet diaphragm and carburetter in alignment. After tightening the jet locking nut, the jet diaphragm must be kept in the same position relative to the carburetter body; the simplest way to do this is to mark one of the corresponding jet diaphragm and carburetter body setscrews holes with a soft pencil. Failure to do this may cause the centralisation to be upset.

Check that the centralisation is correct by noting if there is any difference in the sound of the piston hitting the jet bridge with the jet in its highest and lowest positions. If there is any difference in the sound, the procedure for centralising the jet will have to be repeated.

If difficulty in centring the jet is encountered after carrying out the above procedure, the jet needle can be

FUEL SYSTEM

lowered slightly in the piston to make the centralising effect more positive. The needle must, however, be restored to the normal position when checking the centralisation.

SETTING THE CARBURETTER MIXTURE CONTROL WARNING LIGHT SWITCH

Remove the dash casing below the control slide by withdrawing the drive screws and on Right hand drive models the screwed bezels of the odometer and clock setting drives. Set the lever on the control slide ¼ (6·350 mm.) from the bottom limit of its travel, when a click will be heard and, utilizing the two nuts on the threaded shank of the switch, position the switch so that the warning light ceases to glow when the ignition is switched "on". Actuate the lever up and down once or twice and make any final adjustments necessary. Replace the components by reversing the removal procedure.

THE FUEL SYSTEM

FUEL PUMP

The Lucas 2FP fuel pump is a complete unit, consisting of a cumulative type centrifugal pump driven by a permanent field electric motor. The unit is fully sealed and is mounted inside the fuel tank.

Fuel is delivered to the carburetters at a pressure approximately 2 lbs per sq. in. (0·14 kg/cm^2) when the pump is running.

Electrically the pump is under the control of the ignition switch and will commence to operate when the ignition is switched "ON".

A 5 amp fuse, located behind the instrument panel in the fuse pack, is incorporated in the electrical circuit as a safety measure in the event of a fault developing in the pump or connections and it is essential if a fuse blows to replace it with one of the same value. Under no circumstances should a higher rated fuse be fitted.

Removal

Disconnect the battery positive terminal. Raise the boot lid and remove the carpet from the boot floor. Unscrew the setscrews retaining the boot floor and remove the two floor panels.

Remove the cover from the cable connector block located in the spare wheel compartment. Withdraw the connectors and disconnect the cables, noting that like colours are connected.

Remove the delivery pipe union from the pump.

Drain the tank by removing the drain plug and filter assembly.

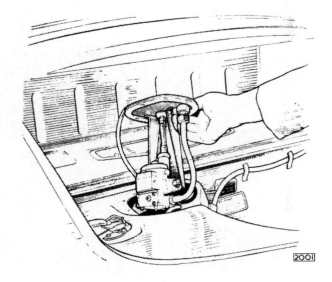

Fig. 14. Removing the fuel pump from the tank

Remove the eight setscrews securing the fuel pump carrier plate to the tank and lift out the fuel pump and the carrier plate, taking care not to damage the joint.

With the pump removed from the tank, disconnect the fuel delivery pipe union from the fuel pump, remove the nut securing the braided cable conduit to the carrier plate, remove the two bolts securing the pump to the carrier plate and withdraw the fuel pump.

Page C.11

FUEL SYSTEM

Refitting

Refitting is the reverse of the removal procedure, but care must be taken when refitting the pump to the tank to ensure that the sealing joint is in a good condition. Renew if damaged. Failure to ensure this will result in an escape of petrol fumes into the car and a petrol leak when the fuel tank is full.

Note: A star washer is provided on one of the petrol-proof grommets on the mounting feet. This washer provides an earthing path from the pump to the mounting bracket via the fixing bolt, so preventing the build-up of electrostatic charges on the pump unit. It is extremely important that the earthing washer is in position when a replacement pump is fitted.

If the inlet pipe has been removed or loosened at the fuel pump, ensure that it is central in the filter sump before finally fitting the pump to the fuel tank.

When the fuel pump is in position, refit the filter and drain plug assembly, by lubricating the filter 'O' ring and feeding the filter onto the fuel inlet pipe.

Ensure that the drain plug cork washer is in good condition and renew if necessary.

A. Cable terminals
B. Armature
C. Gauze flame trap
D. Relief valve
E. Impeller
F. Anti-static earthing washer
G. Commutator brushes

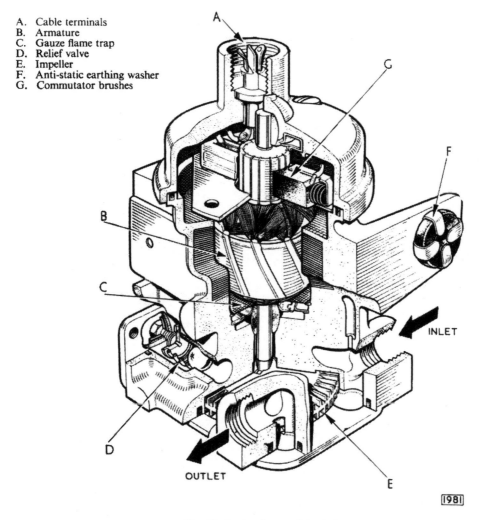

Fig. 15. Sectioned view of the fuel pump

Page C.12

FUEL SYSTEM

SERVICING INSTRUCTIONS

Complaint—Fuel Starvation

(i) Check the level of the fuel in the tank. Replenish if necessary.

(ii) Check the fuse, located behind the instrument panel. If after replacement fuse blows again, check for a short circuit in (a) feed cable or (b) pump unit.
Replace unit or repair cable as required.

(iii) If fuse has not blown, locate cable connectors contained in rubber anti-flash block located in the spare wheel compartment, by which the pump is connected to the battery supply.
Check voltage and current available at the terminal ends with the ignition switched "ON" by using a first grade voltmeter and an ammeter.
The voltage should be 12 volts and the current should not exceed 1·8 amperes.

(iv) If no voltage is shown, check that the fault is not due to a broken or an intermittent connection in the switch, feed or earth. Repair as necessary.

(v) If no current or an excessive current measurement is shown, this will be indicative that the pump is faulty.
Fit replacement pump unit.

Complaint—Fuel Flooding

First check that the needle valves in the carburetters are clean and unworn. If these are satisfactory, check the delivery pressure by connecting a pressure gauge to the fuel line at the carburetter end.

With a voltage of 12 volts (i.e., with the ignition switched on) applied to the pump this pressure should be 2—2½ lbs/sq. in. (·14—·17 kg/cm^2).

If it is higher than 2½ lbs/sq. in (·17 kg/cm^2) the setting of the relief valve (a screw and locknut on the pump cover plate) should be adjusted to reduce the pressure to 2 lbs/sq. in. (·14 kg/cm^2).

It will be necessary to remove the fuel pump from the tank, as outlined on page 11, to carry out this adjustment and it will be more convenient to complete the operation on the test bench with the pump submerged in a receptacle containing sufficient clean paraffin (kerosene) and the pump cables connected to a fully charged battery. To reduce the pressure turn the setting screw in an anti-clockwise direction. It is important to re-tighten the locknut fully after adjustment. When refitting the fuel pump to the tank after adjustment, examine the sealing joint and renew if damaged.

Note: When testing the fuel pump for fuel pressure, the **black** cable on the pump must always be connected to the **positive** battery terminal.

Warning

When bench testing the fuel pump, EXTINGUISH all naked lights or flames in the vicinity and do **not** allow the cables to spark when making connections. To obviate this connect a switch in the test cable circuit and switch "OFF" when connecting the pump to the battery.

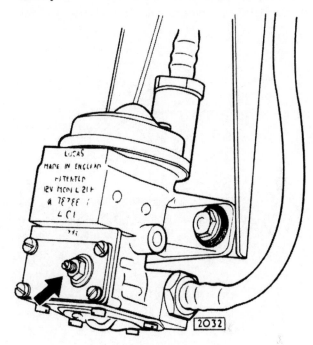

Fig. 16. Fuel pump relief valve adjusting screw

FUEL TANK

Removal

Disconnect the battery positive terminal. Remove the luggage compartment floor covering and the two floor panels (6 setscrews).

Drain the tank by removing the drain plug and filter assembly (4, Fig. 17).

Remove the filter sump from the tank.

Remove the fuel pump cables from the terminal block, noting that like colours are connected and remove the terminal block. Remove the fuel pipe banjo bolt (27) and fibre washers (28) from the top of the fuel pump mounting bracket (17). Remove the fuel gauge wires. Slacken the filler pipe clips (11) push the rubber pipe (10) up the filler neck.

Slacken the breather clip (8) and remove the

Page C.13

FUEL SYSTEM

breather pipe (7) from the filler neck. Remove the four nuts securing the boot lock and remove the boot lock. Slacken the clips and remove the boot channel drain tube. Remove the four nuts and the body strengthening plate from the back of the tank mounting bracket (22) inside the rear suspension aperture. Remove the three pointed end mounting bolts with the flat and spring washers, two from the front of the tank and one from the rear. Remove the tank mounting bracket. Withdraw the fuel tank.

Refitting

Refitting is the reverse of the removal procedure but ensure that the rubber seal is in place between the filter sump (2) and the body. Replace the fuel gauge cables with the white and green cable to the terminal marked "W" and the green and black to the terminal marked "T". The black cable should be fitted to the earth terminal on the element housing.

Note: To ensure that the cable connectors are correctly attached to the blade terminals on the fuel gauge tank unit, slide back the insulating sleeve from the cable connector to expose the terminal end. Push home fully onto the blade and slide the insulating sleeve forward to cover the joint.

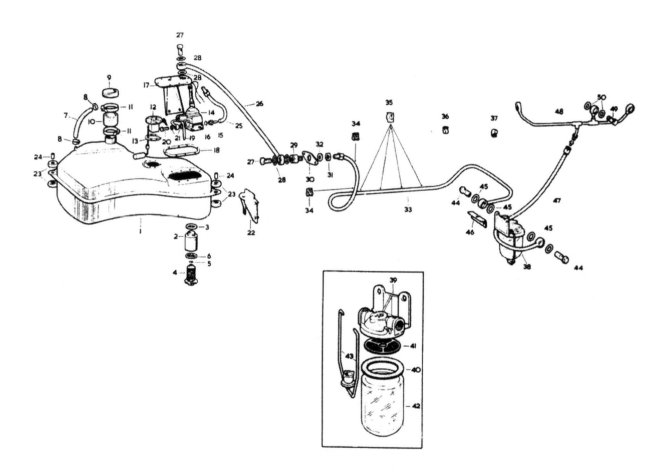

Fig. 17. The fuel system

1. Fuel tank	11. Clip	21. Washer	31. Nut	41. Filter gauze
2. Sump assembly	12. Tank element	22. Mounting bracket	32. Brass washer	42. Bowl
3. Washer	13. Gasket	23. Rubber pad	33. Pipe	43. Retaining strap
4. Filter and drain plug assembly	14. Fuel pump	24. Distance piece	34. Clip	44. Banjo bolt
5. 'O' ring	15. Union	25. Pipe	35. Clip	45. Fibre washer
6. Washer	16. Fibre washer	26. Pipe	36. Clip	46. Mounting bracket
7. Hose	17. Mounting bracket	27. Banjo bolt	37. Clip	47. Pipe
8. Clip	18. Gasket	28. Fibre washer	38. Filter assembly	48. Feed pipe
9. Filler cap	19. Pipe	29. Connector	39. Filter casting	49. Banjo bolt
10. Hose	20. Banjo bolt	30. Mounting plate	40. Sealing washer	50. Fibre washer

Page C.14

FUEL SYSTEM

FUEL TANK GAUGE UNIT

Removal

Disconnect the battery positive terminal.

Remove the luggage compartment floor covering and the two floor panels, six setscrews, to expose the fuel tank gauge unit. Disconnect the three cables. Remove the six setscrews and twelve copper washers attaching the unit to the fuel tank. The seal can be broken by a sharp tap on one side of the unit. Withdraw the unit, taking care not to damage the float arm.

Refitting

The existing gasket should be scraped away from the boss on the fuel tank, taking care that none falls into the tank. Apply a suitable sealing compound to both sides of the new gasket, which should be positioned on the fuel tank boss with the holes in line. Insert the element into the tank so that the float is towards the rear of the car. Replace the six screws and twelve washers and tighten securely. Attach the white and green cable to the terminal marked "W", and the green and black cable to the terminal marked "T".

Note: To ensure that the cable connectors are correctly attached to the blade terminals on the fuel tank unit, slide back the insulating sleeve from the cable connector to expose terminal end. Push the connector home fully on to the blade and slide the insulating sleeve forward to cover the joint.

Attach the earth wire connector to the terminal at the element housing. Refit the boot floor panels and the floor covering. Reconnect the battery positive terminal.

SUPPLEMENTARY INFORMATION TO SECTION C "CARBURETTERS AND FUEL SYSTEM"

"Delrin" Float Chamber Needle

Commencing at Engine Number RA.2464 the carburetters are fitted with "Delrin" float chamber needles. This needle has a body of white plastic material and incorporates a spring-loaded pin to overcome needle "flutter" due to engine rock when idling, causing slow flooding with consequent rough slow running or stalling.

In conjunction with the introduction of this new type of needle, the seat, float lever fork and float chamber lid are also modified.

The new type of float chamber lid assembly is interchangeable with the previous type as a complete assembly. The new type needle and seat can be used to replace the previous type provided the original lever fork is retained. These float lever forks are NOT interchangeable and must be kept to their respective lids. The old type needle and seat must not be fitted to the new type of float chamber lid which can be identified by the embossed AUD 2283 or 2284 on the inside of the lid.

Improved Fuel Pump

	Commencing Chassis Numbers	
	R.H. Drive	L.H. Drive
"E" Type Open 2 Seater	850786	880619
Fixed Head Coupe	861386	889510

An improved fuel pump is introduced at the above chassis numbers which delivers petrol at a pressure of 3–3½ lb./sq. in. (·2–·25 kg/sq. cm.).

This pump is only interchangeable with its predecessor if "Delrin" float chamber needles and seats are fitted.

Page C.15

SECTION D

COOLING SYSTEM
3·8 "E" TYPE
GRAND TOURING MODELS

INDEX

	Page
Data	D.4
Routine Maintenance	
Checking radiator water level	D.4
Care of the cooling system	D.4
Frost Precautions	
Anti-freeze	D.5
Engine heater	D.5
Radiator	D.5
Removal	D.6
Refitting	D.6
Radiator Cowl	
Removal	D.6
Refitting	D.6
Radiator Header Tank	
Removal	D.7
Refitting	D.7

Page D.1/2

INDEX (continued)

	Page
Fan Motor	
Removal	D.7
Refitting	D.7
Fan Motor Relay	
Removal	D.7
Refitting	D.8
Water Pump Belt	
Adjustment	D.8
Removal	D.8
Refitting	D.8
Thermostat	
Removal	D.8
Checking	D.8
Refitting	D.8
Fan Thermostatic Switch	
Removal	D.9
Refitting	D.10
Water Pump	
Removal	D.10
Dismantling	D.10
Checking	D.11
Re-assembly	D.11
Refitting	D.12
Water Temperature Gauge	D.12

COOLING SYSTEM

Water circulation is assisted by an impeller type pump mounted on the front cover of the engine, the system being pressurised and thermostatically controlled. Water is circulated from the right hand side of the cross-flow radiator by the water pump and flows through the cylinder block and cylinder head water passages to the separate radiator header tank via the inlet manifold water jacket and is then returned to the radiator. A fan, driven by an electric motor is thermostatically controlled by a switch mounted in the radiator header tank. The fan operates when the temperature of the engine coolant rises above approximately 80°C.; when the temperature of the coolant falls to 72°C. the fan electric motor automatically cuts out.

DATA

	Imperial pints	U.S. pints	Litres
Total capacity—including heater	32	38½	18.18
Coolant pump—type		Centrifugal	
—drive		Belt	
Coolant pump belt—angle of 'V'		36°	
Coolant pump to engine speed ratio		0.9 : 1	
Cooling system control		Thermostat	
Thermostat Data		See page D9	
Radiator type		Cross flow—10 fins/inch (4 fins/cm.)	

Radiator cap:
    Make and type .. .. A.C.—relief valve
    Release pressure .. .. 4 lbs. per sq. in. (0.28 kg./cm.²)
    Release depression .. .. ½ lb. (0.23 kg.)

ROUTINE MAINTENANCE

DAILY

Checking Radiator Water Level

Every day, check the level of the water in the radiator and, if necessary, top up to the bottom of the filler neck.

Use water that is as soft as is procurable; hard water produces scale which in time will affect the cooling efficiency of the system.

PERIODICALLY

Care of the Cooling System

The entire cooling system should occasionally be flushed out to remove sediment. To do this, open the radiator block and cylinder block drain taps and insert a water hose into the radiator filler neck. Allow the water to flow through the system, with the engine running at a fast idle speed (1,000 r.p.m.) to cause circulation, until the water runs clear.

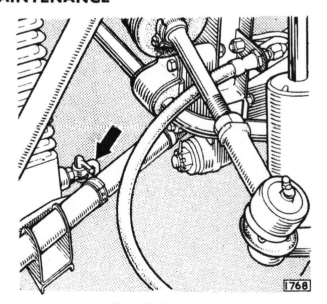

Fig. 1 Radiator drain tap

Page D.4

COOLING SYSTEM

Since deposits in the water will in time cause fouling of the surfaces of the cooling system with consequent impaired efficiency it is desirable to retard this tendency as much as possible by using water that is as nearly neutral (soft) as is available. One of the approved brands of water inhibitor may be used with advantage to obviate the creation of deposits in the system.

When **refilling the cooling system** open the heater control tap by placing the temperature control on the facia water panel in the **hot** position. Check the radiator water level after running the engine and top up if necessary.

FROST PRECAUTIONS

Anti-Freeze—Important

During the winter months it is strongly recommended that an anti-freeze compound with an inhibited Ethylene Glycol base is used in the proportions laid down by the anti-freeze manufacturers. It should be remembered that if anti-freeze is not used it is possible, owing to the action of the thermostat, for the radiator to "freeze-up" whilst the car is being driven, even though the water in the radiator was not frozen when the engine was started.

Before adding anti-freeze solution the cooling system should be cleaned by flushing.

The cylinder head gasket must be in good condition and the cylinder head nuts pulled down correctly, since if the solution leaks into the crankcase a mixture will be formed with the engine oil which is likely to cause blockage of the oil ways with consequent damage to working parts. Check the tightness of all water hose connections, water pump and manifold joints. To ensure satisfactory mixing, measure the recommended proportions of water and anti-freeze solution in a separate container and fill the system from this container, rather than add the solution direct to the cooling system.

When filling the cooling system, open the heater control tap by placing the temperature control on the facia in the "HOT" position. Check the radiator water level after running the engine and top up if necessary. If topping up is necessary during the period in which anti-freeze solution is in use, this topping up must be carried out using anti-freeze solution or the degree of protection provided may be lost. Topping

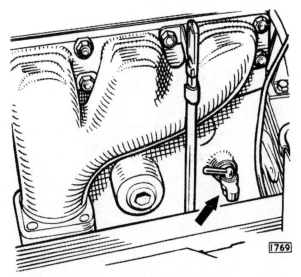

Fig. 2. Cylinder block drain tap

up with water will dilute the mixture possibly to an extent where damage by frost will occur.

Engine Heater

Provision is made on the right-hand side of the cylinder block for the fitment of an American standard engine heater element No. 7, manufactured by James B. Carter Ltd., Electrical, Heating and Manufacturing Division, Winnipeg, Manitoba, Canada, or George Bray & Co. Ltd., Leicester Place, Blackman Lane, Leeds 2, England.

RADIATOR

The aluminium radiator is of the cross flow type having 10 cooling fins per inch (4 fins/cm.). It is pressurised by means of a filler cap incorporated in the separate radiator header tank. The filler cap incorporates a pressure relief valve which is designed to hold a pressure of up to 4 pounds per square inch (0.28 kg./cm.^{2}) above atmospheric pressure inside the system. When the pressure rises above four pounds the spring loaded valve lifts off its seat and the excess pressure escapes via the overflow pipe. As the water temperature falls again a small valve incorporated in the centre of the pressure valve unit, opens and restores atmospheric pressure should a depression be caused by a fall in the temperature of the water.

By raising the pressure inside the cooling system the boiling point of the water is raised approximately

Page D.5

COOLING SYSTEM

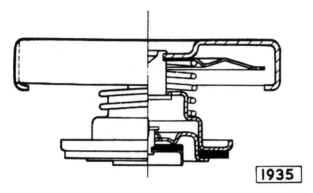

Fig. 3. Sectioned view of radiator filler cap

six degrees thus reducing the risk of water loss from boiling.

Removal

Unscrew the radiator cap and drain the radiator and cylinder block.

Slacken the hose clip securing the water hose from the header tank to the top of the radiator.

Slacken the two hose clips securing the water hose from the water pump to the bottom of the radiator.

Remove the two self-locking nuts and bolts securing the radiator steady brackets to the header tank support bracket.

Remove the two self-locking nuts securing the shield on the front sub frame to the radiator mounting brackets.

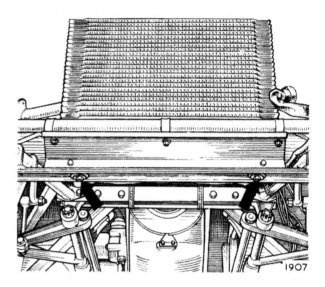

Fig. 5. Nuts securing radiator to sub frame

Remove the two self-locking nuts, washers and mounting rubbers securing the bottom of the radiator to the sub frame.

Remove the radiator and collect the other mounting rubbers and spacers taking care not to damage the radiator or the fan blades.

Refitting

Refitting is the reverse of the removal procedure.

RADIATOR COWL

Removal

Remove the radiator as described on this page.

Remove the two self-locking nuts and plain washers securing the cowl to the bottom of the radiator.

Remove the two self-locking nuts securing the radiator steady brackets to the top of the radiator.

Collect the spacers between the radiator and brackets.

Remove the radiator cowl and sealing rubber.

Remove the two nuts, bolts and washers securing the radiator cowl to the radiator steady brackets and collect the spacers.

Refitting

Refitting is the reverse of the removal procedure.

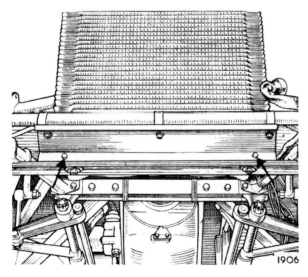

Fig. 4. Bolts securing the sub frame shield to the radiator mounting brackets

Page D.6

COOLING SYSTEM

Fig. 6. *Bolts securing radiator cowl*

Disconnect the two electrical connections from the thermostat fan control in the header tank.

Remove the two nuts and bolts securing the header tank to the mounting bracket.

Disconnect the radiator header tank overflow pipe and remove the header tank.

Refitting

The electrical connections on the black/red wire should be attached to the centre connector on the thermostatic fan control in the header tank. Fit the black wire to the earth connection. Refitting is the reverse of the removal procedure.

FAN MOTOR
Removal

Disconnect the positive lead on the battery.

Remove the four self-locking nuts and plain washers securing the fan motor to the front sub assembly.

Remove the electric motor from its mountings and disconnect the two electrical connections.

Withdraw the electric motor and fan blades from the right hand side between the radiator and frame assembly.

Refitting

Refitting is the reverse of the removal procedure.

FAN MOTOR RELAY (Early cars only)
Removal

Disconnect the positive lead on the battery.

Remove the three electrical cables from the fan

RADIATOR HEADER TANK
Removal

Unscrew the radiator cap and drain the water from the radiator and cylinder block.

Slacken the hose clip securing the water hose from the cylinder head to the header tank.

Slacken the hose clip securing the water hose from the radiator to the header tank.

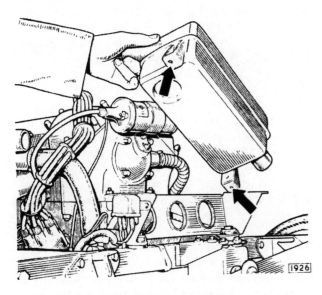

Fig. 7. *Radiator header tank mounting points*

Fig. 8. *Showing position of fan relay. Inset shows relay connections*

Page D.7

COOLING SYSTEM

motor relay on the radiator header tank support bracket.

Remove the two nuts and bolts securing the relay to the side of the support bracket.

Refitting

Refitting is the reverse of the removal procedure but care must be taken to fit the electrical cables to the correct terminals on the fan relay. The black/red cables should be connected to terminal W.1, the green cable to C.2 and the black/green cable to C.1, see Fig. 8.

WATER PUMP BELT
Adjustment—(Early Cars only)

Slacken the two bolts and nuts underneath the dynamo and also the adjusting link bolt. Pull the dynamo outwards until the belt can be flexed approximately $\frac{1}{2}''$ (12.7 cm.) either way, midway between the water pump and dynamo pulleys. Tighten the adjusting link bolt and the dynamo mounting bolts.

Note: Slackness of the belt will cause slip with the possible result of a squealing noise from the belt or a reduced charging rate from the dynamo. Too much tension will create undue wear of the belt, pulleys, water pump and dynamo bearings.

Fig. 9. To adjust the belt tension slacken the three dynamo mounting bolts and move the dynamo into the desired position. Adjustment is not necessary on cars fitted with a jockey pulley.

Removal

Release the belt tension by slackening the two bolts and nuts underneath the dynamo and also the adjusting link bolt. Swing the dynamo towards the engine until the belt can be removed from the pulley. Remove the belt from the crankshaft and water pump pulleys.

Refitting

Refitting is the reverse of the removal but it is important that the belt is not stretched over the pulleys by any means other than by hand. If a tool is used to lever the belt on or off, the endless cords in the belt may be broken.

THERMOSTAT

This is a valve incorporated in the cooling system which restricts the flow of coolant through the radiator until the engine has reached its operating temperature, thus providing rapid warming up of the engine and in cold weather an early supply of warm air to the interior of the car via the heater. When the engine temperature rises to a pre-determined figure (see "Thermostat Data") the thermostat valve commences to open and allows the water to circulate through the radiator. The flow of water increases as the temperature rises until the valve is fully open. Included in the system is a water by-pass utilizing a slot in the thermostat housing integral with the water outlet pipe; this allows the coolant to by-pass the radiator until the thermostat opening temperature is attained.

Removal

Drain sufficient water from the system to allow the level to fall below the thermostat by operating the drain tap situated at the bottom left-hand side of the radiator block. Slacken the hose clip and remove the top water hose from the elbow pipe on the thermostat housing. Remove the two nuts and spring washers securing the water outlet elbow and remove elbow. Lift out the thermostat, noting the gasket between the elbow pipe and thermostat housing.

Checking

Thoroughly clean the thermostat and check that the small hole in the valve is clear. Check the thermostat for correct operation by immersing in a container of cold water together with a thermometer and stirrer. Heat the water, keeping it well stirred and observe if the characteristics of the thermostat are in agreement with the data given under "Thermostat Temperatures".

Refitting

Refitting is the reverse of the removal procedure. Always fit a new gasket between the elbow pipe and the thermostat housing. Ensure that the recess in the thermostat housing and all machined faces are clean.

Page D.8

COOLING SYSTEM

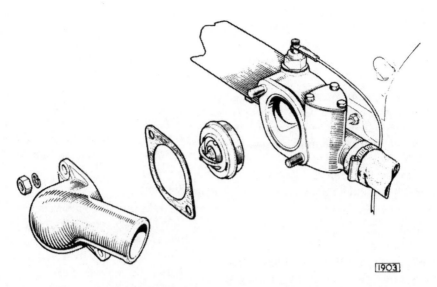

Fig. 10. Exploded view of thermostat and housing

Thermostat Data

Jaguar part number	Start Operating Temperature	Fully Open Temperature	Remarks
C.12867/2 or C.20766/2	165°F. (73·9°C) 159°F. (70·5°C)	187°F. (86·1°C) 168°F. (75·5°C)	

FAN THERMOSTATIC SWITCH

The fan thermostatic switch is situated in the separate radiator header tank. When the water reaches a temperature of approximately 80°C. the thermostatic switch operates and automatically starts the fan motor. The fan motor will continue to run until the temperature of the coolant has fallen to approximately 72°C.

Removal

Disconnect the two electrical connections from the thermostatic switch.

Remove the three securing setscrews and washers.

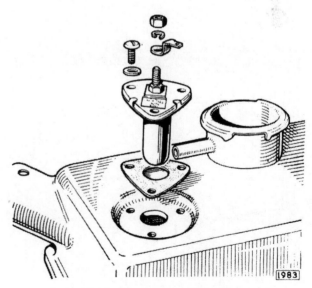

Fig. 11. Exploded view of fan thermostatic switch

Page D.9

COOLING SYSTEM

Withdraw the thermostatic switch and remove the cork gasket.

Refitting

Refitting is the reverse of the removal procedure, but a new cork gasket must be fitted when the switch is replaced. Fit the connection on the red/black wire to the centre connector on the thermostatic switch. Fit the black wire to the earth connection. If any water has escaped during the removal of the switch the radiator header tank should be topped up to the correct level.

WATER PUMP

The water pump (Fig. 12.) is of the centrifugal vane impeller type, the impeller being mounted on a steel spindle which in turn runs in a double row of ball bearings. These are sealed at their ends to exclude all dirt and to retain the lubricant. The main seal on the pump spindle is located in the pump housing by a metal cover and the carbon face maintains a constant pressure on the impeller by means of a thrust spring inside the seal. A hole drilled in the top of the casting acts as an air vent and lead into an annular groove in the casting into which stray water is directed by means of a rubber thrower on the pump spindle. A drain hole at the bottom of the groove leads away any water and prevents seepage into the bearing.

Removal

Drain the cooling system.

Detach the two hose connections to the radiator header tank by unscrewing the hose clips.

Disconnect the two electrical connections to the thermostat fan switch.

Remove the two nuts and bolts securing the header tank and the right hand radiator steady bracket to the header tank support bracket.

Remove the nut and bolt securing the left hand radiator steady bracket.

Detach the three electrical connections from the fan relay fitted on the right hand side of the header tank support bracket.

Remove the two bolts, large washers and mounting rubbers securing the header tank bracket to the sub frame.

Remove the header tank support bracket and collect the remaining rubbers.

Disconnect the three water hoses from the water pump by unscrewing the hose clips.

Slacken off the dynamo bolts and push the dynamo towards the engine, remove the dynamo water pump drive belt. Remove the four setscrews and spring washers securing the water pump pulley to the pulley carrier.

Unscrew the six set bolts, three nuts and spring washers securing the water pump to the engine timing chain cover.

Note the gasket between the pump and timing cover. Withdraw the water pump.

Dismantling

Remove the water pump pulley hub by means of a suitable extractor as shown in Fig. 13. Slacken locknut and remove Allen head locating screw.

Withdraw the spindle and impeller assembly from the pump casting. This assembly must not be pushed out by means of the shaft or the bearing will be damaged. A tube measuring $1\frac{3}{32}''$ (27.77 mm.) outside diameter and $\frac{31}{32}''$ (24.61 mm.) inside diameter

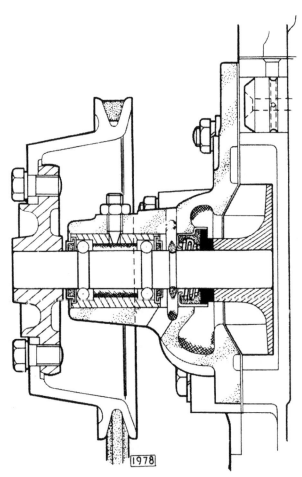

Fig. 12. Sectioned view of water pump

Page D.10

COOLING SYSTEM

must be used to push out the assembly from the front of the pump.

Press out the spindle from the impeller as shown in Fig. 14. and remove seal and rubber thrower. The spindle and bearing assembly cannot be dismantled any further.

Checking

Thoroughly clean all parts of the pump except the spindle and bearing assembly in a suitable cleaning solvent.

Note: The bearing is a permanently sealed and lubricated assembly and therefore must not be washed in the solvent.

Inspect the bearing for excessive end play and remove any burrs, rust or scale from the shaft with fine emery paper, taking the precaution of covering the bearing with a cloth, to prevent emery dust from entering the bearing. If there are any signs of wear or corrosion in the bearing bore or on the face in front of the impeller the housing should be renewed.

Re-assembly

Install the shaft and bearing assembly into the pump body from the rear and line up the location hole

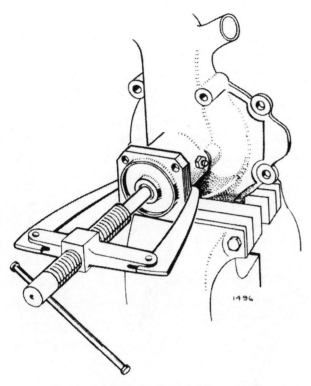

Fig. 13. Withdrawing the fan hub from spindle

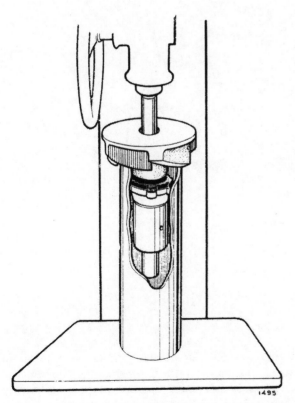

Fig. 14. Removing water pump impeller from pump spindle

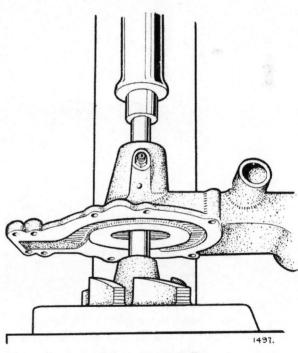

Fig. 15. Fitting impeller

Page D.11

COOLING SYSTEM

in the bearing with the tapped hole in the body. Fit locating screw and locknut. Place the rubber thrower in its groove on the spindle in front of the seal. Coat the outside of the brass seal housing with a suitable water resistant jointing compound and fit into the recess in the pump casting. Push the seal into its housing with the carbon face towards the rear of the pump. Ensure that the seal is seated properly.

Press on impeller as shown in Fig. 15. until the rear face of the impeller is flush with the end of the spindle. In a similar manner press the water pump pulley on to the spindle until it is flush with the end.

Refitting

Refitting is the reverse of the removal procedure although care should be taken to renew the water pump to timing cover gasket, lightly smearing with grease before fitting. When refitting the electrical connections for the fan relay refer to "Fan motor relay-fitting" on page D8. When refitting the fan thermostat electrical connections refer to "Fan thermostatic switch-refitting" on page D10. Adjust the dynamo/water pump belt as described on page D8.

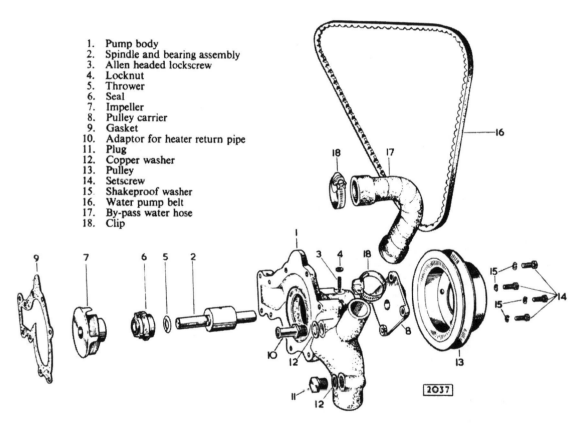

1. Pump body
2. Spindle and bearing assembly
3. Allen headed lockscrew
4. Locknut
5. Thrower
6. Seal
7. Impeller
8. Pulley carrier
9. Gasket
10. Adaptor for heater return pipe
11. Plug
12. Copper washer
13. Pulley
14. Setscrew
15. Shakeproof washer
16. Water pump belt
17. By-pass water hose
18. Clip

Fig. 16. Exploded view of water pump

WATER TEMPERATURE GAUGE

The indicator head is attached to the instrument panel and operates on a thermal principle using a bi-metal strip surrounded by a heater winding. The transmitter unit is mounted in the inlet manifold water jacket adjacent to the thermostat. For the full description and fault analysis of this instrument refer to Section P "Electrical and Instruments".

Page D.12

SUPPLEMENTARY INFORMATION TO SECTION D "COOLING SYSTEM"

Automatic Fan Belt Tensioner

Commencing at engine number R.1845 to R.9999 and RA.1001 and onwards a spring-loaded jockey pulley is fitted on the right hand side of the engine. This pulley maintains the correct tension on the fan belt without the need for periodic adjustment. If it should become necessary to replace the fan belt, slacken the dynamo mounting nuts; press the dynamo towards the engine. Press against the spring tension of the jockey pulley and remove the fan belt.

At engine number RA.1100 and onwards, a modified jockey pulley carrier and stop were fitted to limit the travel of the pulley. These parts may be fitted in pairs to replace those already in use.

Introduction of 9 lb. Pressure Cap

	Commencing Chassis Numbers	
	R.H. Drive	L.H. Drive
"E" Type Fixed Head Coupe	861091	888241
Open 2 Seater	850657	879044

At the above chassis numbers and subsequently, all "E" type cars were fitted with a 9 lb./sq.in. (0.63 kg/sq.cm.) radiator pressure cap, a modified header tank and engine to header tank hose. This hose is of the straight type compared to its predecessor which was convolute.

It is IMPORTANT that these parts are fitted in car sets and under no circumstances should the 9 lb./sq. in. (0.63 kg/sq. cm.) pressure cap be fitted to cars prior to those quoted above unless the new header tank and hose are also fitted.

Deletion of Fan Motor Relay

	Commencing Chassis Numbers	
	R.H. Drive	L.H. Drive
"E" Type Open 2 Seater	850274	878021
Fixed Head Coupe	861187	886749

Commencing at the above chassis numbers the fan motor relay is deleted from production cars and a modified forward harness is used.

The new harness may be fitted in place of the earlier type harness and relay.

Page D.–s.1

SECTION E

CLUTCH
3·8 "E" TYPE
GRAND TOURING MODELS

INDEX

	Page
Clutch—(Early Cars)	
Description	E.4
Data	E.5
Routine Maintenance:	
Clutch fluid level	E.5
Clutch pedal free travel	E.6
Recommended Hydraulic Fluids	E.6
Hydraulic System—General Instructions	E.6
Bleeding the System	E.8
Flushing the System	E.8

Page E.1/2

INDEX *(Continued)*

	Page
Removing and Refitting a Flexible Hose	E.8

The Master Cylinder:

Removal	E.10
Renewing the Master Cylinder Seals	E.10
Master Cylinder Push-rod—Free Travel	E.10

The Slave Cylinder:

Removal	E.11
Dismantling	E.11
Assembling	E.11
Refitting	E.11

The Clutch Unit	E.12
General Instructions	E.13
Clutch Cover Assembly	E.13
Release Bearing	E.13
Condition of Clutch Facings	E.13
Alignment	E.14
Pedal Adjustment	E.14
Removal of Clutch	E.14
Dismantling	E.14

INDEX (Continued)

	Page
Assembling	E.15
Adjusting the Release Levers	E.15
1. Using a Borg and Beck Gauge Plate	E.16
2. Using the Churchill fixture	E.16
3. Using the actual Driven Plate	E.17
Refitting	E.18
Data for Clutch Lever Tip Setting	E.19
Fault Finding	E.20

Clutch—(Later Cars)

	Page
Description	E.23
Data	E.23
The Clutch Unit:	
Description	E.24
General Instructions	E.24
Removal	E.25
Dismantling	E.25
Assembling	E.25
Refitting	E.26
Condition of Clutch Facings	E.26
Pedal Adjustment	E.27

CLUTCH
(Early Cars)

DESCRIPTION

The clutch is of the single dry plate type and consists of a spring loaded driven plate assembly, a cover assembly and a graphite release bearing. The operating mechanism consists of a pendant-type foot pedal, coupled by a push rod to an independent master cylinder. This is connected by piping and a flexible hose to a slave cylinder mounted on the clutch housing. Depressing the clutch pedal moves the piston in the master cylinder and imparts thrust to the slave cylinder piston which in turn, operates the graphite release bearing by means of a push rod and operating fork. The bearing is forced against the clutch release lever plate which causes the release levers to withdraw the pressure plate and thus release the clutch driven plate.

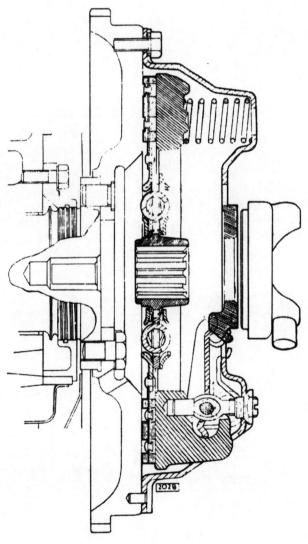

Fig. 1. Sectional view of clutch.

CLUTCH

DATA

	Normal Touring Use	Racing and Competition Use
Make	Borg and Beck	Borg and Beck
Model	10 A6—G	10 A6—G
Outside diameter	9.84″—9.87″ (231 mm.—232 mm.)	9.84″—9.87″ (231 mm.—232 mm.)
Inside diameter	6.12″—6.13″ (153 mm.—154 mm.)	6.12″—6.13″ (153 mm.—154 mm.)
Type	Single dry plate	Single dry plate
Clutch release bearing	Graphite	Graphite
Operation	Hydraulic	Hydraulic
Clutch thrust springs—number	12	12
—colour	Violet	Violet
—free length	2.68″ (68 mm.)	2.68″ (68 mm.)
Driven plate—type	Borglite	Arcuate
—facings	Wound yarn	Wound yarn cemented
Driven plate damper springs—number	6	6
—colour	Brown/Cream	Buff

ROUTINE MAINTENANCE

WEEKLY
Clutch Fluid Level
Right-hand drive cars

The fluid reservoir for the hydraulically operated clutch is situated on the bulkhead (adjacent to the twin brake reservoirs), on the driver's side, and it is important that the fluid does not fall below the level marked "Fluid Level".

Left-hand drive cars

The fluid reservoir for the hydraulically operated clutch is situated on the front frame assembly, adjacent to the twin brake reservoirs and exhaust manifold. It is important that the fluid does not fall below the level marked "Fluid Level".

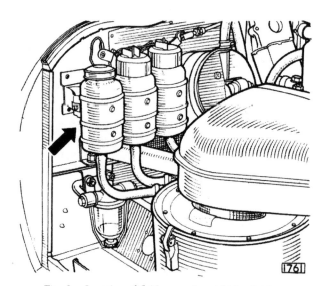

Fig. 2. Location of fluid reservoir—right-hand drive.

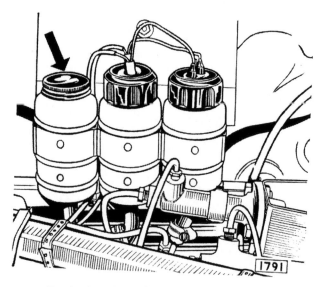

Fig. 3. Location of fluid reservoir—left-hand drive.

Page E.6

CLUTCH

EVERY 2,500 MILES (4,000 KM.)

Clutch Free Travel

Normal Road Use

There should be $\frac{1}{16}''$ (1.6 mm.) free travel measured on the operating rod between the slave cylinder and clutch withdrawal lever.

This free travel is most easily felt, after removal of pedal return spring, by moving operating rod towards slave cylinder and then returning towards withdrawal lever to fullest extent. Adjustment is effected by slackening the locknut, and turning the operating rod. Screwing the rod into the knuckle joint will increase the free travel; screwing the rod out will decrease the free travel. Always replace return spring after adjustment.

Racing

There should be as much free travel of the operating rod, between the slave cylinder and the clutch withdrawal lever, as is possible to obtain without grating of the gears being experienced when engaging first gear.

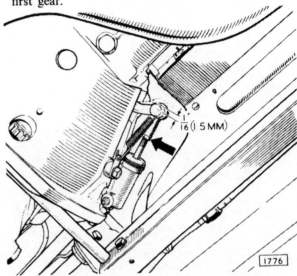

Fig. 4. Adjustment of clutch operating rod

Recommended Hydraulic Fluids

Preferred Fluid

Castrol/Girling Crimson Clutch/Brake Fluid (S.A.E.70 R3).

In countries where the above fluids are unobtainable use only a recognised brake fluid guaranteed to conform to the SAE specification 70 R.3.

In the event of deterioration of the rubber seals and

Alternative Fluids

Lockheed Super Heavy Duty Brake Fluid.

hoses due to the use of an incorrect fluid, all the seals and hoses must be replaced and the system thoroughly flushed and refilled with one of the above fluids. (See "Flushing the System").

HYDRAULIC SYSTEM—GENERAL INSTRUCTIONS

Should it be found necessary to dismantle any part of the clutch system (that is, master cylinder or slave cylinder), the operation must be carried out under conditions of scrupulous cleanliness. Clean the mud and grease off the unit before removal from the vehicle and dismantle on a bench covered with a sheet of clean paper. Do not swill a complete unit, after removal from the vehicle, in paraffin, petrol or trichlorethylene (trike) as this would ruin the rubber parts and, on dismantling, give a misleading impression of their original condition. Do not handle the internal parts, particularly rubbers, with dirty hands. Place all metal parts in a tray of clean brake fluid to soak; afterwards dry off with a clean fluffless cloth, and lay out in order on a sheet of clean paper. Rubber parts should be carefully examined and if there is any sign of swelling or perishing they should be renewed; in any case it is usually good policy to renew **all** rubbers. The main castings may be swilled in any of the normal cleaning fluids but all traces of the cleaner must be dried out before assembly.

All internal parts should be dipped in clean brake fluid and assembled wet, as the fluid acts as a lubricant. When assembling the rubber parts use the fingers only.

Page E.7

CLUTCH

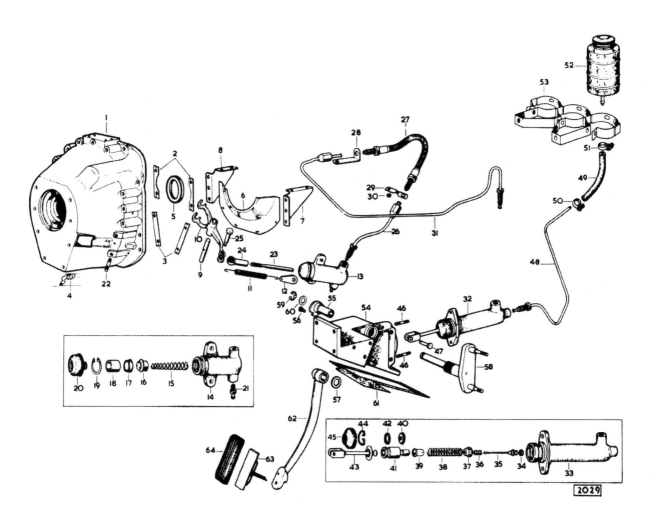

Fig. 5. Clutch operating system.

1. Clutch housing	22. Stud	44. Circlip
2. Locking plate	23. Operating rod	45. Dust cover
3. Locking plate	24. Adjuster assembly	46. Stud
4. Timing aperture cover	25. Pivot pin	47. Clevis pin
5. Oil seal	26. Hydraulic pipe	48. Hydraulic pipe
6. Cover plate	27. Flexible hydraulic pipe	49. Flexible pipe
7. Support bracket	28. Bracket	50. Hose clip
8. Support bracket	29. Bracket	51. Hose clip
9. Shaft	30. Distance piece	52. Reservoir
10. Operating fork	31. Hydraulic pipe	53. Mounting bracket
11. Return spring	32. Master cylinder	54. Clutch pedal housing
12. Anchor plate	33. Master cylinder body	55. Bush
13. Slave cylinder	34. Seal	56. Setscrew
14. Slave cylinder body	35. Valve	57. Fibre washer
15. Spring	36. Spring	58. Pedal shaft
16. Cup filler	37. Spring support	59. Circlip
17. Seal	38. Main spring	60. Washer
18. Piston	39. Spring support	61. Gasket
19. Circlip	40. Cup seal	62. Pedal
20. Rubber dust cover	41. Piston	63. Pedal pad
21. Bleeder screw	42. Static seal	64. Pedal pad cover
	43. Push rod	

Page E.8

CLUTCH

BLEEDING THE SYSTEM

"Bleeding" the clutch hydraulic system (expelling air) is not a routine maintenance operation and should only be necessary when a portion of the hydraulic system has been disconnected or if the level of the fluid in the reservoir has been allowed to fall. The presence of air in the hydraulic system may result in difficulty in engaging gear owing to the clutch not disengaging fully.

The procedure is as follows:—

Fill up the master cylinder reservoir with brake fluid exercising great care to prevent the entry of dirt. Attach a rubber bleed tube to the nipple on the slave cylinder on the right-hand side of the clutch housing and allow the tube to hang in a clean glass jar partly filled with brake fluid. Unscrew the nipple one complete turn. Depress the clutch pedal slowly, **tighten the bleeder nipple before the pedal reaches the end of its travel** and allow the pedal to return unassisted.

Repeat the above procedure, closing the bleed nipple at each stroke, until the fluid issuing from the tube is entirely free of air, care being taken that the reservoir is replenished **frequently** during this operation, for should the level be allowed to drop appreciably air will enter the system.

On completion, top up the master cylinder reservoir to the line marked "Fluid Level".

Do not on any account use the fluid which has been bled through the system to replenish the reservoir as it will have become aerated. Always use fresh fluid straight from the tin.

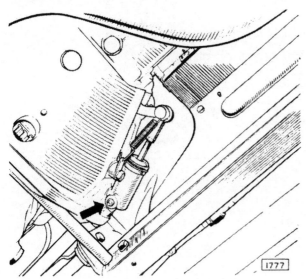

Fig. 6. Position of clutch bleed nipple.

FLUSHING THE SYSTEM

Should the fluid in the system become thick or "gummy" after many years in service, or after a vehicle has been laid up for some considerable time, the system should be drained, flushed and re-filled. It is recommended that this should be carried out once every five years.

Pump all fluid out of the hydraulic system through the bleeder screw of the clutch slave cylinder. To the bleeder screw on the slave cylinder connect one end of a rubber tube, and allow the other end to fall into a container, slacken the screw one complete turn and pump the clutch pedal by depressing it quickly and allowing it to return without assistance; repeat, with a pause in between each operation, until no more fluid is expelled. Discard the fluid extracted.

Fill the supply tank with industrial methylated spirit and flush the system as described above. Keep the supply tank replenished until at least a quart of spirit has passed through the bleeder screw.

Remove the master cylinder and pour off any remaining spirit. Refit the master cylinder, re-fill with clean brake fluid and "bleed" the system.

NOTE: If the system has been contaminated by the use of mineral oil, etc., the above process will not prove effective. It is recommended that the various units, including the pipe lines, be dismantled and thoroughly cleaned and that all rubber parts, including flexible hoses, be renewed. The contaminated fluid should be destroyed immediately.

REMOVAL AND REFITTING A FLEXIBLE HOSE

In some cases, the cause of faulty clutch may be traced to a choked flexible hose. Do not attempt to clear the obstruction by any means except air pressure, otherwise the hose may be damaged. If the obstruction cannot be cleared the hose must be replaced by a new one.

Removal

To renew a flexible hose, adopt the following procedure:—

Unscrew the tube nut from the hose union, then unscrew the locknut and withdraw the hose from the bracket. Disconnect the hose at the other end.

CLUTCH

Refitting

When re-fitting a hose, first ensure that it is not twisted or "kinked" (this is MOST IMPORTANT) then pass the hose union through the bracket and, whilst holding the union with a spanner to prevent the hose from turning, fit the locknut and the shakeproof washer; connect up the pipe by screwing on the tube-nut.

THE MASTER CYLINDER

The master cylinder is mechanically linked to the clutch pedal and provides the hydraulic pressure necessary to operate the clutch. The components of the master cylinder are contained within the bore of a body which at its closed end has two 90° opposed integral pipe connection bosses. Integrally formed around the opposite end of the cylinder is a flange provided with two holes for the master cylinder attachment bolts. In the unloaded condition a spring loaded piston, carrying two seals (see Fig. 7) is held against the underside of a circlip retained dished washer at the head of the cylinder. A hemispherically ended push-rod seats in a similarly formed recess at the head of the piston. A fork end on the outer end of the push-rod provides for attachment to the pedal. A rubber dust excluder, the lip of which seats in a groove, shrouds the head of the master cylinder to prevent the intrusion of foreign matter.

A cylindrical spring support locates around the inner end of the piston and a small drilling in the end of the support is engaged by the stem of a valve. The larger diameter head of the valve locates in a central blind bore in the piston. The valve passes through the bore of a vented spring support and interposed between the spring support and an integral flange formed on the valve is a small coiled spring. A lipped rubber seal registers in a groove around the end of the valve. This assembly forms a recuperation valve which controls fluid flow to and from the reservoir.

When the foot pedal is in the OFF position the master cylinder is fully extended and the valve is held clear of the base of the cylinder by the action of the main spring. In this condition the master cylinder is in fluid communication with the reservoir, thus permitting recuperation of any fluid loss sustained, particularly during the bleeding operation.

When a load is applied to the foot pedal the piston moves down the cylinder against the compression of the main spring. Immediately this movement is in excess of the valve clearance the valve closes under the influence of its spring and isolates the reservoir. Further loading of the pedal results in the discharge of fluid under pressure from the outlet connection, via the pipe lines to the clutch slave cylinder.

Removal of the load from the pedal reverses the sequence, the action of the main spring returns the master cylinder to the extended position.

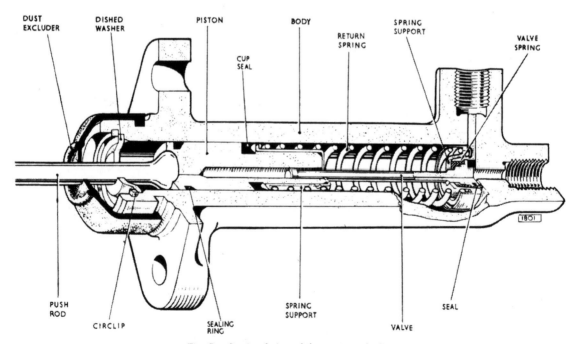

Fig. 7. Sectional view of the master cylinder.

Page E.10

CLUTCH

Removal

Drain the clutch fluid reservoir and detach the inlet and outlet pipes from the clutch master cylinder, by unscrewing the two union nuts. Detach the master cylinder push-rod from the clutch pedal from inside the car by removing the split pin and withdrawing the clevis pin. Remove the clutch master cylinder from the housing situated inside the engine compartment by removing two nuts.

Renewing the Master Cylinder Seals

Ease the dust excluder clear of the head of the master cylinder.

With suitable pliers remove the circlip; this will release the push rod complete with dished washer.

Withdraw the piston and remove both seals.

Withdraw the valve assembly complete with springs and supports. Remove the seal from the end of the valve.

Lubricate the new seals and the bore of the cylinder with brake fluid, fit the seal to the end of the valve ensuring that the lip registers in the groove. Fit the seals in their grooves around the piston.

Insert the piston into the spring support, ensuring that the head of the valve engages the piston bore.

Lubricate the piston with Castrol Rubber Grease H. 95/59 and slide the complete assembly into the cylinder body taking particular care not to damage or twist the seals. The use of a fitting sleeve is advised.

Position the push-rod and depress the piston sufficiently to allow the dished washer to seat on the shoulder at the head of the cylinder. Fit the circlip and check that it fully engages the groove.

Fill the dust excluder with clean Castrol H.95/59 Rubber Grease.

Reseat the dust excluder around the head of the master cylinder.

Master Cylinder Push-rod—Free Travel

To ensure that this piston returns to the fully extended position clearance is provided between the enlarged head of the push-rod, the piston and dished washer. As this washer also forms the return stop for the clutch pedal, no means of adjustment is necessary.

Refitting

Secure the master cylinder to the vehicle by fitting the fixing nuts at the flange. Connect the pipes to the inlet and outlet connections, the push rod to the pedal, and bleed the system. Check for leaks by depressing the clutch pedal once or twice and examining all hydraulic connections.

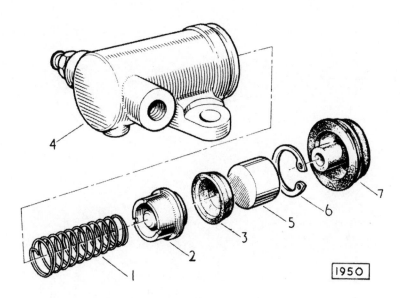

Fig. 8. Exploded view of the clutch slave cylinder.

Page E.11

CLUTCH

THE SLAVE CYLINDER

The clutch slave cylinder consists of a body (4 Fig. 8) which incorporates two threaded connections and is bored to accommodate a piston (5) against the inner face of which a rubber cup (3) is loaded by a cup filler (2) and a spring (1); the travel of the piston is limited by a circlip (6) fitted in a groove at the end of the bore. A rubber boot (7) through which a push-rod passes, is fitted on to the body to prevent the intrusion of dirt or moisture.

One of the connections in the body receives a pipe from the clutch master cylinder, whilst the other is fitted with a bleeder screw; the connection for the pipe is parallel to the mounting flange on the body.

Removal

To remove from the vehicle, disconnect the pipe, detach the rubber boot from the body and remove the fixing screws; leave the push-rod attached to the vehicle. If the boot is not being renewed it may be left on the push-rod.

Dismantling

Remove the circlip (6) from the end of the bore and apply a **low** air pressure to the open connection to expel the piston (5) and the other parts; remove the bleeder screw.

Assembling

Prior to assembly, smear all internal parts and the bore of the body with Rubberlube.

Fit the spring (1) in the cup filler (2) and insert these parts, spring uppermost, into the bore of the body (4). Follow up with the cup (3), lip leading, taking care not to turn back or buckle the lip; then insert the piston (5), flat face innermost, and fit the circlip (6) into the groove at the end of the bore.

Refitting

Fit the rubber boot (7) on the push-rod, if removed previously, and offer up the slave cylinder to the vehicle, with the push-rod entering the bore. Secure the cylinder with the fixing screws and stretch the large end of the boot into the groove on the body. Fit into their respective connections the bleeder screw and the pipe from the clutch master cylinder.

"Bleed" the clutch as described on page E.9.

CLUTCH

THE CLUTCH UNIT
(Early Cars)

The driven plate assembly (14, Fig. 9) is of the flexible centre type, in which a splined hub is indirectly attached to a disc and transmits the power and overrun through a number of coil springs held in position by shrouds.

The cover assembly consists of a pressed steel cover (1) and a cast iron pressure plate (3) loaded by thrust springs (2). Mounted on the pressure plate are release levers (4), which pivot on floating pins (9), retained by eye bolts (8). Adjustment nuts (10) are screwed on to the eye bolts and secured by staking. Struts (7) are interposed between lugs on the pressure plate and the outer end of the release levers. Anti-rattle springs (11) restrain the release levers and retainer springs (6) connect the release lever plate (5) to the levers.

The graphite release bearing (12) is shrunk into a bearing cup which is mounted on the throw-out forks and held by the release bearing retainer springs (13).

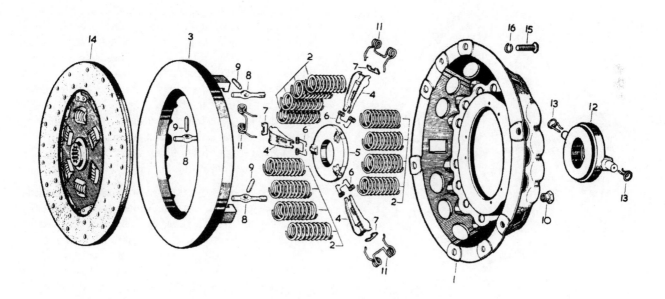

1. Cover
2. Thrust spring
3. Pressure plate
4. Release lever
5. Release lever plate
6. Release lever retainer
7. Release lever strut
8. Release lever eyebolt
9. Eyebolt pin
10. Adjustment nut
11. Anti-rattle spring
12. Release bearing and cup assembly
13. Release bearing retainer
14. Driven plate assembly
15. Securing bolt
16. Spring washer

Fig. 9. Exploded view of the clutch assembly.

Page E.13

CLUTCH

GENERAL INSTRUCTIONS

When overhauling the clutch the following instructions should be noted and carried out:—

Clutch Cover Assembly

Before dismantling the clutch, suitably mark the following parts so that they can be re-assembled in the same relative positions to each other to preserve the balance and adjustment; clutch cover, lugs on the pressure plate and the release levers.

When re-assembling make sure that the markings coincide and, if new parts have been fitted which would affect the adjustment, carefully set the release levers (see page E.16).

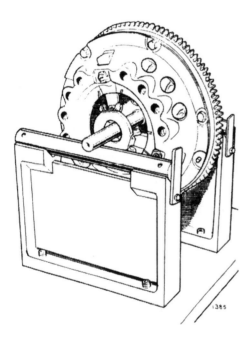

Fig. 10. Clutch and flywheel balance.

If a new pressure plate has been fitted, it is essential that the complete cover assembly should be re-balanced, for which reason it is not a practical proposition where special equipment is not available.

Before assembly, thoroughly clean all parts and renew those which show appreciable wear. A very slight smear of grease such as Lockheed Expander Lubricant or Duckham's Keenol K.O.12 should be applied to the release lever pins, contact faces of the struts, eyebolts seats in the clutch cover, drive lug sides on the pressure plate and the plain end of the eyebolts.

Release Bearing

If the graphite release bearing ring is badly worn it should be replaced by a complete bearing assembly.

CONDITION OF CLUTCH FACINGS

The possibility of further use of the friction facings of the clutch is sometimes raised, because they have a polished appearance after considerable service. It is natural to assume that a rough surface will give a higher frictional value against slipping, but this is not correct.

Since the introduction of non-metallic facings of the moulded asbestos type, in service, a polished surface is a common experience, but it must not be confused with a glazed surface which is sometimes encountered due to conditions discussed below.

The ideal smooth or polished condition will provide a normal contact, but a glazed surface may be due to a film or a condition introduced, which entirely alters the frictional value of the facings. These two conditions might be simply illustrated by the comparison between a polished wood, and a varnished surface. In the former the contact is still made by the original material, whereas in the latter instance, a film of dried varnish is interposed between the contact surfaces.

The following notes are issued with a view to giving useful information on this subject:—

(a) After the clutch has been in use for some little time, under perfect conditions (that is, with the clutch facings working on true and polished or ground surfaces of correct material, without the presence of oil, and with only that amount of slip which the clutch provides for under normal conditions) then the surface of the facings assumes a high polish, through which the grain of the material can be clearly seen. This polished facing is of mid-brown colour and is then in a perfect condition.

(b) Should oil in small quantities gain access to the clutch in such a manner as to come in contact with the facings it will burn off, due to the heat generated by slip which occurs under normal starting conditions. The burning off of this small amount of lubricant, has the effect of gradually darkening the facings, but, provided the polish on the facings remains such that the grain of the material can be clearly distinguished, it has very little effect on clutch performance.

(c) Should increased quantities of oil or grease

CLUTCH

obtain access to the facings, one or two conditions, or a combination of the two, may arise, depending upon the nature of oil, etc.

(1) The oil may burn off and leave on the surface facings a carbon deposit which assumes a high glaze and causes slip. This is a very definite, though very thin deposit, and in general it hides the grain of the material.

(2) The oil may partially burn and leave a resinous deposit on the facings, which frequently produces a fierce clutch, and may also cause a "spinning" clutch due to a tendency of the facings to adhere to the flywheel or pressure plate face.

(3) There may be a combination of (1) and (2) conditions, which is likely to produce a judder during clutch engagement.

(d) Still greater quantities of oil produce a black soaked appearance of the facings, and the effect may be slip, fierceness, or judder in engagement, etc., according to the conditions. If the conditions under (c) or (d) are experienced, the clutch driven plate should be replaced by one fitted with new facings, the cause of the presence of the oil removed and the clutch and flywheel face thoroughly cleaned.

ALIGNMENT

Faulty alignment will cause excessive wear of the splines in the hub of the driven plate, and eventually fracture the steel disc around the hub centre as a result of "swash action" produced by axial movement of the splined shaft.

PEDAL ADJUSTMENT

This adjustment is most important and the instructions given should be carefully followed; faulty adjustment falls under two headings:—

(a) Insufficient free (or unloaded) pedal travel may cause a partly slipping clutch condition which becomes aggravated as additional wear takes place on the facings, and this can result in a slipping clutch leading to burning out unless corrected. Over-travel of effective pedal movement only imposes undue internal strain and causes excessive bearing wear.

(b) Too much free pedal movement results in inadequate release movement of the bearing and may produce a spinning plate condition that is, dragging clutch rendering clean changes impossible.

REMOVAL

To remove the clutch, the engine and gearbox must first be removed (refer to Section B "Engine").

Slacken the clutch mounting screws a turn at a time by diagonal selection until the thrust spring pressure is released. Remove the set screws and withdraw the complete clutch assembly from the flywheel. Remove the driven plate assembly and take care to maintain the driven plate faces in a clean condition. Observe that the clutch and flywheel are balanced as an assembly. This location is indicated by balance marks 'B' stamped on the clutch and flywheel (Fig. 21).

DISMANTLING

Before dismantling, mark all the major components.

To dismantle the clutch, either bolt the assembly to the baseplate of the Churchill fixture, to a spare flywheel, or place the clutch on the bed of a press with blocks under the pressure plate in such a manner that the cover is free to move downwards when pressure is applied.

Having compressed the clutch in one of these various ways, unscrew the nuts (Fig. 11), (considerable torque is initially necessary in order to break off the squeezed-in portion of each nut), and slowly release the clamping pressure. Lift the cover and the thrust springs off the pressure plate and remove the release lever mechanism. Fig. 12 shows the method whereby the strut is disen-

Fig. 11. Removal of the adjustment nuts.

Page E.15

CLUTCH

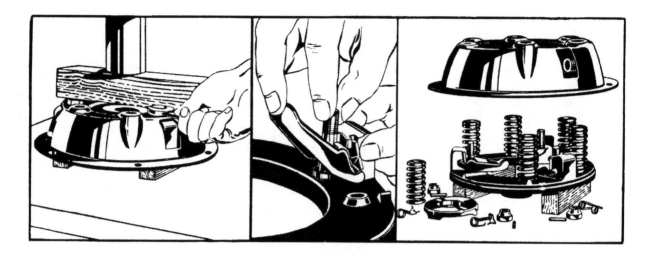

Fig. 12. Dismantling the clutch assembly using a ram press.

gaged from the lever, after which the threaded end of the eye-bolt and the inner end of the lever are held as close together as possible to enable the shank of the eyebolt to clear the hole in the pressure plate.

ASSEMBLING

It is essential that all major components be returned to their original positions if the balance of the assembly is to be preserved.

Fit a pin (9, Fig. 9) into an eyebolt (8) and locate the parts within a release lever (4). Hold the threaded end of the eyebolt and the inner end of the lever as close together as possible and, with the other hand, engage the strut (7) within the slots in a lug on the pressure plate, with the other end of the strut push outwards towards the periphery of the plate. Offer up the lever assembly, first engaging the eyebolt shank within the hole in the plate, then locate the strut within the groove in the lever. Fit the remaining levers in the same way, not forgetting to lubricate all contact faces.

Place the pressure plate on the baseplate of the Churchill fixture, on a spare flywheel, or on blocks on the bed of a press and position the thrust springs (2) on the bosses of the plate. Having arranged all the springs, and after ensuring that the anti-rattle springs (11) are fixed within the cover, rest the cover on the springs, carefully aligning the pressure plate lugs with the cover slots. If the Churchill fixture or a spare flywheel is being used, move the clutch to align the holes in the cover flange with the tapped holes in the flywheel or baseplate and then clamp the cover down with the fixing screws, turning them a little at a time to avoid distortion. If a press is being used, arrange a block across the cover and compress the assembly. Then screw the adjusting nuts (10) into an approximately correct position.

The release levers must now be set to the correct height, adopting any of the three methods elsewhere described after which the adjusting nuts should be locked by punching them into the eyebolt slots. After setting the levers, fit the release lever plate.

ADJUSTING THE RELEASE LEVERS

To ensure satisfactory operation, correct adjustment of the release levers is essential. In service, the original adjustment made by the makers never needs attention and re-adjustment is only necessary if the clutch has been dismantled.

To facilitate adjustment of the release levers the gauge plates once produced by the clutch manufacturer can be utilized. As numerous Traders still possess these plates details as to their identification are given on Page E.20.

An alternative method of lever adjustment is to use the universal fixture known as the No. 99 manufactured by V. L. Churchill & Co. Ltd., which caters for the $6\frac{1}{4}"$—$11"$ clutch.

Finally, where neither a gauge plate nor Churchill tool is available the levers may be set using the actual driven plate as a gauge and these three methods are described below.

Page E.16

CLUTCH

Fig. 13. The gauge plate.

(1) Using a Borg & Beck gauge plate (Fig. 14)

(a) Mount the clutch on the actual or a spare flywheel (1, Fig. 14) or alternatively clamp it down to a flat surface, with the gauge plate (4) occupying the position normally taken by the driven plate. The ground lands of the gauge plate should each be located under a release lever (5).

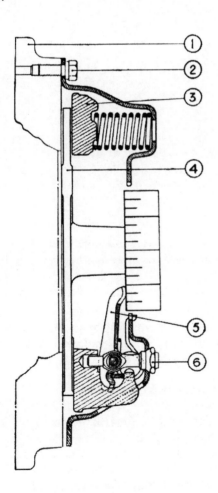

Fig. 14. Release lever adjustment.

(b) Adjust the levers by turning the eyebolt nuts (6) until the levers are just in contact with a short straight edge resting upon the boss of the gauge plate.

(c) Having made a preliminary setting some attempt must be made to operate the clutch several times in order to settle the mechanism. Normally, this operation can be carried out in a drilling machine or light press having a suitable adaptor, arranged to bear upon the lever tips.

(d) Carry out a further check and re-adjust if necessary.

(2) Using the Churchill Fixture

This tool, which is illustrated in Fig. 15 provides for the accurate adjustment of the levers; additionally, it affords a convenient fixture upon which to dismantle and assemble the unit. A device is included to operate the clutch and thereby to settle the working parts after assembly. To use the tool, adopt the following procedure, which also indicates the additional operations when dismantling and assembling the clutch.

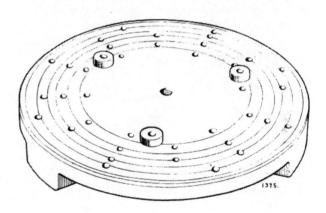

Fig. 15. The special base plate for clutch adjustment.

Remove from the box the gauge finger, the pillar and the actuator, as shown in Fig. 16 and consult the code card to determine the reference of the adaptor and the spacers appropriate to the clutch which is being serviced.

Rest the base plate on a flat surface, wipe it clean and place the spacers upon it in the positions quoted on the code card, as in Fig. 15. Place the clutch on the spacers, aligning it with the appropriate tapped holes in the base, arranging

Page E.17

CLUTCH

it so that the release levers are as close to the spacers as possible.

Screw the actuator into the centre hole in the base plate and press the handle down to clamp the clutch. Then screw the set bolts provided firmly into the tapped holes in the baseplate using a speed brace; remove the actuator.

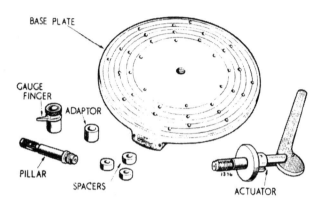

Fig. 16. The base plate and accessories.

Remove the adjusting nuts (Fig. 11) and gradually unscrew the set bolts to relieve the load of the thrust springs (Fig. 17). Lift the cover off the clutch and carry out whatever additional dismantling may be desired.

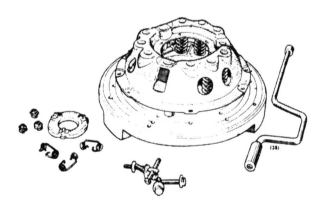

Fig. 17. Removing the clutch cover assembly.

After carrying out the necessary servicing of the clutch components, re-assemble the parts on the clutch pressure plate, place the cover upon it and transfer the assembly to the base plate, resting on the spacers and aligned correctly.

Carefully bolt the cover to the base plate and screw the adjusting nuts on to the eyebolts until flush with the tops of the latter. Screw the actuator into the base (Fig. 18) and pump the handle a dozen times to settle the clutch mechanism. Remove the actuator. Screw the pillar firmly into the base and place upon it the appropriate adaptor, recessed face downwards, and the gauge finger.

Turn the adjusting nuts until the finger just touches the release levers, pressing downwards on the finger assembly to ensure that it is bearing squarely on the adaptor (Fig. 19).

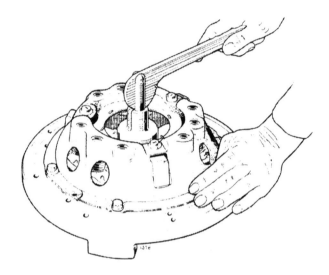

Fig. 18. Screwing the actuator into the base plate.

Remove the finger adaptor and pillar, replace the actuator and operate the clutch a further dozen times. Replace the pillar and check the lever setting, making any final correction.

Finally, lock the adjusting nuts. The cylindrical portion of the nut must be peened into the slot in the eyebolt, using a blunt chisel and hammer.

(3) Using the Actual Driven Plate

This method of setting the levers is not highly accurate and should only be resorted to when neither a gauge plate nor Churchill Fixture is available. The drawback to this method lies in the fact that although the driven plate is produced to close limits, it is difficult to ensure absolute

Page E.18

CLUTCH

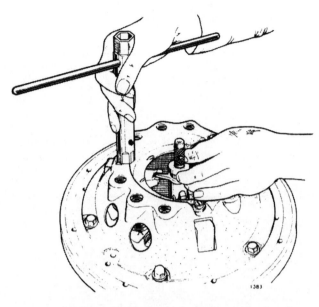

Fig. 19. Using finger assembly to adjust the release levers.

that the 'B' stamped adjacent to one of the dowel holes coincides with the 'B' stamped on the periphery of the flywheel (Fig. 21). Do not remove the dummy shaft until all the setscrews are securely tightened, otherwise the driven plate will come off centre and difficulty will be met in engaging the constant pinion shaft into the bush in the rear end of the crankshaft.

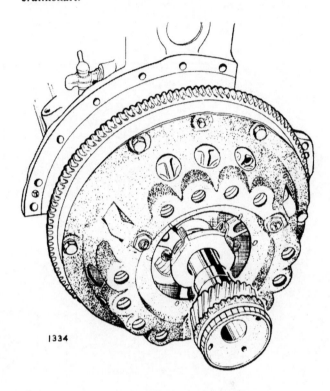

Fig. 20. Centralising the driven plate.

parallelism. Although the error in the plate is small, it is magnified some five-fold at the lever tip due to the lever ratio.

The method to be adopted is as follows:—

(a) Mount the clutch on the flywheel with the driven plate in its normal position or clamp the assembly to any flat surface having a hole within it to accommodate the boss of the driven plate.

(b) Consult the chart on page E.20 to ascertain the height of the lever tip from the flywheel and adjust the levers until this dimension is achieved.

(c) Having made a preliminary setting slacken the clamping pressure, turn the driven plate through a right angle, re-clamp the cover and check the levers again as a safeguard against any lack of truth in the driven plate.

REFITTING

Place the driven plate on the flywheel taking care that the larger part of the splined hub faces the gearbox. Centralise the plate on the flywheel by means of the dummy shaft (a constant pinion shaft may be used for this purpose, Fig. 20). Secure the cover assembly with the six setscrews and spring washers, tightening the screws a turn at a time by diagonal selection. Ensure

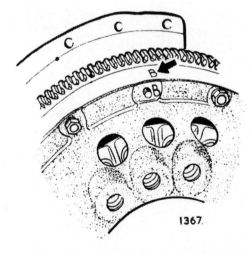

Fig. 21. Balance marks on the clutch and flywheel.

Page E.19

CLUTCH

DATA FOR CLUTCH LEVER TIP SETTING

Clutch Model	Driven Plate	Gauge Plate Part No.	Lever tip height from flywheel face Dimension "A"	Gauge Plate Land Thickness Dimension "C"	Gauge Plate Dia.	Remarks
10"	Borglite or Arcuate	CG14322	1·955" (49·65 mm.)	0·330" (8·381 mm.)	8·375" (212·7mm.)	Dimension "A" 2·45" (62·23mm.) if taken with Release Lever Plate in position.

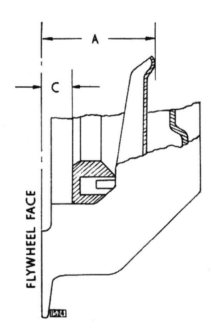

Fig. 22. Dimensions for clutch lever tip setting.

Page E.20

CLUTCH

FAULT FINDING

SYMPTOM	CAUSE	REMEDY
Drag or Spin	(a) Oil or grease on the driven plate facings	Fit new facings. Isolate clutch from possible ingress of oil or grease.
	(b) Misalignment between the engine and splined clutch shaft	Check over and correct the alignment
	(c) Air in clutch system	"Bleed" system
	(d) Bad external leak between the clutch master cylinder and the slave cylinder	Renew pipe and unions
	(e) Excessive clearance between the release bearing and the release lever plate	Adjust to $\frac{1}{16}''$ (1.58 mm.) clearance
	(f) Warped or damaged pressure plate or clutch cover	Renew defective part
	(g) Driven plate hub binding on splined shaft	Clean up splines and lubricate with small quantity of high melting point grease such as Duckham's Keenol
	(h) Distorted driven plate due to the weight of the gearbox being allowed to hang on clutch plate during assembly	Fit new driven plate assembly using a jack to take overhanging weight of the gearbox
	(i) Broken facings of driven plate	Fit new facings, or replace plate
	(j) Dirt or foreign matter in the clutch	Dismantle clutch from flywheel and clean the unit, see that all working parts are free CAUTION: Never use petrol or paraffin for cleaning out clutch
Fierceness or Snatch	(a) Oil or grease on driven plate facings	Fit new facings and ensure isolation of clutch from possible ingress of oil or grease
	(b) Misalignment	Check over and correct alignment
	(c) Worn out driven plate facings	New facings required
Slip	(a) Oil or grease on driven plate facings	Fit new facings and eliminate cause of foreign presence
	(b) Failure to adjust at clutch slave cylinder to compensate for loss of release bearing clearance consequent upon wear of the driven plate facings $\frac{1}{16}''$ (1.58 mm.) clearance is necessary between the release bearing and the release lever plate)	Adjust push rod as necessary
	(c) Seized piston in clutch slave cylinder	Renew parts as necessary
	(d) Master cylinder piston sticking	Free off the piston

Page E.21

CLUTCH

FAULT FINDING

SYMPTOM	CAUSE	REMEDY
Judder	(a) Oil, grease or foreign matter on driven plate facings	Fit new facings or driven plate
	(b) Misalignment	Check over and correct alignment
	(c) Pressure plate out of parallel with flywheel face in excess of the permissible tolerance	Re-adjust levers in plane, and, if necessary, fit new eyebolts
	(d) Contact area of friction facings not evenly distributed. Note that friction facing surface will not show 100% contact until the clutch has been in use for some time, but the contact actually showing should be evenly distributed round the friction facings	This may be due to distortion, if so fit new driven plate assembly
	(e) Bent splined shaft or buckled driven plate	Fit new shaft or driven plate assembly
Rattle	(a) Damaged driven plate, broken springs, etc.,	Fit new parts as necessary
	(b) Worn parts in release mechanism	
	(c) Excessive backlash in transmission	
	(d) Wear in transmission bearings	
	(e) Bent or worn splined shaft	
	(f) Graphite release bearing loose on throw out fork	
Tick or Knock	Hub splines worn due to misalignment	Check and correct alignment then fit new driven plate
Fracture of Driven Plate	(a) Misalignment distorts the plate and causes it to break or tear round the hub or at segment necks	Check and correct alignment and introduce new driven plate
	(b) If the gearbox during assembly be allowed to hang with the shaft in the hub, the driven plate may be distorted, leading to drag, metal fatigue and breakage	Fit new driven plate assembly and ensure satisfactory re-assembly
Abnormal Facing Wear	Usually produced by overloading and by the excessive slip starting associated with overloading	In the hands of the operator

Page E.22

CLUTCH
(Later Cars)

DESCRIPTION

A diaphragm spring clutch is fitted consisting of a spring assembly held flexibly in the lugs of the pressure plate by a spring retaining ring and pivotting on a fulcrum formed by the rims of the clutch cover and the driving plate. Depressing the clutch pedal actuates the release ring causing a corresponding depression of the diaphragm. The lever action of the spring pulls the pressure plate from the driven plate, thus freeing the clutch.

E-23

DATA

	Normal Touring Use
Make	Laycock
Model	10″ Diaphragm Spring
Release Ring Travel	·400″ (10 mm)
Clutch Release Bearing	Graphite
Operation	Hydraulic

Page E.23

CLUTCH

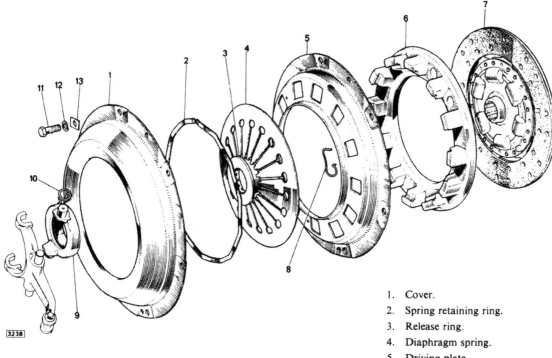

Fig. 23. Exploded view of the clutch unit.

1. Cover.
2. Spring retaining ring.
3. Release ring.
4. Diaphragm spring.
5. Driving plate.
6. Pressure plate.
7. Driven plate assembly.
8. Clip.
9. Release bearing.
10. Clip.
11. Bolt.
12. Spring washer.
13. Balance weight.

THE CLUTCH UNIT
(Later Cars)

DESCRIPTION (Fig. 23)

The driven plate assembly (7) is of the flexible centre type, in which a splined hub is indirectly attached to a disc and transmits the power and overrun through a number of coil springs held in position by shrouds.

The cover assembly consists of a pressed steel cover (1) and a cast iron pressure plate (6) loaded by a spring assembly.

The spring assembly consists of a diaphragm spring (4) and a release ring (3) which is held flexibly in the lugs of the pressure plate by means of a spring retaining ring (2).

Balancing of the clutch assembly is effected by drilling holes in the loose cover plate.

A graphite release bearing (9) is shrunk into a bearing cup which is mounted on the throw-out forks and held by the release bearing retainer springs.

GENERAL INSTRUCTIONS

To enable the balance of the assembly to be preserved after dismantling there are corresponding paint marks on the cover plate and driving plate. In addition, there are corresponding reference numbers stamped in the flanges of the cover and driving plate.

When reassembling ensure that the markings co-incide and that, when refitting the clutch to the flywheel, the letter "B" stamped adjacent to one of the dowel holes co-incides with the "B" stamped on the edge of the flywheel.

Page E.24

CLUTCH

HYDRAULIC SYSTEM—GENERAL INSTRUCTIONS

Should it be found necessary to dismantle any part of the clutch system (that is, master cylinder or slave cylinder), the operation must be carried out under conditions of scrupulous cleanliness. Clean the mud and grease off the unit before removal from the vehicle and dismantle on a bench covered with a sheet of clean paper. Do not swill a complete unit, after removal from the vehicle, in paraffin, petrol or trichlorethylene (trike) as this would ruin the rubber parts and, on dismantling, give a misleading impression of their original condition. Do not handle the internal parts, particularly rubbers, with dirty hands. Place all metal parts in a tray of clean brake fluid to soak; afterwards dry off with a clean fluffless cloth, and lay out in order on a sheet of clean paper. Rubber parts should be carefully examined and if there is any sign of swelling or perishing they should be renewed; in any case it is usually good policy to renew all rubbers. The main castings may be swilled in any of the normal cleaning fluids but all traces of the cleaner must be dried out before assembly.

All internal parts should be dipped in clean brake fluid and assembled wet, as the fluid acts as a lubricant. When assembling the rubber parts use the fingers only.

BLEEDING THE SYSTEM

"Bleeding" the clutch hydraulic system (expelling the air) is not a routine maintenance operation and should only be necessary when a portion of the hydraulic system has been disconnected or if the level of the fluid in the reservoir has been allowed to fall. The presence of air in the hydraulic system may result in difficulty in engaging gear owing to the clutch not disengaging fully.

The procedure is as follows:—

Fill up the reservoir with a brake fluid exercising great care to prevent the entry of dirt. Attach a rubber bleed tube to the nipple on the slave cylinder on the right-hand side of the clutch housing and allow the tube to hang in a clean glass jar partly filled with brake fluid. Unscrew the nipple one complete turn. Depress the clutch pedal slowly, tighten the bleeder nipple before the pedal reaches the end of its travel and allow the pedal to return unassisted.

Repeat the above procedure, closing the bleed nipple at each stroke, until all the fluid issuing from the tube is entirely free of air, care being taken that

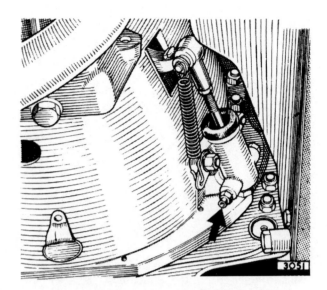

Fig. 26. Clutch slave cylinder bleed nipple.
(Slave cylinder fitted to early cars shown)

the reservoir is replenished **frequently** during this operation, for should it be allowed to become empty more air will enter.

On completion, top up the reservoir to the bottom of the filler neck.

Do not on any account use the fluid which has been bled through the system to replenish the reservoir as it will have become aerated. Always use fresh fluid straight from the tin. Use only the recommended fluid.

FLUSHING THE SYSTEM

Should the fluid in the system become thick or "gummy" after many years in service, or after a vehicle has been laid up for some considerable time, the system should be drained, flushed and re-filled. It is recommended that this should be carried out once every five years.

Pump all fluid out of the hydraulic system through the bleeder screw of the clutch slave cylinder. To the bleeder screw on the slave cylinder connect one end of a rubber tube, and allow the other end to fall into a container, slacken the screw one complete turn and pump the clutch pedal by depressing it quickly and allowing it to return without assistance; repeat, with a pause in between each operation, until no more fluid is expelled. Discard the fluid extracted.

Fill the supply tank with industrial methylated spirit and flush the system as described above. Keep the supply tank replenished until at least a quart of spirit has passed through the bleeder screw.

Page E.25

CLUTCH

When fitting the retaining ring ensure that the 12 crowns fit into the grooves in the 12 pressure plate lugs and that the 11 depressions of the undulations are fitted so as to press on the spring, that is, with the cranked ends of the rings uppermost. It is most important that the retaining ring fits the full depth of the groove in the lugs.

REFITTING

Place the driven plate on the flywheel taking care that the larger part of the splined hub faces the gearbox. Centralise the plate on the flywheel by means of a dummy shaft (a constant pinion shaft may be used for this purpose). Secure the cover assembly with the six setscrews and spring washers, tightening the screws a turn at a time by diagonal selection. Ensure that the "B" stamped adjacent to one of the dowel holes coincides with the "B" stamped on the periphery of the flywheel.

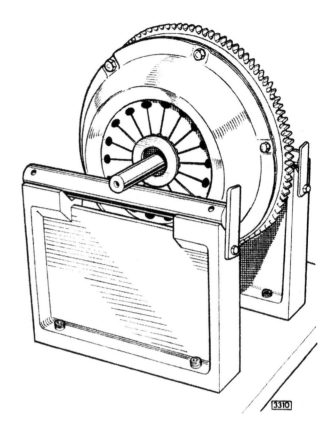

Fig. 27. Checking clutch and flywheel balance.

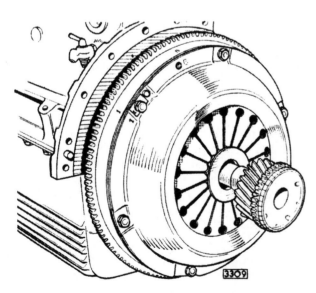

Fig. 26. Centralising the clutch plate on the flywheel by means of a dummy shaft.

CONDITION OF CLUTCH FACINGS

The possibility of further use of the friction facings of the clutch is sometimes raised, because they have a polished appearance after considerable service. It is natural to assume that a rough surface will give a higher fractional value against slipping, but this is not correct.

Since the introduction of non-metallic facings of the moulded asbestos type, in service, a polished surface is a common experience, but it must not be confused with a glazed surface which is sometimes encountered due to the conditions discussed below.

The ideal smooth or polished condition will provide a normal contact, but a glazed surface may be due to a film or a condition introduced, which entirely alters the frictional value of the facings. These two conditions might be simply illustrated by the comparison between a polished wood, and a varnished surface. In the former the contact is still made by the original material, whereas in the latter instance, a film of dried varnish is interposed between the contact surfaces.

Page E.26

CLUTCH

The following notes are issued with a view to giving useful information on this subject:—

(a) After the clutch has been in use for some little time, under perfect conditions (that is, with the clutch facings working on true and polished or ground surfaces of correct material, without the presence of oil, and with only that amount of slip which the clutch provides for under normal conditions) then the surface of the facings assumes a high polish, through which the grain of the material can be clearly seen. This polished facing is of mid-brown colour and is then in a perfect condition.

(b) Should oil in small quantities gain access to the clutch in such a manner as to come in contact with the facings it will burn off, due to the heat generated by slip which occurs under normal starting conditions. The burning off of this small amount of lubricant, has the effect of gradually darkening the facings, but, provided the polish on the facings remains such that the grain of the material can be clearly distinguished, it has very little effect on clutch performance.

(c) Should increased quantities of oil or grease obtain access to the facings, one or two conditions, or a combination of the two, may arise, depending upon the nature of oil, etc.

(1) The oil may burn off and leave on the surface facings a carbon deposit which assumes a high glaze and causes slip. This is a very definite, though very thin deposit, and in general it hides the grain of the material.

(2) The oil may partially burn and leave a resinous deposit on the facings, which frequently produces a fierce clutch, and may also cause a "spinning", clutch due to a tendency of the facings to adhere to the flywheel or pressure plate face.

(3) There may be a combination of (1) and (2) conditions, which is likely to produce a judder during clutch engagement.

(d) Still greater quantities of oil produce a black soaked appearance of the facings, and the effect may be slip, fierceness, or judder in engagement, etc., according to the conditions. If the conditions under (c) or (d) are experienced, the clutch driven plate should be replaced by one fitted with new facings, the cause of the presence of the oil removed and the clutch and flywheel face thoroughly cleaned.

PEDAL ADJUSTMENT

This adjustment is most important and the instructions given should be carefully followed; faulty adjustment falls under two headings:—

(a) Insufficient free (or unloaded) pedal travel may cause a partly slipping clutch condition which becomes aggravated as additional wear takes place on the facings, and this can result in a slipping clutch leading to burning out unless corrected. Over-travel of effective pedal movement only imposes undue internal strain and causes excessive bearing wear.

(b) Too much free pedal movement results in inadequate release movement of the bearing and may produce a spinning plate condition that is, dragging clutch rendering clean changes impossible.

CLUTCH

THE SLAVE CYLINDER (Fig. 28)

The clutch slave cylinder consists of a body (4) which incorporates two threaded connections and is bored to accommodate a piston (5) against the inner face of which a rubber cup (3) is loaded by a cup filler (2) and a spring (1); the travel of the piston is limited by a circlip (6) fitted in a groove at the end of the bore. A rubber boot (7) through which a push-rod passes, is fitted on to the body to prevent the intrusion of dirt or moisture.

One of the connections in the body receives a pipe from the clutch master cylinder, whilst the other is fitted with a bleeder screw; the connection for the pipe is parallel to the mounting flange on the body.

Removal

To remove from the vehicle, disconnect the pipe, detach the rubber boot from the body and remove the fixing screws; leave the push-rod attached to the vehicle. If the boot is not being renewed it may be left on the push-rod.

Dismantling

Remove the circlip (6) from the end of the bore and apply a **low** air pressure to the open connection to expel the piston (5) and the other parts; remove the bleeder screw.

Assembling

Prior to assembly, smear all internal parts and the bore of the body with Rubberlube.

Fit the spring (1) in the cup filler (2) and insert these parts, spring uppermost, into the bore of the body (4). Follow up with the cup (3), lip leading, taking care not to turn back or buckle the lip; then insert the piston (5), flat face innermost, and fit the circlip (6) into the groove at the end of the bore.

Refitting

Fit the rubber boot (7) on the push-rod, if removed previously, and offer up the slave cylinder to the vehicle, with the push-rod entering the bore. Secure the cylinder with the fixing screws and stretch the large end of the boot into the groove on the body. Fit into their respective connections the bleeder screw and the pipe from the clutch master cylinder.

"Bleed" the clutch as described on page E.9.

Note:

On later cars an hydrostatically operated slave cylinder is fitted and, as normal clearance is automatically compensated, no clearance adjustment is required.

The new slave cylinder can be identified by the absence of a return spring.

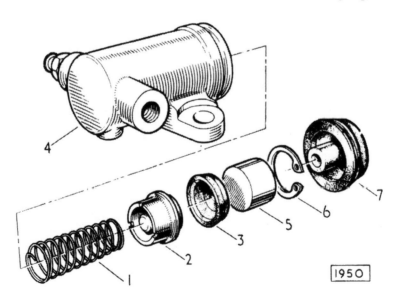

1. Spring.
2. Cup filler.
3. Rubber cup.
4. Body.
5. Piston.
6. Circlip.
7. Rubber boot.

Fig. 28. The clutch slave cylinder.

Page E.28

SECTION F
GEARBOX

3·8 "E" TYPE
GRAND TOURING MODELS

INDEX

	Page
Gearbox Ratio Data	F.3
Data	F.3
Routine Maintenance	F.4
Recommended Lubricants	F.4
Gearbox—To remove and refit	F.7
Gearbox—To dismantle	F.7
Dismantling the mainshaft	F.8
Dismantling the constant pinion shaft	F.9
Gearbox—To re-assemble	
Checking layshaft end float	F.10
Assembling the mainshaft	F.10
Assembling the 2nd gear synchro assembly	F.11
Fitting the 2nd gear assembly to the mainshaft	F.11
Assembling the 3rd/top synchro assembly	F.12
Fitting the 3rd/top synchro assembly to the mainshaft	F.12
Assembling the constant pinion shaft	F.13
Assembling the gears to the casing	F.14
Fitting the top cover	F.14
Fitting the rear cover	F.14
Fitting the clutch housing	F.14

GEARBOX

The gearbox is of the four-speed type with synchromesh on the second, third and top gears; these gears are of single helical form and are in constant mesh. The first and reverse gears have spur teeth which slide into mesh.

	Gearbox Ratios	Overall Ratios			
Gearbox prefix	EB	EB	EB	EB	EB
Gearbox suffix	JS	JS	JS	JS	JS
First and Reverse	3·377 : 1	11·177 : 1	9·894 : 1	10·367 : 1	11·954 : 1
Second	1·86 : 1	6·156 : 1	5·449 : 1	5·710 : 1	6·584 : 1
Third	1·283 : 1	4·246 : 1	3·759 : 1	3·938 : 1	4·541 : 1
Top	1 : 1	3·31 : 1	2·93 : 1	3·07 : 1	3·54 : 1
Axle ratios		3·31 : 1	2·93 : 1	3·07 : 1	3·54 : 1

Ordering Spare Parts

It is essential when ordering spare parts for an individual gearbox, to quote the prefix and suffix letters in addition to the gearbox number.

The gearbox number is stamped on a lug situated at the left-hand rear corner of the gearbox casing and on the top cover.

DATA

Second gear end-float on mainshaft—·002" to ·004" (·05 to ·10 mm.)
Third gear end-float on mainshaft—·002" to ·004" (·05 to ·10 mm.)
Layshaft end-float on countershaft—·002" to ·004" (·05 to ·10 mm.)

Page F.3

GEARBOX

ROUTINE MAINTENANCE

EVERY 2,500 MILES (4,000 KM.)
Gearbox Oil Level

Check the level of the oil in the gearbox with the car standing on level ground. A combined level and filler plug is fitted on the left-hand side of the gearbox.

Clean off any dirt from around the plug before removing it.

The level of the oil should be to the bottom of the filler and level plug hole.

The filler plug is accessible from inside the car through an aperture in the left-hand vertical face of the gearbox cowl. To obtain access to the plug remove the seat cushion, slide the seat rearwards to the full extent; lift the front carpet and roll forward to expose the two snap fasteners retaining the gearbox cowl trim panel to the floor.

Release the snap fasteners and raise the panel.

Remove the front aperture cover now exposed and insert a tubular wrench through the aperture to remove the plug.

In the interests of cleanliness always cover the carpets before carrying out any lubrication operations.

EVERY 10,000 MILES (16,000 KM.)
Changing the Gearbox Oil

The draining of the gearbox should be carried out at the end of a run when the oil is hot and therefore will flow more freely. The drain plug is situated at the front end of the gearbox casing.

After all the oil has drained replace the drain plug and refill the gearbox with the recommended grade of oil through the combined filler and level plug hole situated on the left-hand side of the gearbox casing; the level should be to the bottom of the hole.

Fig. 1. Showing the gearbox filler and level plug access hole.

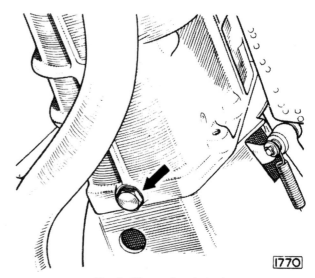

Fig. 2. The gearbox drain plug.

Recommended Lubricants

Mobil	Castrol	Shell	Esso	B.P.	Duckham	Regent Caltex/Texaco
Mobiloil A	Castrol XL	Shell X100 30	Esso Motor Oil 20W/30	Energol SAE 30	NOL 30	Havoline 30

Page F.4

GEARBOX

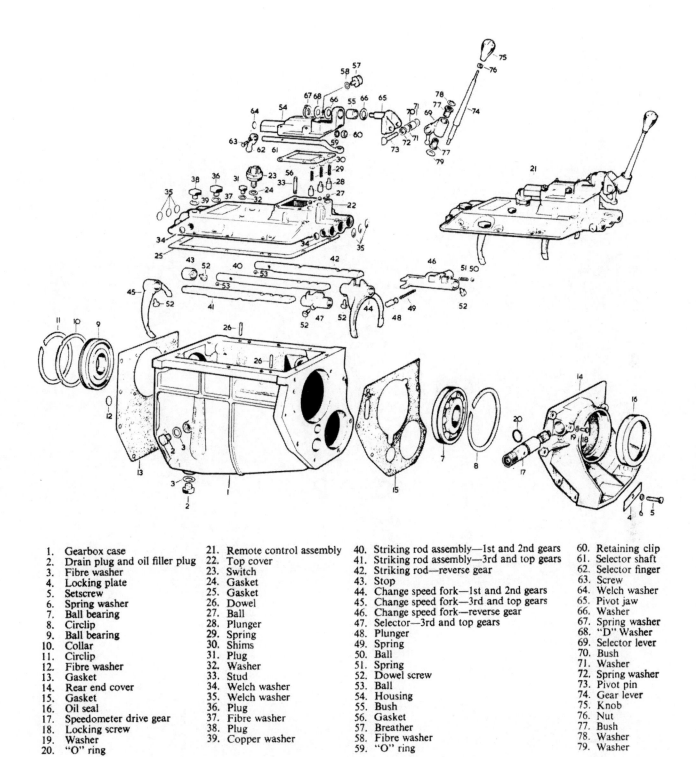

1. Gearbox case
2. Drain plug and oil filler plug
3. Fibre washer
4. Locking plate
5. Setscrew
6. Spring washer
7. Ball bearing
8. Circlip
9. Ball bearing
10. Collar
11. Circlip
12. Fibre washer
13. Gasket
14. Rear end cover
15. Gasket
16. Oil seal
17. Speedometer drive gear
18. Locking screw
19. Washer
20. "O" ring
21. Remote control assembly
22. Top cover
23. Switch
24. Gasket
25. Gasket
26. Dowel
27. Ball
28. Plunger
29. Spring
30. Shims
31. Plug
32. Washer
33. Stud
34. Welch washer
35. Welch washer
36. Plug
37. Fibre washer
38. Plug
39. Copper washer
40. Striking rod assembly—1st and 2nd gears
41. Striking rod assembly—3rd and top gears
42. Striking rod—reverse gear
43. Stop
44. Change speed fork—1st and 2nd gears
45. Change speed fork—3rd and top gears
46. Change speed fork—reverse gear
47. Selector—3rd and top gears
48. Plunger
49. Spring
50. Ball
51. Spring
52. Dowel screw
53. Ball
54. Housing
55. Bush
56. Gasket
57. Breather
58. Fibre washer
59. "O" ring
60. Retaining clip
61. Selector shaft
62. Selector finger
63. Screw
64. Welch washer
65. Pivot jaw
66. Washer
67. Spring washer
68. "D" Washer
69. Selector lever
70. Bush
71. Washer
72. Spring washer
73. Pivot pin
74. Gear lever
75. Knob
76. Nut
77. Bush
78. Washer
79. Washer

Fig. 3. Exploded view of the gearbox casing and top cover.

Page F.5

GEARBOX

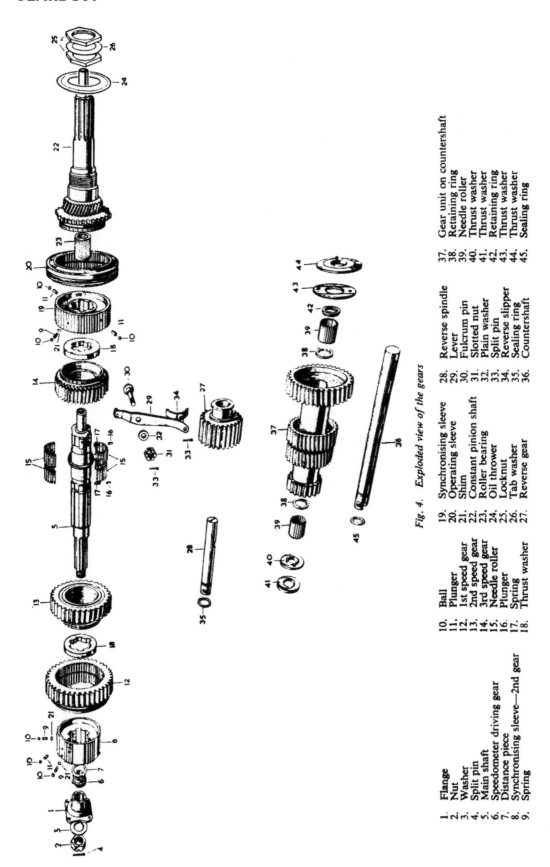

Fig. 4. Exploded view of the gears

1. Flange
2. Nut
3. Washer
4. Split pin
5. Main shaft
6. Speedometer driving gear
7. Distance piece
8. Synchronising sleeve—2nd gear
9. Spring
10. Ball
11. Plunger
12. 1st speed gear
13. 2nd speed gear
14. 3rd speed gear
15. Needle roller
16. Plunger
17. Spring
18. Thrust washer
19. Synchronising sleeve
20. Operating sleeve
21. Shim
22. Constant pinion shaft
23. Roller bearing
24. Oil thrower
25. Locknut
26. Tab washer
27. Reverse gear
28. Reverse spindle
29. Lever
30. Fulcrum pin
31. Slotted nut
32. Plain washer
33. Split pin
34. Reverse slipper
35. Sealing ring
36. Countershaft
37. Gear unit on countershaft
38. Retaining ring
39. Needle roller
40. Thrust washer
41. Retaining ring
42. Thrust washer
43. Thrust washer
44. Thrust washer
45. Sealing ring

Page F.6

GEARBOX

GEARBOX—TO REMOVE AND REFIT

In order to remove the gearbox it is necessary to remove the gearbox and engine as an assembly as described in Section B "Engine".

GEARBOX—TO DISMANTLE

Drain the gearbox by removing plug and fibre washer situated at base of the casing. Place gearbox in neutral and remove the ten setscrews with spring washers securing the top cover. Lift off top cover noting that this is located by two dowels fitted in the gearbox case. Remove and scrap the gasket.

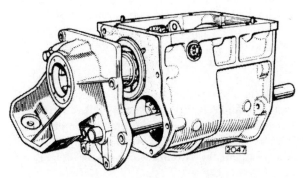

Fig. 6. Showing the removal of the rear cover: note the dummy countershaft inserted at the front end of the casing

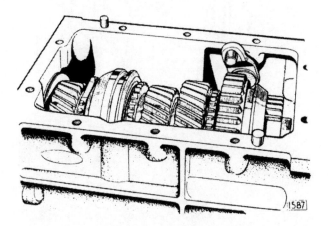

Fig. 5. The top cover removed showing the layout of the mainshaft gears

Remove the clutch slave cylinder from the clutch housing. Detach the spring clips and remove the clutch release bearing. Release the locknut and remove the allan headed screw securing the clutch fork to shaft. Withdraw shaft downwards and remove fork. From inside the clutch housing remove the locking wire from the two bolts and tap back the tabs on the locking washers. Unscrew the eight bolts and remove the clutch housing.

Remove the speedometer cable drive attachment from the speedometer driven gear by rotating the knurled thumb nut in an anti-clockwise direction. Remove the locking screw retaining the speedometer driven gear bush in the end cover. Withdraw the driven gear and bearing.

Remove the fibre blank from the front end of the layshaft.

Engage top and first gear. Extract the split pin, remove the nut and plain washer retaining the universal joint flange to the mainshaft and withdraw the flange from the splines on the shaft.

Remove the seven setscrews securing the rear end cover to the gearbox casing. (Do not disturb the layshaft/reverse idler locking plate). Withdraw the end cover complete with shafts, at the same time inserting a dummy countershaft into the countershaft bore at the front of the gearbox casing (see Fig. 6). The dummy shaft and countershaft must be kept in contact until the countershaft is clear of the casing.

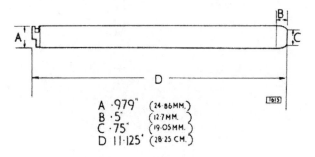

A ·979" (24·86MM.)
B ·5" (12·7MM.)
C ·75" (19·05MM.)
D 11·125" (28·25CM.)

Fig. 7. The dummy countershaft

Page F.7

GEARBOX

Withdraw the speedo drive gear. Withdraw the dummy countershaft allowing the layshaft gear unit to drop into the bottom of the casing.

Rotate the constant pinion shaft until the two cutaway portions of the driving gear are facing the top and bottom of the casing. Tap the mainshaft to the front to knock the constant pinion shaft with ball bearing forward out of the case (see Fig. 8). Remove the constant pinion shaft and withdraw the roller bearing from the shaft spigot. Continue to tap mainshaft forward until free of the rear bearing. Tap the bearing rearward out of the casing.

Push the reverse gear forward out of engagement to clear the mainshaft first speed gear. Lift the front end of the mainshaft upwards and remove complete with all mainshaft gears forward out of the casing leaving the layshaft in the bottom of the casing (see Fig. 9).

Draw reverse wheel rearwards as far as it will go to clear layshaft first speed gear. Lift out layshaft gear unit observing inner and outer thrust washers fitted at each end of the gears. Take care not to lose any needles which are located at each end of the gear unit.

Push reverse gear back into the case and remove through top. Note bush which is a press fit in reverse gear.

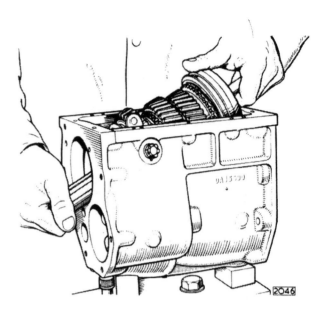

Fig. 9. Removing the mainshaft from the gearbox casing.

DISMANTLING THE MAINSHAFT

Withdraw the top/third gear operating and synchronising sleeves forward off the shaft. Press the operating sleeve off the synchronising sleeve and remove the six synchronising balls and springs. Remove the interlock plungers and balls from the synchro sleeve.

Withdraw the second gear synchronising sleeve complete with first speed gear rearwards off the shaft. Press the first speed gear off the synchronising sleeve and remove the six synchronising balls and springs.

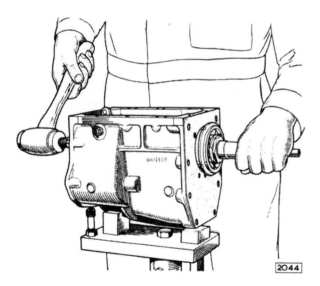

Fig. 8. The constant pinion shaft is removed by tapping the mainshaft forward.

Page F.8

GEARBOX

Remove the interlock ball and plunger from the synchro sleeve.

Press in the plunger locking the third speed gear thrust washer (see Fig. 10) and rotate washer until splines line up, when washer can be withdrawn. Remove the washer forward off shaft followed by third speed gear, taking care not to lose any needles which will emerge as the gear is removed. Remove the spring and plunger.

Press in the plunger locking the second speed gear thrust washer (see Fig. 11) and rotate washer until splines line up, when washer can be withdrawn. Remove the washer rearwards off shaft followed by second speed gear, taking care not to lose any needles which will emerge as the gear is removed. Remove the spring and plunger.

DISMANTLING THE CONSTANT PINION SHAFT

Knock back tab washer securing locknuts and remove locknuts (right-hand thread). Withdraw the bearing from the shaft and remove the oil thrower.

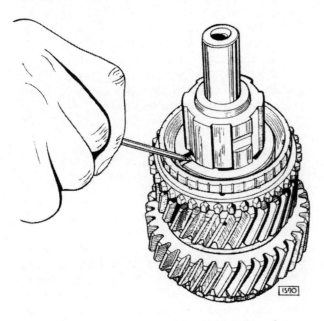

Fig. 10. Depressing the 3rd speed thrust washer locking plunger.

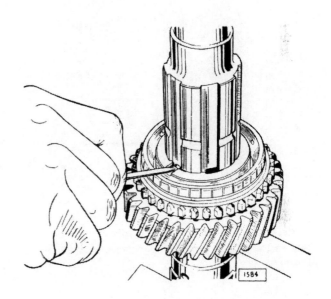

Fig. 11. Depressing the 2nd speed thrust washer plunger.

Page F.9

GEARBOX

GEARBOX—TO RE-ASSEMBLE

CHECKING LAYSHAFT END-FLOAT

Check the clearance between bronze thrust washer and the casing at rear of layshaft (see Fig. 12). The end-float should be .002" to .004" (.05 to .10 mm.). Thrust washers are available in thicknesses of .152", 1.56", .159", .162" and .164" (3.86, 3.96, 4.04, 4.11 and 4.17 mm.) to provide a means of adjusting the end-float.

Note: The gearbox must not be gripped in a vice when checking the end float otherwise a false reading will be obtained.

Remove dummy countershaft and insert a thin rod in its place.

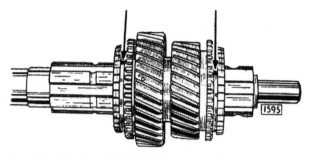

Fig. 13. Showing the holes through which the thrust washer locking plungers are depressed

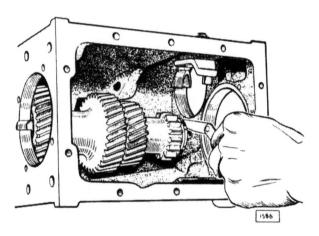

Fig. 12. Checking the layshaft end float

Place bushed reverse gear in slipper and draw gear rearwards as far as possible to give clearance for fitting layshaft gear unit.

ASSEMBLING THE MAINSHAFT

Fit the needle rollers (41 off) behind the shoulder on the mainshaft and slide the second speed gear, synchronising cone to rear, on to rollers. Apply grease to the needle rollers to facilitate assembly. Fit the second speed thrust washer spring and plunger into plunger hole. Slide thrust washer up shaft and over splines. Align large hole in synchro cone and with a steel pin compress plunger and rotate thrust washer into locked position with cutaway in line with plunger. Check the end-float of the second gear on the mainshaft by inserting a feeler gauge between the thrust washer and the shoulder on the mainshaft. The clearance should be .002" to .004" (.05 to .10 mm.). Thrust washers are available in the following thicknesses to enable the end-float to be adjusted:

.471"/.472"—(11.96/11.99 mm.)
.473"/.474"—(12.01/12.03 mm.)
.475"/.476"—(12.06/12.09 mm.)

Fit the needle rollers (41 off) in front of the shoulder on the mainshaft and slide the third speed gear, synchronising cone to front, on to rollers. Apply grease to the needle rollers to facilitate assembly. Fit the third speed thrust washer spring and plunger into plunger hole. Slide thrust washer down shaft and over splines. Align large hole in synchro cone and with a steel pin compress plunger and rotate thrust washer into locked position with cutaway in line with plunger. Check the end-float of the third gear on the mainshaft by inserting a feeler gauge between the thrust washer and the shoulder on the mainshaft. The clearance should be .002" to .004" (.05 to .10 mm.). Thrust

Page F.10

GEARBOX

washers are available in the following thicknesses to enable the end-float to be adjusted:

.471″/.472″—(11.96/11.99 mm.)
.473″/.474″—(12.01/12.03 mm.)
.475″/.476″—(12.06/12.09 mm.)

ASSEMBLING THE 2nd GEAR SYNCHRO ASSEMBLY

Fit the springs and balls (and shims if fitted) to the six blind holes in the synchro sleeve. Fit the 1st speed gear to the 2nd speed synchronising sleeve with the relieved tooth of the internal splines in the gear in line with the stop in the sleeve (see Fig. 14). Compress the springs by means of a hose clip or by inserting the assembly endwise in a vice and slowly closing the jaws. Slide the operating sleeve over the synchronising sleeve until the balls can be heard and felt to engage the neutral position groove.

It should require 62 to 68 lbs. (28 to 31 kg.) pressure to disengage the synchronising sleeve from the neutral position in the operating sleeve. In the absence of the necessary equipment to check this pressure, grip the operating sleeve in the palms of the hands and press the synchronising sleeve with the fingers until it disengages from the neutral position; it should require firm finger pressure before disengaging. Shims can be fitted underneath the springs to adjust the pressure of the balls against the operating sleeve.

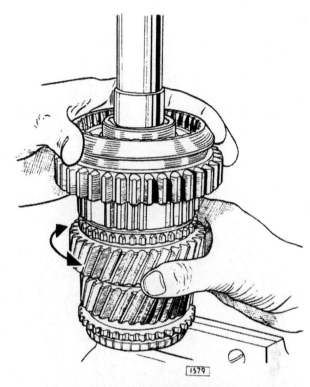

Fig. 15. With the 1st gear engaged and slight downward pressure on the synchro assembly the 2nd gear assembly should be free to rotate

FITTING THE 2nd GEAR ASSEMBLY TO THE MAINSHAFT

Fit the 1st speed gear/2nd speed synchro assembly to the mainshaft (any spline) and check that the synchro sleeve slides freely on the mainshaft, when the ball and plunger is not fitted. If it does not, try the sleeve on different splines on the mainshaft and check for burrs at the end of the splines.

Remove the synchro assembly from the mainshaft, fit the ball and plunger and refit to the same spline on the mainshaft.

Check the interlock plunger as follows:—
Slide the outer operating sleeve into the first gear position as shown in Fig. 15.

With slight downward pressure on the synchro assembly the 2nd speed gear should rotate freely without any tendency for the synchro cones to rub.

If the synchro cones are felt to rub, a longer plunger should be fitted to the synchro sleeve. Plungers are available in the following lengths:—
·490″, ·495″, and ·500″ (12·4, 12·52 and 12·65 mm.).

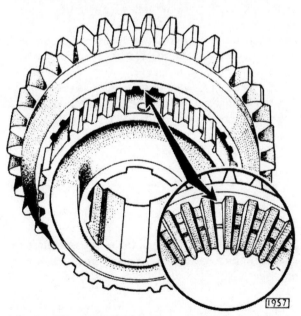

Fig. 14. When fitting the 1st speed gear to the 2nd speed synchro-sleeve the relieved tooth on the internal splines must be in line with the stop pin in the sleeve

Page F.11

GEARBOX

ASSEMBLING THE 3rd/TOP SYNCHRO ASSEMBLY

Fit the springs and balls (and shims if fitted) to the six blind holes in the inner synchronising sleeve. Fit the wide chamfer end of the operating sleeve to the large boss end of inner synchronising sleeve (see Fig. 16) with the two relieved teeth in operating sleeve in

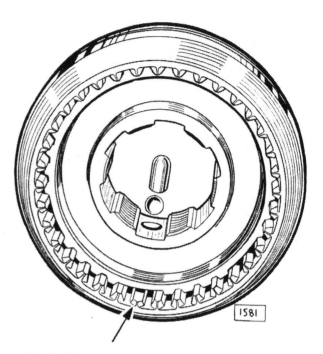

Fig. 17. The relieved tooth in the operating sleeve must be in line with the interlock holes in the synchro sleeve.

FITTING THE 3rd/TOP SYNCHRO ASSEMBLY TO THE MAINSHAFT

Fit the interlock balls and plungers, balls first, to the holes in the synchronising sleeve.

When fitting the 3rd speed/top gear synchro assembly to the mainshaft note the following points:—

(a) There are two transverse grooves on the mainshaft splines which take the 3rd/top synchro assembly and the relieved tooth at the wide chamfer end of the outer operating sleeve must be in line with the **foremost** groove in the mainshaft (Fig. 18). Failure to observe this procedure will result in the locking plungers engaging the wrong grooves thereby preventing full engagement of top and third gears.

(b) The wide chamfer end of the outer operating sleeve must be facing forward, that is, towards the constant pinion shaft end of the gearbox.

The inner sleeve must slide freely on the mainshaft, when the balls and plungers are not fitted. If it does not, check for burrs at the ends of the splines.

Fit the two balls and plungers to the holes in the inner synchro sleeve and refit the synchro assembly to the mainshaft observing points 'a' and 'b' above.

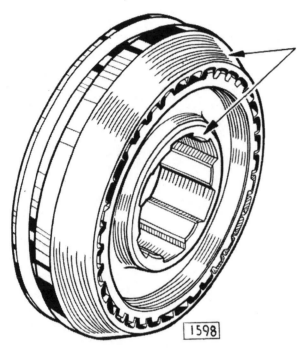

Fig. 16. The wide chamfer end of the operating sleeve must be fitted to the same side as the large boss on the synchro sleeve.

line with the two ball and plunger holes in the synchronising sleeve (see Fig. 17). Compress the springs by means of a hose clip or by inserting the assembly endwise in a vice and slowly closing the jaws. Slide the operating sleeve over the synchronising sleeve until the balls can be heard and felt to engage the neutral position groove.

It should require 52 to 58 lbs. (24 to 26 kg.) pressure to disengage the synchronising sleeve from the neutral position in the operating sleeve. In the absence of the necessary equipment to check this pressure, grip the operating sleeve in the palms of the hands and press the synchronising sleeve with the fingers until it disengages from the neutral position; it should require firm finger pressure before disengaging. Shims can be fitted underneath the springs and balls to adjust the pressure of the balls against the operating sleeve.

Page F.12

GEARBOX

Lift and lower the synchro assembly; it should be possible to move the assembly approximately $\frac{3}{16}$" (4.5 mm.) without any drag being felt. Also with slight downward pressure exerted on the synchro assembly the 3rd speed gear should be free to rotate without any tendency for the synchro cones to rub.

If it is found that the synchro assembly does not move freely a shorter top gear plunger should be fitted. If the 3rd gear synchro cones are felt to rub a longer top gear plunger should be fitted; looking at the wide chamfer end of the outer operating sleeve, the top gear plunger is one in line with the relieved tooth in the operating sleeve.

Plungers are available in the following lengths:— ·490", ·495" and ·500" (12·4, 12·52 and 12·65 mm.).

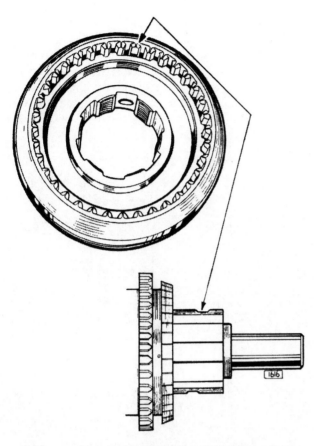

Fig. 18. The relieved tooth at the wide chamfer end of the outer operating sleeve must be in line with the foremost groove in the mainshaft.

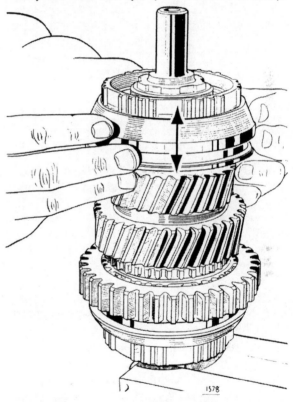

Fig. 19. Checking the 3rd speed interlock plunger with the 3rd speed engaged there should be approximately $\frac{3}{32}$" (2.5 m/m) movement without drag.

Check the interlock plungers as follows:—

Slide the 3rd/top operating sleeve over the 3rd speed gear dogs as shown in Fig. 19. With the 3rd gear engaged lift and lower the synchro assembly; it should be possible to move the assembly approximately $\frac{3}{32}$" (2·5 mm.) without any drag being felt. If it is found that the synchro assembly does not move freely a shorter 3rd speed plunger should be fitted; looking at the wide chamfer end of the outer operating sleeve this is the plunger that is not opposite the relieved tooth in the operating sleeve.

Plungers are available in the following lengths:— ·490", ·495" and ·500" (12·4, 12·52 and 12·65 mm.).

Next slide the operating sleeve into the top gear position as shown in Fig. 20.

ASSEMBLING THE CONSTANT PINION SHAFT

Fit the oil thrower followed by ball bearing on to shaft with circlip and collar fitted to outer track of bearing. Screw on nut (right-hand thread) and fit tab washer and locknut. Fit the roller race into the shaft spigot bore.

Page F.13

GEARBOX

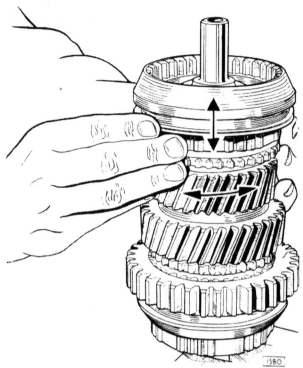

Fig. 20. Checking the 4th (top) interlock plunger. With the top gear engaged there should be approximately $\frac{3}{16}$" (4.76 m/m) axial movement without any drag. With the top gear still engaged and with a slight downward pressure exerted on the synchro assembly the 3rd speed gear should be free to rotate

ASSEMBLING THE GEARS TO THE CASING

Enter the mainshaft through the top of the casing and pass to the rear through bearing hole in case. Fit a new gasket to the front face of casing. Offer up the constant pinion shaft at the front of the case with cutaway portions of toothed driving member facing the top and bottom of the casing. Tap the constant pinion shaft to the rear until the collar and circlip on the bearing butt against the casing. Holding the constant pinion shaft in position tap in the rear bearing complete with circlip.

Lift the layshaft cluster into mesh with the thin rod and insert a dummy countershaft through the countershaft bore in front face of the casing (see Fig. 21).

Engage top and first gears. Fit the Woodruff key and speedo drive gear to the mainshaft. Fit the tab washer and locknut and secure. Place gearbox in neutral.

Fit the clutch operating fork and insert shaft. Fit the locking screw and locknut. Fit the release bearing and spring clips. Engage slave cylinder with operating rod and slide on to studs. Fit the spring anchor plate to lower stud and secure with the nuts. Fit the return spring.

FITTING THE TOP COVER

Fit a new gasket on to top face of case. Offer up the top cover, noting that this is located by two dowels and secure in position with ten setscrews and spring washers. (Two long screws at rear and two short screws at front.) Fit the gearbox drain plug and fibre washer.

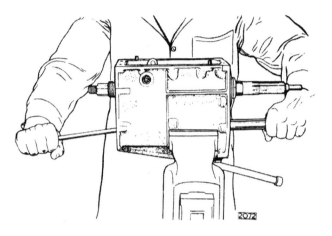

Fig. 21. Lifting the layshaft into mesh and inserting the dummy countershaft

FITTING THE REAR COVER

Fit a new gasket to the rear face of the gearbox casing. Offer up the rear end cover complete with counter and reverse shafts and tap into position, driving the dummy countershaft forward out of the casing. Secure the rear cover with seven setscrews and spring washers.

Fit a new fibre washer at the front end of the countershaft. Fit the speedo driven gear and bearing to the rear cover.

Refit the speedometer cable drive attachment to the speedometer driven gear. Care must be taken to ensure that the square drive shaft protruding from the unit has entered into the gearbox drive correctly before tightening the nut.

FITTING THE CLUTCH HOUSING

Fit a new oil seal into the clutch housing, lip of oil seal facing the gearbox.

Fit the clutch housing and secure with the eight bolts and three tab washers and locking wire.

Page F.14

GEARBOX

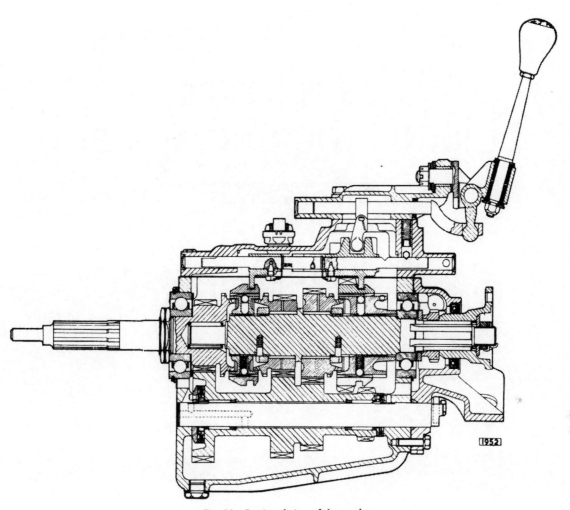

Fig. 22. Sectioned view of the gearbox

Page F.15

SECTION G

PROPELLER SHAFTS
3·8 "E" TYPE
GRAND TOURING MODELS

INDEX

	Page
Description	G.3
Routine Maintenance	
Universal joints	G.3
Sliding spline	G.3
Recommended lubricants	G.4
Propeller Shaft	
Removal	G.5
Refitting	G.5
The Universal Joints	
Examine and check for wear	G.5
To dismantle	G.5
Assembling	G.7

Page G.1/2

PROPELLER SHAFTS

DESCRIPTION

A Hardy Spicer propeller shaft of the open type with needle roller universal joints and a sliding spline at the front end is fitted.

ROUTINE MAINTENANCE

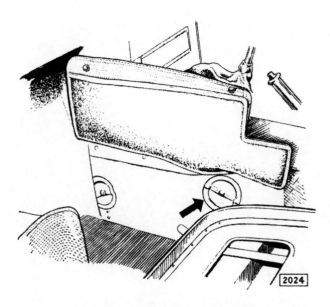

Fig. 1. Access hole to front universal joint

The grease nipple for the front universal joint is only accessible from inside the car through an aperture in the left-hand vertical face of the gearbox cowl.

To obtain access to the nipple remove the seat cushion, slide the seat rearwards to the full extent; lift the front carpet and roll forward to expose the two snap fasteners retaining the gearbox cowl trim panel to the floor.

Release the snap fasteners and raise the panel.

Remove the rear metal or rubber aperture cover now exposed and insert the grease gun through the aperture to grease the universal joint.

It may be necessary to move the car slightly in order to bring the nipple to the required position.

In the interests of cleanliness always cover the carpets before carrying out lubrication.

EVERY 2,500 MILES (4,000 KM.)

Universal Joints (Early cars)

The propeller shaft is fitted with two needle roller bearing universal joints which should be lubricated with the recommended grade of grease.

The grease nipple for the rear end of the propeller shaft is accessible from underneath the car.

Sliding Spline (Early cars)

The front end of the propeller shaft is fitted with a sliding joint which should be lubricated with the recommended grade of grease through the nipple situated at the rear of the universal joint yoke. The grease nipple is accessible through a hole in the gearbox cowl as described in the lubrication of propeller shaft universal joints.

Note: Later cars are fitted with "sealed for life" universal joints and sliding spline which do not require periodic lubrication.

Page G.3

PROPELLER SHAFTS

Recommended Lubricants

Mobil	Castrol	Shell	Esso	B.P.	Duckham	Regent Caltex/Texaco
Mobilgrease MP	Castrolease LM	Retinax A	Multi-purpose Grease H	Energrease L.2	LB.10	Marfak All purpose

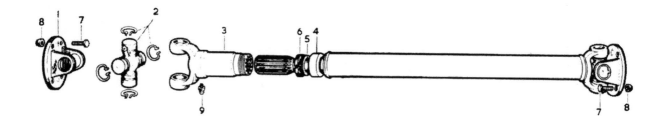

1. Flange yoke
2. Journal assembly
3. Sleeve yoke
4. Dust Cap
5. Steel Washer
6. Cork Washer
7. Bolt
8. Self-locking nut
9. Grease nipple

Fig. 2. Exploded view of propeller shaft assembly

Page G.4

PROPELLER SHAFTS

PROPELLER SHAFT

Removal

To remove the propeller shaft it is necessary to remove either the engine or the rear suspension.

As the removal of the rear suspension is the simpler and quicker operation it is recommended that it should be removed in preference to the engine. Refer to Section K, "Rear Suspension".

Remove the two seat cushions. Remove the four nuts securing each seat to the seat slides and withdraw the seats.

Remove the two screws, on each side of the radio control panel, which secure the ash tray.

Remove the two screws which secure each side of the radio control panel casing to the brackets under the instrument panel. Remove the radio control panel casing.

Remove the three setscrews securing the propeller shaft tunnel cover to the body.

Place the gear lever as far forward as possible and pull the handbrake into the ON position. Unscrew the gear lever knob and lock nut.

Slide the propeller shaft tunnel cover over the gear and handbrake levers and remove the tunnel cover.

Remove the twelve screws and washers securing the plastic gear box cowl to the body. Remove the gearbox cowl.

Remove the four self-locking nuts securing the propeller shaft flange to the gearbox flange.

Withdraw the propeller shaft from the rear of the propeller shaft tunnel.

Refitting

Refitting is the reverse of the removal procedure but it is essential to bleed the rear brakes after refitting the rear suspension. Refer to Section L "Brakes".

THE UNIVERSAL JOINTS

Examine and check for Wear

The parts most likely to show signs of wear after long usage are the bearing races and spider journals. Should looseness in the fit of these parts, load markings or distortion be observed they should be renewed as a unit as worn needle bearings used with a new spider journal or new needle bearings with a worn spider journal will wear more rapidly, making another replacement necessary in a short time.

It is essential that the bearing races are a light drive fit in the yoke trunnion.

In the rare event of wear having taken place in the yoke cross holes, the holes will have become oval and the yokes must be removed.

The other parts likely to show signs of wear are the splined sleeve yoke and splined shaft. A total of ·004" (·1 mm.) circumferential movement, measured on the outside diameter of the spline, should not be exceeded. If wear has taken place above this limit the complete propeller shaft should be replaced.

To Dismantle

To remove the sliding joint from the splined shaft, unscrew the dust cap and pull back the cork washer.

Clean the paint and dirt from the rings and top of bearing races. Remove all the snap rings by pinching together with a pair of pliers and prising out with a screwdriver. If a ring does not snap out of its groove readily, lightly tap end of bearing race to relieve the pressure against the ring.

Hold the joint in the hand and with a soft nosed hammer tap the yoke lug as shown in Fig. 3.

The top bearing will gradually emerge and can finally be removed with the fingers, (see Fig. 4).

If necessary, tap the bearing race from inside with a small diameter bar, taking care not to damage the bearing race. (see Fig. 5).

Repeat this operation for the opposite bearing. The splined sleeve yoke or flange yoke can now be removed. Rest the two exposed trunnions on wood or

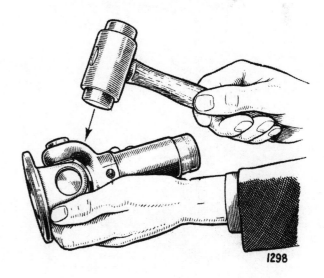

Fig. 3. Tapping the yoke to remove the bearing

Page G.5

PROPELLER SHAFTS

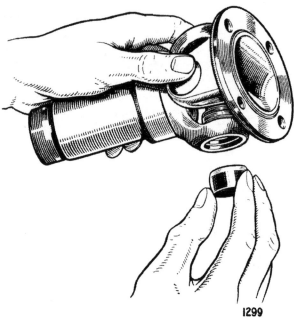

Fig. 4. *Withdrawing the bearing from the universal joint.*

Fig. 6. *Separating the universal joint yokes.*

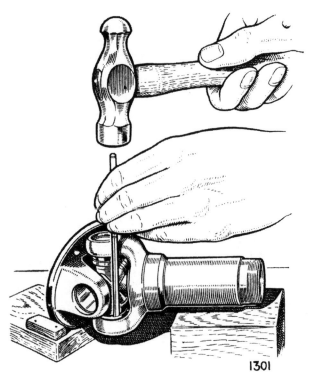

Fig. 5. *Tapping out a bearing with a small diameter bar.*

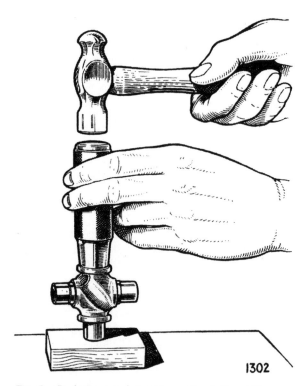

Fig. 7. *Replacing a gasket retainer with a hollow drift.*

Page G.6

PROPELLER SHAFTS

lead blocks, then tap yoke with a soft nosed hammer to remove the two remaining bearing races. Wash all parts in petrol.

Assembling

Insert journal in yoke holes and using a soft round drift with a flat face about $\frac{1}{32}''$ (·8 mm.) smaller in diameter than the hole in the yoke, tap the bearing into position. Repeat this operation for the other three bearings. Fit new snap rings and ensure that they are correctly located in their grooves. If joint appears to bind tap lightly with a wooden mallet, to relieve any pressure of the bearings on the end of the journal. **When replacing the sliding joint it must be refitted with its fixed yoke in line with the fixed yoke at the end of the propeller shaft tube. Arrows are stamped on the two parts to facilitate alignment. (See Fig. 8).**

Should any difficulty be encountered when assembling the needle rollers in the housing, smear the wall of the race with vaseline. It is advisable to install new cork gaskets and gasket retainers on the spider assembly, using a tubular drift as shown in Fig. 7.

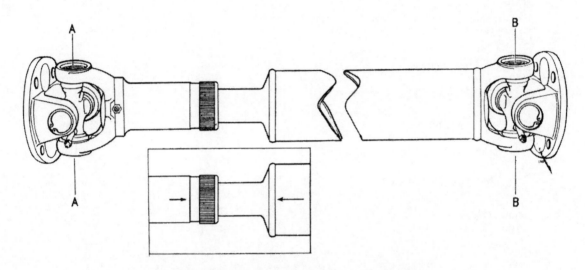

Fig. 8. When refitting the sliding joint to the drive shaft it is ESSENTIAL the yokes A and B are in the same plane. The inset shows the arrows that are stamped on the two parts to facilitate alignment.

SECTION H

REAR AXLE

INDEX

	Page
Description	H.4
Data	H.4
Axle Ratios	H.5
Routine Maintenance	
Checking the oil level	H.8
Half shaft lubrication	H.8
Rear wheel bearing lubrication	H.8
Changing the rear axle oil	H.9
Rear Axle	
Removal	H.9
Refitting	H.10
Half Shaft	
Removal	H.10
Refitting	H.10
The Universal Joints	
Checking for wear	H.11
Dismantling	H.11
Assembling	H.12
The Rear Hub	
Removal	H.12
Dismantling	H.13
Refitting	H.13
Hub bearing end float	H.13

INDEX (Continued)

	Page
The Differential Unit	
Description	H.14
Principle of operation	H.14
Power flow in forward driving	H.16
Power flow in turns	H.16
Power flow with poor traction	H.16
Action on rough roads	H.16
Removing the Differential Assembly from the Carrier	H.16
Dismantling the Differential Assembly	
Removing the pinion	H.16
Removing the output shafts	H.17
Dismantling the output shafts	H.17
Refitting the output shafts	H.18
Dismantling the differential unit	H.18
Assembling the differential unit	H.20
Checking the differential unit for wear	H.20
Pinion adjustment	H.20
Differential bearing and drive gear adjustment	H.23
Final assembly	H.24
Tooth contact	
Ideal contact	H.25
High tooth contact	H.25
Low tooth contact	H.25
Toe contact	H.25
Heel contact	H.25
Backlash	H.25
Gear and pinion movement	H.26

REAR AXLE

DESCRIPTION

The rear axle unit (Fig. 1) is the Salisbury 4.HU type. It is mounted independently from the hubs and road wheels and is fitted with the Thornton "Powr-Lok" differential unit. Short drive shafts with universal joints at each end are coupled to the axle output shafts. These output shafts also provide a mounting for the discs of the inboard mounted disc brakes.

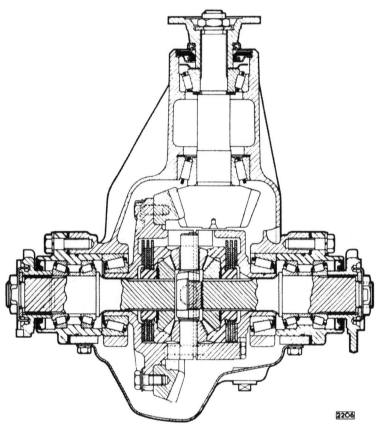

Fig. 1. Sectioned view of the axle unit

DATA

Output Shaft End Float	·001" to ·003" (·02 to ·07 mm.)
Differential Bearing Preload	·006 to ·010" (·15 to ·25 mm.) shim allowance
Pinion Bearing Preload	8 to 12 lbs/in. (·09 to ·14 kg/m.)
Backlash	As etched on drive gear—minimum ·004" (·10 mm.)

Page H.4

REAR AXLE

Tightening Torque
 —Drive Gear Bolts . . . . 70 to 80 lb.ft. (9·7 to 11·1 kgm.)
 —Differential Bearing Cap Bolts . . . . 60 to 65 lb.ft. (8·3 to 9·0 kgm.)
 —Pinion Nut . . . . 120 to 130 lb.ft. (16·6 to 18·0 kgm.)
 Thornton "Powr-Lok" Differential Bolts . . . . . . 40 to 45 lb.ft. (5·5 to 6·2 kgm.)

Axle Ratios

3·07 : 1 (43 × 14)
3·31 : 1 (43 × 13)
3·54 : 1 (46 × 13)

The axle gear ratio is stamped on a tag attached to the assembly by one of the rear cover screws. The axle serial number is stamped on the underside of the gear carrier housing.

Reconditioning Scheme (Great Britain only).
Although full servicing instructions for the rear axle are given in this section it is recommended that, wherever possible, advantage is taken of the factory reconditioning scheme particularly in view of the intricate adjustments and the number of special tools required.

Reconditioned axles are supplied on an exchange basis and comprise an axle complete less half shafts, hubs and brake details; rear axles for overhaul should therefore be returned in this condition.

Recommended Lubricants

	Mobil	Castrol	Shell	Esso	B.P.	Duckham	Regent Caltex/ Texaco
Rear Axle	Mobilube GX 90	Castrol Hypoy	Spirax 90 EP	Esso Gear Oil GP 90/140	Gear Oil SAE 90 EP	Hypoid 90	Multigear Lubricant EP 90
Rear wheel bearings	Mobilgrease MP	Castrolease LM	Retinax A	Esso Multi-purpose Grease H	Energrease L2	LB10	Marfak All purpose

Capacities

Imperial pints	U.S. pints	Litres
2¾	3¼	1·6

Page H.5

REAR AXLE

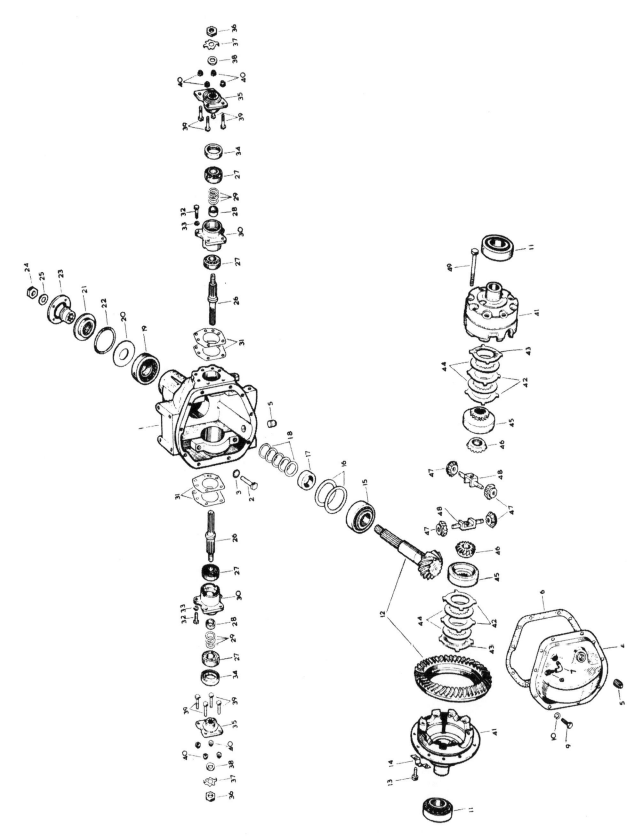

Fig. 2. Exploded view of the axle unit

Page H.6

REAR AXLE

1. Gear carrier
2. Setscrew
3. Lockwasher
4. Cover
5. Plug
6. Gasket
7. Elbow
8. Breather
9. Setscrew
10. Spring washer
11. Roller bearing
12. Crown wheel and pinion
13. Setscrew
14. Locking plate
15. Roller bearing
16. Shim
17. Distance piece
18. Shim
19. Roller bearing
20. Oil thrower
21. Oil seal
22. Gasket
23. Companion flange
24. Nut
25. Washer
26. Output shaft
27. Roller bearing
28. Distance piece
29. Shim
30. Bearing housing
31. Shim
32. Bolt
33. Spring washer
34. Oil seal
35. Flange
36. Nut
37. Tab washer
38. Washer
39. Bolt
40. Self locking nut
41. Differential case
42. Flat friction plate
43. Dished friction plate
44. Friction plate
45. Side gear ring
46. Side gear
47. Pinion mate gear
48. Shaft
49. Bolt

Page H.7

REAR AXLE

ROUTINE MAINTENANCE

EVERY 2,500 MILES (4,000 km.)

Checking Rear Axle Oil Level

Check the level of the oil in the rear axle with the car standing on level ground.

A combined filler and level plug is fitted in the rear of the axle casing accessible from underneath the car. Clean off any dirt from around the plug before removing it.

The level of the oil should be to the bottom of the filler and level plug hole; use only HYPOID oil of the correct grade and since different brands may not mix satisfactorily, draining and refilling is preferable to replenishing if the brand of oil in the axle is unknown.

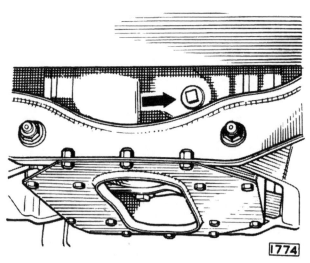

Fig. 3. Rear axle level and filler (the exhaust pipes have been removed for illustrative purposes)

Rear Axle Half Shafts (Early cars)

The two rear axle half shafts are fitted with needle roller bearing universal joints which should be lubricated with the recommended grade of grease through the nipples provided. One nipple is situated at each joint.

Note: Later cars are fitted with "Sealed for life" universal joints which do not require periodic lubrication.

EVERY 10,000 MILES (16,000 km.)

Rear Wheel Bearings

A hole in the hub bearing housing for lubrication of the wheel bearings is accessible after removal of the wheel. Clean off the area around the dust cap to ensure that no dirt enters the hub. Prise out the cap and inject the recommended grade of grease through the hole until no more will enter. If a pressure gun is used take care not to build-up pressure in the hub as the grease may escape past the oil seal. Refit the dust cap.

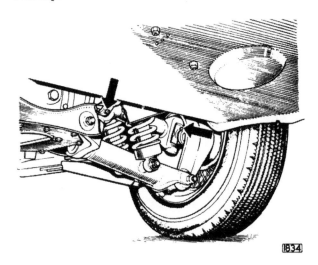

Fig. 4. Half shaft universal joint grease nipples

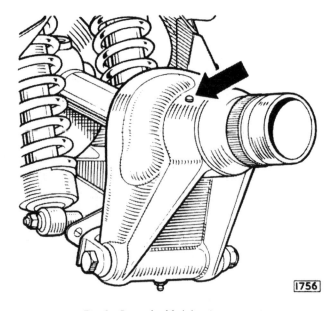

Fig. 5. Rear wheel hub bearing grease cap

Page H.8

REAR AXLE

Changing the Rear Axle Oil

The draining of the rear axle should be carried out at the end of a run when the oil is hot and will therefore flow more freely. The drain plug is situated in the base of the differential casing.

After the oil has drained, replace the drain plug and refill the rear axle with the recommended grade of oil after removal of the combined filler and level plug situated in the rear cover.

The level of the oil should be to the bottom of the filler and level plug hole when the car is standing on level ground.

Use only HYPOID oil of the correct grade.

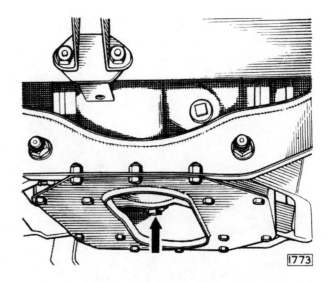

Fig. 6. Rear axle drain plug. (The exhaust pipes have been removed for illustrative purposes)

THE AXLE UNIT

Removal

The following removal and refitting operations are described assuming the rear suspension is removed from the car. If it is possible for the operations to be carried out with the rear suspension in position on the car the fact will be noted in the text.

Remove the rear suspension assembly from the car (as described in section K "Rear Suspension"). Invert the suspension assembly on a bench and remove the 14 bolts securing the tie plate. Remove the tie plate and disconnect the four hydraulic damper and spring units. Remove the four self locking nuts securing the half shaft universal joint to the brake disc and axle output shaft flange. Owing to heat dissipation from the brake disc, it is most important that the locknuts fitted on the output shaft flange studs are of the metal and not nylon locking type. Withdraw the half shaft from the bolts noting the number of camber shims. Remove one self-locking nut from the inner wishbone fulcrum shaft and drift out the shaft. Remove the hub, halfshaft, wishbone and radius arm assembly and repeat the procedure at the other side. Remove the two bolts securing the handbrake

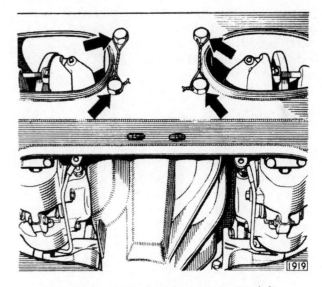

Fig. 7. Showing the top axle casing mounting bolts

compensator linkage and withdraw the compensator. Disconnect the hydraulic feed pipes at the brake calipers. Turn the suspension assembly over and remove the locking wire from the four differential

Page H.9

REAR AXLE

carrier mounting bolts. Unscrew the mounting bolts and remove the cross beam from the differential carrier by tilting forward over the nose of the pinion.

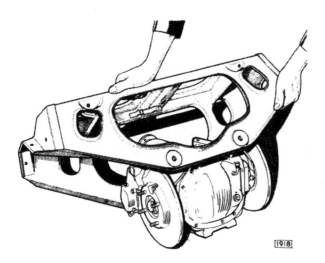

Fig. 8. *Removing the cross beam from the axle unit*

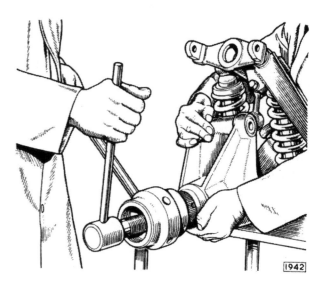

Fig. 9. *Withdrawing the rear hub and carrier from the half shaft using Churchill Tool No. J7*

Refitting

Refitting is the reverse of the removal procedure, it should be noted however, that the inner wishbone fulcrum shaft self-locking nut should be tightened to a torque of 55 lb.ft. (7·6 kgm.). The four differential carrier mounting bolts on the top of the cross beam should be tightened to a torque of 75 lbs/ft. (10·4 kg/m.).

HALF SHAFT

Removal

Remove the lower wishbone outer fulcrum shaft (as described in Section K "Rear Suspension").

Withdraw the split pin and remove the castellated nut and plain washer. Using the extractor, Tool No. J.7 (See "Special Tools" Page H.27) withdraw the hub and hub carrier from the splined end of the half shaft, retaining the inner oil seal track and the end-float spacer. (Early Cars were fitted with shims in addition to the spacer). Remove the front hydraulic damper and spring unit (as described in Section K "Rear Suspension"). Remove the four steel type self-locking nuts securing the half shaft inner universal joint to the axle output shaft flange and inboard brake disc.

Withdraw the half shaft from the bolts noting the number of camber shims. (Fig. 10).

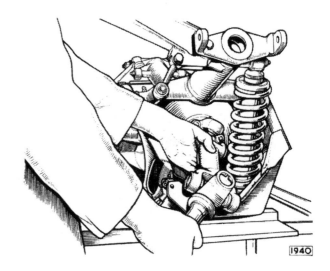

Fig. 10. *Withdrawing the half shaft*

Refitting

Replace the camber shims, place the halfshaft inner universal joint over the four bolts and fit the four locknuts. Refit the front hydraulic damper and spring unit (as described in Section K "Rear Suspension"). Bring the hub carrier into line with the splined end of the half shaft. Place the inner oil seal track, the end-float spacer on to the shaft and introduce the shaft into the hub. Align the split pin hole in the halfshaft with the hole in the hub, locate

Page H.10

REAR AXLE

the splines and feed the splined shaft into the hub. When the threaded end emerges sufficiently refit the washer and castellated nut and draw the shaft into position by tightening the nut to 140 lb.ft. (19·3 kgm.) torque. Replace the split pin. Refit the lower wishbone outer fulcrum shaft as described in Section K "Rear Suspension". If the halfshaft has been renewed it will be necessary to check the camber of the wheels as described in Section K "Rear Suspension".

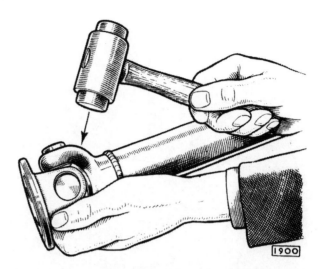

Fig. 12. Tapping the yoke to remove the bearing

Dismantling

Clean the paint and dirt from the rings and top of bearing races. Remove all the snap rings by pinching together with a pair of pliers and prising out with a screwdriver. If a ring does not snap out of its groove readily lightly tap the end of the bearing race to relieve the pressure against the ring.

Hold the joint in the hand and with a soft nosed hammer tap the yoke lug as shown in Fig. 12.

The bearing will gradually emerge and can finally be removed with the fingers (see Fig. 13).

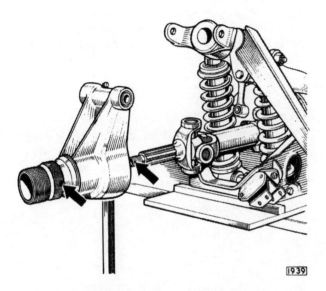

Fig. 11. When assembling the hub to the half shaft ensure that the split pin hole and access hole are in alignment

THE UNIVERSAL JOINTS

Examine and Check for Wear

The part most likely to show wear after long usage are the bearing races and spider journals. Should looseness in the fit of these parts, load markings or distortion be observed, they should be renewed as a unit, as worn needle bearings used with a new spider journal or new needle bearings with a worn spider journal will wear more rapidly, making another replacement necessary in a short time.

It is essential that the bearing races are a light drive fit in the yoke trunnion.

In the rare event of wear having taken place in the yoke cross holes, the holes will have become oval and the yokes must be removed.

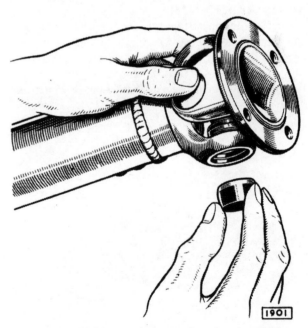

Fig. 13. Withdrawing the bearing from the universal joint

Page H.11

REAR AXLE

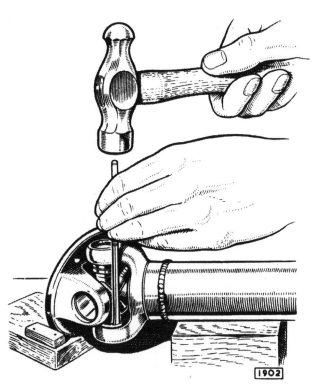

Fig. 14. Tapping out a bearing with a small diameter bar

If necessary tap the bearing race from inside with a small diameter bar taking care not to damage the bearing race (see Fig. 14).

Repeat the operation for the opposite bearing. The flange yoke can now be removed. Rest the two exposed trunnions on wood or lead blocks, then tap the yoke with a soft nosed hammer to remove the two remaining bearing races. Wash all parts in petrol.

Fig. 15. Separating the universal joint yokes

Page H.12

Assembling

Insert the journal in the yoke holes and using a soft round drift with a flat face $\frac{1}{32}''$ ($\cdot$8 mm.) smaller in diameter than the hole in the yoke, tap the bearings into position. Repeat this operation for the other three bearings. Fit new snap rings and ensure that they are correctly located in their grooves. If the joint appears to bind, tap lightly with a wooden mallet to relieve any pressure of the bearings on the end of the journal.

Should any difficulty be encountered when assembling the needle rollers in the housing, smear the wall of the race with vaseline. It is advisable to install new cork gaskets and gasket retainers on the spider assembly using a tubular drift.

THE REAR HUB

Removal

Remove the halfshaft from the hub as described under "Halfshaft Removal". Remove the outer wishbone fork from the hub carrier as described under "Wishbone Removal" (Section K "Rear

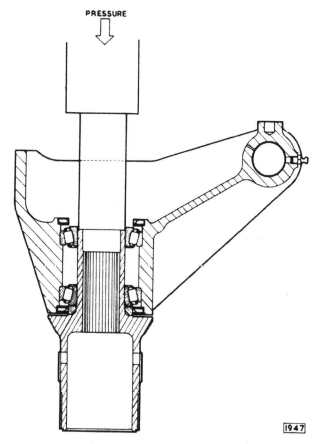

Fig. 16. Pressing the hub from the hub carrier

REAR AXLE

Suspension"). Remove the hub carrier and the hub.

To Dismantle

Invert the hub carrier so that the inner hub bearing is at the top and press out the hub (Fig. 16) with the outer bearing inner race and the outer oil seal track in place, discarding the outer oil seal. Prise out the inner oil seal and remove the inner bearing inner race. Drift out the outer races of the inner and outer bearings if necessary. Withdraw the outer bearing inner race with a suitable extractor.

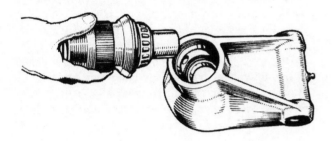

Fig. 17. Removing the hub from the hub carrier

Refitting

If new bearings are to be fitted, press new inner and outer bearing outer races into the hub carrier ensuring that they seat correctly in their recesses.

With the hub carrier held so that the outer bearing will be at the top, place the outer bearing inner race in position and press the outer oil seal into its recess. Press the hub with the outer oil seal track in position into the outer bearing inner race until the hub is fully home.

Hub Bearing End Float

Hold the hub and hub carrier vertically in a vice with the inner end of the hub uppermost. Place the Special Collar (Tool No. J.15) on the hub. Place the inner bearing inner race on the hub and press the race onto the hub until the inner face is flush with the special collar. This will provide end float bearings. The end float should then be measured with a dial test indicator. A spacer should then be fitted in place of the Special Collar to give end float of ·002"—·006" (·051—·152 mm.). Spacers are supplied in thicknesses of ·109"—·143" (2·77—3·63 mm.) in steps of ·003" (·076 mm.) and are lettered A—M (less letter I) as shown in next coloumn.

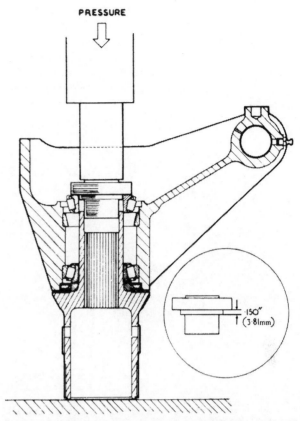

Fig. 18. Pressing in the hub inner bearing inner race using the special collar (Churchill Tool No. J.15)

Spacer Letter	Thickness	
	inches	mm.
A	·109	2·77
B	·112	2·85
C	·115	2·92
D	·118	3·00
E	·121	3·07
F	·124	3·15
G	·127	3·23
H	·130	3·30
J	·133	3·38
K	·136	3·45
L	·139	3·53
M	·142	3·61
P	·145	3·68
Q	·148	3·75
R	·151	3·87

For example, assume the end float measured to be ·025" (·64 mm.). Subtract the nominal end float of ·004" (·10 mm.) from the measured end float giving ·021" (·53 mm.). Since the Special Collar is ·150" (3·81 mm.) thick, the thickness of the spacer to be fitted will be ·150"—·021" i.e. ·129" (3·28 mm.). The

Page H.13

REAR AXLE

nearest spacer is ·130" (3·30 mm.) so a letter H spacer should be fitted in place of the special collar.

The inner oil seal should now be fitted.

When the half shaft splined end has been fitted to the hub as described in "Half shaft—Refitting" and tightened up, the end float should be checked, using the dial indicator.

Fig. 19. Checking the hub bearing end float with a dial test indicator (Churchill Tool No. J.13). The hub must be tapped inwards before taking a reading

THE DIFFERENTIAL UNIT

The Thornton "Powr-Lok" limited slip differential is fitted as standard.

Warning

When a car is equipped with a Thornton "Powr-Lok" differential the engine must NOT be run with the car in gear and one wheel off the ground otherwise, owing to the action of the differential, the car may drive itself off the jack or stand.

If it is desired to turn the transmission by running the engine with the car in gear **both** wheels must be jacked up clear of the ground.

DESCRIPTION

The limited slip differential has two pinion shafts with two mates to each shaft. The pinion shafts are mounted at right angles to each other but do not make contact at their intersection. Double ramps with flat surfaces at each end of the pinion shafts, mate with similar ramps in the differential case. Clearance in the differential case permits slight peripheral movement at the ends of the pinion shafts.

When a driving force is applied to the differential case, the pinion shafts, pinion mates and differential side gears splined to the axle shafts, rotate as a unit. Resistance to turning at the wheel forces the pinion shafts to slide up the differential case ramps, pushing the pinion shafts and side gears apart. As the pinion shafts move apart they apply load to the clutch plates thus restricting turning between the axle shafts and the differential case. Both axle shafts have now become clutched to the differential case to a varying degree dependent upon the amount of torque transmitted. This in effect locks the axle shafts to the differential case, in the normal straight ahead driving position, which reduces spinning of either rear wheel should it leave the road or encounter poor traction such as ice, snow, sand, loose gravel or oil patches.

Due to the lateral movement of the pinion shafts in the differential case, a little more backlash may be apparent in a limited slip rear axle. Slight chatter may also occur when one wheel is on a slippery surface, this is due to surge torque.

PRINCIPLE OF OPERATION

The conventional differential divides the load equally between both driving wheels. In this connection, it should be remembered that the conventional

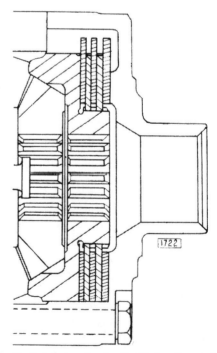

Fig. 20. Sectioned view showing the friction discs and plates

Page H.14

REAR AXLE

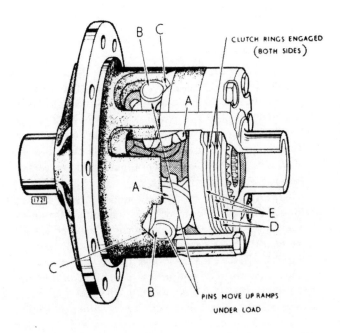

Fig. 21

on the curve has a further distance to travel. With the outer gear over-running and the inner gear fixed, the pinion mates A (see Fig. 22) are caused to rotate, but inasmuch as they are restricted by the fixed gear, they first must move pinion mate shafts B back down the cam surface C relieving the thrust loads on the plate clutches E. Thus when turning the corner, the differential, for all practical purposes, is similar to a conventional differential and the wheels are free to rotate at different speeds.

On straight driving, the clutches are engaged and thus prevent momentary spinning of the wheels when leaving the road or when encountering poor traction. In turning a corner, the load is relieved from the clutch surface so that wear is reduced to a minimum.

differential will always drive the wheel which is easiest to turn. This is a definite disadvantage under adverse conditions of driving where the traction of one wheel is limited.

The main purpose of the limited slip differential is to overcome this limit-action. Many times the torque of the slipping wheel is provided to the driving wheel, thus permitting improved operation under all conditions of driving. The torque is transmitted from the differential case to the cross pins and differential pinions to the side gears in the same manner as torque is applied in the conventional differential.

The driving forces moves the cross pins B, Fig. 21 up the ramp of the cam surfaces C, applying a load to the clutch rings D and restricts turning of the differential through the friction clutches E. This provides a torque ratio between the axle shafts which is based on the amount of friction in the differential and the amount of load that is being applied to the differential.

When turning a corner, this process is in effect partially reversed. The differential gears become a planetary set, with the gear on the inside of the curve becoming the fixed gear of the planetary. The outer gear of the planetary over-runs as the outside wheel

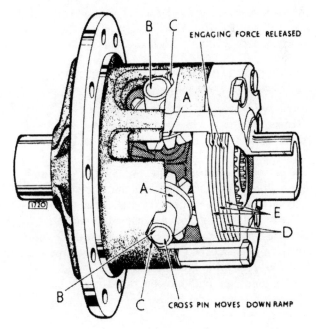

Fig. 22

Page H.15

REAR AXLE

POWER FLOW IN FORWARD DRIVING

Under normal starting and operating conditions the torque or power flow in both the limited slip and conventional type differential is transmitted equally to each axle shaft and wheel. However, when sudden patches of ice, loose gravel or oil are encountered, the limited slip differential will not permit the wheel with the lesser traction to spin, gain momentum and swerve the car when a dry surface is regained.

POWER FLOW IN TURNS

In turning, the limited slip differential gives normal differential action and permits the outer wheel to turn faster than the inner wheel. At the same time the differential applies the major driving force to the inside rear wheel, improving stability and cornering.

POWER FLOW WITH POOR TRACTION

When traction conditions under the rear wheels are dissimilar, the driving force with an ordinary differential is limited by the wheel with the poorer traction. Typically, in this situation, the wheel with the poorer traction spins and the vehicle remains immobile. The limited slip differential enables the wheel with the better traction to apply the major driving force to the road.

ACTION ON ROUGH ROADS

Bumps do not adversely affect wheel action when wheels are controlled by the limited slip differential. The free wheel does not spin and gain momentum. There is no sudden wheel stoppage to cause car swerve or tyre scuffing and wheel hop is reduced.

REMOVING THE DIFFERENTIAL ASSEMBLY FROM THE CARRIER

Remove the axle as described on page H.9.

Knock up the locking tabs and unscrew the brake caliper mounting bolts (on early cars locking wire was used).

Remove the caliper noting the number of small round shims between the caliper and the shims and differential carrier. Remove the brake disc.

Drain the lubricant from the gear carrier and remove the gear carrier rear cover. Flush out the unit thoroughly so that the parts can be carefully inspected.

Unscrew the five bolts securing the output shaft bearing housing. Withdraw the output shaft, bearing housing, bearings and adjustment shims noting the number of preload shims.

Repeat for the other drive shaft. Remove the two bolts holding each differential bearing cap and withdraw the differential unit.

Remove the Pinion

Remove the pinion nut and washer. Withdraw the universal joint companion flange with a suitable puller. PRESS the pinion out of the outer bearing. It is important that the pinion should be pressed out, not driven out, to prevent damage to the outer bearing. The pinion having been pressed from its outer bearing may now be removed from the differential casing.
Note: Keep all shims intact.

Remove the pinion oil seal together with the oil slinger and outer bearing cone. Examine the outer bearing for wear and if replacement is required extract the bearing outer race using Tool No. SL.12 shown in Fig. 23. If the correct tool is not available and the bearing cup is to be scrapped it is possible to drive out the cup, the shoulder locating the bearing being recessed to facilitate this operation. Remove the pinion inner bearing outer race as shown in Fig. 23 using Tool No. SL.12, if the bearing requires replacement or

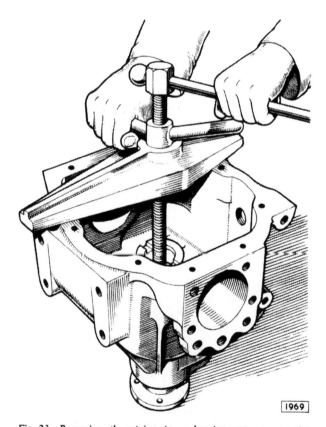

Fig. 23. Removing the pinion inner bearing outer race using Churchill tool SL. 12 with adaptor SL. 12 AB-4

Page H.16

REAR AXLE

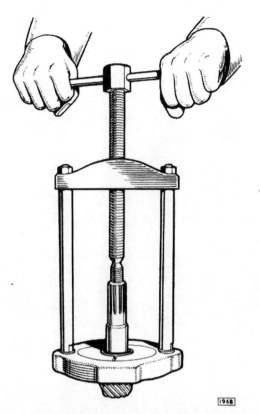

Fig. 24. Withdrawing the pinion inner bearing inner race using Churchill tool SL. 14 with adaptor SL. 11 P/AB-2

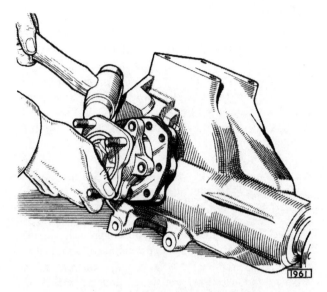

Fig. 25. Removing an output shaft assembly

adjustment of the pinion setting is to be undertaken. Take care of the shims fitted between the bearing cup and the housing abutment face. If the inner bearing is to be replaced it may be driven out but the correct service tool should be used when the bearing is removed in order to carry out pinion setting adjustment.

THE OUTPUT SHAFTS

Removal

Remove the brake caliper and disc as previously described.

Unscrew the five bolts securing the output shaft bearing housings, bearings and adjustment shims, noting the number of pre-load shims.

To Dismantle

Unlock the tab washer and remove the nut, tab washer and plain washer. Press the output shaft with the inner bearing inner race, spacing sleeve and endfloat shims in position through the flange and bearing housing. If it is necessary to replace the bearings, remove the endfloat shims and spacing collar, and using a suitable extractor withdraw the inner bearing inner race from the shaft. Drift out the inner bearing outer race and using a suitable sized tube on the outer race, press out the complete outer bearing and the oil seal. If it is necessary to reset the output shaft endfloat, withdraw the oil seal and the outer bearing inner race.

Fig. 26. Exploded view of an output shaft assembly

Page H.17

REAR AXLE

Refitting

Press in the new inner and outer bearing outer races ensuring that they are fully home in the recesses. The races must be fitted so that the bearings will be opposed. Press the inner bearing inner face on to the shaft ensuring that it is fully home against the shoulder and that the race is fitted the correct way round. Fit the spacing sleever and the endfloat shims. Fit the output shaft into the bearing housing and place the outer bearing inner race on the shaft from the opposite end. Do not fit the oil seal at this stage. Fit the output shaft flange with the plain washer and a new tab washer, fit the nut and tighten.

Check the endfloat with a dial gauge, this should be ·001″—·003″ (·025—·076 mm.). Should adjustment be necessary remove the flange nut, tab and plain washers and withdraw the flange and outer bearing inner race. Add or remove shims to obtain the correct clearance. Adding shims increases the endfloat and removing shims decreases it. When the correct endfloat is obtained replace the outer bearing inner race and press a new oil seal into position, flush with the casing and with the lip inwards. Refit the flange and the plain and tab washers ensuring that the two tags on the tab washer locate in the holes on the flange. Tighten the nut and turn one or more tabs up securing the nut. Ensure that these tabs lie as flat on the nut as possible.

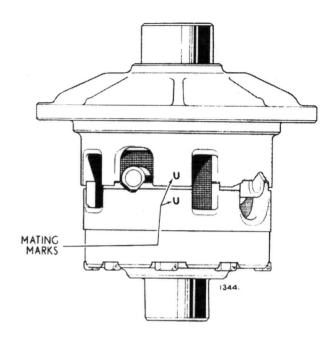

Fig. 27. Alignment marks on the differential case

DISMANTLING THE DIFFERENTIAL UNIT

In the absence of any mating or alignment marks as shown in Fig. 27, scribe a line across the two half casings to facilitate assembly.

Remove the eight bolts (9, Fig. 29) securing the two halves of the differential casing.

Split the casing and remove the clutch discs (3) and plates (2 and 4) from one side.

Remove the differential side gear ring (5).

Remove the pinion side gear (6) and the pinion mate cross shafts (7) complete with the pinion mate gears.

Separate the cross shafts (10).

Remove the remaining side gear and the side gear ring.

Extract the remaining clutch discs and plates.

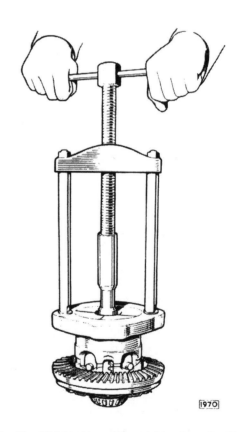

Fig. 28. Withdrawing a differential bearing using Churchill tool SL. 14 with adaptor SL. 11 D/A-5

Page H.18

REAR AXLE

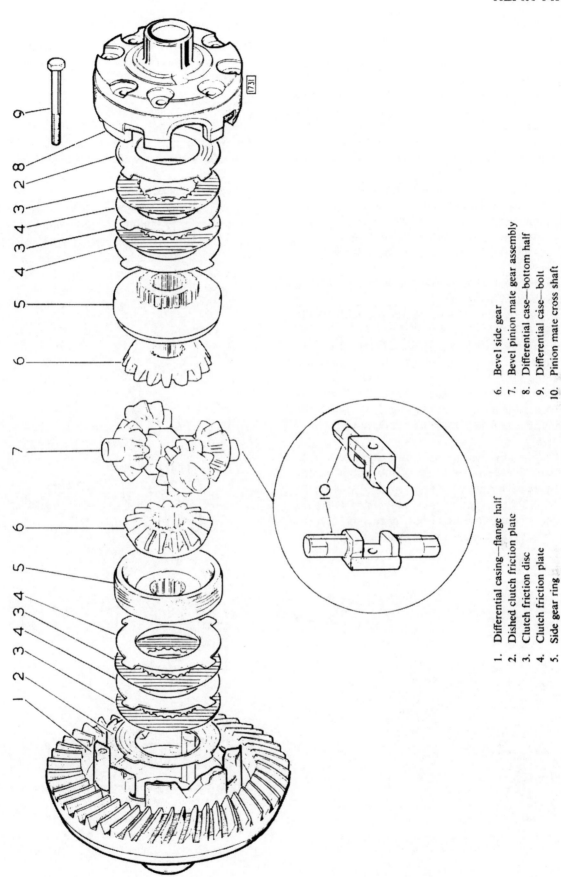

Fig. 29. *Exploded view of the Thornton "Powr-Lok" differential*

1. Differential casing—flange half
2. Dished clutch friction plate
3. Clutch friction disc
4. Clutch friction plate
5. Side gear ring
6. Bevel side gear
7. Bevel pinion mate gear assembly
8. Differential case—bottom half
9. Differential case—bolt
10. Pinion mate cross shaft

Page H.19

REAR AXLE

REASSEMBLING

Refit the clutch plates and discs alternately into the flange half of the casing.

Fit the two "Belleville" clutch plates (i.e. curved plates) so that the convex side is against the diff. casing (see Fig. 29).

Fit the side gear ring so that the serrations on the gear mesh with the serrations in the two clutch discs.

Place one of the side gears into the recess of the side gear ring so the splines in both align.

Fit the cross shafts together.

Refit the pinion mate cross shafts complete with pinion mate gears ensuring that the ramps on the shafts coincide with the mating ramps in the differential case.

Assemble the remaining side gear and side gear ring so the splines in both align.

Refit the remaining clutch plates and discs to the side gear ring.

Offer up the button half of the differential case to the flange half in accordance with the identification marks and position the tongues of the clutch friction plates so they align with the grooves in the differential case. Assemble the button half to the flange half of the differential case with eight bolts but do not tighten at this juncture.

Check the alignment of the splines in the side gear rings and side gears by inserting two output shafts, then tighten the eight bolts to a torque of 35–45 lb.ft. (4·8 to 6·2 kgm.) while the output shafts are in position. Failure to observe this instruction, particularly with the differential unit having the dished clutch friction plates, will render it difficult or impossible to enter the output shafts after the eight bolts have been tightened.

CHECKING FOR WEAR

With one output shaft and the drive pinion locked, the other output shaft must not turn radially more than $\frac{3}{4}''$ (19 mm.) measured on a 6" (152 mm.) radius.

Pinion Adjustment

Re-install the pinion outer bearing outer race with Tool No. SL.12. Re-install the pinion inner bearing outer race with the original adjusting shims positioning same. Press the inner bearing inner race on the pinion, using an arbor press and a length of tube, contacting the inner race only and not the roller retainer.

The hypoid drive pinion should be correctly adjusted

Fig. 31. Replacing the pinion inner bearing outer race using Churchill tool SL. 12 with adaptor SL. 12 AB-4

before attempting further assembly, the greatest care being taken to ensure accuracy.

The correct pinion setting is marked on the ground end of the pinion as shown in Fig. 32. The marked assembly serial number at the top is also marked on

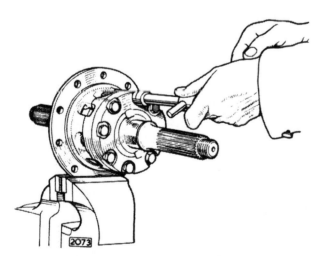

Fig. 30. Tightening the differential casing bolts with the output shafts in position

Page H.20

REAR AXLE

A. Pinion Drop .. .. 1·5" (38·1 mm.)
B. Zero Cone Setting .. .. 2·625" (66·67 mm.)
C. Mounting Distance .. .. 4·312" (108·52 mm.)
D. Centre Line to Bearing Housing .. .. 5·495" (139·57 mm.)
                                                                                                                        to   to
                                                                                                                          5·505" (139·83 mm.)

Fig. 32. Pinion setting marks

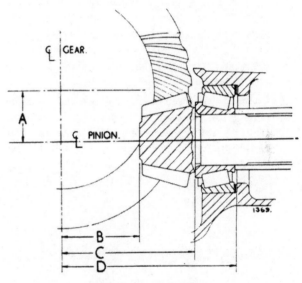

Fig. 33. Pinion setting distances

the drive gear, and care should be taken to keep similarly marked gears and pinions in their matched sets, as each pair is lapped together before despatch to the factory. The letter on the left is a production code letter and has no significance relative to assembly or servicing of any axle. The letter and figure on the right refer to the tolerance on offset or pinion drop dimension "A" in Fig. 33 which is stamped on the cover facing of the gear carrier housing.

The number at the bottom gives the cone setting distance of the pinion and may be Zero (0). Plus (+) or Minus (—). When correctly adjusted, a pinion marked Zero will be at the zero cone setting distance, dimension "B" in Fig. 33 from the centre line of the gear to the face on the small end of the pinion; a pinion marked Plus two (+2) should be adjusted to the nominal (or Zero) cone setting plus ·002" (·051 mm.), and a pinion marked Minus two (—2) to the cone setting distance minus ·002" (·051 mm.).

The zero cone setting distance ("B" Fig. 33) is given above.

Thus for a pinion marked Minus two (—2) the distance from the centre of the drive gear to the face of the pinion should be 2·623" (66·6 mm.) (that is, 2·625" —·002") (66·7—·05 mm.) and for a pinion marked Plus three (+3) the cone setting distance should be 2·628" (66·75 mm.).

When the pinion bearing cups have been installed in the gear carrier, with the original pinion inner bearing adjusting shims, as described in the first paragraph of this section, proceed with pinion as follows:—

(1) Place the pinion, with the inner bearing cone assembled, in the gear carrier.

(2) Turn the carrier over and support the pinion with a suitable block of wood for convenience before attempting further assembly.

(3) Install the pinion bearing spacer if fitted on the unit under repair.

(4) Install the original outer bearing shims on the pinion shank so that they seat on the spacer or a shoulder on the shank, according to the construction of the unit.

Page H.21

REAR AXLE

(5) Fit pinion outer bearing inner race, companion flange, washer and nut only, omitting the oil slinger and oil seal assembly, and tighten the nut.

(6) Check the pinion cone setting distance by means of the gauge, Tool No. SL.3P, as shown in Fig. 34. The procedure for using the gauge is:—

(a) Adjust the bracket carrying the dial indicator to suit the model being serviced, then set the dial indicator to zero with the setting block on a surface plate, using the 4 HA setting.

(b) Place the dial indicator assembly on the fixed spindle of the gauge body.

(c) Fit the fixed spindle of the gauge body into the centre in the pinion head, slide the movable spindle into position, locating in the centre in the pinion shank with the gauge body underneath the gear carrier, and lock the spindle with the screw provided.

(d) Check the pinion setting by taking a dial indicator reading on the differential bore with the bracket assembly seated on the ground face on the end of the pinion. The correct reading will be the minimum obtained; that is, when the indicator spindle is at the bottom of the bore. Slight movement of the assembly will enable the correct reading to be easily ascertained. The dial indicator shows the deviation of the pinion setting from the zero cone setting and it is important to note the direction of any such deviation as well as the magnitude.

(7) If the pinion setting is incorrect it is necessary to dismantle the pinion assembly and remove the pinion inner bearing outer race. Add or remove shims as required from the pack locating the bearing outer race and re-install the shim pack and the bearing outer race. The adjusting shims are available in thickness of ·003″, ·005″ and ·010″ (·076, ·127 and ·254 mm.). Then carry out the operations (1) to (6) detailed on page H.21.

(8) When the correct pinion setting has been obtained, check the pinion bearing preload, which should afford a slight drag or resistance to turning, there being no end play of the pinion. The correct preload for the pinion bearings gives a torque figure as listed in "Data" on page H.4. Less

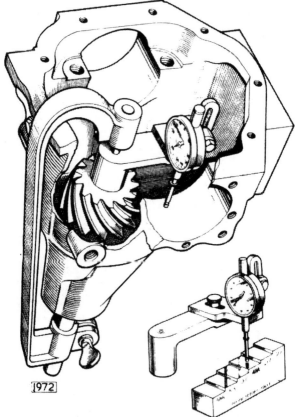

Fig. 34. Checking the pinion cone setting using Churchill Tool SL 3 P (with the later type of magnetic dial gauge post the clamp fixture is not required)

Page H.22

REAR AXLE

than the correct range will result in excessive deflection of the pinion under load, whilst too much preload will lead to pitting and failure of the bearings. To rectify the preload, adjust the shim pack between the outer bearing inner race and the pinion shank spacer, but do not touch the shims behind the inner bearing outer race, which control the position of the pinion. Remove the shims to increase preload and add shims to decrease preload.

Installation of pinion oil seal assembly and oil slinger is usually effected after fitting differential assembly, see operations (1), (2) and (3) under "Final Assembly" on page H.24.

Differential Bearing, Preload and Drive Gear Adjustment

(1) Install the pinion less the oil seal and slinger in the differential carrier. Fit the differential assembly.

(2) Fit the differential bearing caps ensuring that the position of the numerals marked on the differential carrier face and the caps correspond as indicated in Fig. 35. Tighten the bolts securing the caps.

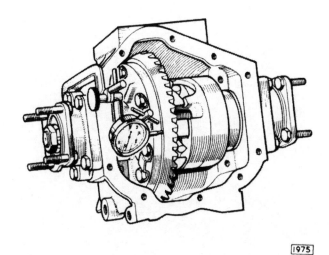

Fig. 36. Checking the drive gear run-out

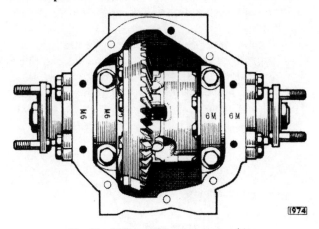

Fig. 35. Differential bearing cap markings

(3) Mount a dial indicator on the gear carrier housing with the button against the back face (as shown in Fig. 36). Turn the pinion by hand and check the run out on the back face which should not exceed ·005" (·13 mm.). If there is excessive run out strip the assembly and rectify by cleaning the surfaces locating the drive gear. Any burrs on these surfaces should be removed.

(4) Install the drive shafts without any shims between the drive shaft bearing housing and the differential carrier. Place three bolts evenly spaced in each bearing housing. Set up a dial indicator on the differential carrier with the button against one of the drive gear teeth as nearly in line with the direction of tooth travel as possible (see Fig. 37).

(5) Move the drive gear by hand to check the backlash which should be etched on the gear. If the backlash is not in accordance with the specification, move the drive gear towards or away from the pinion as necessary by tightening the bolts in the drive gear bearing housing on one side of the differential carrier and slackening them on the other.

(6) When the correct backlash has been obtained, check the gap on each side of the carrier between the drive gear bearing housing and the carrier with a set of feeler gauges.

(7) Note the gap, having checked around the circumference of the housing to ensure that the gap is equal. Subtract ·003" (·076 mm.) from the width of the gap on each side to give the correct preload. Install shims on each side to the requisite amount, the shims being in ·003", ·005", ·010" and ·030" (·076, ·127, ·254 and ·762 mm.) thicknesses.

For example, assume the backlash etched on the drive gear to be ·007" (·178 mm.) when this figure

Page H.23

REAR AXLE

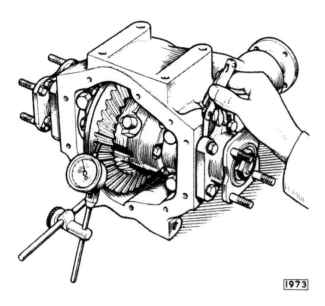

Fig. 37. Checking the backlash and the drive gear location

is obtained as described above, the gap on one side is ·054″ (1·37 mm.) and ·046″ (1·17 mm.) on the other. Then the amount of shims to be fitted are ·054″—·003″ i.e. ·051″ (1·30 mm.) to one side and ·043″ (1·09 mm.) to the other.

Finally fit the output shaft and shims to the differential carrier and tighten up the five bolts on each side.

Final Assembly

To complete the rebuilding of the unit:—

(1) Remove the drive pinion nut, washer and companion flange.
(2) Install the oil slinger, and then fit the pinion oil seal assembly, using Tool No. SL4P/B, as shown in Fig. 38. Place the oil seal with the dust excluder flange uppermost (not omitting the oil seal gasket used with the metal case type seal on later models), fit the installation collar, Tool No. SL.4P/B, and then tighten down the pinion nut and washer to drive the assembly home. Remove the installation collar.
(3) Fit the companion flange, washer and pinion nut, tighten securely.
(4) Fit the rear cover gasket, renewing it if required, and rear cover, securing same with set bolts and lock washers, not omitting the ratio and "Powr Lok" tags which are attached by the set bolts.
(5) Check that the drain plug is securely tightened, then fill with the appropriate quantity of one of the hypoid lubricants recommended on page H.5
(6) Replace the filler plug and check that the cover set bolts are tight.
(7) Check for oil leaks at the cover, pinion oil seal and where the differential cap bolt holes break through.
(8) Replace brake disc and caliper, centralising the caliper by means of the round shims (as described in Section L "Brakes"). Fit new tab washers, tighten the mounting bolts to 55 lb.ft. (7·6 kgm.) torque and secure the bolt heads with the tab washers.

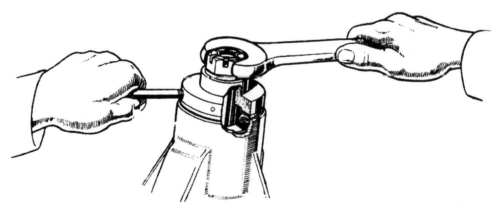

Fig. 38. Fitting the pinion oil seal using Churchill tool SL 4 P/B

Page H.24

REAR AXLE

TOOTH CONTACT

After setting the backlash to the required figure, use a small brush to paint eight or ten of the drive gear teeth with a stiff mixture of marking raddle, used sparingly, or engineers blue may be used if preferred. Move the painted gear teeth in mesh with the pinion until a good impression of the tooth contact is obtained. The resulting impression should be similar to Fig. A in Fig. 39.

The illustrations referred to in this section are those shown in Fig. 39 which indicates the tooth bearing impression as seen on the drive gear.

> The HEEL is the large or outer end of the tooth.
> The TOE is the small or inner end of the tooth.
> The FACE top or addendum is the upper portion of the tooth profile.
> The FLANK or dedendum is the lower portion of the tooth profile.
> The DRIVE side of the drive gear tooth is CONVEX.
> The COAST side of the drive gear tooth is CONCAVE.

(a) Ideal Contact

Fig. A shows the ideal tooth bearing impression on the drive and coast sides of the gear teeth. The area of contact is evenly distributed over the working depth of the tooth profile and is located nearer to the toe (small end) than the heel (large end). This type of contact permits the tooth bearing to spread towards the heel under operating conditions when allowance must be made for deflection.

(b) High Tooth Contact

In Fig. B it will be observed that the tooth contact is heavy on the drive gear face or addendum, that is, high tooth contact. To rectify this condition, move the pinion deeper into mesh, that is, reduce the pinion inner race setting distance, by adding shims between the pinion inner bearing outer race and the housing and adding the same thickness of preload shims between the pinion bearing spacer, or the shoulder of the pinion shank and outer bearing inner race. This correction has a tendency to move the tooth bearing towards the toe on drive and heel on coast, and it may therefore be necessary after making this change to adjust the drive gear as described in paragraphs (d) and (e).

(c) Low Tooth Contact

In Fig. C it will be observed that the tooth contact is heavy on the drive gear flank or dedendum, that is, low tooth contact. This is the opposite condition from that shown in (b) and is therefore corrected by moving the pinion out of mesh, that is, increase the pinion inner race setting distance by removing shims from between the pinion inner bearing outer race and housing, and removing the same thickness of preload shims from between the pinion bearing spacer or the shoulder on the pinion shank and the outer bearing inner race. The correction has a tendency to move the tooth bearing towards the heel on drive and toe on coast, and it may therefore be necessary after making this change to adjust the drive gear as described in (d) and (e).

(d) Toe Contact

Fig. D shows an example of toe contact which occurs when the bearing is concentrated at the small end of the tooth. To rectify this condition, move the drive gear out of mesh, that is, increase backlash, by transferring shims to the drive gear side of the differential from the opposite side.

(e) Heel Contact

Fig. E shows an example of heel contact which is indicated by the concentration of the bearing at the large end of the tooth. To rectify this condition move the drive gear closer into mesh, that is reduce backlash, by removing shims from the drive gear side of the differential and adding an equal thickness of shims to the opposite side.

Note: It is most important to remember when making this adjustment to correct a heel bearing that sufficient backlash for satisfactory operation must be maintained. If there is insufficient backlash the gears will at least be noisy and have a greatly reduced life, whilst scoring of the tooth profile and breakage may result. Therefore, always maintain a minimum backlash requirement of ·004" (·10 mm.).

Backlash

When adjusting backlash always move the drive gear as adjustment of this member has more direct influence on backlash, it being necessary to move the pinion considerably to alter the backlash a small amount—·005" (·13 mm.) movement on pinion will generally alter backlash ·001" (·025 mm.).

Page H.25

REAR AXLE

Gear and Pinion Movement

Moving the gear out of mesh moves the tooth contact towards the heel and raises it slightly towards the top of the tooth.

Moving the pinion out of mesh raises the tooth contact on the face of the tooth and slightly towards the heel on drive, and towards the toe on coast.

	TOOTH CONTACT (DRIVE GEAR)	CONDITION	REMEDY
A	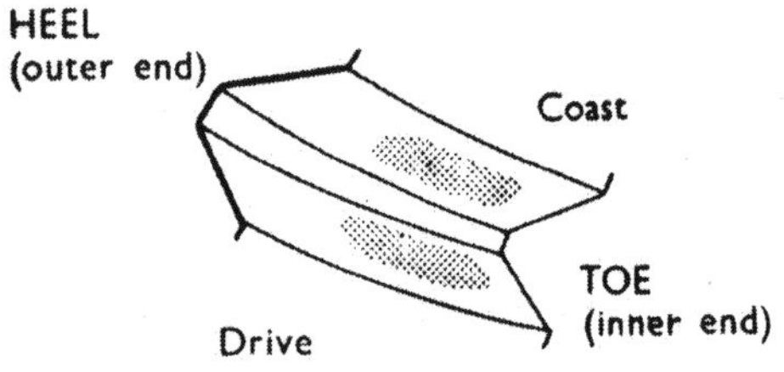	IDEAL TOOTH CONTACT Evenly spread over profile, nearer toe than heel.	o —— o
B	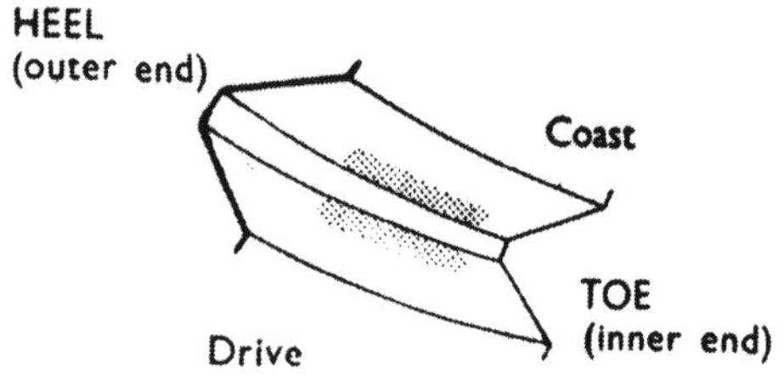	HIGH TOOTH CONTACT Heavy on the top of the drive gear tooth profile.	Move the DRIVE PINION DEEPER INTO MESH. i.e., REDUCE the pinion cone setting.
C	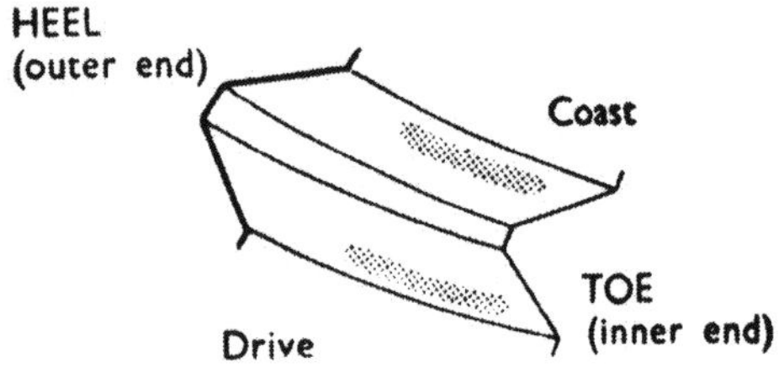	LOW TOOTH CONTACT Heavy in the root of the drive gear tooth profile.	Move the DRIVE PINION OUT OF MESH. i.e., INCREASE the pinion cone setting.
D	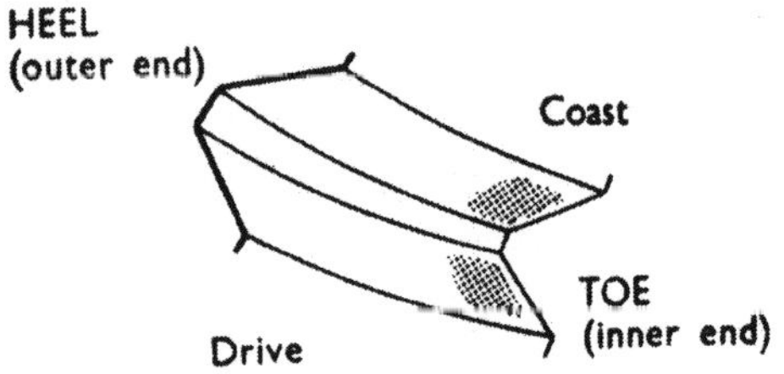	TOE CONTACT Hard on the small end of the drive gear tooth.	Move the DRIVE GEAR OUT OF MESH. i.e., INCREASE backlash.
E	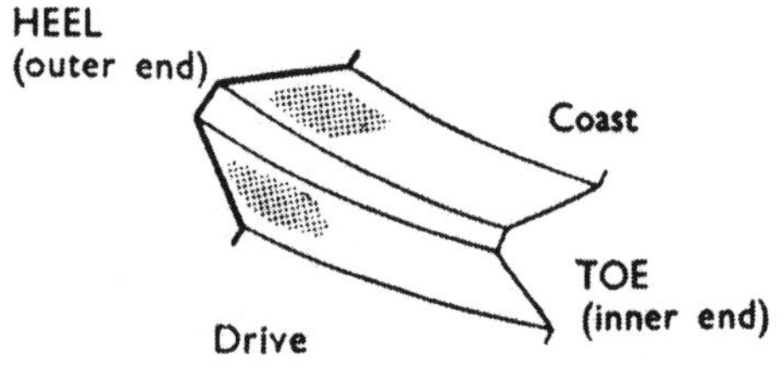	HEEL CONTACT Hard on the large end of the drive gear tooth.	Move the DRIVE GEAR INTO MESH. i.e., DECREASE backlash but maintain minimum backlash as given in "Data"

Fig. 39. Tooth contact indication (contact markings on the drive gear)

Page H.26

REAR AXLE

SPECIAL TOOLS

Description	Tool Number
Multi-purpose Hand Press	SL14
Used in conjunction with the following adaptors:—	
Pinion Bearing Inner Races	
Removing / Replacing } adaptor	SL11 P/AB – 2
Differential Bearing	
Removing adaptor	SL11 D/A – 5
Differential Bearing	
Replacing—Universal Handle	SL 2 DB
Used with adaptor	SL 2 D/B – 2
Main tool and Ring	SL 12
Used in conjunction with the following adaptor:—	
Pinion Bearing Outer Races	
Removing / Replacing } adaptor	SL12 AB – 4
Hub End Float Special Collar	J.15
Hub End Float Dial Test Indicator	J.13
Pinion Cone Setting Gauge	SL 3 P
Rear Hub Extractor	J.7
Pinion Oil Seal Installation Collar	SL 4 P/B

Page H.27

SECTION I

STEERING
3·8 "E" TYPE
GRAND TOURING MODELS

INDEX

	Page
Description	I.3
Data	I.3
Routine Maintenance:	
Steering housing	I.4
Steering tie rods	I.4
Recommended Lubricants	I.4
Steering Housing	
Removal	I.6
Dismantling	I.6
Examining for wear	I.6
Re-assembling	I.7
Refitting	I.8
Steering Wheel:	
Removal	I.8
Refitting	I.8
Upper Steering Column:	
Removal	I.8
Dismantling	I.8
Re-assembling	I.10
Refitting	I.10
Lower Steering Column:	
Removal	I.10
Universal joints	I.10
Refitting	I.10

STEERING

INDEX (continued)

	Page
Tie Rod Ball Joints:	
Removal	I.10
Refitting	I.10
Front Wheel Alignment	I.12
Steering Arm:	
Removal	I.12
Refitting	I.12
Accidental Damage	I.12

STEERING

DESCRIPTION

The steering gear is of the high efficiency rack and pinion type in which motion is transmitted from the inner column through the pinion to the steering rack. Tie-rods operating the steering arms are attached to each end of the steering rack by ball joints enclosed in rubber bellows.

The steering rack assembly is attached as a complete unit to the front cross member of the forward chassis frame between the front of the engine and the radiator. The steering column engages the splined end of the pinion shaft to which it is secured by a clamp bolt.

DATA

Type	Rack and Pinion
Number of turns—lock to lock	2½
Turning circle	37' 0" (11·27 m.)
Diameter of steering wheel	16" (40·5 cm.)
Front wheel alignment	$\frac{1}{16}$"–$\frac{1}{8}$" (1·6—3·2 mm.) toe-in

STEERING

ROUTINE MAINTENANCE

EVERY 2,500 MILES (4,000 KM.)

Steering Housing

The rack and pinion steering housing is attached to the front cross member of the forward chassis frame.

A grease nipple for the lubrication of the rack and pinion assembly is accessible from underneath the front of the car from the driver's side.

Do not over lubricate the steering housing to the extent where the bellows at the ends of the housing become distended.

Check that the clips at the ends of the bellows are fully tightened; otherwise the grease will escape from the housing.

Steering Tie-Rods

Lubricate the ball joints of the two steering tie-rods with the recommended lubricant. When carrying out this operation examine the rubber seals at the bottom of the ball bearing housing to see if they have become displaced or split. In this event they should be repositioned or replaced as any dirt or water that enters the joint will cause premature wear.

Do not over lubricate the ball joints to the extent where grease escapes from the rubber seal.

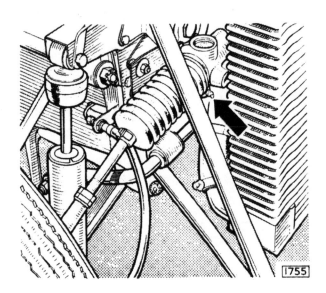

Fig. 1. Steering housing grease nipple (right-hand drive illustrated).

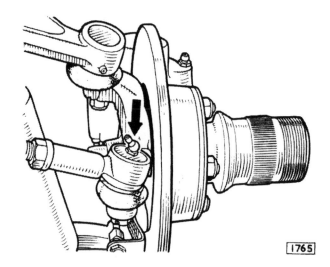

Fig. 2. Steering tie rod grease nipple.

Recommended Lubricants

	Mobil	Castrol	Shell	Esso	BP	Duckham	Regent Caltex/Texaco
Steering Housing Steering Tie-rods	Mobilgrease MP	Castrolease LM	Retinax A	Esso Multi-purpose Grease H	Energrease 2	LB10	Marfak All purpose

Page 1.4

STEERING

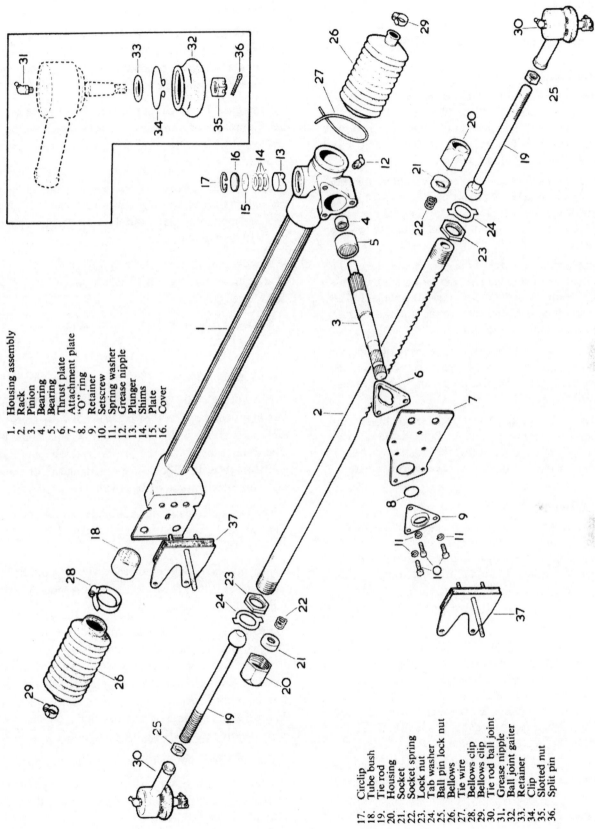

1. Housing assembly
2. Rack
3. Pinion
4. Bearing
5. Bearing
6. Thrust plate
7. Attachment plate
8. "O" ring
9. Retainer
10. Setscrew
11. Spring washer
12. Grease nipple
13. Plunger
14. Shims
15. Plate
16. Cover
17. Circlip
18. Tube bush
19. Tie rod
20. Housing
21. Socket
22. Socket spring
23. Lock nut
24. Tab washer
25. Ball pin lock nut
26. Bellows
27. Tie wire
28. Bellows clip
29. Bellows clip
30. Tie rod ball joint
31. Grease nipple
32. Ball joint gaiter
33. Retainer
34. Clip
35. Slotted nut
36. Split pin

Fig. 3 Exploded view of rack and pinion housing.

Page I.5

STEERING

STEERING HOUSING

The steering housing unit cannot be removed until after the the removal of radiator and fixings.

Radiator-Removal

Slacken off but do not remove the road wheels; the hub caps are marked "Right (Off) Side" and "Left (Near) Side" and the direction of rotation to remove, that is, clockwise for the right hand side and anti-clockwise for the left hand side.

Place the jack under the lower wishbone fulcrum front support bracket and raise the car until the wheel is clear of the ground. Place a stand under the wishbone fulcrum rear support bracket. Repeat for the other side.

Complete the removal of the roadwheels. DO NOT place the jack or stands under the forward frame cross tubes. Drain the radiator, conserve the coolant if an anti-freeze is in use, and remove the top and bottom water hoses from the radiator.

Remove the two bolts and nuts securing the radiator top support brackets to the header tank mounting; remove the two bottom fixing nuts and rubber mounting washers.

Release radiator duct panel from bottom of radiator by removing the two setscrews. Lift out the radiator matrix; care must be taken that the radiator fan blades are not damaged during the removal of the radiator.

Steering Housing-Removal

Remove split pin and nut from both steering tie-rod ball joints and drift out the tie-rod ends from the steering arms, into which they are a taper fit, by tapping on the side face of the steering arms.

Turn steering until Allen screw in steering column lower joint is accessible. Insert Allen key and remove screw.

From the steering housing side remove the two inner self-locking nuts and the central bolt with attached self-locking nut securing the housing to the rubber/steel bonded mounting. Remove the top and bottom outer self-locking nuts, and withdraw the bolts noting the two spacer tubes fitted between the mounting bracket and the frame.

Repeat for the opposite (rack tube) side, but note that on the two outer fixings the spacer tubes have been omitted and replaced with two adjuster lock nuts.

Withdraw steering housing.

Dismantling

Unlock the ball end retaining nuts on the steering tie-rods and remove the ball end assemblies. Release the clips securing the bellows to the rack housing and tie-rods and remove the bellows.

Bend back the locking plate tabs on the two tie-rod ball housings and remove tie-rods and housings. Remove the housing lock nuts, sockets and socket springs.

Remove the three setscrews retaining pinion shaft oil seal retainer to housing, remove retainer "O" ring, housing mounting plate and bearing retainer plate, and withdraw pinion shaft.

Remove circlip, disc, "Belleville" washer, shims, and plunger from housing. Care must be taken not to lose any of the shims.

Withdraw the rack from the housing.

Examining for Wear

Thoroughly clean, dry and examine all parts of the assembly; components showing signs of wear must be replaced with new parts. Particular attention should be paid to the condition of the bellows and the outer ball joint rubber seals. Should they be damaged or show signs of deterioration replace with new.

The two outer ball joint assemblies cannot be dismantled and if worn must be replaced with new components.

Carefully examine the tide-rod ball seats and housings, replace with new parts if excessive wear is evident.

Examine the bush fitted in the end of the rack tubing and renew if worn. The bush can be drifted out, after the removal of the rack, from the opposite end by means of a long drift inserted through the steering housing. Press the new bush into the rack tubing using a shouldered, polished mandrel of the same diameter as the shaft which is to fit into the bearing until the visible end of the bearing is flush with the rack tubing.

The bush must not be opened out after refitting.

Before refitting the bush allow it to thoroughly soak in clean engine oil; this will allow the pores of the bush to be filled with lubricant.

Page I.6

STEERING

Re-assembling

Apply a generous coating of grease to the rack and insert the rack into the housing. Grease and re-assemble the pinion shaft, bearing retaining plate, housing mounting plate, "O" ring and oil seal retainer into the housing. Refit the three setscrews and lock washers and fully tighten. Renew oil seal if damaged.

Insert the steering damper plunger, disc washer and circlip to the housing, but do not fit the "Belleville" washer or shims at this stage.

Attach a clock gauge, mounted on a suitable bracket, to the steering housing with the stem in contact with the centre of the disc, Fig. 4. Apply a downward pressure to the disc to ensure that the pinion is fully engaged with the rack at its lowest point; set the gauge to zero.

Apply an upward pressure to the rack to eliminate all end float in plunger.
Note: Disc will now be in contact with the circlip.

Note the new reading on the clock gauge. Measure the thickness of the "Belleville" washer and subtract this figure from the reading obtained on the gauge.

Select suitable shims to give a final end float figure of ·006″—·010″ (·15 mm.—·25 mm.).

Remove gauge and fit shims, "Belleville" washer, disc washer and circlip.

Screw the tie-rod ball housing lock nuts onto the rack ends to the limit of the threads and refit the locking washers. Renew the washer if the tabs are broken.

Refit the socket spring, socket, tie-rod and ball housing. Tighten the ball housing until no end float is felt in the tie-rod. Advance the locknut to meet the housing. Refit the tie-rod joint and locknut.

Attach a spring balance to the outer ball joint and adjust the ball housing until the tie-rod will articulate under a load of 7 lbs (3·18 kg.) applied at the spring balance, Fig 5. Fully tighten the housing locknut and secure in position by bending the tabs of the washer. Apply a generous coating of grease to the ball housings.

Remove tie rod ball joints with locknuts.

Refit bellows and tighten clips.

The larger end of the bellows attached to the housing is secured by means of locking wire and not by the normal type of clip.

Screw on each tie rod ball joint an equal number of turns.

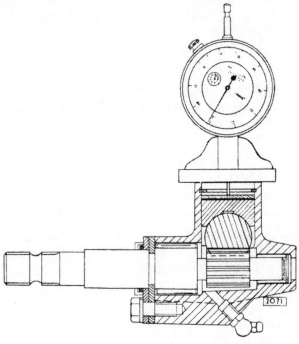

Fig. 4. Method of checking end float in plunger.

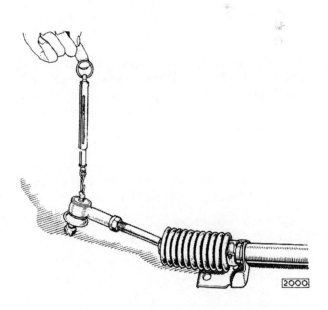

Fig. 5. Checking adjustment of ball housing.

Page I.7

STEERING

Refitting

Refitting is the reverse of the removal procedure but care must be taken that the spacer tubes on the steering housing side are refitted and the nuts fully tightened, also that the mounting on the opposite (rack tube) side is adjusted correctly. To adjust proceed as follows.

Slacken off the four nuts on the two outer bolts, fully tighten the two inner and the single central fixings.

Tighten the two self-locking nuts securing the two outer bolts until the flat washers under the bolt heads can just be rotated with the fingers. Turn the inner lock nuts towards the outer nuts and fully tighten.

After refitting adjust the front wheel alignment as described on page I.12.

STEERING WHEEL

Removal

Remove the three grub screws from the steering wheel hub and withdraw horn push assembly. Release the locknut and remove the hexagon nut securing the steering wheel to the inner column shaft. Extract the flat washer. Exert a sudden pressure behind the steering wheel and withdraw it from the splines on the inner column.

Collect the two halves of the split cone.

Refitting

Slacken the steering column adjuster nut, fully extend the sliding portion of the inner column and lock the steering column adjuster. Refit the two halves of the split cone, making sure that the narrowest part of the cone is towards the top of the column. To retain the cone in position while fitting the steering wheel place a small quantity of grease around the groove in the column and embed the two halves of the cone.

Slide the steering wheel on to the column shaft splines with the central spoke in the 6-o'clock position when the road wheels are pointing straight ahead. Push the wheel fully home on to the split cone.

Fit the flat washer, and nut and fully tighten. Fit locknut and tighten.

Note: When fitting the locknut, secure by using a ring or tubular spanner and do not over-tighten.
Refit the horn push with the head of the Jaguar upright.

UPPER STEERING COLUMN

Removal

Detach the earth lead from the battery. Disconnect the wires leading from the flasher switch at the multi-connector located behind the side facia panel and detach the horn push cable from the connector at the lower end of the upper column tube.

Remove the upper pinch bolt, nut and spring washer retaining the upper column to the universal joint. Remove the two locknuts, bolts and washers securing the column to the lower support bracket. Remove the bolt, nut, washer and spacer tube securing the column to the upper support bracket located behind the side facia panel.

Withdraw column assembly from the splines in the universal joint, and remove the steering column.

Dismantling

Remove the steering wheel as described under "Steering Wheel – Removal".

Remove the inner half of the flashing indicator switch cover by withdrawing towards the centre of the car. Cover is retained in position by means of spring clips. Remove the two screws, washers and clamp retaining the switch to the column and detach the switch.

Unscrew the telescopic adjustment nut from the inner column and withdraw complete with the collet from the inner column shaft splines. The collet is attached to the adjustment nut by means of a circlip.

Withdraw the brass horn button contact rod complete with the spring and insulating bush from the centre of the steering column inner.

Remove the indicator switch striker after unscrewing the two setscrews retaining the striker to the inner column.

Unscrew the stop button now exposed and withdraw the splined shaft from the inner column.

Remove the screw, nut and washer holding the earth contact to the bracket on the outer column.

Remove the bolt and nut holding the slip ring contact to the contact holders.

From the lower end tap the inner column out of the outer tube.

Compress and remove the spring clip and washer retaining felt bush in the upper end of the column and remove the bush.

Page I.8

STEERING

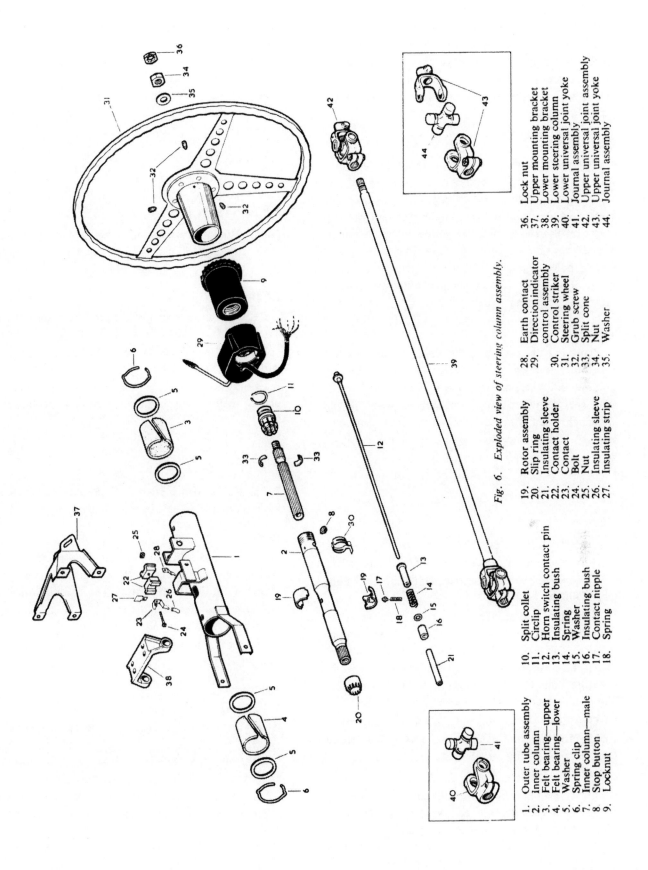

Fig. 6. Exploded view of steering column assembly.

1. Outer tube assembly
2. Inner column
3. Felt bearing—upper
4. Felt bearing—lower
5. Washer
6. Spring clip
7. Inner column—male
8. Stop button
9. Locknut
10. Split collet
11. Circlip
12. Horn switch contact pin
13. Insulating bush
14. Spring
15. Washer
16. Insulating bush
17. Contact nipple
18. Spring
19. Rotor assembly
20. Slip ring
21. Insulating sleeve
22. Contact holder
23. Contact
24. Bolt
25. Nut
26. Insulating sleeve
27. Insulating strip
28. Earth contact
29. Direction indicator control assembly
30. Control striker
31. Steering wheel
32. Grub screw
33. Split cone
34. Nut
35. Washer
36. Lock nut
37. Upper mounting bracket
38. Lower mounting bracket
39. Lower steering column
40. Lower universal joint yoke
41. Journal assembly
42. Upper universal joint assembly
43. Upper universal joint yoke
44. Journal assembly

Page 1.9

STEERING

Repeat for the lower end of the column.

To remove the slip ring if damaged or worn, prise up the slotted end and withdraw from the splined end of the column taking care of the contact and spring which will now be exposed in the lower half of the rubber insulator.

Re-assembling

Re-assembling is the reverse of the dismantling procedure. Renew the felt bushes if worn. When refitting the indicator switch striker care must be taken to ensure that it is adjusted centrally between the two trip levers on the switch with the wheels in the "straight-ahead" position. Adjustment is provided by means of slotted holes under the two fixing screws. Failure to ensure this will result in unequal automatic cancellation of the flasher switch.

Refitting

Refitting is the reverse of the removal procedure but care must be taken to ensure that the upper steering column and wheel are in the central position (that is, with the central spoke of the steering wheel in the 6-o'clock position) with the road wheels "straight-ahead" before engaging the splines in the universal joint.

LOWER STEERING COLUMN

Removal

Disconnect the battery earth lead.

Drain and remove radiator as described on page I.6.

Remove the bonnet as described in Section N, "Body and Exhaust System"

Turn the steering wheel until the lower pinch bolt securing the upper universal joint to the lower steering column is accessible. Remove the bolt, nut and washer. Turn the steering further until the Allen screw in the steering column lower universal joint is accessible. Insert a correct size Allen key and remove screw.

Disconnect the flashing indicator switch cables from the multi-connector, located behind the facia panel, and disconnect the horn cable from the slip ring connector.

From the upper steering column remove the two lower support bolts, nuts and washers and the upper support bolt, nut and washer and spacer tube.

Lower the steering column and withdraw from the splines on the lower steering column shaft. Withdraw lower column shaft in an upward direction until the lower universal joint is clear of the splines on the pinion shaft. The steering column can now be withdrawn through front frame cross members above anti-roll bar.

Universal Joints

Carefully examine universal joints for wear and renew if necessary. For servicing instructions on universal joints see under "Propeller Shafts (Section G)".

Note: No lubrication points are provided in these universal joints, the bearings being pre-packed with grease on assembly.

Refitting

Refitting is the reverse of the removal procedure but care should be taken to ensure that the upper steering column and wheel are in the central position (that is, with the central spoke of the steering wheel in the 6-o'clock position) with the road wheels "straight-ahead" before engaging the splines in the universal joint.

TIE ROD BALL JOINT

The tie rod ball joints cannot be dismantled and if worn a complete new assembly must be fitted.

Removal

Remove the split pin and nut securing the tie-rod ball to the steering arm. Drift out the ball pin which is a taper fit in the steering arm. The ball pin taper will be more easily freed if the sides of the steering arm are tapped with a copper mallet.

Slacken the locknut on the tie-rod and unscrew the ball joint.

Refitting

Refitting is the reverse of the removal procedure. Screw on each tie rod ball joint an equal number of turns.

Lubricate the ball joints with the recommended lubricant. Do not over lubricate the ball joints to the extent where grease escapes from the rubber seal.

Re-set the front wheel alignment as described on page I.12.

Page I.10

STEERING

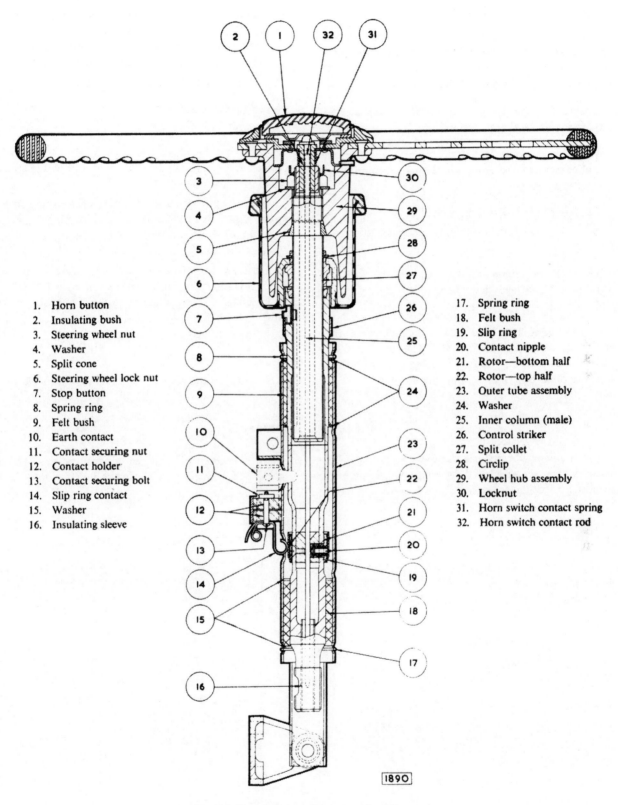

1. Horn button
2. Insulating bush
3. Steering wheel nut
4. Washer
5. Split cone
6. Steering wheel lock nut
7. Stop button
8. Spring ring
9. Felt bush
10. Earth contact
11. Contact securing nut
12. Contact holder
13. Contact securing bolt
14. Slip ring contact
15. Washer
16. Insulating sleeve
17. Spring ring
18. Felt bush
19. Slip ring
20. Contact nipple
21. Rotor—bottom half
22. Rotor—top half
23. Outer tube assembly
24. Washer
25. Inner column (male)
26. Control striker
27. Split collet
28. Circlip
29. Wheel hub assembly
30. Locknut
31. Horn switch contact spring
32. Horn switch contact rod

Fig. 7. Sectioned view of upper steering column.

Page I.11

STEERING

FRONT WHEEL ALIGNMENT

Check that the car is full of petrol, oil and water. If not, additional weight must be added to compensate for, say, a low level of petrol (the weight of 10 gallons of petrol is approximately 80 lbs—36·0 kg.).

Ensure that the tyre pressures are correct and that the car is standing on a level surface.

With the wheels in the straight ahead position check the alignment of the front wheels with an approved track setting gauge.

The front wheel alignment should be:—

$\frac{1}{16}''$—$\frac{1}{8}''$ (1·6—3·2 mm.) toe-in (measured at the wheel rim).

Re-check the alignment after pushing the car forward until the wheels have turned half a revolution (180°).

If adjustment is required slacken the locknuts at the end of each steering tie rod; also slacken the outer (small) clips securing the rack housing rubber bellows to avoid distortion after turning the tie rods.

Turn the tie rods by **equal amounts** in the necessary direction until the alignment of the front wheels is correct. Tighten the locknuts and re-check the alignment. Finally ensure that the rubber bellows are not twisted, and tighten the two clips.

STEERING ARM

Removal

Raise car by placing jack under front suspension lower wishbone fulcrum shaft bearing bracket and remove road wheel.

Remove split pin and nut from steering tie rod ball joint and drift out the tie rod ends from the steering arm, in which it is a taper fit, by tapping on the side face of the steering arm.

Unscrew the self-locking nut and remove the bolt and spring washer attaching the steering arm to the stub axle carrier. Remove the self-locking nut securing the stub axle shaft to the carrier, when the steering arm can be removed.

Refitting

Refitting is the reverse of the removal procedure.

ACCIDENTAL DAMAGE

The following dimensioned drawing is provided to assist in assessing accidental damage. A steering arm suspected of being damaged should be removed from the car, cleaned off and the dimensions checked and compared with those given in the illustration.

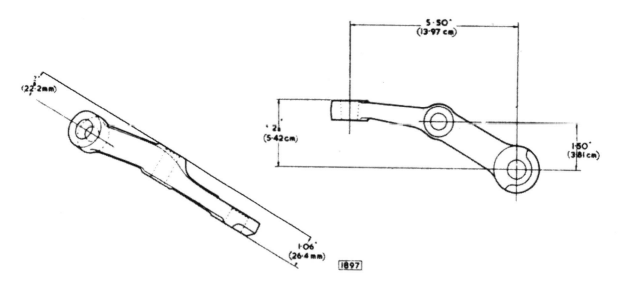

Fig. 8. The steering arm.

Page I.12

SECTION J
FRONT SUSPENSION

3·8 "E" TYPE
GRAND TOURING MODELS

INDEX

	Page
Description	J.4
Data	J.5
Routine Maintenance	J.5
Wishbones and anti-roll bar	J.5
Front hydraulic damper	J.5
Wheel swivels	J.5
Wheel bearings	J.6
Recommended lubricants	J.6
Front Suspension Assembly	
Dismantling	J.7
Upper Wishbone	
Removal	J.7
Refitting	J.7
Removal of fulcrum shaft	J.10
Adjustment of the ball joint	J.10
Renewing the rubber/steel bushes	J.10
Lower Wishbone	
Removal	J.10
Refitting	J.11
Renewing the rubber/steel bushes	J.11
Lower Wishbone Ball Joint	
Dismantling	J.11
Re-assembling	J.12
Adjustment of the ball joint	J.12
Stub Axle Carrier	
Removal	J.12
Refitting	J.13

INDEX (continued)

	Page
Wheel Hubs	
Removal	J.13
Dismantling	J.13
Refitting	J.13
Bearing end-float adjustment	J.14
Hydraulic Dampers	
Removal	J.14
Refitting	J.14
Anti-roll Bar	
Removal	J.15
Renewing the link arm bushes	J.15
Refitting	J.15
Torsion Bar Adjustment	J.15
Castor Angle	
Adjustment	J.17
Camber Angle	
Adjustment	J.18
Accidental Damage	J.19

Page J.3

FRONT SUSPENSION

DESCRIPTION

The right and left hand front suspension units are comprised of the upper and lower wishbones to which are attached the stub axle carriers, the torsion bars and the hydraulic dampers.

The torsion bars are attached at their forward end to the lower wishbones and at the rear end to brackets secured to the chassis frame.

Each torsion bar is controlled by a telescopic direct acting hydraulic damper.

The top of each damper is attached to brackets formed on the forward chassis assembly; the bottom of the damper being bolted to the lower wishbone.

The upper wishbone is a one piece forging secured to the threaded fulcrum shaft by means of pinch bolts through clamps formed on the wishbone inner mounting. The fulcrum shaft is mounted on two rubber/steel bonded bushes.

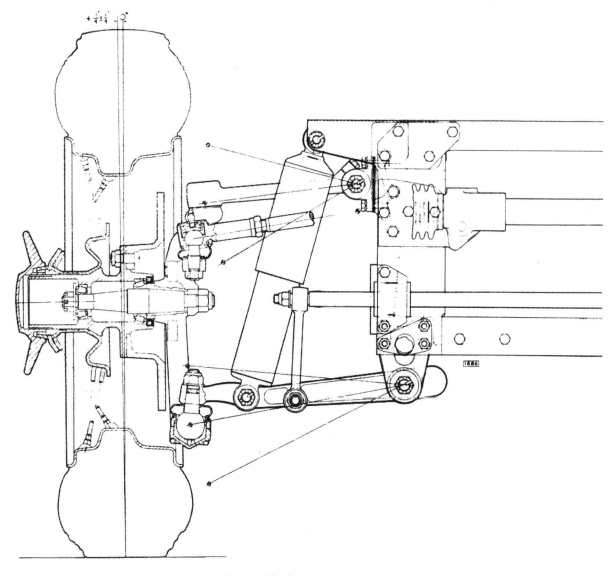

Fig. 1. The front suspension assembly.

Page J.4

FRONT SUSPENSION

The outer ends of the wishbone carry the upper wishbone ball joint which is in turn secured to the hub carrier by the tapered shank of the ball pin and a locknut.

The lower wishbone is a two piece assembly the inner ends of which are mounted at the fulcrum shaft end on rubber/steel bonded bushes.

The outer end of the lower wishbone is secured to the lower wishbone ball joint by the tapered shank of the ball pin and a locknut.

An anti-roll bar fitted between the lower wishbones is attached to the chassis front member by rubber insulated brackets.

The wheel hubs are supported on two tapered roller bearings, of which the inner races fit on a shaft located in a tapered hole bored in the stub axle carrier.

DATA

Type	Independent torsion bars
Dampers	Telescopic hydraulic
Castor Angle	$2° \pm \frac{1}{2}°$ positive
Camber Angle	$\frac{1}{4}° \pm \frac{1}{2}°$ positive
Swivel inclination	$4°$

ROUTINE MAINTENANCE

Wishbones and Anti-Roll Bar

The front suspension wishbone levers and the anti-roll bar are supported on rubber bushes which do not require any attention.

Front Hydraulic Dampers

The front hydraulic dampers are of the telescopic type, and no replenishment with fluid is necessary or provided for.

EVERY 2,500 MILES (4,000 KM.)
Wheel Swivels

Lubricate the nipples (four per car) fitted to the top and bottom of the wheel swivels. The nipples are accessible from underneath the front of the car. Lack of lubrication at these points may cause stiff steering.

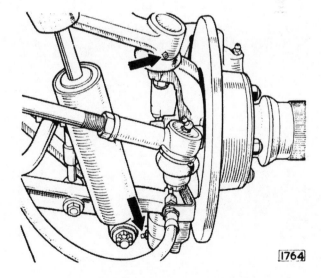

Fig. 2. The steering swivel grease nipples.

Page J.5

FRONT SUSPENSION

EVERY 10,000 MILES (16,000 KM.)

Wheel Bearings

Removal of the wheels will expose a grease nipple in the wheel bearing hubs. Lubricate sparingly with the recommended grade of lubricant. Always thoroughly clean the grease nipple before applying grease gun. An indication that sufficient grease has been applied is by the escape of grease past the outer hub bearing which can be observed through the bore of the splined hub.

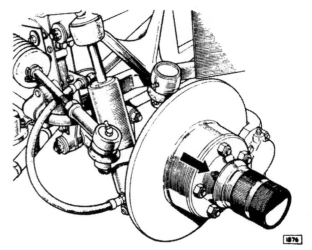

Fig. 3. The front wheel bearing grease nipple.

Recommended Lubricants

Component	Mobil	Castrol	Shell	Esso	B.P.	Duckham	Regent Caltex/Texaco
Front Wheel Bearings	Mobilgrease MP	Castrolease LM	Retinax A	Esso Multi-purpose Grease H	Energrease L2	LB10	Marfak All purpose
Wheel Swivels	Mobilgrease MP	Castrolease LM	Retinax A	Esso Multi-purpose Grease H	Energrease L2	LB10	Marfak All purpose

Page J.6

FRONT SUSPENSION

FRONT SUSPENSION ASSEMBLY— DISMANTLING

It is not advisable to attempt to remove the right hand and left hand front suspension assemblies as complete units. The various components should be removed as separate items.
To dismantle proceed as follows.

UPPER WISHBONE

Removal

Slacken off, but do not remove the hub caps from the road wheels; the hub caps are marked "RIGHT (OFF) SIDE" and "LEFT (NEAR) SIDE" and the direction of rotation to remove, that is, clockwise for the right hand side and anti-clockwise for the left hand side.

Place the jack under the lower wishbone fulcrum support bracket and raise the car until the wheels are clear of the ground.

Place a stand under the wishbone fulcrum rear support bracket.

Complete the removal of the road wheels.

Do NOT place the jack or stands under the forward frame cross tubes.

Remove the self-locking nut and drift out the upper wishbone ball joint from the stub axle carrier, into which it is a taper fit, by tapping on the side face of the carrier adjacent to the pin.

Remove the two bolts, nuts and lock washers retaining the fulcrum shaft rear carrier bracket to the chassis frame.

Identify and remove the shims fitted between the bracket and the chassis frame, and the stiffener plate located behind the two nuts on the inner face of the frame member.

Note: DO NOT confuse the shims with this stiffener plate when refitting the bracket.

Remove the three setscrews and lock washers retaining the fulcrum shaft front carrier bracket to the chassis frame.

Identify and remove the shims fitted between the bracket and the chassis frame.

Remove the upper wishbone.

Extract the split pins and unscrew the nuts retaining the brackets to the fulcrum shaft. Withdraw the brackets and rubber bushes. Note the relative positions of the shims removed from the front and rear brackets as these control the camber angle.

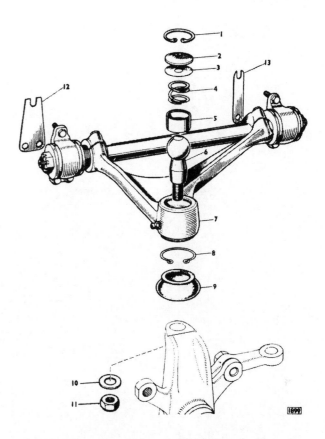

Fig. 4. The upper wishbone and ball pin.

1. Circlip	6. Ball pin	11. Nut
2. Top cover	7. Upper wishbone	12. Camber shims (front carrier bracket)
3. Shims	8. Circlip	
4. Socket spring	9. Rubber gaiter	13. Camber shims (rear carrier bracket)
5. Ball pin socket	10. Washer	

Note: When carrying out the above operation do not allow the flexible brake hose to become extended. Tie up the axle carrier to the frame member.

Refitting

The refitting of the upper wishbone assembly is the reverse of the removal procedure, but the slotted nuts at each end of the fulcrum shafts must not be tightened until the upper wishbone assembly has been fitted and the full weight of the car is on the suspension. Omitting to carry out this procedure will result in undue torsional loading of the rubber bushes with possible premature failure.

Note: Check the ball joint rubber gaiter (9). Replace if worn or damaged.

Check the castor and camber angles after refitting upper wishbone as described on pages J.17 and J.18.

Page J.7

FRONT SUSPENSION

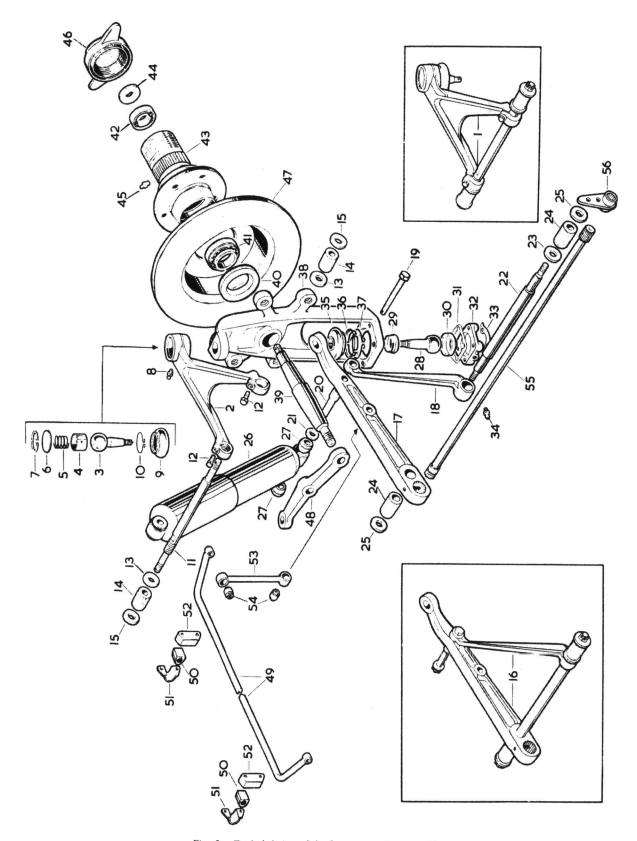

Fig. 5. Exploded view of the front suspension assembly.

FRONT SUSPENSION

1. Upper wishbone assembly (Right-hand)
2. Upper wishbone (Right-hand)
3. Upper wishbone ball pin
4. Ball pin socket
5. Spring
6. Top cover
7. Circlip
8. Grease nipple
9. Rubber gaiter
10. Clip
11. Upper wishbone fulcrum shaft
12. Pinch bolt
13. Distance washer
14. Rubber bush (Upper wishbone)
15. Special washer
16. Lower wishbone assembly (Right-hand)
17. Lower wishbone lever (Right-hand front)
18. Lower wishbone lever (Right-hand rear)
19. Bolt
20. Sleeve
21. Washer
22. Lower wishbone fulcrum shaft
23. Distance washer
24. Rubber bush (lower wishbone)
25. Special washer
26. Shock absorber (front)
27. Shock absorber (bottom bush)
28. Lower wishbone ball pin
29. Ball pin spigot
30. Morganite socket
31. Shims
32. Lower ball pin cap
33. Tab washers
34. Grease nipple
35. Rubber gaiter
36. Gaiter retainer
37. Clip
38. Stub axle carrier
39. Stub axle
40. Oil seal
41. Inner bearing
42. Outer bearing
43. Front hub (Right-hand)
44. "D" washer
45. Grease nipple
46. Hub cap
47. Brake disc
48. Steering arm
49. Anti-roll bar
50. Rubber bush
51. Bracket
52. Distance piece
53. Anti-roll bar link
54. Rubber bush
55. Torsion bar
56. Bracket—torsion bar (rear end)

Page J.9

FRONT SUSPENSION

IMPORTANT

It is essential that the top wishbone ball pin is not allowed to come into hard contact with the sides of the ball socket. When testing the movement of the ball in its socket, move the ball only in the direction of the elongation.

If the top wishbone is removed complete with the stub axle carrier the assembly must not be held by the top wishbone and the axle carrier allowed to swing on the ball pin.

Removal of the Fulcrum Shaft

Release the two clamp screws locking wishbone to fulcrum shaft. Turn shaft in a clockwise direction, looking from the rear, until the threaded portion of the shaft is clear of the wishbone. Withdraw the shaft through the wishbone arms.

Adjustment of the Ball Joint

The correct clearance of the ball pin in its socket is ·004″ (·10 mm.).

Shims for the adjustment of the ball joint are now available in ·004″ (·10 mm.) thicknesses.

To adjust the ball pin clearance to the correct figure, Fig. 4, remove the circlip (1), cover plate (2) and spring (4) from the ball joint. Clean thoroughly all the component parts.

Fit shims (3) between cover plate (2) and upper ball socket (5) until the ball is tight in its sockets when the cover plate and circlip are refitted without the spring.

Remove shims to the value of ·004″ (·10 mm.) and re-assemble ball joint complete with the spring, when it should be possible to move the ball pin by hand.

Finally lubricate with the recommended lubricant.

Note: Shims should not be added to take up excessive wear in the ball pin and sockets; if these parts are badly worn replacements must be fitted.

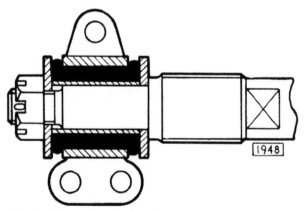

Fig. 6. Section through one of the upper wishbone rubber/steel bushed mounting brackets.

Page J.10

Renewing the Rubber/Steel Bushes

Drift or press out the bush from the bracket. Press the new bush into the bracket ensuring that the bush projects from each side of the bracket by an equal amount. Fitting of the bush will be facilitated if a lubricant made up of twelve parts of water to one part of liquid soap is used.

LOWER WISHBONE

Removal

Slacken off but do not remove the hub caps from the road wheels; the hub caps are marked "RIGHT (OFF) SIDE" and "LEFT (NEAR) SIDE" and the direction of rotation to remove, that is, clockwise for the right hand and anti-clockwise for the left hand side. Make up a block of hard wood to fit into the frame lower cross tube section as shown in Fig. 7.

Remove the cable harness band clips from the cross tube and insert the block of wood under the cross tube; place the jack under the wooden block and raise the car until the road wheels are clear of the ground.

Place stands under the blocks at the two outer ends of the cross tube adjacent to the lower wishbone fulcrum pivots. Complete the removal of the road wheels. Do NOT place the jack or stands under the frame cross tube without the wooden block inserted.

Disconnect the hydraulic brake pipe from the frame connection, remove the brake pipe carrier brackets and blank off the connector to prevent ingress of dirt or loss of fluid.

Remove the split pin and nut from the steering tie rod ball joint and drift out the tie rod end from its tapered seating in the steering arm by tapping on the side face of the steering arm adjacent to the ball pin.

Disconnect the upper wishbone ball joint as described on page J.7. If it is not required to remove the upper wishbone completely for servicing raise the wishbone to its full extent and tie to the frame.

Disconnect the lower wishbone ball joint by removing the self-locking nut and drifting out the ball pin from its tapered seating in the lower wishbone. Remove the axle carrier complete with the brake caliper and disc. Place the jack under the lower suspension arm and raise the jack to take up the weight of the car.

Note: Do not lift the car off the stands.

Remove the self locking nut retaining the anti-roll bar to the lower suspension arm.

FRONT SUSPENSION

Remove the split pin and nuts retaining the telescopic damper to the frame and the wishbone, extract the upper mounting bolt and withdraw the damper.

Lower and remove the jack. Unscrew the two bolts and lock washers securing the torsion bar rear adjuster lever to the frame and slide the lever forward until it is clear of the torsion bar splines.

Remove the locking bolt from the torsion bar front

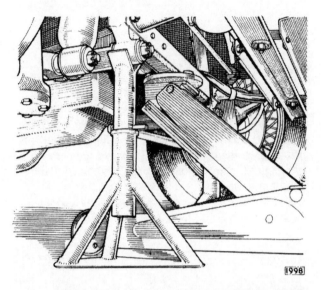

Fig. 7. Showing the front of the car jacked up under the front cross member; note the piece of hardwood which must first be inserted in the member. If only one front wheel is to be raised, the jack can be placed at the front end of the lower wishbone fulcrum shaft at the point where the stand is in position.

mounting. Slide the torsion bar rearwards until the front splines are clear of the wishbone and withdraw in a forward direction.

Remove the two bolts and washers retaining the fulcrum shaft rear carrier to the chassis frame.

Remove the four bolts, nuts and washers retaining the fulcrum shaft front carrier bracket to the chassis frame. Extract the split pin and remove the nuts from the lower wishbone shaft. Withdraw the brackets and rubber bushes.

Refitting

Refitting of the lower wishbone assembly is the reverse of the removal procedure, but it will be necessary to reset the torsion bar as described under "Torsion Bar—Adjustment" page J.15. Check the lower wishbone ball joint for clearance as described under "Lower Wishbone Ball Joint".

Examine the ball joint rubber gaiter. Replace if worn or damaged.

The slotted nuts at each side of the fulcrum shaft must not be tightened until the complete front suspension assembly has been fitted and the full weight of the car is on the suspension. Omitting to carry out this procedure will result in undue torsional loading of the rubber bushes with possible premature failure.

It will be necessary to re-bleed the front hydraulic brakes after refitting the lower wishbone assembly as described in Section L "Brakes".

Renewing the Rubber/Steel Bushes

Drift or press out the bush from the bracket. Press the new bush into the bracket so that the bush projects from each side of the bracket by an equal amount. Fitting of the bush will be facilitated if a lubricant made up of twelve parts of water to one of liquid soap is used.

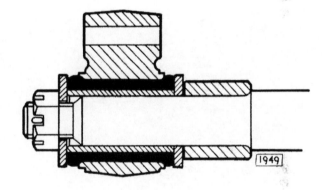

Fig. 8. Section through one of the lower wishbone rubber/steel bushed mounting brackets.

LOWER WISHBONE BALL JOINT

Dismantling

Release the wire clip (4, Fig. 9) and remove the rubber gaiter (3).

Tap back the tab washers (11) and unscrew the four setscrews (12) securing the ball pin cap (9) to the stub axle carrier.

Remove the cap (9), shims (8), ball pin socket (7), and ball pin (6).

Page J.11

FRONT SUSPENSION

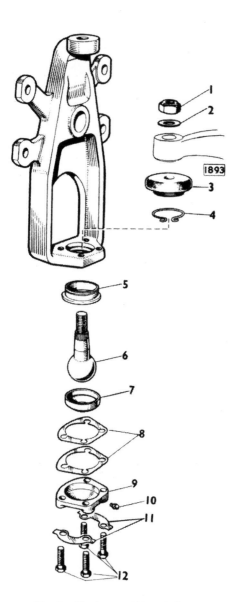

Fig. 9. *The lower wishbone ball joint.*

1. Nut
2. Washer
3. Rubber gaiter
4. Circlip
5. Spigot
6. Ball pin
7. Socket
8. Shims
9. Ball pin cap
10. Grease nipple
11. Tab washers
12. Setscrews

Re-assembling

Re-assembling is the reverse of the dismantling procedure but, if necessary, re-shim the ball joint to obtain the correct clearance of ·004″ to ·006″(·10 mm. to ·15 mm.).

Note: Shims should not be removed to take up excessive wear in the ball pin and sockets; if these parts are badly worn, replacements should be fitted.

Adjustment of the Ball Joint

The correct clearance of the ball pin in its socket is ·004″ to ·006″(·10mm. to·15mm.). Shims for adjustment of the ball joint are available in ·002″ (·05 mm.) and ·004″ (·10 mm.) thicknesses. To adjust the ball pin clearance to the correct figure, remove the shims one by one until, with the ball cap fully tightened, the ball is tight in its sockets. Fit shims to the value of ·004″ to ·006″ (·10 mm. to ·15 mm.) which should enable the shank of the ball pin to be moved by hand.

STUB AXLE CARRIER

Removal

Jack up the car and remove the road wheels as described under "Upper Wishbone—Removal" Page J.7.

Disconnect the hydraulic brake pipe from the frame connection, remove the brake pipe carrier and blank off the connector to prevent ingress of dirt and loss of fluid.

Remove the self-locking nut and plain washer securing the upper wishbone ball joint to the stub axle carrier. Drift out the ball from its tapered seating, by tapping on the side face of the carrier adjacent to the pin.

Raise the wishbone to its full extent and tie back to frame.

Remove the split pin and nut from the steering tie rod ball joint and drift out the tie rod end from its tapered seating by tapping on the side face of the carrier adjacent to the pin.

Remove the self-locking nut and plain washer securing the lower wishbone ball joint to the stub axle

Page J.12

FRONT SUSPENSION

carrier. Drift out the ball pin from its tapered seating by tapping on the side face of the lower wishbone adjacent to the ball pin.

Remove the axle carrier.

Refitting

Refitting is the reverse of the removal procedure. It will be necessary to bleed the front hydraulic brakes system after refitting the axle carrier and suspension arms as described in Section L "Brakes".

WHEEL HUBS

Removal

Jack up the car and remove the road wheel. Disconnect the flexible hydraulic brake pipe from the frame connection and blank off the connector to prevent the ingress of dirt and loss of fluid.

Remove the locking wire from the two brake caliper mounting bolts and unscrew the bolts noting the shims fitted between the caliper and the mounting plate. Remove the caliper. Remove the split pin, (2, Fig. 10), retaining the hub nut; holes are provided in the side of the hub through which the split pin can be withdrawn. Remove the slotted nut (1) and plain washer (3) from the end of the stub axle shaft. The hub can now be withdrawn by hand.

Dismantling

Extract the oil seal (8). Withdraw the inner races of the taper roller bearings (7). Examine bearing for wear. If new bearings are to be fitted the outer races can be drifted out from the hub.

Refitting

Refitting is the reverse of the removal procedure but it will be necessary to re-lubricate the bearings as detailed in "Routine Maintenance" at the beginning of this section and adjust the end float of the hub bearings as described in the following paragraph.

When refitting the brake caliper care should be taken to ensure that the correct clearances are maintained between the inner faces of the caliper and each face of the brake disc. For method of checking the clearance and tolerance permissible refer to Section L "Brakes". Re-bleed the hydraulic brakes after refitting as described in Section L "Brakes".

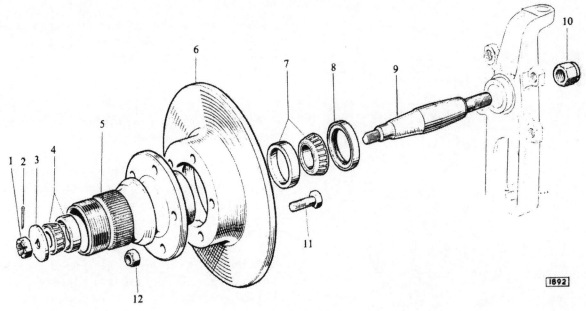

Fig. 10. The front hub.

1. Nut
2. Split pin
3. "D" washer
4. Outer bearing
5. Wheel hub
6. Brake disc
7. Inner bearing
8. Oil seal
9. Stub axle
10. Stub axle securing nut
11. Brake disc securing bolt
12. Nut

Page J.13

FRONT SUSPENSION

Bearing End-float Adjustment

The correct end float of the wheel bearings is ·003" to ·005" (·07 mm. to ·13 mm.). It is particularly important that the end float does not exceed ·005" (·13 mm.) otherwise the brakes may tend to drag and not function correctly.

The wheel bearing end float can be measured with a dial indicator gauge, mounted with the plunger against the hub. If a gauge is not available proceed as follows:

Tighten the end nut until there is no end float, that is, when rotation of the hub feels slightly "sticky".

Slacken back the hub nut between one and two flats depending on the split pin hole relative to the slots in the nut.

HYDRAULIC DAMPERS

The telescopic hydraulic dampers are of the sealed type with no provision for adjustment or "topping-up" with fluid, therefore, in the event of a damper being unserviceable a replacement damper must be fitted.

Before fitting a damper to the car it is advisable to carry out the following procedure to "bleed" any air from the pressure chamber that may have accumulated due to the damper having been stored in a horizontal position.

Hold the damper in its normal vertical position with the shroud uppermost and make several short strokes (not exceeding more than half-way) until there is no lost motion and finish by extending the damper to its full extent once or twice. Do not extend the damper fully until several short strokes have been made first. After the operation of "bleeding" the hydraulic dampers should be kept in their normal upright position until they are fitted to the car.

IMPORTANT

If the hydraulic damper is to be removed do not allow the suspension unit to drop lower than the normal rebound position, otherwise the top ball joint may "neck" in its housing.

Support the outer end of the lower wishbone before removing the damper.

Removal

Jack up the car under the lower wishbone at a point adjacent to the damper lower mounting until the wheels are clear of the ground.

Remove the road wheel.

Remove the split pin and nut from the damper top and bottom mounting bolts.

Remove the top mounting bolt, withdraw the damper from the bottom mounting and remove from the car.

Refitting

Refitting is the reverse of the removal procedure, but the slotted nuts should not be tightened until the full weight of the car is on the suspension. Omitting to carry out this procedure will result in undue torsional loading of the rubber bushes with possible ultimate failure.

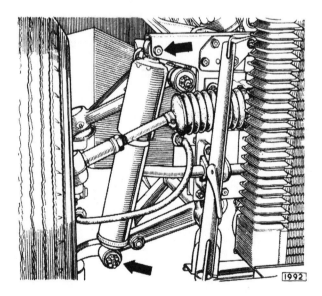

Fig. 11. The hydraulic damper attachment points.

FRONT SUSPENSION

ANTI-ROLL BAR

Removal

Remove the four bolts, nuts and washers from the anti-roll bar support brackets (51, Fig. 5) on the chassis member. Withdraw the two distance pieces.

Remove the self-locking nuts and withdraw the two bolts attaching the arm to the lower wishbone. To separate the anti-roll bar (49) from the link arm (53), remove the self-locking nuts and the washers and withdraw the two bolts. The anti-roll bar bracket rubbers are split to enable them to be removed from the anti-roll bar.

Renewing the Link Arm Bushes

Drift or press out the bushes from the link arm upper and lower eyes. Press the new bush into the eye ensuring the bush projects from each side by an equal amount. The fitting of the bush will be facilitated if a lubricant made up of twelve parts of water to one part of liquid soap is used.

Refitting

Refitting is the reverse of the removal procedure. It is most important when attaching the support bracket to the frame member and also when tightening the self-locking nuts on the link arm attachment bolts to have the full weight of the car on the suspension. Omitting to carry out this procedure will result in undue torsional loading of the rubber bushes with possible premature failure.

TORSION BAR—ADJUSTMENT

Checking

Check that the car is full of petrol, oil and water. If not additional weight must be added to compensate for, say, a low level of petrol (the weight of 10 gallons of petrol is approximately 80 lbs. (36·0 kg.)). Before any check on torsion bar setting is made the car must be placed on a perfectly level surface, wheels in the straight ahead position and tyre pressures correctly adjusted to:

  Front 23 lbs. per sq. in. (1·62 kg./cm.²)
  Rear 25 lbs. per sq. in. (1·76 kg./cm.²)

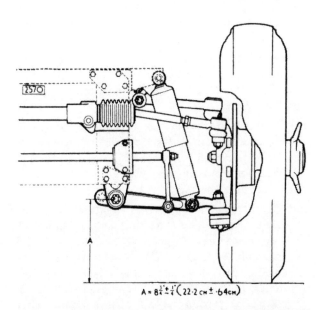

Fig. 12. Showing the method of checking the standing height.

Roll car forward three lengths.

With the torsion bar correctly adjusted the measurement A should be $8\frac{3}{4}'' \pm \frac{1}{4}''$ (22·2 ± ·64 cm.).

Adjustment

If adjustment is necessary proceed as follows.

Jack up the car and place stands under the lower wishbone fulcrum support bracket.

Note: DO NOT place jack or stand immediately under the forward frame tubes.

Remove the road wheels.

Disconnect the upper wishbone ball joint from the stub axle carrier, as described on page J.7.

Disconnect the steering tie-rod ball joint from the stub axle carrier as described on page J.12.

Disconnect the anti-roll bar as described on page J.15.

Place the jack under the lower wishbone at a point adjacent to the damper lower mounting. Raise jack but do not lift the car off the stands.

Remove the split pins and slacken the nuts retaining the lower wishbone rubber mountings.

Remove the hydraulic damper as described on page J.14. Lower the jack.

Remove the two bolts and nuts securing the torsion bar rear adjuster lever to the frame. Fit setting gauge, with two holes drilled at $17\frac{13}{16}''$ (45·24 cm.) centres to damper mounting points to position lower wishbone.

Page J.15

FRONT SUSPENSION

Note: The setting gauge can be easily made using Fig. 13 as a reference.

The two holes in the torsion bar rear adjuster lever and the corresponding holes in the frame should now be in line. If holes are not in line adjustment must be made as follows:

(i) Note which way lever requires to be rotated to bring holes in line. Mark position of the lever on shaft, remove by sliding off the splines, turn in direction required, and locate on fresh splines. Check lever position.

(ii) Repeat operation if further adjustment is necessary. It should be noted that the rear end of the torsion bar has 25 splines whereas the front end has only 24 splines. This permits the bar to be used as its own vernier and allows for a very fine adjustment. If this very fine adjustment is necessary slide torsion bar out of front splines after first removing the locking bolt.

Turn in direction required and engage fresh splines.

If position of lever is now correct refit rear bolts and nuts, also front locking bolt and nut and fully tighten.

Remove the setting gauge and locate damper on lower mounting.

Raise jack until damper upper retaining bolt will pass through bracket and damper eye. Refit nuts but do not tighten. Refit top wishbone steering tie-rod and anti-roll bar.

Repeat operation to left hand side.

Refit road wheels, jack up car, remove stands and lower car.

Tighten damper securing nuts and insert split pins. Tighten lower wishbone fulcrum shaft nuts and insert split pins. Tighten nuts securing anti-roll bar.

Roll car forward three lengths and re-check standing height of car which should now be as shown in Fig. 12.

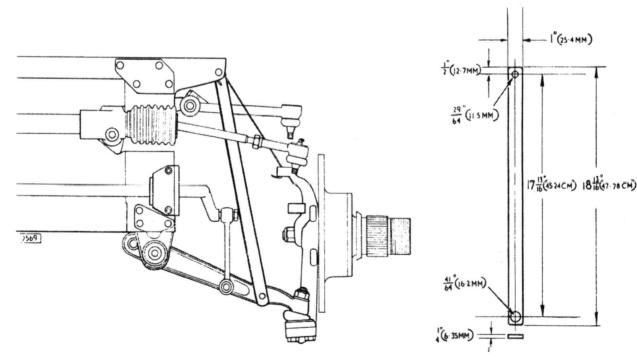

Fig. 13. The torsion bar setting gauge.

Page J.16

FRONT SUSPENSION

CASTOR ANGLE—ADJUSTMENT

Special links must be used when setting the castor angle of the front wheels. Dimensions for making the links are given in Fig. 15. The links, which fit over the top and bottom shock absorber mountings, hold the suspension in the mid-laden position.

Set the rear suspension in the mid-laden position utilising the setting links as described in Section K "Rear Suspension".

Using an approved gauge check the castor angle.

Castor angle $2° \pm \frac{1}{2}°$ positive.

Note: The castor angle for each wheel must not vary by more than $\frac{1}{2}°$.

Adjustment is effected by rotating the round threaded shaft on the front suspension upper wishbone bracket.

Remove the split pins and release the nuts situated at the rear and front of the fulcrum shaft and release the wishbone clamping bolts. The shaft may now be turned with a spanner placed on the two flats provided on the shaft.

Note: It is essential that the split pins be removed and the nuts released from the shaft otherwise a strain will be placed on the rubber mounting bushes.

To increase positive castor angle rotate the shaft anti-clockwise (viewed from the front of the car).

To decrease positive castor angle rotate the shaft clockwise. After adjustment retighten the clamp bolts.

The slotted nuts situated at the front and rear of the fulcrum shaft should not be tightened until the full weight of the car is on the suspension. Omitting to carry out this procedure will result in undue torsional loading of the rubber bushes with possible ultimate failure. Refit split pins.

The front of the car should be jacked up when turning the wheels from lock to lock during checking.

If any adjustment is made to the castor angle, the front wheel alignment should be checked and if necessary reset as described in Section I "Steering".

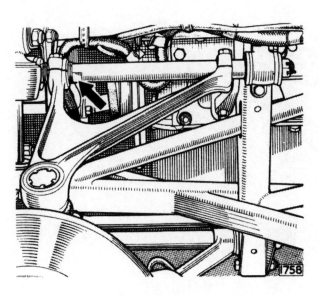

Fig. 14. The castor angle is adjusted by rotating the shaft indicated by the arrow.

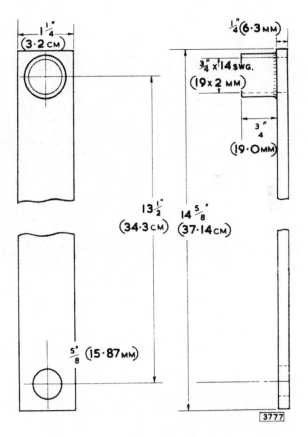

Fig. 15. Dimensions for front suspension setting links.

Page J.17

FRONT SUSPENSION

CAMBER ANGLE—ADJUSTMENT

When setting the camber angle of the front wheels the front and rear suspensions must be locked in the mid-laden position as detailed under the heading "Castor Angle—Adjustment".

Ensure that the tyre pressures are correct and that the car is standing on a level surface. Camber angle $\frac{1}{4}°\pm\frac{1}{4}°$ positive. The camber for each wheel must not vary by more than $\frac{1}{2}°$.

Line up the front wheel being checked parallel to the centre line of the car.

Using an approved gauge check the camber angle. Rotate the wheel being checked through 180° and re-check.

Adjustment is effected by removing or adding shims to the front suspension top wishbone bracket at two points, namely, the front and rear of the bracket.

The top holes in both front and rear shims are slotted and the bolts need only be slackened off to remove or add shims. The bottom holes are not slotted and it is necessary to remove bracket fixing bolts completely.

Inserting shims increases positive camber angle; removing shims increases negative camber angle or decreases positive camber angle. Remove or add an equal thickness of shims from each position otherwise the castor angle will be affected.

It should be noted the $\frac{1}{16}$" (1·6 mm.) of shimming will alter the camber by approximately $\frac{1}{4}°$.

Check the other front wheel in a similar manner. If any adjustment is made to the camber angle the front wheel alignment should be checked and if necessary be re-set as described in Section I "Steering".

Fig. 16. The camber angle is adjusted by means of shims indicated by the arrows. Remove or add an equal thickness of shims from each position.

Page J.18

FRONT SUSPENSION

ACCIDENTAL DAMAGE

The following dimensional drawings are provided to assist in assessing accidental damage. A component suspected of being damaged should be removed from the car, cleaned off, the dimensions checked and compared with those given in the appropriate illustration.

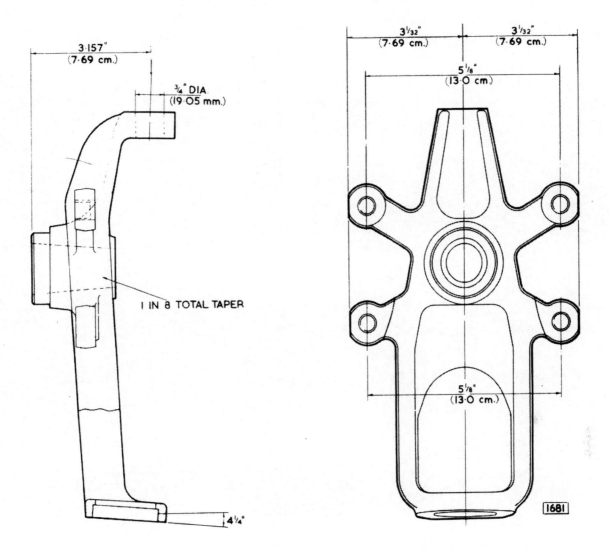

Fig. 17. The stub axle carrier.

Page J.19

FRONT SUSPENSION

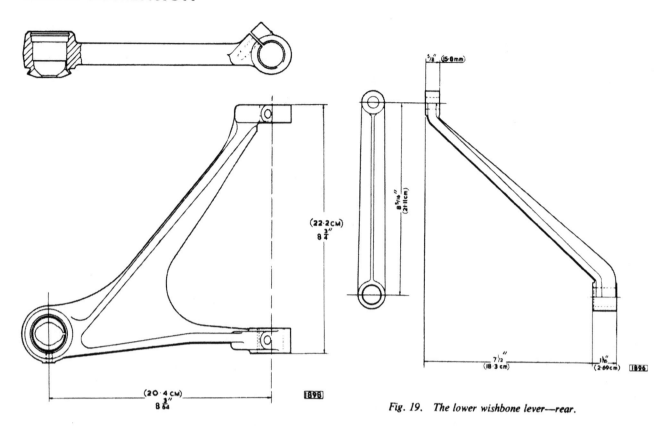

Fig. 18. The upper wishbone.

Fig. 19. The lower wishbone lever—rear.

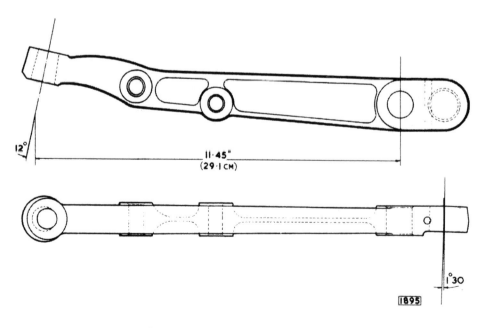

Fig. 20. The lower wishbone lever—front.

Page J.20

SECTION K
REAR SUSPENSION

3·8 "E" TYPE
GRAND TOURING MODELS

INDEX

	Page
Description	K.3
Data	K.4
Special Tools	K.4
Routine Maintenance	
Recommended lubricants	K.4
Rear Suspension	
Removal	K.5
Refitting	K.5
Road Spring and Hydraulic Damper Assembly	
Removal	K.6
Refitting	K.6
Hydraulic Dampers	
Removal	K.7
Refitting	K.7
Radius Arm	
Removal	K.7
Refitting	K.7
Wishbone	
Removal	K.10
Refitting	K.11
Wishbone Outer Pivot	
Removal	K.12
Dismantling	K.12
Re-assembly	K.12
Bearing adjustment	K.13
Refitting	K.14
Inner Fulcrum Wishbone Mounting Bracket	
Removal	K.14
Refitting	K.14
Rear Wheel Camber—Adjustment	K.15

REAR SUSPENSION

Description

The rear wheels are located in a transverse plane by two tubular links of which the top link is the half shafts universally jointed at each end. The lower link is pivoted at the wheel carrier and at the crossbeam adjacent to the differential casing. To provide maximum rigidity in a longitudinal plane the pivot bearings at both ends of the lower link are widely spaced. The suspension medium is provided by four coil springs enclosing telescopic hydraulic dampers, two being mounted on either side of the differential casing. The complete assembly is carried in a fabricated steel crossbeam. The crossbeam is attached to the body by four "Vee" rubber blocks and is located by radius arms. The radius arm pivots are rubber bushes mounted on each side of the car between the lower link and a mounting point on the body structure.

An anti-roll bar fitted between the two lower wishbones, is attached to the underframe side members by rubber insulated brackets.

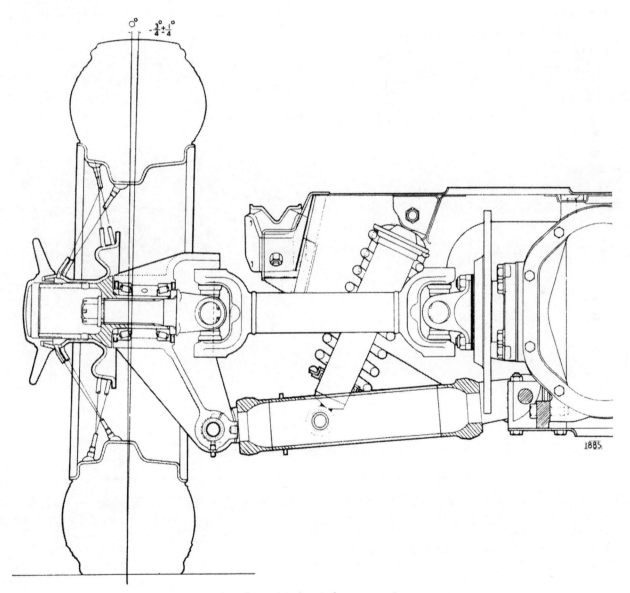

Fig. 1. Sectioned view of rear suspension

Page K.3

REAR SUSPENSION

DATA

	Early Cars	Later Cars
Rear Road Spring		
Free length (approx.)	10·1″ (25·65 cm.)	10·5″ (26·67 cm.)
Number of coils (approx.)	9⅜	10
Wire diameter	·432″ (11·0 mm.)	
Identification colour	—	Red
Dampers	Telescopic	
Road Wheel Movement from mid laden position		
Full bump	3⅛″	
Full rebound	3⅛″	
Track	50¼″	
Rear Wheel Camber	¾° ± ¼° negative	

Special tools	Churchill Tool No.
Rear road spring removal tool (used in conjunction with SL.14)	J.11A
Dummy shaft for wishbone fulcrum points (2 off)	J.14
Rear camber setting links	J.25

ROUTINE MAINTENANCE

EVERY 5,000 MILES (8,000KM.)

Wishbones

Lubricate the wishbone lever pivots. Three grease nipples are provided on each wishbone, see Fig. 2.

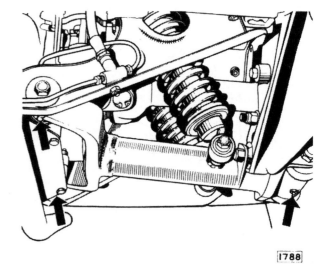

Fig. 2. Outer and inner pivot bearing grease nipples

Recommended Lubricants

	Mobil	Castrol	Shell	Esso	B.P.	Duckham	Regent Caltex/Texaco
Wishbone Pivots	Mobilgrease MP	Castrolease LM	Retinax A	Esso Multi-purpose Grease H	Energrease L2	LB10	Marfak All purpose

Page K.4

REAR SUSPENSION

REAR SUSPENSION

Removal

Slacken the two clamp bolts which secure the tail pipes to the silencers.

Remove the two nuts, bolts and washers securing the exhaust tail pipes to the centre mounting point under the rear of the body.

Withdraw the exhaust tail pipes.

Detach the radius arms at the front end.

Place a stout piece of wood approximately 9" 9" × 1" (22·8 cm. × 22·8 cm. × 25·4 mm.) between the rear suspension tie plate and the jack.

Jack up the rear of the car and place two chassis stands of equal height under the body forward of the radius arm mounting posts. Place blocks of wood between the chassis stands and the body to avoid damage.

Remove the rear road wheels.

Leaving the jack in position under the differential tie plate remove the two self locking nuts and bolts securing the anti-roll bar links to the roll bar.

Disconnect the flexible brake pipe at the connection on the body.

Remove the split pin, washer and clevis pin securing the handbrake cable to the handbrake caliper actuating levers mounted on the suspension cross beam.

Slacken the locknut and screw the outer handbrake cable screw out of the adjuster block.

Remove the four bolts and self locking nuts securing the mounting rubbers at the front of the cross beam to the body frame. Note carefully the number and location of the packing shims between the mounting rubbers and body frame. Remove the six self locking nuts and four bolts securing the rear mounting rubbers to the cross beam.

Remove the four self locking nuts and bolts securing the propeller shaft to the differential pinion flange.

Lower the rear suspension unit on the jack and withdraw the unit from under the car as shown in Fig. 3.

Refitting

Refitting is the reverse of the removal procedure.

Check all mounting rubbers for deterioration.

Bleed the braking system as described in Section L. "Brakes".

If the radius arms have been removed the rear

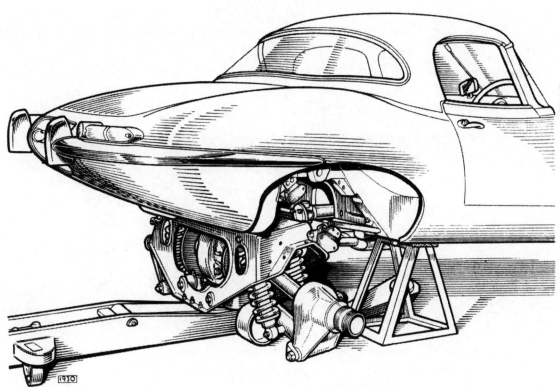

Fig. 3. Removal of the rear suspension assembly from the car

Page K.5

REAR SUSPENSION

suspension should be at the normal riding height before tightening the radius arm securing nuts on the rear suspension wishbone. Refit the radius arms as described on page K.7.

If the rear suspension mounting rubbers have been removed it is essential that the rubbers are refitted with the cut-away flange towards the suspension unit as shown in Fig 4.

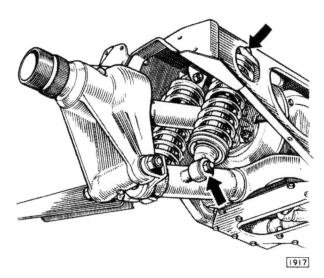

Fig. 5. Hydraulic damper mounting points

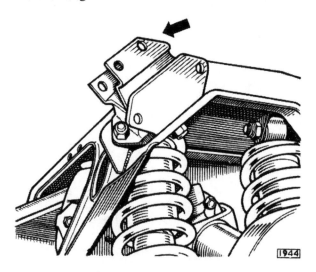

Fig. 4. Showing the correct position of the rear suspension mounting rubber

each hydraulic damper to the cross beam.

Withdraw the hydraulic damper and road spring assembly.

Refitting

Refitting is the reverse of the removal procedure.

IMPORTANT

The following removal and refitting operations are described assuming the rear suspension is removed from the car. If it is possible for the operations to be carried out with the rear suspension in position on the car the fact will be noted in the text.

ROAD SPRING AND HYDRAULIC DAMPER ASSEMBLY

Removal

The road spring and hydraulic damper assembly may be removed from the car with the rear suspension assembly in position.

Remove the two self locking nuts and washers securing the two hydraulic dampers to the wishbone.

Support the appropriate wishbone and drift out the hydraulic damper mounting pin, Fig. 6.

Remove the self locking nut and bolt securing

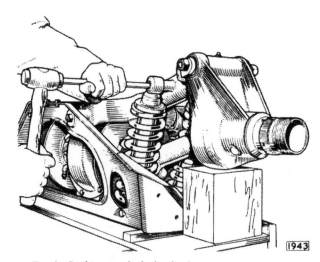

Fig. 6. Drifting out the hydraulic damper mounting pin

Page K.6

REAR SUSPENSION

HYDRAULIC DAMPERS

The telescopic hydraulic dampers are of the sealed type with no provision for adjustment or "topping-up" with fluid. Therefore, in the event of a damper becoming unserviceable a replacement must be fitted.

Before fitting a damper to a car it is advisable to carry out the following procedure to "bleed" any air from the pressure chamber that may have accumulated due to the damper having been stored in the horizontal position. Hold the damper in its normal vertical position with the shroud uppermost and make several short strokes (not extending more than half way) until there is no lost motion. Finish by extending the damper to its full length once or twice. Do not extend the damper fully until several short strokes have been made first. After the operation of "bleeding", the hydraulic dampers should be kept in their normal upright position until they are fitted to the car.

Removal

Remove the road spring and hydraulic damper as described on page K.6.

Utilizing a suitable press, Fig. 7, compress the road

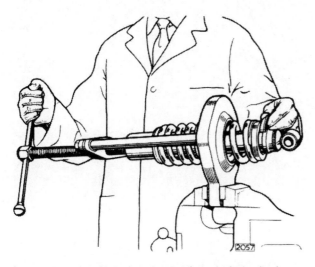

Fig. 7. Removing the rear road spring from the hydraulic damper with Churchill tool J.11 in conjunction with SL.14

spring until the split collet can be removed from under the road spring retaining pad.

Carefully release the pressure on the road spring and withdraw the hydraulic damper.

On early cars an aluminium pad was fitted to either end of the spring. The pad fitted to the shrouded end of the damper was recessed to receive the shroud.

Refitting

Compress the road spring, utilizing Churchill tool No. J.11 and SL.14, sufficiently to allow the hydraulic damper to be passed through the road spring and spring pad and the split collet placed into position, see Fig. 7. Ensure that the split collet and spring pad are seating correctly. Release the pressure on the road spring.

On early cars fit the machined recessed aluminium pad to the shrouded end of the damper. Compress the road spring and pass the damper through the spring and fit the other aluminium pad and secure with the split collet. Release the pressure on the road spring.

Refit the road spring and hydraulic damper assembly as described on page K.6.

RADIUS ARM

Removal

Remove the locking wire from the radius arm safety strap and securing bolt.

Unscrew the two self locking nuts securing the safety strap to the body floor.

Remove the radius arm securing bolt and spring washer and remove the safety strap.

Withdraw the radius arm from the mounting post on the body.

Remove the self locking nut and bolt securing the anti-roll bar to the radius arm.

Remove one of the self locking nuts securing the hub bearing assembly fulcrum shaft to the wishbone.

Drift out the fulcrum shaft from the wishbone and hub assembly as described on page K.12.

Remove the self locking nut and bolt securing the radius arm to the wishbone and remove the radius arm.

Examine the radius arm mounting rubbers for deterioration.

Refitting

Refitting is the reverse of the removal procedure.

When replacing the large radius arm body mounting rubber, the two holes should be in the longitudinal position in the radius arm as shown in Fig. 9.

The rubbers on the wishbone mounted end of the radius arm can be pressed out. Ensure that the rubbers are refitted with an equal amount of space showing on each side of the radius arm.

When refitting the hub bearing assembly shaft refer to page K.14.

Page K.7

REAR SUSPENSION

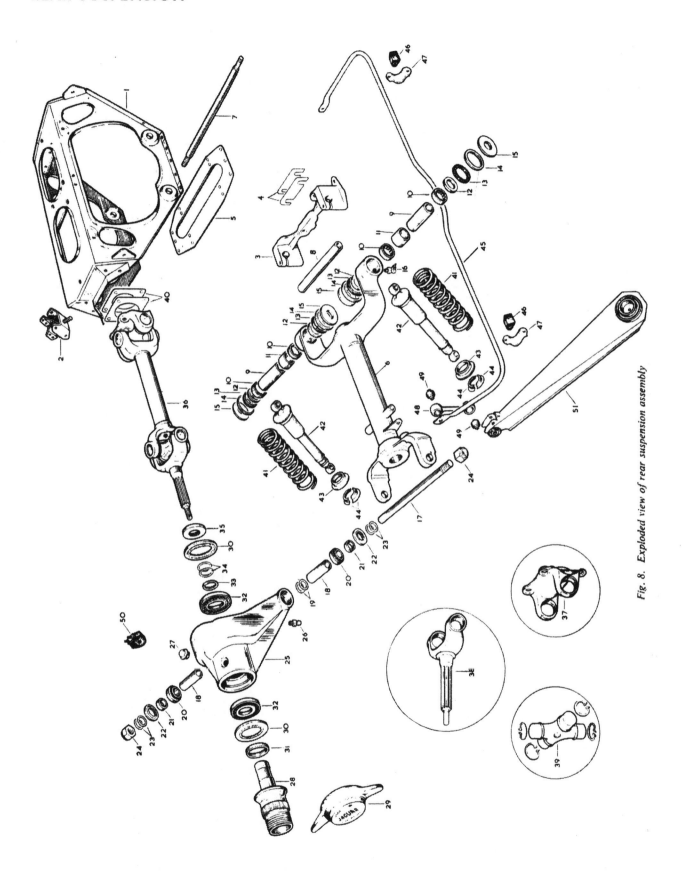

Fig. 8. Exploded view of rear suspension assembly

Page K.8

REAR SUSPENSION

1. Rear suspension cross member.
2. Rubber mounting.
3. Inner fulcrum mounting bracket.
4. Shims.
5. Tie plate
6. Wishbone.
7. Inner fulcrum shaft.
8. Distance tube.
9. Bearing tube.
10. Needle bearings.
11. Spacing collar.
12. Inner thrust washer.
13. Sealing ring.
14. Sealing ring retainer.
15. Outer thrust washer.
16. Grease nipple.
17. Outer fulcrum shaft.
18. Distance tube.
19. Shims.
20. Bearing.
21. Oil seal track.
22. Oil seal.
23. Shims.
24. Self locking nut.
25. Hub carrier.
26. Grease nipple.
27. Grease retaining cap.
28. Rear hub.
29. Hub cap.
30. Oil seal.
31. Oil seal track.
32. Outer bearing.
33. Spacer.
34. Shims (early cars only).
35. Oil seal track.
36. Half shaft.
37. Flange yoke.
38. Splined yoke.
39. Journal assembly.
40. Shim.
41. Coil spring.
42. Shock absorber.
43. Seat.
44. Retaining collet.
45. Anti-roll bar.
46. Rubber bush.
47. Bracket.
48. Link.
49. Rubber bush.
50. Bump stop.
51. Radius arm.

Page K.9

REAR SUSPENSION

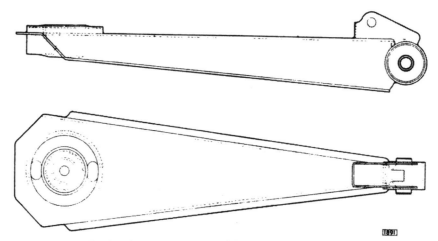

Fig. 9. Showing the position of the mounting rubbers in the radius arm

Refit the safety strap into position, refit the spring washer and radius arm securing bolt.

Refit the two bolts and nuts securing the safety strap to the body.

Tighten the radius arm securing bolt to 46 lb.ft. (6·36 kgm.) and pass the locking wire through the hole in the head of the bolt and secure round the safety strap.

WISHBONE

Removal

Remove the hydraulic dampers from the appropriate wishbone as described on page K.6.

Remove the six self locking nuts and bolts securing the tie plate to the cross beam.

Remove the eight self locking nuts and bolts securing the tie plate to the inner fulcrum wishbone mounting brackets and remove the tie plate.

Remove one of the self locking nuts securing the hub

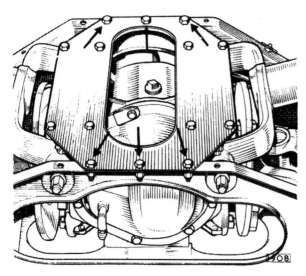

Fig. 10. Showing the six bolts which secure the tie plate to the cross beam

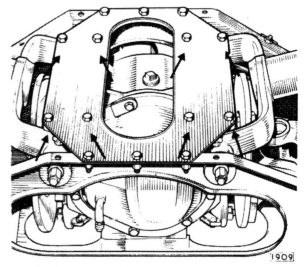

Fig. 11. Showing the eight bolts which secure the tie plate to the inner fulcrum mounting bracket

bearing assembly fulcrum shaft to the wishbone and drift out the fulcrum shaft, see Fig. 16.

Separate the hub carrier from the wishbone. If any shims are fitted between the wishbone and hub assembly note the amount and position of the shims as it is essential to replace the exact amount in the correct

REAR SUSPENSION

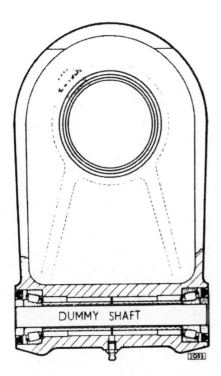

Fig. 12. *Showing the dummy shaft in position in the hub carrier*

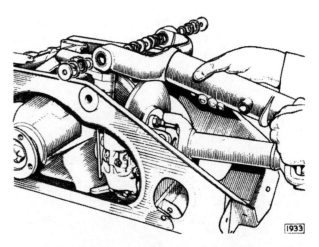

Fig. 13. *Showing the wishbone inner fork and components*

position. To facilitate refitting slide a dummy fulcrum shaft Churchill tool No. J.14 through the hub carrier.

Place a piece of sticky tape over each of the hub carrier assembly oil seal tracks to prevent them becoming displaced.

Remove the self locking nut securing the radius arm to the wishbone. Withdraw the special thin headed bolt and remove the radius arm from the wishbone.

Remove the self locking nut securing the wishbone fulcrum shaft to the cross beam.

Drift the inner fulcrum shaft out of the wishbone and inner fulcrum mounting bracket.

Withdraw the wishbone assembly and collect the four outer thrust washers, inner thrust washers, oil seals and oil seal retainers.

Examine the oil seals for deterioration.

Remove the two bearing tubes.

There is no need to remove the spacer fitted between the inner fulcrum mounting bracket unless the mounting bracket is to be replaced. To remove the spacer, tap out of position. To remove the needle rollers gently tap the needle cages out of the wishbone using a suitable drift. Remove the needle roller spacer.

Refitting

If the needle rollers have been removed from the larger fork of the wishbone lever press one roller cage into position, with the engraving on the roller cage facing outwards.

Insert the roller spacing tube and press in the other roller cage.

Repeat for the other side.

Insert the bearing tubes. Smear the four outer thrust washers, inner thrust washers, oil seals and oil seal retainers with grease and place into position on the wishbone, see Fig. 13.

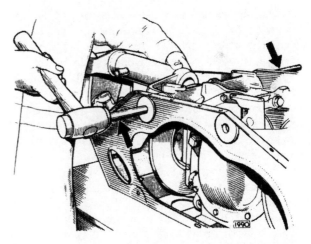

Fig. 14. *Tapping the dummy shafts into position at the wishbone inner fulcrum*

Offer up the wishbone to the inner fulcrum mounting bracket with the radius arm mounting bracket towards the front of the car. Align the holes and spacers. Press a dummy shaft Churchill tool No. J.14 through each side of the cross beam and wishbone.

Page K.11

REAR SUSPENSION

The dummy shafts locate the wishbone, thrust washers, cross beam and inner fulcrum mounting bracket and facilitate refitting of the fulcrum shaft.

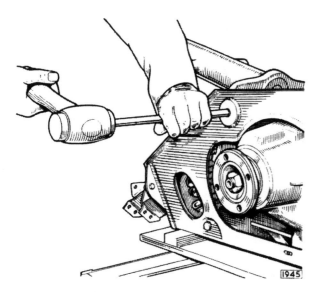

Fig. 15. Drifting the inner fulcrum shaft into position and displacing the dummy shafts

Smear the fulcrum shaft with grease and gently tap the shaft through the cross beam, wishbone and inner fulcrum mounting bracket. As the fulcrum shaft is tapped into position the short dummy shafts will be displaced from the opposite side. It will be found advantageous to keep a slight amount of pressure exerted on the dummy shafts as they emerge from the cross beam. This will reduce the tendency for the dummy shafts to be knocked out of position and allow a spacer or thrust washer to be displaced. If a washer or spacer becomes displaced it will be necessary to remove the fulcrum shaft, dummy shafts and wishbone and then repeat the operation.

When the fulcrum shaft is in position tighten the two self locking nuts to 55 lb.ft. (7·60 kgm.) with a torque wrench.

Refit the eight bolts and self locking nuts securing the tie plate to the inner fulcrum wishbone mounting bracket.

Refit the six bolts and self locking nuts securing the tie plate to the cross beam.

Refit the radius arm to the wishbone as described on page K.7.

Remove the two pieces of sticky tape holding the oil seal tracks in position.

Offer up the wishbone to the hub assembly.

Using a dummy shaft, Churchill tool No. J.14, line up the wishbone hub assembly oil seal tracks and spacers. Smear the fulcrum shaft with grease and gently tap the fulcrum shaft into position and displace the dummy shaft.

It will be found advantageous to apply a small amount of pressure on the locating bar against the fulcrum shaft to prevent the bar being knocked out of position and allowing a spacer to be displaced. If a spacer is displaced it may be necessary to repeat the operation.

Slide the fulcrum shaft through the wishbone and hub carrier. Using feeler gauges check the amount of clearance between the hub carrier and the wishbone lever, see Fig. 19. If necessary fit sufficient shims between the hub carrier and the wishbone to centralize the hub carrier. Tighten the nuts on the fulcrum shaft to 55 lb.ft. (7·60 kgm.).

Check the rear suspension camber angle as described on page K.15.

Refit the hydraulic dampers as described on page K.7.

Refit the rear suspension as described on page K.5.

Re-lubricate the wishbone fulcrum shafts as described in "Routine Maintenance" at the beginning of this section.

WISHBONE OUTER PIVOT

Removal

Support the hub carrier and wishbone.

Remove one of the self locking nuts securing the outer fulcrum shaft.

Drift out the fulcrum shaft, Fig. 16, and collect the shims, if any, between the hub carrier and the wishbone.

Separate the hub carrier and wishbone.

Dismantling

Remove the oil seal track and prise out the oil seals.

Remove the inner races of the tapered roller bearings, spacers and shims.

Re-assembly

Refit the inner races for the tapered roller bearings.

Fit the spacers and a known quantity of shims, this

Page K.12

REAR SUSPENSION

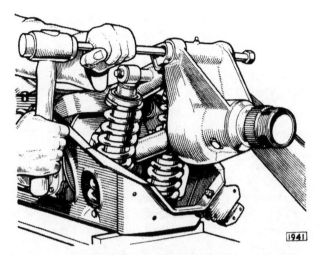

Fig. 16. Drifting out the wishbone outer fulcrum shaft

is necessary to obtain the correct bearing adjustment as described in the following paragraphs.

Fit the tapered roller bearings and oil seal tracks.

Bearing Adjustment

If it is necessary to adjust the tapered roller bearings it will be necessary to extract the hub from the rear axle half shaft as described in Section H "Rear Axle"

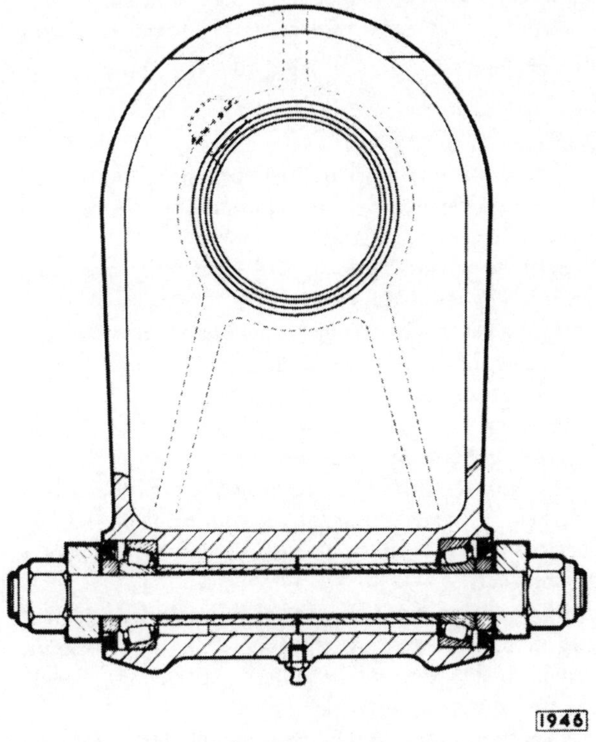

Bearing adjustment is effected by shims fitted between the two fulcrum shaft spacer tubes. The correct bearing adjustment is ·000″—·002″ (·00 mm.—·05 mm.) pre-load.

Shims are available in sizes of ·004″ (·101 mm.) and ·007″ (·17 mm.) thick and 1⅛″ (28·67 mm.) diameter.

A simple jig should be made consisting of a piece of plate steel approximately 7″ × 4″ × ⅜″ (17·7cm. × 10·1cm. × 9·5mm.). Drill and tap a hole suitable to receive the outer fulcrum shaft. Place the steel plate in a vice and screw the fulcrum shaft into the plate and slide an oil seal track onto the shaft. Place the assembly into position on the fulcrum shaft minus the oil seals and with an excess of shims, of a known quantity, between the spacers. Place an inner wishbone fork outer thrust washer onto the fulcrum shaft so that it abuts the oil seal track. Fill the remaining space on the shaft with washers and secure with a nut. Tighten the nut to 55 lb.ft. (7·60 kgm.). Press the hub carrier assembly towards the steel plate using a slight twisting motion to settle the rollers onto the bearing surface. Maintain a steady pressure against the hub carrier and using a feeler gauge measure the amount of clearance between the large diameter washer and the machined face of the hub carrier.

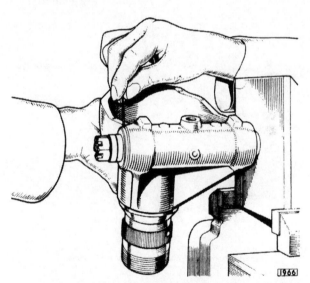

Fig. 18. Measuring the amount of clearance between the hub carrier and large washer to determine the end float in the bearings

Pull the hub carrier assembly towards the large diameter washer slightly rotating the carrier to settle the rollers onto the bearing surface. Maintain a steady pressure against the hub carrier and using feeler gauges measure the amount of clearance between the large diameter washer and the machined face of the hub carrier.

Fig. 17. Section through hub carrier and wishbone showing outer fulcrum shaft in position

Page K.13

REAR SUSPENSION

Subtract the one measurement from the other which gives the amount of end float present in the bearings.

Remove sufficient shims to obtain a reading of $000''-\cdot 002''$ ($\cdot 00$ mm.—$\cdot 05$ mm.) preload.

Example:—

Correct preload $\cdot 000''-\cdot 002''$ ($\cdot 00$ mm.—$\cdot 05$ mm.) Mean $\cdot 001''$ ($\cdot 02$ mm.)

Assume the bearing end float to be $\cdot 010''$ ($\cdot 25$ mm.) Therefore $\cdot 010''+\cdot 001''=\cdot 011''$ ($\cdot 25$ mm.$+\cdot 02$ mm. $=\cdot 27$ mm.) to be removed to give correct preload.

Refit the hub carrier to the half shaft as described in Section H "Rear Axle".

Fit new oil seals with the lips inwards and place the fulcrum shaft into position in the hub carrier.

Offer up the hub carrier to the wishbone. Chase the dummy shaft through the wishbone with the fulcrum shaft.

Using feeler gauges measure the gap between the oil seal track and the wishbone. Shims of $\cdot 004''$ ($\cdot 101$ mm.) thickness by $\frac{7}{8}''$ (22·2 mm.) diameter should be used.

Repeat for the other end and shim as necessary to centralize the hub carrier in the wishbone fork. The above procedure is to prevent the wishbone fork ends

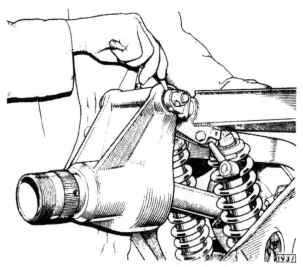

Fig. 19. Using feeler gauges to measure the clearance between the hub carrier oil seal tracks and wishbone fork

from closing inwards. Tighten the nuts on the fulcrum shaft to 55 lb.ft. (7·60 kgm.).

Refitting

To facilitate refitting, slide a dummy shaft Churchill tool No. J.14 through the hub carrier before offering up the wishbone to the hub carrier.

Refitting is the reverse of the removal procedure.

Page K.14

Re-lubricate the bearings as described in "Routine Maintenance" at the beginning of the section.

INNER FULCRUM WISHBONE MOUNTING BRACKET

Removal

Remove the eight bolts and self locking nuts securing the tie plate to the inner fulcrum wishbone mounting bracket.

Remove the six bolts and self locking nuts securing the tie plate to the cross beam.

Remove one self locking nut and drift out the inner fulcrum shaft.

Withdraw the forks of the wishbone from between the cross beam and inner fulcrum wishbone mounting bracket.

Collect the oil seal retainers, oil seals, inner and outer thrust washers and bearing tubes.

Remove the lock wire from the two setscrews which secure the inner fulcrum wishbone mounting bracket to the differential unit.

Remove the spacer between the inner fulcrum mounting bracket

Remove the two setscrews and note the amount of shims between the bracket and the differential.

Remove the inner fulcrum wishbone mounting bracket.

Refitting

If only one inner fulcrum wishbone mounting bracket is removed, replace the same amount of shims between the differential casing and the bracket.

Shims are available in sizes of $\cdot 005''$ ($\cdot 127$ mm.) and $\cdot 007''$ ($\cdot 177$ mm.) thickness.

If, however, both the inner fulcrum wishbone mounting brackets have been removed or replaced, it will be necessary to re-shim the brackets.

Hold the inner fulcrum wishbone mounting bracket in position between the cross beam.

Insert the fulcrum shaft through the cross beam and bracket. Screw the inner fulcrum bracket securing setscrews in two or three threads, enough to locate the bracket.

Insert the required amount of shims and tighten the two setscrews securing the inner fulcrum wishbone mounting bracket to the differential casing. Secure the two setscrews with locking wire.

Tap the spacer, fitted between the inner fulcrum mounting bracket lugs, into position.

REAR SUSPENSION

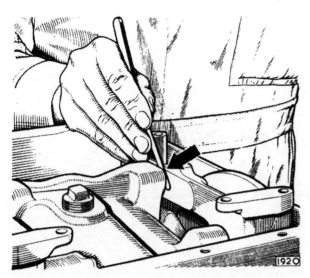

Fig. 20. Measuring the clearance between the inner fulcrum mounting bracket and the differential casing

Withdraw the inner fulcrum shaft from the cross beam and fulcrum bracket.

Offer up the wishbone to the inner fulcrum mounting bracket complete with bearing tubes, needle roller bearing and spacers, inner and outer thrust washers, oil seals and oil seal retainers. Ensure that the radius arm mounting bracket is towards the front of the car.

Align the holes and spacers. Press a dummy shaft through each side of the cross beam and wishbone.

The dummy shafts locate the wishbone, spacers, cross beam and inner fulcrum mounting bracket and facilitate refitting of the fulcrum shaft.

Smear the fulcrum shaft with grease and gently tap the shaft through the cross beam, wishbone and inner fulcrum mounting bracket. As the fulcrum is tapped into position the short dummy shafts will be displaced from the opposite side. It will be found advantageous to keep a slight amount of pressure exerted on the dummy shafts as they emerge from the cross beam. This will reduce the tendency for the dummy shafts to be knocked out of position and allow a spacer or thrust washer to be displaced. If a washer or spacer becomes displaced it will be necessary to remove the fulcrum shaft, dummy shafts and wishbone and then repeat the operation.

When the fulcrum shaft is in position tighten the two self locking nuts to 55 lb.ft. (7·60 kgm.) with a torque wrench.

Refit the eight bolts and self locking nuts securing the tie plate to the inner fulcrum wishbone mounting bracket.

Refit the six bolts and self locking nuts securing the tie plate to the cross beam.

Refit the rear suspension unit as described on page K.5.

REAR WHEEL CAMBER ANGLE—ADJUSTMENT

To check the camber angle of the rear suspension it is necessary for the car's wheels to be on a flat surface and for the tyre pressures to be correct.

Owing to the variations in the camber angle with different suspension heights it is necessary to lock the rear suspension in the mid-laden position by means of two setting links as shown in Fig. 22. To fit the setting links, hook one end in the lower hole of the rear mounting and depress the body until the other end can be slid over the hub carrier fulcrum nut. Repeat for the other side.

Remove the self locking nut and washer securing the forward road spring and hydraulic damper assembly to the wishbone mounting pin. Drift the mounting pin through the wishbone until the assembly is free from the pin.

Check the camber of the rear wheels, using a recommended gauge, by placing the gauge against each rear tyre in turn as shown in Fig. 21. The correct reading is $-\frac{3}{4}° \pm \frac{1}{4}°$. If the reading is incorrect it will be necessary to add or subtract shims between the half shaft and the brake disc. One shim ·020″ (·5 mm.) will alter the rear camber angle by approximately $\frac{1}{4}°$.

Jack up the car on the appropriate side and remove the rear road wheel.

Remove the self locking nut and washer securing the forward road spring and hydraulic damper assembly to the wishbone mounting pin. Drift the mounting pin through the wishbone until the assembly is free from the pin.

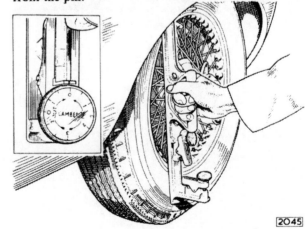

Fig. 21. Checking the rear wheel camber angle

Page K.15

REAR SUSPENSION

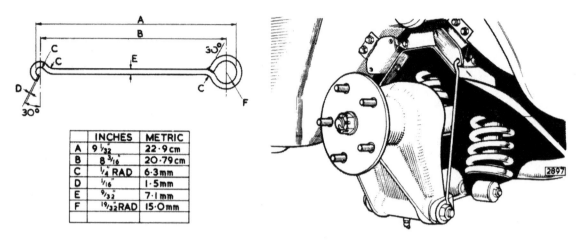

	INCHES	METRIC
A	9 1/32"	22·9 cm
B	8 3/16"	20·79 cm
C	1/4 RAD	6·3 mm
D	1/16"	1·5 mm
E	9/32"	7·1 mm
F	19/32 RAD	15·0 mm

Fig. 22. When checking the rear camber angle the rear suspension must be retained in the mid-laden position by means of the setting links (Churchill Tool No. J.25).

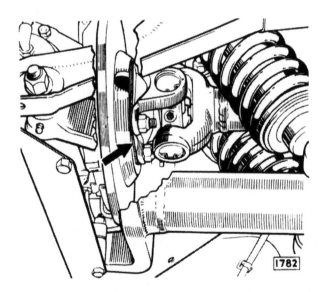

Fig. 23. The rear wheel camber angle is adjusted by means of shims indicated by the arrow

Remove the self locking nut and bolt securing the top of the road spring and hydraulic damper assembly to the cross beam and remove the assembly.

Unscrew the four self locking nuts securing the half shaft and the camber shims to the brake disc. Pull the hub and half shaft away from the shims sufficiently to clear the disc mounting studs. Remove or add shims as necessary.

Offer up the half shaft to the four disc mounting studs and secure with four self locking nuts. Offer up the forward road spring and hydraulic damper assembly to the cross beam and secure with a bolt and self locking nut.

Align the hydraulic damper and road spring assembly bottom mounting with the mounting pin in the wishbone and drift the pin through the assembly. Replace the plain washer and secure with a self locking nut.

Replace the rear road wheel and secure with the hub cap.

Re-check the camber angle.

Warning: After completing the adjustment, do not omit to remove the setting links from the suspension.

Page K.16

SECTION L
BRAKES

3·8 "E" TYPE
GRAND TOURING MODELS

INDEX

	Page
Description	L.4
Data	L.4
Retractor Operation—early type	L.5
Retractor Operation—later type	L.6
Handbrake Operation	L.6
Routine Maintenance	
Brake fluid level	L.7
Brake fluid warning light	L.8
Footbrake adjustment	L.8
Handbrake adjustment	L.8
Brake pedal bearing lubrication	L.9
Examining the friction pads for wear	L.9
Renewing the friction pads	L.9
Recommended brake fluids	L.9
Bleeding the Brake System	L.10
Brake Overhaul—Precautions	L.10
The Master Cylinders	
Removal	L.11
Renewing the master cylinder seals	L.12
Free travel of the master cylinder push rods	L.13
The Front Calipers	
Removal	L.13
Refitting	L.13
The Rear Calipers	
Removal	L.13
Refitting	L.14
The Front Brake Discs	
Removal	L.14
Refitting	L.14

INDEX (Continued)

	Page
The Rear Brake Discs	
Removal	L.16
Refitting	L.16
Brake Disc "Run-out"	L.16
Renewing the Friction Pads	L.16
Renewing the Brake Piston Seals—Early Type	L.18
Renewing the Brake Piston Seals—Later Type	L.18
The Handbrake—(Early Cars)	
Description	L.19
Renewing the handbrake friction pads	L.19
Friction pad carriers—removal	L.19
Friction pad carriers—dismantling	L.19
Friction pad carriers—assembling	L.19
Friction pad carriers—refitting	L.20
Removal and refitting the handbrake cable	L.21
The Brake Fluid Level and Handbrake Warning Light	
Description	L.21
Handbrake warning light switch—setting	L.21
The Vacuum Reservoir and Check Valve	
Description	L.22
Removal and refitting	L.22
The Brake/Clutch Pedal Box Assembly	
Removal and Refitting	L.23
Dismantling the brake linkage	L.23
Re-assembling the brake linkage	L.25
The Bellows Type Vacuum Servo	
Description	L.25
Operation	L.26
Servicing the unit	L.29
Detaching the unit	L.30
Dismantling the unit	L.30
Re-assembling the unit	L.31
Self-adjusting Handbrakes—(Later Cars)	
Description	L.31
	L.32

THE BRAKING SYSTEM

DESCRIPTION

The front wheel brake units are comprised of a hub mounted disc rotating with the wheel and a braking unit rigidly attached to each suspension member. The rear brake units are mounted inboard adjacent to the differential case. The braking unit is rigidly attached to the differential case. The brake unit consists of a caliper which straddles the disc and houses a pair of rectangular friction pad assemblies, each comprising a pad and a securing plate. These assemblies locate between a keep plate bolted to the caliper bridge and two support plates accommodated in slots in the caliper jaw. Cylinder blocks bolted to the outer faces of the caliper accommodate piston assemblies which are keyed to the friction pad assemblies. A spigot formed on the outer face of each piston locates in the bore of a backing plate with an integral boss grooved to accommodate the collar of a flexible rubber dust seal. The outer rim of the seal engages a groove around the block face and so protects the assembly from intrusion of moisture and foreign matter. A piston seal is located between the piston inner face and a plate secured by peen locked screws. (On later cars incorporating the revised retraction arrangement, a one piece piston is fitted).

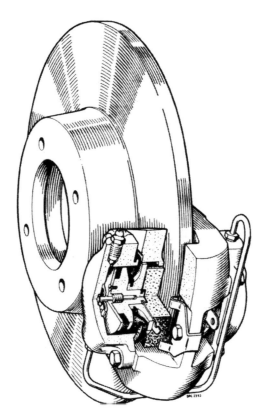

Fig. 1. Sectional view of a front disc brake

DATA

Make	Dunlop
Type	Bridge type caliper with quick change pads
Brake disc diameter—front	11″ (27·9 cm)
rear	10″ (25·4 cm)

Page L.4

BRAKES

Master cylinder bore diameter	$\frac{5}{8}''$ (15·87 mm)
Master cylinder stroke (upper—rear brakes)	1" (25·4 mm)
(lower—front brakes)	$1\frac{3}{8}''$ (34·92 mm)
Brake cylinder bore diameter —front	$2\frac{1}{8}''$ (53·97 mm)
—rear	$1\frac{3}{4}''$ (44·45 mm)
Servo unit type	Dunlop bellows type vacuum servo
Main friction pad material	Mintex M.59*
Handbrake friction pad material	Mintex M.34

Special Tools
Piston re-setting lever ... ... Part Number 7840

*Early cars fitted with M.40 or M.33 pads.

Retractor Operation (early type)

A counterbore in the piston accommodates a retractor bush which tightly grips the stem of a retractor pin. This pin forms part of an assembly which is peened into the base of the cylinder bore. The assembly comprises a retractor stop bush, two spring washers, a dished cap and the retractor pin; it functions as a return spring and maintains a "brake-off" working clearance of approximately 0·008/0·010" (·20–·25 mm) between the pads and the disc throughout the life of the pads.

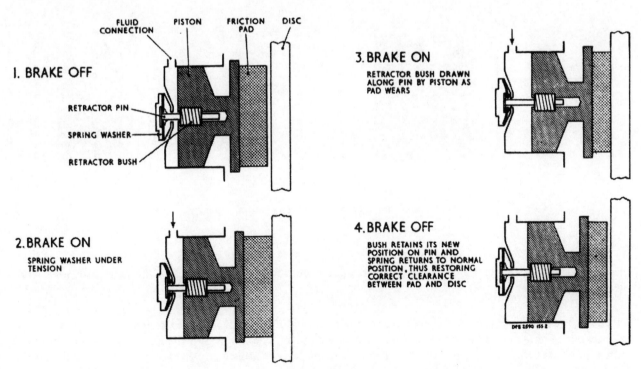

Fig. 2. Operation of the self-adjusting mechanism—early type

Page L.5

BRAKES

Retractor Operation (later type)

The retractor unit (see Fig. 3) comprises the retractor pin pressed into the cylinder block and the retractor bush, washer, return spring and spring retainer peened into the piston.

When the brakes are applied the piston moves the friction pad towards the disc. The retractor bush grips the pin holding the spring retainer and the return spring against the washer. The piston in moving the distance between the pad and disc compresses the return spring and when the brakes are released the return spring expands maintaining an equal clearance between the pad and disc.

When the pad wears and has not made contact with the disc by the time that the washer has fully compressed the return spring, the washer will move the retractor bush down the pin until the pad contacts the disc. The retractor bush stop in this new position and when the brakes are released the return spring expands allowing the pads to maintain the normal "brakes off" clearance of approximately ·008″—·010″ (·20-·25 mm) as before.

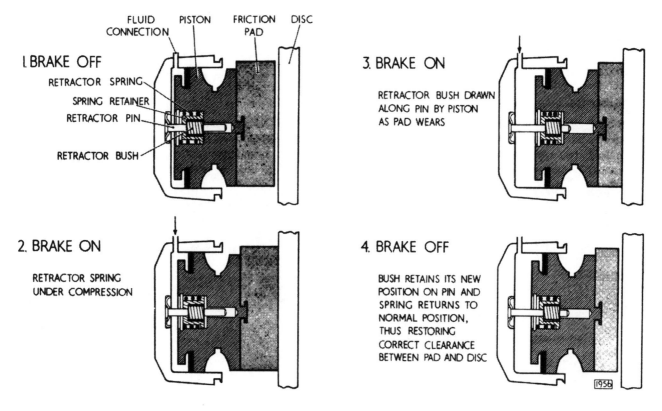

Fig. 3. Operation of the self-adjusting mechanism—later type

Handbrake

The mechanical handbrake units are mounted on and above the caliper bodies of the rear roadwheels brake by means of pivot bolts.

Each handbrake unit consists of two carriers, one each side of the brake disc and attached to the inside face of each carrier by means of a special headed bolt is a friction pad. The free end of the inner pad carrier is equipped with a pivot seat to which the forked end of the operating lever is attached. A trunnion is also mounted within the forked end of the operating lever and carries the threaded end of the adjuster bolt on the end of which is a self-

Page L.6

BRAKES

locking nut. Located on the shank of the adjuster bolt and in a counterbore in the inside face of the inner pad carrier is the operating lever return spring held under load by a nut retained by a spring plate riveted to the inside face of the inner carrier. The adjuster bolt passes through the outer pad carrier and its hemispherically shaped head seats in a suitable recess in the outer carrier.

ROUTINE MAINTENANCE

WEEKLY

Brake Fluid Level

On right-hand drive cars the fluid reservoirs (two) for the hydraulic brakes are attached to the bulkhead on the driver's side. The left-hand reservoir (nearest to centre line of car) supplies the rear brakes and the right-hand supplies the front brakes.

On left-hand drive cars the fluid reservoirs (two) for the hydraulic brakes are attached to the front frame assembly adjacent to the exhaust manifold. The forward reservoir supplies the rear brakes, the rear reservoir supplies the front brakes.

At the recommended intervals check the level of fluid in the reservoir and top up if necessary to the level mark, above fixing strap, marked "Fluid Level" using only the correct specification of Brake Fluid.

Do NOT overfill.

The level can be plainly seen through the plastic reservoir container.

First, disconnect the two electrical cables from the "snap-on" terminals. Unscrew the filler cap and "top-up" if necessary to the recommended level. Insert the combined filler cap and float slowly into the reservoir to allow for displacement of fluid and screw down the cap. Wipe off any fluid from the top of the cap and connect the cables to either of the two terminals.

Note: A further indication that the fluid level is becoming low is provided by an indicator pin situated between the two terminals.

First press down the pin and allow it to return to its normal position; if the pin can then be lifted with the thumb and forefinger the reservoir requires topping-up immediately.

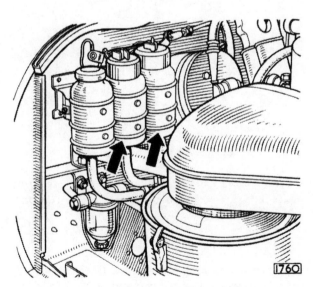

Fig. 4. Fluid reservoirs—Right hand drive

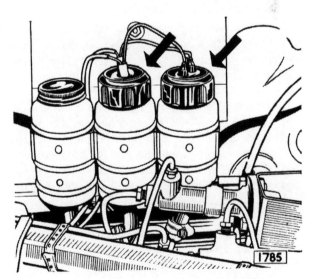

Fig. 5. Fluid reservoirs—Left hand drive

Page L.7

BRAKES

Brake Fluid Level Warning Light

A warning light (marked "Brake Fluid—Handbrake") situated on the facia behind the steering wheel, serves to indicate if the level in one or both of the brake fluid reservoirs has become low, provided the ignition is "on". As the warning light is also illuminated when the handbrake is applied, the handbrake must be fully released before it is assumed that the fluid level is low. If with the ignition "on" and the handbrake fully released the warning light is illuminated, the brake fluid must be "topped-up" immediately.

As the warning light is illuminated when the handbrake is applied and the ignition is "on" a two-fold purpose is served. Firstly, to avoid the possibility of driving away with the handbrake applied. Secondly, as a check that the warning light bulb has not "blown"; if on first starting up the car with the handbrake fully applied, the warning light does not become illuminated the bulb should be changed immediately.

Note: If it is found that the fluid level falls rapidly indicating a leak from the system, the car should be taken immediately to the nearest Jaguar Dealer for examination.

EVERY 2,500 MILES (4,000 km)

Footbrake Adjustment

Both the front and rear wheel brakes are so designed that no manual adjustment to compensate for brake friction pad wear is necessary as this automatically takes place when the footbrake is applied.

Handbrake Adjustment (Early Cars)

The mechanically operated handbrakes are attached to the rear caliper bodies but form an independent mechanically actuated system carrying their own friction pads and individual adjustment.

To adjust the handbrakes to compensate for friction pad wear which will be indicated by excessive handbrake lever travel, carry out the following procedure. Remove the carpet from the luggage compartment floor by lifting the snap fasteners and rolling carpet away. Remove rear axle cover now exposed by unscrewing the seven screws retaining cover to floor.

Insert a ·004" (·10 mm) feeler gauge between the face of one handbrake pad and the disc and screw in the adjuster bolt (using the special key provided in the tool kit) until the feeler gauge is just nipped.

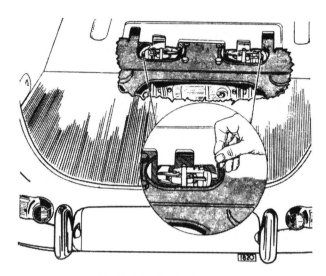

Fig. 6. Handbrake adjustment

Withdraw the feeler gauge and check the disc for free rotation. Repeat for the other side.

If, after carrying out the above adjustment, satisfactory travel of the handbrake lever is not obtained, the handbrake cable should be adjusted as follows:

Fig. 7. Handbrake cable adjustment

Screw in the handbrake adjuster bolt at each rear brake until the handbrake pads are in hard contact with the brake discs.

Fully release the handbrake lever.

Slacken the locknut securing the threaded adaptor to the compensator at the rear end of the handbrake

Page L.8

BRAKES

cable. Screw out the adaptor until there is no slack in the cable; it is, however, important to ensure that the cable is not under tension. Tighten the locknut and reset handbrake pad clearance with a .004″ (.10mm) feeler gauge as described above.

EVERY 5,000 MILES (8,000 KM.)

Brake Pedal Bearing

The brake pedal bearing should be lubricated with engine oil (Fig. 8).

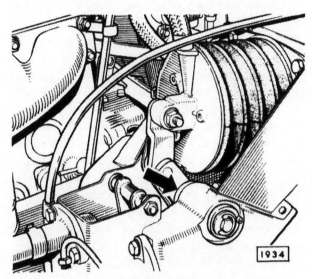

Fig. 8. Pedal bearing lubrication

Friction Pads—Examination for wear

At the recommended intervals, or if a loss of braking efficiency is noticed, the brake friction pads (2 per brake) should be examined for wear; the ends of the pads can be easily observed through the apertures in the brake caliper. When the friction pads have worn down to a thickness of approximately ¼″ (7 mm) they need renewing.

Friction Pads—Renewal

To remove the friction pads, unscrew the nut from the bolt attaching the friction pad retainer to the caliper and extract the bolt. Withdraw the pad retainer.

Insert a hooked implement through the hole in the metal tag attached to the friction pad and withdraw the pad by pulling on the tag.

Fig. 9. Location of rear brake calipers

To enable the new friction pads to be fitted it will be necessary to force the pistons back into the cylinder blocks by means of the special tool (Part number 7840).

Insert the new friction pads into the caliper ensuring that the slot in the metal plate attached to each pad engages with the button in the centre of the piston.

Finally, refit the friction pad retainer and secure with the bolt and nut. Apply the footbrake a few times to operate the self-adjusting mechanism, so that normal travel of the pedal is obtained.

When all the new friction pads have been fitted, top up the supply tank to the recommended level.

RECOMMENDED BRAKE FLUIDS

Preferred Fluid

Castrol/Girling Crimson Clutch/Brake Fluid.
(S.A.E. 70 R3)

Alternative Brake Fluids

Recognised brands of brake fluid conforming to Specification S.A.E. 70 R3 such as:

Lockheed Super Heavy Duty Brake Fluid.

In the event of deterioration of the rubber seals and hoses due to the use of an incorrect fluid, all the seals and hoses must be replaced and the system thoroughly flushed and refilled with one of the above fluids.

Page L.9

BRAKES

BLEEDING THE BRAKE SYSTEM

The following procedure should be adopted either for initial priming of the system or to bleed in service if air has been permitted to enter the system. This latter condition may occur if connections are not maintained, properly tightened, or if the master cylinder periodic level check is neglected. During the bleeding operation it is important that the level in the reservoir is kept topped up to avoid drawing air into the system. It is recommended that new fluid be used for this purpose.

Check that all connections are tightened and all bleed screws closed. Fill the reservoir with brake fluid of the correct specification. Attach the bleeder tube to the bleed screw on the near side rear brake and immerse the open end of the tube in a small quantity of brake fluid contained in a clean glass jar. Slacken the bleed screw and operate the brake pedal slowly backwards and forwards through its full stroke until fluid pumped into the jar is reasonably free from air bubbles. Keep the pedal depressed and close the bleed screw. Release the pedal.

Repeat for offside rear brake.

Repeat for front brakes.

Repeat the complete bleeding sequence until the brake fluid pumped into the jar is completely free from air bubbles.

Lock all bleed screws and finally regulate the fluid level in the reservoir. Apply normal working load on the brake pedal for a period of two or three minutes and examine the entire system for leaks.

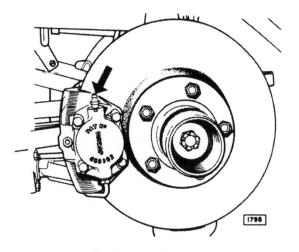

Fig. 10. Brake bleed nipple

BRAKE OVERHAUL—PRECAUTIONS

The complete brake system is designed to require the minimum of attention and providing the hydraulic fluid in the reservoir is not allowed to fall below the recommended level no defects should normally occur. Fluid loss must be supplemented by periodically topping up the reservoir with fluid of the same specification of that in the system.

The inclusion of air in a system of this type will be indicated by sluggish response of the brakes and spongy action of the brake pedal. This condition may be due to air induction at a loose joint or at a reservoir in which the fluid has been allowed to fall to a very low level. These defects must be immediately remedied and the complete system bled. Similarly, bleeding the system is equally essential following any servicing operation involving the disconnecting of part or whole of the hydraulic system.

The following instructions detail the procedure for renewal of component parts and for complete overhaul of the disc brakes, handbrakes and master cylinders. The units should be thoroughly cleaned externally before dismantling. Brake fluid should be used for cleaning internal components, and, except where otherwise stated in these notes, the use of petrol, paraffin or chemical grease solvents should be avoided as they may be detrimental to the rubber components. Throughout the dismantling and assembling operation it is essential that the work bench be maintained in a clean condition and that the components are not handled with dirty or greasy hands. The precision parts should be handled with extreme care and should be carefully placed away from tools or other equipment likely to cause damage. After cleaning, all components should be dried with lint-free rag.

When it is not the intention to renew the rubber components, they must be carefully examined for serviceability. There must be no evidence of defects such as perishing, excessive swelling, cutting or twisting, and where doubt exists comparison with new parts may prove to be of some assistance in making an assessment of their condition. The flexible pipes must show no signs of deterioration or damage and the bores should be cleaned with a jet of compressed air. No attempt should be made to clear blockage by probing as this may result in damage to the lining and serious restriction to fluid flow. Partially or totally blocked flexible pipes should always be renewed.

Page L.10

BRAKES

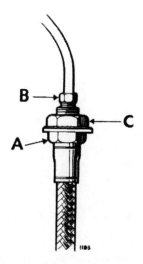

Fig. 11. Flexible hose connection. Hold hexagon "A" with spanner when removing or refitting locknut "C"

When removing or refitting a flexible pipe, the end sleeve hexagon (A, Fig. 11) should be held with the appropriate spanner to prevent the pipe from twisting. A twisted pipe will prove detrimental to efficient brake operation.

THE MASTER CYLINDERS

The master cylinders are mechanically linked to the footbrake pedal and, at a ratio proportional to the load applied, provide the hydraulic pressure necessary to operate the brakes. The components of the master cylinders are contained within the bore of a body which at its closed end has two 90° opposed integral pipe connection bosses. Integrally formed around the opposite end of the cylinder is a flange provided with two holes for the master cylinder attachment bolts. In the unloaded condition a spring loaded piston, carrying two seals (see Fig. 12) is held against the underside of a circlip retained dished washer at the head of the cylinder. A hemispherically ended push-rod seats in a similarly formed recess at the head of the piston. A fork end on the outer end of the push-rod provides for attachment to the pedal. A rubber dust excluder, the lip of which seats in a groove, shrouds the head of the master cylinder to prevent the intrusion of foreign matter.

A cylindrical spring support locates around the inner end of the piston and a small drilling in the end of the support is engaged by the stem of a valve. The larger diameter head of the valve locates in a central blind bore in the piston. The valve passes through the bore of a vented spring support and interposed between the spring support and an integral flange formed on the valve is a small coiled spring. A lipped rubber seal registers in a groove around the end of the valve. This assembly forms a recuperation valve which controls fluid flow to and from the reservoir.

When the foot pedal is in the OFF position the master cylinder is fully extended and the valve is held clear of the base of the cylinder by the action of the main spring. In this condition the master cylinder is in fluid communication with the reservoir, thus permitting recuperation of any fluid loss sustained, particularly during the bleeding operation of the brake system.

When a load is applied to the foot pedal the piston moves down the cylinder against the compression of the main spring. Immediately this movement is in excess of the valve clearance the valve closes under the influence of its spring and isolates the reservoir. Further loading of the pedal results in the discharge of fluid under pressure from the outlet connection, via the pipe lines to the brake system.

Removal of the load from the pedal reverses the sequence, the action of the main spring returns the master cylinder to the extended position.

Removal

Unscrew and withdraw the pipe unions from the ends of the master cylinders. Plug the holes to prevent the ingress of dirt or loss of fluid.

Remove the two bolts and locknuts from the top master cylinder flange.

Slacken the locknut on the top master cylinder push rod. Unscrew the push rod from the yoke and remove the master cylinder. Remove the two bolts and locknuts from the lower master cylinder flange.

Pull the lower master cylinder forward as far as possible. Remove the split pin and withdraw the clevis pin.

Remove the master cylinder.

Refitting is the reverse of the removal procedure.

Adjust the push rod on the top master cylinder to give $\frac{1}{16}$" (1·58 mm) free play—this, by means of the balance lever will give $\frac{1}{32}$" (·794 mm) free play to each master cylinder.

Tighten the locknut at the top master cylinder push rod. Bleed the braking system throughout.

BRAKES

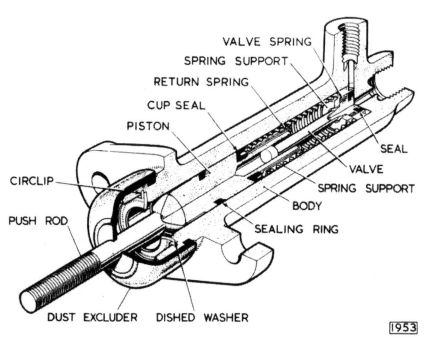

Fig. 12. Sectioned view of a master cylinder

Renewing the Master Cylinder Seals

Ease the dust excluder clear of the head of the master cylinder.

With suitable pliers remove the circlip; this will release the push rod complete with dished washer.

Withdraw the piston and remove both seals.

Withdraw the valve assembly complete with spring and supports. Remove the seal from the end of the valve.

Lubricate the new seals and the bore of the cylinder with brake fluid, fit the seal to the end of the valve ensuring that the lip registers in the groove. Fit the seals in their grooves around the piston.

Insert the valve head into the slotted hole in the spring support. Insert the piston into the other end of the spring support and centralise the valve head in the piston bore. Lubricate the piston with Castrol Rubber Grease H95/59. The piston, valve, main spring and spring supports must be inserted into the cylinder bore as a complete assembly.

Do not assemble the valve, main spring and spring supports into the cylinder bore without the piston.

Care should be taken when inserting the piston not to damage or twist the seals. The use of the fitting sleeve supplied with the master cylinder reconditioning kit is recommended.

Position the push-rod and depress the piston sufficiently to allow the dished washer to seat on the shoulder at the head of the cylinder. Fit the circlip and check that it fully engages the groove.

Fill the dust excluder with clean Castrol H95/59 Rubber Grease.

Reseat the dust excluder around the head of the master cylinder.

Page L.12

BRAKES

Free Travel of Master Cylinder Push-rods

When the brake pedal is in the "off" position, it is necessary that the pistons in the master cylinders are allowed to return to the fully extended position, otherwise pressure may build up in the system causing the brakes to drag or remain on.

To set the push-rods to the correct clearance, slacken the locknut at the top master cylinder push-rod and adjust the push-rod to give $\frac{1}{16}''$ (1·58 mm) free travel—this by means of the balance lever will give $\frac{1}{32}''$ (·794 mm) free travel to each master cylinder. Tighten the locknut at the top master cylinder push rod.

FRONT CALIPERS

Removal

In order to remove the front calipers, jack up the car and remove the road wheel. Disconnect the fluid feed pipe and plug the hole in the caliper. Discard the locking wire from the mounting bolts. Remove the caliper, noting the number of round shims fitted.

Refitting

Locate the caliper body (complete with the cylinder assemblies) in position and secure with two bolts.

Check the gap between each side of the caliper and the disc, both at the top and bottom of the caliper. The difference should not exceed ·010" (·25 mm) and round shims may be fitted between the caliper and the mounting plate to centralise the caliper body. Lockwire the mounting bolts.

If not already fitted, fit the bridge pipe connecting the two cylinder assemblies. Connect the supply pipe to the cylinder body and ensure that it is correctly secured.

Bleed the brakes as described on page L.10.

Important: It is essential that the bridge pipe is fitted with the "hairpin" bend to the inboard cylinder block, that is, furthest from the road wheel (see Fig. 1). The bridge pipe carries a rubber identification sleeve marked "Inner Top".

REAR CALIPERS

Removal

The rear suspension unit must be removed in order to withdraw the rear calipers.

Proceed as described in Section K "Rear Suspension" and support the suspension unit under its centre.

Withdraw the split pin and remove the clevis pin joining the compensator linkage to the handbrake operating lever.

Remove the hydraulic feed pipe at the three-way union.

Remove the friction pads from the caliper as described on page L.16.

Remove the front hydraulic damper and road spring unit (as described in Section K "Rear Suspension") and remove the four self locking nuts from the halfshaft inner universal joint.

Withdraw the joint from the bolts and allow the hub carrier to move outwards—support the carrier in this position.

Note the number of camber shims between the universal joint flange and the brake disc.

Knock back the locking tabs and remove the pivot bolts securing the handbrake pad carriers to the caliper and the retractor plate. Withdraw the handbrake pad carriers from the aperture at the rear of the cross member.

Remove the keep plate on the caliper and using a hooked implement withdraw both brake pads.

Rotate the disc until the holes in the disc line up with the caliper mounting bolts.

Knock back the locking tabs (on early cars locking wire was used) and remove the mounting bolts.

Note the number of small circular shims fitted to the caliper mounting bolts between the caliper and the axle casing (Fig. 13)

The caliper can now be removed from the aperture at the front of the cross member.

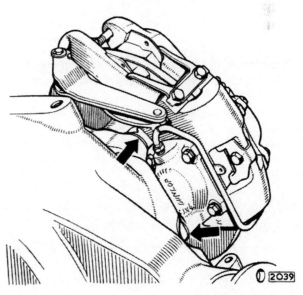

Fig. 13. Location of the rear brake caliper adjustment shims

Page L.13

BRAKES

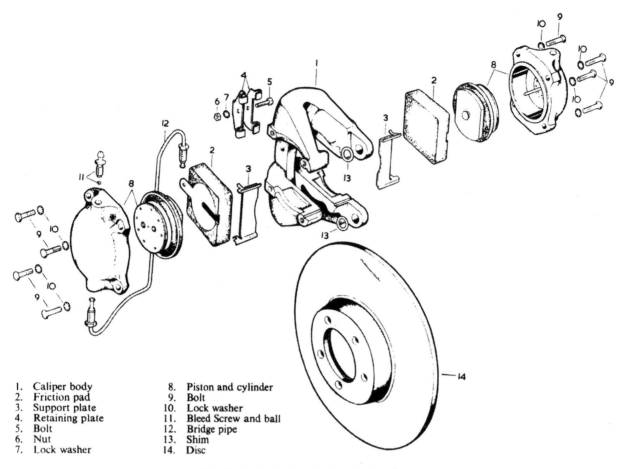

1. Caliper body
2. Friction pad
3. Support plate
4. Retaining plate
5. Bolt
6. Nut
7. Lock washer
8. Piston and cylinder
9. Bolt
10. Lock washer
11. Bleed Screw and ball
12. Bridge pipe
13. Shim
14. Disc

Fig. 14. Exploded view of a front brake caliper

Refitting

Refitting is the reverse of the removal procedure. The correct number of camber shims should be fitted.

When the halfshaft has been refitted check the caliper for centralisation as described in refitting the front calipers. Fit the fluid supply pipe and the bridge pipe if necessary. Bleed the braking system (as described on page L.10).

THE FRONT BRAKE DISCS

Removal

Jack up the car and remove the road wheel. Disconnect the flexible hydraulic pipe from the frame connection and plug the connector to prevent ingress of dirt and loss of fluid.

Discard the locking wire and remove the two caliper mounting bolts noting the number of round shims fitted between the caliper and mounting plate. Remove the caliper.

Remove the hub (as described in Section J "Front Suspension").

Remove the five self locking nuts and bolts securing the disc to the hub and remove the disc.

Refitting

Refitting is the reverse of the removal procedure. The hub bearing endfloat should be set (as described in Section J "Front Suspension") and the caliper fitted and centralised as described previously (page L.13). Reconnect the brakes and bleed the braking system (as described on page L.10).

Page L.14

BRAKES

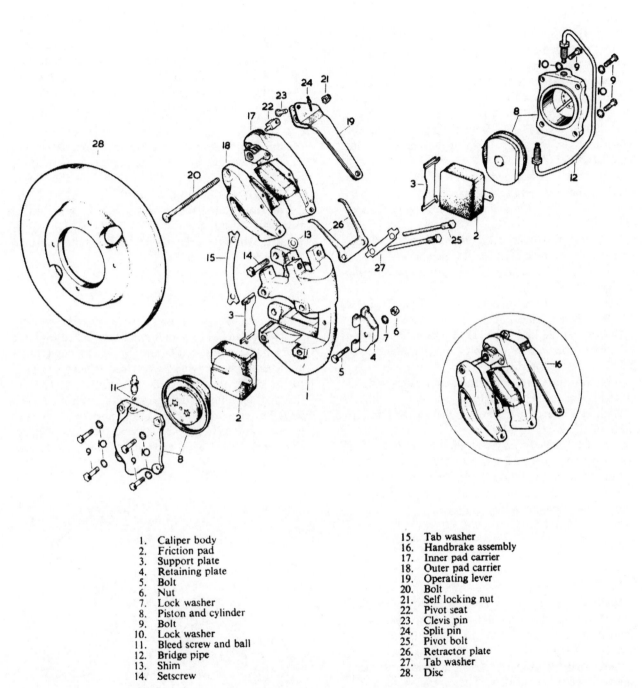

1. Caliper body
2. Friction pad
3. Support plate
4. Retaining plate
5. Bolt
6. Nut
7. Lock washer
8. Piston and cylinder
9. Bolt
10. Lock washer
11. Bleed screw and ball
12. Bridge pipe
13. Shim
14. Setscrew
15. Tab washer
16. Handbrake assembly
17. Inner pad carrier
18. Outer pad carrier
19. Operating lever
20. Bolt
21. Self locking nut
22. Pivot seat
23. Clevis pin
24. Split pin
25. Pivot bolt
26. Retractor plate
27. Tab washer
28. Disc

Fig. 15. Exploded view of a rear brake caliper

Page L.15

BRAKES

THE REAR BRAKE DISCS

Removal

Remove the rear suspension unit (as described in Section K "Rear Suspension").

Invert the suspension and remove the two hydraulic damper and road spring units (as described in Section K "Rear Suspension").

Remove the four steel type self locking nuts securing the halfshaft inner universal joint and brake disc to the axle output shaft flange.

Withdraw the halfshaft from the bolts, noting the number of camber shims between the universal joint and the brake disc.

Knock back the tabs and unscrew the two pivot bolts securing the hand brake pad carriers to the caliper. Remove the pivot bolts and the retractor plate (Fig. 15).

Withdraw the handbrake pad carriers from the aperture at the rear of the cross members.

Knock back the tabs at the caliper mounting bolts (on earlier cars locking wire was used).

Remove the keeper plate on the caliper and using a hooked implement, withdraw both brake pads.

Disconnect the brake fluid feed pipe at the caliper.

Unscrew the mounting bolts through the access holes in the brake disc.

Withdraw the bolts, noting the number and position of the round caliper centralizing shims.

Withdraw the caliper through the aperture at the front of the cross member.

Tap the halfshaft universal joint and brake disc securing bolts back as far as possible.

Lift the lower wishbone, hub carrier and halfshaft assembly upwards until the brake disc can be withdrawn from the mounting bolts.

Refitting

Refitting the brake discs is the reverse of the removal procedure. The securing bolts must be knocked back against the drive shaft flange when the new disc has been fitted.

Care must be taken to refit the caliper centralizing shims in the same position. The centralization of the caliper should be checked (as described in "Refitting the Calipers") when the halfshaft has been refitted.

Refit the rear suspension (as described in Section K "Rear Suspension").

Bleed the brakes as described on page L.10.

Page L.16

BRAKE DISC "RUN-OUT"

Check the brake discs for "run-out" by clamping a dial test indicator to the stub axle carrier for the front discs and the cross member for the rear discs. Clamp the indicator so that the button bears on the face of the disc. "Run-out" should not exceed ·006″ (·15 mm) gauge reading. Manufacturing tolerances on the disc should maintain this truth and in the event of "run-out" exceeding this value, the components should be examined for damage.

Note: It is most important that the endfloat of the front hubs and the rear axle output shafts is within the stated limits otherwise the brakes may not function correctly.

The front hub endfloat adjustment is described in Section J "Front Suspension". The endfloat adjustment of the rear axle output shafts is described in Section H "Rear Axle".

RENEWING THE FRICTION PADS

Brake adjustment is automatic during the wearing life of the pads. The pads should be checked for wear every 5,000 miles (8,000 km) by visual observation and measurement; when wear has reduced the pads to the minimum permissible thickness of $\frac{1}{4}″$ (7 mm) the pad assemblies (complete with securing plates) must be renewed. If checking is neglected the need to renew the pads will be indicated by a loss of brake efficiency. The friction pads fitted have been selected

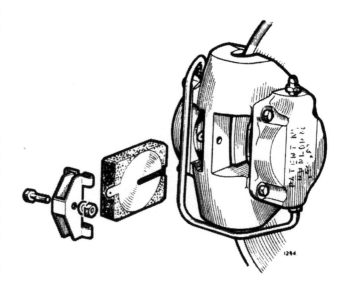

Fig. 16. Friction pad removal

BRAKES

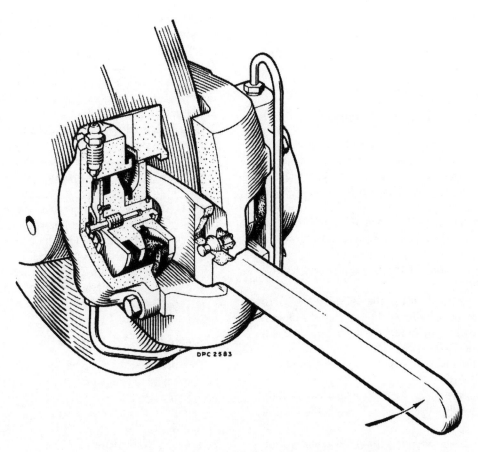

Fig. 17. Resetting the pistons with the special tool (Part No. 7840)

as a result of intensive development, and it is essential at all times to use only factory approved material. To fit the new friction pad assemblies proceed as follows:

Remove the nut, washer and bolt securing the keep plate and withdraw the plate.

With a suitable hooked implement engaged in the hole in the lug of the securing plate withdraw the defective pad assemblies.

Thoroughly clean the backing plate, dust seal and the surrounding area of the caliper.

With the aid of the special tool, press in the piston assemblies to the base of the cylinder bores as shown in Fig. 17.

Note: Before doing this, it is advisable to half empty the brake supply tank, otherwise forcing back the friction pads will eject fluid from the tank with possible damage to the paintwork. When all the new friction pads have been fitted, top up the supply tank to the recommended level.

Insert the forked end of the piston resetting lever into the space between the caliper bridge and one of the piston backing plates, with the fork astride the projecting piston spigot and its convex face bearing on the piston backing plate. Locate the spigot end of the lever pin in the keep plate bolt hole in the bridge. Pivot the lever about the pin to force the piston to the

Page L.17

BRAKES

base of its cylinder. Insert the new friction pad assembly.

Replace the keep plate and secure it with the bolt, washer and nut.

Renewing the Brake Piston Seals—Early Type

Leakage past the piston seals will be denoted by a fall in level of the fluid reservoir or by spongy pedal travel. It is recommended that the dust seal be renewed when fitting a new piston seal. Proceed as follows:

Remove the caliper as described on page L.13.

Withdraw the brake pads as described in the previous paragraphs.

Disconnect and blank off the supply pipe and remove the bridge pipe.

Remove the bolts securing the cylinder blocks to the caliper and withdraw the cylinder blocks. Thoroughly clean the blocks externally before proceeding with further dismantling.

Disengage the dust seal from the groove around the cylinder block face.

Connect the cylinder block to a source of fluid supply and apply pressure to eject the piston assembly.

Remove the screws securing the plate to the piston, lift off the plate and piston seal, withdraw the retractor bush from within the piston bore. Carefully cut away and discard the dust seal.

Support the backing plate on a bush of sufficient bore diameter to just accommodate the piston. With a suitable tubular distance piece placed against the end of the piston spigot and located around the shouldered head, press out this piston from the backing plate. Care must be taken during this operation to avoid damaging the piston.

Engage the collar of a new dust seal with the lip on the backing plate avoiding harmful stretching.

Locate the backing plate on the piston spigot and, with the piston suitably supported, press the backing plate fully home.

Insert the retractor bush into the bore of the piston. Lightly lubricate a new piston seal with brake fluid, and fit it to the piston face. Attach and secure the plate with the screws and peen lock the screws.

Check that the piston and the cylinder bore are thoroughly clean and show no signs of damage. Locate the piston assembly on the end of the retractor pin. With the aid of a hand press slowly apply an even pressure to the backing plate and press the assembly into the cylinder bore. During this operation ensure the piston assembly is in correct alignment in relation to the cylinder bore, and that the piston seal does not become twisted or trapped as it enters. Engage the outer rim of the dust seal in the groove around the cylinder block face. Ensure that the two support plates are in position.

Re-assemble the cylinder blocks to the caliper. Fit the bridge pipes ensuring that they are correctly positioned. Remove the blank and reconnect the supply pipe. Bleed the hydraulic system.

Important: It is essential that the bridge pipe is fitted with the "hairpin" bend end to the inboard cylinder block, that is, furthest from the road wheel (see Fig. 1). The bridge pipe carries a rubber identification sleeve marked "Inner Top".

Renewing the Brake Piston Seals—Later Type

The later type cylinder blocks may be distinguished by the letter "C" cast into the block body at the inlet union hole.

Remove the caliper as described on page L.13.

Withdraw the brake pads as described under "Renewing the Friction Pads".

Disconnect and blank off the supply pipe and remove the bridge pipe.

Remove the bolts securing the cylinder blocks to the caliper and withdraw the cylinder blocks. Thoroughly clean the blocks externally before proceeding with further dismantling.

Disengage the dust seal from the groove around the cylinder block face.

Connect the cylinder block to a source of fluid supply and apply pressure to eject the piston assembly.

Using a blunt screwdriver carefully push out and remove the piston seal and the dust seal. It is impossible to strip the piston down further.

Check that the piston and cylinder bore are thoroughly clean and show no signs of damage.

When replacing the piston and dust seals, first lightly lubricate with brake fluid, then place on the piston using the fingers only. Locate the retractor pin in the retractor bush in the piston, then with even pressure press the piston assembly into the cylinder bore. During this operation ensure the piston assembly is in correct alignment in relation to the cylinder bore and that the piston seal does not become twisted or trapped as it enters. Engage the outer rim of the dust

Page L.18

BRAKES

seal in the groove around the cylinder block face. Ensure that the two support plates are in position.

Re-assemble the cylinder blocks to the caliper. Fit the bridge pipes, ensuring that they are correctly positioned. Connect the supply pipe and bleed the hydraulic system (as described on page L.10).

Important: It is essential that the bridge pipe is fitted with the "hairpin" bend end to the inboard cylinder block, that is, furthest from the road wheel (see Fig. 1). The bridge pipe carries a rubber identification sleeve marked "Inner Top".

THE HANDBRAKE—(Early Cars)

Description (Fig. 15)

The mechanical handbrake units are mounted on and above the caliper bodies of the rear brakes by means of pivot bolts and forked retraction plates.

Each handbrake unit consists of two carriers, one each side of the brake disc and attached to the inside face of each carrier by means of a special headed bolt is a friction pad. The free end of the inner pad carrier is equipped with a pivot seat to which the forked end of the operating lever is attached. A trunnion is also mounted within the forked end of the operating lever and carries the threaded end of the adjuster bolt on the end of which is a self-locking nut. Located on the shank of the adjuster bolt and in a counterbore in the inside face of the inner pad carrier is the operating lever return spring held under load by a nut retained by a spring plate riveted to the inside face of the inner carrier. The adjuster bolt passes through the outer pad carrier and its hemispherically shaped head seats in a suitable recess in the outer carrier.

The handbrake units require periodical adjustment and a hexagonal recess for this purpose is provided in the head of the adjuster bolt.

Handbrake Friction Pads—Renewing

With the friction pad carriers removed withdraw the friction pad by slackening the nuts in the outer face of each carrier and utilizing a hooked tool in the drilling of the friction pad securing plate. Insert two friction pad assemblies into the friction pad carriers, short face upwards, ensuring each pad securing plate locates the head of the retaining bolt protruding through the inside face of the pad carriers and secure by tightening the nuts on the outside faces. Repeat with the second handbrake. Refit the handbrake friction pad carriers as previously described and reset the handbrake as described under "Routine Maintenance" (page L.8).

Friction Pad Carriers—Removal

With the car on a ramp, disconnect the handbrake compensator linkage from the handbrake operating lever at the front of the rear suspension assembly by discarding the split pin and withdrawing the clevis pin. Lift the locking tabs and remove the pivot bolts and retraction plate. Remove the friction pad carriers from the caliper bridge by moving them rearwards around the disc and withdrawing from the rear of the rear suspension assembly. Repeat with the second handbrake.

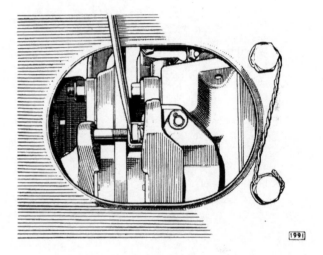

Fig. 18. Preloading the handbrake caliper return spring with a screwdriver

Friction Pad Carriers—Dismantling

Separate the friction pad carriers by withdrawing the adjuster bolt, exercising care to control the run of the self-locking nut in the forked end of the operating lever. Detach the pivot seat from the forked end of the operating lever by discarding the split pin and withdrawing the clevis pin. Do not attempt to remove the spring or squared nut, if either are damaged the pad carrier should be renewed. The pressings of the operating lever are spot welded together with the trunnion block in position, thus it cannot be removed.

Friction Pad Carriers—Assembling (Fig. 15)

Before re-assembling the friction pad carriers, ensure that the trunnion block has complete freedom of movement in the forked end of the operating lever. Ensure that the pin of the pivot seat is a sliding fit

Page L.19

BRAKES

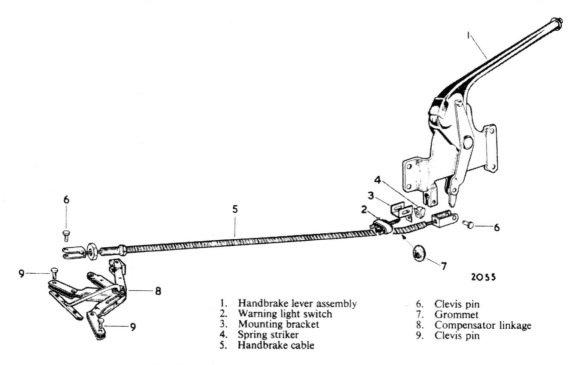

1. Handbrake lever assembly
2. Warning light switch
3. Mounting bracket
4. Spring striker
5. Handbrake cable
6. Clevis pin
7. Grommet
8. Compensator linkage
9. Clevis pin

Fig. 19. Exploded view of the handbrake actuating mechanism

in the drilling at the extreme end of the friction pad carrier. The pivot seat must also be a sliding fit between the forked ends of the operating lever. The clevis pin must be a sliding fit both through the eye of the pivot and through the holes in the forked ends of the operating lever.

Assemble the operating lever and the pivot seat but do not fit this assembly to the inner pad carrier at this stage.

Pass the adjusting bolt through the outer pad carrier and screw into the retaining nut and spring. Fit the operating lever and pivot assembly to the inner pad carrier and screw in adjusting bolt until it comes flush with the outer face of the trunnion block. The spring should then be preloaded by inserting the blade of a screwdriver between the retaining nut and the cage (Fig. 18). The adjusting screw should then be screwed out until the end again becomes flush with the outer face of the trunnion block. Place the self locking nut

on the trunnion block and screw the adjuster bolt into it ensuring that it engages the self locking nut with the first thread. When the adjuster bolt becomes flush with the second face of the self locking nut, withdraw the preloading screwdriver.

Refitting

Refitting is the reverse of the removal procedure but particular attention must be given to the following points.

That the locking plates under the heads of the two pad carrier bolts are replaced with new ones, even though the second pair of locking tags have still to be used.

The fork shaped retraction plates should be reset by lifting the locking tabs, slackening and tightening the pivot bolts and locking the bolt heads by turning up the tabs on the locking plate.

Page L.20

BRAKES

Removal and Refitting the Handbrake Cable

Remove the split pin and withdraw the clevis pin from the inner cable fork at the compensator linkage. Release the locknut and unscrew the outer cable from the retaining block, remove the spring holding the cable away from the propeller shaft.

Remove the four nuts securing each seat to the seat slides and withdraw the seats. Remove the two screws one each side of the radio control panel which secure the ashtray. Remove the two screws which secure each side of the radio control panel to the brackets under the instrument panel. Withdraw the radio control panel casing and remove the three setscrews securing the propeller shaft tunnel cover to the body. Place the gear lever as far forward as possible and

Fig. 20. Brake fluid level and handbrake warning light

pull the handbrake into the "on" position. Unscrew the gear lever knob and locknut. Slide the propeller shaft tunnel cover over the gear and handbrake levers and remove the tunnel cover.

Remove the split pin and withdraw the clevis pin from the forkend at the handbrake lever. Slacken the pinch bolt and remove the outer cable from the retaining block. Remove the grommet and withdraw the handbrake cables from the rear end of the propeller shaft tunnel.

Refitting is the reverse of the removal procedure. It should be noted, however, that when fitting the outer cable to the retaining block at the compensator, the cable must be screwed in with the longer end of the block facing towards the front of the car.

Screw the adaptor in until there is no slack in the cable then tighten the locknut and check the handbrake pad clearance with a ·004" (·10 mm.) feeler gauge as described in "Routine Maintenance" (page L.8).

THE BRAKE FLUID LEVEL AND HANDBRAKE WARNING LIGHT

Description (Fig. 20)

The brake fluid level and handbrake warning light, situated in the side facia panel, will indicate after the ignition has been switched on whether the brake fluid in the reservoir is at a low level or the handbrake has not reached the fully off position. This is effected by three switches, one in the top of each of the fluid reservoirs and a third on the handbrake lever, being in circuit with a single warning lamp which is included in the ignition circuit.

When the ignition is switched on and while the handbrake remains applied, the warning light will glow but will become extinguished when the handbrake is fully released with the brake fluid in the reservoir at a high level.

Should the warning light continue to glow after the handbrake has been fully released, it indicates that the brake fluid in the reservoir is at a very low level and the cause must be immediately determined and eliminated. Should the brake fluid be at a high level, the cause of the handbrake remaining on must be investigated.

Handbrake Warning Light Switch—Setting

A bracket mounted interrupter switch is attached to the handbrake outer cable retaining block on the propeller shaft tunnel below the handbrake lever assembly. An extension of the handbrake lever contacts a spring steel lever which depresses the plunger of the interrupter switch when the handbrake is in the "off" position. It is necessary to remove the propeller shaft tunnel cover as described under "Removal of the Handbrake Cable" to examine the interrupter switch.

Should the warning light fail to extinguish when the handbrake is in the fully "off" position, and the brake fluid levels in the reservoirs are correct, check that the spring steel lever is contacting the interrupter switch correctly before examining the leads for short circuiting.

Examine the handbrake for full travel and the spring steel bracket for misalignment. Apply the handbrake and switch on the ignition, when the warning light should glow. If the warning light fails to glow when the handbrake is applied and the ignition is switched on, before checking the warning light bulb ensure that the spring steel lever is clearing the interrupter switch plunger. If it is not doing so, bend the lever away from the plunger or renew as necessary.

Page L.21

BRAKES

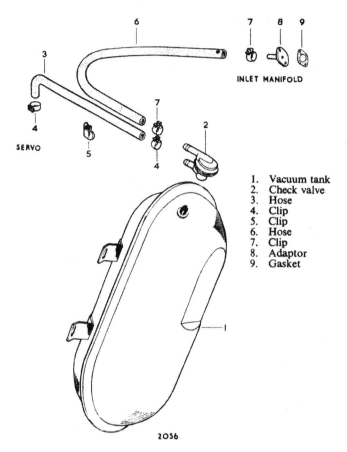

1. Vacuum tank
2. Check valve
3. Hose
4. Clip
5. Clip
6. Hose
7. Clip
8. Adaptor
9. Gasket

Fig. 21. Exploded view of the vacuum reservoir and system

THE VACUUM RESERVOIR AND CHECK VALVE
Description (Fig. 21)

The vacuum reservoir is incorporated in the vacuum line between the inlet manifold and vacuum servo unit. It is located on the bulkhead on the offside of the engine below the carburetter trumpets. Its purpose is to provide a reserve of vacuum in the event of braking being required after the engine has stalled.

A vacuum check valve is fitted at the top end of the front face of the vacuum reservoir, with the topmost connection communicating with the inlet manifold, while the second connection communicates directly with the vacuum port of the vacuum servo unit, thus any reduction of pressure inside the reservoir is conveyed to the vacuum servo unit.

Included in the inlet port of the check valve is a flat rubber spring-loaded valve and when there is a depression in the inlet manifold the valve is drawn away from its seat against its spring loading, thus the interior of the reservoir becomes exhausted. When the depression in the reservoir becomes equal to that of the inlet manifold, the valve spring will return the valve to its seat, thus maintaining the highest possible degree of vacuum in the reservoir.

Removal and Refitting

Detach the tray below the vacuum reservoir by removing four drive screws and two nuts and setscrews. Slacken the clips and remove the two pipes from the check valve. Remove the four setscrews holding the vacuum reservoir to the bulkhead and remove the reservoir from below. Unscrew the check valve from the top of the vacuum reservoir when necessary.

Refitting is the reverse of the removal procedure but particular attention must be given to the following points:

(i) That the rubber hose from the vacuum servo unit is attached to the pipe of the check valve

Page L.22

BRAKES

having the two grooves in its body; it is also the pipe nearest the screwed connection.

(ii) That the rubber hose from the inlet manifold is attached to the pipe of the check valve having two annular ribs in its body; this pipe is moulded into the centre of the check valve cap.

THE BRAKE/CLUTCH PEDAL BOX ASSEMBLY

Removal and Refitting

Remove the air cleaner elbow and the carburetter trumpets, also slacken the rear carburetter float chamber banjo nut and bend the petrol feed pipe towards the float chamber, and remove throttle rods from the bell crank above the servo bellows, on right hand drive models only.

Remove servo vacuum pipe and clips.

Drain the brake and clutch fluid reservoirs, remove fluid inlet pipes from the brake and clutch master cylinders and plug the holes.

Remove the brake fluid warning light wires and remove the brake and clutch fluid reservoirs.

Remove the fluid outlet pipes from the brake and clutch master cylinder and plug the holes. Remove the brake master cylinders as described on page L.11.

From inside the car remove brake and clutch pedal pads, remove dash casing (as described in Section N "Body and Exhaust").

Remove six self-locking nuts and one plain nut and shakeproof washer holding the servo assembly to the bulkhead.

Compress the servo bellows by hand, lift the servo assembly and remove from the car by twisting the unit approximately 90° clockwise to allow the pedals to pass through the hole in the bulkhead.

Fig. 22. *Removing the Brake/Clutch pedal box assembly*

Remove the bulkhead rubber seal.

Remove the four nuts and one setscrew fastening the brake master cylinder mounting bracket.

Remove the self-locking nut from the serrated pin and remove the conical spring and retaining washer.

Remove the pinch bolt from the brake pedal lever.

Remove the circlip and washer from the pedal shaft.

Remove the vacuum checkpoint from the front valve housing.

Remove the brake master cylinder support bracket with linkage and pedal shaft assembly from the pedal housing. Retain the fibre washer from between the brake and clutch pedals.

Remove the throttle bell crank bracket on right hand drive models by removing the four self locking nuts.

Remove the brake vacuum servo assembly.

Refitting is the reverse of the removal procedure, ensure that the rubber seal is in place over the exhausting tube between the servo bellows and the bulkhead.

When refitting the securing nuts inside the car ensure that the plain nut and shakeproof washer go on the short stud at the front centre.

When replacing the fluid reservoirs ensure that the brake fluid warning light wires are fitted with one feed wire (red and green) and one earth wire (black) to each reservoir cap.

Ensure that the petrol feed pipe is clear of the rear float chamber before tightening the banjo union nut.

Ensure that all clevis pins enter freely and without force, failure to do this may prevent the system operating in the "poised position".

Bleed the brake and clutch hydraulic systems.

Dismantling the Brake Linkage

Remove the bulkhead rubber seal (42, Fig. 23).

Remove the four nuts and one setscrew fastening the brake master cylinder mounting bracket (35).

Remove the self-locking nut from the serrated pin (21) and remove the conical spring (24) and retaining washer.

Remove the pinch bolt from the brake pedal lever (1).

Remove the circlip (9) and washer (10) from the pedal shaft.

Remove the vacuum check point union from the front valve housing.

Remove the brake master cylinder support bracket with linkages and pedal shaft assembly from the pedal housing (4). Retain the fibre washer from between the brake and clutch pedals.

Page L.23

BRAKES

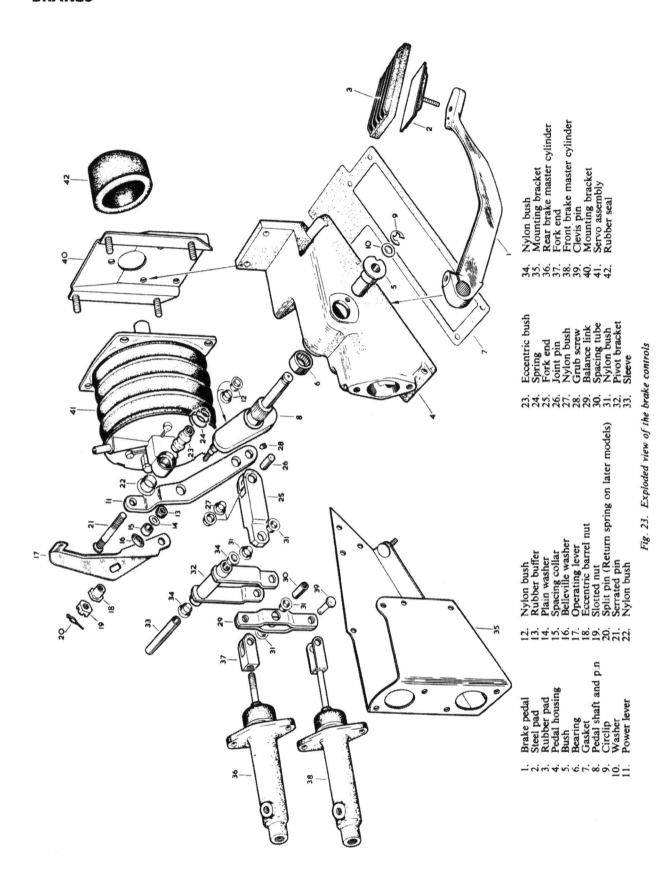

Fig. 23. Exploded view of the brake controls

1. Brake pedal
2. Steel pad
3. Rubber pad
4. Pedal housing
5. Bush
6. Bearing
7. Gasket
8. Pedal shaft and pin
9. Circlip
10. Washer
11. Power lever
12. Nylon bush
13. Rubber buffer
14. Plain washer
15. Spacing collar
16. Belleville washer
17. Operating lever
18. Eccentric barrel nut
19. Slotted nut
20. Split pin (Return spring on later models)
21. Serrated pin
22. Nylon bush
23. Eccentric bush
24. Spring
25. Fork end
26. Joint pin
27. Nylon bush
28. Grub screw
29. Balance link
30. Spacing tube
31. Nylon bush
32. Pivot bracket
33. Sleeve
34. Nylon bush
35. Mounting bracket
36. Rear brake master cylinder
37. Fork end
38. Front brake master cylinder
39. Clevis pin
40. Mounting bracket
41. Servo assembly
42. Rubber seal

Page L.24

BRAKES

Remove the throttle bell crank bracket on right hand drive models.

Remove the four self-locking nuts and remove the brake vacuum servo assembly (41).

Remove the setscrews and brass bush (5) from the pedal housing. Remove the split pin and withdraw the clevis pin. Remove the clutch master cylinder and withdraw the clutch pedal. The caged needle roller bearing (6) should be pressed out and replaced if necessary.

Remove the self-locking nut and withdraw the bolt from the pivot bracket (32). Remove the brake master cylinder support bracket.

Remove the servo operating arm return spring (20). Remove the castellated nut (19) and eccentric barrel nut (18).

Remove the self-locking nut and flat washer from the lower pedal shaft stud and withdraw the servo operating arm (17).

Remove the bellville washer (16), spacing collar (15), chamfered washer (14) and rubber buffer (13).

Remove the power lever (11), from the pedal shaft and pin assembly (8) and renew the nylon bushes (12) if necessary.

Remove the steel bush (23) and nylon bush (22), press out the serrated pin if necessary.

Remove the self locking nut and bolt attaching the two-way fork (25) to the pivot bracket and balance link and dismantle.

Remove the upper brake master cylinder fork end (37) by removing the split pin and withdrawing the clevis pin.

Press out the spacing sleeves (30) and (33) from the compensator fork and lever and renew the nylon bushes (31) and (34) if necessary.

Remove the grub screw (28) and press out the joint pin (26) from the two-way fork. Renew the nylon bushes (27) and (31) if necessary.

Remove the four plain nuts and shakeproof washers and remove the servo mounting bracket (40) from the pedal block.

Reassembly

Re-assembly is the reverse of the dismantling procedure, ensure that all linkages are very free especially the balance link and the servo operating arm.

When replacing the pedal shaft assembly on the pedal lever ensure that the pedal pad is lined up with the clutch pedal pad and also that the brake pedal lever does not foul the pedal box on full stroke.

Ensure that the fibre washer is in place between the brake and clutch pedal.

Reset the air valve operation with the eccentric barrel nut as described under "Servicing the Unit—Valve adjusting eccentric out of adjustment" (page L.29).

BELLOWS TYPE VACUUM SERVO

Description

The power unit consists of an air-vacuum bellows which expands or contracts as the air pressure is varied by the introduction of vacuum or atmosphere. One end of the assembly is connected to the dash unit and the other end to the power unit and pedals. A reserve tank is incorporated in the system to give an increased number of pedal applications. The valves which control the air pressure are located in the valve housing, and are actuated by the movement of the brake pedal. As the pedal is depressed the air valve is closed, the vacuum valve is opened and air is evacuated out of the bellows by the depression in the inlet manifold causing the bellows to contract. This in turn exerts a pull on the power lever in proportion to the pedal pressure applied by the driver; it thus provides the power assistance to the driver in depressing the pedal and applying the brakes.

It is therefore a "pedal-assistance" type unit operating in conjunction with the conventional hydraulic brake system.

In the event of no assistance as with the loss of vacuum, the hydraulic brakes can still be applied in the normal manner. Lifting the pedal pressure closes the vacuum and opens the air valve to the atmosphere, so destroying the vacuum and releasing the brakes. If the pedal pressure is partially applied and then held, both valves are closed and the vacuum remains constant until pedal is further depressed or released completely. This is known as the "poised position".

BRAKES

Operation

1. Brakes are in the "off" position. The bellows are fully extended and filled with air admitted through the air filter and air valve, which is open.

   The vacuum valve is closed, sealing the bellows from the vacuum supply. It will be noted, however, that vacuum is being applied at all times against the vacuum valve, so that any opening of the valve will immediately begin to exhaust air from the bellows.

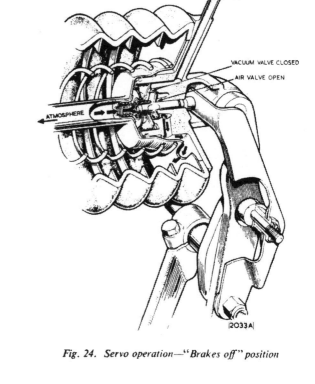

Fig. 24. Servo operation—"Brakes off" position

2. Applying the Brakes. When the brake pedal is operated, the operating lever applies pressure to the cap of the air valve in the servo. This overcomes the air valve spring and closes the air valve. Continued movement opens the vacuum valve and admits vacuum to the bellows, causing it to contract. Because of its linkage to the power lever, the assisted movement is transmitted to the push rod of the master cylinder to apply the brakes. The applying movement of the bellows tends to carry the air valve button away from the trigger; thus the continuing exhaust of air from the bellows will only occur with greater pressure of the brake pedal.

   When pedal pressure is held, both valves immediately close and the servo remains poised until pressure is again increased or released.

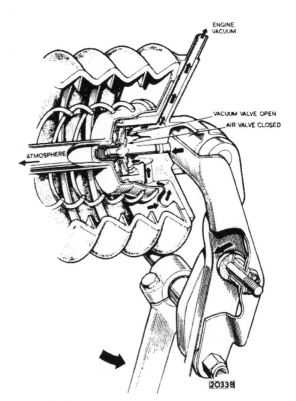

Fig. 25. Servo operation—Applying the brakes

Page L.26

BRAKES

3. **Brakes Fully Applied.** When the brakes are fully applied, and the servo is giving maximum assistance, any extra pedal pressure results in still greater increase of pressure to the master cylinder, through the combination of the pedal and power lever, acting as one through the eccentric, fully compressing the rubber collar, as shown in the diagram. The full assistance of the power unit is maintained during the increase.

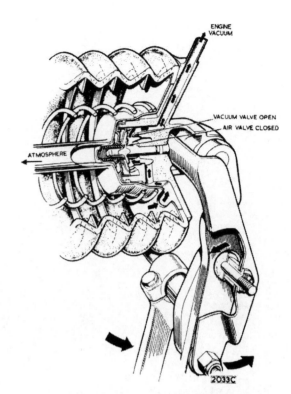

Fig 26. Servo operation—Brakes fully applied

4. **Releasing the Brake.** When the brakes are released, the trigger moves away from the air button, the vacuum is closed by spring tension and the air valve is re-opened. Air again enters the bellows, causing it to expand. At any point during the release the driver may hold the brakes, and the unit will immediately become poised. On complete release the servo regains the position shown in Fig. L.24.

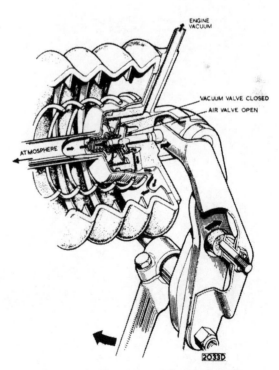

Fig. 27. Servo operation—Releasing the brakes

Page L.27

BRAKES

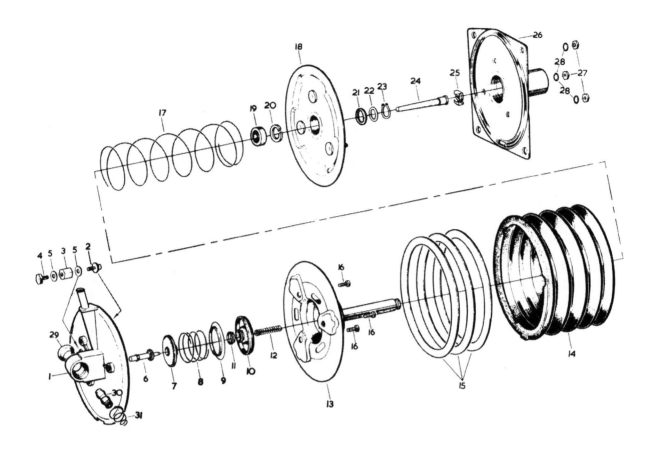

1. Valve housing
2. Nipple
3. Adaptor
4. Plug
5. Gasket
6. Air valve
7. Vacuum valve
8. Return spring
9. Balancing washer
10. Balancing diaphragm
11. Retainer
12. Control spring
13. Retainer sleeve
14. Bellows
15. Support ring
16. Bolt
17. Main return spring
18. Mounting hub
19. Seal
20. Guide sleeve
21. Rubber buffer
22. Stop washer
23. Circlip
24. Air filter
25. Baffle
26. Mounting plate
27. Nut
28. Lock washer
29. Nylon bush
30. Eccentric bush
31. Spring

Fig. 28. Exploded view of the vacuum servo

Page L.28

BRAKES

SERVICING THE UNIT

Symptom: Hard pedal; power assistance not operating.

Cause (1) Blocked, kinked or leaking vacuum line.

Remedy: Remove the rubber vacuum hose from the power unit and with the engine running check the vacuum source. Check that the valve unit in the reserve tank is operating correctly, replace if faulty, Fig. 21. Check that the hoses are not blocked, kinked or loosely connected. Replace or repair as necessary.

Note: Any vacuum leaks in the system can usually be located easily when the engine is running by a hissing sound.

Cause (2) Vacuum leaks in the unit.

Remedy: With the engine running and brake pedal pressure applied listen at the unit for a hissing sound indicating a vacuum leak. Locate and correct. If it is necessary to remove the unit see separate note on removal routine to be followed.

Cause (3) Valve adjusting eccentric out of adjustment.

(i) Connect a vacuum gauge (reading 0–30 ins. (0–76.20 cm.) of mercury) to the union on the valve housing, Fig. 29.

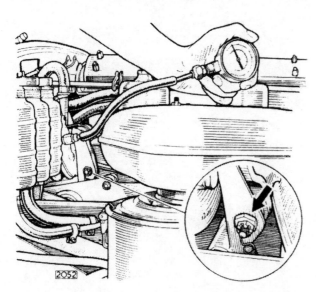

Fig. 29. Checking the servo with a vacuum gauge, Churchill tool No. J12—Gauge and Adaptor J12/1—Pipe and Adaptor, inset shows eccentric adjusting nut

Note: On early models the union will be found on the back mounting plate and on later models on the front auto-valve housing plate.

(ii) Run the engine and apply normal full pressure to the brake pedal. Gauge should now register 20 inches (50.8 cm.) of mercury. If no or only partial vacuum is registered it will be necessary to adjust the valve eccentric as follows:

Remedy: Remove the return spring and release the locknut, apply a spanner to the hexagon on the eccentric bush, Fig. 29, turn until required vacuum is obtained. This will be when the air valve is closed and the vacuum valve fully open. Tighten the locknut, apply the brakes and check. Release the pedal pressure completely; vacuum valve should now be closed and the air valve open with the gauge registering zero. Brakes should be free. If the brakes are not free this indicates that the vacuum valve is not completely closed. Release the locknut and adjust the eccentric in the opposite direction until the gauge registers zero and the brakes are free.

Note: Do not adjust the eccentric more than is necessary.
Re-tighten the locknut, and fit the return spring. Recheck by applying the brakes with the engine running. If the adjustment is now correct switch off the engine, remove the gauge and close the union.

Symptom: Slow return of the brake pedal.

Cause: Choked air filter.

Remedy: Remove the bellows unit from the car (for procedure see note headed "The Brake/Clutch Pedal Box Assembly Removal").
Hold the Pedal Box Assembly in a vice, use soft jaws and do not grip tightly, collapse the bellows by hand. This will expose the end of the air inlet tube with circlip attached. Remove the air intake baffle. The filter will now lift out, clean and dry thoroughly. Refit and replace the unit in the car, Fig. 22.

Page L.29

BRAKES

To Detach Bellows from Assembly

Proceed as described under "Dismantling the Brake Linkage" until the servo unit can be withdrawn.

Dismantling

Clamp the servo unit in a soft-jawed vice. Remove the three setscrews and shakeproof washers and remove the mounting bracket (26, Fig. 28).

Remove the air filter retaining baffle (25), withdraw the air filter (24).

Hold the mounting hub (18) down against the return spring (17). Remove the circlip (23), washer (22) and discard the rubber washer (21).

Holding the mounting hub down, remove the lip of the bellows (14) from the hub.

Remove the mounting hub and the return spring.

Remove the three self-locking setscrews, remove the guide sleeve (13) and bellows from the valve housing. Withdraw the guide sleeve from the bellows.

Remove the air valve control spring (12), the valve balancing diaphragm (10), the vacuum valve spring (8), vacuum valve assembly (7) and air valve assembly (6) from the valve housing (1).

Remove the valve balancing washer (9) and retainer (11) from the valve balancing diaphragm.

Discard the valve balancing diaphragm, the vacuum valve assembly and the air valve assembly.

Clean all metal parts except the mounting hub assembly but including the air filter in alcohol or other oil free solvent and dry with compressed air.

Important: The leather seal in the mounting hub assembly is filled with a silicone lubricant which must not be removed. If necessary the mounting hub should be cleaned with a dry cloth only.

Clean the bellows if necessary by washing in a mild soap and water solution after removing the three support rings.

Rinse in clean water and dry with compressed air.

Inspect all parts for wear or damage. All worn or damaged parts must be replaced.

If the vacuum valve seat in the vacuum valve housing is damaged the valve housing must be replaced.

Replace all parts listed below whether they show damage or not.

Replace the following:

	No. off	Fig. No.	
Rubber washer	1	(28)	21
Air valve assembly comprising:	1		6
Air valve			
Air valve buffer			
Air valve cup			
Air valve cap			
Vacuum valve assembly	1		7
Valve balancing diaphragm	1		10

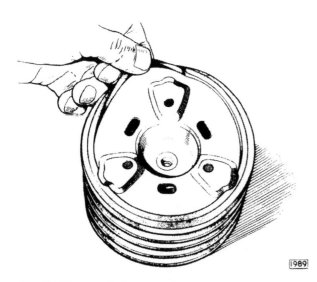

Fig. 30. Lining up the bosses on the retainer sleeve with the recesses on the bellows

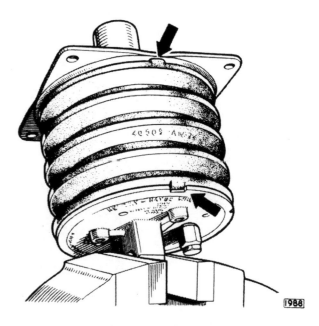

Fig. 31. Lining up the bosses on the bellows with the recesses on the mounting plate and valve housing

Page L.30

BRAKES

Reassembly

Re-assembly is the reverse of the dismantling procedure. It should be noted that when fitting the guide sleeve to the bellows the three bosses in the guide sleeve must line up with the three recesses in the bellows (Fig. 30). When fitting the valve housing to the bellows the boss on the bellows must line up with the recess in the valve housing. Similarly the cut-out in the mounting plate must line up with the boss on the bellows when attaching it to the mounting hub (Fig. 31).

When the unit is assembled, test the air valve by pressing the air valve cap down with the flat of a screwdriver. Two definite stages of movement should be felt and the valve should snap back readily.

SELF-ADJUSTING HANDBRAKES

Description

The self-adjusting handbrakes fitted to later models are attached to the rear brake caliper bodies but form an independent mechanically actuated system carrying its own friction pads. The handbrakes are self-adjusting to compensate for friction pad wear and automatically provide the necessary clearance between the brake discs and the friction pads.

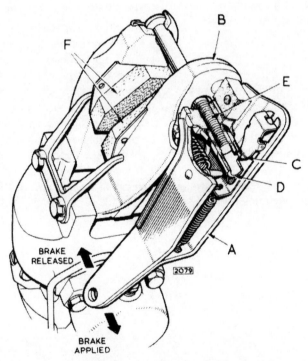

Fig. 32. Sectioned view of the adjusting mechanism.

Page L.31

BRAKES

Operation

When the handbrake lever in the car is operated, the operating lever (A, Fig. 1) is moved away from the friction pad carrier (B) and draws the friction pads (F) together. Under normal conditions, when the lever is released the pawl (C) in the adjusting mechanism returns to its normal position, thus the normal running clearance between the brake disc and the friction pads is maintained.

In the event of there being increased clearance, the pawl will turn the ratchet nut (D) on the bolt thread drawing the adjuster bolt (E) inwards and bringing the friction pads closer to the brake disc until the normal running clearance is restored.

Removal

With the car on a ramp, disconnect the handbrake compensator linkage from the handbrake operating lever at the front of the rear suspension assembly by discarding the split pin and withdrawing the clevis pin. Lift the locking tabs and remove the pivot bolts and retraction plate. Remove the friction pad carriers from the caliper bridge by moving them rearwards round the disc and withdrawing from the rear of the rear suspension assembly. Repeat for the second handbrake.

Dismantling

Remove the cover securing bolt, discard the split pin and withdraw the pivot clevis pin. Remove the dust cover and remove the split pin from the screwdriver slot in the adjusting bolt. Unscrew the adjusting bolt from the ratchet nut and withdraw the nut and bolt. Detach the pawl return spring and withdraw the pawl over the locating dowel. Detach the operating lever return spring and remove the operating lever and lower cover plate.

Assembling

Assembly is the reverse of the dismantling procedure.

Refitting

Refitting is the reverse of the removal procedure but the handbrake should be set as follows:—

With the split pin removed from the screwdriver slot in the adjusting bolt, screw the bolt in or out until there is a distance of $\frac{7}{16}''$ (11·1 mm.) between the friction pads, that is, the thickness of the disc plus $\frac{1}{16}''$ (1·5 mm.).

Refit the split pin and refit the caliper to the car.

Pull and release the operating lever at the caliper repeatedly when the ratchet will be heard to "click-over". Repeat the operation until the ratchet will not operate which will indicate that the correct clearance is maintained between the disc and the friction pads.

Reconnect the handbrake compensator linkage to the operating levers and check the cable adjustment as follows:—

Handbrake Cable Adjustment

Fully release the handbrake lever in the car. Slacken the locknut at the rear end of the handbrake cable.

Adjust the length of the cable by screwing out the threaded adaptor to a point just short of where the handbrake operating levers at the calipers start to move. Check the adjustment by pressing each operating lever at the same time towards the caliper; if any appreciable movement of the compensator linkage takes place the cable is too tight.

When correctly adjusted a certain amount of slackness will be apparent in the cable; no attempt should be made to place the cable under tension or the handbrakes may bind.

Page L.32

SECTION M
WHEELS AND TYRES

3·8 "E" TYPE
GRAND TOURING MODELS

INDEX

	Page
Description	M.3
Data	M.3
Inflation Pressures for Touring	M.3
Wheels and Tyres for Racing	M.3
Inflation Pressures for Racing	M.4
Tyres—General Information	M.4
Inflation Pressures	M.4
Valve Cores and Caps	M.5
Tyre Examination	M.5
Tyre and Wheel Balance	M.5
Static Balance	M.5
Dynamic Balance	M.5
Changing Position of Tyres	M.5
Wire Spoke Wheels	M.6
Description	M.6
Removal and Dismantling	M.6
Rebuilding	M.6
Trueing	M.7
Lateral Correction	M.7
Radial Correction	M.7
Checking for Dish	M.8
Adjustment for Dish	M.8

WHEELS AND TYRES

DESCRIPTION
Conventional tyres and tubes are fitted to "E" Type cars with wire spoke wheels as standard.

DATA

Road Wheels

Type and Make	Dunlop—72 wire spoke
Fixing	Centre lock, knock on hub cap
Rim Section	5K
Rim diameter	15" (381 mm.)
Number of spokes	72

Tyres

Make	Dunlop
Type	Conventional tyre and tube (RS.5)
Size	6·40 × 15"

Inflation Pressures (Dunlop R.S.5)

	Front	Rear
Normal use up to maximum speed of 130 m.p.h. (210 k.p.h.)	23 lbs sq. in. (1·62 kg. cm.^{2})	25 lbs sq. in. (1·76 kg. cm.^{2})
For sustained high speeds and maximum performance	30 lbs sq. in. (2·11 kg. cm.^{2})	35 lbs sq. in. (2·46 kg. cm.^{2})

Inflation Pressures (Dunlop SP.41 HR 185 × 15)

For speeds up to 125 m.p.h. (200 k.p.h.)	32 lbs/sq. in.	32 lbs sq. in.
For speeds up to maximum	40 lbs/sq. in.	40 lbs sq. in.

Note: The Dunlop SP.41 Tyre must not be used on "E" Type Cars unless the maximum speed is restricted to 125 m.p.h. (200 k.p.h.). Pressures should be checked when the tyres are cold, such as after standing overnight, and not when they have attained normal running temperatures.

WHEELS AND TYRES FOR RACING

Note that chrome-plated wheels are not recommended for use on cars which will be participating in serious competition work.

If it is desired to use 6·50 × 15 Road Racing tyres on the rear wheels for competition purposes, these tyres must be fitted to special wheels (Part No. C.18922) having a wider rim section and revised spoking, which maintains the normal clearance between the tyre and the wheel arch panel and results in the rear track being increased.

These special rear wheels must in no circumstances be used in the front wheel position on the "E" Type car. Also note that, since these wheels are recommended only for competition use, they will not be supplied chromium plated. Special rear wheels (Part No. C.18922) will be supplied only as spares and NOT as part of the specification of a new car.

It is recommended that, prior to and following participation in competition events covering racing or rallies, the wheels are checked to ensure that they are in an undamaged condition, are running true and that the spokes are correctly tensioned.

TYRES

The Dunlop Road Speed RS.5 tyres which are standard equipment on the "E" Type model give the best all round results for road use. It is not desirable that Road Racing tyres should be fitted to cars which will be used only on the road.

Racing Tyres

6·00 × 15 Dunlop R.5 Road Racing tyres should be fitted if "E" Type cars are being raced. If it is desired to fit larger section rear tyres to reduce the possibility of wheel spin under full power acceleration or to adjust the gear ratio, 6·50 × 15 Dunlop R.5. Road Racing tyres can be fitted, but only if these tyres are fitted on the special rear wheels described above.

Note that it is not desirable that cars should be run under normal touring conditions using Dunlop R.5 Road Racing tyres since, although these tyres give the best handling qualities under racing conditions, they do not have the same qualities for touring purposes as the Road Speed tyre, in addition to which the tyre walls are more liable to damage through "kerbing".

Page M.3

WHEELS AND TYRES

Tyre Pressures for Racing

Recommended tyre pressures for racing purposes are:—

    45 p.s.i. front and rear, Cold
    (3·2 kg/cm^2)

Dependent upon temperature and maximum speed conditions these pressures should be raised to:—

    50 p.s.i. front and rear, Cold
    (3·5 kg/cm^2)

The minimum tyre pressures for Dunlop R.5 Road Racing tyres if used for normal touring purposes are:—

    30 p.s.i. front and rear, Cold
    (2·1 kg/cm^2)

TYRES—GENERAL INFORMATION

TYRES

Dunlop tyres (RS.5) Road Speed tyres have been specially designed for cars with the high speed range of the Jaguar "E" Type class.

When replacing worn or damaged tyres and tubes it is essential that tyres with exactly the same characteristics are fitted.

Due to the high speed performance capabilities of the Jaguar "E" Type it is important that no attempt is made to repair damaged or punctured tyres.

All tyres which are suspect in any way should be submitted to the tyre manufacturers for their examination and report. The importance of maintaining **all** tyres in perfect condition cannot be too highly stressed.

Inflation Pressures

It is important to maintain the tyre pressures at the correct figures, incorrect pressures will affect the steering, riding comfort, and tyre wear.

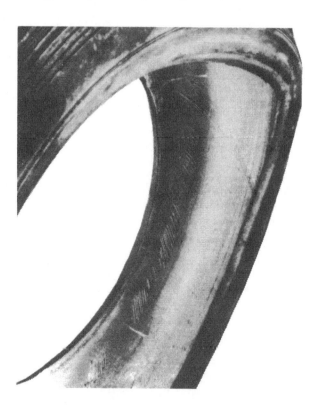

Fig. 2. Running deflated has destroyed this cover

Check the inflation pressures when the tyres are cold and not when they have attained their normal running temperature; tyre pressures increase with driving and any such increase should be ignored.

Always ensure that the valve caps are fitted to the end of the valve as they prevent the ingress of dirt and form a secondary seal to the valve core.

Fig. 1. Excessive tyre distortion from persistent under-inflation causes rapid wear on the shoulders and leaves the centre standing proud

Page M.4

WHEELS AND TYRES

Valve Cores and Caps

Valve cores are inexpensive and it is a wise precaution to renew them periodically.

Valve caps should always be fitted and renewed when the rubber seatings have become damaged after constant use.

Tyre Examination

Examine tyres periodically for flints, nails, etc., which may have become embedded in the tread. These should be removed with a blunt screwdriver or a similar instrument.

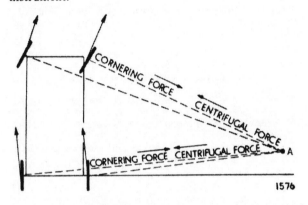

Fig. 3. Slip when cornering increases tyre wear

TYRE AND WHEEL BALANCE

Static Balance

In the interests of smooth riding, precise steering and the avoidance of high speed "tramp" or "wheel hop" all Dunlop tyres are balance checked to predetermined limits.

To ensure the best degree of tyre balance the covers are marked with white spots on one bead, and these indicate the lightest part of the cover. Tubes are marked on the base with black spots at the heaviest point. By fitting the tyre so that the marks on the cover bead exactly coincide with the marks on the tube, a high degree of tyre balance is achieved. When using tubes which do not have the coloured spots it is usually advantageous to fit the covers so that the white spots are at the valve position.

The original degree of balance is not necessarily maintained and it may be affected by uneven tread wear, by cover and tube repairs, by tyre removal and refitting or by wheel damage and eccentricity. The car may also become more sensitive to unbalance due to normal wear of moving parts.

If roughness or high speed steering troubles develop, and mechanical investigation fails to disclose a possible cause, wheel and tyre balance should be suspected.

A Tyre Balancing Machine is marketed by the Dunlop Company to enable Service Stations to deal with such cases.

Warning

If balancing equipment is used which dynamically balances the road wheels on the car, the following precaution should be observed.

In the case of the rear wheels always jack **both** wheels off the ground otherwise damage may be caused to the differential.

This is doubly important in the case of the "E" Type which is fitted with a Thornton "Powr-Lok" differential as in addition to possible damage to the differential, the car may drive itself off the jack or stand.

Dynamic Balance

Static unbalance can be measured when the tyre and wheel assembly is stationary. There is another form known as dynamic unbalance which can be detected only when the assembly is revolving.

There may be no heavy spot—that is, there may be no natural tendency for the assembly to rotate about its centre due to gravity—but the weight may be unevenly distributed each side of the tyre centre line. Laterally eccentric wheels give the same effect. During rotation the off set weight distribution sets up a totating couple which tends to steer the wheel to right and left alternately.

Dynamic unbalance of tyre and wheel assemblies can be measured on the Dunlop Tyre Balancing Machine and suitable corrections made when cars show sensitivity to this form of unbalance. Where it is clear that a damaged wheel is the primary cause of severe unbalance it is advisable for the wheel to be replaced.

TYRE REPLACEMENT AND WHEEL INTERCHANGING

When replacement of the rear tyres becomes necessary, fit new tyres to the existing rear wheels and, after balancing, fit these wheels to the front wheel positions on the car, fitting the existing front wheel and tyre assemblies (which should have useful tread life left) to the rear wheel positions on the car.

If at the time this operation is carried out the tyre of the spare wheel is in new condition, it can be fitted to one of the front wheel positions in preference to replacing one of the original rear tyres, which wheel and tyre then become the spare.

Note: Due to the change in the steering characteristics which can be introduced by fitting to the front wheel positions wheels and tyres which have been used on the rear wheel positions, interchanging of part worn tyres from rear to front wheel positions is not recommended.

Page M.5

WHEELS AND TYRES

WIRE SPOKE WHEELS
REPAIR AND ADJUSTMENT

DESCRIPTION

Dunlop 72 Cross-spoked wire wheels are fitted as standard to the Jaguar "E" Type and the following instructions are issued to assist in the repair and adjustment of the road wheels in the event of damage due to accident or from any other cause.

Cross-spoking refers to the spoke pattern, where the spokes radiate from the well of the wheel rim to the nose or outer edge of the hub shell, and from the tyre seat of the rim to the flanged or inner end of the shell (Fig. 5).

REMOVAL AND DISMANTLING

Detach wheel from car and remove tyre complete from wheel rim.

Remove spoke nipples and detach spokes from rim and centre.

Check wheel rims and centre; renew if damaged beyond normal repair.

Examine spokes and renew as necessary.

REBUILDING

Place the wheel centre and the rim on a flat surface with the valve hole upwards in the 6-o'clock position.

Note: All spoking operations commence in this position, and the valve hole is always the starting point for all rebuilding operations.

With the valve hole in the 6-o'clock position, fit one A, B, C, and D spoke to produce the pattern as shown in Fig. 4.

Having established the correct pattern remove the A and B spokes and proceed as follows:—

(1) Attach the D spoke to the rim, and screw up the nipple finger tight; leave the C spoke loosely fitted without a nipple attached.

(2) Attach all the D spokes with the nipples finger tight.

(3) Insert all the C spokes through the hub shell without nipples.

(4) Attach all the B spokes as paragraph 2 above.

(5) Attach all the A spokes as paragraph 2 above.

(6) Attach the nipples and finger tighten all C spokes.

(7) Tighten the two C spokes and the two D spokes on each side of the valve hole until the ends of the spokes are just below the slot in the nipple heads.

(8) Tighten the four C and D spokes diametrically opposed to the valve hole (12-o'clock position).

(9) Mark around the wheel until all the C and D spokes are similarly tightened.

(10) Follow with all A and B spokes as in paragraphs 7, 8 and 9 above.

(11) Work around the wheel with a spoke spanner and tighten all nipples until some resistance is felt. Diametrically opposed spokes should be tightened in sequence.

The wheel is now ready for trueing and adjustment.

TRUEING

Wheels can be out of truth in a lateral or radial direction, or in a combination of both.

As a general rule, lateral out of truth should be corrected first.

The wheel to be trued must be mounted on a free-running trueing stand before any adjustment can be carried out.

Page M.6

WHEELS AND TYRES

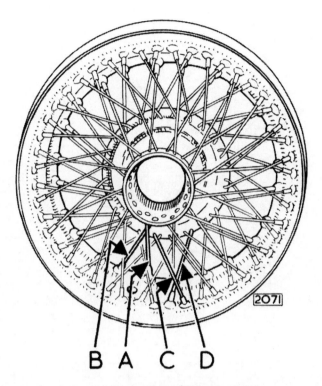

Fig. 4. Showing the spoking arrangement

Lateral Correction

Mount the wheel on the trueing stand. Spin the wheel, and holding a piece of chalk near the wall of the rim flange, mark any high spots. Tighten the A and B spokes in the region of the chalk marks and slacken the C and D spokes in the area.

Note: Throughout the trueing operations, no spoke should be tightened to such an extent that it is impossible to tighten it further without risk of damage. If any spoke is as tight as it will go, all the other spokes should be slackened.

Radial Correction

When lateral out of truth has been corrected, spin the wheel on the trueing stand, and, with the chalk, mark the high spots on the horizontal tyre seat. Tighten **all** spokes in the region of the chalk marks, or if the spokes are on the limit of tightness, slacken all the remaining spokes.

CHECKING FOR "DISH"

The term "dish" defines the lateral dimension from the inner face of the flanges of the wheel centre to the inner edge of the wheel rim. To check "dish" place straight edge across the inner edge of the wheel rim and measure the distance to the inner face of the wheel centre flange (Fig. 5). This dimension should be $3\frac{7}{16}'' \pm \frac{1}{16}''$ (87·3 mm. ± 1·58 mm.).

Adjustment for "Dish"

If the "dish" is in excess of the correct dimension $3\frac{7}{16}'' \pm \frac{1}{16}''$ (87·3 mm. ± 1·58 mm.) tighten all A and B spokes, and slacken all C and D spokes by a similar amount.

When the "dish" dimension is less than the given tolerance slacken all A and B spokes and tighten all C and D spokes by a similar amount.

Page M.7

WHEELS AND TYRES

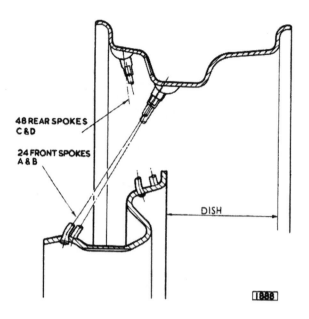

Fig. 5. Location for measuring the dish and the "A," "B," "C," and "D" spokes

It will be necessary after completing the "dish" adjustments to repeat the lateral and radial trueing procedure until the wheel is not more than ·060" (1·5 mm.) out of truth in either direction.

It is important that after the wheel trueing operation is completed that all spokes should be tensioned uniformly, and to a reasonably high degree.

Correct tension can be closely estimated from the high pitched note emitted when the spokes are lightly tapped with a small hammer.

If a spoke nipple spanner of the torque recording type is used, a normal torque figure should be in the order of 60 lb.in. (0·7 kgm.).

Page M.8

SECTION N
BODY & EXHAUST SYSTEM

3·8 "E" TYPE
GRAND TOURING MODELS

INDEX

BODY

	Page
Side Facia Panel	
Removal (Left hand drive)	N.5
Removal (Right hand drive)	N.5
Refitting (Left hand drive)	N.6
Refitting (Right hand drive)	N.6
Glovebox	
Removal	N.6
Refitting	N.7
Top Facia Panel	
Removal	N.7
Refitting	N.7
Bonnet	
To open (Early cars)	N.7
To open (Later cars)	N.7
Removal	N.8
Refitting	N.8
Adjustment (Early cars)	N.9
Adjustment (Later cars)	N.9
Accidental damage	N.9
Motif Bar	
Removal	N.9
Refitting	N.9
Bonnet Side Panel	
Removal	N.9
Refitting	N.9
Bonnet Centre Section	
Removal	N.10
Refitting	N.10
Air Vent Grille	N.10
Bonnet Safety Catch	N.10

INDEX (Continued)

The Under Panel
- Removal N.10
- Refitting N.10

Chrome Strips on the Bonnet
- Removal N.13
- Refitting N.13

Front Bumper
- Removal N.13
- Refitting N.13

Front Bumper Over-riders
- Removal N.13
- Refitting N.13

Rear Bumpers
- Removal N.14
- Refitting N.14

Rear Bumper Over-riders
- Removal N.14
- Refitting N.14

Luggage Compartment Lid and Hinges
- Removal N.15
- Refitting N.15
- Luggage compartment lock adjustment (Open 2-seater) N.15
- Luggage compartment lock adjustment (Fixed Head Coupe) .. N.15

Petrol Filler Lid
- Removal N.16
- Refitting N.16

Windscreen
- Removal (Open 2-seater) N.17
- Refitting N.17
- Removal (Fixed Head Coupe) N.18
- Refitting N.18

Rear Window Glass
- Removal (Fixed Head Coupe) N.19
- Refitting N.19
- Removal (Detachable hard top) N.19

INDEX (Continued)

Doors and Hinges
    Removal N.19
    Refitting N.20

Door Trim Casings
    Removal N.20
    Refitting N.20

Door Window Glass and Frame
    Removal (Door window glass) N.20
    Removal (Door window frame) N.20
    Refitting N.21

Window Regulator
    Removal N.21
    Refitting N.21

Door Window Outer Seal
    Removal N.21
    Removal of Chrome door finisher (Open 2-seater) N.21
    Refitting N.21

Seats and Runners
    Removal N.21
    Refitting N.21

No Draught Ventilators
    Removal N.22
    Refitting N.22

Removal and Refitting of Door Lock Mechanism N.22

Accidental Damage N.25

EXHAUST SYSTEM

Exhaust System
    Removal N.39
    Refitting N.39

BODY AND EXHAUST SYSTEM

BODY

SIDE FACIA PANEL

Removal (Left-hand Drive)

Disconnect the positive lead on the battery. Unscrew the two chrome bezels securing the speedometer trip and the time clock control cables to the under scuttle casings.

Remove the two under scuttle casings by unscrewing the drive screws and withdrawing the casings away from the retaining clips.

Withdraw the headlamp, ignition and fuel warning lights from the rear of the speedometer. Disconnect the speedometer drive cable from the rear of the speedometer.

Remove the upper steering column top fixing bolt and nut securing the column to the support bracket, noting the distance tube between the bracket side flanges.

Release the upper steering column lower mounting bolts and nuts.

Disconnect the flasher switch cables from the multi-snap connector attached to the harness and located behind the facia panel. Lower the column and allow the steering wheel to rest on the driver's seat.

Remove the two thumb screws securing the centre instrument panel to the body and allow the panel to rest in the horizontal position. Remove the three slotted setscrews and lockwashers retaining the side facia panel to the centre instrument panel support brackets.

Remove the headlamp dipper switch from the side facia panel by removing the chrome ring nut securing the switch to the facia and withdrawing the switch lever through the panel.

Remove the two nuts and washers at the rear of the side facia panel securing the panel to the bracket attached to the body adjacent to the door hinge post.

Detach the panel.

Release the two setscrews securing the two heater control inner cables to the control levers and withdraw the cables.

Withdraw the two instrument illumination bulb holders from the speedometer.

Withdraw the two instrument illumination bulb holders from the revolution counter. Withdraw the two flasher indicator warning light bulb holders from the indicator light unit.

Disconnect the clock connection at the snap connector.

Remove the two cables from the brake fluid warning light.

Disconnect the two "Lucar" connectors from the rear of the revolution counter.

Remove the side facia panel.

Removal (Right-hand Drive)

Disconnect the positive lead on the battery. Unscrew the two chrome bezels securing the speedometer trip and the time clock control cables to the under scuttle casings. Remove the under scuttle casings by unscrewing the drive screws and withdrawing the casings away from the retaining clips.

Withdraw the headlamp, ignition and fuel warning lights from the rear of the speedometer.

Disconnect the drive cable from the rear of the speedometer.

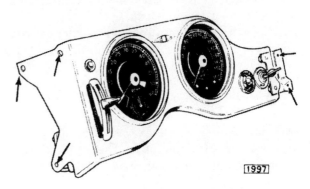

Fig. 1. Location of the side facia panel attachment points (Right hand drive).

Remove the upper steering column top fixing bolt and nut securing the column to the support bracket noting the distance tube between the bracket side flanges.

Release the upper steering column lower mounting bolts and nuts.

Disconnect the flasher switch cables from the multi-snap connector attached to the harness and located behind the facia panel.

Page N.5

BODY

Lower the column and allow the steering wheel to rest on the driver's seat.

Remove the two thumb screws securing the centre instrument panel to the body and allow the panel to rest in the horizontal position.

Remove the three slotted setscrews and lock washers retaining the side facia panel to the instrument panel support bracket. Remove the bolt, nut and washer retaining the mixture control bracket to the centre panel support bracket.

Remove the headlamp dipper switch from the side facia panel by removing the chrome ring nut securing the switch to the facia and withdrawing the switch lever through the panel. Remove the two nuts and washers at the rear of the side facia panel securing the panel to the bracket attached to the body adjacent to the door hinge post.

Detach the panel.

Release the setscrew securing the mixture control inner cable to the control lever and withdraw the cable; remove the mixture control warning light bulb holder and disconnect the two cables from the warning light switch. Withdraw the two instrument illumination bulb holders from the speedometer.

Withdraw the two instrument illumination bulb holders from the revolution counter and disconnect the two "Lucar" connectors from the rear of the instrument.

Withdraw the flasher indicator warning light bulb holders from the indicator light unit.

Disconnect the two cables from the brake fluid warning light.

Disconnect the clock connection at the snap connector.

Remove the side facia panel.

Refitting

Refitting is the reverse of the removal procedure, but particular attention must be paid to the following points.

When refitting the headlamp dipper switch note that the terminals with the Blue/Yellow and Blue/Green cables attached are uppermost and that the flat on the switch stem is registering correctly with the flat in the mounting hole.

Insert the flasher warning lights into their correct sockets; that is, with the warning light attached to the black/white cable in the right-hand indicator bulb holder and the black/red cable in the left-hand side.

Left-hand Drive

Reconnect the two heater control cables ensuring that the full movement of the lever marked "HOT and COLD" with the water control tap and the lever marked "OFF AIR-ON" with the air control flap is maintained. For full instructions on adjustment see Section O, "Car Heating and Ventilating Equipment".

Right-hand drive

Reconnect the mixture control cable ensuring that the full movement of the control lever and the lever on the carburetter is maintained.

To adjust the control pass the cable through the boss on the lever, place the lever in the "COLD" position and position the lever on the carburetter towards the rear of the engine. Tighten the setscrew securing the cable control wire and recheck.

Refit the steering column and adjust for rake.

Reconnect the flasher indicator switch cables to the multi-snap connector using the wiring diagram as a reference.

GLOVEBOX

Removal

Disconnect the positive lead on the battery. Remove the under scuttle casing by unscrewing the drive screws and withdrawing casing away from the retaining clips.

Remove the two thumb screws securing the centre instrument panel to the body and allow the panel to rest in the horizontal position.

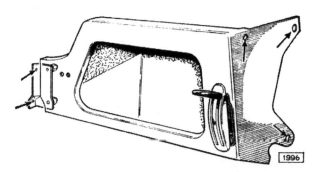

Fig. 2. Location of the glovebox attachment points (Right hand drive).

Remove the grab handle by removing the two setscrews from the hidden face of the glovebox, and on early cars, one screw from the base of the screen pillar exposed after lifting the draught rubber and pulling away the trim welt; on later models remove one setscrew securing the handle to the bracket located at the base of the pillar.

BODY

Remove the three slotted setscrews and lock washers retaining the glovebox to the centre instrument panel support bracket.

Remove the two nuts and lock washers at the rear of the glovebox securing the glovebox to the bracket attached to the body adjacent to the door hinge post.

On right-hand drive cars disconnect the heater controls as detailed on page 5. (Side facia, Removal—Left-hand drive).

On left-hand drive cars disconnect mixture control warning light and switch as detailed on page 5. (Side facia, Removal—Right-hand drive).

Refitting

Refitting is the reverse of the removal procedure, but particular attention must be paid to maintaining full movement of the heater control on right-hand drive cars as detailed on page 6 (Side facia, Refitting), and the mixture control on left-hand drive cars as detailed on page 6 (Side facia, Refitting).

TOP FACIA PANEL

Removal

Disconnect the positive lead on the battery.

Remove all under scuttle casings by unscrewing the drive screws and withdrawing casings away from the retaining clips. Remove central console panel by removing the four large round headed setscrews attaching console to the body brackets. Withdraw console away from facia. If a radio is fitted to the car, withdraw control head complete with the console after detaching the aerial and power cables.

Remove the thumb screws securing the centre instrument panel to the body and allow the panel to rest in the horizontal position. Remove the two $\frac{3}{16}''$ nuts, lockwashers and plain washers securing the top facia panel to the brackets attached to the centre panel supports. Remove the two outer fixing nuts and washers securing the panel to the brackets attached to the body side panel below the screen pillars.

Withdraw the three flexible demister conduit pipes from the rubber elbow connections attached to the bulkhead below the instrument panel.

Disconnect the two cables attached to the map light.

Remove the top facia panel complete with the demister nozzles and pipes.

Refitting

Refitting is the reverse of the removal procedure. Utilizing the slotted holes in the brackets adjust the forward edge of the facia to the screen frame.

BONNET

To Open (Early cars)

To open the bonnet insert the "T" handle provided into the lock and on the right-hand side turn the key clockwise and on the left-hand side turn the key anti-clockwise.

This will release the bonnet which will now be retained by the safety catch.

Insert the fingers under the rear edge of the bonnet and press in the safety catch.

To Open (Later cars)

To open turn the two small levers located on the right and left-hand door hinge posts anti-clockwise and pull to full extent. This will release the bonnet which will now be retained by the safety catch.

Insert the fingers under the rear edge of the bonnet and press in the safety catch.

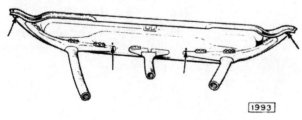

Fig. 3. Location of the top facia panel attachment points.

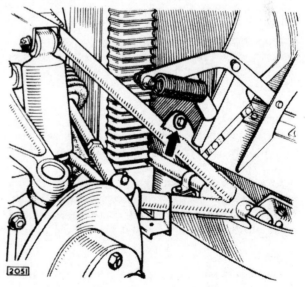

Fig. 4. The bonnet spring mechanism pivot points.

BODY

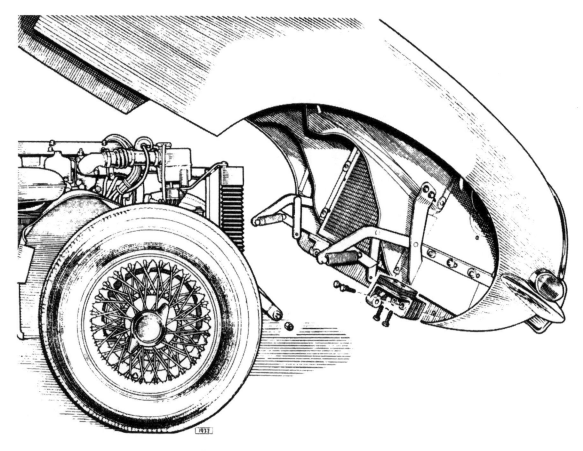

Fig. 5. The bonnet hinge mountings.

Removal

Disconnect the multi-pin socket from the left-hand side of the bonnet.

Mark the position of the hinges on the bonnet to facilitate refitting.

Remove the two self-locking nuts and washers securing the bonnet hinges to the front sub-frame mounting pin (Fig. 5).

Remove the two pivot pins and nuts securing the helper spring mechanism to the sub-frame (Fig. 4).

Supporting the bonnet, remove the four setscrews and washers securing the left-hand hinge to the bonnet (Fig. 5).

Remove the hinge noting the amount and location of the packing pieces between the hinge and the bonnet.

Still supporting the bonnet slide the right-hand hinge off the mounting pin and remove the bonnet.

Refitting

Refitting is the reverse of the removal procedure. The multi-pin electrical socket will only fit into the plug one way and therefore it is essential to mate the socket correctly with the pins (Fig. 6).

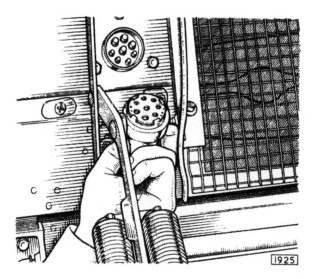

Fig. 6. Location of the multi-pin socket connections.

Page N.8

BODY

Adjustment (Early cars)

To ensure locking of the bonnet, adjustment is provided by means of packing pieces inserted under the bonnet lock plate attached by two screws to the body. Remove or add packing pieces until lock pawl retains bonnet firmly when locked.

Adjustment (Later cars)

To ensure secure locking of the bonnet, adjustment is provided by means of rubber buffers attached to the adjustable spigot pins. To adjust bonnet release spigot pin locknut, turn the pin until the lock pawl retains bonnet firmly when locked. Re-tighten the locknut.

Accidental Damage

The bonnet is composed of eleven main components each of which is replaceable if damaged. The components are listed below:

1. Bonnet side panel (Right-hand side).
2. Bonnet side panel (Left-hand side).
3. Bonnet centre panel.
4. Front under panel.
5. Front diaphragm (Right-hand side).
6. Front diaphragm (Left-hand side).
7. Rear diaphragm (Right-hand side).
8. Rear diaphragm (Left-hand side).
9. Valance (Right-hand side).
10. Valance (Left-hand side).
11. Air duct lower.

MOTIF BAR

Removal

To remove the motif bar from the bonnet orifice remove the two hexagon headed setscrews securing the bar to the two front bumpers. These setscrews are accessible from the rear of the bumper extension pieces and require the use of a $\frac{7}{16}''$ A.F. socket wrench; preferably of the rachet type.

Refitting

Refitting is the reverse of the removal procedure.

BONNET SIDE PANEL

Removal

Remove the bonnet as detailed on page 8.
Remove the glass headlamps cover and duct as detailed on page 13. "FRONT Bumper—Removal".
Remove the front bumper as detailed on page 13.
Remove the side/flasher lamp after detaching the

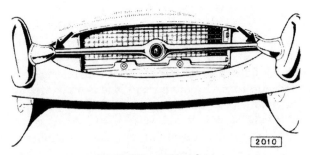

Fig. 7. The motif bar fixings.

cover by removing the three fixing screws and disconnecting the two attached cables from the snap connectors in the headlamp nacelle.

Remove the 5 bolts, nuts, plain and lock washers from the bottom flange securing the side panel to the under panel and the two bolts, nuts, plain and lock washers securing the side panel to the centre panel.

Remove the four bolts, nuts, plain and lock washers securing the side panel to the headlamp mounting diaphragm and the nine bolts, nuts and washers attaching the side panel to the centre panel along the crown line of the side panel.

Remove the five bolts, nuts and washers securing the side panel to the engine valance panel and the four bolts, nuts and washers attaching the side panel to the rear diaphragm.

Straighten the brass tabs of the two chromium beading strips, nine clips will be found on the long strip and two on the smaller one. Remove the closing plate attaching the side panel to the centre panel at the rear by withdrawing the four setscrews and washers. Remove the panel.

Refitting

Refitting is the reverse of the removal procedure.

Care must be taken during assembly to ensure that the edge lines of the centre section and the side panel are flush when bolted together. Failure to maintain this will prevent the chrome strip from fitting neatly to the bonnet.

Refit chrome strips as detailed on page 13. After assembly generously coat all under wing joints with a good quality sealing compound.

BODY

BONNET CENTRE SECTION

Removal

Remove the bonnet as detailed on page 8.

Remove both glass head lamp covers and ducts as detailed on page 13. "Front Bumper—Removal".

Remove both front bumpers as detailed on page 13. and motif bar as detailed on page 9.

Remove the radiator stone guard after unscrewing the eight cross headed drive screws and the two bolts and nuts securing the guard to the bonnet, withdraw the guard from the bottom noting the felt sealing strip at the top edge. From the right-hand side remove the ten cross headed drive screws and washers attaching the centre section to the valance and three screws from the rear diaphragm. From inside the headlamp nacelle remove the two bolts and nuts and washers from the vertical flange attaching the side panel to the centre section and the two bolts, nuts and washers securing the section to the under panel. Straighten the brass tabs of the two chrome beading strips and remove the nine bolts, nuts and washers securing the centre section to the side panel along the crown line.

Remove the beading strips and the closing plate connecting the centre section to the side panel at the rear after withdrawing the four setscrews and washers.

Repeat the operation to the left-hand side.

Refitting

Refitting is the reverse of the removal procedure. Care must be taken during assembly to ensure that the edge lines of the centre section and the side panel are flush when bolted together. Failure to maintain this will prevent the chrome strips from fitting neatly to the bonnet.

Refit chrome strips as detailed on page 13. When refitting the radiator stone guard ensure that the felt sealing strip is in good condition. Renew if necessary.

After assembly generously coat all under wing joints with a good quality sealing compound.

AIR VENT GRILLE

The chromium plated grille located at the rear of the centre section of the bonnet can be detached after removing the two bolts and nuts from the bottom edge and the two spring steel nut fasteners from the top fixing pegs.

Utilize the external flats of a $\frac{7}{16}''$ A.F. tubular spanner to remove the nut fasteners.

Refitting is the reverse of the removal procedure.

BONNET SAFETY CATCH

Remove the bonnet safety catch by unscrewing the four setscrews and washers. When refitting adjust the catch utilizing the slotted holes so that the lever will retain the bonnet when the locks are released, but will, when pressed, allow the bonnet to be fully opened.

THE UNDER PANEL

Removal

Remove the bonnet as detailed on page 8.

Remove both headlamp covers and ducts as detailed on page 13, under "Front Bumper—Removal". Remove both front bumpers as detailed on page 13, and the motif bar as detailed on page 9.

Remove the radiator stone guard after unscrewing the eight cross headed drive screws and the two bolts and nuts securing the guard to the bonnet. Withdraw the guard from the bottom noting the felt sealing strip at the top edge.

From the right-hand side of the bonnet remove the bolts, nuts and washers located in the headlamp nacelle securing the under panel to the centre section and side panel.

Remove the bottom hinge bracket after withdrawing four setscrews and lock washers. Note the quantity of spacer shims fitted.

Mark the position of the bonnet spring bracket for reference when refitting, remove the spring bracket after withdrawing four setscrews.

Remove the five cross headed drive screws retaining the under panel to the head lamp mounting diaphragm and the two cross headed drive screws securing the under panel to the valance.

Repeat the sequence for the left-hand side.

Remove the three cross headed drive screws and the two bolts, nuts and washers attaching the under panel to the orifice lower panel.

Remove the lower panel.

Refitting

Refitting is the reverse of the removal procedure. After assembly generously coat all under wing joints with a good quality sealing compound.

BODY

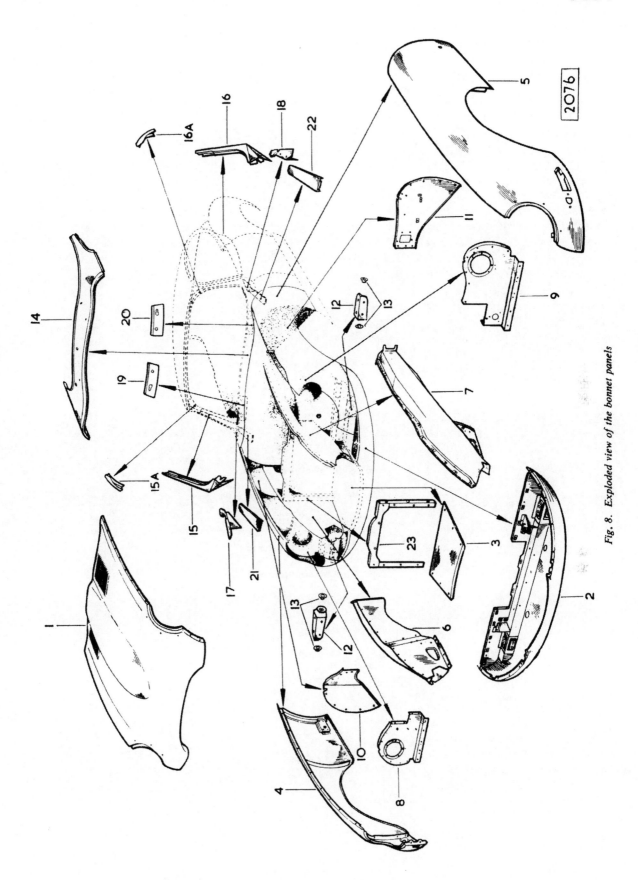

Fig. 8. Exploded view of the bonnet panels

Page N.11

BODY

1. Centre section
2. Under panel
3. Lower air duct
4. Side panel—right hand
5. Side panel—left hand
6. Valance—right hand
7. Valance—left hand
8. Front diaphragm—right hand
9. Front diaphragm—left hand
10. Rear diaphragm—right hand
11. Rear diaphragm—left hand
12. Bonnet hinge
13. Nylon bush
14. Scuttle top panel
15. Windscreen pillar—right hand
16. Windscreen pillar—left hand
15a. Reinforcement channel
16a. Reinforcement channel
17. Filler panel
18. Filler panel
19. Corner panel
20. Corner panel
21. Closing panel
22. Closing panel
23. Stoneguard mounting frame

BODY

CHROME STRIPS ON BONNET

Removal

The chrome strips along the crown line of the bonnet are secured with clips.

To remove, release the bolts and nuts retaining the centre section to the bonnet side panel, Straighten the prongs of the clips and withdraw the chrome strips.

Refitting

Refitting is the reverse of the removal procedure. Re-bend the clips after re-tightening all the under wing flange bolts.

After re-assembly generously coat all under wing joints with a good quality sealing compound.

FRONT BUMPER

Removal

The front bumper is comprised of two sections (right and left-hand) linked by the motif bar. Removal of either section is identical.

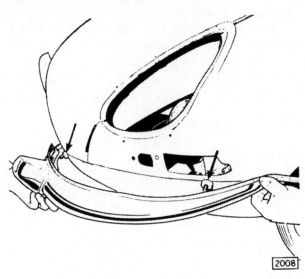

Fig. 9. Removing the left hand front bumper.

To gain access to the bumper fixing bolts it is necessary to remove the glass headlamp cover. Remove the six screws holding the cover retaining the ring to the wing. Remove the ring and rubber seal now exposed.

Remove the glass cover. Remove the three setscrews securing the headlamp duct to the diaphragm panel and withdraw duct forward through nacelle. Remove the setscrew retaining the motif bar to the bumper (See Motif Bar—Removal) and unscrew the two $\frac{3}{8}''$ U.N.F. setscrews, located in the wing nacelle, securing the bumper to the wing.

Detach the bumper and beading.

The curved extension attached to the bumper at its inner end can be removed by withdrawing the two setscrews.

Refitting

Refitting is the reverse of the removal procedure. When refitting ensure that the beading is replaced between the bumper and extension and also between the bumper and the body.

FRONT BUMPER OVER-RIDERS

Removal

Remove the front bumper (See "Front Bumper—Removal").

Remove the nut, plain and lock washer securing the over-rider to the bumper.

Remove the over-rider and beading.

Refitting

When refitting replace the beading between the over-rider and bumper. Refitting is the reverse of the removal procedure.

Page N.13

BODY

NEAR BUMPERS

Removal

The rear bumper is comprised of two sections (right-hand and left-hand). Removal varies only in respect of the components it is necessary to remove to gain access to the fixing screws.

Right-hand Bumper

Remove the section of the boot floor over the spare wheel by raising the forward edge until the peg attached to the floor board clears the spring clip. Slide the floorboard forward and remove.

Remove the spare wheel by unscrewing the centre fixing nut.

Remove the side trim casing after unscrewing the three chrome drive screws.

Remove the three bumper retaining setscrews. The forward screw is located within the wheel arch, the remaining two being accessible from the boot interior.

Refitting is the reverse of the removal procedure. When refitting ensure that the rubber beading is replaced, between the bumper and the body.

Left-hand Bumper

Disconnect the positive lead on the battery. Remove the floor board covering spare wheel, and remove the spare wheel. Remove the floor board covering the petrol tank by unscrewing the countersunk screws. Disconnect the two cables from the petrol tank gauge unit.

Remove the cover from the rubber junction block located in the spare wheel compartment and disconnect petrol pump cables.

Disconnect the petrol pipe from the petrol tank and tie up union to boot lid hinge to prevent loss of petrol. Note the two fibre washers. Remove the side trim casing after unscrewing the three chrome drive screws.

Release the clips and remove the petrol filler hose.

Remove the three setscrews from the petrol tank mounting and remove the petrol tank.

Remove the three bumper retaining setscrews. The forward screw is located within the wheel arch, the remaining two being accessible from the boot interior.

Refitting

Refitting is the reverse of the removal procedure. Always ensure that the rubber beading is replaced between the bumper and body. When re-connecting the petrol pipe, note that the two fibre washers are replaced one to each side of the banjo connection.

Reconnect the tank unit and the petrol pump, using wiring diagram as a reference.

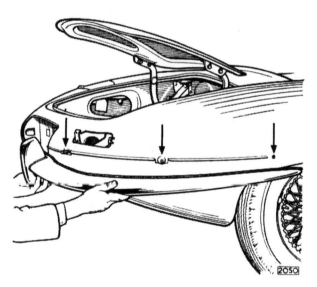

Fig. 10. Removing the right hand rear bumper

REAR BUMPER OVER-RIDERS

Removal

Remove the rear bumper (See "Rear Bumper—Removal").

Remove the nut, plain and lock washer securing the over-rider to the bumper.

Remove the over-rider and beading.

Refitting

When refitting replace the beading between the over-rider and the bumper. Refitting is the reverse of the removal procedure.

Page N.14

BODY

LUGGAGE COMPARTMENT LID AND HINGES

Removal

Raise the luggage compartment lid and on the Fixed head coupe retain in position by lowering the stay.

The lid on the Open 2-seater is retained in the open position by the action of helper springs.

Mark the position of the hinges on the lid. Remove the four setscrews, plain and lock washers and remove the lid.

Mark the position of the hinges on the body and remove the four setscrews, nuts and lock washers securing the hinge to the body.

Refitting

Refitting is the reverse of the removal procedure.

Luggage Compartment Lock Adjustment—Open 2-seater

Slacken the four setscrews securing the luggage compartment lid striker to the luggage compartment lid (see Fig 11). Move the striker in the elongated holes until the lock operates correctly and does not rattle. Tighten the retaining screws.

Luggage Compartment Lock Adjustment—Fixed Head Coupe

Slacken the two cross-headed screws in the lock striker and the two nuts securing the striker to the lid.

Move the striker in the elongated holes until the lock and the safety catch operate correctly and do not rattle. Tighten the retaining setscrews.

Further adjustment is provided if required by the four slotted holes in the lock attached to the body panel.

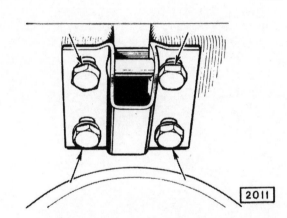

Fig. 11. Location of the screws for adjustment of the luggage compartment lid striker (Open 2-seater)

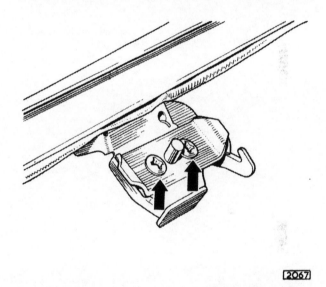

Fig. 12. Location of the screws for adjustment of the luggage compartment lid striker (Fixed Head Coupe)

Page N.15

BODY

Fig. 13. Showing the adjustment for the luggage compartment lid striker bracket (Fixed Head Coupe).

PETROL FILLER LID

Removal

Remove the return spring. Unscrew the two setscrews and washers securing the lid and hinge to the inner wall of the petrol filler cap compartment.

Remove the two setscrews and washers securing the lid to the hinge.

Refitting

Refitting is the reverse of the removal procedure.

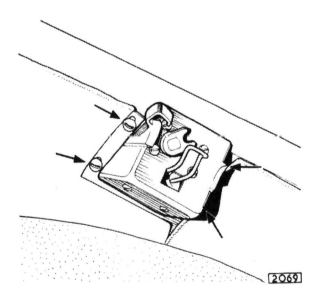

Fig. 14. The adjustment screws for the luggage compartment lock (Fixed Head Coupe).

BODY

WINDSCREEN

Removal—Open 2-seater

On Open 2-seater models it is necessary to detach the windscreen stay from the bracket attached centrally to the top screen frame by withdrawing the two slotted setscrews.

Remove the two chrome screen pillar cappings from the screen pillars by extracting the two cross-headed screws from each capping.

Note: The two screws have different heads and must be replaced in the same holes when refitting the screen.

Remove the screen pillar trim welts by withdrawing away from the flange on the pillars. The welt is retained in position by spring clips.

Using a No. 35 drill remove the two "Pop" rivets now exposed, retaining the chrome finisher to each screen pillar. Prise away the finisher from the screen rubber.

Prise off the chrome finisher from the bottom of the windscreen rubber. Extract one end of the screen rubber insert and withdraw completely. Run a suitable thin bladed tool around the windscreen to break the seal between the rubber and the windscreen aperture flange.

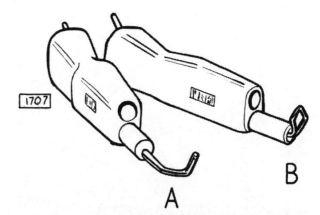

Fig. 16. The two special tools used when refitting the windscreen.

Fig. 15. Removing the windscreen.

Strike the glass with the flat of the hand from the inside of the car, starting in one corner and working towards the bottom.

Repeat this process around the complete windscreen. Withdraw the screen.

Remove the windscreen top frame by inserting a thin flat bladed tool between the sealer and the glass to break the seal and gently prise away the frame. Do not use undue force when removing the frame.

Refitting

Remove the old sealer from the windscreen flange. Examine the screen rubber for cuts.

If the windscreen was not broken by a projectile the windscreen aperture flange should be examined for a bump in the metal. If this is found the bump should be filed away otherwise the glass may break again.

The rubber should be attached to the windscreen aperture with the flat side of the rubber towards the rear.

Fig. 17. Using the special tool ("A") Fig. 16, for lifting the windscreen rubber over the glass

Page N.17

BODY

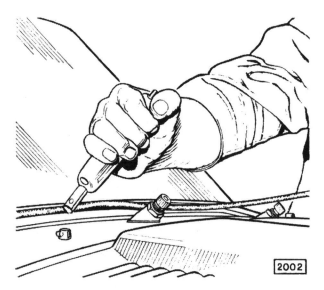

Fig. 18. Using the special tool ("B") Fig. 16. for inserting the rubber sealing strip in the windscreen sealing rubber

Using the special tool (A, Fig. 16) insert the screen into the rubber along the bottom edge first. DO NOT fit one end and then try to fit the other. Using the special tool (B, Fig. 16) insert the rubber sealing strip with the rounded wide edge to the outside.

Using a pressure gun filled with a sealing compound and fitted with a copper nozzle (so that the glass will not be scratched) apply the nozzle of the gun between the metal body flange and the rubber. Repeat the operation between the glass and the rubber. Remove excess sealing compound with a cloth soaked in white spirit. DO NOT USE THINNERS as this will damage the paintwork.

Fit the chrome strip on top of the windscreen rubber and bend to suit contour if necessary. Coat the inside of the strip with a layer of Bostik 1251 and allow to become tacky.

Place the chrome strip on the rubber over the sealing strip and with the special tool (A) lip the rubber over the chrome finisher.

Refit the windscreen top frame. Always use a new length of sealing strip and do not apply undue force when refitting. If difficulty is experienced when fitting frame lubricate sealing strip and glass with a liquid soap solution.

Coat the inside face of the screen pillar finisher with Bostik 1251 and allow to become tacky.

Note: It is only necessary to apply the Bostik to that portion of the finisher which comes into contact with the screen rubber.

Place the finisher on the rubber over the sealing strip and with the special tool (A) lip the rubber over the finisher. Secure the finisher to the screen pillar with two "Pop" rivets inserted in the original holes.

Refit the chrome screen pillar cappings to the screen pillars.

It is essential that the flat countersunk screw is fitted to the inside face of the screen pillar capping and the raised screw is fitted to the top face.

Failure to ensure this will prevent the hood from fitting correctly to the screen frame.

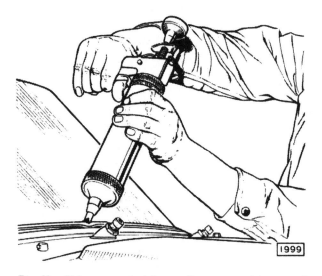

Fig. 19. Using a gun to inject sealing compound between the surround and the glass

Removal—Fixed Head Coupe

Prise off the two screen pillar chrome finishers from the windscreen rubber and repeat with the upper and lower finishers. Extract one end of the rubber insert and withdraw completely.

Run a suitable thin bladed tool around the windscreen to break the seal between the rubber and the windscreen aperture flange.

Strike the glass with the flat of the hand from inside the car, starting in one corner and working towards the bottom.

Repeat this process around the complete windscreen. Withdraw the windscreen.

Refitting

Remove the old sealer from the windscreen flange.

The procedure for refitting and re-sealing the glass is similar to the instructions given for the Open 2-seater (page 17).

Page N.18

BODY

Fit the upper chrome strip on top of the windscreen rubber and bend to suit contour if necessary. Coat the inside of the strip with Bostik 1251 and allow to become tacky. Place the chrome strip on the rubber over the rubber sealing strip and with a hook (A, Fig. 16) lip the rubber over the finisher. Repeat the operation with the lower chrome strip. Refit the two screen pillar chrome finishers. Coat the inside of the finisher with Bostik and lip the rubber over using the same tool. The screen pillar finishers will overlap the upper and lower finishers at the two ends.

REAR WINDOW GLASS

Removal—Fixed Head Coupe

Prise away the chrome finisher strip from the outside of the rubber.

Extract one end of the rubber insert and withdraw completely.

Run a suitable thin bladed tool around the glass to break the seal between the rubber and the glass aperture flange.

Strike the glass with the flat of the hand from inside the car, starting in one corner and working towards the bottom.

Repeat this process around the complete glass.

Withdraw the glass.

Fig. 20. Removal of the rear glass (Fixed Head Coupe)

Refitting

Remove the old sealer from the glass flange.

The procedure for refitting and re-sealing of the rear glass is similar to the instructions given for fitting the windscreen (page 17).

Fit the chrome strip on top of the rubber and bend to suit contour if necessary. Coat the inside of the strip with Bostik 4251 and allow to become tacky.

Place the strip on the rubber and using tool (A, Fig. 16) lip the rubber over the finisher.

Removal—Detachable Hard Top

The rear light on the detachable hard top is made from a clear plastic material which will not break under ordinary circumstances. If however, the rear light becomes badly scratched it may be renewed by proceeding as for windscreen removal and refitting on page 17.

Care must be taken when removing the excess sealing compound that the rear light is not scratched. Always use a very soft cloth soaked in white spirit. DO NOT use thinners.

DOORS AND HINGES

Removal

Mark the position of the hinges on the door hinge pillar.

Remove the eight bolts securing the hinges to the pillar.

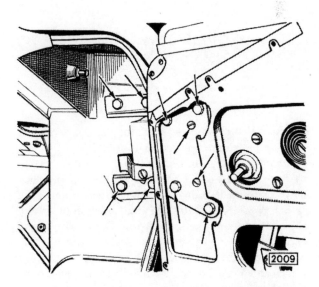

Fig. 21. Location of the screws securing the door hinges

Page N.19

BODY

To remove the hinges from the door remove the door trim casings (see "Door Trim Casings").

Remove the four setscrews and lock washers and the two drive screws attaching the hinges to the door panel.

Refitting

Refitting is the reverse of the removal procedure.

DOOR TRIM CASINGS

Removal

Remove the door handle by inserting a screwdriver between the handle and the spring cap and press the cap inwards (see Fig. 22). This will expose the retaining pin which should be tapped out. The handle, spring clip and escutcheon can now be removed.

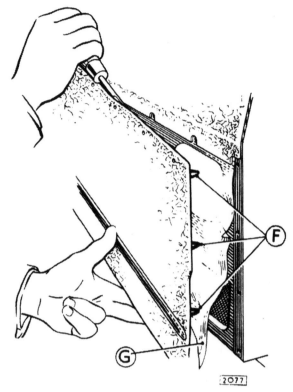

Fig. 23. Removing the door trim casing F—spring clips. G—plastic cover.

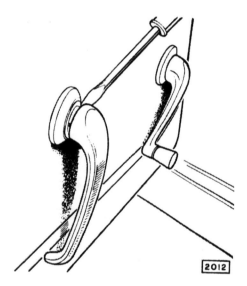

Fig. 22. Location of the interior door lock handle retaining pin.

Remove the window regulator handle which is secured in the same way as the door handle.

Remove the top chrome strip from the door casing by inserting a screwdriver under the strip at the door hinge end and levering strip away from its retaining spring clip. Repeat for the remaining four spring clips. Remove the chrome strip. Detach the spring clips by removing the five drive screws. Insert a thin bladed screwdriver between the casing and the door frame and prise off the casing which is secured by twenty-one clips.

Refitting

Refitting is the reverse of the removal procedure.

DOOR WINDOW GLASS AND FRAME

Removal—Door Window Glass

Remove the door trim casing as previously described.
Pull off the clear plastic sheet which is stuck to the door frame with upholstery solution.

Remove the six screws and washers retaining the closing strip to the top of the door frame. Wind the window down until the roller on the window regulator is accessible through the lower aperture in the door inner panel, and unscrew the regulator stop pin located in the channel, Fig. 24.

Raise the window until the regulator channel is above the door panel. Ease the regulator slide from the channel and withdraw the glass.

Removal—Door Window Frame

Remove the door window glass as described previously. Remove the three drive screws securing the

Page N.20

BODY

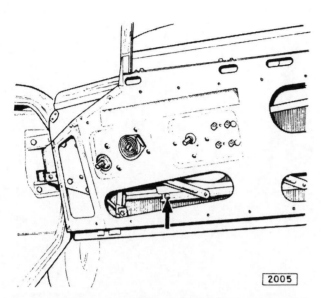

Fig. 24. Location of the window regulator stop pin.

frame to the top of the door panel. Note the location of the spacing shims fitted between the frame and the door panel.

Remove the two nuts and washers securing the glass frame to the two brackets located on the door lower panel.

Withdraw the frame.

Refitting

The refitting of the door window glass and frame is the reverse of the removal procedure.

WINDOW REGULATOR

Removal

Remove the door casing, glass and frame as described on page 20.

Remove the four nuts and lock washers securing the regulator to the door frame.

Remove the four screws and lock washers securing the window regulator spring to the door frame.

Lower the regulator mechanism within the door frame and withdraw through the aperture at the bottom of the door panel.

Refitting

Refitting is the reverse of the removal procedure.

DOOR WINDOW OUTER SEAL

Removal

Remove the door casing and glass as described on page 20.

Remove the five screws and outer seal retaining strip securing seal to the door panel. Detach outer seal.

Removal of the chromium door finisher—Open 2-Seater

The chromium finisher fitted to the top of the door panel can be removed after the removal of the outer seal by extracting the two screws, located in the front and rear faces of the finisher.

Refitting

Refitting is the reverse of the removal procedure.

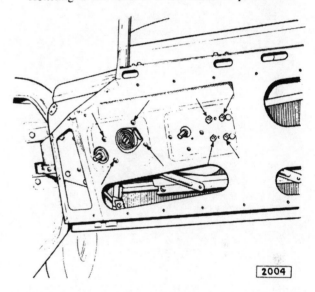

Fig. 25. Location of the nuts and screws securing the window regulator to the door panel.

SEATS AND RUNNERS

Removal

Remove the cushions from the front seats. Remove the four nuts and washers securing each seat pan to the runners and lift off the seat.

If it is required to remove the seat runners slide the runners rearwards and remove the two setscrews securing the front of the runners to the body floor.

Slide the runners forward and remove the two setscrews retaining the rear of the runners to the floor.

Refitting

Refitting is the reverse of the removal procedure.

Page N.21

BODY

NO DRAUGHT VENTILATOR (Fixed Head Coupe)

Removal

Remove the two screws securing the N.D.V. catch arm bracket to the body, accessible from inside the car.
Open the N.D.V.
Remove the five screws securing the N.D.V. light hinge to the frame post.

Refitting

Refitting is the reverse of the removal procedure.

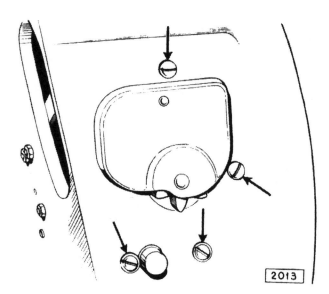

Fig. 27. Location of the screws retaining the door lock in position

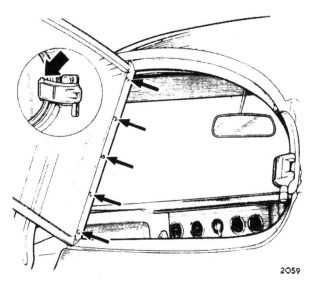

Fig. 26. The screws retaining the N.D.V. hinge. Inset shows the catch retaining pin

is accessible through the aperture in the inner door panel. The lock is detached from the door by removing the four screws (L). To take the lock out of the door it must be pressed inwards and downwards slightly and passed around the window channel which is immediately behind it.

Removing Outside Push Button Handle

This is retained by two nuts (M, Fig. 30).

REMOVAL AND REPLACEMENT OF DOOR LOCK MECHANISM

Remove the door trim casing as described on page 20.

Detaching Remote Control Connecting Link

The lock and remote control units are joined by a connecting link which can be detached to enable either unit to be moved independently.
The link is secured to the dowel (H, Fig. 30) on the lock lever by a circlip.

Removing the Lock Unit

First release the spring (I, Fig. 30) holding the bottom of the outside handle extendable link (J) to the dowel (K) on the lock intermediate lever. This

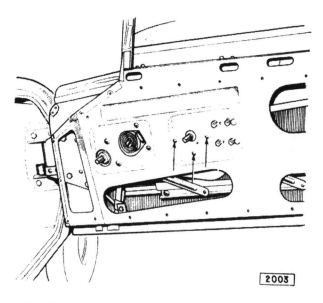

Fig. 28. Location of the remote control retaining screws

Page N.22

BODY

Removing the Remote Control Unit

Remove the three screws (N, Fig. 30) and take the remote control with its connecting link out through the aperture in the inner door panel.

Removing the Striker Unit

Do not disturb the three fixing screws (O) unless it is necessary to make adjustments or fit a new replacement.

Fitting the Lock Unit

The lock unit is inserted through the aperture in the inner door panel, passed around the window channel and lifted slightly until it projects through its cut-out in the door shut face. The four securing screws (L, Fig. 30) (with shakeproof washers) should then be fitted and tightened.

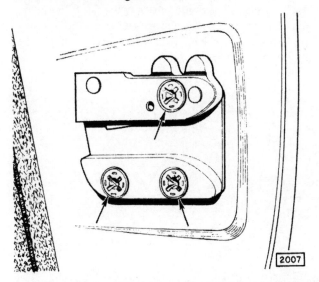

Fig. 29. Location of the door striker plate securing screws

Locating the Remote Control Unit

The remote control *must* be fitted in the locked position. For this reason it is supplied pinned in the locked position as shown at (P, Fig. 30).

Insert the remote control unit through the aperture in the inner door panel and position it so that its spindle and pin (P) project through their respective holes in the panel.

At this stage *loosely* fit the three securing screws (N) (with shakeproof washers).

Attaching the Remote Control Connecting Link

The connecting link is fitted to the dowel (H, Fig. 30) on the lock operating lever and retained by a circlip. A waved washer is interposed between the lever and link and a plain washer is fitted under the circlip.

Aligning the Remote Control Unit

Move the remote control *towards* the lock until the operating lever is in *contact* with the lock case as illustrated and tighten the three securing screws (N, Fig. 30).

Important: In certain cases it may be necessary to elongate the four holes in the inner door panel to achieve this condition.

Fitting the Outside Push Button Handle

The plunger housings on the outside handles are stamped LH or RH (left hand or right hand). The appropriate handle should be held in position on the door panel and the clearance between the push button plunger (Q, Fig. 30) and the lock contactor (R) checked through the aperture in the inner door panel. Do not check the clearance by depressing the push button as this may be deceptive. The clearance should be $\frac{1}{32}''$

However, before making any adjustments turn the operating lever (S) to the unlocked position so that depression of the push button moves the plunger through its housing. Only in this position, release the locknut (T), screw the plunger bolt (Q) in or out as required and re-tighten the locknut *before* releasing the push button.

Before finally fitting the handle attach the extendable link (J) to the operating lever (S) and retain with a circlip.

The operating lever should then be turned to the locked position, i.e. until the location holes in the operating lever and plunger housing are in line. To maintain the operating lever in this position insert a short length of $\frac{1}{8}''$ diameter rod (U) (suitably cranked for easy removal after assembly) through the locating holes illustrated. Manoeuvre the rod and the extendable link (J) through the handle aperture so that they hang down inside the door, then the handle fixing nuts (M) (with plain and shakeproof washers) can be fitted and tightened.

Connecting Push Button Mechanism to the Lock Nut

Ensure that the remote control cam is pinned (P, Fig. 30) in the locked position. It will be observed that one of the three holes in the bottom of the extendable link (J) can be aligned with the dowel (K) on the lock intermediate lever. The extendable link is simply pressed

BODY

on, being automatically retained by the spring (I). Finally withdraw the cranked rod (U) and the pin (P).

Fitting and Adjusting Strike Unit

Attach the striker loosely by means of its three screws (O, Fig 30) which pass through the door pillar into an adjustable tapping plate.

Positioning is carried out by a process of trial and error until the door can be closed easily but without rattling and no lifting or dropping of the door is apparent. Ensure that the securing screws are finally tightened.

Important: The strike must be retained in the horizontal plane relative to the door axis.

Master Check for Correct Alignment

Fit an inside handle *vertically downwards* on the remote control spindle. Turn the handle *forward* into the locked position. It will automatically return to the central position when released. Close the door while holding the push button in the *fully depressed* position. The door will remain locked although the push button may be freely depressed.

Insert the key in the slot in the push button and turn in the appropriate direction. Push button control will then be restored and the door can be opened.

After turning the key automatically return to the *vertical* position when it can be removed.

Important: The key must be removed from the locking device before closing a door in the locked position.

Lubrication

Before fitting the door casing ensure that any moving parts are adequately greased, using a protective grease such as "Astrolan".

After assembly introduce a few drops of thin machine oil into the oil hole (V, Fig. 30) provided on top of each lock case and into the private lock key slots. These items should be lubricated once a month.

Important: The private lock cylinders must not under any circumstances be lubricated with grease.

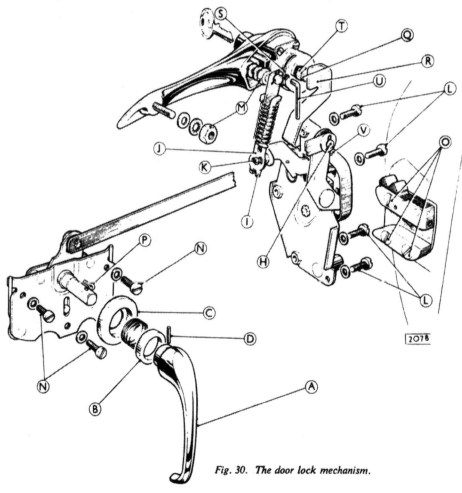

Fig. 30. The door lock mechanism.

Page N.24

BODY AND EXHAUST SYSTEM

ACCIDENTAL DAMAGE

The repair of the stressed steel body of monocoque construction together with the front and rear sub-frame assemblies varies in some degree depending on the extent of the damage to that of separate body and chassis construction or integral construction bodies.

Superficial damage can be affected in a similar manner to that employed on "all steel" bodies, which is familiar to all body repairers.

Repairs to rectify extensive damage affecting the front and rear sub-assemblies and also to the body must be carried out so that when the repair is completed the main mounting points for the engine and front and rear suspension units are in correct relationship to each other.

Important

It is most important, when accidental damage has been sustained at the front frame, that the appropriate sub-frame assembly should be replaced. NO ATTEMPT SHOULD BE MADE TO WELD OR BRAZE REPLACEMENT TUBES INTO THESE ASSEMBLIES NOR SHOULD HEAT IN ANY FORM BE APPLIED IN AN EFFORT TO STRAIGHTEN THEM.

CHECKING THE BODY AND FRONT FRAME FOR ALIGNMENT

Checking for distortion in the horizontal plane

The plan view of the body (see Fig. 1) provides the important dimensions for checking distortion in the body and front-frame. These dimensions can be measured actually on the underside of the body or by dropping perpendiculars from the points indicated by means of a plumb bob onto a clean level floor. If the latter method is used, the area below each point should be chalked over and the position at which the plumb bob touches the floor marked with a pencilled cross.

Checking for distortion in the vertical plane

For checking the body and sub-frames for distortion in the vertical plane, the side elevation gives the details of the important dimensions from a datum line

Page N.25

BODY AND EXHAUST SYSTEM

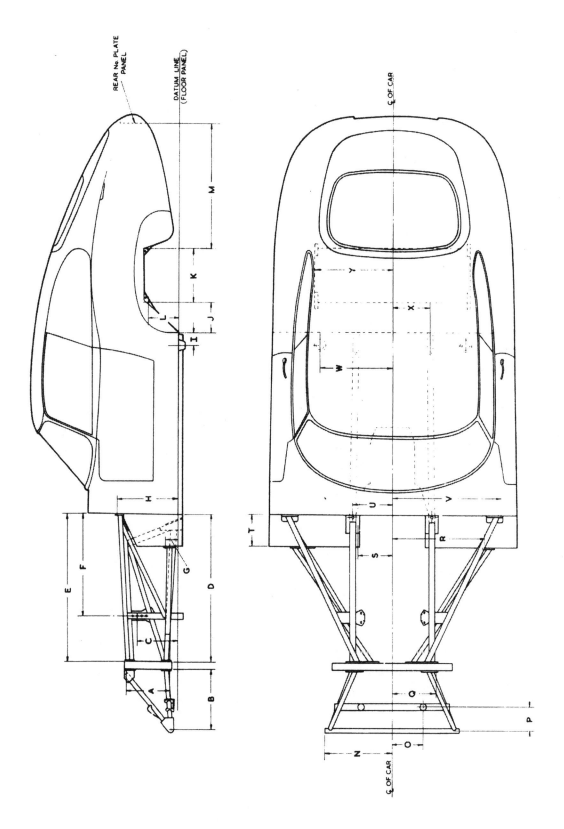

Fig. 1. *Body and front frame alignment diagram*

Page N.26

BODY AND EXHAUST SYSTEM

KEY TO ALIGNMENT DIAGRAM

Symbol	Measurement taken from	Dimension	
A	Upper hole of top mounting flange to lower hole of lower mounting flange on front sub frame.	10¾"	27·31 cm.
B	Centre line front tube to rear mounting flange front sub frame	14 5/32"	35·96 cm.
C	Second from top hole on engine mounting post to datum line	10⅛"	25·72 cm.
D	Lower rear face of front cross member to front bulkhead face	35 25/32"	90·88 cm.
E	Upper rear face of front cross member to front bulkhead face	35¾"	90·81 cm.
F	Centre line of holes on engine mounting post to front bulkhead face	25⅛"	63·82 cm.
G	Top hole in outer side member to datum line	3⅛"	7·94 cm.
H	Top hole in upper outrigger flange to datum line	15⅛"	38·42 cm.
I	Centre of radius arm mounting post to front face of rear wheel aperture	2 15/16"	7·46 cm.
J	Centre line of front lower mounting hole to front face of rear wheel aperture	9 23/32"	24·69 cm.
K	Centre line of front lower mounting hole to centre line of rear mounting hole	10 37/64"	26·87 cm.
L	Centre line of lower mounting hole to datum line	8 7/64"	20·60 cm.
M	Centre line of rear lower mounting hole to rear number plate panel	31 43/64"	80·45 cm.
N	Outer face of front sub frame front tube to centre line of car	16 15/16"	43·02 cm.
O	Centre line of radiator mounting hole to centre line of car	8¼"	20·95 cm.
P	Centre line of front frame front tube to centre line of radiator mounting hole	5¾"	14·61 cm.
Q	Top outer hole of front sub frame upper mounting flange to centre line of car	10 11/16"	27·15 cm.
R	Inner hole of outer side member mounting flange to centre line of car	22⅞"	58·10 cm.

Page N.27

BODY AND EXHAUST SYSTEM

KEY TO ALIGNMENT DIAGRAM (continued)

Symbol	Measurement taken from	Dimension	
S	Inner hole of top side member mounting flange to centre line of car	8⅝"	21·91 cm.
T	Rear mounting flange face of outer side member to front bulkhead face	8"	20·32 cm.
U	Centre line of hole in front of body underframe side members to centre line of car	9 25/32"	24·84 cm.
V	Outer hole of outrigger mounting bracket to centre line of car	26⅝"	67·63 cm.
W	Centre line of radius arm mounting post to centre line of car	18"	45·72 cm.
X	Centre line of hole in rear of body underframe side members to centre line of car	9⅛"	23·18 cm.
Y	Outside face of rear suspension mounting points to centre line of car	18 9/16"	47·15 cm.

Page N.28

BODY AND EXHAUST SYSTEM

REPLACEMENT BODY PANELS

Where the existing panels or members are badly damaged and it is not possible to effect a satisfactory repair in position, the affected panels will have to be cut out and replacement panels welded in their place.

It will frequently be found advantageous to use only a part of a given panel so that the welded joint can be made in a more accessible position. Great care must, of course, be taken when cutting the mating portions of the panel to ensure that perfect matching is obtained.

Any unused portions of replacement panels should be retained as it will often be found that they can be used for some future repair job.

Where a replacement panel to be fitted forms part of an aperture such as for a door or the luggage boot lid, an undamaged door or lid should be temporarily hinged in position and used as a template to assist location while a replacement panel is clamped and welded in position.

Before any dismantling takes place after accidental damage a check of the body alignment should be carried out.

THE FRONT FRAME

The front frame assembly is fabricated from square section steel tubing and is bolted to the main body structure.

To facilitate repair and reduce the cost of replacement in the event of damage, the frame is a built up unit consisting of two triangular side members and a deep front cross member.

Disconnect the torsion bars before removing the sub frame from the body. (See Section J "Front Suspension"). The plan and side elevation views given on page 4 provide all the important dimensions necessary for checking both the side-members and the front cross-member.

WELDING METHODS

The following are the principal methods of welding used in the assembly of the body and underframe panels. The instructions given below for breaking the different types of welds should be adhered to when removing a damaged panel as this will facilitate the assembly of the new panel.

Spot Welding

This type of welding is used for the joining of two or more overlapping panels and consists of passing electric current of high amperage through the panels by means of two copper electrodes.

This results in complete fusion of the metal between the electrodes forming a "spot" weld which is frequently repeated along the length of the panels to be joined. Spot welds can easily be recognised by slight indentation of the metal.

Lap joints on the outer body panels which are spot welded together are usually lead filled and in this case it will be necessary to direct the flame of an oxy-acetylene torch on to the lead so that the filling can be melted and wiped off by means of a piece of cloth.

Breaking Spot Welds

Spot welds cannot be broken satisfactorily other than by drilling; any attempt to separate the panels by using a chisel will result in the tearing of the metal in the vicinity of the spot welds.

Use a $\frac{3}{16}''$ (4·7 mm.) diameter drill and carefully drill out each weld. There is no necessity to drill completely through both panels; if the "spot" is drilled out of one of the panels the weld can be completely broken by inserting a sharp thin chisel between the two panels and tapping lightly with a hammer.

Where possible, drill the spot welds completely out of the panel that is to be left in position on the body. This will allow the new panel to be joined to the mating panel on the body by gas welding through the holes in the overlapping flange. (This does not apply if spot welding equipment is available).

If this is not possible, and the holes have to be drilled out in the damaged panel, new holes can be drilled in the replacement panel and the same type of weld effected.

Gas Welding

This type of welding is carried out by means of oxy-acetylene equipment and is used for the joining of overlapping panels or the butt welding of the edges of two panels.

Breaking Gas Welds

Gas welds may be broken either by means of a sharp chisel or by cutting through with a hacksaw; welding can be removed by grinding with a pointed emery wheel.

Page N.29

BODY AND EXHAUST SYSTEM

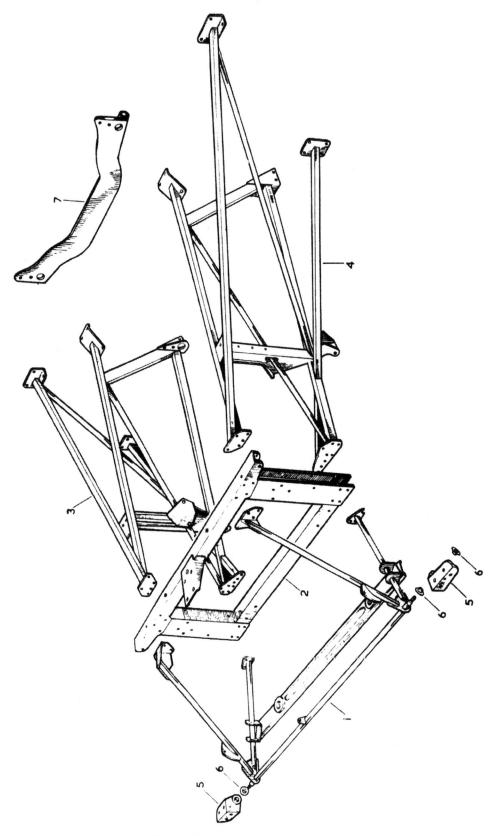

Fig. 2. Exploded view of the front frame assembly

BODY AND EXHAUST SYSTEM

1. Front sub frame assembly
2. Front cross member assembly
3. Right-hand side member assembly
4. Left-hand side member assembly
5. Bonnet hinge bracket
6. Nylon bush
7. Torsion bar anchor bracket reaction plate

BODY AND EXHAUST SYSTEM

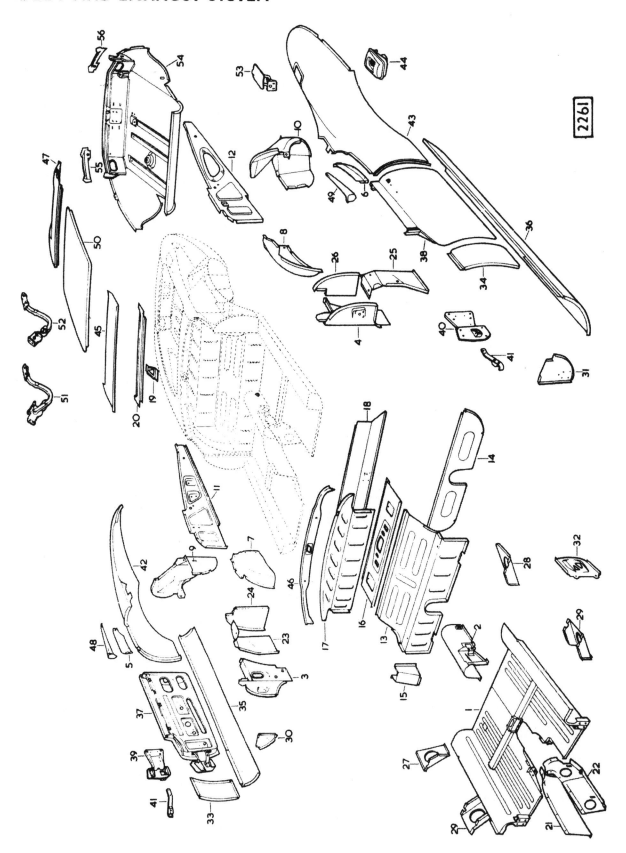

Fig. 3. Body panels (Open 2 seater)

Page N.32

BODY AND EXHAUST SYSTEM

1. Floor assembly
2. Tunnel assembly
3. Shut pillar (right-hand side)
4. Shut pillar (left-hand side)
5. Support panel (right-hand rear quarter)
6. Support panel (left-hand rear quarter)
7. Wheel arch panel (right-hand forward)
8. Wheel arch panel (left-hand forward)
9. Wheel arch panel (right-hand rear)
10. Wheel arch panel (left-hand rear)
11. Valance (behind right-hand wheel arch)
12. Valance (behind left-hand wheel arch)
13. Floor panel (rear)
14. Cross member (rear floor)
15. Stiffener bracket (sides of rear cross member)
16. Top panel (above rear floor)
17. Rear bulkhead panel assembly
18. Panel assembly (front of spare wheel compartment)
19. Shield (interior light)
20. Panel (reinforcing tonneau)
21. Gearbox panel (right-hand)
22. Gearbox panel (left hand)
23. Reinforcement panel (right-hand shut pillar)
24. Closing panel (right-hand shut pillar)
25. Reinforcement panel (left-hand shut pillar)
26. Closing panel (left-hand shut pillar)
27. Reinforcement panel (right-hand sill, rear)
28. Reinforcement panel (left-hand sill, rear)
29. Reinforcement panel (left and right-hand sill, front)
30. Closing panel (right-hand sill, front)
31. Closing panel (left-hand sill, front)
32. Reinforcement panel (left-hand dash)
33. Exterior panel (right-hand dash)
34. Exterior panel (left-hand dash)
35. Sill outer panel (right-hand)
36. Sill outer panel (left-hand)
37. Door shell (right-hand)
38. Door shell (left-hand)
39. Hinge (right-hand)
40. Hinge (left-hand)
41. Check arm (both doors)
42. Rear wing panel (right-hand)
43. Rear wing panel (left-hand)
44. Petrol filler box
45. Tonneau top panel
46. Support panel (tonneau top panel)
47. Tonneau rear panel
48. Top quarter panel (right-hand)
49. Top quarter panel (left-hand)
50. Boot lid shell
51. Boot lid hinge (right-hand)
52. Boot lid hinge (left-hand)
53. Petrol filler box lid
54. Lower rear panel
55. Filler panel (right-hand stop/tail lamp)
56. Filler panel (left-hand stop/tail lamp)

Page N.33

BODY AND EXHAUST SYSTEM

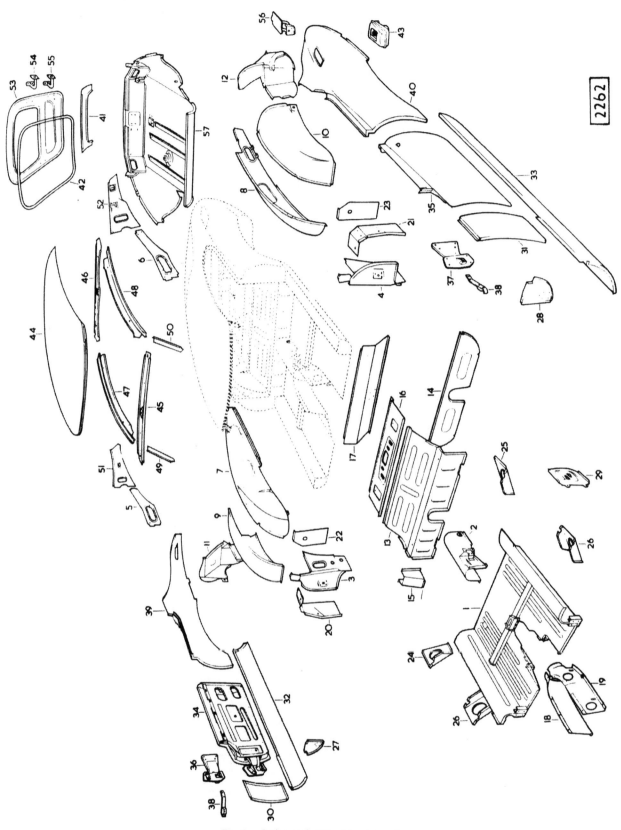

Fig. 4. Body panels (Fixed head coupe)

Page N.34

BODY AND EXHAUST SYSTEM

1. Floor assembly
2. Tunnel assembly
3. Shut pillar (right-hand side)
4. Shut pillar (left-hand side)
5. Support panel (right-hand)
6. Support panel (left-hand)
7. Wheel arch (right-hand inner)
8. Wheel arch (left-hand inner)
9. Wheel arch (right-hand outer)
10. Wheel arch (left-hand outer)
11. Wheel arch (right-hand rear)
12. Wheel arch (left-hand rear)
13. Floor panel (rear)
14. Cross member (rear floor)
15. Stiffener bracket
16. Top panel assembly
17. Rear bulkhead panel assembly
18. Gearbox panel (right-hand)
19. Gearbox panel (left-hand)
20. Reinforcement panel (right-hand shut pillar)
21. Reinforcement panel (left-hand shut pillar)
22. Closing panel (right-hand shut pillar)
23. Closing panel (left-hand shut pillar)
24. Panel (right-hand sill, rear)
25. Panel (left-hand sill, rear)
26. Panel (left and right-hand sill, front)
27. Closing panel (right-hand sill, front)
28. Closing panel (left-hand sill, front)
29. Panel (left-hand dash)
30. Exterior panel (right-hand dash side)
31. Exterior panel (left-hand dash side)
32. Sill outer panel (right-hand)
33. Sill outer panel (left-hand)
34. Door shell (right-hand)
35. Door shell (left-hand)
36. Hinge assembly (right-hand door)
37. Hinge assembly (left-hand door)
38. Check arm (left and right-hand doors)
39. Rear wing panel (right-hand)
40. Rear wing panel (left-hand)
41. Tail panel
42. Gutter (boot lid aperture)
43. Petrol filler box
44. Roof panel
45. Windscreen header panel
46. Reinforcement rail (rear)
47. Cantrail panel (right-hand)
48. Cantrail panel (left-hand)
49. Bead extension (right-hand cantrail)
50. Bead extension (left-hand cantrail)
51. Support panel (right-hand)
52. Support panel (left-hand)
53. Boot lid shell
54. Upper hinge (boot lid)
55. Lower hinge (boot lid)
56. Petrol filler box lid
57. Lower panel (rear)

Page N.35

BODY AND EXHAUST SYSTEM

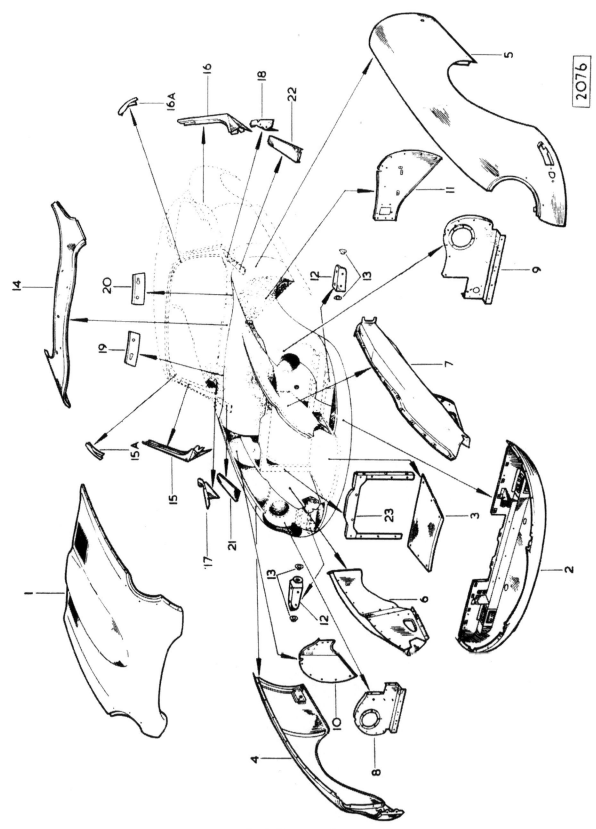

Fig. 5. Bonnet panels

BODY AND EXHAUST SYSTEM

1. Centre section
2. Under panel
3. Lower air duct
4. Side panel—right hand
5. Side panel—left hand
6. Valance—right hand
7. Valance—left-hand
8. Front diaphragm—right-hand
9. Front diaphragm—left-hand
10. Rear diaphragm—right-hand
11. Rear diaphragm—left-hand
12. Bonnet hinge
13. Nylon Bush
14. Scuttle top panel
15. Windscreen pillar—right-hand
16. Windscreen pillar—left-hand
15a. Reinforcement channel
16a. Reinforcement channel
17. Filler panel
18. Filler panel
19. Corner panel
20. Corner panel
21. Closing panel
22. Closing panel
23. Stoneguard mounting frame

Page N.37

EXHAUST SYSTEM

EXHAUST SYSTEM

Removal

Release the two clips securing the muffler boxes to the rear of the silencer assembly.

Remove the bolt, nut and washer securing the mufflers.

Remove the bolt, nut and washer securing the mufflers to the rubber mounting bracket. Withdraw the mufflers.

Release the two clips securing the silencers to the two down pipes.

Remove the four bolts, nuts and washers securing the silencer assembly to the rubber mounting brackets attached to the body.

Lower the silencer assembly and withdraw from the down pipes.

Remove the four nuts and washers securing each down pipe to the exhaust manifolds and withdraw the down pipes.

Collect the sealing rings which are between the exhaust manifolds and the down pipe.

To remove the rubber mounting brackets securing the silencers, remove the nuts and washer securing the bracket to the body. To remove the muffler mounting bracket unscrew the two bolts and nuts securing the bracket to the body attachment.

Collect the rings which are between the exhaust manifolds and the down pipes.

Refitting

Renew the rings when refitting the exhaust down pipes to the manifold.

Refitting is the reverse of the removal procedure.

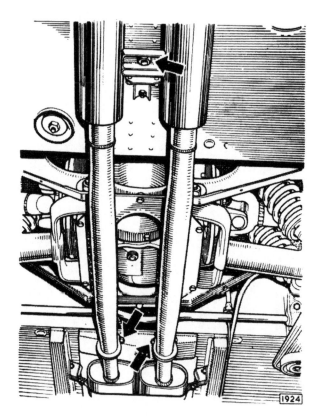

Fig. 31. The attachment points for the exhaust tail pipes.

EXHAUST SYSTEM

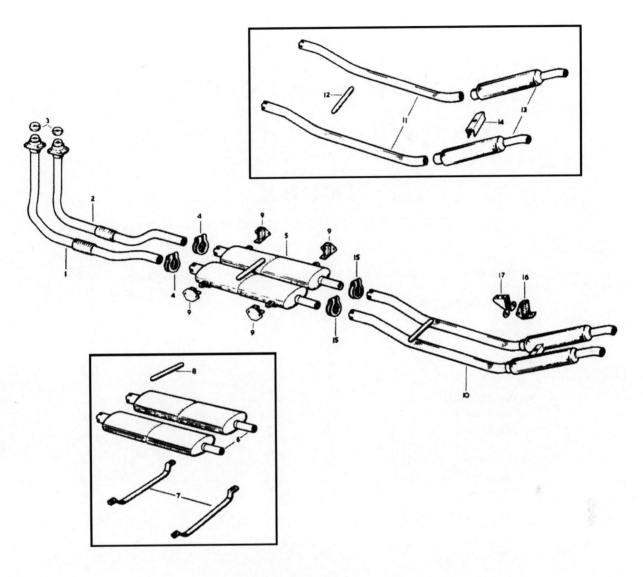

Fig. 32. Exploded view of the exhaust system

1. Front down pipe assembly
2. Rear down pipe assembly
3. Gasket
4. Clip
5. Twin silencer assembly
6. Silencer
7. Mounting strap
8. Stiffener
9. Rubber mounting
10. Tail pipe assembly
11. Tail pipe
12. Strap
13. Muffler box
14. Mounting bracket
15. Clip
16. Rubber mounting
17. Bracket

Page N.40

SECTION O
HEATING & WINDSCREEN WASHING EQUIPMENT

INDEX

CAR HEATING AND VENTILATING SYSTEM

	Page
Air Control	O.3
Temperature Control	O.3
Air Distribution	O.3
Fan Switch	O.3
Heater Unit	
Removal	O.6
Refitting	O.6
Heater Water Control Tap	
Removal	O.6
Refitting	O.6
Fan Motor	
Removal	O.7
Refitting	O.7

WINDSCREEN WASHING EQUIPMENT

	Page
Operation	O.7
Warning	O.7
Filling-up	O.7
Cold Weather	O.8
Adjusting the Jets	O.8
Jet Nozzles	
Cleaning	O.8
Lubrication	O.8

HEATING AND WINDSCREEN WASHING EQUIPMENT

CAR HEATING AND VENTILATING SYSTEM

The car heating and ventilating equipment consists of a heating element and an electrically driven fan mounted on the engine side of the bulkhead. Air from the heater unit is conducted:

(a) To a built-in duct fitted with two doors situated behind the instrument panel.

(b) To vents at the bottom of the windscreen to provide demisting and defrosting.

The amount of fresh air can be controlled at the will of the driver and is introduced into the system by operating the "Air" control lever and switching on the fan.

AIR CONTROL

The air control (marked "OFF-AIR-ON") controls the amount of fresh air passing through the heater element; when this control is placed in the "OFF" position the supply of air is completely cut off.

Placed in the "ON" position the maximum amount of air passes through the heater element. By placing the control in intermediate positions varying amounts of air may be obtained.

TEMPERATURE CONTROL

The temperature control (marked "HOT-COLD") situated on the facia panel operates a valve which controls the amount of hot water passing through the heater element; when this control is placed in the "COLD" position the supply of hot water to the element is completely cut off so that the cold air only can be admitted for ventilating the car in hot weather.

Placed in the "HOT" position the maximum amount of hot water passes through the heater element. By placing the control in intermediate positions varying degrees of heat may be obtained.

AIR DISTRIBUTION

The proportion of air directed to the windscreen or the interior of the car can be controlled by the position of the two doors situated under the duct behind the instrument panel.

With the heater doors fully closed the maximum amount of air will be directed to the windscreen for rapid demisting or defrosting.

With the heater doors fully open, air will be directed into the car interior and to a lesser degree to the windscreen.

FAN SWITCH

The heater fan for the car heating and ventilating system considerably increases the flow of air through the system and is controlled by a three-position switch (marked "Fan").

Lift the switch to the second position for slow speed and to the third position for maximum speed, whichever is required.

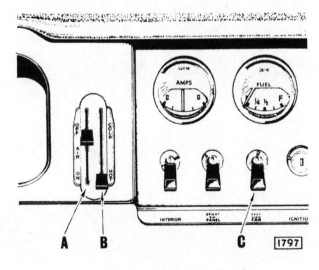

Fig. 1. Heating and ventilating controls.
   A Air control lever.
   B Temperature control lever.
   C Fan switch.

Page O.3

HEATING AND WINDSCREEN WASHING EQUIPMENT

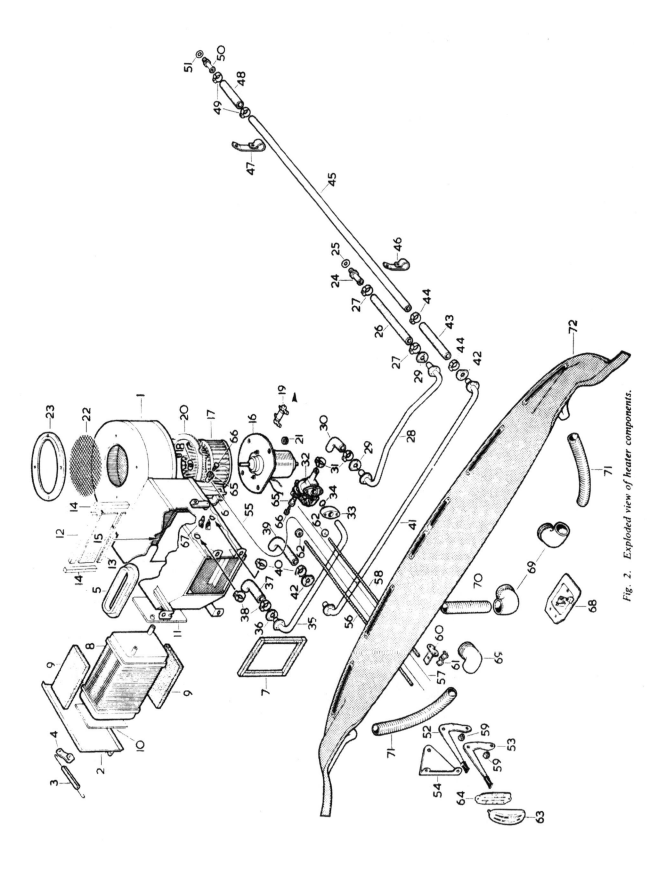

Fig. 2. Exploded view of heater components.

Page O.4

WINDSCREEN WASHING AND HEATING EQUIPMENT

1. Heater case.
2. Side panel.
3. Spring.
4. Flap lever.
5. Air release duct.
6. Mounting bracket.
7. Seal between heater case and dash.
8. Water radiator for heater.
9. Felt seal.
10. Seal.
11. Seal.
12. Seal on air control flap.
13. Seal on heater case.
14. Seal on air control flap.
15. Air release duct seal.
16. Fan motor.
17. Fan.
18. Spire nut.
19. Electrical resistance.
20. Sealing ring.
21. Grommet.
22. Wire mesh.
23. Wire mesh securing ring.
24. Manifold heater pipe adaptor.
25. Copper washer.
26. Water hose.
27. Hose clip.
28. Water feed pipe.
29. Feed pipe flange.
30. Water hose elbow.
31. Hose clip.
32. Water control tap.
33. Control tap mounting block.
34. Sealing ring.
35. Feed pipe from water control tap.
36. Feed pipe securing flange.
37. Water hose elbow.
38. Hose clip.
39. Water hose elbow.
40. Hose clip.
41. Water return pipe.
42. Securing flange.
43. Water hose.
44. Hose clip.
45. Water return pipe.
46. Return pipe mounting clip.
47. Return pipe mounting clip.
48. Water hose.
49. Hose clips.
50. Water pump adaptor.
51. Copper washer.
52. Air flap control lever.
53. Water control tap lever.
54. Control lever support bracket.
55. Air flap control cable.
56. Conduit for air flap control cable.
57. Water tap control cable.
58. Conduit for water tap control cable.
59. Control cable retaining clip.
60. Cable abutment clamp bracket.
61. Abutment clamp.
62. Grommet.
63. Control lever escutcheon.
64. Plate.
65. Inner control cable trunnions.
66. Setscrew.
67. Abutment clamp.
68. Heater doors.
69. Rubber elbow.
70. Demister hose.
71. Demister hose.
72. Screen rail.

Page O.5

WINDSCREEN WASHING AND HEATING EQUIPMENT

Operation of the fan is required mainly when the car is stationary or running at a slow speed. At higher road speeds it will be found possible to dispense with the fan as air will be forced through the system due to the passage of the car through the air.

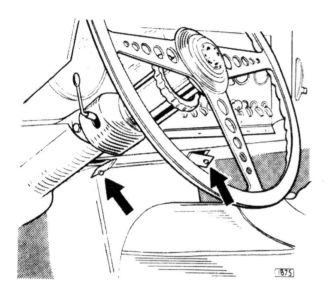

Fig. 3. Heater duct doors.

COLD WEATHER

To obtain fresh air heating, demisting and defrosting:
(a) Set fresh air control to DESIRED POSITION.
(b) Set temperature control to DESIRED POSITION.
(c) Switch ON fan at required speed.
(d) OPEN heater doors.

To obtain rapid demisting and defrosting:
(a) Turn fresh air control to FULLY ON.
(b) Set temperature control to HOT.
(c) Switch ON fan—fast position.
(d) CLOSE heater doors.

HOT WEATHER

To obtain ventilation and demisting:
(a) Set fresh air control to DESIRED POSITION.
(b) Set temperature control to COLD.
(c) Switch ON fan at required speed.
(d) OPEN heater doors.

To obtain rapid demisting:
(a) Set fresh air control to FULLY ON.
(b) Set temperature to COLD.
(c) Switch ON fan—fast position.
(d) CLOSE heater doors.

Page O.6

WARNING

There is the possibility that fumes may be drawn into the car from the atmosphere when travelling in dense traffic and in such conditions it is advisable to close the heater air control and switch off the fan.

HEATER

Removal

Raise the bonnet and drain the radiator and cylinder block.

Disconnect the positive battery terminal.

Slacken the two jubilee clips securing the heater hoses to the heater body.

Slacken the pinch bolt securing the heater air control flap cable to the lever.

Slacken the pinch bolt securing the conduit casing to the heater body and remove the cable.

Disconnect the three electrical wires for the fan at the snap connector.

Remove the four bolts, plain and serrated washers securing the heater body to the scuttle.

Remove the two screws securing the heater body bracket to the sub frame.

Remove the heater.

Refitting

Refitting is the reverse of the removal procedure. Ensure that the rubber seal attached to the heater body outlet to car aperture is in the correct position. Move the heater flap operating lever on the side facia panel into the "ON" position. Move the heater flap lever into the fully forward position (A Fig. 4) and pass the control cable through the two securing points. Tighten the two pinch bolts securing the control cable and conduit casing.

HEATER WATER CONTROL TAP

Removal

Slacken the pinch bolt securing the water tap control lever to the cable.

Slacken the pinch bolt securing the cable conduit casing to the water tap.

Slacken the jubilee clip securing the rubber hose to the water tap.

Remove the two bolts and spring washers which secure the water tap, rubber sealing washer and distance piece to the scuttle.

Withdraw the heater tap from the heater pipe.

Refitting

Refitting is the reverse of the removal procedure. Ensure that the sealing rubber is fitted into the machined

HEATING AND WINDSCREEN WASHING EQUIPMENT

faces of the water control tap and distance piece. Move the water control tap operating lever into the "HOT" open position. Move the lever on the water control tap fully forward into the "HOT" position (B Fig. 4) and pass the control cable through the securing points. Tighten the two pinch bolts securing the control cable and conduit casing.

FAN MOTOR

Removal

Remove the heater as described on page O.6.

Remove the three bolts, serrated washers and rubber seals securing the fan to the heater.

Withdraw the fan motor and fan from the heater body.

Withdraw the small spring clip securing the fan to fan motor spindle.

Refitting

Refitting is the reverse of the removal procedure. Ensure the small fan retaining spring clip is replaced.

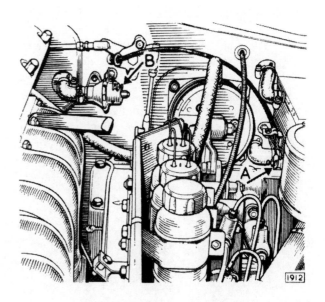

Fig. 4   Setting the heater operating levers.

WINDSCREEN WASHING EQUIPMENT

The windscreen washer is electrically operated and comprises a glass water container mounted in the engine compartment which is connected to jets at the base of the windscreen. Water is delivered to the jets by an electrically driven pump incorporated in the water container.

OPERATION

The windscreen washer should be used in conjunction with the windscreen wipers to remove foreign matter that settles on the windscreen.

Lift the switch lever (marked "Washer") and release immediately when the washer should operate at once and continue to function for approximately seven seconds. Allow a lapse of time before operating the switch for a second time.

WARNING

If the washer does not function immediately check that there is water in the container.

The motor will be damaged if the switch is held closed for more than one or two seconds if the water in the container is frozen.

The washer should not be used under freezing conditions as the fine jets of water spread over the windscreen by the blades will tend to freeze up.

In the summer the washer should be used freely to remove insects before they dry and harden on the screen.

FILLING-UP

The water should be absolutely CLEAN. If possible use SOFT water for filling the container, but if this is not obtainable and hard water has to be used, frequent operation and occasional attention to the nozzle outlet holes will be amply repaid in preventing the formation of unwelcome deposits.

The correct water level is up to the bottom of the container neck. Do not overfill, or unnecessary splashing may result. Always replace the rubber filler cover correctly after filling, pressing it fully home.

It is not possible to empty the container completely with the pump. **Refilling is necessary when the water level has fallen so that the top of the auxiliary reservoir**

Page O.7

HEATING AND WINDSCREEN WASHING EQUIPMENT

is uncovered. About 30 full operations will be obtained from one filling.

When using the washer, an indication of the need to refill the container is given by the behaviour of the unit. The time taken for the auxiliary reservoir to refill increases as the water level in the container falls.

As soon as the water level has fallen to the top of the auxiliary reservoir, the amount of water delivered to the windscreen will decrease with successive operations and the time the unit runs will, in proportion, become less.

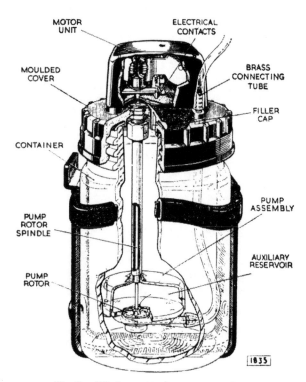

Fig. 5. Windscreen washer water container.

If the water level is allowed to fall still further, until it is down to the bottom of the auxiliary reservoir, the automatic action will cease and water will be delivered to the windscreen only as long as the switch is operated. This will continue until the water level has fallen to the inlet orifices, when the pump will be above the water level and no water will be available for delivery to the windscreen.

Do not continue to operate the switch after the available water has been used up, otherwise damage may be caused to the unit.

Refilling the container will restore normal operation of the unit.

Page O.8

COLD WEATHER

To avoid damage by frost, add denatured alcohol (methylated spirits) as follows:

The underside of the rubber filler cover will be found to form a measure. Two measures of denatured alcohol should be added per container of water. USE NO OTHER ADDITIVES WHATSOEVER.

ADJUSTING THE JETS

With a screwdrive turn each nozzle in the jet holder until the jets of water strike the windscreen in the area swept by the wiper blades. It may be necessary to adjust the nozzles slightly after a trial on the road due to the jets of water being deflected by the airstream.

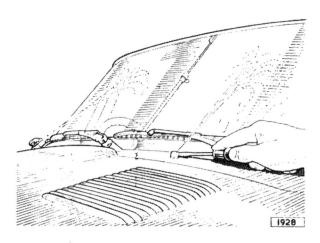

Fig. 6. Adjusting the windscreen washer jets.

JET NOZZLES

Cleaning

To clear a blocked jet nozzle completely unscrew the nozzle from the jet holder. Clear the small orifices with a thin piece of wire or blow out with compressed air; operate the washer with the nozzle removed. Allow the water to flush through the jet holder and then replace the nozzle.

LUBRICATION

If, after lengthy service, the motor is found to be running slowly, unscrew the moulded cover from the container and apply one or two drops only of thin machine oil to the felt pad situated in the gap between the cover and the motor unit. Do not over-lubricate or excess oil may find its way into the water container when the cover is refitted, with consequent smearing of the windscreen.

SECTION P

ELECTRICAL AND INSTRUMENTS

3·8 "E" TYPE
GRAND TOURING MODELS

INDEX

Battery

	Page
Data	P.6
Routine Maintenance	P.6
Removal	P.6
Refitting	P.6
Persistent Low State of Charge	P.6
Recharging from an External Supply	P.8
Preparing New Unfilled, Uncharged Batteries For Service	P.8
Preparing New "Dry Charged" Batteries For Service	P.9

Distributor

Removal	P.10
Refitting	
Ignition Timing	P.10
Routine Maintenance	P.10
Data	P.11
Servicing	
Dismantling	P.12
Bearing replacement	P.12
Reassembly	P.12
Distributor Test Data	P.13

Flasher Units

Information	P.14

Fuse Unit

Information	P.15

Generator Model C42
(Fitted to later "E" Type models)

Removal	P.16
Refitting	P.16
General	P.16
Routine Maintenance	P.16
Performance Data	P.17
Servicing	P.17

Generator Model C45 PVS-6
(Fitted to early "E" Type models)

Removal	P.21
Refitting	P.21
Brushgear Inspection	P.21
Commutator End Bearing	P.21

Page P.2

INDEX *(continued)*

Horns

	Page
Adjustment	P.22

Lamps

Light Bulbs	P.23
Headlamps	
Bulb replacement	P.24
Headlamp setting	P.24
Sidelamp Bulb—Replacement	P.25
Front Flasher Bulb—Replacement	P.25
Rear Flasher Bulb—Replacement	P.26
Rear/Brake Bulb—Replacement	P.26
Number Plate Lamp Bulb—Replacement	P.26
Interior—Luggage Lamp Bulb—Replacement	P.26
Reverse Lamp Bulb—Replacement	P.26

RB 310 Current and Voltage Regulator

Checking Continuity Between Battery and Control Box	P.27
Voltage Regulator Adjustment	P.27
Current Regulator Adjustment	P.28
Cleaning Regulator Contacts	P.28
Cut-out Adjustment	P.28
Cleaning Cut-out Contacts	P.29

RB 340 Current and Voltage Regulator

Checking Continuity Between Battery and Control Box	P.30
Voltage Regulator Adjustment	P.30
Current Regulator Adjustment	P.30
Cleaning Regulator Contacts	P.30
Cut-out Adjustment	P.30
Cleaning Cut-out Contacts	P.30

Starter Motor

Removal	P.31
Refitting	P.31
General	P.31
Routine Maintenance	P.31
Performance Data	P.32
Servicing	P.32

Starter Drive

General	P.35
Routine Maintenance	P.35
Dismantling and Reassembling	P.36

Page P.3

INDEX (continued)

Windscreen Wiper

	Page
Removal of Wiper Motor	P.36
Refitting	P.37
Removal of Windscreen Wiper Spindle Housings	P.37
Refitting	P.37
Data	P.38
Description	P.38
Maintenance	P.38
Fault Diagnosis	P.39
Testing	P.40

Miscellaneous

Electric Clock
- Removal .. P.41
- Adjustment .. P.41
- Refitting .. P.41

The Brake Fluid and Handbrake Warning Light
- Renewing the bulb .. P.42

Mixture Control Warning Light
- Renewing the bulb .. P.42
- Setting the carburetter mixture control warning light switch .. P.42

Flashing Indicator Control
- Removal .. P.42
- Refitting .. P.42

Flashing Indicator Warning Light Bulb
- Replacement .. P.42

The Wiring Diagram .. P.43

Instruments

Dash Casings
- Removal .. P.44
- Refitting .. P.44

The Instrument Panel
- Opening .. P.44
- Removal .. P.44
- Refitting .. P.44
- Closing .. P.44

The Glove Box
- Removal .. P.44
- Refitting .. P.44

INDEX (continued)

Instruments (continued)

	Page
The Side Facia Panel	
Removal	P.45
Refitting	P.45
The Speedometer and Odometer	
Removal	P.45
Refitting	P.45
The Revolution Counter and Clock	
Removal	P.46
Refitting	P.46
Testing	P.46
Revolution Counter Drive	
Removing	P.46
Refitting	P.46
The Removal of Instrument Panel Components	
The ignition switch	P.46
The cigar lighter	P.46
The starter push switch	P.47
The head and side light switch	P.47
The tumbler light switches	P.47
The ammeter and oil pressure gauges	P.47
The fuel and water temperature gauges	P.48
The voltage regulator	P.48
Renewing the switch indicator strip bulbs	P.48
The Engine Temperature, Fuel Tank Contents and Oil Pressure Gauges	
Description	P.49
Operation of the engine temperature gauge	P.49
Operation of the fuel tank contents gauge	P.49
Analysis of faults	P.50
Operation of the oil pressure gauge	P.51
Speedometer Cable	
Removal	P.53
Refitting	P.53
Speedometer Cable—General Instructions	P.53
Speedometer—General Instructions	P.54

Page P.5

ELECTRICAL AND INSTRUMENTS

BATTERY

The Lucas FRV11/7A battery is of the semi-linkless type, the short cell inter-connectors being partially exposed to enable testing of the individual cells to be carried out with a heavy discharge tester.

DATA

Battery type	FRV11/7A
Voltage	12
Number of plates per cell	11
Capacity at 10-hour rate	55 ampere hours
Capacity at 20-hour rate	60 ampere hours

ROUTINE MAINTENANCE

Wipe away any foreign matter or moisture from the top of the battery, and ensure that the connections and the fixings are clean and tight.

About once a month, or more frequently in hot weather, examine the level of the electrolyte in the cells. If necessary add distilled water to bring the electrolyte just level with the separator guards, which can be seen when the vent plugs are removed.

The use of a Lucas battery filler will be found helpful in this topping-up process as it ensures that the correct electrolyte level is obtained automatically and also prevents distilled water from being spilled over the battery top.

Distilled water should always be used for topping-up. In an emergency however, clean soft rain water collected in an earthenware container may be used.

Note: Never use a naked light when examining a battery, as the mixture of oxygen and hydrogen given off by the battery when on charge, and to a lesser extent when standing idle, can be dangerously explosive.

REMOVAL

Unscrew the two wing nuts retaining the battery strap; remove the fixing rods and strap. Disconnect terminals and lift out battery from cradle.

REFITTING

Refitting is the reverse of the removal procedure. Before refitting the cable connectors, clean the terminals and coat with petroleum jelly.

PERSISTENT LOW STATE OF CHARGE

First consider the conditions under which the battery is used. If the battery is subjected to long periods of discharge without suitable opportunities for recharging, a low state of charge can be expected. A fault in the generator or regulator, or neglect of the battery during a period of low or zero mileage may also be responsible for the trouble.

Vent Plugs

See that the ventilating holes in each vent plug are clear.

ELECTRICAL AND INSTRUMENTS

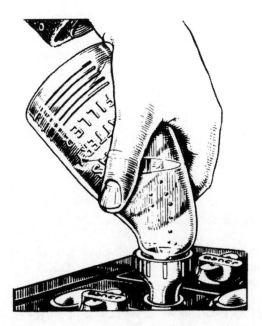

Fig. 1. Lucas battery filler.

The readings given by each cell should be approximately the same. If one cell differs appreciably from the others, an internal fault in the cell is indicated.

The appearance of the electrolyte drawn into the hydrometer when taking a reading gives a useful indication of the state of the plates. If the electrolyte is very dirty, or contains small particles in suspension, it is possible that the plates are in a bad condition.

The specific gravity of the electrolyte varies with the temperature, therefore, for convenience in comparing specific gravities, this is always corrected to 60°F., which is adopted as a reference temperature. The method of correction is as follows:—

For every 5°F. below 60°F. deduct ·002 from the observed reading to obtain the true specific gravity at 60°F.

For every 5°F. above 60°F. add ·002 to the observed reading to obtain the true specific gravity at 60°F.

The temperature must be that indicated by a thermometer actually immersed in the electrolyte, and not the air temperature.

Compare the specific gravity of the electrolyte with the values given in the table and so ascertain the state of charge of the battery.

If the battery is in a discharged state, it should be recharged, either on the vehicle by a period of daytime running or on the bench from an external supply, as described under "Recharging From An External Supply".

Level of Electrolyte

The surface of the electrolyte should be just level with the tops of the separators guards. If necessary, top up with distilled water. Any loss of acid from spilling or spraying (as opposed to the normal loss of water by evaporation) should be made good by dilute acid of the same specific gravity as that already in the cell.

Cleanliness

See that the top of the battery is free from dirt or moisture which might provide a discharge path. Ensure that the battery connections are clean and tight.

Hydrometer Tests

Measure the specific gravity of the acid in each cell in turn with a hydrometer. To avoid misleading readings, do not take hydrometer readings immediately after topping-up.

Discharge Test

A heavy discharge tester consists of a voltmeter, 2 or 3 volts full scale, across which is connected a shunt resistance capable of carrying a current of several hundred amperes. Pointed prongs are provided for making contact with the inter-cell connectors.

Press the contact prongs against the exposed positive and negative terminals of each cell. A good cell will maintain a reading of 1·2—1·5 volts, depending on the state of charge, for at least 6 seconds. If, however, the reading rapidly falls off, the cell is probably faulty and a new plate assembly may have to be fitted.

Page P.7

ELECTRICAL AND INSTRUMENTS

State	Home and climates with shade temperature ordinarily below 80°F (26·6°C). Specific gravity of electrolyte (corrected to 60°F)	Climates with shade temperature frequently over 80°F (26·6°C). Specific gravity of electrolyte (corrected to 60°F)
Fully charged	1·270—1·290	1·210—1·230
About half discharged	1·190—1·210	1·130—1·150
Completely discharged	1·110—1·130	1·050—1·070

RECHARGING FROM AN EXTERNAL SUPPLY

If the battery tests indicate that the battery is merely discharged, and is otherwise in a good condition, it should be recharged, either on the vehicle by a period of daytime running or on the bench from an external supply.

If the latter, the battery should be charged at 5·5 amperes until the specific gravity and voltage show no increase over three successive hourly readings. During the charge the electrolyte must be kept level with the tops of the separator guards by the addition of distilled water.

A battery that shows a general falling-off in efficiency common to all cells, will often respond to the process known as "cycling". This process consists of fully charging the battery as described above and then discharging it by connecting to a lamp board, or other load, taking a current of 5 amperes. The battery should be capable of providing this current for at least 7 hours before it is fully discharged, as indicated by the voltage of each cell falling to 1·8. If the battery discharges in a shorter time, repeat the "cycle" of charge and discharge.

PREPARING NEW UNFILLED, UNCHARGED BATTERIES (MODEL FRV11/7A) FOR SERVICE

Preparation of Electrolyte

Batteries should not be filled with acid until required for initial charging.

Electrolyte of the specific gravity required is prepared by mixing distilled water and concentrated sulphuric acid, usually of 1·835 specific gravity. The mixing must be carried out either in a lead-lined tank or in a suitable glass or earthenware vessel. Slowly add the acid to the water, stirring with a glass rod. **Never add the water to the acid,** as the resulting chemical reaction causes violent and dangerous spurting of the concentrated acid. The correct specific gravity for the filling acid and approximate proportions of acid and water are indicated in the following table;

Specific Gravity of Filling Acid (corrected to 60°F.)	
Home and Climates with shade temperature ordinarily below 80°F (26·6°C)	Climates with shade temperatures frequently above 80°F (26·6°C)
1·270	1·210
Add 1 part by volume of acid (1·835 S.G.) to 2·8 parts of distilled water to mix this electrolyte	Add 1 part by volume of acid (1·835 S.G.) to 4 parts of distilled water to mix this electrolyte
Quantity of electrolyte required per cell 1¼ pints approximately (720 cc.)	

Page P.8

ELECTRICAL AND INSTRUMENTS

Heat is produced by the mixture of acid and water, and the electrolyte should be allowed to cool before taking hydrometer readings—unless a thermometer is used to measure the actual temperature, and a correction applied to the reading before pouring the electrolyte into the battery.

Filling the Battery

The temperature of the acid, battery and filling-in must not be below 32°F.

Carefully break the seals in the filling holes and fill each cell to the level of the separator guard with electrolyte of the appropriate specific gravity. Allow the battery to stand for twelve hours, in order to dissipate the heat generated by the chemical action of the acid on the plates and separators. Restore levels by adding more acid of the same specific gravity and then proceed with the initial charge.

Initial Charge Rate

Charge at a rate of 3·5 amps until the voltage and specific gravity readings show no increase over five successive hourly readings. This may take up to 80 hours, depending on the length of time the battery has been stored before charging.

Keep the current constant by varying the series resistance of the circuit or the generator output.

This charge should not be broken by long rest periods. If, however, the temperature of any cell rises above the permissible maximum (that is, 100°F. for batteries filled with 1·270 S.G. acids, 120°F. for those with 1·210 S.G. acid), the charge must be interrupted until the temperature has fallen at least 10°F., below that figure. Throughout the charge, the electrolyte must be kept level with the top of the separator guards by the addition of acid solution of the same specific gravity as the original filling-in acid, until the specific gravity and voltage readings have remained constant for five successive hourly readings. If the charge is continued beyond that point, top up with distilled water.

At the end of the charge carefully check the specific gravity in each cell to ensure that, when corrected to 60°F., it lies within the specified fully-charged limits.

If any cell requires adjustment, some of the electrolyte must be siphoned off and replaced either by distilled water or by acid of the strength originally used for filling-in, depending on whether the specific gravity is too high or too low. Continue the charge for an hour or so to ensure adequate mixing of the electrolyte and again check the specific gravity readings. If necessary, repeat the adjustment process until the desired reading is obtained in each cell. Finally, allow the battery to cool, and siphon off any electrolyte above the tops of the separator guards.

PREPARING NEW "DRY-CHARGED" BATTERIES (MODEL FRVZ11/7A) FOR SERVICE

Filling the Cells

Carefully break the seals in the filling holes and fill each cell with correct specific gravity acid as shown in the table on page P.8 to the top of the separator guards in one operation. The temperatures of the filling room, battery and acid should be maintained at between 60°F. and 100°F. If the battery has been stored in a cool place, it should be allowed to warm up to room temperature before filling.

Freshening Charge

Batteries filled in this way are up to 90% charged and capable of giving a starting discharge one hour after filling. When time permits, however, a short freshening charge will ensure that the battery is fully charged.

Such a freshening charge should be 5 amperes for not more than 4 hours.

During the charge the electrolyte must be kept level with the top of the separators by the addition of distilled water. Check the specific gravity of the electrolyte at the end of the charge; if 1·270 acid was used to fill the battery, the specific gravity should now be between 1·270 and 1·290; if 1·210 acid, between 1·210 and 1·230.

Maintenance in Service

After filling, a dry-charged battery needs only the attention normally given to all lead-acid type batteries.

Page P.9

ELECTRICAL AND INSTRUMENTS

DISTRIBUTOR

REMOVAL

Spring back the clips and remove the distributor cap.

Disconnect the low tension wire from the distributor.

Disconnect the vacuum pipe by unscrewing the union nut at the vacuum advance unit.

Remove distributor clamping plate retaining setscrew and withdraw distributor.

REFITTING

If the distributor clamping plate pinch bolt has not been slackened during removal of distributor refitting will be the reverse of the removal procedure. Enter the distributor into the cylinder block with the vacuum advance unit connection facing the cylinder block.

Rotate the rotor arm until the driving dog engages with the distributor drive shaft.

If the distributor clamping plate pinch bolt has been slackened during removal of distributor it will be necessary to reset the ignition timing as follows:—

Ignition Timing

Set the micrometer adjustment in the centre of the scale.

Connect the low tension wire to the terminal on the distributor body.

Enter the distributor into the cylinder block with the vacuum advance unit connection facing the cylinder block.

Rotate the rotor arm until the driving dog engages with the distributor drive shaft.

Rotate the engine until the rotor arm approaches the No. 6 (front) cylinder segment in the distributor cap.

Slowly rotate the engine until the ignition timing scale on the crankshaft damper is the appropriate number of degrees before the pointer on the sump. (See Data).

Connect a 12 volt test lamp with one lead to the distributor terminal (or the CB terminal of the ignition coil) and the other to a good earth.

Slowly rotate the distributor body until the points are just breaking, that is, when the lamp lights up.

Tighten the distributor plate pinch bolt.

A maximum of six clicks on the vernier adjustment from this setting, to either advance or retard, is allowed.

ROUTINE MAINTENANCE

Distributor Contact Breaker Points

Every 2,500 miles (500 miles with new contact set) check the gap between the contact points with feeler gauges when the points are fully opened by one of the cams on the distributor shaft. A combined screwdriver and feeler gauge is provided in the tool kit.

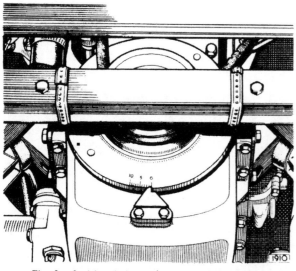

Fig. 2. Ignition timing scale on crankshaft damper.

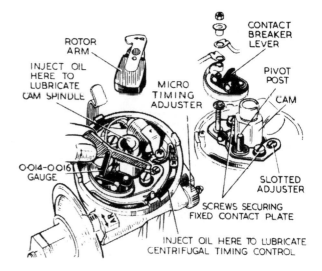

Fig. 3. Checking point gap and lubrication points.

Page P.10

ELECTRICAL AND INSTRUMENTS

The correct gap is ·014″—·016″ (·36—·41 mm.).

If the gap is incorrect, slacken the two screws securing the fixed contact plate and turn the eccentric-headed adjustment screw in its slot until the required gap is obtained. Tighten the securing screws and recheck the gap. (Fig. 3).

Lubrication—Every 2,500 miles

Remove the moulded cover and withdraw the rotor arm. A tight rotor arm can be withdrawn by using a suitable pair of levers carefully applied at opposite points below the rotor moulding—never against the metal electrode.

Important: Do not allow oil or grease on or near the contacts when carrying out the following lubrication.

Cam Bearing

To lubricate the cam bearing, inject a few drops of thin machine oil into the rotor arm spindle (Fig. 3). Do not remove or slacken the screw located inside the spindle—a space is provided beneath the screwhead to allow the lubricant to reach the cam bearing.

Cam

Lightly smear the faces of the cam (Fig. 3) with Mobilgrease No. 2 or with clean engine oil.

Centrifugal Timing Control

Inject a few drops of thin machine oil through a convenient aperture in the contact breaker base plate.

Cleaning

Clean the moulded cover inside and outside with a soft dry cloth. Pay particular attention to spaces between the terminals. Check that the small carbon brush inside the moulding can move freely in its holder.

Whilst the rotor arm is removed, examine the contact breaker. Rough, burned or blackened contacts can be cleaned with fine carborundum stone or emery cloth. After cleaning remove any grease or metallic dust with a petrol moistened cloth.

Contact cleaning is facilitated by removing the lever to which the moving contact is attached. To do this, remove the nut, insulating piece and electrical connections from the post to which the contact breaker spring is anchored. The contact breaker lever can then be lifted off the pivot post and the spring from the anchor post.

After cleaning and trimming the contacts, smear the pivot post (Fig. 3) with Ragosine Molybdenised Non-creep Oil or with Mobilgrease No. 2. Re-assemble the contact breaker and check the setting.

Refit the rotor arm, carefully locating its moulded projection in the spindle keyway and pushing it on as far as it will go.

Refit the moulded cover and spring the two side clips into position.

DATA

Ignition Distributor Type	DMBZ.6A
8 to 1 Compression Ratio	40617A
9 to 1 Compression Ratio	40617A
Cam dwell angle	35° ± 2°
Contact breaker gap	0·014″—0·016″ (0·36—0·41 mm.)
Contact breaker spring tension (Measured at free contact)	18—24 ozs. (512—682 gms.)

IGNITION TIMING

8 to 1 Compression Ratio	9° BTDC
9 to 1 Compression Ratio	10° BTDC

Page P.11

ELECTRICAL AND INSTRUMENTS

SERVICING

Dismantling

When dismantling, note carefully the position in which the various components are fitted in order to simplify their re-assembly

Bearing Replacement

The ball bearing at the upper end of the shank can be removed with a shouldered mandrel locating on the inner journal of the bearing.

When fitting a new ball bearing, the shouldered mandrel must locate on both inner and outer journals of the bearing.

The bearing bush at the lower end of the shank can be driven out with a suitable punch.

A bearing bush may be prepared for fitting by allowing it to stand completely immersed in medium viscosity (S.A.E. 30—40) engine oil for at least 24 hours. In cases of extreme urgency, this period of soaking may be shortened by heating the oil to 100°C. for 2 hours and then allowing to cool before removing the bush.

The bush is pressed into the shank with a shouldered mandrel. The mandrel should be hardened and polished and approximately 0·0005" greater in diameter than the distributor shaft. To prevent subsequent withdrawal of the bush with the mandrel, a stripping washer should be fitted between the shoulder of the mandrel and the bush.

Under no circumstances should the bush be overbored by reamering or by any other means, since this will impair the porosity and therefore the lubricating quality of the bush.

Re-assembly

When re-assembling, Ragosine molybdenised non-creep oil or (failing this) clean engine oil, should be smeared on the shaft and, more lightly, on the contact breaker bearing plate.

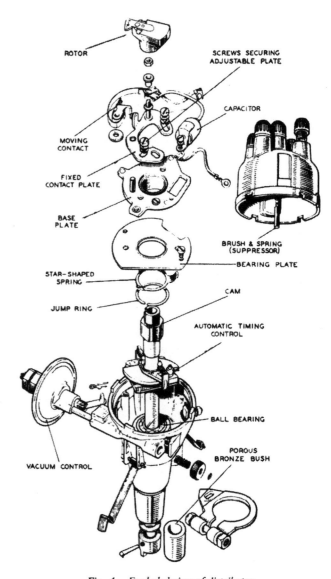

Fig. 4. Exploded view of distributor.

Page P.12

ELECTRICAL AND INSTRUMENTS

IGNITION DISTRIBUTOR TEST DATA

Distributor Type	Lucas Service Number	Lucas Vacuum Unit Number	VACUUM TIMING ADVANCE TESTS The distributor must be run immediately below the speed at which the centrifugal advance begins to function to obviate the possibility of an incorrect reading being registered.			Lucas Advance Springs Number	CENTRIFUGAL TIMING ADVANCE TESTS Mount distributor in centrifugal advance test rig and set to spark at zero degrees at 100 r.p.m.				
			Vacuum in inches of mercury and advance in degrees		No advance in timing below-ins. of mercury		Accelerate to-RPM and note advance in degrees		Decelerate to-RPM and note advance in degrees		No advance in timing below-RPM
			Inches	Degrees			RPM	Degrees	RPM	Degrees	
DMBZ 6A	40617A	54410415	20 13 9 7½ 6	7—9 6—8½ 2½—5½ 0—3 0—½	4½	54410416	2,000	12	1,500 1,300 850 650 450	10—12 9—11 7—9 3½—6½ 0—2½	325

Auto advance weights Lucas number 410033/S.   One inch of mercury = 0·0345 kg/cm²

Page P.13

ELECTRICAL AND INSTRUMENTS

FLASHER UNITS

The flasher unit is housed in a cylindrical container plugged into a base block which is a part of the main wiring harness, and is attached to the bulkhead behind the facia on the driver's side.

The electrical contact is made by means of three blades, extending from the base of the unit. These blades are offset to prevent any possibility of a wrong connection being made.

The automatic operation of the flasher lamps is controlled by means of a switch, contained in the flasher unit, being operated automatically by the alternative heating and cooling of an actuating wire; also incorporated is a small relay to flash the indicator warning lights when the system is functioning correctly. Failure of either of these lights to flash will indicate a fault.

In the event of trouble occurring the following procedure should be followed:—

(i) Check bulbs for broken filaments.

(ii) Refer to the wiring diagram and check all flasher circuit connections.

(iii) Switch on the ignition and check with a voltmeter that flasher unit terminal 'B' is at 12 volts, with respect to earth.

(iv) Connect together flasher unit terminals 'B' and 'L' and operate the direction indicator switch. If the flasher lamps now light the flasher unit is defective and must be replaced.

(v) If after the above checks the bulb still does not light a fault is indicated in the flasher switch which is best checked by substitution.

Note: It is important that only bulbs of the correct wattage rating (that is, 21 watts) are used in the flasher lamps.

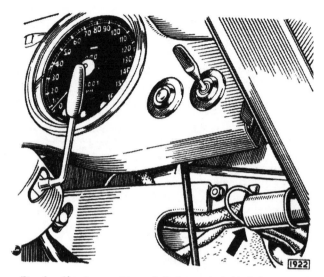

Fig. 5. Showing position of flasher unit behind facia panel.

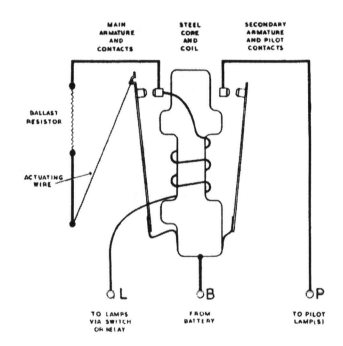

Fig. 6. Flasher unit circuit diagram.

Page P.14

ELECTRICAL AND INSTRUMENTS

FUSE UNITS

Four Model 4 FJ fuse units, each carrying two live glass cartridge type fuses and two spares, are incorporated in the electrical system and are located behind the instrument panel.

Access to the fuses is obtained by removing the two instrument panel retaining screws (top left-hand and top right-hand corners).

The instrument panel will then hinge downwards exposing the fuses and the fuse indicator panel. The circuits controlled by individual fuses are shown on the indicator panel and it is essential that the blown fuse is replaced by one of the correct value.

Only one end of the spare fuses is visible and they are retained in position by a small spring clip. Always replace the spare fuse as soon as possible.

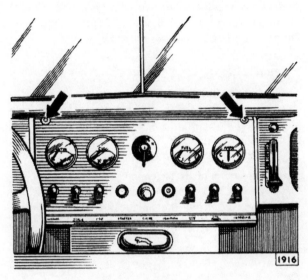

Fig. 7. The instrument panel, the two arrows indicate the securing screws.

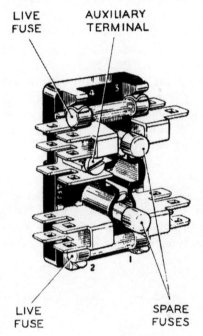

Fig. 8. The Model 4J fuse unit.

Page P.15

ELECTRICAL AND INSTRUMENTS

GENERATOR—MODEL C42
(Fitted to later "E" Type models)

REMOVAL

Disconnect the cables from the two terminals at the rear of the dynamo noting that they are of different sizes.

Remove the nut and bolt securing the adjusting link to the dynamo.

Remove the two nuts and bolts securing the dynamo to the mounting bracket when the dynamo can be lifted out.

Remove the dynamo belt.

REFITTING

Refitting is the reverse of the removal procedure. When the dynamo belt has been refitted move the dynamo to a position where it is possible to depress the belt about $\frac{1}{2}$" (12 mm.) midway between the water pump and dynamo pulleys.

1. GENERAL

The generator is a shunt-wound, two-pole, two-brush machine, arranged to work in conjunction with Lucas regulator unit model RB340. A fan, integral with the driving pulley, draws cooling air through the generator, inlet and outlet holes being provided in the end brackets of the unit.

The output of the generator is controlled by the regulator unit and is dependent on the state of charge of the battery and the loading of the electrical equipment in use. When the battery is in a low state of charge, the generator gives a high output, whereas if the battery is fully charged, the generator gives only sufficient output to keep the battery in good condition without any possibility of over-charging. An increase in output is given to balance the current taken by lamps and other accessories when in use.

2. ROUTINE MAINTENANCE

(a) **Lubrication**

Every 5,000 miles, inject a few drops of high quality viscosity (S.A.E. 30) engine oil into the hole marked "OIL" at the end of the C.E. bracket bearing housing.

(b) **Inspection of Brushgear**

Every 24,000 miles the generator should be removed from the engine and the brushgear checked as detailed in paragraph 4c.

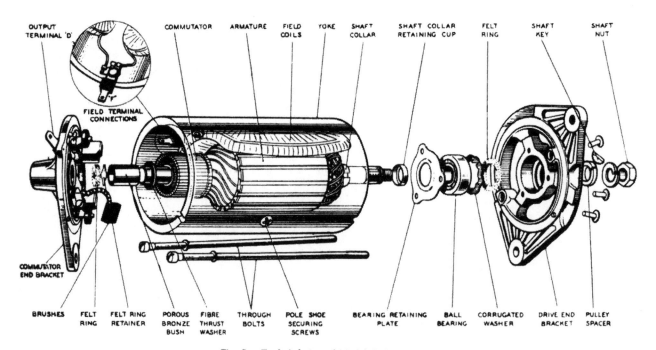

Fig. 9. Exploded view of Model C42 generator.

Page P.16

ELECTRICAL AND INSTRUMENTS

(c) **Belt Adjustment**

Occasionally inspect the generator driving belt and, if necessary, adjust to take up any undue slackness by turning the generator on its mounting. Care should be taken to avoid overtightening the belt, the tension needed being just enough to drive without slipping. See that the machine is properly aligned, otherwise undue strain will be thrown on the generator bearings.

3. PERFORMANCE DATA

Cutting-in Speed	1,250 r.p.m. (max.) at 13·0 generator volts.
Maximum Output	30 amps at 2,200 r.p.m. (max.) at 13·5 generator volts.
Field Resistance	4·5 ohms.

4. SERVICING

(a) **Testing in position to Locate Fault in Charging Circuit**

In the event of a fault in the charging circuit, adopt the following procedure to locate the cause of the trouble.

i. Inspect the driving belt and adjust if necessary (see Paragraph 2c).
ii. Check the connections on the commutator end bracket. The larger connector carries the main generator output, the smaller connector the field current.
iii. Pull off the connectors from the terminal blades of the generator and connect the two terminal blades with a short length of wire.
iv. Start the engine and set to run at normal idling speed.
v. Clip the negative lead of a moving coil type voltmeter, calibrated 0—20 volts, to one generator terminal and the positive lead to a good earthing point on the yoke.
vi. Gradually increase the engine speed, when the voltmeter reading should rise rapidly and without fluctuation. Do not allow the voltmeter reading to reach 20 volts and do not race the engine in an attempt to increase the voltage.

It is sufficient to run the generator up to a speed of 1,000 r.p.m.

If the voltage does not rise rapidly and without fluctuation the unit must be dismantled (see Paragraph 4b) for internal examination.

Excessive sparking at the commutator in the above test indicates a defective armature which should be replaced.

NOTE: If a radio suppression capacitor is fitted between the output terminal and earth, disconnect this capacitor and re-test the generator before dismantling. If a reading is now given on the voltmeter, the capacitor is defective and must be replaced.

If the generator is in good order, remove the link from between the terminals and restore the original connections.

(b) **To Dismantle**

i. Take off the driving pulley.
ii. Unscrew and withdraw the two through bolts.
iii. Withdraw the commutator end bracket from the yoke.
iv. Lift the driving end bracket and armature from the yoke. Take care not to lose the fibre thrust washer or collar from the commutator end of the shaft.
v. The driving end bracket, which on removal from the yoke has withdrawn with it the armature and armature shaft ball bearing, need not be separated from the shaft unless the bearing is suspected and requires examination, or the armature is to be replaced; in this event the armature should be removed from the end bracket by means of a hand press, having first removed the shaft key.

(c) **Brushgear** (Checking with yoke removed)

i. Lift the brushes up into the brush boxes and secure them in that position by positioning the brush spring at the side of the brush.
ii. Fit the commutator end bracket over the commutator and release the brushes.
iii. Hold back each of the brush springs and move the brush by pulling gently on its flexible connector. If the movement is sluggish, remove the brush from its holder and ease the sides by lightly polishing on a smooth file. Always refit

Page P.17

ELECTRICAL AND INSTRUMENTS

brushes in their original positions. If the brushes are badly worn, new brushes must be fitted and bedded to the commutator. The minimum permissible length of brush is ¼".

iv. Test the brush spring tension utilizing a spring balance. The tension needed to just lift the spring from contact with the brush with a new spring and a new brush is 33 ozs. but with a brush worn to ¼" it may reduce to 16 ozs. Both pressures should be measured. Renew any brush spring when the tension falls below these values.

(d) **Commutator**

A commutator in good condition will be smooth and free from pits or burned spots.

Clean the commutator with a petrol-moistened cloth. If this is ineffective carefully polish with a strip of fine glass paper while rotating the armature. To remedy a badly worn commutator, first rough turn the commutator and then undercut the insulator between the segments to a depth of $\frac{1}{32}$". Finally, take a light skim with a very sharp (preferably diamond-tipped) tool. If a non-diamond tipped tool is used for machining, the commutator should be lightly polished with a very fine glass paper. Emery cloth must not be used on the commutator. Finally clean away any dust.

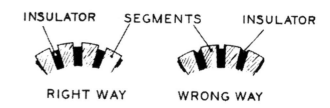

Fig. 11. Showing the correct and incorrect way of undercutting the commutator insulation.

(e) **Armature**

Indication of an open-circuited armature winding will be given by burnt commutator segments. If armature testing facilities are not available, an armature can be checked by substitution. To separate the armature shaft from the drive end bracket, press the shaft out of the drive end bracket bearing.

When fitting the new armature, support the inner journal of the ball bearing, using a mild steel tube of suitable diameter, whilst pressing the armature shaft firmly home (see also paragraph 4h).

(f) **Field Coils**

Measure the resistance of the field coils, without removing them from the generator yoke, by means of an ohm meter connected between the field terminal and the yoke. Field resistance is 4·5 ohms.

If an ohm meter is not available, connect a 12 volt d.c. supply between the field terminal and generator yoke with an ammeter in series. The ammeter reading should be approximately 2·7 amperes. Zero reading on the ammeter or an "Infinity" ohm meter indicates an open circuit in the field winding.

If the current reading is much more than 2·7 amperes, or the ohm meter reading much below 4·5 ohms, it is an indication that the insulation of one of the field coils has broken down.

In either event, unless a substitute generator is available, the field coils must be replaced. To do this, carry out the procedure outlined below:

i. Drill out the rivet securing the field coil terminal assembly to the yoke and remove the insulating sleeve from the terminal block to protect it from the heat of soldering. Unsolder the terminal blade and earthing eyelet.

Fig. 10. Undercutting the commutator insulation.

Page P.18

ELECTRICAL AND INSTRUMENTS

ii. Remove the insulation piece which is provided to prevent the junction of the field coils from contacting with the yoke.

iii. Mark the yoke and pole shoes so that the latter can be refitted in their original positions.

iv. Unscrew the two pole shoe retaining screws by means of a wheel-operated screwdriver.

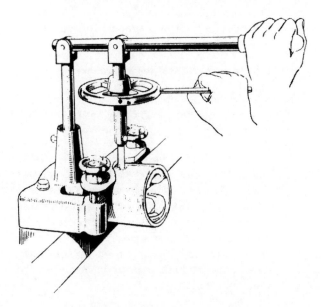

Fig. 12. Tightening the pole shoe retaining screws.

v. Draw the pole shoes and coils out of the yoke and lift off the coils.

vi. Fit the new field coils over the pole shoes and place them in position inside the yoke. Take care to ensure that the taping of the field coils is not trapped between the pole shoes and the yoke.

vii. Locate the pole shoes and field coils by lightly tightening the fixing screws.

viii. Fully tighten the screws by means of the wheel-operated screwdriver.

ix. Solder the original terminal blade and earthing eyelet to the appropriate coil ends.

x. Refit the insulating sleeve and re-rivet the terminal assembly to the yoke.

xi. Refit the insulation piece behind the junction of the two coils.

(g) **Bearings**

Bearings which are worn to such an extent that they will allow side movement of the armature shaft must be replaced.

To replace the bearing bush in a commutator end bracket, proceed as follows:—

i. Remove the old bearing bush from the end bracket. The bearing can be withdrawn with a suitable extractor or by screwing a ⅝" tap into the bush for a few turns and pulling out the bush with the tap. Screw the tap squarely into the bush to avoid damage to the bracket.

ii. Withdraw and clean the felt retainer and felt ring.

iii. Insert the felt ring and felt ring retainer in the bearing housing, then press the new bearing bush into the end bracket, using a self-extracting tool as illustrated, the fitting pin or mandrel portion being of 0·5924" diameter and highly polished. To withdraw the pin after pressing

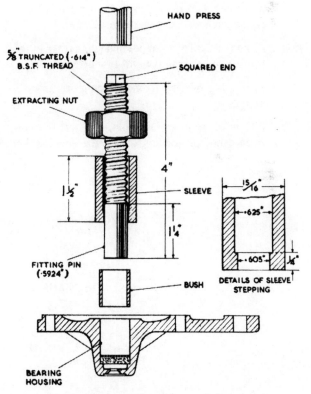

Fig. 13. Method of fitting the porous bronze bush.

Page P.19

ELECTRICAL AND INSTRUMENTS

the bush fully home, turn the nut against the sleeve while gripping the squared end of the fitting pin.

Porous bronze bushes must not be opened out after fitting, or the porosity of the bush may be impaired.

Note: Before fitting the new bearing bush, it should be allowed to stand for 24 hours completely immersed in a good grade S.A.E. 30 engine oil; this will allow the pores of the bush to be filled with lubricant.

The ball bearing at the driving end is replaced as follows:—

i. Drill out the rivets which secure the bearing retaining plate to the end bracket and remove the plate.

ii. Press the bearing out of the end bracket. Remove and clean the corrugated washer and felt ring.

iii. Before fitting the replacement bearing, see that it is clean and pack it with high melting point grease such as Energrease RBB3.

iv. Place the felt ring and corrugated washer in the bearing housing in the end bracket.

v. Locate the bearing in the housing and press it home.

vi. Fit the bearing retaining plate. Insert the new rivets from the pulley side of the end bracket and open the rivets by means of a punch to secure the plate rigidly in position.

(h) To Re-assemble

i. Fit the drive end bracket to the armature shaft. The inner journal of the bearing must be supported by a tube, approximately 4" long $\frac{1}{8}$" thick and internal diameter $\frac{5}{8}$". Do not use the drive end bracket as a support for the bearing whilst fitting an armature.

ii. Fit the yoke to the drive end bracket.

iii. Lift the brushes up into the brush boxes and secure them in that position by positioning each brush spring at the side of its brush.

iv. Fit the fibre thrust washer on the shaft. Fit the commutator end bracket to the yoke, so that the dowel on the bracket locates with the groove on the yoke. Take care not to trap the brush connector pigtails. Insert a thin screwdriver through the ventilator apertures adjacent to the brush boxes and carefully lever up the spring arms until the bushes locate correctly on the commutator.

v. Refit the two through bolts, pulley spacer and shaft key.

vi. After reassembly, lubricate the commutator end bearing (see Paragraph 2a).

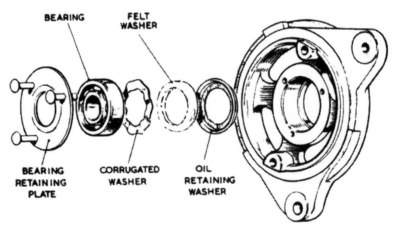

Fig. 14. Exploded view of drive and bearing.

Page P.20

ELECTRICAL AND INSTRUMENTS

GENERATOR—MODEL C.45 PVS-6

(Fitted to early "E" Type models)

REMOVAL

Disconnect the cables from the two terminals at the rear of the dynamo noting that they are of different sizes.

Remove the nut and bolt securing the adjusting link to the dynamo.

Remove the two nuts and bolts securing the dynamo to the mounting bracket when the dynamo can be lifted out.

Remove the dynamo belt.

REFITTING

Refitting is the reverse of the removal procedure. When the dynamo belt has been refitted, move the dynamo to a position where it is possible to depress the belt about $\frac{1}{2}''$ (12 mm.) midway between water pump and dynamo pulleys.

While the generator has different dimensions and performance from Model C42 previously described, its construction is similar, and the same servicing procedure applied in general. The essential differences between the two generators concern:—

(i) Performance.
(ii) Brushgear inspection.
(iii) Commutator end bearing.

PERFORMANCE

Cutting-in Speed	1,300 (max.) r.p.m. at 13·0 generator volts.
Maximum Output	25 amperes at 2,050 (max.) r.p.m. at 13·5 generator volts.
Field Resistance	6·0 ohms.

BRUSHGEAR INSPECTION

The yoke is provided with "windows" and a band cover. The instructions given for model C42 under paragraph 4c (i-iii) need not, therefore, be followed in order to gain access to the brushes for inspection and spring testing—it being only necessary to slacken a single clamping screw and release the band cover.

Minimum permissible brush length is $\frac{11}{32}''$. Brush spring tesion 28 ozs. with new brush, 20 ozs. with brush worn to $\frac{11}{32}''$.

Fig. 15. Testing the brush spring tension.

COMMUTATOR END BEARING

A ball bearing is fitted at the commutator end of the armature shaft. Details are shown in the illustration. The bearing is secured to the shaft by a thrust screw and can be withdrawn with an extractor after the screw has been removed.

When replacing a defective bearing see that the new bearing is clean and packed with high melting point grease. It must be pressed home against the shoulder on the shaft and secured with the thrust screw.

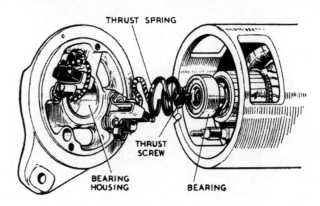

Fig. 16. Showing the end plate removed.

Page P.21

ELECTRICAL AND INSTRUMENTS

HORNS

It is important to keep the horn mounting bolts tight and to maintain rigid the mountings of any units fitted near the horns. Electrical connections and cables should be checked occasionally and rectified as required.

REMOVAL

Remove the six screws securing the headlight rim, remove the rim, rubber seal and headlight glass. Remove the three screws securing the headlight duct to the diaphragm panel and withdraw the duct forwards through the headlight glass aperture. The horn may now be seen through the aperture. Remove the two securing nuts and bolts, remove the cover from the horn by unscrewing the central screw and detach the wires. The horn may now be withdrawn.

ADJUSTMENT

Adjustment is effected after removal of the domed cover by means of the fixed contact screw.

Connect a 0—20 first grade moving coil ammeter in series with horn. Release contact locknut and adjust contact until horn will pass 13—15 amperes at 12 volts. Retighten locknut and check.

Refit domed cover.

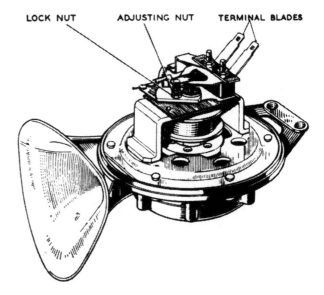

Fig. 17. The horn adjustment screw.

Page P.22

ELECTRICAL AND INSTRUMENTS

LAMPS

LIGHT BULBS

LAMP	LUCAS BULB NUMBER	VOLTS	WATTS	APPLICATION
Head (Yellow)	416 417 410 411	12 12 12 12 Sealed beam unit	60/40 60/40 45/40 45/40	Home and R.H. Drive Export L.H. Drive Export European Continental France U.S.A. and Canada
Side	989	12	6	
Front and Rear Flashing Indicators	382	12	21	
Rear/Brake	380	12	21/6	
Interior Lights	382 989	12 12	21 6	Open 2 Seater Fixed Head Coupe
Map Light	989	12	6	
Instrument Illumination Headlamp Warning Light Ignition Warning Light Fuel Level Warning Light Handbrake/Brake Fluid Warning Light Mixture Control Warning Light	987	12	2·2	
Switch Indicator Strip Flashing Indicator Warning Light	281	12	2	

Page P.23

ELECTRICAL AND INSTRUMENTS

HEADLAMPS

The headlamps comprise two Lucas light units with pre-focus double-filament bulbs (excepting U.S.A., export models, which are provided with an adaptor to accept American Sealed Beam Units) front rims and dust excluding rubber rings.

Since the spread of light and its position on the kerbside in the dipped position is a function of lensing and bulb design, special light units and bulbs are fitted to suit lighting regulations of the country in which a car is used. Special care should therefore be taken when replacing a bulb to see that the correct replacement is fitted.

Bulb Replacement

Remove the six screws holding glass headlamp cover retaining ring to wing. Remove ring and rubber ring now exposed. Remove glass cover.

Release the three cross-headed screws retaining headlamp glass and reflector unit rim and remove rim by turning in an anti-clockwise direction.

Note: It is not necessary to remove screws completely.

Light unit can now be withdrawn.

Remove plug with attached cables from unit. Release bulb retaining spring clips and withdraw bulb.

Replace with bulb of correct type. When re-assembling note that a groove in the bulb plate must register with a raised portion on the bulb retainer.

Replace spring clips and refit light unit assembly.

Refit retaining ring by turning in a clockwise direction and tighten the three cross-headed screws.

Note: Do not turn the two slotted screws or the setting of the headlamp will be upset.

Refit glass cover and retaining ring with rubber seal.

Headlamp Setting

The headlamps should be set so that when the car is carrying its normal load the driving beams are projected parallel with each other and parallel with the ground (see Fig.19).

When setting remove glass cover retaining ring rubber seal and glass cover. Cover one lamp whilst adjusting the other.

The setting of the beams are adjusted by the two slotted screws, one being located at the bottom centre and the other one at centre right-hand side. The bottom screw is for vertical adjustment, the side screw being for horizontal. After adjustment replace glass cover and retaining ring with rubber seal.

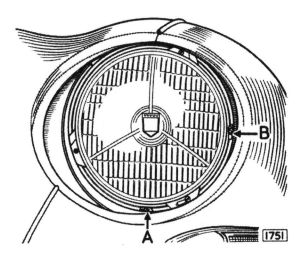

Fig. 19. Adjustment of the screw 'A' will alter the headlamp beam in the vertical plane; adjustment of the screw 'B' will alter the beam in the horizontal plane.

Fig. 18. Headlamp bulb removal.

ELECTRICAL AND INSTRUMENTS

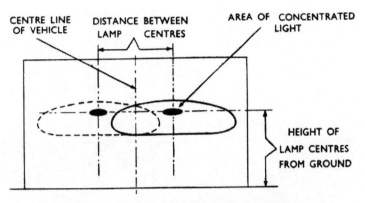

(A) FRONT OF VEHICLE TO BE SQUARE WITH SCREEN

(B) VEHICLE TO BE LOADED AND STANDING ON LEVEL GROUND

(C) RECOMMENDED DISTANCE FOR SETTING IS AT LEAST 25FT.

(D) FOR EASE OF SETTING ONE HEADLAMP SHOULD BE COVERED

Fig. 20. Headlamp beam setting.

Sidelamp Bulb—Replacement

Remove the three screws retaining the lamp glass and remove glass. The sidelamp bulb is the inner one of the two exposed and is removed by pressing inwards and turning anti-clockwise.

Front Flasher Bulb—Replacement

Proceed as for the sidelamp bulb. The flasher bulb is the outer one of the two exposed.

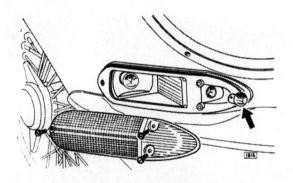

Fig. 21. Sidelamp bulb removal.

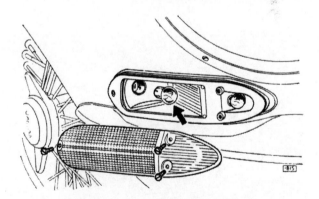

Fig. 22. Front flasher bulb removal.

ELECTRICAL AND INSTRUMENTS

Rear/Brake Bulb—Replacement

Remove the two screws retaining the lamp glass and remove glass. The rear/brake bulb is the inner one of the two bulbs exposed and is removed by pressing inwards and turning anti-clockwise. When fitting a replacement bulb note that the pins are offset.

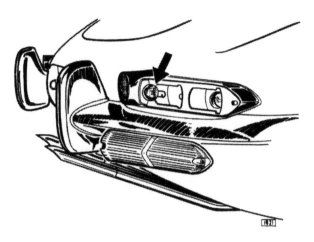

Fig. 23. Rear/brake bulb removal.

Rear Flasher Bulb—Replacement

Proceed as for rear/brake bulb. The flasher bulb is the outer one of the two exposed.

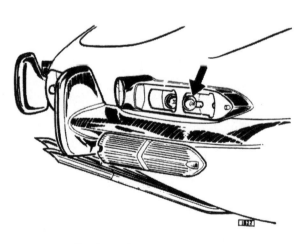

Fig. 24. Rear flasher bulb removal.

Number Plate Lamp Bulb—Replacement

Remove the fixing screw retaining rim and lamp glass and detach glass rim and gasket. Remove bulb by pressing inwards and turning anti-clockwise.

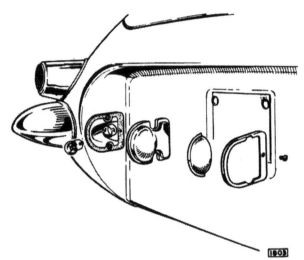

Fig. 25. Number plate lamp bulb removal.

Interior—Luggage Lamp Bulb—Replacement

The interior—luggage lamp bulb is retained in a holder accessible when the boot lid is raised. To remove bulb from its holder press inwards and turn anti-clockwise.

Reverse Lamp Bulb—Replacement

Remove the two screws retaining the lamp glass and detach the glass and gasket. Remove the bulb by pressing and rotating in an anti-clockwise direction.

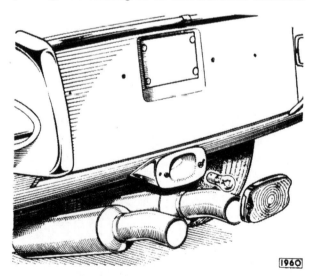

Fig. 26. Reverse lamp bulb removal.

Page P.26

ELECTRICAL AND INSTRUMENTS

RB 310 CURRENT—VOLTAGE REGULATOR

(Fitted to early "E" type models to control generator C45 PV-6)

(a) CHECKING CONTINUITY BETWEEN BATTERY AND CONTROL BOX

If the generator and battery are in order, disconnect the cables from control box terminal blades 'B' and connect them to the negative terminal of a good quality 0—20 moving coil voltmeter.

Connect the positive terminal of the voltmeter to an earthing point on the chassis. If the meter registers battery voltage, i.e. 12 volts, the wiring is in order and the control box settings should be checked.

If there is no reading, re-connect the cables to terminal blades 'B' and examine the wiring between battery, ammeter, and control box for defective cables or loose connections.

(b) VOLTAGE REGULATOR ADJUSTMENT

The regulator is carefully set during manufacture and, in general, it should not be necessary to make further adjustment. However, if the battery fails to keep in a charged condition or if the generator output does not fall when the battery is fully charged, the setting should be checked and, if necessary, corrected.

It is important to check before altering the regulator setting that the low state of charge of the battery is not due to a defective battery or to slipping of the generator belt. Only a good quality MOVING COIL VOLTMETER (0—20 volts) must be used when checking the regulator. The open circuit setting can be checked without removing the cover from the control box.

Disconnect the cables from the control box terminal blades 'B' and join the ignition and battery feeds together using a suitable "jumper lead".

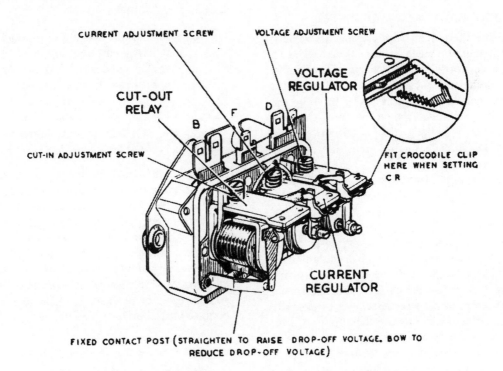

Fig. 27. *The RB.310 control box showing the position of the three spring loaded adjusting screws.*

Page P.27

ELECTRICAL AND INSTRUMENTS

Connect the voltmeter to control box terminal 'D' and a good earthing point.

The regulator should be at ambient temperature, i.e. as measured in its immediate vicinity, and adjustment should be completed within thirty seconds, otherwise heating of the shunt coil by the energising current may cause false settings to be made.

Run the engine up until the generator speed reaches 3,000 r.p.m. (2,000 engine r.p.m.) when the open circuit voltage reading should lie within the following limits:—

Ambient Temperature	Open Circuit Voltage Setting
10°C. (50°F.)	15·1—15·7
20°C. (68°F.)	14·9—15·5
30°C. (86°F.)	14·7—15·3
40°C. (104°F.)	14·5—15·1

If the voltmeter reading is outside the specified limits rotate the voltage regulator adjusting screw, which is adjacent to the 'D' terminal, clockwise, to raise the setting or anti-clockwise to reduce the setting. Check the setting by switching off the engine, restarting and then raising the generator speed to 3,000 r.p.m. (2,000 engine r.p.m.) and make any final adjustment.

(c) CURRENT REGULATOR ADJUSTMENT

When setting the current regulator on the vehicle, the generator must be made to develop its full rated output, regardless of the state of charge of the battery at the time of setting. The voltage regulator must therefore be rendered inoperative. To do this, the voltage regulator contact should be short-circuited with a crocodile or bulldog clip placed between the insulated fixed contact bracket and the voltage regulator frame.

Disconnect the cables from terminal blades 'B' and, using a suitable "jumper lead" connect a 0—40 first grade moving coil ammeter between these cables and terminal blades 'B'.

Start the engine and run the generator at about 4,000 r.p.m. (2,700 engine r.p.m.) when the ammeter should read 24—26 amperes. If the ammeter is outside the specified limit rotate the current adjusting screw, which is the centre of the three, clockwise to raise the setting or anti-clockwise to reduce the setting. Check the setting by switching off the engine, restarting and then raising the generator speed to 4,000 r.p.m. (2,700 r.p.m.) and make any final adjustment.

Restore the original connections.

(d) CLEANING REGULATOR CONTACTS

After long periods of service it may be found necessary to clean the contacts of the voltage and current regulators. These may be cleaned with silicon carbide paper, fine carborundum stone or fine emery cloth. All traces of metal dust or other foreign matter must be removed with methylated spirits (denatured alcohol).

(e) CUT-OUT ADJUSTMENT

If the regulator is correctly set but the battery is still not being charged, the cut-out may be out of adjustment.

i. **Method of Setting Cut-in Voltage**

   Partially withdraw the Lucar cable connector from control box terminal blade 'D'.
   Connect a first-grade 0—20 volt moving coil voltmeter between the exposed portion of terminal blade 'D' and a good earthing point, taking care not to short-circuit terminal 'D' to the base.
   Start the engine and slowly increase the speed, while observing the voltmeter pointer. The voltage should rise steadily and then drop slightly at the instant of contact closure. The cut-in voltage is that which is indicated immediately before the pointer drops back. It should lie between the limits 12·7—13·3 volts.

   Note: Should the instant of contact closure be indeterminate and difficult to ascertain, due to the cut-in and battery voltages being approximately equal, switch on the headlamps in order to depress the battery voltage. Repeat the rising voltage check, when a definite drop should be observed as contacts close.

If the cut-in voltage occurs outside the above limits, an adjustment must be made by rotating the cut-out adjusting screw, which is adjacent to the 'B' terminal blades, a fraction at a time clockwise to raise the setting or anti-clockwise to reduce the setting. Test after each adjustment by increasing the engine speed and note the voltmeter reading at the instant of contact closure. Electrical settings of the cut-out, like the voltage regulator, must be effected as quickly as possible because of temperature rise effects.

ii. **Method of Setting Drop-off Voltage**

   Withdraw the cables from control box terminal blades 'B' and (to provide a battery feed to the

Page P.28

ELECTRICAL AND INSTRUMENTS

ignition coil) connect them together with a suitable "jumper lead".

Connect a first-grade 0—20 volt moving coil voltmeter between one of the terminal blades 'B' and a good earthing point.

Start the engine and run it up to above cut-in speed.

Slowly decelerate and observe the voltmeter pointer.

Opening of the contacts, indicated by the voltmeter pointer dropping to zero, should occur between the limits 9·5—11·0 volts. If it does not, the spring force exerted by the moving contact blade must be adjusted by altering the height of the fixed contact.

To do this, carefully straighten the legs of the fixed contact post to raise the drop-off voltage or bow them to reduce it. Repeat the test and, if necessary, re-adjust until the armature releases at the specified voltage.

(f) CLEANING CUT-OUT CONTACTS

After long periods of service it may be found necessary to clean the cut-out contacts. These may be cleaned with fine glass paper. All traces of metal dust or other foreign matter must be removed with methylated spirits.

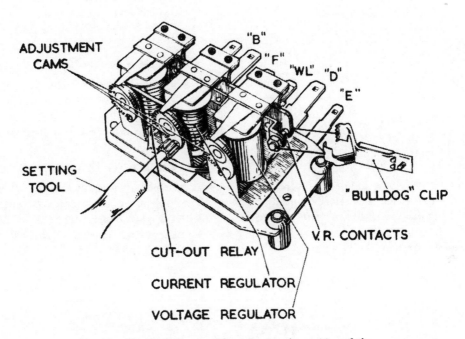

Fig. 28. The RB.340 control box showing the position of the three cam adjusters.

Page P.29

ELECTRICAL AND INSTRUMENTS

RB. 340 CURRENT—VOLTAGE REGULATOR

(Fitted to later "E" type models to control generator model C42)

(a) CHECKING CONTINUITY BETWEEN BATTERY AND CONTROL BOX
Instructions as given for model RB.310.

(b) VOLTAGE REGULATOR ADJUSTMENT
Instructions as given for model RB.310 except for actual setting procedure and voltage limits which are as follows:—

Using a suitable tool, turn the voltage adjustment cam until the correct setting is obtained—turning the tool clockwise to raise the setting or anti-clockwise to lower it.

Ambient Temperature	Open Circuit Voltage Setting
10°C. (50°F.)	15·0—15·6
20°C. (68°F.)	14·8—15·4
30°C. (86°F.)	14·6—15·2
40°C. (104°F.)	14·4—15·0

(c) CURRENT REGULATOR ADJUSTMENT
Instructions as given for model RB.310 except for actual setting procedure and current limits, which are as follows:—

Using a suitable tool, turn the current adjustment cam until the correct setting is obtained—turning the tool clockwise to raise the setting or anti-clockwise to lower.

Current Regulator Setting 30 ± 1½ amperes.

(d) CLEANING REGULATOR CONTACTS
Instructions as given for model RB.310.

(e) CUT-OUT ADJUSTMENT
Instructions as given for model RB.310 except as follows:—

i. **Method of Setting Cut-in Voltage**

Using a suitable tool, turn the cut-out relay adjustment cam until the correct setting is obtained—turning the tool clockwise to raise the setting or anti-clockwise to lower it.
Cut-in Voltage Setting 12·6—13·4 volts.

ii. **Method of Setting Drop-off Voltage**

between the limits 9·25—11·25 volts.
To do this, carefully bend the fixed contact bracket. Closing the gap will raise the drop-off voltage. Opening the gap will reduce the drop-off voltage.

(f) CLEANING CUT-OUT CONTACTS
After long periods of service it may be found necessary to clean the cut-out contacts. These may be cleaned with fine glass paper. All traces of metal dust or other foreign matter must be removed with methylated spirits (denatured alcohol).

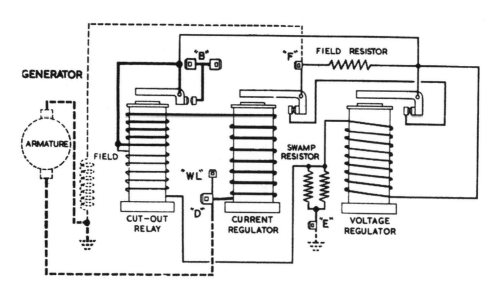

Fig. 29. The circuit diagram of the RB.340 control box.

Page P.30

ELECTRICAL AND INSTRUMENTS

STARTER MOTOR

REMOVAL

Detach the earth lead from the battery. Disconnect the cable from the terminal at the end of the starter motor.

Release the clips and detach the two rubber hose pipes from the brake servo vacuum situated on the bulkhead above the starter motor (Note hose pipe connections for later fitting).

Remove the four nuts and washers retaining vacuum tank to bulkhead and remove tank.

Remove the two nuts from the rear ends of the starter motor securing bolts. Support starter motor from below by hand and withdraw both bolts.

Withdraw starter motor through chassis frame.

REFITTING

Refitting is the reverse of the removal procedure. Care must be taken when reconnecting to ensure that the vacuum tank hoses are fitted to the correct unions. Refer to Section L "Brakes" before making connections.

1. GENERAL

The electric starting motor is a four-pole, four-brush machine having an extended shaft which carries the engine engagement gear, or starter drive as it is more usually named. The diameter of the yoke is $4\frac{1}{2}"$.

The starting motor is of similar construction to the generator except that heavier copper wire is used in the construction of the armature and field coils. The field coils are series parallel connected between the field terminal and the insulated pair of brushes.

2. ROUTINE MAINTENANCE

The only maintenance normally required by the starting motor is the occasional checking of brush-gear and commutator. About every 10,000 miles, remove the metal band cover. Check that the brushes move freely in their holders by holding back the brush springs and pulling gently on the flexible connectors. If a brush is inclined to stick, remove it from its holder and clean its sides with a petrol moistened cloth. Be careful to replace brushes in their original positions in order to retain "bedding". Brushes which have worn so that they will not "bed" properly on the commutator or have worn less than $\frac{5}{16}"$ in length must be renewed.

The commutator should be clean, free from oil or dirt and should have a polished appearance. If it is dirty, clean it by pressing a fine dry cloth against it while the starter is turned by hand by means of a spanner applied to the squared extension of the shaft. Access to the squared shaft is gained by removing the thimble-shaped metal cover. If the commutator is very dirty moisten the cloth with petrol.

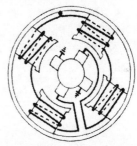

Fig. 30. Showing the internal connections of the starter motor.

Page P.31

ELECTRICAL AND INSTRUMENTS

3. PERFORMANCE DATA

Model	M 45 G
Lock Torque	22 lbs/ft. with 430–450 amperes at 7·8–7·4 volts.
Torque at 1,000 r.p.m.	8·3 lbs/ft. with 200–220 amperes at 10·2–9·8 volts.
Light running current	45 amperes at 5,800–6,800 r.p.m.

4. SERVICING

(a) TESTING IN POSITION

Check that the battery is fully charged and terminals are clean and tight. Recharge if necessary.

(i) Switch on the lamps and operate the starter control. If the lights go dim, but the starter motor is not heard to operate, an indication is given that the current is flowing through the starting motor windings but that the armature is not rotating for some reason; possibly the pinion is meshing permanently with the geared ring on the flywheel. In this case the starting motor must be removed from the engine for examination.

(ii) Should the lamps retain their full brilliance when the starter switch is operated, check the circuit for continuity from battery to starting motor via the starter switch, and examine the connections at these units. If the supply voltage is found to be applied to the starting motor when the switch is operated, an internal fault in the motor is indicated and the unit must be removed from the engine for examination.

(iii) Sluggish or slow action of the starting motor is usually due to a loose connection causing a high resistance in the motor circuit. Check as described above.

(iv) If the motor is heard to operate, but does not crank the engine, indication is given of damage to the drive.

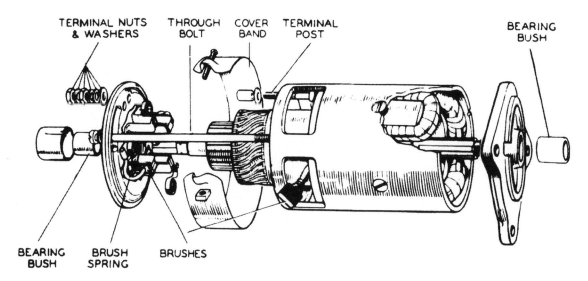

Fig. 31. Exploded view of the starter motor.

Page P.32

ELECTRICAL AND INSTRUMENTS

(b) BENCH TESTING AND EXAMINATION OF BRUSHGEAR AND COMMUTATOR

(i) Remove the starting motor from the engine, as described on page P.31.

(ii) After removing the starting motor from the engine secure the body in a vice and test by connecting it with heavy gauge cables to a battery of the appropriate voltage. One cable must be connected to the starter terminal and the other held against the body or end bracket. Under these light load conditions, the starter should run at a very high speed (see Paragraph 3) without excessive noise and without excessive sparking at the commutator.

(iii) If the operation of the starting motor is unsatisfactory, remove the cover band and examine the brushes and commutator. Hold back each of the brush springs and move the brush by pulling gently on its flexible connector. If the movement is sluggish, remove the brush from its holder and ease the sides by lightly polishing on a smooth file. Always replace brushes in their original positions. If the brushes are worn so that they will not bear on the commutator, or if the brush flexible is exposed on the running face, they must be replaced (see paragraph 4d).

Check the tension of the brush springs with a spring scale. The correct tension is 30—40 ozs. New springs should be fitted if the tension is low.

If the commutator is blackened or dirty, clean it by holding a petrol-moistened cloth against it while the armature is rotated.

(iv) Re-test the starter as described under (ii). If the operation is still unsatisfactory, the unit can be dismantled for detailed inspection and testing as follows:—

(c) TO DISMANTLE

(i) Remove the cover band, hold back the brush springs and lift the brushes from their holders.

(ii) Remove the nuts from the terminal post which protrudes from the commutator end bracket.

(iii) Unscrew the two through bolts from the commutator end bracket. Remove the commutator end bracket from the yoke.

(iv) Remove the driving end bracket complete with armature and drive from the starting motor yoke. If it is necessary to remove the armature from the driving end bracket, it can be done by means of a hand press after the drive has been dismantled.

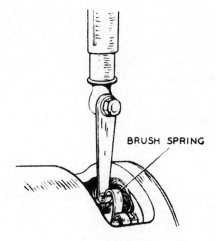

Fig. 33. Testing the brush spring tension.

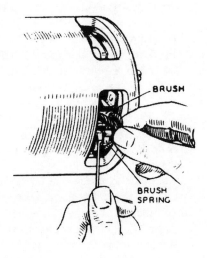

Fig. 32. Checking the brush gear.

Page P.33

ELECTRICAL AND INSTRUMENTS

(d) REPLACEMENT OF BRUSHES

If the brushes are worn to less than $\frac{5}{16}$" in length, they must be replaced.

Two of the brushes are connected to terminal eyelets attached to the brush boxes on the commutator end bracket and two are connected to the field coils.

The flexible connectors must be removed by unsoldering and the connectors of the new brushes

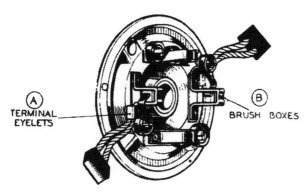

Fig. 31. The commutator end bracket brush connections.

secured in their place by soldering. The new brushes are preformed so that the bedding to the commutator is unnecessary.

(e) COMMUTATOR

A commutator in good condition will be smooth and free from pits and burned spots. Clean the commutator with a petrol-moistened cloth. If this is ineffective, carefully polish with a strip of fine glass paper, while rotating the armature. To remedy a badly worn commutator, dismantle the starter drive and remove the armature from the end bracket. Now mount the armature in a lathe, rotate at a high speed and take a light cut with a very sharp tool. Do not remove any more metal than is necessary. Finally polish with very fine glass paper.

The insulators between the commutator segments MUST NOT BE UNDERCUT.

(f) ARMATURE

Examination of the armature may reveal the cause of failure, e.g., conductors lifted from the commutator due to the starter motor being engaged while the engine is running and causing the armature to be rotated at an excessive speed. A damaged armature must always be replaced—no attempts should be made to machine the armature core or to true a distorted armature shaft.

(g) FIELD COILS

(i) Test the field coils for continuity by connecting a 12-volt test lamp between the starting motor terminal and to each field brush in turn.

(ii) Lighting of the lamp does not necessarily mean that the field coils are in order, as it is possible that one of them may be earthed to a pole-shoe or to the yoke. This may be checked with a 110-volt test lamp, the test leads being connected between the starting motor terminal and a clean part of the yoke. If the lamp lights, defective insulation of the field coils or of the terminal post is indicated. In this event, see that the insulating band is in position and examine the field coils and terminal connections for any obvious point of contact with the yoke. If from the above tests the coils are shown to be open-circuited or earthed and the point of contact cannot be readily located and rectified, either the complete starting motor or the field coils must be replaced. If the field coils are to be replaced, follow the procedure outlined below, using a wheel-operated screwdriver.

Remove the insulation piece wich is provided to prevent the intercoil connectors from contacting with the yoke.

Mark the yoke and pole shoes so that the latter can be refitted in their original positions. Unscrew the four pole shoe retaining screws with the wheel-operated screwdriver.

Draw the pole shoes and coils out of the yoke and lift off the coils. Fit the new field coils over the pole shoes and place them in position inside the yoke.

Take care to ensure that the taping of the field coils is not trapped between the pole shoes and the yoke.

Locate the pole shoes and field coils by lightly tightening the fixing screw. Fully tighten the screws with the wheel-operated screwdriver. Replace the insulation piece between the field coil connections and the yoke.

Page P.34

ELECTRICAL AND INSTRUMENTS

(h) BEARINGS

Bearings which are worn to such an extent that they will allow excessive side-play of the armature shaft must be replaced. To replace the bearing bushes proceed as follows:—

(i) Press the bearing bush out of the end bracket.

(ii) Press the new bearing bush into the end bracket using a shouldered, highly polished mandrel of the same diameter as the shaft which is to fit in the bearing. Porous bronze bushes must not be opened out after fitting, or the porosity of the bush may be impaired.

Note: Before fitting a new porous bronze bearing bush it must be completely immersed for 24 hours in clean thin engine oil.

(j) REASSEMBLY

The re-assembly of the starting motor is a reversal of the dismantling procedure.

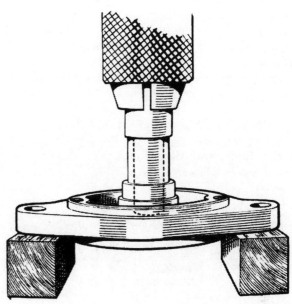

Fig. 35. Method of fitting bush.

STARTER DRIVE

1. GENERAL

The pinion is mounted on a threaded sleeve which is carried on splines on the armature shaft, the sleeve being arranged so that it can move along the shaft against a compression spring so as to reduce the shock loading at the moment engagement takes place.

When the starter switch is operated, the shaft and screwed sleeve rotate, and owing to the inertia of the pinion the screwed sleeve turns inside the pinion causing the latter to move along the sleeve into engagement with the flywheel ring. The starter will then turn the engine.

As soon as the engine fires and commences to run under its own power, the flywheel will be driven faster by the engine than by the starter. This will cause the pinion to be screwed back along the sleeve and so thrown out of mesh with the flywheel teeth. In this manner the drive safeguards the starter against damage due to being driven at high speeds by the engine.

A pinion restraining spring is fitted over the starter shaft to prevent the pinion being vibrated into contact with the flywheel when the engine is running.

2. ROUTINE MAINTENANCE

If any difficulty is experienced with the starting motor not meshing correctly with the flywheel, it may be that the drive requires cleaning. The pinion should move freely on the screwed sleeve; if there is any dirt

Page P.35

ELECTRICAL AND INSTRUMENTS

or other foreign matter on the sleeve it must be washed off with paraffin.

In the event of the pinion becoming jammed in mesh with the flywheel, it can usually be freed by turning the starter motor armature by means of a spanner applied to the shaft extension at the commutator end.

This is accessible by removing the cap which is a push fit.

3. DISMANTLING AND REASSEMBLY

Having removed the armature as described in the section dealing with starting motors the drive can be dismantled as follows:—

Remove the split pin (A) from the shaft nut (B) at the end of the starter drive. Hold the squared starter shaft extension at the commutator end by means of a spanner and unscrew shaft nut (B). Lift off the main spring (C), washer (D), screwed sleeve with pinion (E), collar (F), pinion restraining spring (G) and restraining spring sleeve (H).

Note: If either the screwed sleeve or pinion are worn or damaged they must be replaced as a pair, not separately.

The reassembly of the drive is a reversal of the dismantling procedure.

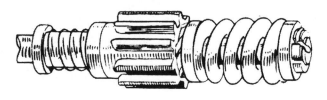

Fig. 36. Showing the starter drive assembled.

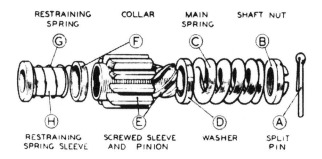

Fig. 37. Exploded view of the starter drive assembly.

WINDSCREEN WIPER

The windscreen wiper assembly consists of a two-speed motor coupled by connecting rods to three wiper spindle bearings. A control cable is attached to the centre spindle bearing mechanism for adjustment of the parking switch. The knurled adjusting knob attached to the cable is accessible in the engine compartment on the bulkhead.

Turning this control will raise or lower the parking limits of wiper arms.

REMOVAL OF WIPER MOTOR

Disconnect the battery earth cable.

Disconnect the ball joint from the throttle control shaft at the pivot bracket and remove bracket by unscrewing the two setscrews.

Release snap connector clip from bulkhead and disconnect cables. Lower the instrument panel after removing the two retaining screws in the top right hand and left hand corners and disconnect the ball

Fig. 38. The windscreen wiper parking adjuster screw.

Page P.36

ELECTRICAL AND INSTRUMENTS

joint from the central windscreen wiper spindle housing. Remove the four setscrews retaining the windscreen wiper motor to bulkhead and withdraw the motor complete with the attached link rod.

REFITTING

Refitting is the reverse of the removal procedure.

Note: It is essential when refitting that the length of the link rod is not altered. Any alteration in the length of this rod will place the windscreen wiper arms out of phase with each other.

When refitting the throttle control pivot bearing bracket, care must be taken that the control rod is central in its bearing. Adjustment is provided by means of the two slotted holes in the bracket.

REMOVAL OF WINDSCREEN WIPER SPINDLE HOUSINGS

The following instructions apply to right-hand drive cars; instructions for left-hand drive models are identical with the exception of the side facia panels which are in this case reversed (i.e.) the instrument facia panel being in each case on the driver's side.

REMOVAL (Right-hand or Left-hand Housings)

Disconnect battery.

Withdraw wiper arms from spindles.

Lower the centre instrument panel after removing the two retaining screws in the top right hand and top left hand corners.

Remove side facia panel (see page P.45) for the removal of right hand spindle housing or remove glove box (see page P.44) for removal of left hand spindle housing.

Disconnect the ball joint from the wiper spindle crank. From outside the car unscrew the large nut securing the spindle housing to the scuttle.

Remove the chrome distance piece and rubber seal.

From inside the car withdraw the spindle housing.

REFITTING

Refitting is the reverse of the removal procedure.

Note: It is essential when refitting that the length of the link rod is not altered. Any alteration in the length of this rod will place the windscreen wiper arms out of phase with each other. If both spindle housings are removed care must be taken to ensure when refitting that the spindle with the longer crank is fitted to the driver's side.

CENTRE HOUSING

Disconnect battery.

Withdraw wiper arm from spindle.

Lower the centre instrument panel after removing the two retaining screws in the top right-hand and top left-hand corners. Remove side facia panel (see page P.45) and glove box (see page P.44).

Disconnect the ball joints from the two outer spindle cranks.

Disconnect the two cables attached to parking switch.

Remove the nut attaching the wiper parking switch control to the engine side of the bulkhead and withdraw the control from inside the car.

From outside the car unscrew the large nut securing the centre housing to the scuttle. Remove the chrome distance piece and rubber seal.

From inside the car withdraw the housing from the scuttle.

Withdraw housing and attached rods through centre aperture in dash panel.

REFITTING

Refitting is the reverse of the removal procedure.

Note: It is essential when refitting that the length of the link rods are not altered. Any alteration in the length of these rods will place the windscreen wiper arms out of phase with each other.

Page P.37

ELECTRICAL AND INSTRUMENTS

DATA

	Normal	High
Wiping Speeds	44—48 Cycles/minute	58—68 Cycles/minute
Operating Currents Arms and Blades removed Motor only	3·0—3·7 amp. 2·5—3·2 amp.	2·2—2·9 amp. 1·7—2·4 amp.
Resistance of Field Coil	8·0—9·5 ohms	
Value of Field Resistor	9·5—11·0 ohms	
Pressure of Blades against Windscreen	11—13 ozs.	

DESCRIPTION

The windscreen wiper is a two-speed, thermostatically protected, self parking, link operated unit.

The link and spindle housing assembly comprises a back plate with the three attached spindle housings, the spindle housings being detachable separately from the assembly.

One control rod operates from the motor to the centre spindle and the remaining two from the centre to the two outer spindles.

The motor is controlled by a switch giving Park, Normal and High speed operation. The higher speed is intended to be used when driving fast through heavy rain or light snow. It should not be used with heavy snow or with a dry or drying windscreen.

If overloaded the motor windings will overheat and cause the thermostat to trip and isolate the motor from the supply. Possible causes include: Packed snow or ice on screen, over-frictional or oil contaminated blades, damaged drive mechanism or spindle units. Provided the obstruction or other cause of excessive heating is removed, normal working resumes automatically when the temperature falls to a safe level.

MAINTENANCE

Efficient wiping is dependent upon having a clean windscreen and wiper blades in good condition.

Use methylated spirits (denatured alcohol) to remove oil, tar spots and other stains from the windscreen. Silicone and wax polishes should not be used for this purpose.

Worn or perished wiper blades are readily removed for replacement.

When necessary, adjustments to the self-parking mechanism can be made by turning the knurled nut located on the bulkhead. Turn the nut only one or two serrations at a time and test the effect of each setting before proceeding.

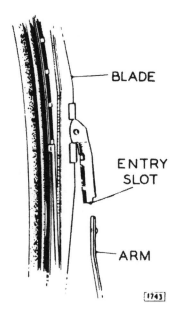

Fig. 39. Wiper blade to arm attachment.

Page P.38

ELECTRICAL AND INSTRUMENTS

FAULT DIAGNOSIS

Poor performance can be electrical or mechanical in origin and not necessarily due to a faulty motor, for example:

Low voltage at the motor due to poor connections or a discharged battery.

Excessive loading on the wiper blades.

Spindles binding in housings.

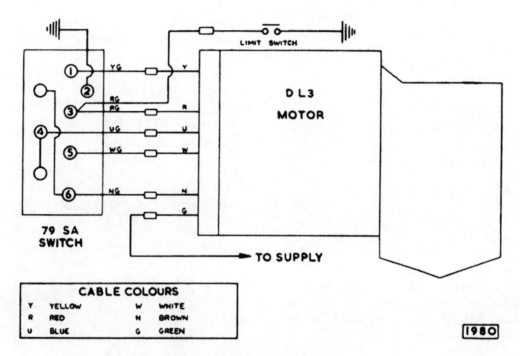

Fig. 40. Wiring connections switch to wiper.

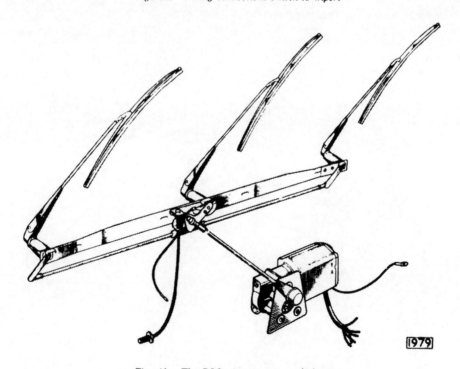

Fig. 41. The DL3 wiper motor and linkage.

Page P.39

ELECTRICAL AND INSTRUMENTS

TESTING

Unless the origin of the fault is apparent, proceed as follows to determine the cause of failure.

Measuring Supply Voltage

Using a first grade moving coil voltmeter, measure the voltage between the motor supply terminal (to which the green cable is connected) and a good earthing point. This should be 11·5 volts with wiper working normally. If the reading is low, check the battery, switch (by substitution), cabling and connections.

Measuring Light Running Current

If the normal terminal voltage is correct, measure the light running current by means of a first grade moving coil ammeter, connected in series with the supply cable.

Remove the windscreen wiper arms and blades.

To Check the "Fast" Speed Current

Using a fully charged 12v battery and two test leads, connect the "GREEN" cable on the wiper motor to the "Negative" battery terminal. Join the "YELLOW" and "RED" cables together and connect to the "Positive" battery terminal. Connect the "BLUE" and "WHITE" cables together. Check the cycles per minute of the wiper spindle.

To Check the "Slow" Speed Current

Connect the "GREEN" cable to the "Negative" battery terminal. Join the "BROWN" and "RED" cables together and connect to the "Positive" battery terminal. Connect the "BLUE AND WHITE" cables together. Check the cycles per minute of the wiper spindle.

The light running current must not exceed:

3·0—3·7 amperes at slow speed—44—48 c.p.m./or r.p.m. of output motor shaft or 2·2—2·9 amperes at fast speed—58—68 c.p.m./or r.p.m. of output motor shaft.

If the current is in excess of these figures change the motor. See DATA chart for other information.

Checking Spindle Housings

Renew seized housings.

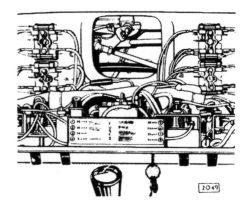

Fig. 43. The central wiper wheel box.

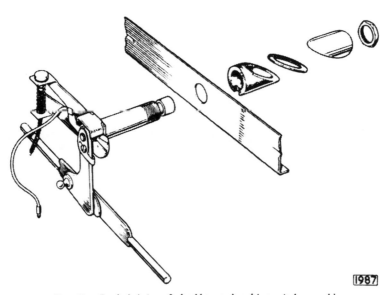

Fig. 42. Exploded view of wheel box and parking switch assembly.

Page P.40

ELECTRICAL AND INSTRUMENTS

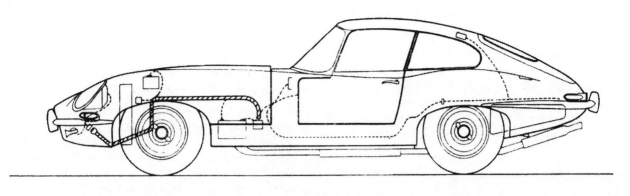

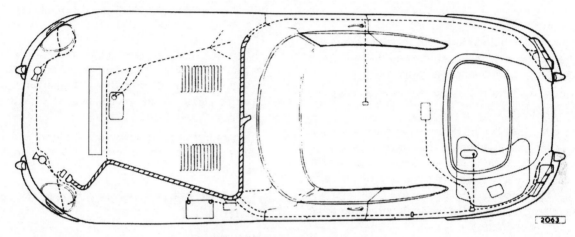

Fig. 44. The layout of wiring harnesses.

MISCELLANEOUS

ELECTRIC CLOCK
Removal

Detach the earth lead from the battery. Remove the revolution counter from the instrument panel as detailed under "Revolution Counter and Clock Removal". Detach the clock from the hidden face of the revolution counter by removing the two nuts. The flexible setting drive can be removed by slackening the knurled nut. Disconnect the cable at the snap connector.

Adjustment

Adjustment is effected by means of a small screw surrounded by a semi-circular seal, located at the back of the instrument.

If the clock is gaining turn the screw towards the minus (—) sign; if the clock is losing turn the screw towards the positive (+) sign.

Note: The action of resetting the hands automatically restarts the clock.

Refitting

Refitting is the reverse of the removal procedure.

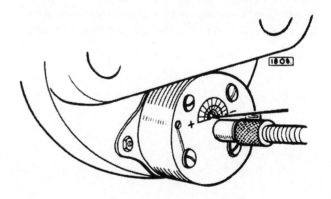

Fig. 45. Adjustment screw for clock.

Page P.41

ELECTRICAL AND INSTRUMENTS

BRAKE FLUID AND HANDBRAKE WARNING LIGHT

Unscrew the bezel of the lamp, exercising care to control the run of the spring loaded bulb beneath. Feed the bulb into the spring-loaded bulb holder, ensure that the red transparent window is retained in the bezel by a small circlip, position the designation plate on the bulb holder and screw on the bezel.

CARBURETTER MIXTURE CONTROL WARNING LIGHT

Renewing the Bulb

Withdraw the bulb holder from the rear of the light unit above the lever quadrant and withdraw the bulb by rotating in an anti-clockwise direction.

Replace the bulb holder and bulb by reversing the removal sequence.

The lamp unit can be removed from the side facia panel after the bulb holder has been removed by unscrewing the body of the unit and withdrawing the red plastic window from the front face of the facia board. The replacement of the lamp unit is the reverse of the removal sequence but the angle terminal bracket must not be omitted.

SETTING THE CARBURETTER MIXTURE CONTROL WARNING LIGHT SWITCH

Set the lever of the carburetter mixture control $\frac{1}{4}"$ (6·350 mm.) from the bottom limit of its travel, when a click will be heard and utilizing the two nuts on the threaded shank of the switch, position the switch so that the warning light ceases to glow when the ignition is switched "on". Actuate the lever up and down once or twice and make any final adjustments necessary.

FLASHING INDICATOR CONTROL

Removal

Detach the earth lead from the battery.

Disconnect the seven cable harness from the snap connectors situated behind the facia panel.

Remove inner half of switch cover by withdrawing towards the centre of the car; cover is retained in position by means of spring clips. Switch and outer half of cover can now be withdrawn after removing the two screws and the clamp retaining the switch to steering column. Detach the outer half of switch cover from switch by removing the two fixing screws.

Refitting

Refitting is the reverse of the removal procedure. Particular attention must be paid to ensure that the switch is positioned correctly on the steering column, that the spigot on the switch is located in the hole drilled in the steering column.

Reconnect cable harness into the multi-snap connector so that similar coloured cables are connected together.

FLASHING DIRECTION INDICATOR WARNING LIGHT BULB

Replacement

Detach the earth lead from the battery. Withdraw one or both of the bulb holders from the rear of the light unit situated between the speedometer and the revolution counter. Remove the bulb from the holder by applying an inward pressure and turning in an anti-clockwise direction.

Refitting is the reverse of the removal sequence. Care must be taken to ensure that the bulb holders are replaced in the correct position, i.e., replace right hand indicator bulb behind right hand arrow.

Page P.42

ELECTRICAL AND INSTRUMENTS

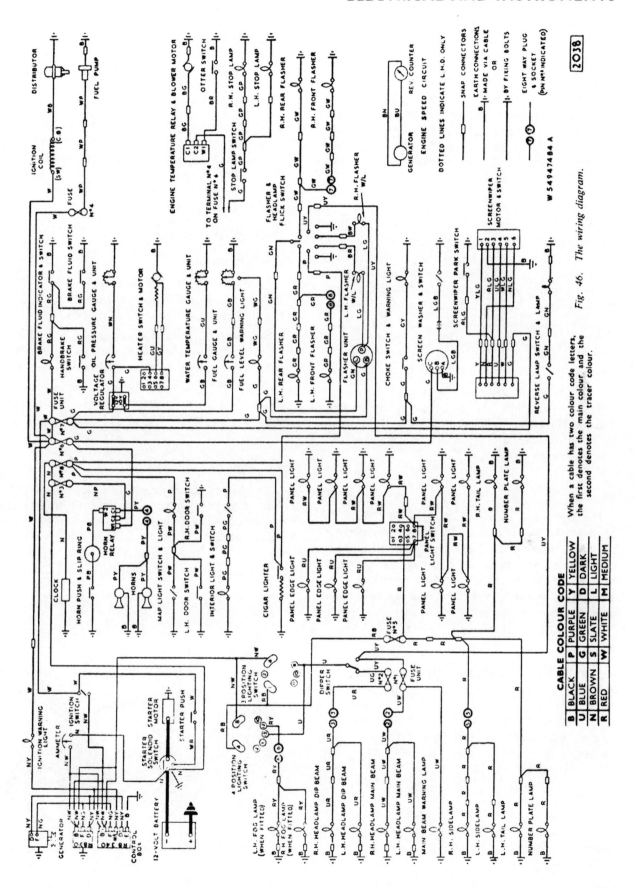

Fig. 46. *The wiring diagram.*

Page P.43

ELECTRICAL AND INSTRUMENTS

THE INSTRUMENTS

DASH CASINGS

Removal

Detach one or both dash casings situated beneath the glove box or side facia panel by withdrawing the drive screws, and in the instance of the dash casing on the steering column side, the screwed bezels of the odometer and clock setting drives.

Refitting

Refitting is the reverse of the removal procedure but in the instance of the dash casing on the steering column side, it will be necessary to attach the odometer and clock setting drives to the casing before attaching the latter to the underside of the instrument panel.

THE INSTRUMENT PANEL

Opening

Detach the earth lead from the battery.
Remove the ignition key and cigar lighter for safe keeping. Hinge the centre instrument panel downwards on its bottom edge, after withdrawing the thumb screws situated in each top corner.

Removal

The ingtrument panel can be removed completely by detaching the earth lead from the battery, identifying and removing the leads from the instruments, cigar lighter and switches, removing the electrical harness and clips from the instrument panel and withdrawing the two hinge pivot bolts from the instrument panel support brackets.

Refitting

Refitting is the reverse of the removal procedure, but particular attention must be given to the following point.
That the leads are refitted in accordance with their colour coding, utilizing the wiring diagram as a reference.

Closing

Closing is the reverse of the opening proceedure but particular attention must be given to the following points:

(i) That the leads are replaced in accordance with their colour coding, utilizing the wiring diagram as a reference.

(ii) That the clips securing the main harness to the instrument panel will in no way foul any of the switch or instrument terminals, otherwise a direct short will occur when the battery is connected.

GLOVEBOX—Removal

Disconnect battery.

Lower the centre instrument panel after removing the two retaining setscrews in the top right hand and top left hand corners.

Remove the three setscrews retaining glove box now exposed. Remove the two nuts retaining glove box to the bracket located on side panel below screen pillar.

Detach glove box and disconnect the heater control cables from heater control quadrant. Remove glove box.

Refitting

Refitting is the reverse of the removal procedure. Care must be taken to ensure that the heater control is connected correctly and full travel of the control maintained.

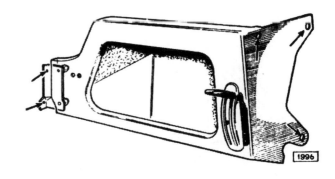

Fig. 47. The glove box showing attachment details.

Page P.44

ELECTRICAL AND INSTRUMENTS

SIDE FACIA PANEL—Removal

Disconnect battery.

Lower the centre instrument panel after removing the two retaining setscrews in the top right hand and top left hand corners. Remove the three setscrews retaining side facia panel now exposed. Remove the two nuts retaining facia panel to the bracket located on side panel below screen pillar.

Disconnect speedo cable from speedometer and the flexible setting cable from the electric clock.

Remove the circular nut retaining dipper switch to panel and remove switch.

Detach facia panel.

Disconnect the brake fluid level warning light cables from the unit and the electric clock cables from the snap connector.

Disconnect the mixture control cable from the mixture control quadrant and detach the warning light unit by withdrawing the bulb holder from the socket. Disconnect the cable from mixture control warning light switch. Disconnect the two cables attached to the revolution counter and remove the ignition main beam and petrol tank warning lights. Detach the two flasher warning light bulbs by withdrawing the bulb holders from the two sockets; withdraw panel illumination bulbs.

Remove facia panel.

Fig. 48. The side facia panel showing attachment details.

Refitting

Refitting is the reverse of the removal procedure. Care must be taken when refitting to ensure that the mixture control cable is connected correctly and the full travel of the control maintained. Replace flasher warning light units in their correct holders. When refitting dipper switch ensure that the two terminals on the switch with the cables coloured blue/yellow and blue/green are uppermost.

THE SPEEDOMETER

Removal

Detach the earth lead from the battery and raise the steering to the highest position. Detach the speedometer from the facia board by removing the two knurled nuts, earth lead and the two retaining pieces.

Withdraw the flexible drive from the centre of the instrument by slackening the knurled sleeve nut.

Remove the speedometer from the facia board; identify and remove the three warning lamps and the two instrument illumination lamps from the hidden face of the instrument. Remove the flexible odometer trip setting drive by slackening the knurled sleeve nut.

Refitting

Refitting is the reverse of the removal procedure but particular attention must be paid to the following points.

(i) That the two instrument illumination lamps are inserted in the apertures at the side of the instrument.

(ii) That the headlamp warning light is inserted in the right hand bottom aperture.

(iii) That the fuel warning light is inserted in the centre bottom aperture.

(iv) That the ignition warning light is inserted in the left hand bottom aperture.

THE REVOLUTION COUNTER AND CLOCK

The revolution counter and clock are of the electrical type and the electrical leads to both are included in the car harness.

The clock is mounted at the bottom of the revolution counter indicator head and to effect its removal it is necessary to remove the revolution counter from the side facia panel.

The revolution counter consists of an A.C. generator fitted to the rear end of the camshaft with an indicator head mounted in the side facia panel.

Page P.45

ELECTRICAL AND INSTRUMENTS

Removal

Detach the earth lead from the battery.

Detach the revolution counter from the facia board by removing the two knurled nuts, earth lead and retaining pieces. Withdraw the revolution counter, remove the two centre leads and the two instrument illumination lamps from the hidden face of the instrument and from the clock at the snap connector.

Detach the flexible clock setting drive by slackening the knurled sleeve nut, and the clock from the revolution counter, by removing the two nuts.

TESTING OPERATION OF REVOLUTION COUNTER

Utilizing an A.C. voltmeter check the current across the terminals of the generator at the rear of the right hand camshaft while the engine is running; as a rough guide it can be assumed that there is one volt output per 100 r.p.m. When electrical current is evident, check the continuity of the two leads by attaching the terminals to the generator and connecting the voltmeter to the opposite ends of the cables after removal from revolution counter. If when running engine continuity is evident, it can be assumed that the instrument is unserviceable and must be exchanged.

THE REVOLUTION COUNTER DRIVE

The revolution counter drive takes the form of a small A.C. electrical generator fitted at the rear R.H. end of the cylinder head where its tongued driving spindle engages a slotted adaptor screwed in the rear end of the inlet camshaft. Leads included in the electrical harness of the car connect with the Lucar tabs pointing upward in the body of the generator and with similar tabs at the rear of the instrument lead in the side facia panel. The Lucar tabs are of the same size and the leads can be fitted either way round.

Removal

Open the engine compartment and detach the earth lead from the battery. Remove the electrical harness from the two Lucar tabs on the A.C. generator on the rear R.H. end of the cylinder head. Detach the A.C. generator from the rear R.H. end of the cylinder head by withdrawing three allen screws and a plate washer, remove the generator in a rearward direction and note the position of the tongued driving spindle.

Refitting

Refitting is the reverse of the removal procedure but particular attention must be given to the following point:

That the tongued driving spindle is positioned in the same attitude as it was when it was removed; whenever difficulty is experienced in engaging the tongued spindle do not apply any force but remove the generator, ascertain the position of the slot in the camshaft with a mirror and set the tongued drive in a similar position.

THE REMOVAL OF THE INSTRUMENT PANEL COMPONENTS

The Ignition Switch

Detach the earth lead from the battery and hinge the instrument panel downward. Identify and remove the leads from the ignition switch. Withdraw the ignition switch from the hidden face of the instrument panel by removing the chrome ring. The lock barrel can be withdrawn by inserting a thin rod through a hole in the body of the switch.

Refitting is the reverse of the removal procedure but particular attention should be given to the following points:

(i) That the number of the ignition key is stamped on the lock barrel.

(ii) That the flat on the thread is positioned toward the right-hand side of the panel.

(iii) That the leads are refitted in accordance to their colour coding, utilizing the wiring diagram as a reference.

Renewing the Cigar Lighter Element

Withdraw the cigar lighter unit from the instrument panel and ensure that it is cold. Place the unit into the palm of the hand, knob first, and hold the sleeve downward against the pressure of the spring with the fingers and unscrew the lighter element and fit a replacement. It must be noted that the spring must not be omitted or tampered with for it ejects the lighter unit when it attains its correct temperature.

Page P.46

ELECTRICAL AND INSTRUMENTS

Cigar Lighter Unit—Removal

Withdraw the cigar lighter unit, detach the earth lead from the battery and hinge the instrument panel downward. Identify and remove the leads from the cigar lighter housing. Withdraw the cigar lighter housing through the face of the instrument panel after removing the nut and 'U' piece from the centre terminal post. It is not wise to dismantle the cigar lighter housing any further, otherwise direct shorting may occur on assembly.

Refitting is the reverse of the removal procedure but particular attention must be given to the following points:

(i) That the centre terminal post is firm and tight.

(ii) That the insulated washer in the 'U' piece is tight and in good condition, a sub-standard fit and poor condition of this washer could cause a direct short.

(iii) That the black lead is attached by its Lucar connection to the tag at the top of the instrument panel and the purple lead from the main harness is attached to the centre terminal post.

The Starter Push Switch

Detach the earth lead from the battery and hinge the instrument panel downward. Identify and remove the leads from the starter push switch. Withdraw the starter push switch through the face of the instrument panel by removing the nut on the hidden face.

The Head and Side Light Switch—Removal

Remove the light switch control lever from the face of the instrument panel by depressing the plunger in the right hand side.

Detach the earth lead from the battery and hinge the instrument panel downward. Identify and remove the leads from the light switch and detach the light switch from the three posts on the hidden face of the instrument panel by removing the three nuts.

The designation plate can be removed from the face of the instrument panel by detaching the nut on the hidden face.

Refitting

Refitting is the reverse of the removal procedure but particular attention must be given to the following points:

(i) That the designation plate is mounted on the face of the instrument panel by allowing the flat on the threaded barrel to locate a flat in the panel.

(ii) That the control lever is pressed on to the rod of the switch protruding through the face of the instrument panel so that the control rod plunger locates a drilling in the hub of the lever, a smear of vaseline on the plunger greatly facilitates this operation.

(iii) That the leads are refitted in accordance to their colour coding utilizing the wiring diagram as a reference.

The Tumbler Type Switches

Detach the earth lead from the battery and hinge the instrument panel downward. Identify and remove the leads from the Lucar tags on the body of the desired switches and withdraw the tumbler switch from the hidden face of the instrument panel by holding the switch lever in a horizontal position and removing the screwed chromium ring from the face of the instrument panel.

Refitting is the reverse of the removal procedure but particular attention must be given to the following points:

(i) That the switch is fitted to the instrument panel so that the flat face of the switch lever is downward.

(ii) That the leads are refitted in accordance to their colour coding and utilizing the wiring diagram as a reference.

The Ammeter and Oil Pressure Gauge—Removal

Detach the earth lead from the battery and hinge the instrument panel downward. Withdraw the illumination bulb holder from the instrument and detach the leads. Remove the two knurled nuts and 'U' clamp.

Withdraw through front face of panel.

Refitting is the reverse of the removal procedure but particular attention must be given to the following points:

(i) That the 'U' piece is fitted so that it will not foul any terminal or bulb holder, one side is cut away for this purpose.

Page P.47

ELECTRICAL AND INSTRUMENTS

(ii) That the leads are refitted in accordance with the colour coding utilizing the wiring diagram as a reference.

The Fuel and Water Temperature Gauges

These instruments are removed and refitted in a similar manner to the ammeter and oil pressure gauges but in this instance only one knurled nut secures the 'U' piece.

The removal and replacement of the fuel gauge tank unit and the water temperature transmitter unit are detailed in the "Fuel System" and "Cooling System" sections respectively.

The Voltage Regulator (Fuel and Water Temperature Gauges)

Removal

Detach the earth lead from the battery and hinge the instrument panel downwards. Identify and remove the leads from the voltage regulator situated at the top right hand side of the instrument panel.

Detach the voltage regulator from the panel by removing one nut.

Refitting

Refitting is the reverse of the removal procedure but particular attention must be given to the following points:

(i) That a good earth is made between the voltage regulator and the panel.

(ii) That the leads are refitted in accordance with the colour coding utilizing the wiring diagram as a reference.

Renewing the Switch Indicator Strip Bulbs

Detach the earth lead from the battery and hinge the instrument panel downwards. Three bulbs are provided, one being in each bottom corner and one at the bottom centre. Withdraw the bulb holder from the socket. Remove the bulb from the holder by applying an inward pressure and rotating 90°. The bulb is replaced by inserting the cap in the holder and rotating 90° until the notches in the bulb holder are located.

Remove the indicator strip, chrome finisher and light filter from the bottom edge of the instrument panel by withdrawing the four screws.

Page P.48

ELECTRICAL AND INSTRUMENTS

THE BI-METAL RESISTANCE INSTRUMENTATION

Engine Temperature, Fuel Tank and Oil Pressure Gauges

DESCRIPTION

The Bi-metal Resistance Instrumentation for engine temperature, petrol tank contents and engine oil pressure consists of a gauge unit fitted in the instrument panel, a transmitter unit fitted in the engine unit or petrol tank and connected together to the battery, the oil pressure gauge being an exception, through a common voltage regulator. The purpose of the latter is to ensure a constant power supply at a predetermined voltage thus avoiding errors due to a low battery voltage. In the instance of the oil pressure gauge this is not quite so critical to supply voltage.

In all systems the gauge unit operates on the thermal principal utilizing a heater winding wound on a bi-metal strip, while the transmitter units of the engine temperature and petrol tank contents gauge are of the resistance type but in both instances the system is voltage sensitive. The transmitter unit of the oil pressure gauge is of the thermal pressure principal utilizing a heater winding wound on a bimetal strip having contact at one end with the second contact mounted on a diaphragm which is sensitive to engine oil pressure.

OPERATION OF THE ENGINE TEMPERATURE GAUGE

The transmitter unit of the engine temperature gauge is fitted in the water outlet pipe of the engine unit and is a variable resistance and consists of a temperature sensitive resistance element contained in a brass bulb. The resistance element is a semi-conductor which has a high negative temperature co-efficient of resistance and its electrical resistance decreases rapidly with an increase in its temperature. As the temperature of the engine unit rises the resistance of the semi-conductor decreases and increases the flow of current through the transmitter similarly a decrease in engine temperature reduces the flow of current.

The gauge unit fitted in the instrument panel consists of a heater winding, connected at one end to the transmitter unit and at the second end to the 'I' terminal of the voltage regulator, wound on a bimetal strip which is linked to the indicator needle. The heater winding and bimetal strip assembly is sensitive to the changes in voltage received from the transmitter unit causing the heater winding to heat or cool in the bimetal strip, resulting in the deflection of the indicator needle over the scale provided. The calibration of the scale is such that the movement of the indicator needle over it is relative to the temperature of the transmitter unit bulb and therefore the temperature of the engine unit.

OPERATION OF THE FUEL TANK GAUGE

The transmitter unit of the petrol gauge is fitted in the petrol tank and is a variable resistance actuated by a float, the arm of which carriers a contact travelling across a resistance housed in the transmitter body. The float arm takes up a position relative to the level of petrol in the tank and thus varies the amount of current passing through the indicator unit.

The gauge unit in the instrument panel consists of a heater winding, connected at one end to the transmitter unit and at the other to the 'I' terminal of the voltage regulator, wound on a bimetal strip which is linked to the indicator needle. The heater winding and bimetal strip assembly is sensitive to the changes in voltage received from the position of the transmitter float, causing the heater winding to heat or cool the bimetal strip, resulting in the deflection of the indicator needle over the scale provided. The calibration of the scale is such that the movement of the indicator needle over it is relative to the position of the transmitter float actuated by the level of the contents in the petrol tank.

Exaggerated indicator needle movement due to petrol swirl in the tank is considerably reduced as there is a delay before current changes from the transmitter unit can heat or cool the bimetal and heater winding assembly in the indicator unit, which in fact causes the deflection of the needle. Similarly the indicator needle will take a few moments to register the contents of the petrol tank when the ignition is first switched on.

Page P.49

ELECTRICAL AND INSTRUMENTS

ANALYSIS OF THE ENGINE TEMPERATURE AND PETROL TANK GAUGE FAULTS

NOTE: THE INSTRUMENT PANEL GAUGES MUST NEVER BE CHECKED BY SHORT-CIRCUITING THE TRANSMITTER UNITS TO EARTH

Symptom	Unit Possibly at Fault	Action
Instrument panel gauge showing a "zero" reading	Voltage regulator	Check that output voltage at terminal 'I' is 10 volts
	Instrument panel gauge	Check for continuity between the gauge terminals with the leads disconnected.
	Transmitter unit in petrol tank or engine unit.	Check for continuity between the terminal and the case with lead disconnected.
	Wiring	Check for continuity between the gauge, the transmitter and the voltage regulator, also that the transmitter unit is earthed.
Instrument panel gauge showing a high/low reading when ignition switched on	Voltage regulator	Check output voltage at terminal 'I' is 10 volts.
	Instrument panel gauge	Check by substituting another instrument panel gauge.
	Transmitter unit in petrol tank or engine	Check by substituting another transmitter unit in petrol tank or engine unit.
	Wiring	Check for leak to earth.
Instrument panel gauge showing a high reading and overheating	Voltage regulator	Check output voltage at terminal 'I' is 10 volts.
	Wiring	Check for short circuits on wiring to each transmitter unit.
Instrument panel gauge showing an intermittent reading	Voltage regulator	Check by substituting another voltage regulator.
	Instrument panel gauge	Check by substituting another instrument panel gauge.
	Transmitter unit in petrol tank or engine unit	Check by substituting another transmitter unit in petrol tank or engine unit.
	Wiring	Check terminals for security, earthing and wiring continuity.

Page P.50

ELECTRICAL AND INSTRUMENTS

OPERATION OF THE OIL PRESSURE GAUGE

The transmitter unit of the oil pressure gauge, fitted in the head of the engine oil filter, is a voltage compensated pressure unit and consists of a diaphragm, a bimetal strip with a heater winding wound thereon, a resistance and a pair of contacts. One contact is attached to the diaphragm while the second is mounted on one end of the bimetal strip, the second end of which is connected through the resistance and the gauge unit to the battery supply; the heater winding is also connected to the battery supply but not through the resistance. Engine oil pressure will close the contacts causing current to flow through the gauge unit, bimetal strip and contacts to earth resulting in the heating of the heater winding which will, after a time, open the contacts.

The gauge unit fitted in the instrument panel consists of a winding, connected at one end to the battery supply and at the second to the transmitter unit wound on to a bimetal strip which is linked to an indicating needle. The heater winding and bimetal strip assembly is sensitive to the continuity changes received from the thermal pressure unit, fitted in the engine oil filter, causing the heater winding to heat or cool the bimetal strip resulting in the deflection of the indicating needle over the scale provided.

The changes in continuity of current from the transmitter unit will vary according to the amount of oil pressure for, as the latter rises, the outward moving diaphragm contact limits the return travel of the bimetal strip contact thus allowing a longer continuity period. This results in a greater heating of the heater winding in the gauge unit and increased deflection of the indicating needle over the scale showing a greater oil pressure.

The opening and closing of the transmitter unit contacts is continuous thus the temperature of the heater winding in the gauge unit is kept within close limits and the calibration of the scale is such that the movement of the indicating needle over it is relative to the opening of the transmitter unit contacts and therefore the oil pressure of the engine is recorded.

ANALYSIS OF THE OIL PRESSURE GAUGE FAULTS

Symptom	Unit Possibly at Fault	Action
Instrument panel gauge showing a "zero" reading	Wiring	Check for continuity between the gauge and the transmitter unit and that the latter is earthed.
	Instrument panel gauge	Check for continuity between the gauge terminals with leads disconnected. If satisfactory replace the transmitter unit.
Instrument panel gauge showing a reading with ignition switched on but engine not running	Transmitter unit on oil filter head	Check by substituting another transmitter unit.
Instrument panel gauge showing a high reading and overheating	Transmitter unit on oil filter head	Check by substituting another transmitter unit.
Instrument panel gauge showing a below "zero" reading with ignition switched off	Instrument panel gauge	Check by substituting another instrument panel gauge.

ELECTRICAL AND INSTRUMENTS

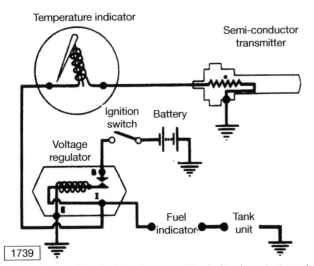

Fig.49 The combined wiring diagram of the fuel tank contents and water termperature guages with the voltage regulator

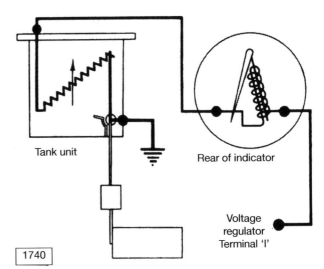

Fig.50 The fuel tank contents and guage circuit

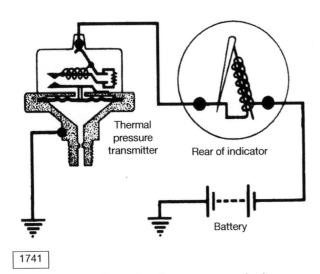

Fig.51 The engine oil pressure guage circuit

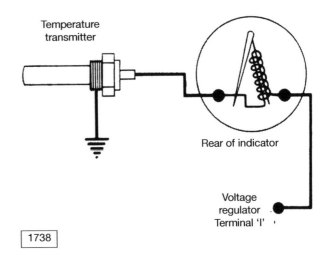

Fig.52 The water temperature guage circuit

Page P.52

ELECTRICAL AND INSTRUMENTS

THE SPEEDOMETER DRIVE CABLE

Removal

Disconnect the flexible drive cable and remove the speedometer from the side instrument facia as previously detailed. Detach the flexible drive cable from the right-angle drive attachment on the gearbox and release it from the retaining clips.

Refitting

Refitting is the reverse of the removal procedure but particular attention must be given to the following points:

(i) That the run of the flexible drive cable is without any sharp bends.

(ii) That the securing clips are so shaped that they only hold the cable in position without crushing it.

SPEEDOMETER CABLE—GENERAL INSTRUCTIONS

Flexible cable condition to a great extent affects performance of speedometers. Poor installation or damage to the flexible drive will show up as apparent faults. It is most important that the flexible drive should be correctly fitted and maintained as illustrated in the following diagrams.

1. **Smooth Run**

   Run of flexible drive must be smooth. Minimum bend radius 6". No bend within 2" of connections.

2. **Securing**

   Avoid sharp bends at clips. If necessary change their positions. Do not allow flexible drive to flap freely. Clip at suitable points.

3. **Securing**

   Avoid crushing flexible drive by over-tightening clip.

4. **Connection**

   Ensure tightness of outer flex connections. They should be finger tight only. It may be necessary to clean thoroughly the point of drive before the connection can be screwed completely home.

5. **Connection of Inner Flexible Shaft**

   Where possible slightly withdraw inner flex and connect outer first. Then slide inner into engagement.

6. **Removal of Inner Shaft**

   Most inner flexes can be removed by disconnecting instrument end and pulling out flex. Broken inner flex will have to be withdrawn from both ends.

7. **Examination of Inner Flexible Shaft**

   Check for kinked inner flexible shaft by rolling on clean flat surface. Kinks will be seen and felt.

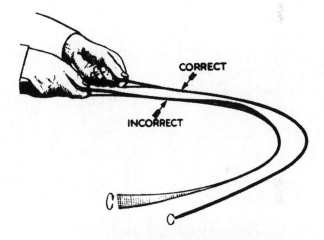

Fig. 53. Checking the inner flex for kinks.

8. **Lubrication Every 10,000 Miles**

   Withdraw inner flexible drive (see paragraph 6). Place blob of grease on end of outer cable and insert flex through it, carrying grease inside. Use Esse T.S.D. 119 or equivalent. Do NOT use oil.

9. **Excessive Lubrication**

   Avoid excessive lubrication. If oil appears in flexible drive, suspect faulty oil-seal at point of drive.

Page P.53

ELECTRICAL AND INSTRUMENTS

10. Inner Shaft Projection

Check ⅜" projection of inner flex beyond outer casing at instrument end. This ensures correct engagement in instrument and point of drive.

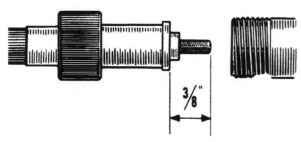

Fig. 54. Showing the amount the inner flex must protrude from outer cable.

11. Concentric Rotation

Check that inner flex rotates in centre of outer cable.

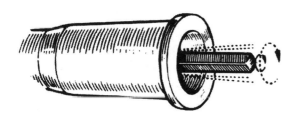

Fig. 55. Checking the inner flex for "run-out."

12. Damaged Inner Shaft

Examine inner flex ends for wear or other damage. Before fitting new flex ensure instrument main spindle is free.

13. Damage Drive End Connections

Examine point of drive for damage or slip on gears in gearbox.

14. Ensuring Correct Drive Fitted

When ordering, state Make, Year and Model of vehicle. State also length of drive required when alternatives are shown.

SPEEDOMETERS—GENERAL INSTRUCTIONS

Speedometer performance is dependent on the flexible drive, and apparent faults in the instrument may be due to some failure of the drive. Before returning a speedometer for service, the flexible drive should be checked, as described in the previous paragraphs. The following diagrams show you how to check the instrument performance.

15. Instrument Not Operating

Flexible drive not properly connected (see paragraph 5). Broken or damaged inner flexible shaft or fault at point of drive (see paragraphs 12 and 13), in which case remove and replace flex (see paragraphs 6 and 8) or rectify point of drive fault. Insufficient engagement of inner shaft (see paragraph 10). Defective instrument—return for service.

16. Instrument Inaccurate

Incorrect speedometer fitted. Check code number.

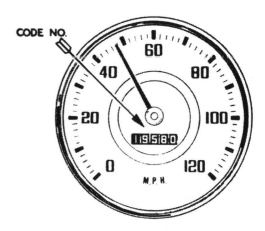

Fig. 56. Showing the code number on the face of the instrument.

Page P.54

ELECTRICAL AND INSTRUMENTS

17. Speedometer Inaccurate
Check tyre pressures. Inaccuracy can be caused by badly worn tyres. Non-standard tyres fitted, apply to Smiths for specially calibrated instrument.

18. Speedometer Inaccurate
Rear axle non-standard. Drive ratio in vehicle gearbox non-standard. A rapid and simple check is obtained by entering in the formula the figures found in the test (see paragraph 19).

$$\frac{1680 N}{R} = \text{T.P.M. No.}$$

Where N = Number of turns made by the inner shaft for 6 turns of rear wheel and R = Radius of rear wheel in inches measured from centre of hub to ground.

Example

Cardboard pointer on inner shaft (see 19) rotates $9\frac{1}{4}$ times as vehicle is pushed forward 6 turns of rear wheel. Rear wheel radius $12\frac{1}{4}''$.

Flex turns per mile:

$$\frac{1680 \times 9\frac{1}{4}}{12\frac{1}{4}} = \frac{15330}{12\frac{1}{4}} = 1251 = \text{T.P.M. No.}$$

19. Gearing Test
Disconnect flexible drive from speedometer. With the gears in neutral, count the number of turns of the inner shaft for six turns of the rear wheels when the vehicle is pushed forward in a straight line. Measure rolling radius of rear wheels—centre of hub to ground. Apply figures in formula (see paragraph 18).

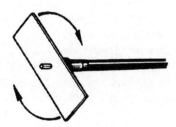

Fig. 57 Cardboard counter on the inner flex for checking the number of turns.

20. Correct Speedometer
Number illustrated should correspond within 25 either way with the number obtained from paragraphs 18 and 19. If it does not, apply to Smiths for specially calibrated instrument, giving details of test and vehicle.

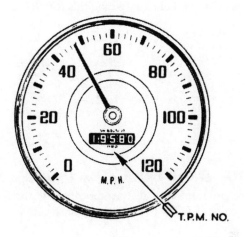

Fig. 58 Showing the turns per mile on the face of the instrument.

21. Pointer Waver
Oiled up instrument. Replace oil seal if necessary, clean and lubricate flexible drive (see paragraph 8). Return instrument for replacement.

22. Pointer Waver
Inner flexible shaft not engaging fully. Check 10, then try 4. Also check 12.

23. Pointer Waver
Kinked or crushed flexible drive. Check 7 and 3. For withdrawal of inner shaft see paragraph 6. Bends of too small radius in flexible drive, check 1.

24. Pointer Waver
If 21, 22 and 23 show no sign of trouble, instrument is probably defective. Return for replacement.

25. Noisy Installation
Tapping noises. Check 5 and 2. Flexible drive damaged. Check 7 and 12 (also see paragraph 6), check lubrication is sufficient. Check 10 and 11.

Page P.55

ELECTRICAL AND INSTRUMENTS

26. Noisy Installation
General high noise level. Withdraw inner shaft (see paragraph 6) and reconnect outer flex. If noise continues at lower level then source of noise is in vehicle point of drive. Fitting new P.V.C. covered flexible drive with nylon bush on inner shaft and instrument with rubber mounted movement should overcome this trouble.

27. Noisy Installation
Regular ticking in time with speedometer decimal distance counter. Return speedometer for replacement.

28. Noisy Installation
Loud screeching noise more prevalent in cold weather return instrument for replacement.

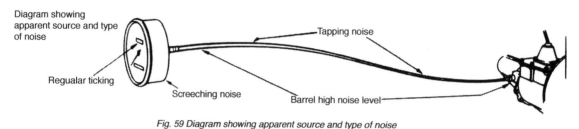

Fig. 59 Diagram showing apparent source and type of noise

RIGHT ANGLE DRIVE ATTACHMENT

No provision is made for lubrication or dismantling this unit. If faulty remove and replace with new unit.

Removal
Detach the speedometer cable from unit.
Remove unit from gearbox by releasing the large thumb nut.

Refitting
Refitting is the reverse of the removal procedure but particular attention must be given that the square drive shaft protruding from the unit has entered into the gearbox drive correctly before tightening nut.

SUPPLEMENTARY INFORMATION TO SECTION P "ELECTRICAL AND INSTRUMENTS"

Introduction of Modified Distributor

Later 3·8 Litre "E" type cars were fitted with a new distributor—the Lucas 22D6. This type of distributor differs from its predecessor—DMBZ6 type—in construction but the advance curves remain the same.

The method of adjusting the contact points differs also.

If the gap is incorrect, slacken, very slightly, the contact plate securing screw and adjust the gap by turning a screwdriver in the slot in the contact plate (clockwise to decrease the gap and anti-clockwise to increase the gap). Tighten the securing screw and re-check the gap.

The correct gap remains the same—·014"–·016" (·36 mm.–·41 mm.).

Electric Time Clock

	Commencing Chassis Numbers	
	R.H. Drive	L.H. Drive
"E" Type Open 2 Seater	850702	879324
Fixed Head Coupe	861169	888543

Commencing at the above chassis numbers the electric clock fitted to the revolution counter dial incorporates a rectifier. This is to reduce fouling of the contact points in the clock.

If at any time the clock is removed for servicing and subsequent bench testing, IT IS MOST IMPORTANT that the feed terminal on the back of the clock is connected to the negative side of the battery and that the outer casing of the clock is positively earthed. Incorrect connection of a rectified clock to the battery will instantly destroy the rectifier.

Page P.56

SUPPLEMENTARY INFORMATION

FOR

4·2 LITRE "E" TYPE AND 2+2 CARS

(SERIES 1)

This Supplement covers the variations between the 4·2 Litre "E" Type, 2+2 cars and the 3·8 Litre versions of the "E" Type. Insert the Supplement at the end of the 3·8 Litre "E" Type Service Manual, Publication No. E.123.

ISSUED BY

JAGUAR CARS LIMITED, COVENTRY, ENGLAND

Telephone	Code	Telegraphic Address
ALLESLEY 2121 (P.B.X.)	BENTLEY'S SECOND	"JAGUAR," COVENTRY. Telex. 31622

Publication No. E123B/3

INDEX TO SECTIONS

SECTION TITLE	SECTION REFERENCE
ENGINE	BX
CARBURETTERS AND FUEL SYSTEM	CX
COOLING SYSTEM	DX
CLUTCH	EX
GEARBOX	FX
AUTOMATIC TRANSMISSION	FFX
FRONT SUSPENSION	JX
BRAKES	LX
BODY AND EXHAUST SYSTEM	NX
HEATING AND WINDSCREEN WASHER	OX
ELECTRICAL AND INSTRUMENTS	PX

SECTION B
ENGINE

DATA

Camshaft

    Permissible end float ..   ·004″ to ·006″ (·10 to ·15 mm.)

Connecting Rod

    Big end—Diameter clearance ..   ·0015″ to ·0033″ (·037 to ·083 mm.)

Crankshaft Main Bearings

    Journal diameter ..   2·750″ to 2·7505″ (69·85 to 69·86 mm.)

    Journal length

        —Front .. $1\frac{9}{16}″$ (39·06 mm.)

        —Centre .. $1\frac{3}{8}″$ + ·001″ (34·37 mm. + ·025 mm.)
                                                       − ·0005″   − ·0125 mm.

        —Rear .. $1\frac{11}{16}″$ (42·86 mm.)

        —Intermediate .. $11\frac{7}{32}″$ ± ·002″ (30·96 mm. ± ·05 mm.)

Cylinder Block

    Bore size for fitting liners ..   3·761″ to 3·762″ (94·03 to 94·05 mm.)

    Outside diameter of liner ..   3·765″ to 3·766″ (94·13 to 94·15 mm.)

    Interference fit ..   ·003″ to ·005″ (·08 to ·13 mm.)

    Overall length of liner ..   6·959″ to 6·979″ (17·39 to 17·45 cm.)

    Outside diameter of lead-in ..   3·758″ to 3·760″ (93·95 to 94·00 mm.)

    Size of bore honed after assembly—cylinder block—Nominal ..   92·07 mm. (3·625″)

Gudgeon Pin

    Length ..   3·00″ (75 mm.)

Page B.X.s.1

ENGINE

Piston and Piston Rings

Gudgeon pin bore	·8571″ to ·8753″ (2·188 to 2·1883 mm.)
Piston rings—Width Compression	·0770″ to ·0780″ (1·97 to 2·00 mm.)
Oil Control	Self expanding (Maxiflex)
Piston rings—Thickness	·151″ to ·158″ (3·775 to 3·95 mm.)
Piston rings—Gap when fitted to cylinder bore	
Oil Control	·015″ to ·033″ (·38 to ·82 mm.)

ROUTINE MAINTENANCE

DAILY
Check the engine oil level.

EVERY 3,000 MILES (5,000 KM.)
Drain the engine sump; renew oil filter element and seal; refill with new oil.

Top up carburetter hydraulic piston dampers.

Check carburetter slow running (700 r.p.m.-all synchromesh gearbox, 500 r.p.m. 2+2 automatic transmission).

Lubricate distributor and check contact points.

Clean, adjust and test sparking plugs.

EVERY 6,000 MILES (10,000 KM.)

Carry out 3,000 mile (5,000 km.) service.

Tune carburetters.
Clean fuel feed line filter.
Adjust top timing chain (if necessary).
Check alternator belt for wear.

EVERY 12,000 MILES (20,000 KM.)
Carry out 3,000 mile and 6,000 mile (5,000 and 10,000 km.) service.

Renew air cleaner element.
Renew sparking plugs.

ENGINE

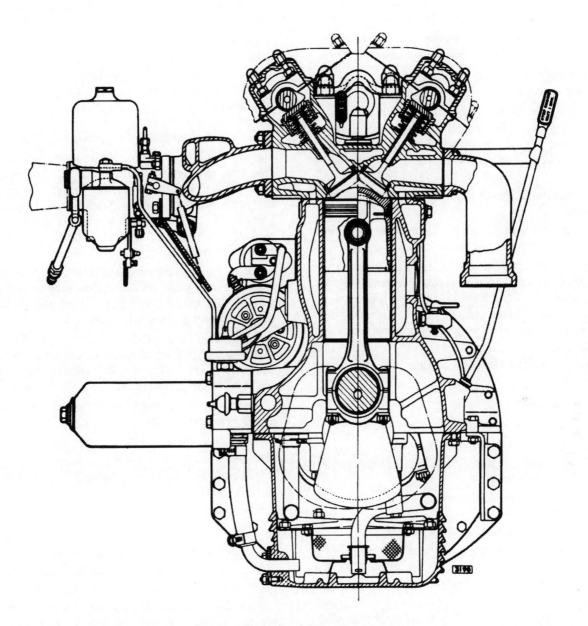

Fig. 1. *Cross sectional view of the engine.*

Page B.X.s.3

ENGINE

ENGINE REMOVAL AND REFITTING

REMOVAL

Remove the bonnet.

Disconnect the battery.

Drain the cooling system and cylinder block; conserve the coolant if antifreeze is in use.

Slacken the clip on the breather pipe; unscrew the two wing nuts and withdraw the top of the air cleaner.

Disconnect the petrol feed pipe from under the centre carburetter.

Slacken the clamps and remove the water hoses from the cylinder head and radiator to the header tank. Slacken the two clamps and withdraw the water pump hose. Remove the heater hoses from the inlet manifold.

Disconnect the brake vacuum hose from the inlet manifold.

Pull off the two Lucar connectors from the fan thermostat control in the header tank.

Remove the two bolts securing the header tank mounting bracket to the front cross member. Remove the two nuts and two bolts securing the header tank straps to the radiator and fan cowl. Withdraw the header tank complete with mounting bracket and straps.

Disconnect the throttle linkage at the rear carburetter.

Disconnect:—

    The two coil leads.

    The water temperature transmitter.

    The battery cable and solenoid switch cable from the starter.

    The main output cables from the alternator (on early cars, note the location of each Lucar connector to ensure correct refitting).

    On early cars, the cable from the switch on the right hand side of the engine block operating the oil pressure/ignition warning light.

    On later cars, the Lucar connector for the 3AW warning light control on the alternator.

    The engine earth strap from the left hand side member.

Withdraw the oil filter canister; catch the escaping oil in a drip tray.

Remove the crankshaft pulley, damper and drive belt. Mark the pulley and damper to facilitate refitting. Remove the ignition timing pointer from the sump.

Remove the revolution counter generator complete with cables.

Remove the four nuts and washers securing each exhaust downpipe from the manifold. Unclip the pipes at the silencers and withdraw the downpipes. Collect the sealing rings between the pipes and the manifold.

On standard transmission cars, proceed as follows:—

Remove the seats. Remove the knob and locknut from the gear lever. Remove two hexagon headed setscrews and two chromium-plated nuts and detach the radio/ash tray console panel from the gearbox tunnel. If a radio is fitted, disconnect the electrical cables from the control head to enable the panel to be completely removed.

On 2+2 cars, raise the central arm-rest; lift out the bottom panel; withdraw five self-tapping screws and remove the central arm-rest. Lift off the trimmed cover panel from the gearbox tunnel.

On all other cars, withdraw two pan-headed screws and two seat belt attachments before lifting off the trimmed cover. Withdraw the self-tapping screws and remove the gearbox cover.

Disconnect the reverse lamp cables from the switch on the gearbox top cover.

Disconnect the speedometer drive cable from the gearbox.

Remove the clutch slave operating cylinder from the clutch cover.

Disconnect the propeller shaft.

On automatic transmission cars, proceed as follows:—

Withdraw the transmission dipstick and unscrew the dipstick tube from the transmission oil pan.

Place the selector lever in L and, from underneath the car, unscrew the nut securing the selector cable adjustable ball joint to the transmission lever. Release the nut securing the outer cable clamp to the abutment bracket.

Remove the speedometer drive cable from the transmission extension housing.

Page B.X.s.4

ENGINE

Disconnect the transmission oil cooler pipes from the right hand side of the radiator block and from the transmission unit. Withdraw the clips and remove the pipes.

Disconnect the kickdown cable at the rear of the cylinder head.

Remove the central arm-rest and lift off the trimmed cover panel from the gearbox tunnel. Withdraw the drive screws securing the cover plate on the transmission tunnel. Disconnect the propeller shaft.

For all models, proceed as follows:—

Remove the nuts securing the torsion bar reaction tie plate on each side and tap the bolts back flush with the face of the tie plate. With the aid of a helper, place a lever between the head of the bolt just released and the torsion bar. Exert pressure on the bolt head to release the tension on the upper bolt. Remove the nut and tap the upper bolt back flush with the face of the tie plate. Withdraw the bolts securing the tie plate on each side to the body underframe channels through the side members. Tap the tie plate off the four bolts.

Note: Failure to relieve the tension on the upper bolts when tapping them back to the tie plate will result in stripping the threads. If this occurs, new bolts must be fitted and the torsion bars re-set.

Support the engine by means of two individual lifting tackles using the hooks provided on the cylinder head. Insert a trolley jack under the transmission (or gearbox) and support the transmission.

Remove the self-locking nut and washer from the engine stabiliser.

Remove the bolts securing the rear engine mounting plate. Remove the bolts from the front engine mountings.

Raise the engine on the lifting tackles and, keeping the combined engine and transmission assembly level, move forwards ensuring that the water pump pulley clears the sub-frame top cross member. Carefully raise the front of the engine and withdraw forwards and upwards.

REFITTING

Refitting is the reverse of the removal procedure. After the unit is in place, it is important that the engine stabiliser is adjusted and that the clutch slave cylinder is mounted correctly.

On automatic transmission 2+2 cars, the kickdown cable must be adjusted and the manual linkage connected in accordance with the instructions given in Section FF.

Page B.X.s.5

ENGINE

THE CYLINDER BLOCK

OVERHAUL

Reboring is normally recommended when the bore wear exceeds ·006″ (·15 mm). Reboring beyond the limit of ·030″ (·76 mm) is not recommended and, when the bores will not clean out at ·030″ (·76 mm), liners and standard size pistons should be fitted.

The worn liners must be pressed out from below utilising the stepped block illustrated.

PISTONS AND GUDGEON PINS

Piston Grades

Grade Identification Letter	To suit cylinder bore size
F	3·6250″ to 3·6253″ (92·075 to 92·0826 mm.)
G	3·6254″ to 3·6257″ (92·0852 to 92·0928 mm.)
H	3·6258″ to 3·6261″ (92·0953 to 92·1029 mm.)
J	3·6262″ to 3·6265″ (92·1055 to 92·1131 mm.)
K	3·6266″ to 3·6269″ (92·1156 to 92·1123 mm.)

Oversize Pistons

Oversize pistons are available in the following sizes: +·010″ (·25 mm.) ±·020″ (·51 mm.) +·030″ (·76 mm.).

There are no selective grades in oversize pistons as grading is necessary purely for factory production methods.

Tapered Periphery Rings

All engine units are fitted with tapered periphery piston rings and these must be fitted the correct way up.

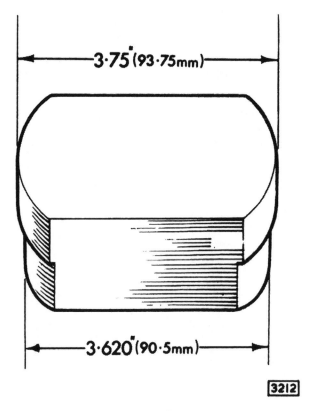

Fig. 2. *Stepped block for cylinder liner removal.*

Page B.X.s.6

ENGINE

The narrowest part of the ring must be fitted uppermost; to assist in identifying the narrowest face a letter "T" or "Top" is marked on the side of the ring to be fitted uppermost.

The Maxiflex oil control ring consists of two steel rails with a spacer between. These rails are held together on assembly with an adhesive. The expander, which is fitted inside the oil control ring, should be assembled with the two lugs positioned in the hole directly above the gudgeon pin bore.

Later engines are fitted with Hepworth and Grandage pistons which have a solid skirt. The oil control rin on these pistons is of similar construction to the Maxiflex ring but the ends of the expander ring (internal ring) are butted together. If the internal ring is fitted to the piston groove with the ehds overlapping, the outer ring assembly cannot be seated properly.

Pistons

Skirt clearance (measured at bottom of skirt at 90 to gudgeon pin pin axis) ·0007" to ·0013" (·018 to ·03 mm.)

Ring gap—when fitted to bore
Top compression ·015" to ·020" (·38 to ·51 mm.)
Lower compression ·010" to ·015" (·254 to ·38 mm.)
Scraper ·015" to ·045" (·38 to 1·143 mm.)
Side clearance in groove ·001" to ·003" (·02 to ·07 mm.)

Gudgeon Pins
Grades (Red) ·8753" to ·8754" (22·23 to 22·24 mm.)
(Green) ·8752" to ·8753" (22·22 to 22·23 mm.)
Clearance in piston ·0001" to ·0003" (·0025 to ·0076 mm.)

Cargraph Treatment—Piston Rings

The chromium plated ring (top compression) is Cargraph treated on the outside diameter to assist in bedding in the chromium surface. This coating is coloured Red for identification purposes and **should not be removed**. Excess oil or grease may be removed with clean paraffin but **rings should not be soaked in any degreasing agent**.

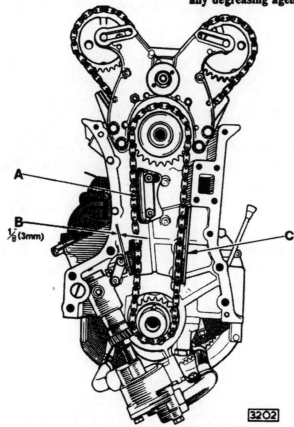

Fig. 3. *The timing gear arrangement.*

When fitting a new lower timing chain, set the intermediate damper (A) in light contact with the chain when there is a ⅛" (3 mm) gap between the rubber slipper and the tensioner body. In the case of a worn chain, the gap (B) may have to be increased to avoid fouling between the chain and the cylinder block. Set the lower damper (C) in light contact with the chain.

Page B.X.s.7

ENGINE

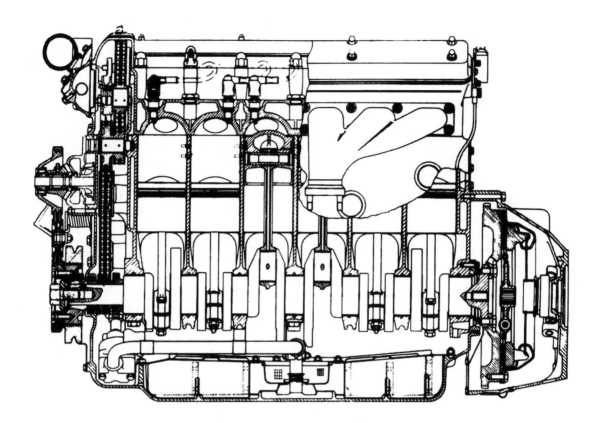

Fig. 4. *Longitudinal section of the engine.*

CYLINDER HEAD GASKET

Commencing at engine number 7E.9210 on 4·2 Litre "E" type cars and 7E.50963, 2+2 cars a new type of cylinder head gasket is fitted. This gasket is thicker than the previous type and is of asbestos compound coated steel-backed construction.

The two faces are treated with a double varnish finish which eliminates the use of any jointing compound when fitting a new gasket to the engine. It is interchangeable with its predecessor but, if difficulty is experienced with a tight top timing chain due to the increased thickness, a replacement idler eccentric shaft (Part No. C.27189) must be fitted.

The cylinder head torque tightening figure has been increased to 58 lb. ft. (8·0 kg/m) with the fitment of the thicker gasket.

It should be noted that certain individual engines prior to those quoted were fitted with the new cylinder head gasket.

OIL SUMP

REMOVAL

Drain the sump; disconnect the oil return pipe and remove the crankshaft damper.

Remove the self-locking nut and washer from the top of the engine stabiliser. Screw down the lower flanged washer to the limit of the stud thread.

Sling the engine from the rear lifting loop and raise the engine approximately 1" (25·4 mm.).

Remove the sump securing screws, lower the front end and withdraw forward.

REFITTING

Refitting is the reverse of the removal procedure but care must be taken to ensure that the rear oil seal is positioned correctly. Adjust the engine stabiliser after refitting.

Check for oil leakage after refilling the sump and running the engine.

Page B.X.s.8

ENGINE

VALVE GUIDES

Later engines will have circlips fitted to the valve guides to ensure positive location in the cylinder head. These valve guides are chamfered at the upper ends and have the outside diameter reduced at the lower end to provide a "lead-in" when fitting.

From engine number 7E.11668 oil seals are fitted to the inlet valve guide—a second groove being machined in the guide above the circlip groove to seat the oil seal.

Checking Valve Guides

Examine the guides for evidence of wear in the bore. The clearance between the valve stem and the guide when new is ·001″–·004″ (·025–·10 mm).

If it is found necessary to renew worn valve guides, they must be fitted in accordance with the following instructions and only genuine factory replacement parts used.

Valve Guide—Replacement

Heat the cylinder head by immersing in boiling water for 30 minutes. With a piloted drift, drive out the old valve guide from the combustion chamber end.

Note: If carbon deposits around the valve guide in the combustion chamber are quite heavy, they should be cleaned off thoroughly before attempting to drive out the old valve guide.

Valve guides when fitted during engine assembly are to the following dimensions and may be fitted in mixed form.
(1) ·501″ to ·502″ (12·70 to 12·725 mm.)
(2) ·503″ to ·504″ (12·776 to 12·801 mm.)

The valve guide (2) will be identified by the machining of one circular groove on the shank of the guide: valve guide (1) will **not** have the groove.

When removing worn guides, care must be taken to identify each individual guide to its particular bore in the cylinder head.

Replacement guides are available in the following sizes and will have identification grooves machined in the shank as noted:—

1st oversize	·503″ to ·504″
(one groove)	(12·776 to 12·801 mm.)
2nd oversize	·506″ to ·507″
(two grooves)	(12·852 to 12·877 mm.)
3rd oversize	·511″ to ·512″
(three grooves)	(12·979 to 13·005 mm.)

Valve guides with one groove should only be fitted as replacements for those originally fitted without a groove: the bore in the cylinder head will not require reaming before fitting.

Guides with two grooves should be used as replacements for those with one groove and guides with three grooves for those with two. Cylinder head bores should be reamed to the following dimensions:—

Valve Guide	Ream to Size
2nd oversize (two grooves)	·505″ $^{+.0005″}_{-.0002″}$ (12.83 mm. $^{+.012 mm.}_{-.005 mm.}$)
3rd oversize (three grooves)	·510″ $^{+.0005″}_{-.0002″}$ (12.95 mm. $^{+.012 mm.}_{-.005 mm.}$)

Coat the valves with graphite grease and fit the circlips. Reheat the cylinder head. With a piloted drift, drive in the valve guide from the top until the circlip registers in the groove machined in the guide bore of the cylinder head. Visually check that the circlip has seated correctly.

Page B.X.s.9

SECTION C

CARBURETTERS AND FUEL SYSTEM

CARBURETTERS

Removal

Drain the cooling system.

Disconnect the battery.

Slacken the hose clip securing the water hose from the inlet manifold to the header tank. Remove the hose.

Disconnect the two electrical connections from the thermostat fan control in the header tank.

Remove the throttle return springs.

Unclip hose connection to breather pipe.

Remove the two butterfly nuts at the carburetter trumpets and remove the air cleaner elbow.

Remove the carburetter trumpet from the carburetters having removed the six nuts and spring washers together with the three gaskets.

Disconnect the throttle linkage at the rear carburetter.

Remove the three banjo union bolts and six fibre washers from the float chambers.

Ensure that the three float chamber filters are not mislaid.

Disconnect the mixture control outer and inner cables.

Remove the suction pipe from the front carburetter.

Disconnect the brown/black cable from the oil pressure switch.

Slacken the clips and disconnect the heater pipes at the water manifold and below the inlet manifold.

On 2+2 cars fitted with automatic transmission, disconnect the kickdown cable at the rear of the cylinder head.

Fig. 1. *Refitting the mixture control rods with the jet levers against the stops.*

CARBURETTERS AND FUEL SYSTEM

Remove the inlet manifold complete with the carburetters and linkage.

Remove the four nuts and spring washers, together with the return spring bracket from each carburetter. Remove all three carburetters together.

If necessary, remove the mixture control linkage from each carburetter by removing the split pins and withdrawing the clevis pins.

Refitting

Refitting is the reverse of the removal procedure except that new gaskets should be fitted to the inlet manifold, to either side of the heat insulating gasket and also to the carburetter trumpet flanges.

Adjust the kickdown cable as detailed on page FF.s.24.

ROUTINE MAINTENANCE
EVERY 3,000 MILES (5,000 KM.)

Lubricate carburetter hydraulic piston damper.

Check carburetter slow running.

EVERY 6,000 MILES (10,000 KM.)

Tune carburetters.

Clean carburetter filters.

Clean fuel feed line filter.

CARBURETTER TUNING

The method of tuning carburetters is identical with that given for 3·8 litre "E" Type cars, however, the idling speed on standard transmission cars should be 700 r.p.m. in order to eliminate any chatter from the constant mesh gears in the all-synchromesh gearbox.

On automatic transmission 2+2 cars, the idling speed should be 500 r.p.m.

INTRODUCTION OF VITON TIPPED NEEDLES

At engine number 7E.2226, Viton tipped needles are fitted to the carburetter float chambers. They are identified by a black rubber tip.

If used as replacements for the previous type needle, **the needle and seat assembly** must be used.

FUEL FEED LINE FILTER

Model	Chassis Number	
	R.H. Drive	L.H. Drive
Open 2 Seater	1E.1905	1E.16057
Fixed Head Coupe	1E.21662	1E.34772
2+2	1E.50143	1E.77701

On cars bearing the above chassis numbers and subsequent, the fuel feed line filter incorporates a renewable fibre filter element. This element should not be cleaned but must be renewed every 12,000 miles. When renewing, the two sealing washers should also be replaced.

If sediment build-up is excessive, the element should be renewed more frequently than stated above.

THE FUEL SYSTEM

THE PETROL PUMP

Description (Fig. 2)

The pump consists of three main assemblies, the main body casting (A); the diaphragm armature and magnet assembly (M) contained within the housing; and the contact breaker assembly housed within the end cap (T2). A non-return valve assembly (C) is affixed to the end cover moulding to assist in the circulation of air through the contact breaker chamber.

The main fuel inlet (B) provides access to an inlet air bottle (I) while access to the main pumping chamber (N) is provided by an inlet valve assembly. This assembly consists of a Melinex valve disc (F) permanently assembled within a pressed-steel cage, held in position by a valve cover (E1).

The outlet from the pumping chamber is provided by an identical valve assembly which operates in the reverse direction. Both inlet and outlet valve assemblies together with the filters are held in position by a clamp plate (H). The valve assemblies may be removed by detaching the clamp plate (H) after removing the self-tapping screws. A filter (E) is provided on the delivery side of the inlet valve assembly. The delivery chamber (O) is bounded by a flexible plastic spring loaded diaphragm (L) contained by the vented cover (P). Sealing of the diaphragm (L) is provided by the rubber sealing ring (L.2).

The magnetic unit consists of an iron coil housing, an iron core (Q), an iron armature (A1) provided with a central spindle (P1) which is permanently united with the diaphragm assembly (L1), a magnet coil (R) and a contact breaker assembly consisting of parts (P2), (U1), (U), (T1) and (V). Between the coil housing and the armature are located eleven spherically edged rollers (S). These rollers locate the armature (A1) centrally within the coil housing and permit freedom of movement in a longitudinal direction.

The contact breaker consists of a bakelite pedestal moulding (T) carrying two rockers (U) and (U1) which are both hinged to the moulding at one end by the rocker spindle (Z). These rockers are interconnected at their top ends by means of two small springs arranged to give a throw-over action. A trunnion (P2) is carried by the inner rocker and the armature spindle (P1) is screwed into this trunnion. The outer rocker (U) is fitted with two tungsten points which contact with corresponding tungsten points which form part of the

Page C.X.s.2

CARBURETTERS AND FUEL SYSTEM

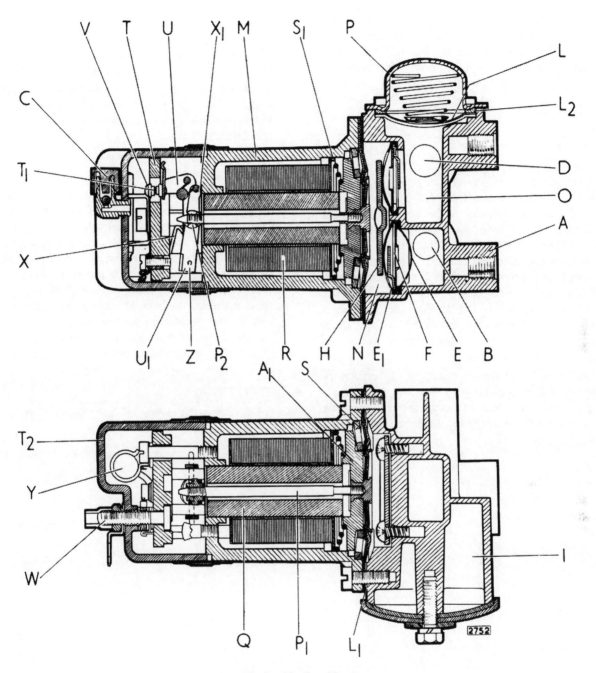

Fig. 2. *The Petrol Pump.*

WARNING: If at any time, it becomes necessary to blow through the fuel feed pipes the outlet pipes must be disconnected from the pumps. Failure to observe this procedure will cause the Melinex valves to be displaced or damaged.

Page C.X.s.3

CARBURETTERS AND FUEL SYSTEM

spring blade (V) connected with one end of the coil. The other end of the coil is connected to a terminal (W) while a short length of flexible wire (X) connecting the outer rocker to one of the screws holding the pedestal moulding onto the coil housing provides an earth return to the body of the pump. It is important that the body of the pump be effectively earthed to the body of the vehicle by means of the earthing terminal provided on the flange of the coil housing.

OPERATION

When the pump is at rest the outer rocker (U) lies in the outer position and the tungsten points are in contact. Current passes from Lucar connector (W) through the coil and back to the blade (V), through the points and to earth, thus energising the coil and attracting the armature (A1). The armature, together with the diaphragm assembly, then retracts thereby sucking petrol through the inlet valve into the pumping chamber (N). When the armature has travelled nearly to the end of its stroke, the throw-over mechanism operates and the outer rocker moves rapidly backwards, thus separating the points and breaking the circuit.

The spring (S1) then reasserts itself forcing the armature and diaphragm away from the coil housing. This action forces petrol through the delivery valve at a rate determined by the requirements of the engine.

As the armature nears the end of its stroke the throw-over mechanism again operates, the tungsten points remake contact and the cycle of operations is repeated.

The spring blade (V) rests against the small projection moulding (T) and it should be set so that, when the points are in contact, it is deflected away from the moulding. The gap at the points should be approximately ·030″ (·75 mm.) when the rocker (U) is manually deflected until it contacts the end face of the coil housing.

REMOVAL

Remove both inlet and outlet pipes from the side of the pump by withdrawing the banjo bolt and washers. Disconnect the electrical feed cable to the pump by unscrewing the knurled knob on the end of the pump. Remove the two self-locking nuts attaching the pump to the bracket and withdraw the two washers from each stud. The pump can now be withdrawn from the bracket leaving the two rubber grommets in position. The rubber grommets in the brackets should be examined for deterioration and replaced if necessary, otherwise excessive petrol pump noise may result.

REFITTING

Refitting is the reverse of the removal procedure.

Page C.X.s.4

CARBURETTERS AND FUEL SYSTEM

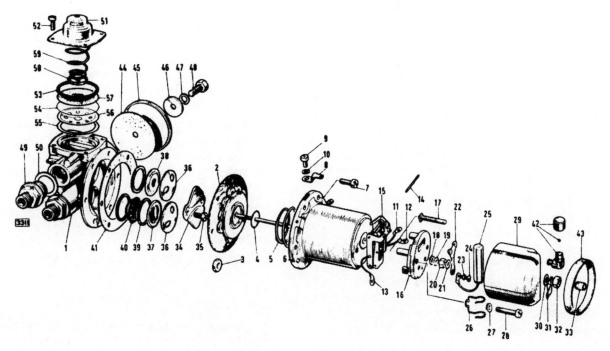

Fig. 3. *Exploded view of the petrol pump.*

1. Pump body.
2. Diaphragm and spindle assembly.
3. Roller—armature centralising.
4. Washer—impact.
5. Spring—armature.
6. Housing—coil.
7. Screw—securing housing—2 B.A.
8. Connector—earth.
9. Screw—4 B.A.
10. Spring washer.
11. Terminal tag.
12. Terminal tag.
13. Earth tag.
14. Rocker pivot pin.
15. Rocker mechanism.
16. Pedestal.
17. Terminal stud.
18. Spring washer.
19. Lead washer.
20. Terminal nut.
21. Washer.
22. Contact blade.
23. Washer.
24. Screw.
25. Condenser.
26. Clip.
27. Spring washer.
28. Screw.
29. End cover.
30. Shakeproof washer.
31. Lucas connector.
32. Nut.
33. Insulating sleeve.
34. Clamp plate.
35. Screw.
36. Valve cap.
37. Inlet valve.
38. Outlet valve.
39. Sealing washer.
40. Filter.
41. Gasket.
42. Vent valve.
43. Sealing band.
44. Joint.
45. Inlet air bottle cover.
46. Dished washer.
47. Spring washer.
48. Screw.
49. Outlet connection.
50. Fibre washer.
51. Cover.
52. Screw.
53. 'O' ring.
54. Diaphragm barrier.
55. Sealing washer.
56. Diaphragm plate.
57. Diaphragm.
58. Spring end cap.
59. Diaphragm spring.

Page C.X.s.5

CARBURETTERS AND FUEL SYSTEM

DISMANTLING (Fig. 3).

Contact Breaker

Remove the insulated sleeve (33), terminal nut (32), and connector (31), together with its shakeproof washer (30). Remove the tape seal (if fitted) and take off the end-cover.

Unscrew the 5 B.A. screw (24) which holds the contact blade (22) to the pedestal (16) and remove the condenser (25) from its clip. This will allow the washer (23), terminal tag (11), and the contact blade to be removed.

Coil housing and diaphragm

Unscrew the coil housing securing screws (7), using a thick-bladed screwdriver to avoid damaging the screw heads.

Remove the earthing screw (9).

The coil housing (6) may now be removed from the body (1). Next remove the diaphragm and spindle assembly (2) by taking hold of the diaphragm and unscrewing it anti-clockwise until the armature spring (5) pushes the diaphragm away from the coil housing. It is advisable to hold the housing over the bench so that the 11 brass rollers (3) will not fall on the floor. The diaphragm and its spindle are serviced as a unit and should not be separated.

Pedestal and rocker

Remove the end-cover seal washer (21), unscrew the terminal nut (20), and remove the lead washer (19). This will have flattened on the terminal tag and thread and is best cut away with cutting pliers or a knife. Unscrew the two 2 B.A. screws (28), holding the pedestal to the coil housing, remove the earth terminal tag (13) together with the condenser clip (26). Tip the pedestal and withdraw the terminal stud (17) from the terminal tag (12). The pedestal (16) may now be removed with the rocker mechanism (15) attached.

Push out the hardened steel pin (14) which holds the rocker mechanism to the pedestal and separate the two.

Body and valves

Unscrew the two Phillips screws (35) securing the valve clamp plate (34), remove the valve caps (36), valves (37) and (38), sealing washers (39) and filter (40).

Page C.X.s.6

Note: Dismantling of the delivery flow-smoothing device should only be undertaken if the operation of it is faulty, and if the necessary equipment for pressure-testing after assembly is available. On this understanding proceed as follows:

Remove the four 4 B.A. screws (52) securing the delivery flow-smoothing device vented cover (51), remove the cover, the diaphragm spring (59), rubber 'O' ring (53), spring cap (58), diaphragm (57), barrier (54), diaphragm plate (56) and sealing washer (55).

Remove the single 2 B.A. screw (48), securing the inlet air bottle cover (45). Remove the cover and gasket (44).

Unscrew the inlet and outlet connections.

INSPECTION

If gum formation has occurred in the fuel used in the pump, the parts in contact with the fuel will have become coated with a substance similar to varnish. This has a strong stale smell and may attack the neoprene diaphragm. Brass and steel parts so affected can be cleaned by being boiled in a 20 per cent. solution of caustic soda, dipped in a strong nitric acid solution and finally washed in boiling water. Light alloy parts must be well soaked in methylated spirits and then cleaned.

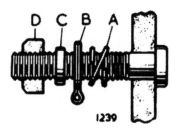

Fig. 4. *The terminal arrangement.*

A—*Double coil spring washer.*
B—*Cable tag.*
C—*Lead washer.*
D—*Countersunk nut.*

CARBURETTERS AND FUEL SYSTEM

Clean the pump and inspect for cracks, damaged joint faces and threads.

Examine the plastic valve assemblies for kinks or damage to the valve plates. They can best be checked by blowing and sucking with the mouth.

Check that the narrow tongue on the valve cage, which is bent over to retain the valve and to prevent it being forced out of position, has not been distorted but allows a valve lift of approximately $\frac{1}{16}$ in. (1·6 mm.).

Examine the delivery flow-smoothing device diaphragm, barrier, plate, spring, and spring cap for damage. If in doubt, renew the diaphragm.

Examine the inlet air bottle cover for damage. Examine the valve recesses in the body for damage and corrosion; if it is impossible to remove the corrosion, or if the recess is pitted, the body must be discarded.

Clean the filter with a brush and examine for fractures, renew if necessary.

Examine the coil lead tag for security and the lead insulation for damage.

Examine the contact breaker points for signs of burning and pitting; if this is evident, the rocker assembly and spring blade must be renewed.

Examine the pedestal for cracks or other damage, in particular to the narrow ridge on the edge of the rectangular hole on which the contact blade rests.

Examine the non-return vent valve in the end-cover for damage, ensure that the small ball valve is free to move.

Examine the diaphragm for signs of deterioration.

Renew the following parts: all fibre and cork washers, gaskets, and 'O' section sealing rings, rollers showing signs of wear on periphery, damaged bolts, and unions.

ASSEMBLY

Pedestal and rocker

Note: The steel pin which secures the rocker mechanism to the pedestal is specially hardened and must not be replaced by other than a genuine S.U. part.

Invert the pedestal and fit the rocker assembly to it by pushing the steel pin (14, Fig. 3) through the small holes in the rockers and pedestal struts. Then position the centre toggle so that, with the inner rocker spindle in tension against the rear of the contact point, the centre toggle spring is above the spindle on which the white rollers run. This positioning is important to

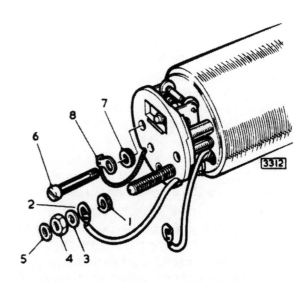

Fig. 5. *Attaching the pedestal to the coil housing.*

obtain the correct "throw over" action; it is also essential that the rockers are perfectly free to swing on the pivot pin and that the arms are not binding on the legs of the pedestal.

If necessary the rockers can be squared up with a pair of thin-nosed pliers.

Assemble the square-headed 2 B.A. terminal stud to the pedestal, the back of which is recessed to take the square head.

Assemble the 2 B.A. spring washer (1) (Fig. 5), and put the terminal stud through the 2 B.A. terminal tag (2), then fit the lead washer (3) and the coned nut (4) with its coned face to the lead washer. (This makes better contact than an ordinary flat washer and nut).

Tighten the 2 B.A. nut and finally add the end-cover seal washer (5).

Assemble the pedestal to the coil housing by fitting the two 2 B.A. pedestal screws (6), ensuring that the spring washer (7) on the left-hand screw (9 o'clock position) is between the pedestal and the earthing tag (8). When a condenser is fitted, its wire clip base is placed under the earthing tag and the spring washer is not required.

Tighten the screws, taking care to prevent the earthing tag (8) from turning, as this will strain or break the earthing flex. Do not tighten the screws or the pedestal will crack.

Do not fit the contact blade at this stage.

Page C.X.s.7

CARBURETTERS AND FUEL SYSTEM

Diaphragm assembly

Place the armature spring into the coil housing with its larger diameter towards the coil (5, Fig. 3).

Before fitting the diaphragm make sure that the impact washer is fitted to the armature. (This is a small neoprene washer that fits in the armature recess). Do not use jointing compound or dope on the diaphragm.

Fit the diaphragm by inserting the spindle in the hole in the coil and screwing it into the threaded trunnion in the centre of the rocker assembly.

Screw in the diaphragm until the rocker will not "throw over"; this must not be confused with jamming the armature on the coil housing internal steps.

Fit the 11 brass centralizing rollers (3, Fig. 3) by turning back the diaphragm edge and dropping the rollers into the coil recess. The pump should be held in the left hand, rocker end downwards, to prevent the rollers from falling out.

Fit the contact blade and adjust the finger settings as described in "Contact gap setting", then carefully remove the contact blade.

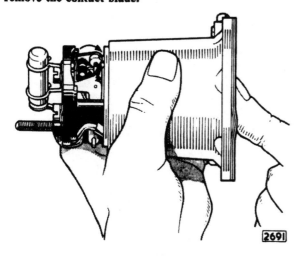

Fig. 6. *Setting the diaphragm.*

Holding the coil housing assembly in the left hand in an approximately horizontal position (see Fig. 6), push the diaphragm spindle in with the thumb of the right hand, pushing firmly but steadily. Unscrew the diaphragm, pressing and releasing with the thumb of the right hand until the rocker just "throws over". Now turn the diaphragm back (unscrew) to the nearest hole and again **4** holes (two-thirds of a complete turn). The diaphragm is now correctly set.

Press the centre of the armature and fit the retaining fork at the back of the rocker assembly.

This is done to prevent the rollers from falling out when the coil housing is placed on the bench prior to fitting the body, and is not intended to stretch the diaphragm before tightening the body screws.

Body components

The valve assemblies are retained internally in the body by a clamp plate secured with self-tapping screws (35, Fig. 3). The inlet valve recess in the body is deeper than the outlet recess to allow for the filter and extra washer. Another feature of these pumps is the incorporation of an air bottle on the inlet and a flow-smoothing device on the delivery side.

The inlet air bottle is a chamber in the body casting blanked off by a simple cover and joint washer held by a single screw. The delivery flow-smoothing device is formed by a perforated metal plate which is in contact with a plastic barrier backed by a rubber diaphragm, all held in position by a spring and end-cap retained by a vented cover. This assembly seals the delivery chamber in the body.

Screw in the inlet and outlet connections with their sealing rings. Assemble the outlet valve components into the outless recess in the following order, first a joint washer, then the valve, tongue side downwards, then the valve cap.

Assemble the inlet valve into the inlet recess as follows: first a joint washer, then the filter, dome side downwards, then another joint washer, followed by the valve assembly, tongue side upwards, then the valve cap.

Take care that both valve assemblies nest down into their respective recesses, place the clamp plate on top, and tighten down firmly to the body with the two screws.

Replace the inlet air bottle cover with its joint washer and tighten down the central screw.

Place the sealing washer in the bottom of the delivery flow-smoothing device recess, follow this with the perforated diaphragm plate, dome side downwards, then the plastic barrier, followed by the rubber diaphragm. Insert the "O" section sealing ring into the recess and ensure that it seats evenly. Place the diaphragm spring, large end towards the vented cover, into the cover, place the spring end-cap on the small end of the spring, pass the assembly tool through the cover, spring, and end cap and turn it through 90° so that tension may be applied to the spring during assembly. Finally fit the spring and cap assembly onto the diaphragm, tighten the four retaining screws, and

Page C.X.s.8

CARBURETTERS AND FUEL SYSTEM

release the assembly tool. The pump should be pressure-tested after disturbance of the flow-smoothing device.

Body attachment

Fit the joint washer to the body, aligning the screw holes.

Offer up the coil housing to the body, ensuring correct seating between them.

Line up the six securing screw holes, making sure that the cast lugs on the coil housing are at the bottom, insert the six 2 B.A. screws finger-tight. Fit the earthing screw with its Lucar connector.

Remove the roller retaining fork before tightening the body securing screws, making sure that the rollers retain their position; a displaced roller will cut the diaphragm. It is not necessary to stretch the diaphragm before tightening the securing screws.

Tighten the securing screws in sequence as they appear diametrically opposite each other.

Contact blade (Fig. 7)

Fit the contact blade and coil lead to the pedestal with the 5 B.A. washer and screw. The condenser should be fitted with the tag placed under the coil lead tag.

Adjust the contact blade so that the points are a little above the points on the rocker when closed, also that when the contact points make or break, one pair of points completely covers the other. As the contact blade is provided with a slot for the attachment screw, some degree of adjustment is possible.

Tighten the contact blade attachment screw when the correct setting is obtained.

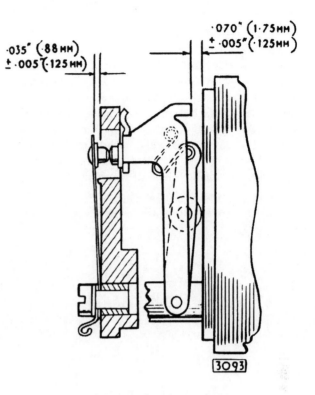

Fig. 7. *Rocker and contact clearances.*

Page C.X.s.9

CARBURETTERS AND FUEL SYSTEM

Contact gap setting

Check that when the outer rocker is pressed onto the coil housing, the contact blade rests on the narrow rib or ridge which projects slightly above the main face of the pedestal. If it does not, slacken the contact blade attachment screw, swing the blade clear of the pedestal, and bend it downwards a sufficient amount so that when repositioned it rests against the rib lightly, over-tensioning of the blade will restrict the travel of the rocker mechanism.

Correct positioning gives a gap of ·035″ ±·005″ (·9±·13 mm.) between the pedestal and tip of spring blade (Fig. 7).

Check the gap between rocker finger and coil housing with a feeler gauge, bending the stop finger, if necessary, to obtain a gap of ·070±·005 in. (1·8±·13 mm.).

End-cover

Tuck all spare cable into position so that it cannot foul the rocker mechanism. Ensure that the end-cover seal washer is in position on the terminal stud, fit the bakelite end-cover and lock washer, secure with the brass nut, fit the terminal tag or connector, and the insulated sleeve.

The pump is now ready for test.

After test, replace the rubber sealing band over the end cover gap and seal with adhesive tape.

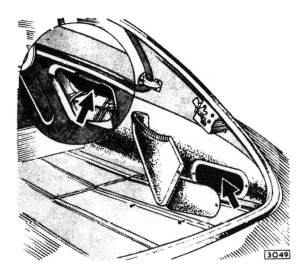

Fig. 8. *The location of the petrol pump.*
(Fixed head coupe).
Inset shows location in open 2-seater model.

SECTION D
COOLING SYSTEM

RADIATOR

The copper radiator is of the cross flow type having 10 cooling fins per inch (4 fins/cm). It is pressurised by means of a filler cap incorporated in the separate radiator header tank. The filler cap incorporates a pressure relief valve which is designed to hold a pressure of up to 7 pounds per square inch (0·49 kg/cm^2) above atmospheric pressure inside the system. When the pressure rises above seven pounds

Fig. 1. *Radiator drain tap.*

the spring loaded valve lifts off its seat and the excess pressure escapes via the overflow pipe. As the water temperature falls again a small valve incorporated in the centre of the pressure valve unit, opens and restores atmospheric pressure should a depression be caused by a fall in the temperature of the water.

By raising the pressure inside the cooling system the boiling point of the water is raised approximately six degrees thus reducing the risk of water loss from boiling.

Removal

Drain the radiator; conserve the coolant if anti-freeze is in use. Remove the three hoses from the top of the radiator and the single hose at the bottom.

On automatic transmission 2+2 cars, disconnect the oil cooler pipes at the right hand side of the radiator block.

Remove the two bolts and nuts securing the radiator top support brackets to the header tank mounting; remove the two bottom fixing nuts and rubber mounting washers.

Release radiator duct panel from the bottom of radiator by removing the two setscrews. Lift out the radiator matrix ensuring that the radiator fan blades are not damaged during the removal of the radiator.

Refitting

Refitting is the reverse of the removal procedure.

RADIATOR COWL (Figs. 2, 3 and 4)

Removal

Remove the radiator as described on page D.s.1.

Remove the two self-locking nuts and plain washers securing the cowl to the bottom of the radiator.

Remove the two self-locking nuts securing the radiator steady brackets to the top of the radiator.

Collect the spacers between the radiator and brackets.

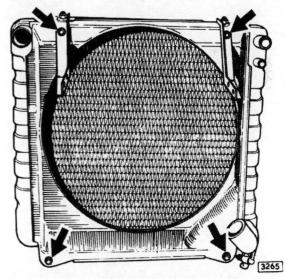

Fig. 2. *Bolts securing the radiator cowl.*

Remove the radiator cowl and sealing rubber.

Remove the two nuts, bolts and washers securing the radiator cowl to the radiator steady brackets and collect the spacers.

Page D.X.s.1

COOLING SYSTEM

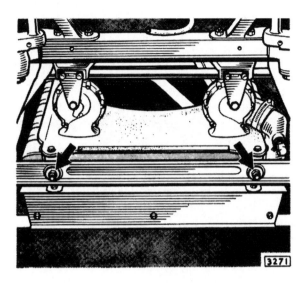

Fig. 3. *Nuts securing radiator to sub-frame.*

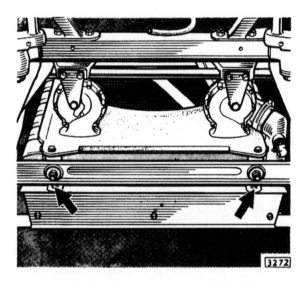

Fig. 4. *Bolts securing radiator cowl to the radiator closing bracket.*

Refitting

Refitting is the reverse of the removal procedure.

RADIATOR HEADER TANK (Fig. 5).

Removal

Drain the coolant from the radiator and cylinder block. Remove the water hose from the cylinder head to the header tank; remove the two water hoses from the header tank to the radiator.

Pull off the two Lucar connections from the thermostatic fan control switch.

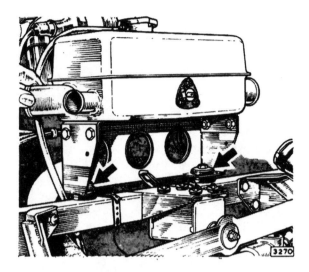

Fig. 5. *Radiator header tank mounting points.*

Withdraw the nuts and bolts securing the radiator tie bars to the radiator and cowl.

Remove the two bolts securing the header tank support bracket to the cross member collecting the rubber mounting pads, distance collars and washers. Withdraw the header tank complete with mounting bracket and radiator tie bars.

Separate the header tank from the mounting bracket and tie bars by removing the four setscrews.

Refitting

The electrical connections on the black/red wire should be attached to the centre connector on the thermostatic fan control in the header tank. Fit the black wire to the earth connection. Refitting is the reverse of the removal procedure.

FAN MOTOR

Removal

Disconnect the negative lead on the battery.

Remove the four self-locking nuts and plain washers securing the fan motor to the front sub-assembly.

Remove the electric motor from its mountings and disconnect the two electrical connections.

Withdraw the electric motor and fan blades from the right-hand side between the radiator and frame assembly.

Refitting

Refitting is the reverse of the removal procedure.

Page D.X.s.2

COOLING SYSTEM

Fig. 6. *Removal of water pump and alternator drive belt.*

WATER PUMP BELT

The drive belt should be examined for wear periodically. Routine adjustment is not necessary as the drive belt is automatically adjusted by means of a spring loaded jockey pulley.

Removal (Fig. 6).

To remove the belt, release the top mounting bolt (B) in the alternator supporting strap (the nut is welded to the support bracket and cannot be turned). Release the bottom mounting nut (C) and swing the alternator inwards on bolt (A) to release the belt.

Refitting

Refitting is the reverse of the removal procedure.

THERMOSTAT (Fig. 7).

This is a valve incorporated in the cooling system which restricts the flow of coolant through the radiator until the engine has reached its operating temperature, thus providing rapid warming up of the engine and, in cold weather, an early supply of warm air to the interior of the car by way of the heater. When the engine temperature rises to a pre-determined figure (see Thermostat Data) the thermostat valve commences to open permitting water to circulate through the radiator. The flow of water increases as the temperature rises until the valve is fully open. Included in the system is a water by-pass utilising a slot in the thermostat housing integral with the water outlet pipe, this allows the coolant to by-pass the radiator until the thermostat opening temperature is attained.

Removal

Drain sufficient water from the system to allow the level to fall below the thermostat by operating the drain tap situated at the bottom left-hand side of the radiator block.

Slacken the hose clip and remove the top water hose from the elbow pipe on the thermostat housing. Remove the two nuts and spring washers securing the water outlet elbow and remove elbow. Lift out the thermostat, noting the gasket between the elbow pipe and thermostat housing.

Checking

Thoroughly clean the thermostat and check that the small hole in the valve is clear. Check the thermostat for correct operation by immersing in a container of cold water together with a thermometer and stirrer.

Heat the water, keeping it well stirred and observe if the characteristics of the thermostat are in agreement with the data given under Thermostat Temperatures.

Refitting

Refitting is the reverse of the removal procedure.

Always fit a new gasket between the elbow pipe and the thermostat housing. Ensure that the recess in the thermostat housing and all machined faces are clear.

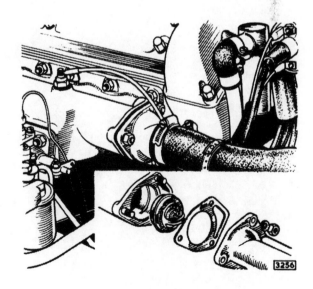

Fig. 7. *Removal of the thermostat.*

Page D.X.s.3

COOLING SYSTEM

FAN THERMOSTATIC SWITCH

The fan thermostatic switch is situated in the separate radiator header tank. When the water reaches a temperature of approximately 86°C. the thermostatic switch operates and automatically starts the fan motor. The fan motor will continue to run until the temperature of the coolant has fallen to approximately 74°C.

Removal

Disconnect the two electrical connections from the thermostatic switch.

Remove the three securing setscrews and washers.

Withdraw the thermostatic switch and remove the cork gasket.

Fig. 8. *Removal of the fan thermostatic switch.*

Refitting

Refitting is the reverse of the removal procedure, but a new cork gasket must be fitted when the switch is replaced. Fit the connection on the red/black wire to the centre connector on the thermostatic switch. Fit the black wire to the earth connection. If any water has escaped during the removal of the switch the radiator header tank should be topped up to the correct level.

1. Impeller.
2. Seal.
3. Thrower.
4. Spindle and bearing assembly.
5. Gasket.
6. Pump body.
7. Allen-headed lockscrew.
8. Locknut.
9. Pulley carrier.
10. Pulley.
11. Spring washer.
12. Setscrew.
13. Drive belt.
14. Adaptor for heater return pipe.
15. Copper washer.

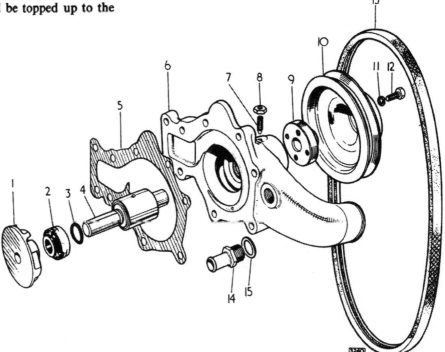

Fig. 9. *Exploded view of the water pump.*

Page D.X.s.4

COOLING SYSTEM

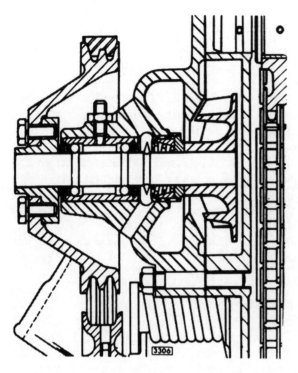

Fig. 10. *Sectioned view of water pump.*

Filling Up

Remove the radiator and expansion tank filler caps. Fill the radiator to the bottom of the filler neck. Replace the filler cap and tighten down fully.

Top up the expansion tank to the half-way mark, refit the cap and tighten down fully.

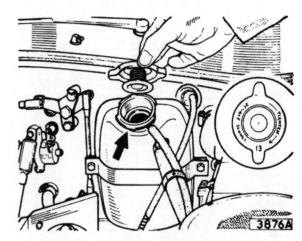

Fig. 11. The expansion tank and pressure cap. (Inset shows the pressure cap fitted to cars with Air-Conditioning Equipment).

Model	Chassis Number	
	R.H. Drive	L.H. Drive
Open 2 Seater	1E.2051	1E.15980–U.S.A. Only
		1E.16010–Other than U.S.A.
Fixed Head Coupe	1E.21807	1E.34583–U.S.A. Only
		1E.34752–Other than U.S.A.
2+2	1E.51213	1E.77709

From the above chassis numbers cars now have a sealed cooling system with a vertical flow radiator and an expansion tank.

The thermostat is retained in a revised housing.

The radiator top tank incorporates a plain (non-pressure) cap, the pressure cap being fitted to the expansion tank mounted on the bulkhead.

PRESSURE CAP RATING

With Standard Equipment	7 lb.
With Air-Conditioning System	13 lb.

Instructions for filling or checking the coolant level in the system differ from those stated for earlier cars fitted with the cross flow radiator, as follows:—

Note: Care must be taken to ensure that the radiator and the expansion tank filler caps are not reversed.

Checking the Coolant Level

IMPORTANT: The coolant level must be checked at the expansion tank and NOT at the radiator top tank.

Check when the system is COLD.

Remove the pressure cap and top up to the half-way mark in the tank.

Replace the pressure cap and tighten down fully.

Refilling the Cooling System— Important

When refilling the cooling system following complete drainage, place the heater temperature control in the "Hot" position to allow the heater circuit to be filled with coolant. Re-check the level after running the engine for a short period.

Page D.X.s.5

COOLING SYSTEM

THE RADIATOR
Removal
Release the filler cap, open the drain tap and drain the cooling system. Conserve the coolant if anti-freeze is in use.

Disconnect the multi-pin socket from the left-hand side of the bonnet.

Remove two bolts, self-locking nuts and washers securing the bonnet linkage to the sub-frame.

Withdraw two hexagon-headed pivot pins and washers securing the bonnet pivot to the sub-frame front lower cross tube, and remove the bonnet assembly.

Release the hose clips and disconnect the top and bottom hoses from the radiator.

Disconnect the oil cooler pipes (2 + 2 automatic transmission cars only) and blank off the unions to prevent loss of oil.

Remove six setscrews securing the cowl to the matrix side brackets. Disconnect the fan thermostat switch cables at the cable junction.

Release the radiator duct panel from the bottom of the matrix by removing two setscrews.

Remove the two bottom fixing nuts and rubber mounting washers.

Lift out the radiator matrix and collect the rubber washers fitted between the bottom tank and the mounting brackets.

NOTE : If air-conditioning equipment is fitted to the car, the condenser unit should be left in position after removal of the two setscrews securing the side support brackets to the matrix.

DO NOT DISTURB THE HOSE CONNECTIONS AT THE CONDENSER UNIT. IT IS DANGEROUS FOR AN UNQUALIFIED PERSON TO ATTEMPT TO DISCONNECT OR REMOVE ANY PART OF THE AIR-CONDITIONING SYSTEM.

Care must be taken when removing the radiator matrix that the fins of the condenser are not damaged.

Refitting
Refitting is the reverse of the removal procedure.

THERMOSTAT
The thermostat differs from that stated on Page D.X.s.3 in respect of the mounting only.
Removal
Drain sufficient coolant from the system to allow the level to fall below the thermostat.

Disconnect the three hoses from the thermostat housing.

Remove three nuts and washers and detach the housing to gain access to the thermostat.
Refitting
Refitting is the reverse of the removal procedure. Renew all gaskets.

To avoid distortion of the flange faces do not over-tighten the nuts.

RADIATOR COWL
Removal
Disconnect the cables from the twin fan motors.

Remove six setscrews securing the cowl to the radiator and remove the cowl complete with fan motors and mounting brackets.
Refitting
Refitting is the reverse of the removal procedure.

FAN MOTORS
Remove the fan cowl as detailed above.

Remove three nuts and setscrews securing each fan mounting bracket to the cowl and detach the bracket assembly.

Remove four nuts and washers securing each motor and detach the motor units from the brackets.

SECTION E
CLUTCH

DESCRIPTION

From the introduction of the 4.2 Litre 'E' Type to chassis No. 7E.13500 (Open 2-seater and Fixed Head Coupe) and 7E.53581 (2+2) a Laycock diaphragm spring clutch is fitted consisting of a spring assembly held flexibly in the lugs of the pressure plate by a spring retaining ring and pivoting on a fulcrum formed by the rims of the clutch cover and the driving plate. Depressing the clutch pedal actuates the release ring causing a corresponding depression of the diaphragm. The lever action of the spring pulls the pressure plate from the driven plate, thus freeing the clutch.

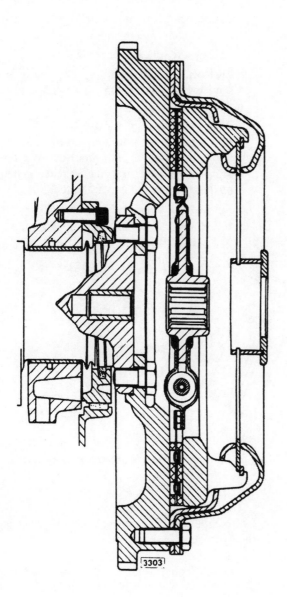

Fig. 1. *Sectioned view of clutch.*

Page E.X.s.1

CLUTCH

DATA

Make	Laycock
Model	10" Diaphragm Spring
Release Ring Travel	·400" (10 mm.)
Clutch Release Bearing	Graphite
Operation	Hydraulic

ROUTINE MAINTENANCE

WEEKLY

Check fluid level in clutch master cylinder reservoir.

EVERY 3,000 MILES (5,000 KM)

Check clutch pedal free travel (not hydrostatic clutch).

HYDROSTATIC CLUTCH SLAVE CYLINDER

Commencing at engine number 7E.4607, an hydrostatic clutch operating slave cylinder is fitted. Normal clutch wear is automatically compensated for by this slave cylinder and no clearance adjustment is necessary.

Fitting

It is important that the operating rod adjustment dimension as shown in Fig. 2 is adhered to when replacing a slave cylinder or clutch unit.

To obtain this dimension proceed as follows:—

1. Extract the clevis pin securing the operating rod to the clutch lever.
2. Release the fork end locknut.
3. Push the clutch operating lever away from the slave cylinder until resistance is felt and retain in this position.
4. Push the operating rod to the limit of its travel into the slave cylinder and adjust the fork end to a dimension of ·75" (19 mm.) between the centre of the fork end and the centre of the clutch operating lever. Tighten the locknut.
5. Release the operating rod and connect the fork end to the lever. Refit the clevis pin.
6. Bleed the hydraulic system in the normal manner.

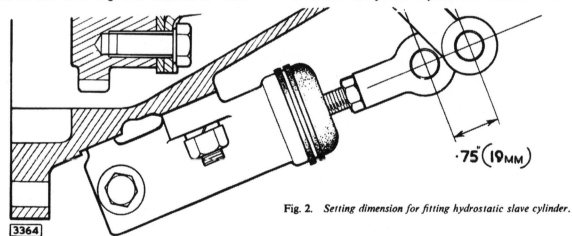

Fig. 2. *Setting dimension for fitting hydrostatic slave cylinder.*

Clutch Slave Cylinder Model	Engine Number
Open 2 Seater and Fixed Head Coupe	7E.18356
2+2	7E.55558

From the above engine numbers, a modified (non-hydrostatic) clutch slave cylinder was fitted replacing the hydrostatic unit.

Normal clutch wear is not automatically adjusted and, should adjustment be required, instructions given on Page E.7 should be followed.

Page E.X.s.2

CLUTCH

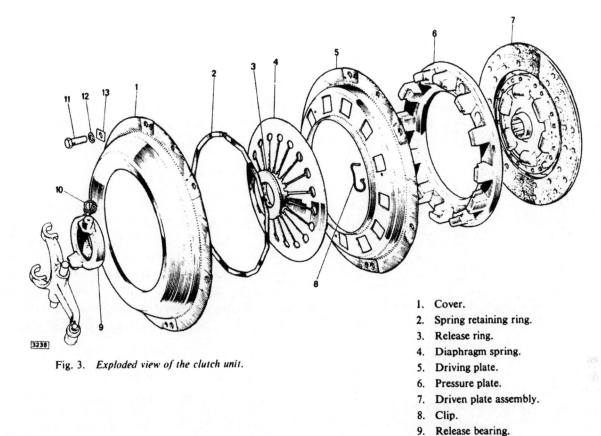

Fig. 3. *Exploded view of the clutch unit.*

1. Cover.
2. Spring retaining ring.
3. Release ring.
4. Diaphragm spring.
5. Driving plate.
6. Pressure plate.
7. Driven plate assembly.
8. Clip.
9. Release bearing.
10. Clip.
11. Bolt.
12. Spring washer.
13. Balance weight.

THE CLUTCH UNIT

DESCRIPTION (Fig. 3)

The driven plate assembly (7) is of the flexible centre type, in which a splined hub is indirectly attached to a disc and transmits the power and overrun through a number of coil springs held in position by shrouds.

The cover assembly consists of a pressed steel cover (1) and a cast iron pressure plate (6) loaded by a spring assembly.

The spring assembly consists of a diaphragm spring (4) and a release ring (3) which is held flexibly in the lugs of the pressure plate by means of a spring retaining ring (2).

Balancing of the clutch assembly is effected by drilling holes in the loose cover plate.

A graphite release bearing (9) is shrunk into a bearing cup which is mounted on the throw-out forks and held by the release bearing retainer springs.

GENERAL INSTRUCTIONS

To enable the balance of the assembly to be preserved after dismantling there are corresponding paint marks on the cover plate and driving plate. In addition, there are corresponding reference numbers stamped in the flanges of the cover and driving plate.

When reassembling ensure that the markings coincide and that, when refitting the clutch to the flywheel, the letter "B" stamped adjacent to one of the dowel holes coincides with the "B" stamped on the edge of the flywheel.

Page E.X.s.3

CLUTCH

Fig. 4. *Balance marks on the clutch and flywheel.*

The clutch is balanced in conjunction with the flywheel by means of loose balance pieces (13) which are fitted under the appropriate securing bolt. Each balance piece must be re-fitted in its original position, the number stamped on the balance weight corresponding to the number stamped on the cover plate. There are three balance weights stamped 1, 2 and 3, the weight stamped 3 being the heaviest.

If the graphite release bearing ring is badly worn it should be replaced by a complete bearing assembly.

Fig. 5. *Identification marks for balance weights.*

Page E.X.s.4

REMOVAL

In order to remove the clutch, the engine and gearbox must be first removed (see Section B—Engine, Removal).

Remove gearbox and clutch housing from engine.

Remove the bolts securing the clutch to the flywheel and withdraw the clutch assembly.

DISMANTLING

Remove the retaining ring and separate the diaphragm spring from the pressure plate.

If the clutch is faulty, it should be replaced as a complete assembly but, should this not be practicable, to dismantle completely proceed as follows:—

1. Mark all parts to ensure that they are assembled in the same relative position.

2. Place the clutch face downwards on the bench.

    (a) Lift off the cover.

    (b) Remove the retaining ring.

    (c) Lift out the diaphragm spring.

    (d) Remove the three spring clips.

    (e) Lift the driving plate off the pressure plate.

3. Examine all parts, paying particular attention to the following points:

    (a) Check for excessive clearance between the pressure plate lugs and the locating apertures in the driving plate.

    (b) Check for heat discoloration, distortion or surface damage at pressure plate face.

    (c) Check for wear on driven plate facings.

    (d) Check for loss of cushion on the spring segments between the facings.

ASSEMBLING

It is essential that all major components be returned to their original positions if the balance of the assembly is to be preserved.

Apply a trace of molydisulphide or zinc based grease to the sides of the pressure plate lugs, fulcrum points for the diaphragm spring on the pressure plate, driving plate and cover and also to the finger-tips where they enter the release tubes.

CLUTCH

When fitting the retaining ring ensure that the 12 crowns fit into the grooves in the 12 pressure plate lugs and that the 11 depressions of the undulations are fitted so as to press on the spring, that is, with the cranked ends of the rings uppermost. It is most important that the retaining ring fits the full depth of the groove in the lugs.

REFITTING

Place the driven plate on the flywheel taking care that the larger part of the splined hub faces the gearbox. Centralise the plate on the flywheel by means of a dummy shaft (a constant pinion shaft may be used for this purpose). Secure the cover assembly with the six setscrews and spring washers, tightening the screws a turn at a time by diagonal selection. Ensure that the "B" stamped adjacent to one of the dowel holes coincides with the "B" stamped on the periphery of the flywheel.

Fig. 7. *Checking clutch and flywheel balance.*

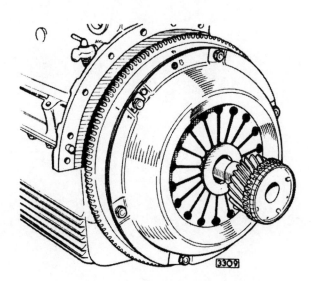

Fig. 6. *Centralising the clutch plate on the flywheel by means of a dummy shaft.*

CONDITION OF CLUTCH FACINGS

The possibility of further use of the friction facings of the clutch is sometimes raised, because they have a polished appearance after considerable service. It is natural to assume that a rough surface will give a higher frictional value against slipping, but this is not correct.

Since the introduction of non-metallic facings of the moulded asbestos type, in service, a polished surface is a common experience, but it must not be confused with a glazed surface which is sometimes encountered due to the conditions discussed below.

The ideal smooth or polished condition will provide a normal contact, but a glazed surface may be due to a film or a condition introduced, which entirely alters the frictional value of the facings. These two conditions might be simply illustrated by the comparison between a polished wood, and a varnished surface. In the former the contact is still made by the original material, whereas in the latter instance, a film of dried varnish is interposed between the contact surfaces.

Page E.X.s.5

CLUTCH

The following notes are issued with a view to giving useful information on this subject:—

(a) After the clutch has been in use for some little time, under perfect conditions (that is, with the clutch facings working on true and polished or ground surfaces of correct material, without the presence of oil, and with only that amount of slip which the clutch provides for under normal conditions) then the surface of the facings assumes a high polish, through which the grain of the material can be clearly seen. This polished facing is of mid-brown colour and is then in a perfect condition.

(b) Should oil in small quantities gain access to the clutch in such a manner as to come in contact with the facings it will burn off, due to the heat generated by slip which occurs under normal starting conditions The burning off of this small amount of lubricant, has the effect of gradually darkening the facings, but, provided the polish on the facings remains such that the grain of the material can be clearly distinguished, it has very little effect on clutch performance.

(c) Should increased quantities of oil or grease obtain access to the facings, one or two conditions, or a combination of the two, may arise, depending upon the nature of oil, etc.

(1) The oil may burn off and leave on the surface facings a carbon deposit which assumes a high glaze and causes slip. This is a very definite, though very thin deposit, and in general it hides the grain of the material.

(2) The oil may partially burn and leave a resinous deposit on the facings, which frequently produces a fierce clutch, and may also cause a "spinning" clutch due to a tendency of the facings to adhere to the flywheel or pressure plate face.

(3) There may be a combination of (1) and (2) conditions, which is likely to produce a judder during clutch engagement.

(d) Still greater quantities of oil produce a black soaked appearance of the facings, and the effect may be slip, fierceness, or judder in engagement, etc., according to the conditions. If the conditions under (c) or (d) are experienced, the clutch driven plate should be replaced by one fitted with new facings, the cause of the presence of the oil removed and the clutch and flywheel face thoroughly cleaned.

PEDAL ADJUSTMENT

On cars **not** fitted with an hydrostatic clutch slave cylinder, adjust the free travel as detailed in Section E.

On cars **fitted** with the hydrostatic clutch slave cylinder, refit the slave cylinder in accordance with instructions given on page E.s.2.

CLUTCH

CLUTCH

DESCRIPTION

Commencing at Chassis No. 7E.13501 ("E" Type), and 7E.53582 (2+2 model) the Laycock Clutch Assembly was replaced by a **Borg and Beck Clutch Assembly Model BB9/412G**.

A diaphragm spring clutch is fitted to all cars equipped with manual transmission.

The diaphragm spring is riveted inside the cover pressing with two fulcrum rings interposed between the shoulders of the rivets and the cover pressing. The diaphragm spring also pivots on these two fulcrum rings. Depressing the clutch pedal actuates the release bearing causing a corresponding deflection of the diaphragm spring thus pulling the pressure plate from the driven plate and freeing the clutch.

DATA

Make	Borg and Beck
Model	BB9/412G
Clutch Release Bearing	Graphite
Operation	Hydraulic
Hydraulic Fluid	Castrol/Girling Crimson Clutch/Brake Fluid

Page E.X.s.7

CLUTCH

THE CLUTCH UNIT

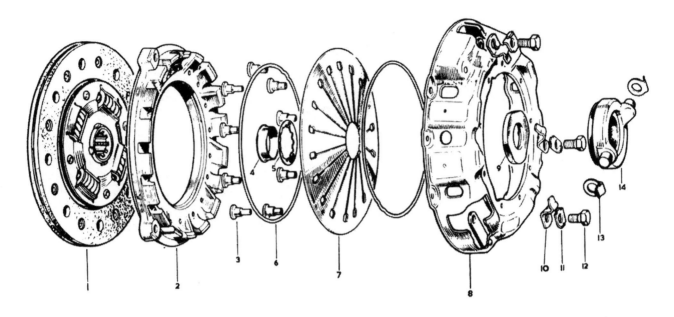

1. Driven plate.
2. Pressure plate.
3. Rivet.
4. Centre sleeve.
5. Belleville washer.
6. Fulcrum ring.
7. Diaphragm spring.
8. Cover pressing.
9. Release plate.
10. Retainer.
11. Tab washer.
12. Setscrew.
13. Retainer.
14. Release bearing.

Fig. 8. *Exploded view of the diaphragm spring clutch.*

SERVICING

The Borg and Beck diaphragm spring clutch is serviced in the U.K. ONLY by fitting an exchange unit which is available from the Works, Spares Division, Coventry. Individual parts are available from the same source for the repair of this clutch in Overseas Markets where exchange units may not be readily available. IT IS ESSENTIAL when overhauling the diaphragm spring clutch, to rigidly observe the service instructions detailed below and particular attention is drawn to the necessary special tools required.

GENERAL INSTRUCTIONS

To enable the balance of the assembly to be preserved after dismantling, there are corresponding paint marks on the cover plate and driving plate. In addition, there are corresponding reference numbers stamped in the flanges of the cover and driving plate.

When reassembling ensure that the markings coincide, and that, when refitting the clutch to the flywheel, the letter "B" stamped adjacent to one of the dowel holes coincides with the "B" stamped on the edge of the flywheel.

The clutch is balanced in conjunction with the flywheel by means of loose balance pieces which are fitted under the appropriate securing bolt. Each balance piece must be refitted in its original position, the number stamped on the balance weight corresponding to the number stamped on the cover plate. There are three balance weights stamped 1, 2 and 3, the weight stamped 3 being the heaviest.

If the graphite release bearing ring is badly worn it should be replaced by a complete bearing assembly.

Page E.X.s.8

CLUTCH

CLUTCH REMOVAL

In order to remove the clutch, the engine and gearbox must first be removed (see Page B.s.4).

Remove gearbox and clutch housing from engine.

Remove the bolts securing the clutch to the flywheel and withdraw the clutch assembly.

Retain any balance weight fitted.

DISMANTLING

Removing Release Plate

The centrally mounted release plate is held in position by a small centre sleeve which passes through the diaphragm spring and belleville washer into the release plate.

To free the plate, collapse the centre sleeve with a hammer and chisel. To avoid any possible damage whilst carrying out this operation, support the release plate in the locating boss of the special tool which should be held firmly in a vice.

Separating the Pressure Plate from Cover Pressing

Knock back the locking tabs and remove the three setscrews securing the pressure plate to the straps

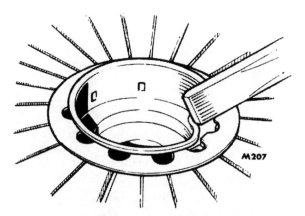

Fig. 10. *Collapsing the centre sleeve with a hammer and chisel.*

riveted to the cover pressing. These straps within the cover pressing must NOT be detached as this is an assembly reduced to its minimum as a spare part.

Dismantling the Cover Assembly

Remove the rivets securing the diaphragm spring and fulcrum rings by machining the shank of the rivets using a spot face cutter.

IT IS ESSENTIAL that the thickness of the cover is not reduced in excess of ·005″ (·127 mm.) at any point. The remaining portions of the rivets may be removed with a standard pin punch.

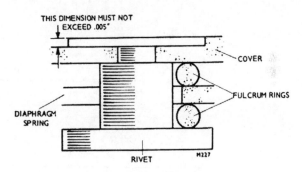

Fig. 11. *Do not reduce the thickness of the cover pressing in excess of ·005″ (·127 mm.).*

REBUILDING

The Cover Assembly

Prior to rebuilding, check the cover pressing for distortion. Bolt the cover firmly to a **flat** surface plate and check that measurements taken at various points from the cover flange to the machined land inside the cover pressing do not vary by more than ·007″ (·2 mm.). If the measurement exceeds this figure the cover must be replaced.

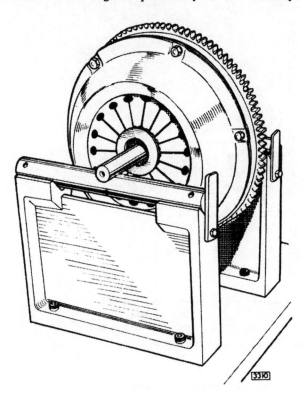

Fig. 9. *Clutch and flywheel balancing.*

Page E.X.s.9

CLUTCH

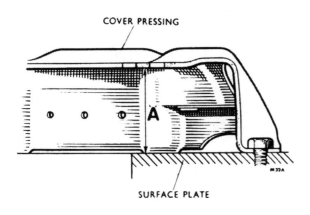

Fig. 12. *The measurement "A" must not vary by more than ·007" (·2 mm.).*

To achieve a satisfactory result when riveting the diaphragm spring into the cover pressing, a special tool must be fabricated to the specifications given in Fig. 13.

All parts except the spring can be made from mild steel. Position the fulcrum ring inside the cover pressing so that the location notches in the fulcrum ring engage a depression between two of the larger diameter holes in the cover pressing.

Place the diaphragm spring on the fulcrum ring inside the cover and line the long slots in the spring with the small holes in the cover pressing. Locate a further fulcrum ring on the diaphragm spring so that the location notches are diametrically opposite the location notches in the first ring. Fit new shouldered rivets, ensuring that the shouldered portion of each seats on the machined land inside the cover.

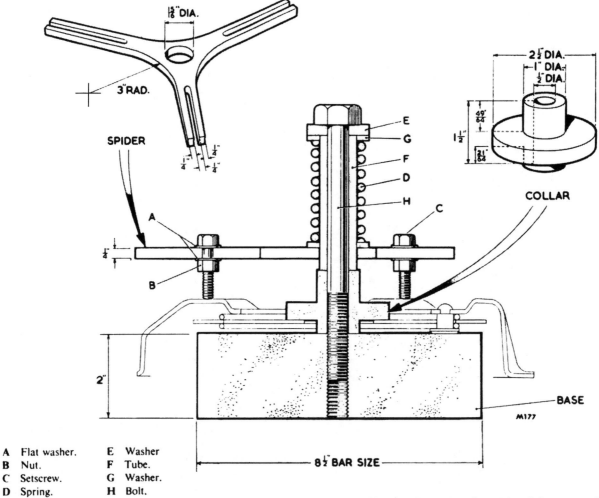

A Flat washer.
B Nut.
C Setscrew.
D Spring.
E Washer.
F Tube.
G Washer.
H Bolt.

Fig. 13. *Dimension of special tool for compressing the diaphragm spring when riveting the spring to cover pressing.*

CLUTCH

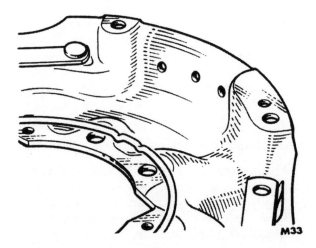

Fig. 14. *Assembly of cover pressing and fulcrum ring.*

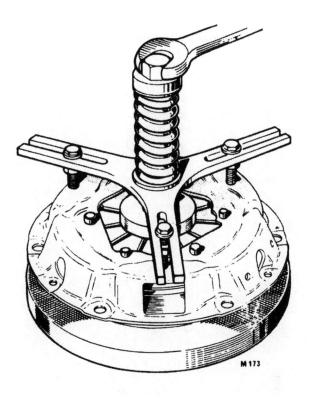

Fig. 16. *Tighten down the large nut so that the diaphragm spring is compressed flat.*

Place the base of the special tool on to the rivet heads. Invert the clutch and base plate.

Fit the collar to the large bolt and fit the large bolt complete with spring, spider and collar into the tapped hole in the base. Position the three setscrews on the spider so that they contact the cover pressing. Tighten down the centre bolt until the diaphragm spring becomes flat and the cover pressing is held firmly by the setscrews.

Rivet securely with a hand punch.

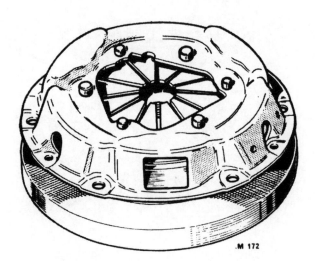

Fig. 15. *Clutch and base plate inverted.*

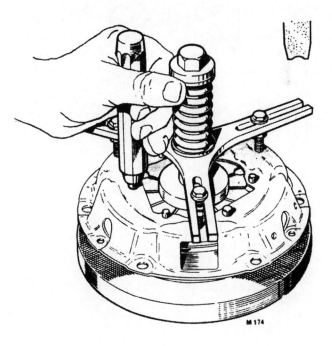

Fig. 17. *Riveting with a hand punch.*

Page E.X.s.11

CLUTCH

Assembling the Pressure Plate to Cover Pressing

Before assembling the pressure plate to the cover pressing, examine the plate for any signs of wear. Should it have been damaged or have excessive scoring, it is strongly recommended that a new plate is fitted. If, however, renewal of the pressure plate is not possible, grinding of the original unit may be undertaken by a competent machinist, bearing in mind that incorrect grinding of the plate may seriously affect the operation of the clutch. IN NO CIRCUMSTANCES MUST THE PRESSURE PLATE BE GROUND TO A THICKNESS OF LESS THAN 1·070″ (27.178 mm.)

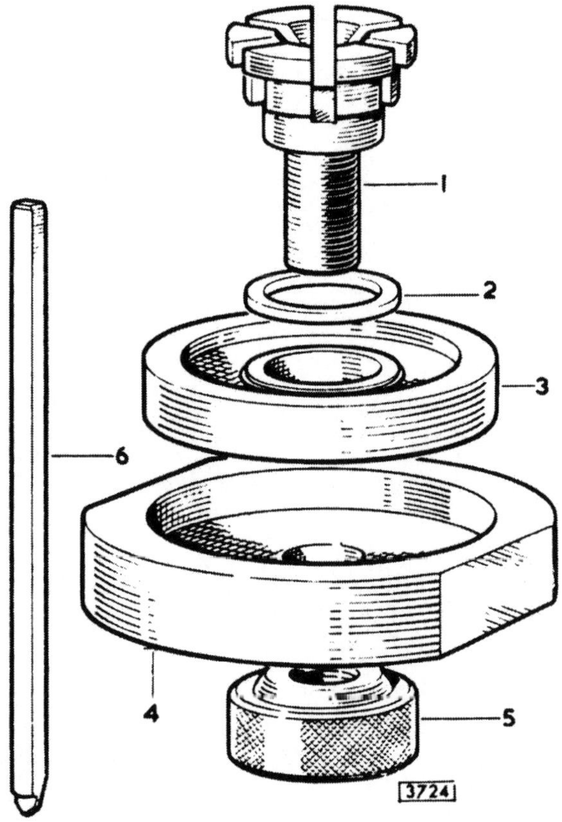

Position the pressure plate inside the cover assembly so that the lugs on the plate engage the slots in the cover pressing. Insert the three setscrews through the straps which are riveted to the cover pressing and lock with the tab washers.

Fitting a New Release Plate

A special tool (Part No. SSC.805) is available from Automotive Products Ltd., Service and Spares Division, Banbury, England, for completion of this operation. Ensure that all parts of the clutch and special tool are clean.

Grip the base of the tool in a vice and place the locating boss into the counterbore of the base plate. Place the release plate, face downwards, into the counterbore of the locating boss.

Apply a little high melting point grease to the tips of the diaphragm spring fingers and position the clutch, pressure plate friction face upwards, on to the release plate.

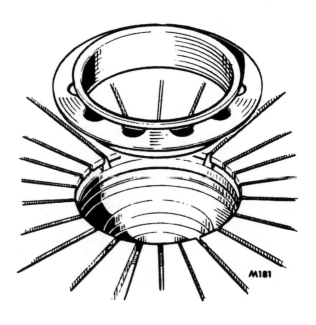

1. Staking guide.
2. Washer.
3. Locating base.
4. Base plate.
5. Knurled nut.
6. Punch.

Fig. 18. *Special Tool (SSC805).*

Fig. 19. *Fitting the sleeve and belleville washer.*

CLUTCH

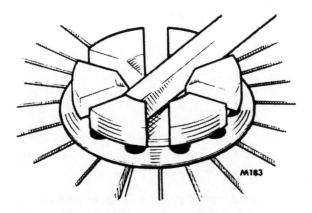

Fig. 20. *Staking the sleeve to the release plate.*

Place the belleville washer, concave surface towards the spring, on to the centre of the diaphragm spring and then push the centre sleeve through the spring into the release plate.

Drop the special washer into the sleeve and insert the staking guide into the centre of the assembly. Fit the knurled nut to the thread on the staking guide, tighten down until the whole assembly is solid. Using the special punch, stake the centre sleeve in six places into the groove in the release plate.

REFITTING

Place the driven plate on the flywheel, taking care that the larger part of the splined hub faces the gear-

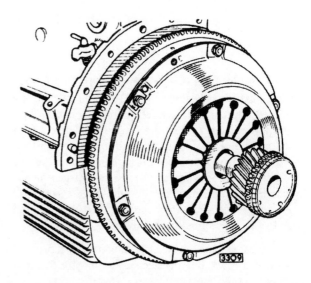

Fig. 21. *Centralising the driven plate on the flywheel by means of a dummy plate.*

box. Centralize the plate on the flywheel by means of the dummy shaft (a constant pinion shaft may be used for this purpose). Secure the cover assembly with the six setscrews and spring washers, tightening the screws a turn at a time by diagonal selection. Ensure that the "B" stamped adjacent to one of the dowel holes coincides with the "B" stamped on the periphery of the flywheel.

Page E.X.s.13

CLUTCH

CONDITION OF CLUTCH FACINGS

The possibility of further use of the friction facings of the clutch is sometimes raised, because they have a polished appearance after considerable service. It is natural to assume that a rough surface will give higher frictional value against slipping, but this not correct. Since the introduction of non-metallic facings of the moulded asbestos type, in service a polished surface is a common experience, but it must not be confused with a glazed surface which is sometimes encountered due to the conditions discussed below.

The ideal smooth or polished condition will provide a normal contact, but a glazed surface may be due to a film or a condition introduced, which entirely alters the frictional value of the facings. These two conditions might be simply illustrated by the comparison between a polished wood and a varnished surface. In the former the contact is still made by the original material whereas, in the latter instance, a film of dried varnish is interposed between the contact surfaces.

The following notes are issued with a view to giving useful information on this subject:—

(a) After the clutch has been in use for some little time under perfect conditions (that is, with the clutch facings working on true and polished or ground surfaces of correct material, without the presence of oil, and with only that amount of slip which the clutch provides for under normal conditions) then the surface of the facings assumes a high polish, through which the grain of the material can be clearly seen. This polished facing is of mid-brown colour and is then in a perfect condition.

(b) Should oil in small quantities gain access to the clutch in such a manner as to come into contact with the facings, it will burn off due to the heat generated by slip which occurs under normal starting conditions. The burning off of the small amount of lubricant has the effect of gradually darkening the facings, but provided the polish on the facings remains such that the grain of the material can be clearly distinguished, it has very little effect on clutch performance.

(c) Should increased quantities of oil or grease obtain access to the facing, one or two conditions, or a combination of the two, may arise, depending upon the nature of oil, etc.

  (i) The oil may burn off and leave on the surface a carbon deposit which assumes a high glaze and causes slip. This is a very definite, though very thin deposit, and in general it hides the grain of the material.

  (ii) The oil may partially burn and leave a resinous deposit on the facings, which frequently produces a fierce clutch, and may also cause a "spinning" clutch due to tendency of the facings to adhere to the flywheel or pressure plate face.

  (iii) There may be a combination of (i) and (ii) conditions which is likely to produce a judder during clutch engagement.

(d) Still greater quantities of oil produces a black soaked appearance to the facings, and the effect may be slip, fierceness, or judder in engagement, etc., according to the conditions. If the conditions under (c) or (d) are experienced, the clutch driven plate should be replaced by one fitted with new facings, the cause of the presence of oil removed and the clutch and flywheel face thoroughly cleaned.

Page E.X.s.14

CLUTCH

FAULT FINDING

SYMPTOM	CAUSE	REMEDY
Drag or Spin	(a) Oil or grease on the driven plate facings.	Fit new facings or replace plate.
	(b) Misalignment between the engine and splined clutch shaft.	Check over and correct the alignment.
	(c) Air in clutch system.	"Bleed" system. Check all unions and pipes.
	(d) Bad external leak between the clutch master cylinder and the slave cylinder.	Renew pipe and unions.
	(e) Warped or damaged pressure plate or clutch cover.	Renew defective part.
	(f) Driven plate hub binding on splined shaft.	Clean up splines and lubricate with small quantity of high melting point grease.
	(g) Distorted driven plate due to the weight of the gearbox being allowed to hang on clutch plate during assembly.	Fit new driven plate assembly using a jack to take overhanging weight of the gearbox.
	(h) Broken facings of driven plate.	Fit new facings, or replace plate.
	(i) Dirt or foreign matter in the clutch.	Dismantle clutch from flywheel and clean the unit; see that all working parts are free. **Caution:** Never use petrol or paraffin for cleaning out clutch.
Fierceness or Snatch	(a) Oil or grease on driven plate facings.	Fit new facings and ensure isolation of clutch from possible ingress of oil or grease.
	(b) Misalignment.	Check over and correct alignment.
	(c) Worn out driven plate facings.	Fit new facings or replace plate.
Slip	(a) Oil or grease on driven plate facings.	Fit new facings and eliminate cause.
	(b) Seized piston in clutch slave cylinder.	Renew parts as necessary.
	(c) Master cylinder piston sticking.	Free off piston.
Judder	(a) Oil, grease or foreign matter on driven plate facings.	Fit new facings or driven plate.
	(b) Misalignment.	Check over and correct alignment.
	(c) Bent splined shaft or buckled driven plate.	Fit new shaft or driven plate assembly.

Page E.X.s.15

CLUTCH

FAULT FINDING (continued)

SYMPTOM	CAUSE	REMEDY
Rattle	(a) Damaged driven plate. (b) Excessive backlash in transmission. (c) Wear in transmission bearings. (d) Bent or worn splined shaft. (e) Release bearing loose on throw out fork.	Fit new parts as necessary.
Tick or Knock	Hub splines worn due to misalignment.	Check and correct alignment then fit new driven plate.
Fracture of Driven Plate	(a) Misalignment distorts the plate and causes it to break or tear round the hub or at segment necks. (b) If the gearbox during assembly be allowed to hang with the shaft in the hub, the driven plate may be distorted, leading to drag, metal fatigue and breakage.	Check and correct alignment and fit new driven plate. Fit new driven plate assembly and ensure satisfactory re-assembly.
Abnormal Facing Wear	Usually produced by over-loading and by excessive clutch slip when starting.	In the hands of the operator.

Page E.X.s.16

SECTION F
GEARBOX

DESCRIPTION

The gearbox is of the four speed type with baulk-ring synchromesh on all forward gears. With the exception of reverse, the detents for the gears are incorporated in the synchro assemblies, the three synchro balls engaging with grooves in the operating sleeve. The detent for reverse gear is a spring loaded ball which engages on a groove in the selector rod.

Two interlock balls and a pin located at the front of selector rods prevent the engagement of two gears at the same time.

The gears are pressure fed at approximately 5 lb. per sq. in. (0·35 kg/cm.^{2}) from a pump driven from the rear of the mainshaft.

DATA

Identification number — EJ 001 onwards

Ratios

1st gear	2·68:1	3rd gear	1·27:1
2nd gear	1·74:1	4th (Top) gear	1·00:1
Reverse	3·08:1		

Identification number	Open 2 seater and F.H. Coupe	KE 101 onwards
	2+2	KJS 101 onwards

Commencing at the above gearbox numbers, the helix angle of the gear teeth has been altered for quietness and the taper dog gear lock altered to prevent possible jumping out of gear. Dismantling and assembling of these units is the same as for previous units.

Ratios

1st gear	2.933:1	3rd gear	1.389:1
2nd gear	1.905:1	4th gear	1.000:1
Reverse	3.378:1		

- 1st gear—end float on mainshaft ·005″ to ·007″ (·13—·18 mm.)
- 2nd gear—end float on mainshaft ·005″ to ·008″ (·13—·20 mm.)
- 3rd gear—end float on mainshaft ·005″ to ·008″ (·13—·20 mm.)
- Countershaft gear unit end float ·004″ to ·006″ (·10—·15 mm.)

ROUTINE MAINTENANCE

EVERY 3,000 MILES (5,000 KM.)

Check gearbox oil level.

EVERY 12,000 MILES (20,000 KM.)

Drain and refill gearbox.

RECOMMENDED LUBRICANTS

Mobilube GX 90	Castrol Hypoy	Spirax 90 E.P.	Esso Gear Oil GP 90/140	Gear Oil SAE 90 E.P.	Hypoid 90	Multigear Lubricant EP90

Page F.X.s.1

GEARBOX

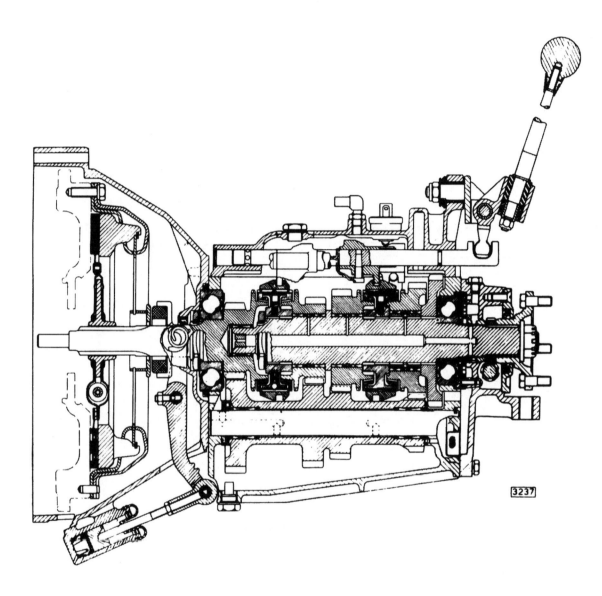

Fig. 1.   *Longitudinal section of clutch and gearbox.*

GEARBOX DISMANTLING

REMOVAL OF CLUTCH HOUSING

Detach the springs and remove the carbon thrust bearing.

Unscrew the two nuts and remove the clutch slave cylinder.

Remove the allen screw, push out the fulcrum pin and detach the clutch fork.

Tap back the locking tabs and break the locking wire and remove the eight setscrews.

Detach the clutch housing.

REMOVAL OF TOP COVER

Place the gear lever in neutral.

Remove the eight setscrews and two nuts and lift off the lid.

REMOVAL OF REAR EXTENSION

Engage first and reverse gears to lock the unit.

Remove the split pin and unscrew the flange nut.

Withdraw the flange.

Remove the four setscrews and detach the rear cover.

Remove the speedometer pinion and bush assembly after unscrewing the retaining bolt.

Withdraw the speedometer driving gear from the mainshaft.

Remove the seven setscrews and withdraw the extension.

Collect the distance piece and oil pump driving pin.

REMOVAL OF OIL PUMP

From the inside face of the rear extension break the staking and remove the three countersunk screws securing the oil pump gear housing. Withdraw the housing by entering two of the securing screws into the tapped holes in the housing; screw in the two screws evenly until the housing is free.

Mark the gears with marking ink so that they can be replaced the same way up in the housing.

REMOVAL OF COUNTERSHAFT

Remove the fibre plug from the front end of the countershaft.

Drive out the countershaft from the front of the casing.

Important:

Ensure that the rear washer (pegged to casing) drops down in a clockwise direction looking from the rear to avoid trapping the washer with the reverse gear when driving the mainshaft forward (see Fig. 2). This is effected by rocking the gearbox casing and moving the reverse lever backwards and forwards, or by pushing the washer down with a piece of wire bent at right angles.

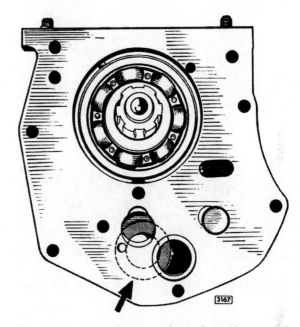

Fig. 2. *Ensure that the rear washer (indicated by arrow) drops down in a clockwise direction.*

REMOVAL OF CONSTANT PINION SHAFT

Rotate the constant pinion shaft until the cutaway portions of the driving gear are facing the top and bottom of the casing otherwise the gear will foul the countershaft.

With the aid of two levers ease the constant pinion shaft and front bearing assembly forward until it can be withdrawn (see Fig. 3).

DISMANTLING THE CONSTANT PINION SHAFT

Remove the roller bearing from inside the constant pinion shaft. On early cars, a spacer was also fitted along with the needle roller bearing.

GEARBOX

Tap back the tab washer and remove the large nut, tab washer and oil thrower.

Tap the shaft sharply against a metal plate to dislodge the bearing.

REMOVAL OF MAINSHAFT

Rotate the mainshaft until one of the cutaway portions in 3rd/Top synchro hub is in line with the countershaft (see Fig. 4), otherwise the hub will foul the constant gear or the countershaft.

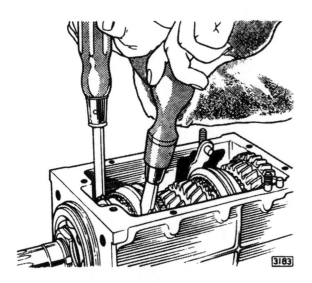

Fig. 3. *With the aid of two levers ease the constant pinion shaft forward.*

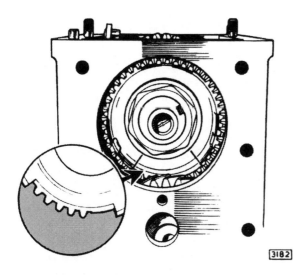

Fig. 4. *Rotate the mainshaft until one of the cutaway portions in 3rd/Top synchro hub is in line with the countershaft.*

Tap or press the mainshaft through the rear bearing ensuring that the reverse gear is kept tight against the first gear (see Fig. 5).

Fig. 5. *Tapping the mainshaft through the rear bearing.*

Remove the rear bearing from the casing and fit a hose clip to the mainshaft to prevent the reverse gear from sliding off (see Fig. 6).

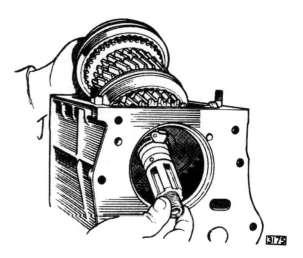

Fig. 6. *Removal of the mainshaft. Note the hose clip fitted to the mainshaft to retain the reverse gear.*

Slacken the reverse lever bolt until the lever can be moved to the rear.

Lift out the mainshaft forward and upward.

Lift out the countershaft gear unit and collect the needle bearings and retaining rings.

Withdraw the reverse idler shaft and lift out the gear.

Page F.X.s.4

GEARBOX

DISMANTLING THE MAINSHAFT

Note: The needle rollers are graded on diameter and must be kept in sets for their respective positions.

Remove the hose clip.

Withdraw the reverse gear from the mainshaft.

Withdraw the 1st gear and collect the 120 needle rollers, spacer and sleeve.

Withdraw the 1st/2nd synchro assembly and collect the two loose synchro-rings.

Withdraw the 2nd speed gear and collect the 106 needle rollers leaving the spacer on the mainshaft.

Tap back the tab washer and remove the large nut retaining the 3rd/Top synchro assembly to the mainshaft.

Withdraw the 3rd/Top synchro assembly from the mainshaft and collect the two loose synchro-rings.

Withdraw the 3rd speed gear and collect the 106 needle rollers and spacer.

DISMANTLING THE SYNCHRO ASSEMBLY

Completely surround the synchro assembly with a cloth and push out the synchro hub from the operating sleeve. Collect the synchro balls and springs, and the thrust members, plungers and springs.

DISMANTLING TOP COVER

Unscrew the self-locking nut and remove the double coil spring, washer, flat washer and fibre washer securing the gear lever to the top cover.

Withdraw the gear lever and collect the remaining fibre washer.

Remove the locking wire and unscrew the selector rod retaining screws.

Withdraw the 3rd/Top selector rods and collect the selector, spacing tube and interlock ball. Note the loose interlock pin at the front end of the 1st/2nd selector rod.

Withdraw the reverse selector rod and collect the reverse fork, stop spring and detent plunger.

Withdraw the 1st/2nd selector rod and collect the fork and short spacer tube.

GEARBOX RE-ASSEMBLING

ASSEMBLING THE SYNCHRO ASSEMBLIES

The assembly procedure for the 1st/2nd and 3rd/Top synchro assemblies is the same.

Note: Although the 3rd/Top and 1st/2nd synchro hubs are similar in appearance they are not identical and to distinguish them a groove is machined on the edge of the 3rd/Top synchro hub (see Fig. 7).

Assemble the synchro hub to the operating sleeve with;

(i) The wide boss of the hub on the opposite side to the wide chamfer end of the sleeve (see Fig. 8).

(ii) The three ball and springs in line with the teeth having three detent grooves (see Fig. 10).

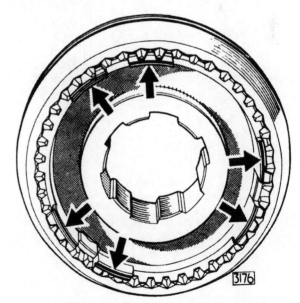

Fig. 7. *Identification grooves—3rd/Top, synchro assembly.*

Page F.X.s.5

GEARBOX

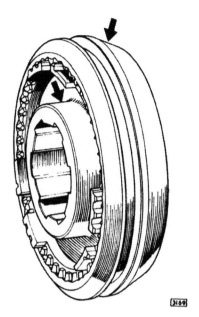

Fig. 8. *Assembly of synchro hub.*

Pack up the synchro hub so that holes for the ball and springs are exactly level with the top of the operating sleeve (see Fig. 11).

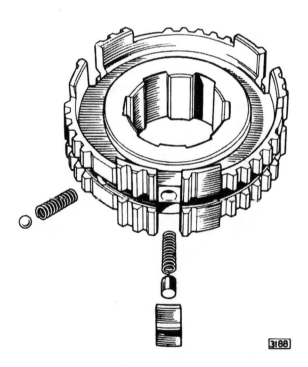

Fig. 9. *Showing the relative positions of the detent ball, plunger and thrust member.*

Fit the three springs, plungers and thrust members to their correct positions with grease; press down the thrust members as far as possible. Fit the three springs and balls to the remaining holes with grease.

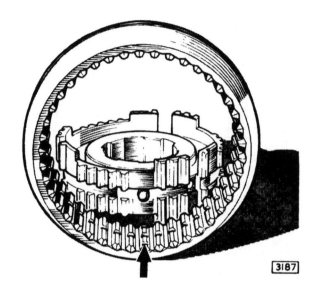

Fig. 10. *Fitting the synchro hub in the sleeve.*

Compress the springs with a large hose clip or a piston ring clamp as shown in Fig. 12 and carefully lift off the synchro assembly from the packing piece.

Depress the hub slightly and push down the thrust members with a screwdriver until they engage the neutral groove in the operating sleeve (see Fig. 13).

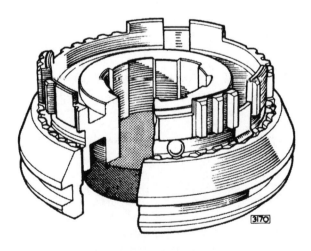

Fig. 11. *Fitting the springs, plungers and thrust members.*

Page F.X.s.6

GEARBOX

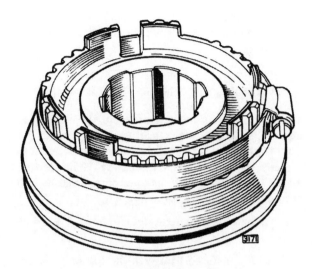

Fig. 12. *Compressing the springs.*

Finally tap the hub down until the balls can be heard and felt to engage the neutral groove (see Fig. 14).

Fig. 13. *Pushing down the thrust members.*

ASSEMBLING THE CLUSTER GEAR

Fit one retaining ring in the front end of the cluster gear. Locate the 29 needle roller bearings with grease and fit the inner thrust washer ensuring that the peg on the washer locates in a groove machined on the face of the cluster gear.

Fit a retaining ring, 29 needle roller bearings and a second retaining ring to the rear end of the cluster gear.

Fig. 14. *Tapping the hub into position.*

CHECKING THE CLUSTER GEAR END FLOAT

Fit the reverse idler gear, lever and idler shaft.

Fit the pegged rear washer to its boss on the casing with grease.

Locate the outer thrust washer to the front of the cluster gear with grease; lower the cluster gear into position carefully. Insert a dummy shaft and check the clearance between the rear thrust washer and the cluster gear. The clearance should be ·004″–·006″ (·10 mm.–15 mm.) and is adjusted by means of the outer thrust washers. This is available in the following selective thicknesses:—

Part Number	Thickness
C.1862/3	·152″ (3·86 mm.)
C.1862	·156″ (3·96 mm.)
C.1862/1	·159″ (4·04 mm.)
C.1862/2	·162″ (4·11 mm.)
C.1862/4	·164″ (4·17 mm.)

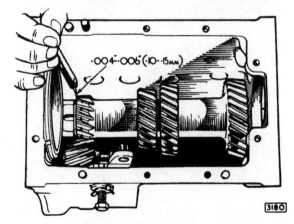

Fig. 15. *Checking the clearance between the rear thrust washer and the countershaft cluster gear.*

Page F.X.s.7

GEARBOX

ASSEMBLING THE CONSTANT PINION SHAFT

Assembling is the reverse of the dismantling procedure but care must be taken to ensure that the bearing is seated squarely on the constant pinion shaft.

ASSEMBLING THE MAINSHAFT

The re-assembly of the mainshaft is the reverse of the dismantling instructions but the following instructions should be noted.

(i) The end float of the gears on the mainshaft is given in "Data" at the beginning of this section and if found to be excessive the end float can only be restored by the fitting of new parts.

(ii) The needle rollers which support the gears on the mainshaft are graded on diameter and rollers of one grade only must be used for an individual gear. The grades are identified by /1, /2, and /3 after the part number.

(iii) The "E" Type constant pinion, countershaft and 3rd speed gear have a groove machined around the periphery of the gear, see Fig. 16. This is to distinguish the "E" Type gears from those fitted to the same type of gearbox on other models which have different ratios.

(iv) Fit a hose clip to prevent the reverse gear from sliding off when assembling the mainshaft to the casing.

ASSEMBLING THE GEARS TO THE CASING

Withdraw the dummy shaft from the cluster gear and, at the same time, substitute a thin rod keeping both the dummy shaft and the rod in contact until the dummy shaft is clear of the casing. The thin rod allows the cluster gear to be lowered sufficiently in the casing for insertion of the mainshaft.

Fit a new paper gasket to the front face of the casing.

Enter the mainshaft through the top of the casing and pass the rear of shaft through the bearing hole.

Enter the constant pinion shaft at the front of the casing with the cutaway portions of the tooth driving member at the top and bottom.

Tap the constant pinion shaft into position and enter the front end of the mainshaft into the spigot bearing of the constant pinion shaft.

Hold the constant pinion shaft in position and with a hollow drift tap the rear bearing into position.

Withdraw the thin rod from the front bore of the cluster gear approximately half way and lever the cluster gear upwards, rotating the mainshaft and constant pinion shaft gently until the cluster gear meshes. Carefully insert the countershaft from the rear and withdraw the rod. Fit the key locating the countershaft in the casing.

REFITTING REAR EXTENSION

Refit the gears to the oil pump the same way as removed, having previously coated the gears and the inside of the pump body with oil. Secure the pump housing with the three countersunk screws and retain by staking.

Fit a new paper gasket to the rear face of the casing.

Fit the distance piece and driving pin to the oil pump in the rear extension.

Offer up the rear extension and secure with the seven screws.

Fit the speedometer driving gear to the mainshaft.

Fig. 16. *Showing the groove which identifies the 'E' type gears.*

GEARBOX

Fit the speedometer driven gear and bush with the hole in the bush in line with the hole in the casing and secure with the retaining bolt.

Fit a new gasket to the rear cover face.

Fit a new oil seal to the rear cover with the lip facing forward.

Fit the rear cover to the extension noting that the setscrew holes are offset.

Fit the four bolts to the companion flange, slide on the flange and secure with flat washer with split pin.

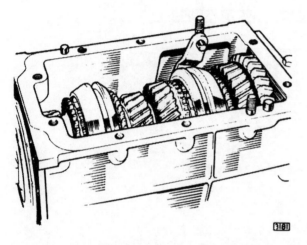

Fig. 17. *Re-assembled gearbox prior to refitting of top cover.*

RE-ASSEMBLING THE TOP COVER
(see Fig. 20).

Re-assembly of the top cover is the reverse of the dismantling instructions. When assembling the selector rods do not omit to fit the interlock balls and pin.

Renew the "O" rings on the selector rods.

To adjust the reverse plunger fit the plunger and spring.

Fit the ball and spring and start the screw and locknut; press the plunger inwards as far as possible and tighten the screw to lock the plunger.

Slowly slacken the screw until the plunger is released and the ball engages with the circular groove in the plunger. Hold the screw and tighten the locknut.

FITTING THE TOP COVER

Fit a new paper gasket.

Ensure that the gearbox and the top cover are in the neutral position.

Ensure that the reverse idler gear is out of mesh with the reverse gear on the mainshaft by pushing the lever rearwards.

Engage the selector forks with the grooves in the synchro assemblies.

Secure the top cover with the nuts and bolts noting that they are of different lengths.

REFITTING THE CLUTCH HOUSING

Refitting the clutch housing is the reverse of the removal procedure.

Fit a new oil seal to the clutch housing with the lip of the seal facing the gearbox. The oil seal has a metal flange and should be pressed in fully.

The two clutch housing securing bolts adjacent to the clutch fork trunnions are secured with locking wire; the remainder are secured with tab washers.

Note: After refitting the gearbox, run the car in top gear as soon as possible to attain the necessary mainshaft speed to prime the oil pump.

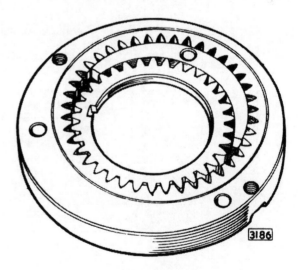

Fig. 18. *The oil pump.*

Page F.X.s.9

GEARBOX

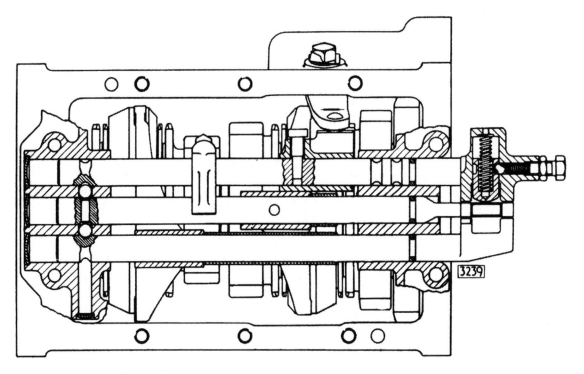

Fig. 19. *Plan view of gearbox showing selector arrangement.*

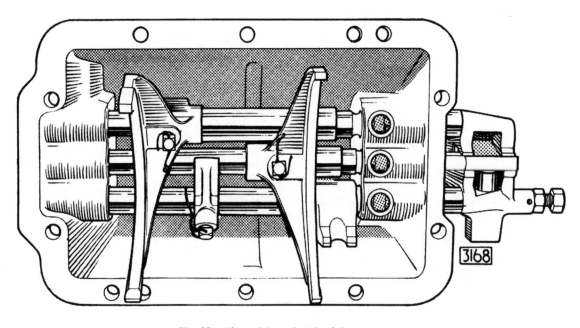

Fig. 20. *View of the underside of the top cover.*

SECTION FF
AUTOMATIC TRANSMISSION

GENERAL DATA

Maximum ratio of torque converter	2·00:1
1st Gear reduction	2·40:1
2nd Gear reduction	1·46:1
3rd Gear reduction	1·00:1
Reverse Gear reduction	2·00:1

AUTOMATIC SHIFT SPEEDS

185 × 15 SP 41 HR Tyres—2·88:1 Final Drive

Selector Position	Throttle Position	Upshifts 1–2	Upshifts 2–3	Downshifts 3–2	Downshifts 3–1	Downshifts 2–1
				M.P.H.		
D1	Minimum	7 – 9	12 – 15	8 – 14	—	4 – 8
D1	Full	38 – 44	66 – 71	23 – 37	—	—
D1	Kickdown	52 – 56	81 – 89	73 – 81	20 – 24	20 – 24
D2	Minimum	—	12 – 15	8 – 14	—	—
D2	Full	—	66 – 71	23 – 37	—	—
D2	Kickdown	—	81 – 89	73 – 81	—	—
L	Zero	—	—	60	—	12 – 20
				K.P.H.		
D1	Minimum	11 – 14	19 – 24	13 – 23	—	6 – 13
D1	Full	61 – 71	106 – 114	37 – 60	—	—
D1	Kickdown	83 – 90	130 – 143	118 – 130	32 – 39	32 – 39
D2	Minimum	—	19 – 24	13 – 23	—	—
D2	Full	—	106 – 114	37 – 60	—	—
D2	Kickdown	—	130 – 143	118 – 130	—	—
L	Zero	—	—	96	—	19 – 32

Page FF.X.s.1

AUTOMATIC TRANSMISSION

AUTOMATIC SHIFT SPEEDS (Continued)

185 × 15 SP 41 HR Tyres – 3·31:1 Final Drive

Selector Position	Throttle Position	Upshifts 1 – 2	Upshifts 2 – 3	Downshifts 3 – 2	Downshifts 3 – 1	Downshifts 2 – 1
				M.P.H.		
	Minimum	6 – 8	11 – 13	7 – 13	—	3 – 7
D1	Full	33 – 40	58 – 62	19 – 33	—	—
	Kickdown	45 – 49	70 – 78	63 – 71	17 – 21	17 – 21
	Minimum	—	11 – 13	7 – 13	—	—
D2	Full	—	58 – 62	19 – 33	—	—
	Kickdown	—	70 – 78	63 – 71	—	—
L	Zero	—	—	60	—	10 – 18
				K.P.H.		
	Minimum	9 – 13	18 – 21	11 – 21	—	5 – 11
D1	Full	53 – 64	93 – 100	31 – 53	—	—
	Kickdown	73 – 80	113 – 126	101 – 114	28 – 34	28 – 34
	Minimum	—	18 – 21	11 – 21	—	—
D2	Full	—	93 – 100	31 – 53	—	—
	Kickdown	—	113 – 126	101 – 114	—	—
L	Zero	—	—	96	—	16 – 29

Note: Shift points are approximate and not absolute values. Reasonable deviations from the above values are permissible.

AUTOMATIC TRANSMISSION

TIGHTENING TORQUE FIGURES

	lb. ft.	kgm.
Front pump to transmission case bolts	17 – 22	2·35 – 3·04
Front servo to transmission case bolts	30 – 35	4·15 – 4·70
Rear servo to transmission case bolts	40 – 45	5·53 – 6·22
Centre support to transmission case bolts	20 – 25	2·76 – 3·46
Upper valve body to lower valve body bolts	4 – 6	0·55 – 0·83
Control valve body to transmission case bolts	8 – 10	1·11 – 1·38
Pressure regulator assembly to transmission case bolts	17 – 22	2·35 – 3·04
Extension assembly to transmission case bolts	28 – 33	3·87 – 4·56
Oil pan to transmission case bolts	10 – 13	1·38 – 1·80
Case assembly—gauge hole plug	10 – 15	1·38 – 2·07
Oil pan drain plug	25 – 30	3·46 – 4·15
Rear band adjusting screw lock nut	35 – 40	4·70 – 5·53
Front band adjusting screw lock nut	20 – 25	2·76 – 3·46
Detent lever attaching nut	35 – 40	4·70 – 5·53
Companion flange nut	90 – 120	12·44 – 16·58
Bearing retainer to extension housing bolts	28 – 33	3·87 – 4·56

	lb. in.	kgm.
Front pump cover attaching screws	25 – 35	0·29 – 0·40
Rear pump cover attaching screws ¼" (6·30 mm.)	50 – 60	0·58 – 0·69
Rear pump attaching screws Nos. 10–24	20 – 30	0·24 – 0·35
Governor inspection cover attaching screws	50 – 60	0·58 – 0·69
Governor valve body to counterweight screws	50 – 60	0·58 – 0·69
Governor valve body cover screws	20 – 30	0·24 – 0·35
Pressure regulator cover attaching screws	20 – 30	0·24 – 0·35
Control valve body screws	20 – 30	0·24 – 0·35
Control valve body plug	10 – 14	0·11 – 0·16
Control valve lower body plug	7 – 15	0·08 – 0·17

Page FF.X.s.3

AUTOMATIC TRANSMISSION

SPECIAL SERVICE TOOLS

Service tools are not available from Borg-Warner Limited. Distributors and Dealers should obtain the following tools illustrated in this manual from Messrs. V. L. Churchill & Co. Ltd., London Road, Daventry, Northants.

Description

Mainshaft end play gauge (CB.W.33).

Rear clutch spring compressor (C.B.W. 37A used with W.G.37).

Hydraulic pressure test gauge equipment (C.B.W. 1A used with adaptor C.B.W.1A-5A).

Spring beam torque wrench (used in conjunction with the following adaptor) (C.B.W.547A-50).

Rear band adjusting adaptor (C.B.W.547A-50-2).

Torque screwdriver (used in conjunction with the following adaptor) (C.B.W.548).

Front band adjusting adaptor (C.B.W.548-2).

Front band setting gauge (C.B.W.34).

Circlip pliers (used with "J" points) (7066).

Bench cradle (C.W.G.35).

Rear clutch piston assembly sleeve (C.W.G.41).

Front clutch piston assembly sleeve (C.W.G.42).

Rear pump discharge tube remover (C.W.G.45).

AUTOMATIC TRANSMISSION

DESCRIPTION AND OPERATION

The Model 8 automatic transmission incorporates a fluid torque converter in place of the usual flywheel and clutch. The converter is coupled to a hydraulically operated planetary gearbox which provides three forward ratios and reverse. All forward ratios are automatically engaged in accordance with accelerator position and car speed.

Overriding control by the driver is available upon demand for engine braking by manual selection of "L".

TORQUE CONVERTER

The feature of using a hydraulic converter in conjunction with a three-speed automatic gearbox provides a means of obtaining a smooth application of engine power to the driving wheels and additional engine torque multiplication to the 1st and 2nd gears of the gearbox.

The converter also provides extreme low-speed flexibility when the gearbox is in 3rd gear and, due to the ability of multiplying engine torque, it provides good acceleration from very low road speed without having to resort to a down-shift in the gearbox.

Torque multiplication from the converter is infinitely variable between the ratios of 2:1 and 1:1. The speed range, during which the torque multiplication can be achieved, is also variable, depending upon the accelerator position.

The hydraulic torque converter for use in conjunction with the automatic gearbox has a mean fluid circuit diameter of 11" (27·9 cm.).

It is of the single-phase, three-element type, comprising an impeller connected to the engine crankshaft, a turbine connected to the input shaft of the gearbox, and a stator mounted on a sprag-type one-way clutch supported on a fixed hub projecting from the gearbox case.

THE GEAR SET

The planetary gear set consists of two sun gears, two sets of pinions, a pinion carrier, and a ring gear. Helical, involute tooth forms are used throughout. Power enters the gear set via the sun gears. In all forward gears power enters through the forward sun gear; in reverse, power enters through the reverse sun gear. Power leaves the gear set by the ring gear. The pinions are used to transmit power from the sun gears to the ring gear. In reverse a single set of pinions is used, which causes the ring gear to rotate in the opposite direction to the sun gear. In forward gears a double set of pinions is used to cause the ring gear to rotate in the same direction as the sun gear. The carrier locates the pinions in their correct positions relative to the sun gears and the ring gear (and also forms a reaction member for certain conditions). The various mechanical ratios of the gear set are obtained by the engagement of hydraulically operated multi-disc clutches and brake bands.

Page FF.X.s.4

AUTOMATIC TRANSMISSION

CLUTCHES

Multi-disc clutches operated by hydraulic pistons connect the converter to the gear set. In all forward gears the front clutch connects the converter to the forward sun gear; for reverse the rear clutch connects the converter to the reverse sun gear.

BANDS

Brake bands, operated by hydraulic servos, hold elements of the gear set stationary to effect an output speed and a torque increase. In Lockup the rear band holds the planet carrier stationary and provides the 1st gear ratio of 2·40:1 and, in reverse, a ratio of 2·00:1. The front band holds the reverse sun gear stationary to provide the 2nd gear ratio of 1·46:1.

ONE-WAY CLUTCH

In D1, a one-way clutch is used in place of the rear band to prevent anti-clockwise rotation of the planet carrier, thus providing the 1st gear ratio of 2·40:1. This one-way clutch, allowing the gear set to freewheel in 1st gear, provides smooth ratio changes from 1st to 2nd, and vice-versa.

Selector Position	Ratio		Applied	Driving	Held
L	Lock-up	1st	Front Clutch Rear Band Sprag Clutch	Forward	Sun	Planet Carrier
D1	Drive One	1st	Front Clutch Sprag Clutch	Forward	Sun	Planet Carrier
L D1 D2	Lock-up Drive One Drive Two	2nd 2nd	Front Clutch Front Band	Forward	Sun	Reverse Sun
D1 D2	Drive One Drive Two	3rd	Front Clutch Rear Clutch	Forward Secondary	Sun Sun	
R	Reverse	Reverse	Rear Clutch Rear Band	Reverse	Sun	Planet Carrier

MECHANICAL POWER FLOW

First Gear (Lockup selected)

The front clutch is applied, connecting the converter to the forward sun gear. The rear band is applied, holding the planet carrier stationary, the gear set providing the reduction of 2·40:1. The reverse sun gear rotates freely in the opposite direction to the forward sun gear.

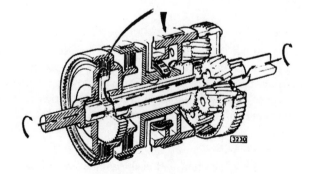

Fig. 1. *Mechanical power flow—1st gear (L) selected.*

Page FF.X.s.5

AUTOMATIC TRANSMISSION

First Gear (Drive 1 selected)

The front clutch is applied, connecting the converter to the forward sun gear. The one-way clutch is in operation, preventing the planet carrier from rotating anti-clockwise; the gear set provides the reduction of 2·40:1. When the vehicle is coasting the one-way clutch over-runs and the gear set freewheels.

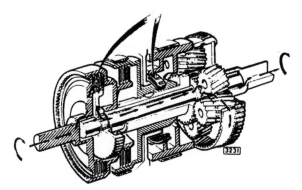

Fig. 2. *Mechanical power flow—1st gear (D) selected.*

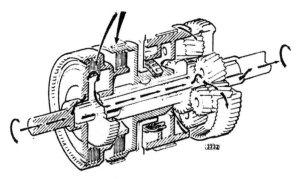

Fig. 3. *Mechanical power flow—2nd gear (L or D2) selected.*

Second Gear (Lockup or Drive 2 selected)

Again the front clutch is applied, connecting the converter to the forward sun gear. The front band is applied, holding the reverse sun gear stationary; the gear set provides the reduction of 1·46:1.

Third Gear

Again the front clutch is applied, connecting the converter to the forward sun gear. The rear clutch is applied, connecting the converter also to the reverse sun gear; thus both sun gears are locked together and the gear set rotates as a unit, providing a ratio of 1:1.

Neutral and Park

In neutral the front and rear clutches are off, and no power is transmitted from the converter to the gear

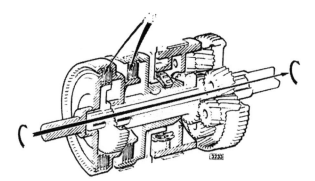

Fig. 4. *Mechanical power flow—3rd gear (D) selected.*

set. The front and rear bands are also released. In "P" the Front Servo Apply and Release and Rear Servo circuits are pressurised while the engine is running, so that the rear band is applied.

Reverse Gear

The rear clutch is applied, connecting the converter to the reverse sun gear. The rear band is applied, holding the planet carrier stationary, the gear set providing the reduction of 2·00:1 in the reverse direction.

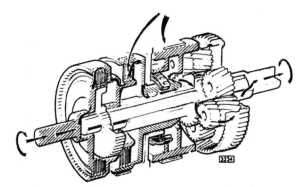

Fig. 5. *Mechanical power flow—Reverse (R) selected.*

THE HYDRAULIC SYSTEM

The hydraulic system contains a front and rear pump, both of the internal/external gear pattern, picking up fluid from the oil pan through a common strainer. Shift control is provided by a centrifugally operated hydraulic governor on the transmission output shaft. This governor works in conjunction with valves in the valve body assembly located in the base of the transmission. These valves regulate fluid pressure and direct it to appropriate transmission components.

The Front Pump

The front pump, driven by the converter impeller,

Page FF.X.s.6

AUTOMATIC TRANSMISSION

is in operation whenever the engine is running. This pump, through the primary and secondary regulator valves supplies the hydraulic requirements of the transmission with the engine running when the vehicle is stationary, as well as at low vehicle speeds before the rear pump becomes effective.

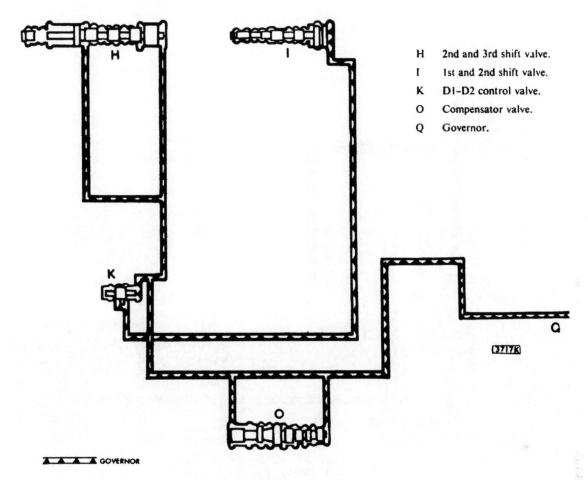

H 2nd and 3rd shift valve.
I 1st and 2nd shift valve.
K D1-D2 control valve.
O Compensator valve.
Q Governor.

Fig. 6. *Governor circuit.*

The Rear Pump

The rear pump is driven by the output shaft of the transmission. It is fully effective at speeds above approximately 20 m.p.h. (32 k.p.h.) and then supplies most of the hydraulic requirements.

If, due to a dead engine, the front pump is inoperative, the rear pump, above approximately 20 m.p.h. (32 k.p.h.) can provide all hydraulic requirements, thus enabling the engine to be started through the transmission.

The Governor

The governor, revolving with the output shaft, is essentially a pressure regulating valve which reduces line pressure to a value which varies with output shaft speed. This variable pressure is utilised in the control system to effect up and down shifts through the 1-2 and 2-3 shift valves. Rotation of the governor at low speeds causes the governor weight and valve to be affected by centrifugal force. The outward force is opposed by an opposite and equal hydraulic force produced by pressure acting on the regulating area of the governor valve. The governor valve is a regulating valve and will attempt to maintain equilibrium. Governor pressure will rise in proportion to the increase in centrifugal force caused by higher output shaft speed.

As rotational speed increases the governor weight moves outward to rest on a stop in the governor body, and can move no further. When this occurs, a spring located between the counter weight and the valve

Page FF.X.s.7

AUTOMATIC TRANSMISSION

becomes effective. The constant force of this spring then combines with the centrifugal force of the governor valve and the total force is opposed by governor pressure. This combination renders governor pressure less sensitive to output shaft speed variations.

It can be seen from the above, that the governor provides two distinct phases of regulation, the first of which is a fast rising pressure for accurate control of the low speed shift points.

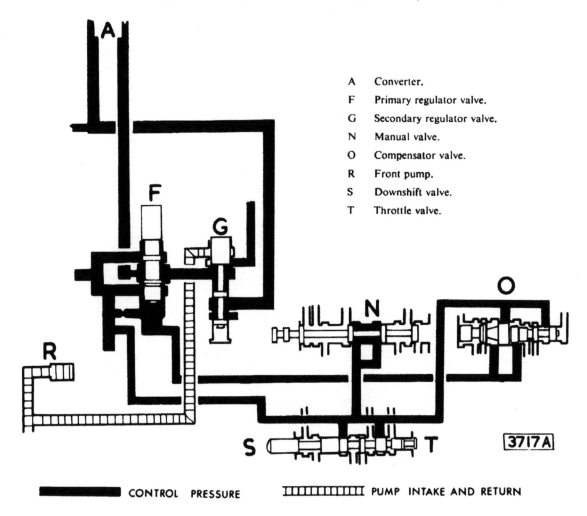

A   Converter.
F   Primary regulator valve.
G   Secondary regulator valve.
N   Manual valve.
O   Compensator valve.
R   Front pump.
S   Downshift valve.
T   Throttle valve.

Fig. 7. *Hydraulic circuit—neutral.*

THE CONTROL SYSTEM

Neutral—Engine Running (see Fig. 7)

When the selector is moved to the neutral position, the manual control valve is positioned so that control pressure cannot pass through the manual valve to the clutches or servos; therefore, the clutches and servos cannot apply. There is no transmission of power through the transmission in the neutral position.

The pressure regulation system, however, is functioning. With the engine running, the front pump is driven and fluid is picked up from the pan by the front pump inlet. Fluid, circulated by the front pump is directed to the control pressure regulator. The primary regulator valve will maintain correct control pressure by expelling the excess fluid to feed the secondary regulator valve. The secondary regulator valve maintains correct pressure for converter feed and lubrication, then forces the excess fluid back to the pump inlet.

Control pressure is directed to the manual control valve, where it is blocked by two lands on the valve. Control pressure is also directed to the throttle valve and the downshift valve and, with the valve closed

Page FF.X.s.8

AUTOMATIC TRANSMISSION

(accelerator at idle position) it is blocked by lands on the valves. Control pressure to the compensator valve is regulated by that valve, and compensating pressure is directed to the primary regulator valve.

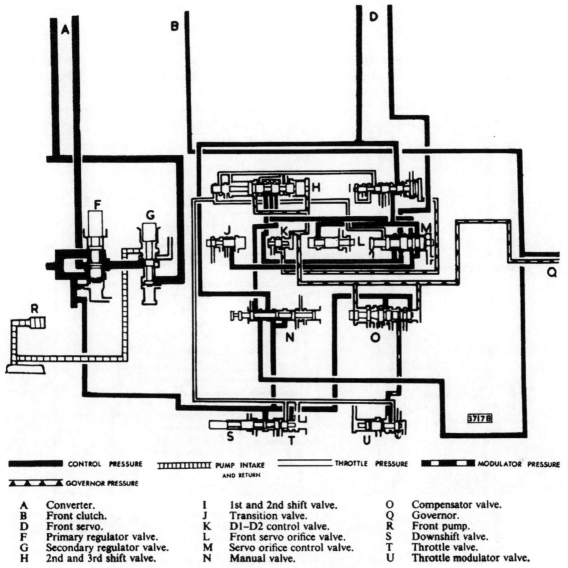

A	Converter.	I	1st and 2nd shift valve.	O	Compensator valve.
B	Front clutch.	J	Transition valve.	Q	Governor.
D	Front servo.	K	D1–D2 control valve.	R	Front pump.
F	Primary regulator valve.	L	Front servo orifice valve.	S	Downshift valve.
G	Secondary regulator valve.	M	Servo orifice control valve.	T	Throttle valve.
H	2nd and 3rd shift valve.	N	Manual valve.	U	Throttle modulator valve.

Fig. 8. *Hydraulic circuit—1st gear (D1 range).*

First Gear, D1 Range (see Fig. 8)

When the selector lever is placed in the D1 position, with the car standing still, and the engine running, the manual control valve is moved to admit control pressure to apply the front clutch.

Control pressure is also directed to the governor, but with the car standing still, the control pressure is blocked at the governor valve.

Control pressure from the manual valve is directed through another passage to the apply side of the front servo and the 1–2 shift valve.

From the 1–2 shift valve pressure then passes to the servo orifice control valve and the front servo release valve where it is blocked.

Control pressure is then directed from the servo orifice control valve via the 2–3 shift valve and again through the control valve to the release side of the front servo.

Pressure is also present at the transition valve where it is blocked.

With pressure on both sides of the front servo piston, the servo is held in a released position. The

Page FF.X.s.9

AUTOMATIC TRANSMISSION

one-way clutch takes the reaction torque on the rear drum, thus eliminating need for rear servo action.

The front pump supplies the pressure to operate the transmission and this pressure is controlled as it was in the neutral position.

When the accelerator is depressed and the car starts to move, centrifugal force, acting on the governor weight and valve, moves the valve to regulate governor pressure, which is directed to the 1–2 shift valve, 2–3 shift valve, and plug, and the compensator valve.

Movement of the accelerator also opens the throttle valve so that throttle pressure is directed to the modulator valve, orifice control valve, and the shift plug on the end of the 2–3 shift valve. Throttle pressure to the modulator valve is re-directed to the compensator valve to increase control pressure.

Throttle pressure to the shift plug on the 2–3 shift valve is reduced, and the reduced pressure is directed to the ends of the 1–2 shift valve and the 2–3 shift valve. This reduced pressure on the shift valves opposes governor pressure.

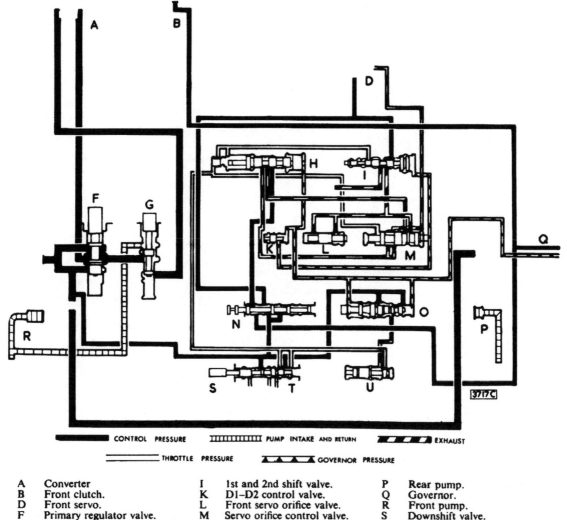

A	Converter	I	1st and 2nd shift valve.	P	Rear pump.
B	Front clutch.	K	D1–D2 control valve.	Q	Governor.
D	Front servo.	L	Front servo orifice valve.	R	Front pump.
F	Primary regulator valve.	M	Servo orifice control valve.	S	Downshift valve.
G	Secondary regulator valve.	O	Compensator valve.	T	Throttle valve.
H	2nd and 3rd shift valve.			U	Throttle modulator valve.

Fig. 9. *Hydraulic circuit—2nd gear (D1 range).*

Second Gear, D1 Range (Fig. 9)

As the car speed increases, the governor pressure builds up until it can overcome the opposite force of the 1–2 shift valve spring and reduced throttle pressure on the end of the valve and so moves the valve. When the 1–2 shift valve moves, control pressure at the valve is shut off and the front servo release pressure is

Page FFX.s.10

AUTOMATIC TRANSMISSION

exhausted, first slowly through a restricting orifice and then fast through the front servo release orifice valve. This leaves the front clutch and the front band applied.

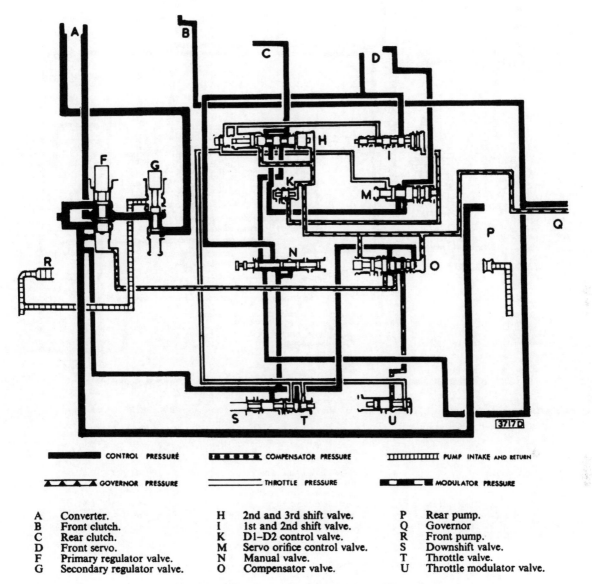

A	Converter.	H	2nd and 3rd shift valve.	P	Rear pump.
B	Front clutch.	I	1st and 2nd shift valve.	Q	Governor.
C	Rear clutch.	K	D1–D2 control valve.	R	Front pump.
D	Front servo.	M	Servo orifice control valve.	S	Downshift valve.
F	Primary regulator valve.	N	Manual valve.	T	Throttle valve.
G	Secondary regulator valve.	O	Compensator valve.	U	Throttle modulator valve.

Fig. 10. *Hydraulic circuit—3rd gear (D1 or D2 range).*

Third Gear, D1 or D2 Range (Fig. 10)

As the car speed continues to increase, the governor pressure also increases until it overcomes the 2–3 shift valve spring and the reduced throttle pressure on the end of the 2–3 shift valve, thus causing the valve to move. When the valve moves, control pressure is admitted to the rear clutch and through the annulus of the servo orifice control valve to the release side of the front servo, thus applying the rear clutch and placing the front servo in the released position. This leaves the front clutch and the rear clutch applied.

As the governor pressure continues to increase, it acts against modulator pressure at the compensator valve to increase compensator pressure and decrease control pressure through the movement of the valve in the primary regulator.

Page FFX.s.11

AUTOMATIC TRANSMISSION

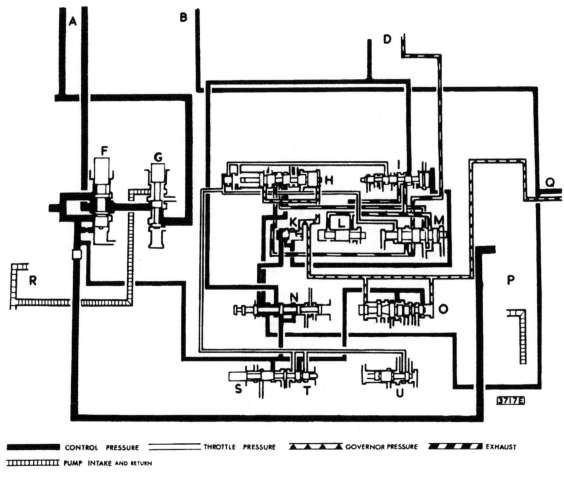

A	Converter.	I	1st and 2nd shift valve.	P	Rear pump
B	Front clutch.	K	D1-D2 control valve.	Q	Governor.
D	Front servo.	L	Front servo orifice valve.	R	Front pump.
F	Primary regulator valve.	M	Servo orifice control valve.	S	Downshift valve.
G	Secondary regulator valve.	N	Manual valve.	T	Throttle valve.
H	2nd and 3rd shift valve.	O	Compensator valve.	U	Throttle modulator valve.

Fig. 11. *Hydraulic circuit—2nd gear (D2 range).*

Second Gear, D2 Range (Fig. 11)

When the selector lever is placed in the D2 (drive) position, with the car standing still and the engine running, control pressure passes through the manual valve to the D1 and D2 control valve, overcomes any governor pressure acting on this valve and passes through the valve to the governor pressure area of the 1-2 shift valve, thus positioning it in the 2nd gear position.

Pressure is exhausted from the release side of the front servo, which results in the front clutch and front band being applied.

All upshifts from 2nd gear ratio direct will be similar to the description of 3rd gear D1 range.

Page FF.X.s.12

AUTOMATIC TRANSMISSION

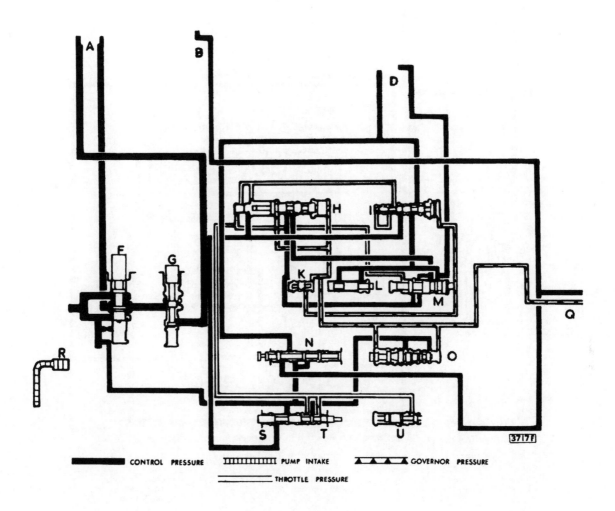

A Converter.	I 1st and 2nd shift valve.	Q Governor.
B Front clutch.	K D1–D2 control valve.	R Front pump.
D Front servo.	L Front servo orifice valve.	S Downshift valve.
F Primary regulator valve.	M Servo orifice control valve.	T Throttle valve.
G Secondary regulator valve.	N Manual valve.	U Throttle modulator valve.
H 2nd and 3rd shift valve.	O Compensator valve.	

Fig. 12. *Hydraulic circuit—2–1 kickdown (D1 range).*

2–1 Kickdown, D1 Range (Fig. 12)

At car speeds up to approximately 20 m.p.h. (32 k.p.h.), after the transmission has shifted from 1st to 2nd or 3rd gear, the transmission can be downshifted to 1st gear by depressing the accelerator pedal beyond the wide open throttle position.

Movement of the accelerator to kickdown position causes the throttle cable to move the downshift valve to allow control pressure to pass through the downshift valve to another land on the 1–2 shift valve. The combination of control pressure and the 1–2 shift valve spring is sufficient to overcome governor pressure and return the valve to the 1st gear position. In this position, control pressure is admitted to the release side of the front servo. This places the front servo in the released position, leaving the front clutch applied and the one-way clutch holding the rear drum.

Page FF.X.s.13

AUTOMATIC TRANSMISSION

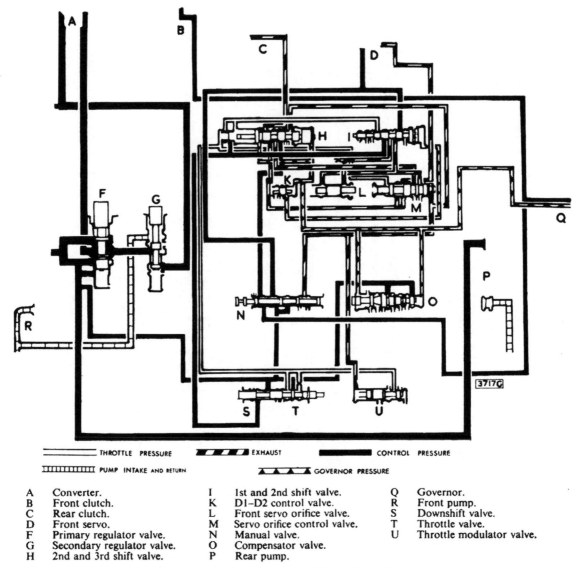

A	Converter.	I	1st and 2nd shift valve.	Q	Governor.
B	Front clutch.	K	D1–D2 control valve.	R	Front pump.
C	Rear clutch.	L	Front servo orifice valve.	S	Downshift valve.
D	Front servo.	M	Servo orifice control valve.	T	Throttle valve.
F	Primary regulator valve.	N	Manual valve.	U	Throttle modulator valve.
G	Secondary regulator valve.	O	Compensator valve.		
H	2nd and 3rd shift valve.	P	Rear pump.		

Fig. 13. *Hydraulic circuit—3–2 kickdown (D1 or D2 range).*

3–2 Kickdown, D1 or D2 Range (Fig. 13)

At car speeds between approximately 22 to 66 m.p.h. (35 to 106 k.p.h.) after the transmission has shifted to 3rd gear, the transmission can be downshifted from 3rd gear to 2nd gear by depressing the accelerator pedal beyond the wide open throttle position.

Movement of the accelerator causes the throttle cable to move the downshift valve to allow control pressure to pass through the downshift valve to the spring end of the 2–3 shift valve. The combination of control pressure at the end on the 2–3 shift valve and 2–3 shift valve springs is sufficient to overcome governor pressure to move the valve. When the valve is in 2nd gear position, control pressure to the rear clutch and through the servo orifice control valve to the release side of the front servo is shut off. The rear clutch circuit exhausts through the exhaust port of the manual control valve, whereas the front servo release circuit exhausts through the 1–2 shift valve, orifice and front servo release orifice valve. This leaves the front clutch and front band applied.

If the accelerator is left in the kickdown position, governor pressure will increase as the car speed increases until the governor pressure is greater than the combined pressures on the 2–3 shift valve, and the transmission will again upshift to 3rd gear.

At speeds above approximately 66 m.p.h. (106 k.p.h.) the governor pressure is so great that the combined pressures on the 2–3 shift valve cannot overcome governor pressure; therefore, there is no kickdown.

Page FFX.s.14

AUTOMATIC TRANSMISSION

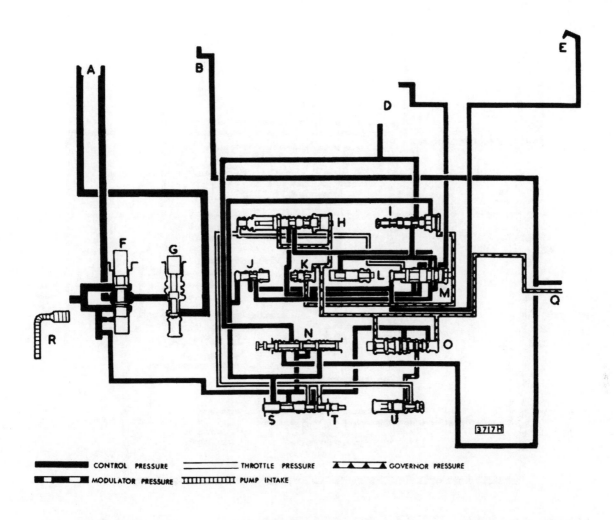

A	Converter.	I	1st and 2nd shift valve.	Q	Governor.
B	Front clutch.	J	Transition valve.	R	Front pump.
D	Front servo.	K	D1–D2 control valve.	S	Downshift valve.
E	Rear servo.	L	Front servo orifice valve.	T	Throttle valve.
F	Primary regulator valve.	M	Servo control valve.	U	Throttle modulator valve.
G	Secondary regulator valve.	N	Manual valve.		
H	2nd and 3rd shift valve.	O	Compensator valve.		

Fig. 14. *Hydraulic circuit—Lockup (1st gear).*

Lockup—First Gear (Fig. 14)

When the selector lever is placed in the Lockup position, the manual control valve is moved to admit through one port, control pressure to the governor feed and to apply the front clutch. Another port supplies both sides of the front servo which is held in the released position and also to the rear servo to apply the rear band through the servo orifice control and transition valves. A third port supplies pressure to move the transition valve and to an additional land on the 1–2 shift valve.

In this position, there is no automatic upshift to a higher gear ratio, since the combination of control pressure on the 1–2 shift valve and the 1–2 shift valve spring is greater than governor pressure acting against the valve, so that the valve cannot move. The combination of control pressure on the 2–3 shift valve and the 2–3 valve spring is also greater than the governor pressure acting against the valve so that the 2–3 shift valve cannot move.

Page FFX.s.15

AUTOMATIC TRANSMISSION

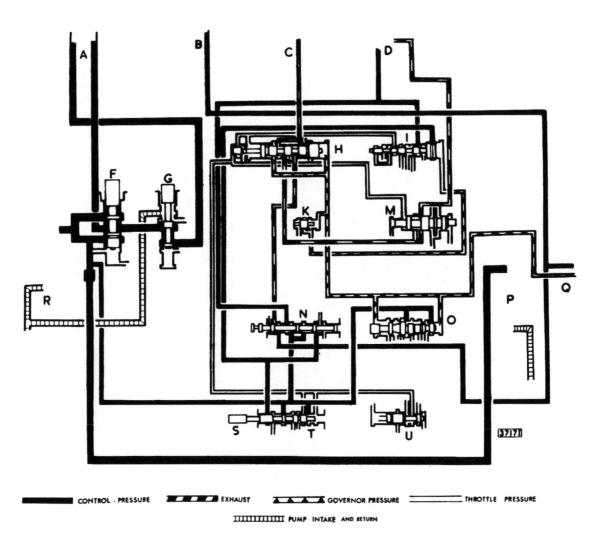

A	Converter.	H	2nd and 3rd shift valve.	P	Rear pump.
B	Front clutch.	I	1st and 2nd shift valve.	Q	Governor.
C	Rear clutch.	K	D1-D2 control valve.	R	Front pump.
D	Front servo.	M	Servo orifice control valve.	S	Downshift valve.
F	Primary regulator valve.	N	Manual valve.	T	Throttle valve.
G	Secondary regulator valve.	O	Compensator valve.	U	Throttle modulator valve.

Fig. 15. *Hydraulic circuit—Lockup (2nd gear).*

Lockup—Second Gear

In L the manual control valve opens to exhaust the rear clutch and front servo release circuit from the 2-3 shift valve. This causes a downshift from 3rd gear whenever L is selected at speed. In this condition, governor pressure will have moved the 1-2 shift valve; the result is that supply to the rear servo through the servo orifice control valve and transition valve is blocked and as front servo release pressure also exhausts through the 2-3 shift valve, the front band will be applied. This band, in conjunction with the front clutch, provides 2nd gear.

Page FF.X.s.16

AUTOMATIC TRANSMISSION

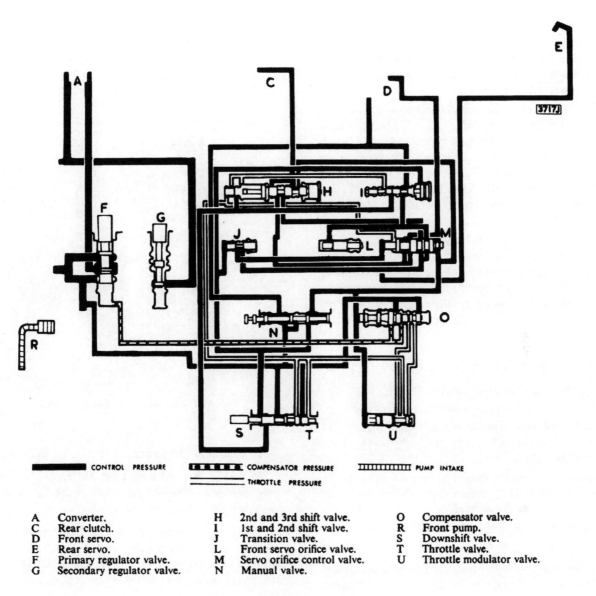

A	Converter.	H	2nd and 3rd shift valve.	O	Compensator valve.
C	Rear clutch.	I	1st and 2nd shift valve.	R	Front pump.
D	Front servo.	J	Transition valve.	S	Downshift valve.
E	Rear servo.	L	Front servo orifice valve.	T	Throttle valve.
F	Primary regulator valve.	M	Servo orifice control valve.	U	Throttle modulator valve.
G	Secondary regulator valve.	N	Manual valve.		

Fig. 16. *Hydraulic circuit—reverse gear.*

Reverse (Fig. 16)

When the selector lever is placed in the reverse position, the manual control valve moves to admit control pressure to the rear clutch, both sides of the front servo and the rear servo. This applies the rear clutch and the rear band.

Control pressure is also directed to the modulator valve to move the valve so when the throttle valve is opened by depressing the accelerator, the throttle pressure passes through the modulator valve to two lands on the compensator valve to reduce compensating pressure, thus increasing control pressure.

High control pressure is desired in reverse, since the reaction forces increase appreciably and higher pressure is required to hold the rear drum.

Page FF.X.s.17

AUTOMATIC TRANSMISSION

MAINTENANCE

It is most IMPORTANT that the following maintenance instructions are closely followed and absolute cleanliness is maintained when topping-up or filling the transmission.

It is **vitally important** when checking the fluid level that no dirt or foreign matter enters the transmission, otherwise trouble will almost certainly arise. Before removing the transmission dipstick, the surrounding area must be cleaned off to prevent dirt from entering the dipstick aperture. When filling the transmission with fluid ensure that the fluid container and funnel are perfectly clean.

In countries where ambient temperatures are unusually high, dust and/or mud must not be allowed to decrease the effective areas of the stoneguards in the converter housing or the slots in the transmission case. Also any foreign matter on the oil pan must be removed as it would act as a temperature insulator.

EVERY 3,000 MILES (5,000 KM.)

Check Transmission Fluid Level

The transmission filler tube is located on the right-hand side of the engine under the bonnet just forward of the bulkhead. Check the fluid level every 3,000 miles (5,000 km.).

Before checking the fluid level, the car should be on level ground and the transmission should be at the normal operating temperature.

Set the handbrake firmly and select P position.

The engine should be at normal idle.

When the engine is running, remove the dipstick, wipe clean and replace in the filler tube in its correct position.

Withdraw immediately and check.

If necessary, add fluid to bring the level to the FULL mark on the dipstick. The difference between FULL and LOW marks on the stick represents approximately 1½ pints (2 U.S. pints or 0·75 litres).

Fig. 17. *Automatic transmission dipstick.*

Be careful not to overfill.

If fluid is checked with transmission cold, a false reading will be obtained and filling to the FULL mark will cause it to be overfilled.

If it is found necessary to add fluid frequently, it will be an indication that there is a leakage in the transmission and it should be investigated immediately to prevent damage to transmission.

Total fluid capacity (including cooler) 16 Imperial pints from dry (19 U.S. pints, 9 litres).

RECOMMENDED AUTOMATIC TRANSMISSION FLUIDS

Mobil	Castrol	Shell	Esso	B.P.	Duckham	Regent Caltex/Texaco
Mobilfluid 200	Castrol T.Q.	Shell Donax T.6	Esso Automatic Transmission Fluid	Automatic Transmission Fluid, Type A	Nolmatic	Teaxamatic Fluid

Page FF.X.s.18

AUTOMATIC TRANSMISSION

If these recommended lubricants are not available, only a transmission fluid conforming to the following specification should be used:—

Automatic Transmission Fluid, Type "A" or Type "A" Suffix "A" (AQ-ATF)

ROAD TEST AND FAULT DIAGNOSIS

TESTING THE CAR

It is important to gain as much information as possible on the precise nature of any fault. In all cases the following road test procedure should be completely carried out, as there may be more than one fault.

Check that the starter will operate only with the selector in "P" and "N" and that the reverse light operates only in "R".

Apply the brakes and, with the engine at normal idling speed, select N-D, N-L, N-R. Transmission engagement should be felt in each position selected.

Check the engine stall speed (see converter diagnosis) with the transmission in "L" and "R". Check for slip or clutch break-away.

Note: Do not stall for longer than 10 seconds, or the transmission will overheat.

With the transmission at normal running temperature, select "D1". Release the brakes and accelerate with minimum throttle opening. Check for 1-2 and 2-3 shifts.

Note: At minimum throttle opening the shifts may be difficult to detect. Confirmation that the transmission is in 3rd gear may be obtained by selecting "L", when a 3-2 downshift will be felt.

At just over 30 m.p.h. (48 k.p.h.), select "N", switch off the ignition and let the car coast. At 30 m.p.h. (48 k.p.h.), switch on the ignition and select "L". The engine should start through the rear wheels, indicating that the rear oil pump of the transmission is operating.

Stop and restart, using full-throttle acceleration, i.e., accelerator at the detent. Check for 1-2 and 2-3 shifts according to the shift speed chart.

At 26 m.p.h. (42 k.p.h.), in 3rd gear, depress the accelerator to full-throttle position. The car should accelerate in 3rd gear and should not downshift to 2nd.

At 30 m.p.h. (48 k.p.h.), in 3rd gear, depress the accelerator to the kick-down position, i.e., through the detent. The transmission should downshift to 2nd gear.

At 18 m.p.h. (29 k.p.h.) in 3rd gear, depress the accelerator to the kick-down position. The transmission should downshift to 1st gear.

Stop and restart, using forced throttle acceleration (i.e., accelerator through the detent). Check for 1-2 and 2-3 shifts according to shift speed chart.

At 40 m.p.h. (64 k.p.h.) in 3rd gear, release the accelerator and select "L". Check for 3-2 downshift and engine braking. Check for inhibited 2-1 downshift and engine braking.

Stop, and with "L" still engaged, release the brakes and, using full throttle, accelerate to 20 m.p.h. (32 k.p.h.). Check for no slip or clutch break-away noise and no up-shifts.

Stop and select "R". Release the brakes and reverse, using full throttle if possible. Check for no slip or clutch break-away noise.

Stop on brakes facing downhill on gradient and select "P". Release the brakes and check that the parking pawl will hold the car. Re-apply brakes before disengaging the parking pawl. Repeat with car facing uphill.

Check that the selector is trapped by the gate in "Park" position.

At 30 m.p.h. (48 k.p.h.), in 3rd gear, D1, coast to a stop. Check roll out shifts for quality and speed in m.p.h. or k.p.h.

The front pump can be checked, with the selector in neutral, by revving the engine between idle and 2,000 r.p.m. A high pitched whine indicates a noisy front pump, a restricted front pump suction line, or a dirty oil screen.

At idle or slightly above idle speed in neutral, a gear whine indicates dragging front clutch plates. A tendency for the car to creep in neutral is a further

Page FF.X.s.19

AUTOMATIC TRANSMISSION

indication of dragging front clutch plates. Check carefully, to avoid confusing this with front pump or engine noises.

PRESSURE TESTS

See "Throttle Cable Adjustment" section and ascertain correct adjustment of throttle cable and engine idle. The pressure gauge is used to check transmission pressures, which should correspond to values given below.

Note: Figures given in table are normal for transmission temperatures from 150° to 185°F. only (65·5°C. to 85°C.).

Selector Position	Control Pressure Idle r.p.m.	Control Pressure Stall r.p.m.
D2	50–60	150–185
D1	50–60	150–185
L	50–60	150–185
R	50–60	190–210
N	55–60	—

Recording stall speed and stall pressures at the time the coverter is being checked will reduce the overall stalling time, which should be kept to a minimum.

Pressures which have been recorded should be analysed as follows: Low pressure indicates leakage in the circuit tested. Low pressure in all selector positions would indicate leakage, faulty pump or incorrect pressure regulation. High pressures, in all selector positions, indicate faulty pressure regulation incorrect cable adjustment or stuck valves.

FAULT DIAGNOSIS

Converter

If the general vehicle performance is below standard, check the engine stall speed with the revolution indicator by applying maximum pressure on the foot brake pedal, selecting lock-up, and fully depressing the accelerator. If the engine stall speed is up to 300 r.p.m. below normal, the engine is not developing its full power.

Inability to start on steep gradients combined with poor acceleration from rest indicates that the converter stator one-way clutch is slipping. This condition permits the stator to rotate in an opposite direction to the turbine and torque multiplication cannot occur. Check the stall speed, and if it is more than 600 r.p.m. below normal the converter assembly must be renewed.

Below standard acceleration in 3rd gear above 30 m.p.h. (48 k.p.h.), combined with a substantially reduced maximum speed, indicates that the stator one-way clutch has locked in the engaged condition. The stator will not rotate with the turbine and impeller, therefore the fluid flywheel phase of the converter performance cannot occur. This condition will also be indicated by excessive overheating of the transmission, although the stall speed will remain normal. The converter assembly must be replaced

Stall speed higher than normal indicates that the converter is not receiving its required fluid supply or that slip is occurring in the clutches of the automatic gearbox.

Note: When checking stall speeds ensure that the transmission is at normal operating temperature. Do not stall for longer than 10 seconds, or the transmission will overheat.

The torque converters are sealed by welding and serviced by replacement only.

The stoneguards in the converter housing must be unobstructed.

Stall Speed Test

This test provides a rapid check on the correct functioning of the converter as well as the gearbox.

The stall speed is the maximum speed at which the engine can drive the torque impeller while the turbine is held stationary. As the stall speed is dependent both on engine and torque converter characteristics, it will vary with the condition of the engine as well as with the condition of the transmission. It will be necessary, therefore, to determine the condition of the engine in order to correctly interpret a low stall speed.

To obtain the stall speed, allow the engine and the transmission to attain normal working temperature, set the handbrake, chock the wheels and apply the footbrake. Select "L" or "R" and fully depress the accelerator. Note the reading on the revolution indicator.

Note: To avoid overheating, the period of stall test must not exceed 10 seconds.

Page FF.X.s.20

AUTOMATIC TRANSMISSION

R.P.M.	Condition Indicated
Under 1,000	Stator freewheel slip
1,600–1,700	Normal
Over 2,100	Slip in the transmission gearbox

Clutch and Band Checks

To determine if a clutch or band has failed, without removing a transmission, check as detailed below.

Refer to the chart on page FF.s.5, showing the clutches and bands applied in each gear position.

Apply the handbrake and start the engine.

Engage each gear ratio and determine if drive is obtained through the component to be checked. If a clutch or band functions in one selector position it is reasonable to assume that the element in question is normal and that trouble lies elsewhere. If the clutch or band is tried in two positions and no drive is obtained in either position, it can be assumed that the element is faulty.

Air Pressure Checks

Air pressure may be used to test various transmission components in the car on the bench. Care should be exercised when air pressure checks are being made to prevent oil blowing on the clothing or into the eyes.

Knowledge of various circuits should be acquired referring to Figs. 6 to 16. It is necessary to remove the valve body to complete these checks.

Apply air pressure to the front clutch passage. A definite thump will indicate engagement. A similar sound should be heard when the rear clutch circuit is tested.

If clutch engagement noise is indefinite it is almost certainly due to damaged piston rings.

Servo action may be watched as air is applied to apply circuits of each servo.

It can be assumed, that if air pressure checks indicate that clutches and servos are being applied normally with air pressure, then the trouble lies in the hydraulic system.

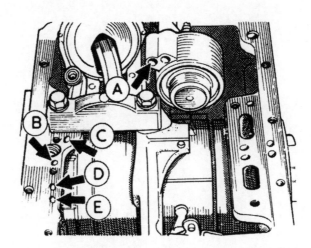

A. *Front servo apply.*
B. *Front clutch.*
C. *Rear servo.*
D. *Rear clutch.*
E. *Governor feed.*

Fig. 18. *Showing pressure passages with valve body removed.*

Page FF.X.s.21

AUTOMATIC TRANSMISSION

FAULT DIAGNOSIS

ENGAGEMENT	In Car	On Bench
Harsh	B, D, c, d	2, 4
Delayed	A, C, D, E, F, a, c, d	b
None	A, C, a, c, d	b, 9, 10, 11, 13
No forward	A, C, a, c, d	B, 1, 4, 7
No reverse	A, C, F, a, c, j, k, h	b, 2, 3, 6
Jumps in forward	C, D, E, F	4, 7, 8
Jumps in reverse	C, D, E	2
No neutral	C, c	2

UPSHIFTS	In Car	On Bench
No. 1–2	C, E, a, c, d, f, g, h, j	b, 5, 17
No. 2–3	C, a, c, d, f, g, h, k, l	b, 3, 17
Shift points too high	B, C, c, d, f, g, h, j, k, l	b
Shift points too low	B, c, f, g, h, l	B

UPSHIFT QUALITY	In Car	On Bench
1–2 slips or runs up	A, B, C, E, a, c, d, f, g, k	b, 1, 5
2–3 slips or runs up	C, a, c, d, f, g, h, k, l	b, 3, 5
1–2 harsh	B, C, E, c, d, f, g, h	1, 7, 8
2–3 harsh	B, C, E, s, d, f	4
1–2 Ties up or grabs	F, c	4, 7, 8
2–3 Ties up or grabs	E, F, C	4

DOWNSHIFTS	In Car	On Bench
No. 2–1	B, C, c, h, j	7
No. 3–2	B, c, h, k	4
Shift points too high	B, C, c, f, h, j, k, l	b
Shift points too low	B, C, c, f, h, j, k, l	b

DOWNSHIFT QUALITY	In Car	On Bench
2–1 Slides		7
3–2 Slides	B, C, E, a, c, d, f, g	b, 3, 5
2–1 Harsh		b, 1, 7
3–2 Harsh	B, E, c, d, f, g, 5	3, 4, 5

REVERSE	In Car	On Bench
Slips or chatters	A, B, F, d, c, g	b, 2, 3, 6

Page FF.X.s.22

AUTOMATIC TRANSMISSION

FAULT DIAGNOSIS (continued)

	In Car	On Bench
LINE PRESSURE		
Low idle pressure	A, C, D, a, c, d	b, 11
High idle pressure	B, c, d, e, f, g	
Low stall pressure	A, B, a, c, d, f, g, h	b, 11
High stall pressure	B, c, d, f, g	
STALL SPEED		
Too low (200 r.p.m. or more)		13
Too high (200 r.p.m. or more)	A, B, C, F, a, c, d, f	b, 1, 3, 6, 7, 9, 13
OTHERS		
No push starts	A, C, E, F, c	12
Transmission overheats	E, F, e	1, 2, 3, 4, 5, 6, 13, 18
Poor acceleration		13
Noisy in neutral	m	2, 4
Noisy in park	m	14
Noisy in all gears	m	2, 4, 14, 16
Noisy during coast (30–20 m.p.h.)		16, 19
Park brake does not hold	C, 15	15

KEY TO THE FAULT DIAGNOSIS CHART

1. **Preliminary Checks in Car**
   - A. Low fluid level.
   - B. Throttle cable incorrectly assembled or adjusted.
   - C. Manual linkage incorrectly assembled or adjusted.
   - D. Engine idle speed.
   - E. Front band adjustment.
   - F. Rear band adjustment.

2. **Hydraulic Faults**
   - (a) Oil tubes missing or broken.
   - (b) Sealing rings missing or broken.
   - (c) Valve body screws missing or not correctly tightened.
   - (d) Primary valve sticking.
   - (e) Secondary valve sticking.
   - (f) Throttle valve sticking.
   - (g) Compensator or modulator valve sticking.
   - (h) Governor valve sticking leaking or incorrectly assembled.
   - (i) Orifice control valve sticking.
   - (j) 1-2 shift valve sticking.
   - (k) 2-3 shift valve sticking.
   - (l) 2-3 shift valve plunger sticking.
   - (m) Regulator.

3. **Mechanical Faults**
   1. Front clutch slipping due to worn plates or faulty parts.
   2. Front clutch seized or plates distorted.
   3. Rear clutch slipping due to worn or faulty parts.
   4. Rear clutch seized or plates distorted.
   5. Front band slipping due to faulty servo, broken or worn band.
   6. Rear band slipping due to faulty servo, broken or worn band.
   7. One-way clutch slipping or incorrectly installed.
   8. One-way clutch seized.
   9. Broken input shaft.
   10. Front pump drive tangs on converter hub broken.
   11. Front pump worn.
   12. Rear pump worn or drive key broken.
   13. Converter blading and/or one-way clutch failed.
   14. Front pump.
   15. Parking linkage.
   16. Planetary assembly.
   17. Fluid distributor sleeve in output shaft.
   18. Oil cooler connections.
   19. Rear pump.

Page FF.X.s.23

AUTOMATIC TRANSMISSION

SERVICE ADJUSTMENTS

THROTTLE/KICKDOWN CABLE ADJUSTMENT

The importance of correct throttle cable adjustment cannot be over-emphasised. The shift quality and correct shift positions are controlled by precise movement of the cable in relation to the carburetter throttle shaft movement.

Preliminary Testing

Test the car on a flat road.

With the selector in the D1 or D2 position and at a minimum throttle opening, the 2-3 upshift should occur at 1,100–1,200 r.p.m.

A "run-up" of 200–400 r.p.m. at the change point indicates LOW pressure.

At full throttle opening, a jerky 2-3 upshift or a sharp 2-1 downshift (in D1 when stopping the car) indicates HIGH pressure.

Install a pressure gauge, 0–200 lb./sq. in. (0–14 kg./sq. cm.) in the line pressure point at the left hand rear face of the transmission unit. Start the engine and allow to reach normal operating temperature.

Select D1 or D2, apply the handbrake firmly and increase the idling speed to exactly 1,250 r.p.m.

The pressure gauge reading should be 72·5 ± 2·5 lb./sq. in. (5·097 ± ·175 kg/cm. sq.).

Adjustment

If road and pressure tests indicate that the throttle/kickdown cable setting is incorrect, adjustment is made at the fork end (see Fig. 20).

Release the fork end locknut, remove the split pin and fork end clevis pin.

To LOWER the pressure, turn the fork end clockwise: to RAISE the pressure, turn anti-clockwise.

Note: One full turn will alter the setting by 9 lb./sq. in. (·63 kg./sq. cm.).

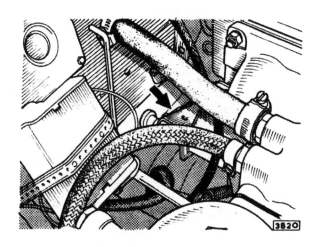

Fig. 20. *The kickdown cable adjustment.*

Slight adjustment only should be necessary; excessive adjustment will result in loss of "kickdown" or an increase in shift speeds.

Refit the fork end joint pin and split pin and tighten the locknut.

Restart the engine and check the pressure at 1,250 r.p.m.

Check that the carburetter butterfly valves are closed at idling speed after adjustment is completed.

If, after repeated attempts to stabilize the change points, the pressure still fluctuates, the throttle/kickdown inner cable may be binding or kinked and the cable should be replaced.

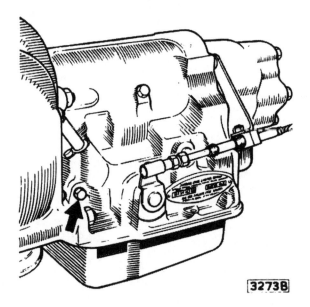

Fig. 19. *The transmission pressure take-off point.*

Page FF.X.s.24

AUTOMATIC TRANSMISSION

Throttle/Kickdown Cable Renewal

Disconnect the cable at the fork end.

Remove the cable retaining clip after withdrawing the setscrew.

Lift the carpets and the underfelts from the gearbox tunnel on the left-hand side.

Remove six drive screws and detach the aperture cover plate now exposed.

Remove the Allen-headed screw and washer retaining the outer cable.

Withdraw the outer cable and locate the spring clip securing the inner cable to the control rod operating the kickdown cam in the transmission unit.

Spring the clip open with a small screw driver and withdraw the inner cable.

Refitting is the reverse of the removal procedure.

Adjust the length of the operating cable to $3\frac{5}{16}''$ (84·1 mm.) between the centre line of the clevis and the end of the outer cable.

Check that the carburetter butterfly valves are closed before commencing adjustments described under the previous heading.

MANUAL LINKAGE ADJUSTMENT

(See Fig. 23)

Remove the transmission tunnel finisher assembly and the carpet at the side of the transmission cover. Remove the rubberised felt and withdraw the setscrews securing the cover plate at the left-hand side of the transmission cover.

Loosen the linkage cable locknut and remove the cable from the transmission lever. Push the transmission lever fully forward to the Lockup detent. Place the gear selector lever in the Lockup position.

Adjust the cable end to fit freely on to the transmission lever. Temporarily re-attach the cable to the lever. Move the gear selector lever through the various positions checking that gating at positions L, D1, R and P does not interfere with the transmission lever setting at the detent positions. The transmission lever must locate the transmission detents positively. Once correct adjustment is established, be sure the linkage cable is secured to the transmission lever and the locknut is tightened.

REMOVAL OF OIL PAN

Prior to front band adjustment or a check of internal parts, the gearbox fluid must be drained and the oil pan removed. When this is done an inspection should be made. A few wear particles in the dregs of the fluid in the pan are normal. An excess of wear particles whether ferrous or non-ferrous, or pieces of band lining material, would indicate that further checking should be done. A new gasket should be used when refitting the pan and the 14 attaching screws torqued to 10–15 lb. ft. (1·38–2·07 kgm.). Always use fresh fluid when refilling.

FRONT BAND ADJUSTMENT

(See Fig. 21)

The front band should be adjusted after the first 1,000 miles (1,600 km.) of operation and at 21,000 mile (35,000 km.) intervals thereafter.

Drain the oil by removing the oil filler connection and remove the oil pan. Loosen the adjusting screw locknut on the servo, apply lever and check that the screws turn freely in the lever. Install a $\frac{1}{4}''$ (6·4 mm.) thick gauge block between the servo piston pin and the servo adjusting screw, then tighten the adjusting screw with a suitable torque wrench or adjusting tool until 10 lb. ins. (0·12 kgm.) is reached. Retighten the adjusting screw locknut to 20–25 lb. ft. (2·76–3·46 kgm.). Remove the $\frac{1}{4}''$ (6·3 mm.) spacer.

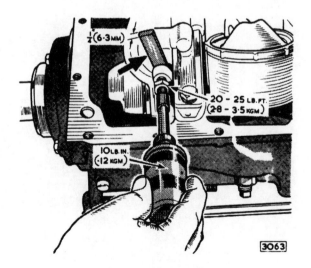

Fig. 21. *Front band adjustment.*

Page FF.X.s.25

AUTOMATIC TRANSMISSION

REAR BAND ADJUSTMENT

The rear band adjustment at the first 1,000 miles (1,600 km.) and at 21,000 miles (35.000 km.) intervals thereafter is made externally. To make the adjustment, first loosen and back off the adjusting screw locknut three or four turns and then make sure that the adjusting screw works freely in the threads in the case. Turn the adjusting screw in with a torque wrench or special tool for this purpose to 10 lb. ft. (1·382 kgm.) torque reading. Back the adjusting screw off 1½ turns exactly, then retighten the locknut to 35–40 lb. ft. (4·84–5·53 kgm.). The adjusting screw is on the right-hand side of the casing and an access hole is provided in the transmission cowl.

GOVERNOR

The governor can be inspected without removal of the oil pan. Remove the inspection cover and gasket. This will expose the governor, but the output shaft may have to be turned to position the governor head at the opening. First check for freedom of the valve by pushing and pulling on the governor weight. If removal of the governor body is desired, take out the two screws which retain it, being careful that they are not dropped inside the extension housing After removal of the body, dismantle it completely and clean all parts. When reassembling the governor, torque the governor body plate screws to 20–30 lb. in. (0·24–0·36 kgm.). When replacing the governor body on to the transmission, torque the screws which retain it to 50–60 lb. in. (0·60–0·72 kgm.). Replace the governor inspection cover, using a new gasket and torque its retaining screws to 50–60 lb. in. (0·60–0·72 kgm.).

It should be noted that if any of the four governor screws mentioned above are loose, the governor will not function correctly.

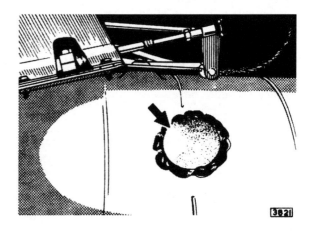

Fig. 22. *Rear band adjustment access point.*

Fig. 23. *Manual selector linkage adjustment.*

Page FF.X.s.26

AUTOMATIC TRANSMISSION

TRANSMISSION UNIT
REMOVAL AND REFITTING

To remove the transmission unit, it is necessary to withdraw the engine and transmission as a complete unit from the car before separating the transmission.

Removal

Disconnect the battery.

Remove the bonnet.

Drain the cooling system and cylinder block. Conserve the coolant if antifreeze is in use.

Slacken the clip on the breather pipe; unscrew the two wing nuts and withdraw the top of the air cleaner.

Disconnect the petrol feed pipe under the centre carburetter.

Slacken the clamps and remove the water hoses from the cylinder head and radiator to the header tank.

Remove the transmission oil cooler pipes from the radiator block.

Remove the heater hoses from the inlet manifold.

Disconnect the brake vacuum pipe.

Pull off the two Lucar connectors from the fan control thermostat in the header tank.

Remove the two bolts securing the header tank mounting bracket to the front cross member. Remove two nuts and two bolts securing the header tank straps to the radiator and fan cowl. Remove the header tank complete with bracket and straps.

Disconnect the throttle linkage at the rear carburetter and the kickdown cable at the rear of the cylinder head.

Disconnect:—

The two coil leads.

The water temperature transmitter cable.

The battery cable and solenoid switch cable from the starter motor.

The oil pressure cable at the top of the oil filter body.

The main harness connector and the Lucar connector for the 3AW warning light control from the alternator.

The engine earth strap from the left-hand side member.

Withdraw the bolt securing the oil filter canister and remove the canister complete with filter. Catch the escaping oil in a drip pan.

Remove the crankshaft pulley; damper and drive belt. Remove the ignition timing pointer from the sump. Mark the pulley and damper to facilitate refitting.

Slacken the two clamps of the water pump hose and withdraw the hose.

Remove the revolution counter generator complete with cables.

Remove the four nuts and washers securing each exhaust downpipe to the manifold. Unclip the pipes at the silencers and withdraw the downpipes. Collect the sealing rings between the downpipes and the manifold.

Withdraw the transmission dipstick and unscrew the dipstick tube from the oil pan.

Place the selector lever in L and withdraw the nut securing the selector cable adjustable ball joint to the transmission lever. Release the nut securing the outer cable clamp to the abutment bracket.

Remove the two lower nuts securing the torsion bar reaction tie plate on each side and tap the bolts back flush with the face of the tie plate. With the aid of a helper, place a lever between the head of the bolt just released and the torsion bar. Exert pressure on the bolt head to relieve the tension on the upper bolt. Remove the nut and tap the upper bolt back flush with the face of the tie plate. Tap the tie plate off the four bolts.

Note: Failure to relieve the tension on the upper bolts when tapping them back against the face of the tie plate will result in stripping the threads. If this occurs, new bolts must be fitted and the torsion bars re-set.

Disconnect the speedometer cable from the rear extension of the transmission unit.

Support the engine by means of two individual lifting tackles using the hooks on the cylinder head. Insert a trolley jack under the transmission and support the unit.

Page FF.X.s.27

AUTOMATIC TRANSMISSION

Remove the self-locking nut and stepped washer from the engine stabiliser.

Remove the bolts securing the rear mounting plate. Disconnect the propeller shaft at the front universal joint.

Remove the bolts from the front engine mountings.

Raise the engine on the lifting tackles and, keeping the unit level, move forwards ensuring that the converter housing clears the torsion bar anchor brackets and that the water pump pulley clears the sub-frame top cross member. Carefully raise the front of the engine and withdraw the complete unit forwards and upwards.

Refitting

Reverse the removal procedure to refit the transmission and engine. IT IS IMPORTANT that the engine stabiliser is adjusted properly and that the kickdown linkage is set correctly when refitting.

TRANSMISSION UNIT
Removal

Disconnect the kickdown linkage at the operating shaft. Drain the oil from the transmission unit. Remove the bolts securing the transmission to the converter housing and withdraw the unit.

TORQUE CONVERTER AND FLYWHEEL
Removal

Withdraw the cover from the front of the converter housing. Remove the starter motor and withdraw the setscrews securing the converter housing to the engine.

Remove the four setscrews, accessible through the starter motor mounting aperture, securing the torque converter to the flywheel. Rotate the engine to gain access to each setscrew in turn.

Remove the setscrews and locking plate securing the flywheel to the crankshaft and withdraw the flywheel.

TRANSMISSION
DISMANTLING AND ASSEMBLY

TRANSMISSION—DISMANTLING

Dismantling should not begin until the transmission exterior and work area have been thoroughly cleaned.

Place the transmission (bottom side up) on a suitable stand or holding fixture.

Remove the oil pan bolts, oil pan and gasket. Remove the oil screen retaining clip, lift off the oil screen from the regulator; lift and remove the screen from the rear pump suction tube. (See Fig. 24).

Use a screwdriver to prise the compensator tube from the valve body and regulator assemblies (Fig. 25).

The control pressure tube should be prised from the valve body, then removed from the regulator (Fig. 26).

Remove the rear pump suction tube by pulling and twisting it at the same time.

Fig. 24. *Removing the screen from the rear suction tube.*

AUTOMATIC TRANSMISSION

Loosen the front and rear servo adjusting screw locknuts and adjusting screws. This will aid in dismantling and later, in assembling, the transmission.

Fig. 25. *Removing the compensator tube.*

Carefully remove the pressure regulator spring retainer. Maintain pressure on the retainer to prevent distortion of the retainer, and sudden release of the springs (Fig. 27).

Fig. 27. *Removing the pressure spring retainer.*

Remove the three valve body attaching capscrews and lockwashers (Fig. 29).

Loosen the front servo to case capscrew and lockwasher approximately $\tfrac{5}{16}''$ (7·94 mm.) (Fig. 30).

Fig. 26. *Removing the line pressure tube.*

Remove springs and spring pilots, but do not remove the regulator valves at this time. The valves will be protected as long as they remain in the regulator body.

Remove the two regulator attaching capscrews and lockwashers, then lift the regulator assembly from the transmission case (Fig. 28).

Fig. 28. *The regulator retaining screws.*

Page FF.X.s.29

AUTOMATIC TRANSMISSION

Place the manual selector lever in park or reverse position. Lift the valve body until the throttle control rod will clear the manual detent lever, then remove the hook from the throttle cam using the index finger or a screwdriver.

Remove the front servo apply and release tubes (Fig. 32).

Remove the front servo bolt and lift the servo from the transmission, catching the servo strut with the index and middle finger of the left hand (Fig. 33).

Fig. 29. *The valve body attaching screws.*

Fig. 31. *Lifting the valve body to clear the front servo.*

Lift the valve body and servo until the valve body will clear the linkage and slide it off the servo apply and release tubes (Fig. 31).

Remove the two rear servo attaching capscrews and lockwashers, then lift the rear servo assembly from the transmission (Fig. 34)

Fig. 30. *Slackening the front servo screws.*

Fig. 32. *Withdrawing the apply and release tubes.*

Page FF.X.s.30

AUTOMATIC TRANSMISSION

Remove the rear band apply and anchor struts.

Remove the rear pump outlet tube, using special extractor tool Part No. CWG.45 (Fig. 35).

Fig. 33. *Removing the front servo.*

Check the end play at this time. Should the end play need correcting it will be done during assembly of the transmission (see Fig. 36). Place an indicator against the end of the input shaft. Prise between the front of the case and the front clutch to move clutch assemblies to their extreme rearward position. Set the indicator to "O". Prise between the planet carrier and the internal gear with a screwdriver to move the clutches to their extreme forward position. Read the end play on the indicator. The allowable limits are 0·008″–0·044″ (0·2–1·1 mm.). It is preferable to have approximately 0·020″ (0·5 mm.). Should correction be necessary, remove the output shaft, extension housing and companion flange as an assembly so that the selective washer can be changed.

Fig. 35. *Removing the rear pump outlet tube (Extractor Tool Part No. CWG45).*

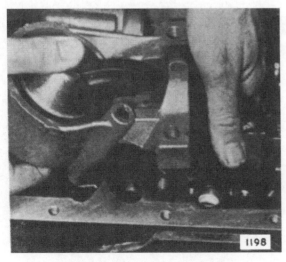

Fig. 34. *Removing the rear servo.*

Fig. 36. *Checking end play.*

Page FF.X.s.31

AUTOMATIC TRANSMISSION

Fig. 37. *Removing the selective thrust washer.*

Selective thrust washers are available in the following thicknesses:

0·061″–0·063″   0·074″–0·076″   0·092″–0·094″
(1·53–1·58 mm.)   (1·85–1·90 mm.)   (2·3–2·35 mm.)

0·067″–0·069″   0·081″–0·083″   0·105″–0·107″
(1·68–1·73 mm.)   (2·03–2·08 mm.)   (2·63–2·68 mm.)

Place the shift selector in park position to hold the output shaft, then remove the companion flange nut, lockwasher, flat washer and flange.

Remove the bearing retainer capscrews, the bearing retainer and the bearing retainer gasket.

Slide the speedometer drive gear off the output shaft.

Remove the governor inspection cover and gasket.

Remove the five extension housing capscrews and remove the output shaft and extension housing assembly.

Fig. 39. *Removing the rear band.*

Remove the two hook type seal rings from the rear of the primary sun gear shaft.

Fig. 38. *Removing the planet carrier.*

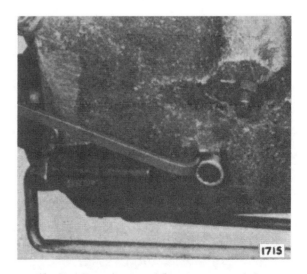

Fig. 40. *Removing one of the centre support bolts.*

Page FF.X.s.32

AUTOMATIC TRANSMISSION

Remove the selective thrust washer from the rear of the planet carrier (Fig. 37).

Pull the planet carrier from the transmission (Fig. 38).

Remove the two centre support bolts; one from each side of the case (Fig. 40).

Fig. 41. *Removing the clutch assemblies.*

Fig. 43. *Removing the attaching setscrew.*

Pull the rear band through the rear opening of the transmission. Hold the two ends of the band together with the left hand while pulling rearward through the rear of the case with the right hand (Fig. 39).

Remove the centre support, push on the end of the input shaft to start the rearward movement of the centre support.

Remove the front and rear clutch assemblies, placing them in a suitable stand for dismantling (Fig. 41). (The planet carrier can be used as a stand for dismantling and assembling the clutches).

Remove the front band (up and out of the case).

Remove the front pump oil seal. Use a seal puller or punch.

Remove the four front pump attaching capscrews and lift off the front pump (Fig. 42).

Remove the front pump oil seal ring from the case.

Front Pump—Dismantling

Remove the stator support attaching screw and remove the stator support (Fig. 43). Mark the top of the internal and external gears with marking ink or a crayon. Lift the gears from the pump body.

Inspect the pump body, the internal and external tooth gears, and stator supports for scores, scratches and excessive wear.

Fig. 42. *Removing the front pump.*

Page FF.X.s.33

AUTOMATIC TRANSMISSION

Minor scratches and scores can be removed with crocus cloth or jewellers' rouge. However, parts showing deep scratches, scores or excessive wear should be replaced. If excessive wear or scoring is observed, replace the complete pump assembly (since the gears and body are carefully matched when built, these parts should not be interchanged or individually replaced).

Front Pump—Assembling

Drive a new seal into the pump body until it bottoms.

Lubricate all pump parts with transmission fluid before assembly. Install the internal and external gears in the pump body with marks previously made in the upward position. Insert the stator support on the pump body and install the retaining screw. Torque the screw to 25–35 lb. in. (0·29–0·40 kgm.). Check the gears for free movement.

Manual Linkage—Dismantling

Pull the retainer clip from the forward end of the linkage rod (Fig. 44). Disconnect the rod from the manual valve detent lever. Release the detent ball and spring by rocking the manual valve lever to the extreme of its travel. The ball will be released with considerable force, but can be caught in a shop towel or even in the hands. Remove the manual lever locknut, the manual detent lever, and then pull the manual control lever from the transmission. Prise the manual lever oil seal from the transmission case with a screwdriver.

Manual Linkage—Assembling

Install a new manual lever oil seal. Assemble the manual control lever through the transmission case boss. Place the manual valve detent lever and locknut on the manual control lever shaft. Rock the manual valve lever to its extreme travel, then install the detent spring. Place the ball in position on the spring, then using the lubrication ball and spring (Fig. 45), rock the manual valve lever back over the ball and spring. Connect the linkage rod and insert the retainer spring clip.

Fig. 45. *Releasing the detent ball.*

Park Linkage—Dismantling

Pull the retainer clip from the rear of the parking brake linkage rod. Disconnect the linkage rod from the torsion lever. Remove the retainer spring from the torsion lever pin and slide the washer with the torsion lever off the pin. Tap the toggle lever rearward to loosen the pin retainer (Fig. 46), then pull the retainer using snap ring pliers (Fig. 47). The toggle lever pin and toggle lever can now be removed. A magnet may be used to pull the parking pawl anchor pin from the transmission case. The parking pawl is now free to be removed.

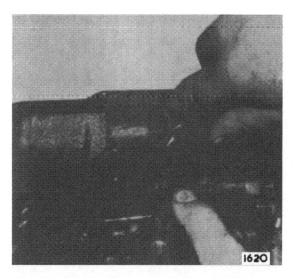

Fig. 44. *Removing the retainer clip from the linkage rod.*

Page FFX.s.34

AUTOMATIC TRANSMISSION

Fig. 46. *Tapping the toggle lever rearwards.*

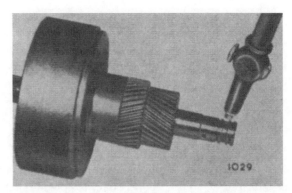

Fig. 48. *Applying compressed air to the clutch feed hole.*

Parking Linkage—Assembling

Assemble the parking pawl and shaft. Use a new toggle lever retainer to assemble the toggle lever and toggle pin. Assemble the torsion lever pin, then the washer, and then place the retainer spring on the torsion lever pin. Connect the linkage rod to the torsion lever and insert the spring clip.

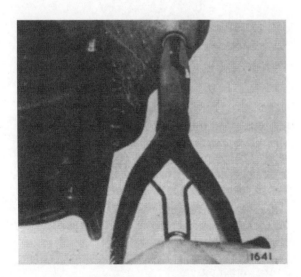

Fig. 47. *Removing the toggle lever pin retainer.*

Clutches—Dismantling

Place the clutch pack in a suitable stand. The planet carrier will work very well for this purpose.

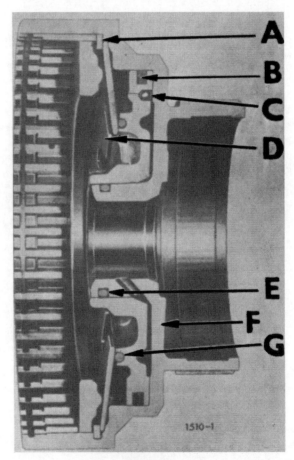

A. Clutch spring ring.
B. Sealing ring.
C. $\tfrac{1}{8}''$ steel ball.
D. Clutch spring.
E. Sealing ring.
F. Cylinder.
G. Piston.

Fig. 49. *Sectioned view of the clutch front drum.*

Page FF X.s.35

AUTOMATIC TRANSMISSION

Lift the complete front clutch assembly from the rear clutch and forward sun gear.

Remove the snap ring and lift the input shaft from the clutch cylinder. (The clutch hub thrust washer may stick to the input shaft).

Lift the clutch hub and thrust washer from the clutch assembly.

Lift the front clutch plates and the pressure plate from the assembly.

Remove the clutch return spring snap ring and then the return spring. It is not necessary to compress the spring to remove the snap ring.

Compressed air applied to the clutch feed hole in the clutch hub will force the piston from the clutch cylinder (Fig. 48).

Remove the rubber seal rings from the clutch hub and clutch piston.

Remove the two front clutch sealing rings from the forward sun gear shaft (Fig. 50).

Use the service tool to compress the clutch return spring, then remove the spring retainer snap ring. Release the spring, but do not permit the spring retainer to catch in the snap ring groove as the spring is being released (Fig. 51).

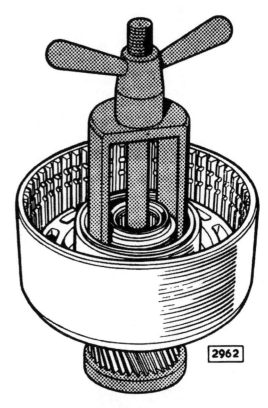

Fig. 51. *Dismantling the clutch using the special tool (Part No. CBW37A).*

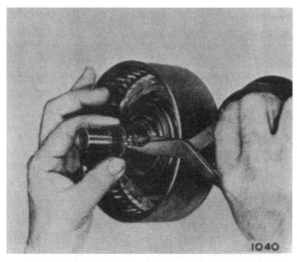

Fig. 50. *Removing the two front clutch sealing rings.*

Remove the thrust washer and thrust plate from the shoulder of the rear clutch hub.

Lift the rear clutch assembly up and off the forward sun gear shaft.

Remove the rear clutch ring.

Remove the clutch pressure plate and the clutch plates.

Replace the forward sun gear shaft in the clutch hub, being careful not to break the cast iron sealing rings. The clutch piston can now be removed from the clutch cylinder by blowing compressed air through the rear clutch passage of the forward sun gear.

Remove the forward sun gear from the clutch cylinder and remove the two rear clutch sealing rings from their grooves in the shaft.

Remove the rubber seal rings from the clutch hub and the clutch piston.

Inspection of Clutches

Inspect all parts for burrs, scratches, cracks and wear. Check all the front clutch plates and the rear clutch friction plates for flatness. Check the rear

Page FF.X.s.36

AUTOMATIC TRANSMISSION

clutch steel plates for proper cone. Lay plates on a flat surface when checking for flatness and cone. Cone should be 0·010" to 0·020" (0·25 to 0·5 mm.). Replace friction plates when wear has progressed so that the grooves are no longer visible. Replace all warped plates. Replace complete set of steel or friction plates in any clutch. Do not replace individual plates (Fig. 52).

Fig. 52. *Checking a clutch plate.*

Inspect the band surfaces of the drum for wear. If only slightly scored the drum may be refaced. Renew if excessive.

Inspect the clutch bushing and the needle bearing for wear and brinelling and for scores. The cast iron sealing rings are normally replaced. If the transmission is being rebuilt and has had little service, the rings may be re-used if they have not worn excessively and are not scratched or distorted.

Inspect the forward sun gear for broken or worn teeth. Inspect all journals and thrust surfaces for scores. Inspect all fluid passages for obstruction or leakage. Inspect the front clutch lubrication valve for freedom (Fig. 53).

Clutches—Assembling

Place the planet carrier on the assembly bench.

Place the forward sun gear in the carrier. Be sure the thrust washer is on the shaft (Fig. 54).

Fig. 54. *Placing the forward sun gear on the carrier.*

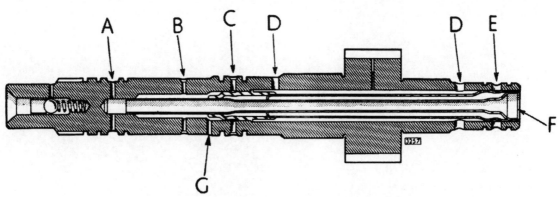

Fig. 53. *Longitudinal section of the forward sun gear showing oil ways.*
A, F—Front clutch: C, E—Rear clutch: B, D, G—lubrication.

Page FF.X.s.37

AUTOMATIC TRANSMISSION

Assemble the rubber "O" ring in its groove on the rear clutch hub (Fig. 55).

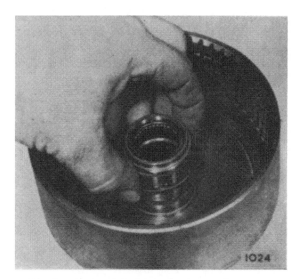

Fig. 55. *Fitting the "O" ring on the rear clutch hub.*

Assemble the square section rubber seal ring in its groove on the rear clutch piston (Fig. 56).

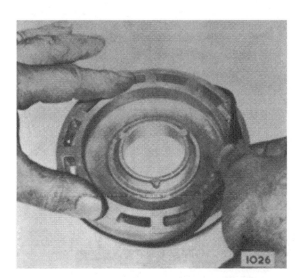

Fig. 56. *Fitting the rear clutch piston sealing ring.*

Assemble the clutch piston in the rear clutch cylinder using Tool Part No. CWG.41 to force it into position. Be sure to lubricate the seal rings so that they will assemble easier.

Place the rear clutch return spring and spring retainer in position on the clutch piston. The rear clutch spring fixture is then used to compress the spring, then the snap ring is assembled in its groove in the clutch.

Fig. 57. *Fitting rear clutch over primary sun gear ring.*

Install the rear clutch cast iron sealing rings in their grooves on the forward sun gear. Be sure that the rings are free in their grooves. Centre each ring in its groove, so that ends do not overlap edges of groove.

Fig. 58. *Fitting a rear clutch steel plate.*

Page FFX.s.38

AUTOMATIC TRANSMISSION

Place the rear clutch piston and cylinder assembly over the forward sun gear and gently slide it down over the sealing rings (Fig. 57).

Install the rear clutch pressure plate.

Install the rear clutch snap ring. This ring has one tanged end (Fig. 60).

Fig. 59. *Fitting a rear clutch friction plate.*

Install a rear clutch steel plate with its concave face up or forward facing in the transmission. Note that these plates are identified by missing teeth on the O.D. and are not interchangeable with front clutch steel plates (Fig. 58).

Fig. 61. *Fitting the sealing rings.*

Install the front clutch cast iron sealing rings in their grooves on the forward sun gear. Centre each ring in its groove so that ends do not overlap edges of the groove (Fig. 61).

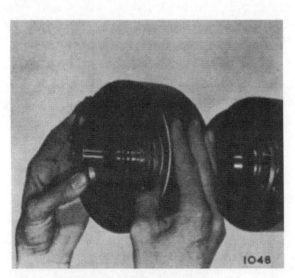

Fig. 60. *Fitting the snap ring.*

Install a rear clutch friction plate, then alternating with first a steel and then a friction plate, complete the clutch pack (Fig. 59).

Fig. 62. *Fitting the front thrust plate.*

Install the front clutch cylinder thrust plate (Fig. 62). Be sure flats on the washer match flats on shaft.

Page FFX.s.39

AUTOMATIC TRANSMISSION

Install the front clutch cylinder thrust washer (Fig. 63).

Fig. 63. *Fitting the front clutch cylinder thrust washer.*

Assemble the front clutch hub "O" ring into its groove in the clutch hub.

Assemble the front clutch piston square section rubber sealing ring in the groove of the clutch piston.

Install the clutch piston into the clutch cylinder after thoroughly lubricating the parts. Press the piston into position using Tool Part No. WG.42.

Fig. 64. *Assembling the front clutch.*

Install the front clutch belleville spring and snap ring. This snap ring is thicker than the other two clutch snap rings and has two tanged ends instead of one.

Fig. 65. *Fitting the front pressure plate.*

Assemble the front clutch assembly over the forward sun gear shaft and into the rear clutch, being careful not to distort or break the cast iron sealing rings. Use a short oscillating movement to engage splines of the rear clutch friction plates (Fig. 64).

Install the front clutch pressure plate (Fig. 65).

Install the front clutch hub, followed by front clutch hub thrust washer (Fig. 66).

Fig. 66. *Fitting the front clutch hub thrust washer.*

Page FF.X.s.40

AUTOMATIC TRANSMISSION

Install a front clutch friction plate over the splines of the hub (Fig. 67). Next, install a front clutch outer plate, meshing splines in the cylinder, alternating as above, complete assembly of plates (Fig. 68).

Centre Support

The centre support is serviced as an assembly. Therefore, there is no dismantling or assembly procedure.

Inspect the support for burrs or distortion, the race bearing surface for scores or scratches.

Fig. 67. *Fitting a front friction plate.*

Assemble the input shaft to the front clutch cylinder.

Assemble the snap ring that holds the input shaft in place (Fig. 69).

Fig. 69. *Fitting the snap ring.*

Fig. 68. *Fitting a front clutch outer plate.*

Place the thrust washer on the input shaft and the clutch assemblies are complete (Fig. 70).

Fig. 70. *Placing the thrust washer in position.*

Pinion Carrier Assembly

The pinion carrier is serviced as an assembly. Therefore, there is no dismantling or assembly procedure.

Page FF.X.s.41

AUTOMATIC TRANSMISSION

Inspect the band surface and the inner and outer bushing for scores. Rotate pinions on their shafts to check for freedom of movement and for worn or broken teeth. Use a feeler gauge to check pinion end play. End play should be 0·010" to 0·020" (·23 to ·5 mm.). Inspect pinion shafts for tightness to the planet carrier.

Sprag Clutch

A sprag-type one-way clutch assembly is incorporated in the planet carrier assembly and is held in place by a snap ring.

When installing the sprag clutch, the flange side of the sprag cage is located down into the outer race of the planet carrier assembly with the copper tension springs towards the centre support.

After the planet carrier and sprag assembly are installed in the case, the planet carrier will freewheel when turned counterclockwise and lock when turned clockwise (from the rear).

Output Shaft

Remove the extension housing and bearing from the output shaft by lifting the housing and tapping the shaft with a heavy plastic hammer.

Remove the bearing spacer washer.

Slide the oil collector and tubes from the shaft.

Remove the four sealing rings.

Remove the governor snap ring, governor and governor drive ball from the output shaft.

Lift the rear pump from the shaft and remove the rear pump drive key.

The snap ring may be removed and the output shaft removed from the ring gear; however, this is not necessary unless replacing one of these parts.

Inspect the output shaft thrust surfaces and journals for scores and the internal gear for broken teeth. Check the ring grooves, splines and gear teeth for burrs, wear or damage. The output shaft is a two-piece assembly and is serviced separately. Inspect the distributor and sleeve mating surfaces for excessive wear and for burrs, scores or leakage.

Governor

Remove the governor body cover plate attaching screws and remove the plate (Fig. 71). Remove the governor body attaching screws, then remove the body from the counter weight. Slide the spring retainer from the stem of governor weight and remove the spring. Remove the valve and weight from the governor body.

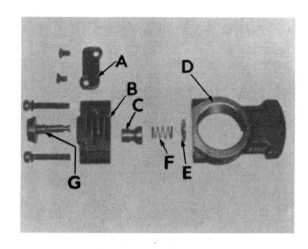

A. Governor body cover plate.
B. Governor body.
C. Valve.
D. Counter weight.
E. Spring retainer.
F. Spring.
G. Weight.

Fig. 71. *Exploded view of the governor.*

Inspect the governor weight, valve and bore for scores. Minor scores may be removed with crocus cloth. Replace the governor valve, weight or body if deeply scored. Check for free movement of the weight and valve in the bore. Inspect all fluid passages in the governor body and counterweight for obstruction. All fluid passages must be clean. Inspect the mating surfaces of the governor body and counterweight for burrs and distortion. Check governor spring retainer washer for burrs. The mating surfaces must be smooth and flat.

Re-install governor body cover plate, torqueing screws to 20–30 lb. in. (0.24 to 0.35 kgm.).

Install the governor valve in the bore of the body. Install the weight in the governor valve. Compress the spring and slide the retainer onto the stem of the weight and release the spring tension. Install the governor body on the counterweight.

Note: Make sure the fluid passages in the body and counterweight are aligned.

Torque the governor body attaching screws to 50–60 lb. in. (0·58 to 0·69 kgm.).

Page FF.X.s.42

AUTOMATIC TRANSMISSION

Rear Pump

Withdraw the five ¼" (6·4 mm.) screws, also the No. 10 U.N.C. screw and remove the cover. Mark the top face of the gears with marking ink or a crayon to assure correct re-installation of gears upon assembly (Fig. 72). Remove the drive and driven gears from the pump body.

Check the pump for free movement of the gears.

Fig. 73. *Replacing the gears.*

Output Shaft and Rear Pump—Assembling

Install the rear pump drive key in the output shaft.

Install rear pump assembly over the shaft.

Install the governor drive ball into the recess in the output shaft, using a spot of petrolatum to hold in place.

Install governor assembly, with plate on the governor body down (facing pump assembly). Install snap ring to lock governor in place (Fig. 74).

Fig. 72. *Marking the top face of the gears.*

Inspect the gear pockets and crescent of the pump body for scores or pitting. Inspect the bushing and drive and driven gear bearing surfaces for scores. Check all fluid passages for obstructions and clean if necessary. Inspect the mating surfaces, gear teeth, pump body and cover for burrs. If any pump parts are defective beyond minor burrs or scores, which cannot be removed with a crocus cloth, replace complete pump as a unit.

Lubricate parts with transmission fluid and replace both gears with the marks facing upward. Install the pump cover, attaching screws and lock-washers. Tighten the ¼" (6·4 mm.) screws to 50–60 lb. in. (0·58 to 0·69 kgm.) torque and the number 10 screw to 20–30 lb. in. (0·24 to 0·35 kgm.) torque (Fig. 73).

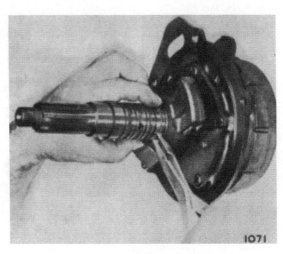

Fig. 74. *Fitting the snap ring.*

Page FF.X.s.43

AUTOMATIC TRANSMISSION

Install the four output shaft sealing rings, making sure they are free in their grooves (Fig. 75).

Install oil collector sleeve and tube assembly. Compress each ring with the fingers and carefully slide the sleeve over them (Fig. 76).

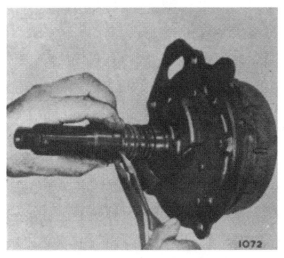

Fig. 75. *Fitting the output shaft sealing ring.*

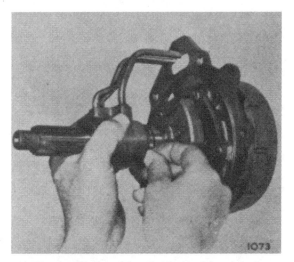

Fig. 76. *Installation of the oil collector sleeve and tube.*

Assemble the bearing spacer washer against the shoulder on the output shaft (Fig. 77).

Fig. 77. *Fitting the bearing spacer washer.*

Page FF.X.s.44

AUTOMATIC TRANSMISSION

Fig. 78. *Exploded view of the front servo.*

Front Servo—Dismantling

Use a small screwdriver to remove the snap ring.

Pull the sleeve and piston from the servo body.

Remove the piston from the servo sleeve.

Remove all sealing rings.

If the servo lever needs attention, it may be removed by first driving the roll pin from the servo and then removing the pivot pin and lever. Use a $\frac{1}{8}''$ (3.1 mm.) drift punch to remove the roll pin.

Inspect the servo parts for cracks, scratches and wear. Check the adjusting screw for freedom in the lever. Check the lever for freedom of movement.

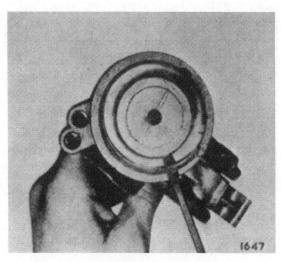

Fig. 79. *Removing the snap ring.*

Page FF.X.s.45

AUTOMATIC TRANSMISSION

Front Servo—Assembling

Assemble the servo lever, pivot pin and the roll pin.

Assemble the sealing rings on the sleeve and piston.

Assemble the piston to the sleeve, place the spring in the piston, and assemble the sleeve, piston and spring into the housing.

Replace the snap ring.

Remove the lever and shaft.

Depress the spring retainer while removing the snap ring.

Remove the servo release spring, piston and rubber "O" ring.

Inspect the servo body for cracks, burrs and obstructed passages and the piston bore and stem for scores. Inspect the actuating lever and shaft for wear and brinnelling.

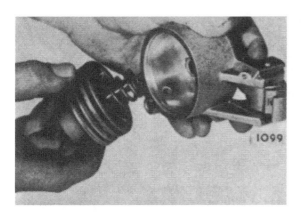

Fig. 80. *Assembling the front servo.*

Rear Servo—Dismantling

Remove the actuating lever roll pin with a ⅛" (3·1 mm.) drift punch.

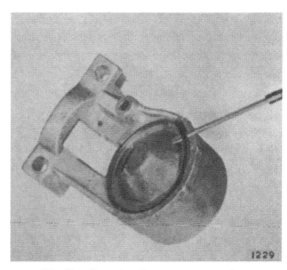

Fig. 82. *Removing the rear servo snap ring.*

Fig. 81. *Removing the rear servo roll pin.*

Fig. 83. *Replacing the roll pin.*

Page FF.X.s.46

AUTOMATIC TRANSMISSION

Rear Servo—Assembling

Lubricate all parts of the servo with transmission fluid before starting assembly.

Install a new "O" ring and then install piston in the servo body.

Install the release spring, retainer and snap ring.

Replace the servo lever, shaft and roll pin.

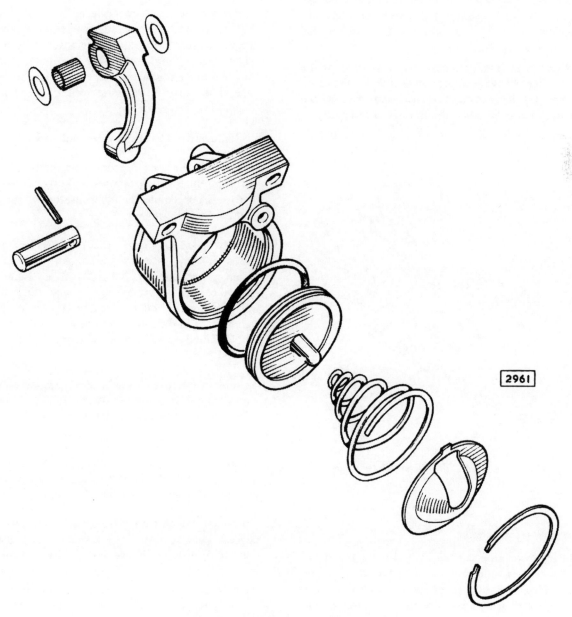

Fig. 84. *Exploded view of the rear servo.*

Page FF.X.s.47

AUTOMATIC TRANSMISSION

Pressure Regulator

Remove the valves from the regulator body. Remove the regulator body cover attaching screws and remove the cover. Remove the separator plate from the regulator body.

Wash all parts thoroughly in cleaning solvent and dry with compressed air. Inspect the regulator body and cover mating surfaces for burrs. Check all fluid passages for obstructions. Inspect the control pressure and converter pressure valves and bores for burrs and scores. Remove all burrs carefully with crocus cloth. Check free movement of the valves in their respective bores. The valves should fall freely into the bores when both the valve and bore are dry. Inspect the valve springs for distortion.

When assembling, be careful to avoid damaging the parts. Replace the separator plate and then the cover on the regulator body. Install and torque the attaching screws to 20–30 lb. in. (0·24–0·35 kgm.).

Insert the valves in the pressure regulator body.

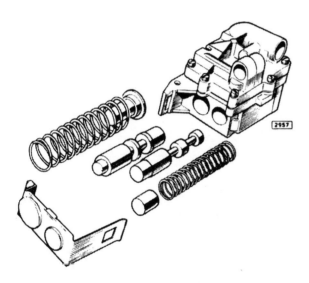

Fig. 85. *Regulator assembly. Valves, springs and retainer shown exploded.*

Valve Body—Dismantling

During dismantling of the control valve assembly, avoid damage to the valve parts and keep the parts clean. Place the valve parts and the assembly on a clean surface while performing the dismantling operation.

Remove the manual valve from the upper valve body.

Remove the four cap screws that retain the valve bodies.

Remove the cover and separator plates from the valve bodies. The body plate is attached to the lower valve body by a cheese head screw and to the upper valve body by a cheese head and a flat head screw. The separator plate and the lower valve body cover are held together by two cheese head screws.

Remove the front upper valve body plate retained by two screws. Remove the compensator valve plug, sleeve, springs and valve. Remove the modulator valve and spring assembly. The outer spring is retained to the modulator valve by a stamped retainer. The spring may be removed by tilting and pressing outward on the retainer.

Remove the downshift valve and spring.

Remove the rear upper valve body plate and throttle return spring retained by three screws to the body. Then remove the compensator cut back valve and the throttle valve.

Remove the four screws that retain the end body to the lower body. Remove the 2–3 shift valve inner and outer springs and the 2–3 shift valve. Remove the orifice control valve and spring and the transition valve spring and valve. Remove the orifice control valve plug and the 2–3 shift valve plug from end body. The end body plate should be removed for cleaning the end body.

Remove the four cheese head screws that retain the lower valve body side plate. Remove the 2–3 governor plug, the D1 and D2 control valve spring and valve.

The rear pump check valve, spring and sleeve generally should not be removed. The sleeve may be removed with snap ring pliers, if necessary.

Remove the end plate from the lower valve body cover. Then remove the 1–2 shift valve and spring and the front servo release orifice valve and spring.

Note: When removing all plates, be sure to hold the plates until screws are removed and release slowly as they are spring loaded.

Page FF.X.s.48

AUTOMATIC TRANSMISSION

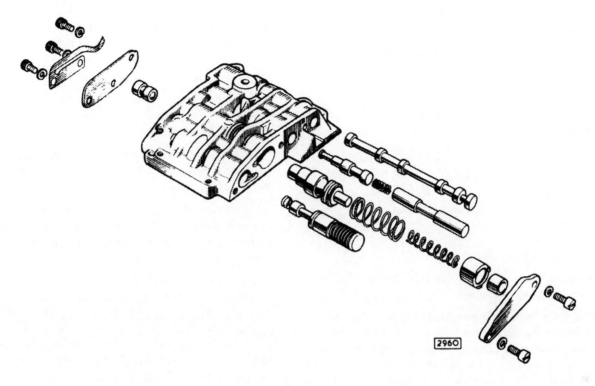

Fig. 86. *Upper valve body exploded—Manual valve, downshift valve, compensator valve and throttle modulator valve exploded.*

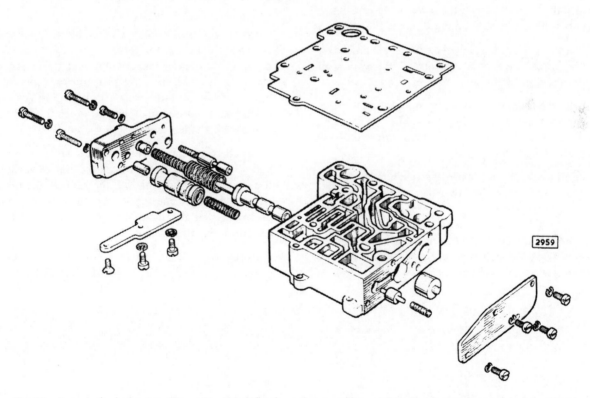

Fig. 87. *Lower valve body—transition valve, 2-3 shift valve and servo orifice control valve (from right to left on left of body), 2nd and 3rd governor plug and D1, D2 control valve exploded on right of body.*

Page FF.X.s.49

AUTOMATIC TRANSMISSION

Inspection

Clean all parts thoroughly in a cleaning solvent, then dry them with compressed air. Inspect all fluid passages for obstructions. Inspect the check valve for free movement. Inspect all mating surfaces for burrs and distortion. Inspect all plugs and valves for burrs and scores.

Note: Crocus cloth can be used to polish the valves and plugs if care is taken to avoid rounding the sharp edges.

Valve Body—Assembling

When assembling the control valve bodies, always use the following procedure:

Install the valve body plate on the upper valve body (retained by one cheese head and one flat head screw). Do not tighten the screws. If the rear pump check valve sleeve, valve and spring were removed from the lower valve body, install them, carefully staking the sleeve in the bore with the smooth end against the valve.

Place the upper body on the lower body and install the cheese head screw, but do not tighten the screw.

Place the lower valve body separator plate and cover on the lower valve body and install the two head screws, leaving them loose.

Install the four cap screws and lockwashers; torque the four screws to 72 lb. in. (·84 kgm.), then tighten the cheese head screws and flat head screw to 20–30 lb. in. (0·23–0·35 kgm.).

Try all valves dry in their respective bores, rotating them to make sure that they are free before final assembly in the valve body. If any sticking or binding occurs, the valve bodies will have to be separated and each surface lapped on crocus cloth, using a surface plate or a glass plate, to ensure against low or high spots or a warped condition.

Note: Lubricate all valves and plugs with automatic transmission fluid before final assembly in their respective bores.

Install the 1–2 shift valve spring and valve in the lower valve body cover. Install the front servo release orifice valve spring and valve and the cover end plate with two cheese head screws.

Install the range control valve and spring, the governor plug, and then install the side plate with four cheese head screws.

Install the orifice control valve spring and valve, the 2–3 shift valve, the 2–3 shift valve inner and outer springs, the transition valve, and spring in the lower valve body.

Replace the end body plate using one flat head and two cheese head screws and torque to 20–30 lb. in. (0·23–0·35 kgm.). Install the orifice control valve plug and the 2–3 shift valve plug in the lower valve body. Install the end body to the lower valve body, guiding the 2–3 shift valve inner spring into the 2–3 shift valve plug. Three long and one short special cheese head screws are used to retain the end body.

Note: Make sure the inner spring is piloted on the 2–3 shift valve plug.

Install the modulator valve and spring assembly. Install the compensator valve, compensator inner and outer springs, compensator plug and sleeve (be sure end of sleeve with the three protrusions is toward the plate and the smooth end to the spring in the upper valve body). Assemble the plate which is retained by two cheese headed screws.

Install the compensator cut-back valve in the rear end of the upper body. Install the rear plate so that the edge of the plate fits into the band of the throttle valve and install one screw to hold the rear plate in place. Install the throttle return spring and install the two remaining cheese headed screws.

Install the manual valve. Torque on all cheese headed screws should be 20–30 lb. in. (0·23 to 0·35 kgm.)

Page FF.X.s.50

AUTOMATIC TRANSMISSION

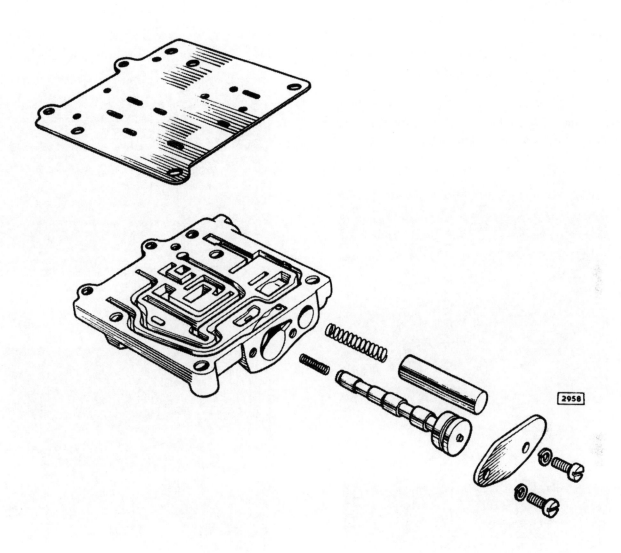

Fig. 88. *Lower valve body cover—front servo release orifice valve and 1-2 shift valve exploded.*

AUTOMATIC TRANSMISSION

TRANSMISSION ASSEMBLING

Lubricate all parts as they are assembled, with the same fluid used for filling the transmission. Petrolatum can be used sparingly to hold gaskets or thrust washers in position during assembly.

Wash the transmission case and dry with compressed air.

Install a new front pump to case gasket, then install the front pump. Torque the four attaching cap screws to 17–22 lb. ft. (2·35 to 3·04 kgm.).

Install the front band through the bottom of the case, positioning the band so that the anchor end is aligned with the anchor in the case.

Install the front clutch, rear clutch and forward sun gear assembly in the case. Handle the clutch assemblies in a manner that will prevent the clutches being pulled apart.

Fig. 91 *Installing the front band.*

Fig. 89. *Fitting a new front pump gasket.*

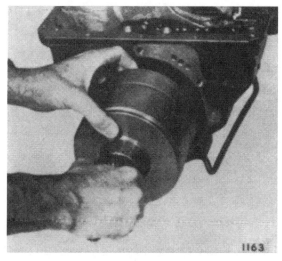

Fig. 92. *Installing the front clutch.*

Install the centre support in the transmission case with the three positioning holes aligned with the holes in the case.

Install the centre support cap screws with the rolled edge of each lockwasher towards the case. Torque to 20–25 lb. ft. (2·76 to 3·46 kgm.).

Fig. 90. *Installing the front pump.*

Page FF.X.s.52

AUTOMATIC TRANSMISSION

Install the rear band through the rear of the case. Be sure that the end with the depression or dimple is placed toward the adjusting screw.

Choose a selective washer to give the correct end play (end play determined during dismantling is used to determine the need for a different thrust washer).

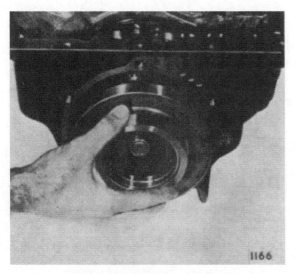

Fig. 93. *Installing the centre support.*

Fig. 95. *Fitting the rear band.*

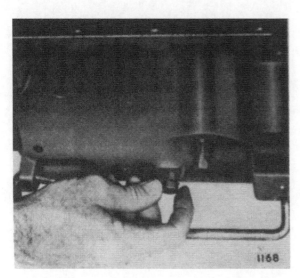

Fig. 94. *Fitting the centre support cap screws.*

Fig. 96. *Fitting the thrust plate and needle bearing.*

Use petrolatum sparingly to hold the forward sun gear thrust plate and needle bearing in the planet carrier, while the carrier is assembled over the sun gear.

Install the hook type seal rings on the rear of the forward sun gear. Check the rings for free movement in their grooves.

Install washer on the rear of the planet carrier.

Use petrolatum to hold the rear pump to case gasket to rear of the case.

Install the ring gear and output shaft assembly. Align the three oil tubes as the assembly is fitted and tap them in position.

Page FF.X.s.53

AUTOMATIC TRANSMISSION

Place the rear pump to extension housing gasket in position, then assemble the extension housing. Torque the five extension housing cap screws to 28–33 lb. ft. (3·87 to 4·56 kgm.).

Install the companion flange, flat washer, lockwasher and nut. Torque the nut to 90–120 lb. ft. (12·44–16·58 kgm.).

Fig. 97. *Assembling the carrier over the sun gear.*

Fig. 99. *Fitting the washer on the rear of planet carrier.*

Fig. 98. *Fitting the sealing rings.*

Fig. 100. *Tapping the output shaft assembly into position.*

Install the bearing snap ring, and then tap the ball bearing into position in the extension housing and on the output shaft (be sure spacer washer is on shaft ahead of bearing).

Slide the speedometer drive gear on the output shaft.

Install rear seal in bearing retainer. Assemble the bearing retainer in its gasket.

Front Servo Installation

Rotate the front band into position so that the anchor end is positioned over the anchor pin in the case.

Position the servo strut with the slotted end aligned with the servo actuating lever, and hold it in position with the middle and index fingers of the left hand.

Page FF.X.s.54

AUTOMATIC TRANSMISSION

Engage the end of the band with the small end of the strut then position the servo over the dowel pin.

Install the attaching cap screw but do not screw it in more than two or three threads at this time.

Rear Servo Installation

Position the servo anchor strut over the adjusting screw, then rotate the rear band to engage this strut. Place the servo actuating lever strut with the notched end to the band and lift the other end with index finger or screwdriver, while locking the servo lever over the strut.

Install the long pointed bolt in the forward servo hole so that it will engage the centre support.

The other shorter bolt is used in the rear position.

Torque the bolts to 40–50 lb. ft. (5·53–6·91 kgm.).

Fig. 101. *Installing the front servo.*

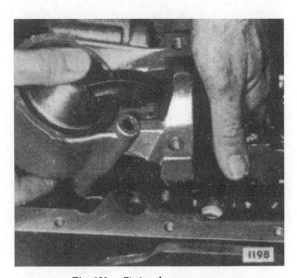

Fig. 103. *Fitting the rear servo.*

Fig. 102. *Engaging the servo anchor strut.*

Valve Body Installation

Place the manual selector in park or reverse position. Carefully align the valve body with the servo tubes and gently slide the valve body further onto the tubes.

The front servo must be pulled up off the dowel to allow easy assembly. Be careful at this point—the servo apply strut may become disengaged from the servo. Before seating the valve body on the case, install the nipple end of the throttle cable into the throttle cam.

Page FF.X.s.55

AUTOMATIC TRANSMISSION

Next, align the manual valve with the inside lever pin and the valve body will then drop into position. Torque the three valve body attaching cap screws to 8–10 lb. in. (0·09–0·12 kgm.).

Replace the control pressure tube, by first assembling the long straight end into the regulator, then rocking the tube downward into the control valve body. If too much resistance is encountered, it will help to loosen the control body attaching cap screws until the tube can be assembled.

Fig. 104. *Fitting the servo tubes.*

Fig. 106. *The valve body in position.*

Fig. 105. *Positioning the valve body.*

Torque the front servo attaching cap screw to 30–35 lb. ft. (4·15–4·84 kgm.) and adjust the front servo.

Fig. 107. *Replacing the control pressure tube.*

AUTOMATIC TRANSMISSION

Pressure Regulator Installation

Assemble the regulator, with the valves in position in their bores, to the case with the attaching cap screws.

Torque cap screws to 17–22 lb. ft. (2·35–3·04 kgm.). Install both springs and guides, then install the spring retainer.

Install the front servo apply and release tubes in the servo.

Install the rear pump inlet and outlet tubes, using new "O" rings.

Replace the compensator tube by aligning one end with the pressure regulator and the other end with the control valve body and then tap it into position.

Assemble the long end of the lubrication tube into the rear pump, then rock the other end into position and tap it into the pressure regulator assembly.

Fig. 108. *The pressure regulator installed.*

Fig. 110. *Fitting the apply and release tubes.*

Fig. 109. *Fitting the pressure regulator springs.*

Fig. 111. *Fitting the lubrication tube.*

AUTOMATIC TRANSMISSION

Replace the front band lubrication tube. Be sure the tube is aligned so that the open end will direct oil onto the front drum surface at the front band gap. Tube should point at approximately the centre of the gap.

Assemble the oil screen assembly onto the rear pump inlet tube and then rock into position over the front pump inlet on the pressure regulator assembly. Hook the screen retainer under the lubrication tube, lay across screen, and snap onto compensator tube.

Install the oil pan gasket, the oil pan and torque the 14 cap screws to 10–20 lb. ft. (1·38–2·76 kgm.).

Adjust the rear band.

Fig. 113. *Fitting the rear pump inlet tube.*

Fig. 112. *Replacing the compensator tube.*

Fig. 114. *View of the Model 8 transmission inverted.*

CONVERTER AND CONVERTER HOUSING

When installing the converter housing, the maximum allowable runout should not exceed 0·010″ (0·25 mm.) for bore or face indicator readings relative to crankshaft centre line; however, it is preferable to have less than 0·006″ (0·015 mm.) reading for both.

When installing the transmission to the converter housing and converter assembly, be certain that the converter lugs are properly aligned with the front pump drive gear, so that the parts will not be damaged by forcing impeller hub drive tangs against the pump drive gear lugs.

Page FF X.s.58

AUTOMATIC TRANSMISSION

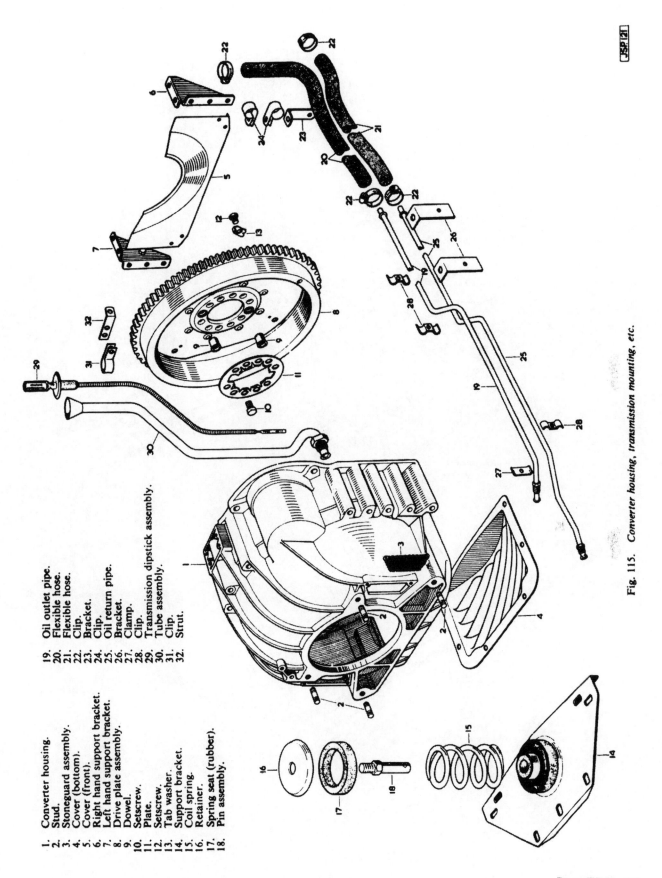

Fig. 115. *Converter housing, transmission mounting, etc.*

1. Converter housing.
2. Stud.
3. Stoneguard assembly.
4. Cover (bottom).
5. Cover (front).
6. Right hand support bracket.
7. Left hand support bracket.
8. Drive plate assembly.
9. Dowel.
10. Setscrew.
11. Plate.
12. Setscrew.
13. Tab washer.
14. Support bracket.
15. Coil spring.
16. Retainer.
17. Spring seat (rubber).
18. Pin assembly.
19. Oil outlet pipe.
20. Flexible hose.
21. Flexible hose.
22. Clip.
23. Bracket.
24. Clip.
25. Oil return pipe.
26. Bracket.
27. Clamp.
28. Clip.
29. Transmission dipstick assembly.
30. Tube assembly.
31. Clip.
32. Strut.

Page FF X.s.59

AUTOMATIC TRANSMISSION

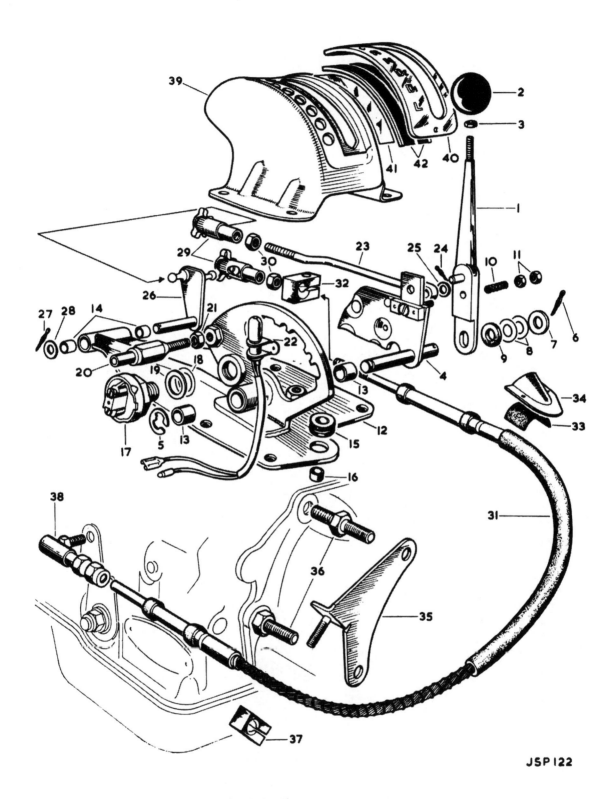

Fig. 116. *The transmission controls.*

Page FF.X.s.60

AUTOMATIC TRANSMISSION

1. Selector lever assembly.
2. Knob.
3. Nut.
4. Cam plate assembly.
5. Circlip.
6. Split pin.
7. Washer.
8. Shim.
9. Washer (rubber).
10. Spring.
11. Nut.
12. Mounting plate and selector gate assembly
13. Bush.
14. Bush.
15. Grommet.
16. Distance tube.
17. Reverse lamp switch.
18. Shim.
19. Shim.
20. Starter cut-out switch.
21. Nut.
22. Lamp assembly.
23. Operating rod assembly.
24. Split pin.
25. Washer.
26. Transfer lever assembly.
27. Split pin.
28. Washer.
29. Ball joint.
30. Nut.
31. Gear control cable assembly.
32. Clamp.
33. Pad.
34. Plate.
35. Abutment bracket.
36. Stud.
37. Clamp.
38. Adjustable ball joint.
39. Cover assembly.
40. Indicator plate.
41. Light filter.
42. Seal.

Page FF.X.s.61

AUTOMATIC TRANSMISSION

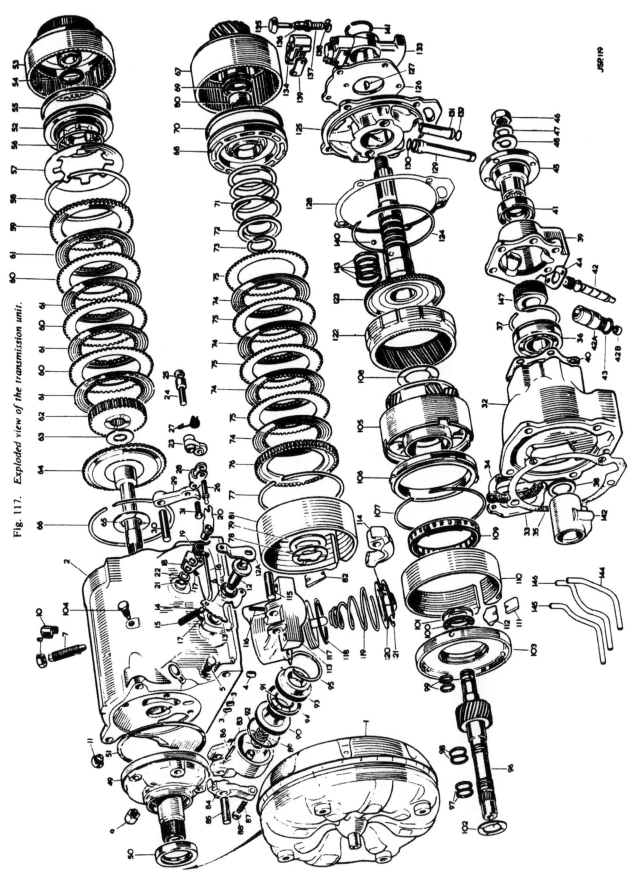

Fig. 117. Exploded view of the transmission unit.

Page FF X.s.62

AUTOMATIC TRANSMISSION

1. Converter assembly.
2. Transmission case assembly.
3. Plug.
4. Dowel.
5. Plug.
6. Oil seal.
7. Screw.
8. Nut.
9. Union.
10. Union.
11. Breather assembly.
12. Manual control shaft assembly.
12A. Selector lever.
13. Lever assembly.
14. ¼" ball.
15. Spring.
16. Link.
17. Clip.
18. Torsion lever.
19. Spring.
20. Forked lever.
21. Clip.
22. Washer.
23. Toggle lever.
24. Toggle pin.
25. Plug.
26. Ball pin.
27. Spring.
28. Link.
29. Pawl.
30. Pivot pin.
31. Pin.
32. Extension case assembly.
33. Cover plate.
34. Gasket.
35. Gasket.
36. Bearing.
37. Snap ring.
38. Spacing washer.
39. Speedometer housing.
40. Gasket.
41. Oil seal assembly.
42. Speedometer driven gear.
42A. Bearing.
42B. Oil seal.
43. "O" ring.
44. Plate.
45. Flange.
46. Nut.
47. Lockwasher.
48. Washer.
49. Front pump assembly.
50. Oil seal assembly.
51. Sealing ring.
52. Piston assembly.
53. Cylinder.
54. Sealing ring (inner).
55. Sealing ring (outer).
56. Split ring.
57. Spring.
58. Snap ring.
59. Pressure plate.
60. Clutch plate (drive).
61. Clutch plate (friction).
62. Hub.
63. Thrust washer (fibre).
64. Input shaft assembly.
65. Thrust washer.
66. Snap ring.
67. Front drum assembly.
68. Piston assembly.
69. Sealing ring (inner).
70. Sealing ring (outer).
71. Spring.
72. Seat.
73. Snap ring.
74. Clutch plate (friction).
75. Clutch plate (drive).
76. Pressure plate.
77. Snap ring.
78. Thrust washer (bronze).
79. Thrust washer (steel).
80. Needle bearing.
81. Brake band.
82. Strut (servo).
83. Body.
84. Lever.
85. Pivot pin.
86. Roll pin.
87. Screw.
88. Nut.
89. Return spring.
90. Piston assembly.
91. "O" ring (small).
92. "O" ring (large).
93. Piston sleeve.
94. Sealing ring.
95. Snap ring.
96. Forward sun gear assembly.
97. Sealing ring.
98. Sealing ring.
99. Sealing ring.
100. Thrust bearing.
101. Race.
102. Thrust washer (bronze).
103. Centre support assembly.
104. Screw.
105. Planetary gears and rear drum assembly.
106. Outer race.
107. Snap ring.
108. Thrust washer.
109. One way clutch assembly.
110. Brake band for rear drum.
111. Strut (servo).
112. Anchor strut.
113. Body assembly.
114. Lever.
115. Shaft.
116. Roll pin.
117. Piston.
118. "O" ring.
119. Return spring.
120. Plate.
121. Snap ring.
122. Ring gear.
123. Mainshaft assembly.
124. Snap ring.
125. Rear pump assembly.
126. Plate.
127. Key.
128. Gasket.
129. Oil inlet tube.
130. "O" ring.
131. Oil outlet tube.
132. "O" ring.
133. Governor assembly.
134. Governor body.
135. Governor weight.
136. Governor valve.
137. Spring.
138. Retainer.
139. Cover plate.
140. ¼" ball.
141. Snap ring.
142. Oil collector sleeve.
143. Piston ring.
144. Oil collector tube (front).
145. Oil collector tube (intermediate).
146. Oil collector tube (rear).
147. Speedometer drive gear.

Page FFX.s.63

SECTION H
REAR AXLE

HALF SHAFTS—UNIVERSAL JOINTS

Model	Chassis Number	
	R.H. Drive	L.H. Drive
Open 2 Seater	1E.1926	1E.16721
Fixed Head Coupe	1E.21669	1E.34851
2+2	1E.51067	1E.77705

Commencing at the above chassis numbers, grease nipples are re-introduced to the universal joints of the rear axle half shafts.

Access to the nipples of the outer joints is gained by removing the plastic sealing plugs from the joint covers. The universal joints should be greased every 3,000 miles (5,000 km.).

SECTION I

STEERING

Model	Chassis Number
	L.H. Drive
Open 2 Seater ..	1E.15980–U.S.A. Only
Open 2 Seater ..	1E.16010–Other than U.S.A.
Fixed Head Coupe	1E.34583–U.S.A. Only
Fixed Head Coupe ..	1E.34752–Other than U.S.A.
2+2	1E.77709

Note: The following details do not apply to any Right Hand Drive cars.

From the above chassis numbers the upper and lower steering columns and mountings are of the collapsible type designed to comply with U.S.A. Federal Safety Regulations.

The collapse points are retained by nylon plugs which will shear on impact, allowing the steering wheel and columns (upper and lower) to move forward.

NO ATTEMPT must be made to repair the units if damaged due to accident.

NEW replacement items MUST be fitted.

UPPER STEERING COLUMN

Description

The upper steering column (inner) is composed of two separate sliding shafts retained to a fixed length by nylon plugs, the outer column being pierced in a lattice form.

The inner shaft assembly is supported in the outer column by two pre-lubricated taper roller bearings.

A gaiter covers the pierced portion of the outer column to seal against the ingress of dirt.

Removal

Disconnect the battery.

Withdraw the self-tapping screws and remove the under-scuttle casing above the steering column.

Disconnect the cables contained in the direction indicator switch harness.

Note the location of the connections for reference when refitting.

Withdraw the ignition key, remove the ring nut and detach the ignition lock from the mounting bracket on the steering column.

Note: If the car is fitted with air-conditioning equipment the switch will be mounted on a bracket attached to the evaporator unit and need not be removed.

Release three grub screws in the steering wheel hub and remove the steering wheel motif.

Remove the locknut, hexagon nut and flat washer and withdraw the steering wheel from the splines on the inner column.

Remove the nut, lockwasher and pinch bolt securing the upper universal joint to the lower steering column.

Remove two nuts and lockwashers securing the upper column lower mounting bracket to the underside of the scuttle.

Remove two bolts, nuts, lockwashers and distance pieces securing the upper mounting bracket to the support bracket on the body.

Withdraw the upper column from the splines on the lower column.

Note: If the steering column has not been damaged by impact, i.e., if the nylon plugs in the inner column or the top mounting bracket have not sheared, excessive force must NOT be used to separate the upper universal joint from the lower column.

Refitting

Refitting is the reverse of the removal procedure.

Set the road wheels in the straight ahead position and check that the bolt holes in the lugs of the upper column universal joint register correctly with the groove machined in the lower column splines. Tighten the pinch bolt to a torque of 16-18 lb. ft. (2.2-2.5 kgm).

IMPORTANT

Excessive force as noted under 'Removal' must not be used when reassembling the universal joint to the column.

UNDER NO CIRCUMSTANCES should a mallet or similar tool be used when engaging the splines in the joint and column.

If the splines will not engage freely, inspect for damage or burrs and remove with a fine file.

NO ATTEMPT must be made to repair any nylon plugs which have sheared due to impact.

Dismantling

Dismantling is confined to removing the steering column adjuster locknut, the splined shaft and the direction indicator switch as detailed on page I.8.

LOWER STEERING COLUMN

Description

The lower steering column comprises two sliding shafts retained to a fixed length by nylon plugs.

Page I.X.s.1

STEERING

Removal

Remove the upper steering column as detailed previously.

Remove the nut, lockwasher and bolt securing the column to the lower universal joint and withdraw the column rearwards through the grommet.

Note : If the steering column has not been damaged by impact, i.e., if the nylon plugs in the column have not sheared, excessive force must NOT be used to separate the column and the lower universal joint.

Refitting

Refitting is the reverse of the removal procedure.

Check that the bolt holes in the universal joint register correctly with the groove machined in the column splines. Tighten the pinch bolt to a torque of 16 - 18 lb. ft. (2.2 - 2.5 kgm).

IMPORTANT

Excessive force as noted under 'Removal' must not be used when reassembling the universal joint to the column. **UNDER NO CIRCUMSTANCES** should a mallet or similar tool be used when engaging the splines in the joint or the column.

If the splines will not engage freely, inspect for damage or burrs and carefully remove with a fine file. NO ATTEMPT must be made to repair any nylon plugs which have sheared due to impact.

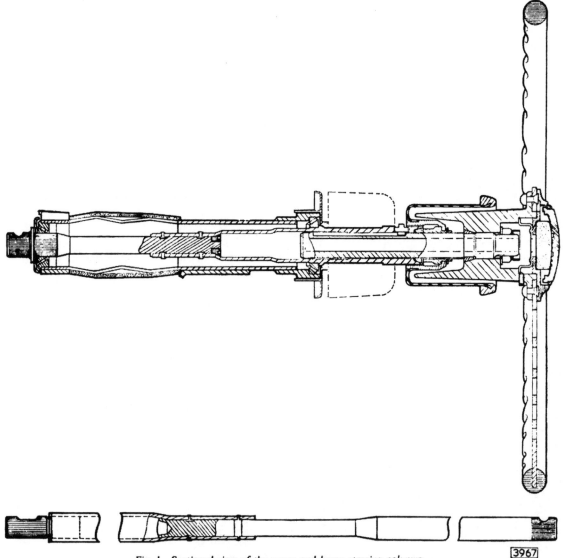

Fig. 1. Sectioned view of the upper and lower steering columns showing the nylon plugs.

Page I.X.s.2

SECTION J

FRONT SUSPENSION

TORSION BAR—CHECKING

Before any check on torsion bar setting is made the car must be placed on a perfectly level surface, wheels in the straight ahead position and the tyre pressures correctly adjusted.

Referring to the illustration overpage, take the measurement "A" from the centre line of each road wheel to the ground. Record the measurement "B" from the centre line of each inner fulcrum of the lower wishbone assembly. Subtract "B" from "A" to give the dimension "C". This should be $3\frac{1}{2}'' \pm \frac{1}{4}''$ (88·9 ± 6·35 mm.) for 4·2 "E" Type cars and $3\frac{3}{4}'' \pm \frac{1}{4}''$ (95·25 ± 6·35 mm.) for 2+2 cars.

If any adjustment is required, this should be carried out in accordance with instructions given in Section J.

The correct dimensions between hole centres for the setting links are as follows:—

4·2 "E" Type F.H.C.	..	$17\frac{13}{16}''$ (45·25 cm.)
Open Sports ..		$17\frac{13}{16}''$ (45·25 cm.)
4·2 "E" Type 2+2 ..	..	$18\frac{1}{4}''$ (46·36 cm.)

4·2 "E" Type F.H.C.	..	1E.35382 (L.H.D.)
Open Sports ..		1E.17532 (L.H.D.)
4·2 "E" Type 2+2 ..	..	1E.50875 (R.H.D.)
		1E.77407 (L.H.D.)

commencing at the above chassis numbers, torsion bars of larger diameter (·780''—·784'' (19·81—19·9 mm.)) are fitted. Cars with these torsion bars require setting links with the following centres between the holes:—

4·2 "E" Type F.H.C. (L.H.D.)	$17\frac{3}{4}''$ (45·1 cm.)
Open Sports (L.H.D.) ..	$17\frac{3}{4}''$ (45·1 cm.)
4·2 "E" Type 2+2 (R.H.D. and L.H.D.) ..	$18''$ (45·7 cm.)

All Air-conditioned 4·2 "E" Type and 2+2 cars have been fitted with the larger diameter torsion bars, the correct distance between the hole centres of the setting link for these cars being as follows:—

4·2 "E" Type F.H.C.	..	$17\frac{3}{4}''$ (45·1 cm.)
Open Sports ..		$17\frac{3}{4}''$ (45·1 cm.)
4·2 "E" Type 2+2	..	$18\frac{1}{8}''$ (48·87 cm.)

Page J.X.s.1

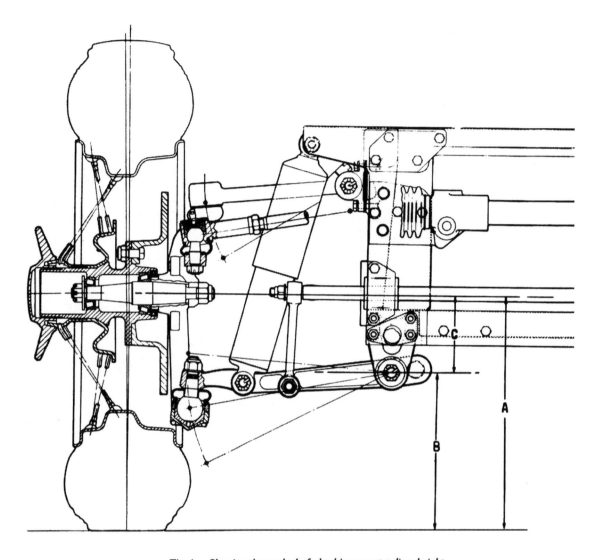

Fig. 1. *Showing the method of checking car standing height.*

SECTION L
BRAKES
DATA

Caliper type	Dunlop bridge type with quick change pads
Brake disc diameter—front	11″ (27·9 cm.)
—rear	10″ (25·4 cm.)
Master cylinder bore diameter	$\frac{7}{8}$″ (22·23 mm.)
Master cylinder stroke	1·30″ (3·3 cm.)
Brake cylinder bore diameter—front	$2\frac{1}{8}$″ (5·39 cm.)
—rear	$1\frac{3}{4}$″ (4·45 cm.)
Servo unit type	Lockheed Dual—line
Main friction pad material	Mintex M.59
Handbrake friction pad material	Mintex M.34

ROUTINE MAINTENANCE

EVERY 3,000 MILES (5,000 KM.)

Check fluid level in master cylinder reservoirs.

EVERY 6,000 MILES (10,000 KM.)

Check brake friction pads for wear.

Page L.X.s.1

BRAKES

DUAL-LINE SERVO BRAKING SYSTEM

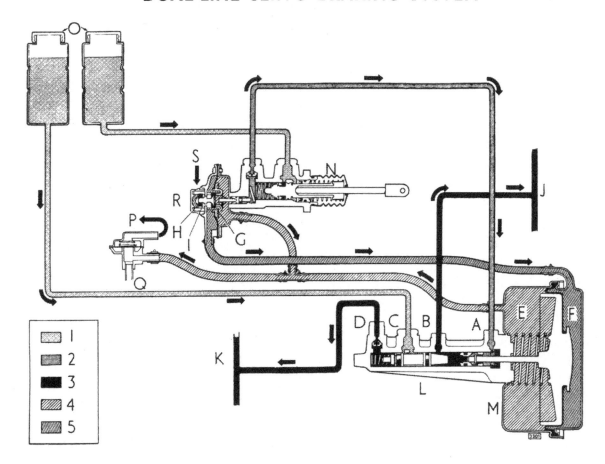

Fig. 1. *Dual-line servo braking system (Early cars).*

Key to Figs. 1 and 2

1. Fluid at feed pressure.
2. Fluid at master cylinder delivery pressure.
3. Fluid at system delivery pressure.
4. Vacuum.
5. Air at atmospheric pressure

A Primary chamber—slave cylinder.
B Outlet port—rear brakes.
C Inlet port—secondary piston.
D Outlet port—front brakes.
E Vacuum.
F Air pressure.
G Diaphragm.
H Filter.
I Air control.
J To rear brakes.
K To front brakes.
L Tandem slave cylinder.
M Vacuum cylinder.
N Master cylinder.
O Fluid reservoirs.
P To manifold.
Q To reservac.
R Reaction valve.

Page L.X.s.2

BRAKES

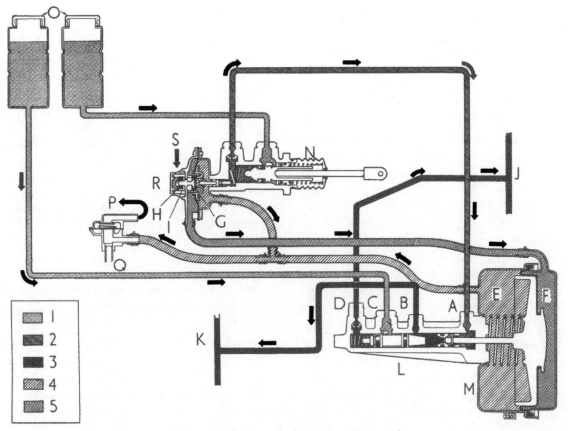

Fig. 2. *Dual-line servo braking system (Later cars)*

DESCRIPTION

The dual-line servo braking system consists of an integral vacuum booster with tandem slave cylinder, a master cylinder combined with a booster reaction valve and two fluid reservoirs.

The master cylinder is of conventional design consisting of a single cast iron cylinder housing a steel, black oxided piston sealed by a single hydraulic cup. This piston is deeply skirted to engage the operating push rod. The smaller intermediate piston, housed in its own bore in the nose of the master cylinder, is actuated by hydraulic pressure generated within the main chamber.

Mounted on the end of the master cylinder, the reaction valve consists of a pair of flow control valves which sequence the flow of air to the booster. Both control valves are operated by the intermediate piston in the master cylinder. A flat plate, interposed between the two master cylinder pistons, enables the intermediate piston to function mechanically in the event of an hydraulic failure.

The booster portion of the integral booster and slave cylinder assembly consists of a pressed steel tank which houses a moulded phenolic resin piston and a rubber rolling diaphragm. A push rod, secured to the piston, extends through the forward face of the tank into the slave cylinder. This push rod provides the principal motive force for the tandem pistons.

On the forward face of the boost tank is mounted the tandem slave cylinder which consists of a single cast iron cylinder housing two pistons in tandem, each piston having its own inlet and outlet port. Either piston will, in the event of a failure, operate independently.

OPERATION (Figs. 1 and 2)

When the system is at rest, both sides of the boost system are continuously exhausted by the engine manifold depression.

As the brake pedal is depressed, the master cylinder

Page L.X.s.3

BRAKES

piston moves along the cylinder building up pressure and forcing fluid out to the primary chamber of the slave cylinder (A). Simultaneously, the intermediate piston, in the end of the master cylinder, closes the diaphragm valve (G) in the reaction valve and, in so doing, isolates the vacuum (E) from the air pressure side (F) of the boost system.

Further progress of the intermediate piston along its bore will crack the air control spool (I) in the reaction valve thus admitting air at atmospheric pressure to the rear of the boost cylinder piston. The air enters the system through a small cylindrical filter (H) on the reaction valve.

The pressure imbalance, created by the admission of air to the pressure side of the boost system, will push the boost piston down the cylinder transmitting a linear force, through the push rod, to the primary piston of the slave cylinder.

Forward motion of the primary piston, supplemented by the output of the master cylinder, transmits hydraulic pressure to the secondary piston (C) and fluid under pressure flows simultaneously from the two output ports (B and D), to the front and rear brakes.

SAFETY FACTORS

In the event of a fluid line failure in the pipe linking the master cylinder to the slave cylinder or the pipe linking the master cylinder to the fluid supply tanks, the reaction valve will be actuated mechanically by the master cylinder piston providing the booster pressure to the front and rear brakes.

A failure in the fluid line coupling the slave cylinder to the front brakes will result in the slave cylinder secondary piston travelling to its fullest extent, down the bore. This has the effect of isolating the front brake line from the rest of the system and permitting normal fluid pressure to build up in the rear brake line.

If a fault exists in the rear brake line, the slave cylinder piston will travel along the bore until it contacts the other piston and the two pistons will then travel along the bore together to apply the front brakes.

Note: On later cars, this process is reversed (See Fig. 2).

In the case of leaks in either the air or vacuum pipes both front and rear brakes may still be applied by the displacement of fluid at master cylinder pressure.

REMOTE SERVO AND SLAVE CYLINDER

Removal

Remove the trim on the floor recess panel on the left-hand side of the car. This will disclose the three nuts securing the remote servo to the bulkhead. Withdraw the three nuts.

Drain the fluid from the system as detailed on page L.10.

Disconnect the four brake pipe unions and the two flexible hoses.

Remove the battery and carrier bracket for the battery tray.

Withdraw the bolt securing the slave cylinder to the mounting bracket on the outer side member. Remove the servo together with the slave cylinder.

Refitting

Refitting the servo is the reverse of the removal procedure. Bleed the system after replenishing with fresh fluid.

MASTER CYLINDER AND REACTION VALVE

Removal

Drain the fluid from the system. Disconnect the two hydraulic pipes from the master cylinder. Disconnect the vacuum hose from the reaction valve.

Remove the clevis pin, which is retained by a split pin, securing the brake pedal to the master cylinder push rod from inside the car. In the case of right-hand drive cars, remove the top of the air cleaner and reaction valve prior to removing the two nuts securing the master cylinder to the mounting.

On left-hand drive cars the master cylinder and reaction valve can be removed as a complete unit.

Refitting

Refitting is the reverse of the removal procedure. Bleed the system after replenishing with fresh fluid.

Page L.X.s.4

BRAKES

SERVICING THE UNIT

General

Prior to dismantling either the remote servo or the master cylinder reaction valve assembly, it is advisable to obtain repair kits containing all the necessary rubber parts required during overhaul. Three separate repair kits are available as follows:—

(a) Remote servo repair kit.

(b) Reaction valve repair kit.

(c) Master cylinder repair kit.

When either of the units have been dismantled the component parts should be washed in denatured alcohol (industrial methylated spirits). Parts that have been washed should be thoroughly dried using a clean lint-free cloth or pressure line and then laid out on clean paper to prevent dirt being assembled into the servo or master cylinder and reaction valve assembly.

Examine all metal parts for damage, with particular reference to those listed below and make renewals where necessary:—

(a) the reaction valve piston and bore.

(b) the master cylinder piston and bore.

(c) the servo slave cylinder pistons and bore.

(d) the servo push rod stem.

If any of the vacuum hose connections have become loose in service these must be rectified prior to reassembly.

The vacuum non-return valve is a sealed unit and, if faulty, it must be replaced by a new assembly.

THE REMOTE SERVO (Fig. 3)

Dismantling

Support the servo slave cylinder in the jaws of a vice, shell uppermost, with specially formed wooden blocks placed either side of the cylinder and against the jaws of the vice.

Fit the cover removal tool (Churchill Tool No. J.31) to the end cover and secure it by fitting the three nuts.

Turn the end cover in an anti-clockwise direction until the indents in the servo shell line-up with the small radii around the periphery of the end cover. At this stage the end cover may be removed from the servo.

Remove the diaphragm (11) from its groove in the diaphragm support (10) and, with the servo removed from the jaws of the vice, apply a gentle pressure to the diaphragm support and shake out the key (12).

The diaphragm support (10) and diaphragm support return spring (8) can then be removed.

Bend down the tabs on the locking plate (16) and remove the locking plate, abutment plate (17) and servo shell (14) from the slave cylinder by unscrewing and removing three screws (15).

Extract the seal (19) and bearing (18) from the mouth of the slave cylinder bore which will permit the removal of the push rod (9) together with the slave cylinder piston assembly.

The push rod may be separated from the piston by sliding back the spring steel clip (6) around the piston and removing the pin (5). It is not necessary to remove the cup (21) from the piston as a new piston together with a cup are contained in the repair kit.

Unscrew and remove the fluid inlet connection (3) and extract the piston stop pin (30) from the base of the inlet fluid port. To facilitate this operation, apply gentle pressure to the secondary piston (4).

Tap the open end of the slave cylinder body with a hide or rubber hammer to remove the secondary piston together with the piston return spring (28) from the bore.

The rubber seal (25) located in the groove adjacent to the heel of the piston may be removed but it is advisable to first remove the spring retainer (26) from the piston head extension before attempting to remove the seal (25) and piston washer (24). Removal of the plastic spring retainer (26) is sometimes difficult but, as a new one is provided in the repair kit, this part can be replaced if damaged.

To remove the trap valve assembly, unscrew and remove the adaptor (1) from the fluid outlet port. If it is necessary to remove the shim-like clip from the body of the trap valve (29) ensure that this part is not distorted in any way.

Page L.X.s.5

BRAKES

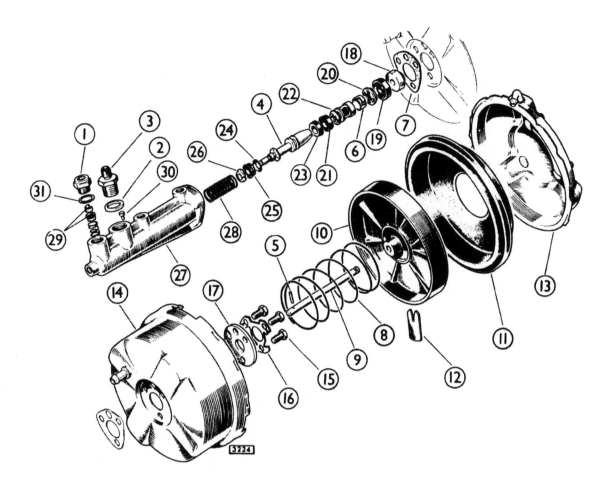

Fig. 3. *Exploded view of the remote servo.*

1. Outlet connection.
2. Gasket.
3. Inlet connection.
4. Piston.
5. Pin.
6. Retaining clip.
7. Gasket.
8. Spring.
9. Push rod.
10. Diaphragm support.
11. Diaphragm.
12. Key.
13. Cover.
14. Vacuum cylinder shell.
15. Screw.
16. Locking plate.
17. Abutment plate.
18. Bearing.
19. Seal.
20. Spacer.
21. Cup.
22. Piston.
23. Cup.
24. Piston washer.
25. Seal.
26. Retainer.
27. Slave cylinder body.
28. Spring.
29. Trap valve.
30. Stop pin.
31. Gasket.

Page L.X.s.6

BRAKES

Assembling

Assemble the trap valve (29) complete with spring and clip into the outlet port and secure it by fitting the fluid outlet adaptor (1) together with the copper gasket (31).

Prior to further assembly, lightly coat the four rubber seals to be replaced in the slave cylinder bore with Lockheed Disc Brake Lubricant.

Locate the piston washer (24) over the piston head extension, convex face towards the piston flange and, using the fingers only, assemble the two rubber seal (23 and 25) onto the piston so that their concave faces oppose each other.

Press the spring retainer (26) onto the piston head extension with both seals in position.

Fit the piston return spring (28) to the secondary piston complete and assemble into the slave cylinder bore, spring leading.

Press the piston assembly down the cylinder bore, using a short length of brass bar, until the drilled piston flange passes the piston stop pin hole.

Insert the piston stop pin (30) into the fluid inlet port and secure it by fitting the inlet adaptor (3) complete with the copper gasket (2). Place the push rod (9) in the primary piston and, with the aid of a small screwdriver, compress the small spring within the piston to enable the pin (5) to be inserted. Prior to fitting the pin retainer (6), it is important to establish that the small coil spring is loaded between the heel of the piston and the pin. Ensure that the pin does not pass through the coils of the spring.

Fit the spring retainer by sliding it into position along the piston ensuring that no corners are left standing proud after assembly.

Using fingers only, fit a new cup (21) into the groove on the piston so that its lip (concave face) faces towards the piston head and assemble the piston into the slave cylinder bore.

Insert the spacer (20), gland seal (19) and plastic bearing (18) into the slave cylinder counterbore leaving the bearing projecting slightly from the mouth of the bore.

Place the gasket (7) in position on the end face of the slave cylinder, using the plastic bearing as a location spigot and fit the vacuum shell (14), abutment plate (17) and locking plate (16).

Insert the three securing screws (15) and tighten down to a torque of 150/170 lb./ins. (1·7–1·9 kgm.).

Bend the tabs on the locking plate against the flats on the three screws.

Locate the diaphragm support return spring (8) centrally inside the vacuum shell, fit the diaphragm support (10) to the push rod and secure it by dropping the key (12) into the slot provided in the diaphragm support.

Stretch the rubber diaphragm (11) into position on the diaphragm support ensuring that the bead around its inside diameter fits snugly into the groove in the diaphragm support.

If the surface of the rubber diaphragm appears wavy or crinkled this indicates that it is not correctly seated. To ease assembly, smear the outside edges of the diaphragm liberally with Lockheed disc brake lubricant.

Fit the end cover using Churchill Tool No. J.31.

Note: As it is possible to fit the end cover in three different positions, ensure that the end cover hose connections line up with the slave cylinder inlet and outlet ports when assembly is complete.

BRAKES

1. Diaphragm.
2. Screw.
3. Shakeproof washer.
4. Gasket.
5. Bolt.
6. Outlet adaptor.
6A. Copper gasket.
7. Trap valve body.
8. Washer.
9. Banjo.
10. Copper gasket.
11. Body.
12. Bearing.
13. Secondary cup.
14. Seal.
15. Piston.

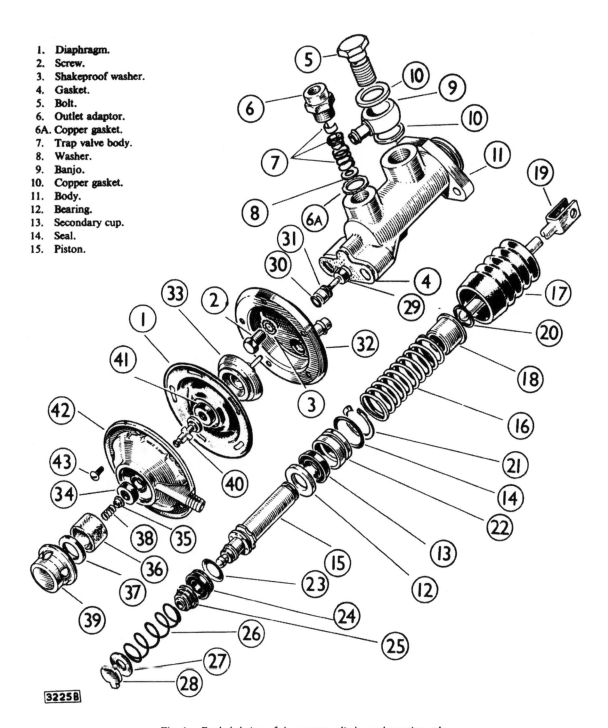

Fig. 4. *Exploded view of the master cylinder and reaction valve.*

16. Return spring.	23. Piston washer.	30. Seal.	37. Sorbo washer.
17. Rubber boot.	24. Main cup.	31. Piston.	38. Spring.
18. Spring retainer.	25. Retainer.	32. Valve housing.	39. Filter cover.
19. Push rod.	26. Spring.	33. Diaphragm support.	40. Valve stem.
20. Spirolox circlip.	27. Retainer.	34. Valve rubber.	41. Valve rubber.
21. Circlip.	28. Lever.	35. Valve cap.	42. Valve cover.
22. Bearing.	29. Seal.	36. Filter.	43. Screw.

Page L.X.s.8

BRAKES

MASTER CYLINDER AND REACTION VALVE

Dismantling (Fig. 4)

Unscrew and remove the fluid outlet adaptor (6) and extract the trap valve assembly (7) from the outlet port.

Remove the rubber boot (17) from the mouth of the cylinder bore, compress the piston return spring (16) and unwind the spirolox circlip (20) from the heel of the piston. The spring retainer (18) and piston return spring (16) can at this stage be removed.

Press the piston (15) down the bore and, with the aid of special circlip pliers (Tool number 7066) extract the circlip (21) from the mouth of the cylinder bore. Care should be taken during this operation not to damage the finely machined cylinder piston.

The piston assembly, complete with nylon bearings and rubber seals, can be withdrawn from the cylinder bore.

Remove the plastic bearing (22), complete with "O" ring (14), secondary cup (13) and rectangular section plastic bearing (12) from the piston by sliding the assembly along the finely machined portion.

Due to the plastic spring retainer (25) being an interference fit onto the piston head extension, this part is likely to become damaged during dismantling. In view of this a new spring retainer is contained in the appropriate repair kit. To remove the spring retainer, hold the piston on a bench, piston head downwards, applying a downwards force to the back face of the spring retainer with a slim open-ended spanner. The piston return spring (26), pressed steel retainer (27) and lever (28) may, at this stage, be withdrawn from the cylinder bore.

Remove the filter cover (39) and collect the filter (36) sorbo washer (37) and spring (38).

Unscrew and remove the five screws securing the valve cover (42), remove the valve cover assembly from the valve housing (32) which can be dismantled further by prising off the snap-on clip securing the valve rubber (34).

The valve stem (40) complete with the other valve rubber (41) can now be withdrawn from the valve housing and the valve rubber removed from the valve stem flange. The reaction valve diaphragm (1) can now be separated from the diaphragm support (33) and, by unscrewing the two hexagon-headed screws (2), the valve housing can be separated from the master cylinder body.

Removal of the valve piston (31) assembly can be effected by inserting a small blunt instrument into the master cylinder fluid outlet port and easing the valve piston assembly along its bore until it can be removed by hand.

Important: No attempt should be made to withdraw the valve piston assembly along its bore by using pliers.

Assembling

Prior to assembly liberally coat all rubber seals and plastic bearings, with the exception of the two valve rubbers, with Lockheed disc brake lubricant.

Holding the master cylinder body at an angle of approximately 25° to the horizontal, insert the lever (28), tab foremost, into the cylinder bore ensuring that, when it reaches the bottom of the bore, the tab on the lever drops into the recessed portion provided.

Place the piston washer (23) on the piston head, convex face towards the piston flange, together with a new main cup (24) and press the plastic spring retainer (25) onto the piston head extension.

Drop the pressed steel spring retainer (27) into the bottom of the bore following up with the piston return spring (26). When these two parts have been assembled it is advisable to recheck the position of the lever.

Press the piston assembly into the cylinder bore and locate the rectangular section plastic bearing (12), secondary cup (13) and bearing (22) together with seal (14) into the mouth of the cylinder bore.

Press the assembly down the bore to its fullest extent and with the aid of the special circlip pliers (Tool number 7006 with "K" points) fit the circlip to retain the internal parts.

Locate the other piston return spring (16) over the heel of the piston together with the pressed steel spring retainer (18), slide the spring retainer down the finely machined portion of the piston against the load of the spring and fit the spirolox circlip (20) into the groove ground around the heel of the piston.

Using the fingers only, stretch a new valve seal (29) and "O" ring into position on the valve piston and insert the assembly into the valve box.

Page L.X.s.9

BRAKES

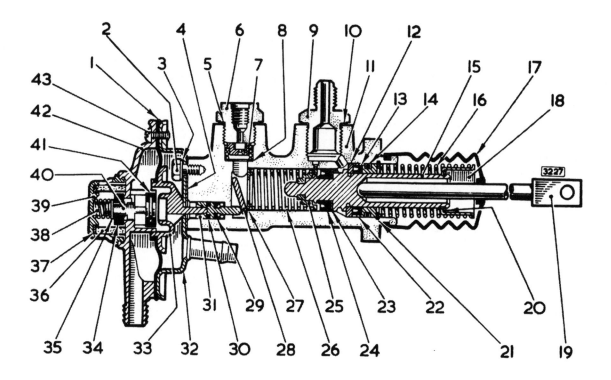

Fig. 5. *Sectioned view of the master cylinder and reaction valve.*

1. Diaphragm.
2. Screw.
3. Shakeproof washer.
4. Gasket.
5. Gasket.
6. Outlet adaptor.
7. Trap valve body.
8. Washer.
9. Inlet adaptor.
10. Copper gasket.
11. Body.
12. Bearing.
13. Secondary cup.
14. Seal.
15. Piston.
16. Return spring.
17. Rubber boot.
18. Spring retainer.
19. Push rod.
20. Spirolox circlip.
21. Circlip.
22. Bearing.
23. Piston washer.
24. Main cup.
25. Retainer.
26. Spring.
27. Retainer.
28. Lever.
29. Seal.
30. Seal.
31. Piston.
32. Valve housing.
33. Diaphragm support.
34. Valve rubber.
35. Valve cap.
36. Filter.
37. Sorbo washer.
38. Spring.
39. Filter cover.
40. Valve stem.
41. Valve rubber.
42. Valve cover.
43. Screw.

Page L.X.s.10

BRAKES

Secure the valve housing to the master cylinder body by fitting the two hexagon headed screws (2) complete with spring washers and tighten each screw to a torque of 160/180 lb./ins. (1·8–2 kgm.). A new gasket should be fitted between the valve housing and the master cylinder body.

Stretch the reaction valve diaphragm onto the diaphragm support through the hole in the valve housing so that it engages the depression in the valve piston.

Using the fingers only, stretch the valve rubber, which is formed with the groove around its inside diameter, onto the valve stem flange, insert the valve stem through the hole in the valve cover and secure it by placing the other valve rubber over the valve stem and fitting the snap-on clip.

The valve cover assembly can now be placed into position on the valve housing ensuring that all the holes line up and that the hose connections are in line with each other at the bottom of the unit. Secure the valve cover assembly by fitting the five self-tapping screws.

Hold the master cylinder in an upright position (valve uppermost) and place the air filter together with the rubber washer in position upon the valve cover with the small spring on the snap-on valve stem clip.

Carefully locate the air filter cover over the air filter and press it firmly home.

If the trap valve assembly has been dismantled; insert the small clip into the trap valve body ensuring that it does not become distorted and locate the spring on the reduced diameter of the trap valve body.

Assemble the trap valve complete (spring innermost) into the master cylinder fluid outlet port.

Place a copper gasket under the head of the fluid outlet adaptor and screw the adaptor into the fluid outlet port. If the fluid inlet adaptor has been removed, this must be replaced in the same manner using a copper gasket under the head.

The master cylinder push rod and convoluted rubber boot can best be fitted during the installation of the assembly.

FRONT CALIPERS

Removal

In order to remove the front calipers, jack up the car and remove the road wheel. Disconnect the fluid feed pipe and plug the hole in the caliper. Discard the locking wire from the mounting bolts. Remove the caliper, noting the number of round shims fitted.

Refitting

Locate the caliper body (complete with the cylinder assemblies) in position and secure with two bolts.

Check the gap (A Fig. 6) between each side of the caliper and the disc, both at the top and bottom of the caliper. The difference should not exceed ·010″ (·25 mm.) and round shims may be fitted between the caliper and the mounting plate to centralise the caliper body. Lockwire the mounting bolts.

If not already fitted, fit the bridge pipe connecting the two cylinder assemblies. Connect the supply pipe to the cylinder body and ensure that it is correctly secured.

Bleed the brakes as described on page L.10.

Important: It is essential that the bridge pipe is fitted with the "hairpin" bend to the inboard cylinder block, that is, furthest from the road wheel. The bridge pipe carries a rubber identification sleeve.

REAR CALIPERS

Removal

The rear suspension unit must be removed in order to withdraw the rear calipers.

Proceed as described in Section K "Rear Suspension" and support the suspension unit under its centre.

Disconnect the handbrake compensator linkage from the handbrake operating levers. Discard the split pins and withdraw the clevis pins.

Lift the locking tabs and remove the pivot bolts together with the retraction plate.

Remove the friction pad carriers from the caliper bridges by moving them rearwards round the discs and withdrawing from the rear of the rear suspension assembly.

Remove the hydraulic feed pipe at the caliper and plug the hole to prevent the entry of dirt.

Remove the friction pads from the caliper as described on page L.32.

Remove the front hydraulic damper and road spring unit (as described in Section K "Rear Suspension") and remove the four self-locking nuts from the half-shaft inner universal joint.

Withdraw the joint from the bolts and allow the hub carrier to move outwards—support the carrier in this position.

Page L.X.s.11

BRAKES

Note the number of small circular shims fitted to the caliper mounting bolts between the caliper and the adaptor plate.

The caliper can now be removed from the aperture at the front of the cross-member.

Refitting

Refitting is the reverse of the removal procedure.

The correct number of camber shims should be fitted.

When the halfshaft has been refitted check the caliper for centralisation as described in refitting the front calipers. Fit the fluid supply pipe and the bridge pipe.

Bleed the braking system.

THE FRONT BRAKE DISCS

Removal

Jack up the car and remove the road wheel. Disconnect the flexible hydraulic pipe from the frame connection and plug the connector to prevent ingress of dirt and loss of fluid.

Discard the locking wire and remove the two caliper mounting bolts noting the number of round shims fitted between the caliper and mounting plate. Remove the caliper.

Remove the hub (as described in Section J "Front Suspension").

THE REAR BRAKE DISCS

Removal

Remove the rear suspension unit (as described in Section K "Rear Suspension").

Invert the suspension and remove the two hydraulic damper and road spring units (as described in Section K "Rear Suspension").

Remove the four steel type self-locking nuts securing the halfshaft inner universal joint and brake disc to the axle output shaft flange.

Withdraw the halfshaft from the bolts, noting the number of camber shins between the universal joint and the brake disc.

Knock back the tabs and unscrew the two pivot bolts securing the hand brake pad carriers to the caliper. Remove the pivot bolts and the retraction plate (Fig. 7).

Withdraw the handbrake pad carriers from the aperture at the rear of the cross members.

Knock back the tabs at the caliper mounting bolts (on earlier cars locking wire was used).

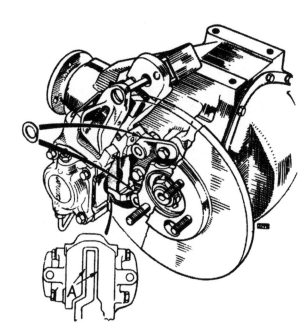

Fig. 6. Location of the shims fitted between the caliper and the adaptor plate.

Remove the keeper plate on the caliper and using a hooked implement, withdraw both brake pads.

Disconnect the brake fluid feed pipe at the caliper.

Unscrew the mounting bolts through the access holes in the brake disc.

Withdraw the bolts, noting the number and position of the round caliper centralizing shims.

Withdraw the caliper through the aperture at the front of the cross member.

Tap the halfshaft universal joint and brake disc securing bolts back as far as possible.

Lift the lower wishbone, hub carrier and halfshaft assembly upwards until the brake disc can be withdrawn from the mounting bolts.

Refitting

Refitting the brake discs is the reverse of the removal procedure. The securing bolts must be knocked back against the drive shaft flange when the new disc has been fitted.

Care must be taken to refit the caliper centralizing shims in the same position. The centralization of the caliper should be checked (as described in "Refitting the Calipers") when the halfshaft has been refitted.

Refit the rear suspension (as described in Section K "Rear Suspension").

Bleed the brakes.

Page L.X.s.12

BRAKES

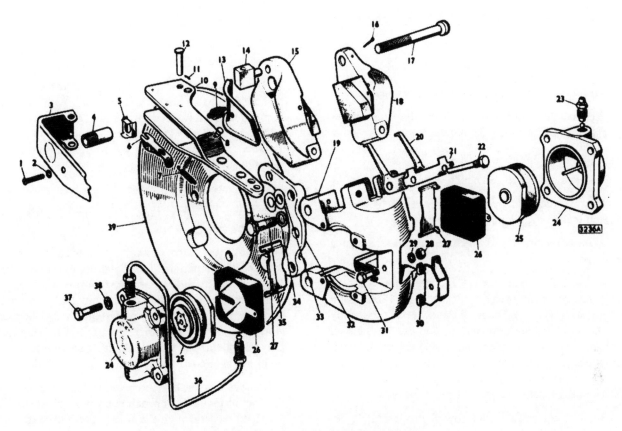

Fig. 7. *Exploded view of a rear brake caliper.*

1. Bolt.
2. Shakeproof Washer.
3. Protection cover assembly (rear).
4. Adjusting nut.
5. Friction spring.
6. Pawl assembly.
7. Tension spring.
8. Anchor pin.
9. Return spring.
10. Operating lever.
11. Split pin.
12. Hinge pin.
13. Protection cover assembly (front)
14. Pivot seat.
15. Inner pad carrier.
16. Split pin.
17. Bolt.
18. Outer pad carrier.
19. Rear caliper.
20. Retraction plate.
21. Tab washer.
22. Bolt.
23. Bleed screw and ball assembly.
24. Brake cylinder.
25. Piston.
26. Friction pad.
27. Support plate.
28. Nut.
29. Lock washer.
30. Retaining plate.
31. Bolt.
32. Locking plate.
33. Shim.
34. Spring washer.
35. Setscrew.
36. Bridge pipe.
37. Bolt.
38. Lock washer.
39. Disc.

Page L.X.s.13

BRAKES

THE BRAKE/CLUTCH PEDAL BOX ASSEMBLY

Removal (L.H. Drive)

Remove the servo vacuum pipe and clips.

Drain the brake and clutch fluid reservoirs.

Remove fluid inlet pipes from the clutch and brake master cylinders. Plug the holes.

Remove the brake fluid warning light wires.

Remove the brake and clutch reservoirs.

Remove the fluid outlet pipes from the brake and clutch master cylinders. Plug all holes.

Remove the brake and clutch pedal pads from inside the car.

Remove the dash casing in accordance with the instructions contained in Section N (Body and Exhaust). The nuts securing the pedal box assembly to the bulkhead are now exposed and can be removed together with two distance pieces and the brake pedal stop plate. Note that there are six self-locking nuts and one plain nut with a shakeproof washer. The plain nut is located on the bottom centre stud.

Remove the brake/clutch pedal box assembly by turning it through approximately 90° to allow the pedals to pass through the hole in the bulkhead.

Removal (R.H. Drive)

Remove the air cleaner elbow and the carburetter trumpets.

Remove the servo vacuum pipe and clips.

Drain the brake and clutch fluid reservoirs.

Remove the fluid inlet pipes from the clutch and brake master cylinders. Plug the holes.

Slacken the rear carburetter float chamber banjo nut and bend the petrol feed pipe towards the float chamber.

Remove the brake fluid warning light wires.

Remove the brake and clutch reservoirs.

Remove the fluid outlet pipes from the brake and clutch master cylinders. Plug all holes.

Remove the five screws securing the reaction valve assembly to the valve housing and withdraw the complete assembly. The valve housing can be removed by unscrewing the two setscrews, together with the shakeproof washers, which secure the housing to the body of the master cylinder.

Remove the throttle bell crank bracket.

Remove the brake and clutch pedal pads from inside the car.

Remove the dash casing in accordance with the instructions contained in Section N (Body and Exhaust). The nuts securing the pedal box assembly to the bulkhead are now exposed and can be removed together with two distance pieces and the brake pedal stop plate. Note that there are six self-locking nuts and one plain nut with a shakeproof washer. The plain nut is located on the bottom centre stud.

Remove the brake/clutch pedal box assembly by turning it through approximately 90° to allow the pedals to pass through the hole in the bulkhead.

Refitting

Refitting is the reverse of the removal procedure.

When refitting the securing nuts inside the car ensure that the plain nut and the shakeproof washer are fitted on the short stud in the bottom centre position.

Ensure that the brake fluid warning light wires are fitted with one feed wire (red and green) and one earth wire (black) to each reservoir cap.

When tightening the banjo union nut ensure that the petrol feed pipe is clear of the rear float chamber.

Bleed the brake and clutch hydraulic systems.

Page L.X.s.14

SECTION N

BODY AND EXHAUST

REAR SEAT (2+2 CARS)

Removal

Lift out the rear seat cushion assembly.

Remove three setscrews and shakeproof washers securing the upper seat squab to the body on each side. To gain access to the foremost setscrew, push the squab forwards to its highest position.

Remove the two nuts, shakeproof and special washers securing the lower seat squab to the seat pan and withdraw the squab. Lift off the trim panel on the propeller shaft tunnel.

Refitting

Refitting is the reverse of the removal procedure.

FRONT BUMPER

	Chassis Number	
	R.H. Drive	L.H. Drive
4·2 "E" Type, Open	1E. 1479	1E. 12580
4·2 "E" Type, Fixed Head Coupe	1E. 21228	1E. 32632

Commencing at the above chassis numbers, front bumpers have been modified to allow access to the mounting studs from underneath the car. The bonnet and front wing assemblies have also been modified and individual parts are not, therefore, interchangeable with their predecessors.

REAR BUMPER

	Chassis Number	
	R.H. Drive	L.H. Drive
4·2 "E" Type, Open	1E.1413	1E.11741
4·2 "E" Type, Fixed Head Coupe	1E.21000	1E.32010

Commencing at the above chassis numbers, the fixings of the rear bumpers are accessible from outside the car.

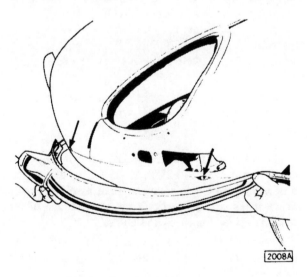

Fig. 1. *Front bumper mounting points.*

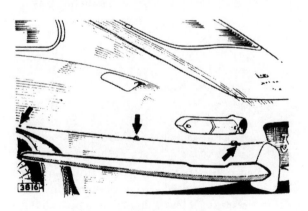

Fig. 2. *Rear bumper mounting points.*

Page N.X.s.1

SECTION O

HEATING AND WINDSCREEN WASHING EQUIPMENT

HEATER (2+2 CARS)

The heater unit fitted to 2+2 cars is identical to that fitted to 4·2 "E" type Open and Fixed Head Coupe cars except for the air distribution controls.

These outlets, situated under the duct behind the instrument panel, are fitted with finger operated direction controls on the facia board. Fully rotating the right-hand knob clockwise and the left-hand knob anti-clockwise will cut off the supply of air to the interior of the car and direct all air to the ducts at the windscreen. Reverse rotation of the knobs will re-direct air progressively from the windscreen to the car interior.

Model	Chassis Number	
	R.H. Drive	L.H. Drive
Open 2 Seater	1E.2037	1E.15980–U.S.A. Only
		1E.16010–Other than U.S.A.
Fixed Head Coupe	1E.21784	1E.34583–U.S.A. Only
		1E.34752–Other than U.S.A.

Commencing at the chassis numbers quoted above, all models are equipped with heater controls similar to those employed on 2+2 cars.

VENT CONTROL CABLES

Removal

Withdraw the parcel tray on each side of the dash by removing four drive screws and four thumb screws.

Release the locknuts securing the outer cables to the vent bracket. Disconnect the cables and collect the loose adaptor. Unscrew the cable from the centre finisher and withdraw the assemblies. A thin spanner will be required to remove the outer casing from the finisher.

Refitting

Reverse the removal procedure to refit the cables.

	Chassis Numbers	
	R.H. Drive	L.H. Drive
4·2 "E" Type, Open 2 Seater	1E.1165	1E.10754
4·2 "E" Type, Fixed Head Coupe	1E.20371	1E.30825

On cars with the above chassis numbers and onwards, Lucas 5SJ windscreen washers are fitted.

The Lucas 5SJ screen-jet is an electrically operated unit comprising a small permanent-magnet motor driving a centrifugal pump through a 3-piece Oldham type coupling. The water container is moulded in high density polythene.

The motor unit is controlled by a switch on the instrument panel and will operate as long as the switch is held in the upward position.

If the washer does not function immediately, check that there is water in the container. The motor will be damaged if the switch is held pressed for more than one or two seconds if the water in the container is frozen.

DATA

Minimum water delivery pressure	4·5 lb./sq. in. (0·32 kg./sq. cm.).
Minimum water delivery per second	3·5 c.c.
Container capacity	2¼ pints (1·1 litres).
Usable quantity of water	2 pints (1 litre).
Diameter of nozzle orifice	0·25"–0·28" (6·3–7 mm.)
Nominal voltage of unit	12
Maximum current consumption	2 amps.
Resistance between commutator segments	2·8–3·1 ohms.

Filling Up

The correct water level is up to the bottom of the container neck. Do not overfill or unnecessary splashing may result. Always replace the filler cover correctly after filling up. It is not possible to empty the container with the pump. Refilling is necessary when the water level has fallen below the level of the pump.

Keep the pump filter clean and the container free from sediment.

Cold Weather

The water container can be given a safe degree of protection down to −28°F. (−33°C.) by the use of proprietary antifreeze solutions such as marketed by Trico or Holts. Instructions regarding the use of the solvent will be found on the container.

Denatured alcohol (methylated spirits) must NOT be used. The use of this chemical will discolour the paintwork.

Page O.X.s.1

HEATING AND WINDSCREEN WASHING EQUIPMENT

SERVICING

Testing in Position

(a) Testing with a voltmeter:-

Connect a suitable direct current voltmeter to the motor terminals observing the polarity as indicated on the moulding housing. Operate the switch. If a low or zero voltage is indicated, the No. 6 fuse, switch and external connections should be checked and corrected as necessary.

If the voltmeter gives a reverse reading, the connections to the motor must be transposed.

If supply voltage is registered at the motor terminals but the unit fails to function, an open-circuit winding or faulty brush gear can be suspected. Dismantle the motor as described under the heading "Dismantling".

(b) Checking the external nozzles and tubes:—

If the motor operates but little or no water is delivered to the screen, the external tubes and nozzles may be blocked.

Remove the external plastic tube from the short connector on the container and, after checking that the connector tube is clear, operate the washer switch. If a jet of water is ejected, check the external tubes and nozzles for damage or blockage.

If no water is ejected, proceed as detailed under "Dismantling".

(c) Testing with an ammeter:—

Connect a suitable direct current ammeter in series with the motor and operate the switch. If the motor does not operate but the current reading exceeds that given in "Data", remove the motor and check that the pump impeller shaft turns freely.

If the shaft is difficult to turn, the water pump unit must be replaced. If the shaft turns freely, the fault lies in the motor which must be dismantled and its component parts inspected.

Dismantling

Disconnect the external tube and the electrical connections and remove the cover from the container. Remove the self-tapping screw which secures the motor to the cover and pull away the motor unit. Take care not to lose the loose intermediate coupling which connects the armature coupling to the pump spindle coupling.

Remove the armature coupling from the armature shaft as follows:—

Hold the armature shaft firmly with a pair of snipe-nosed pliers and, using a second pair of pliers, draw off the armature coupling.

Remove the two self-tapping screws from the bearing plate. The bearing plate and rubber gasket can now be removed. Remove the two terminal screws. The terminal nuts and brushes can now be removed and the armature withdrawn. Take care not to lose the bearing washer which fits loosely on the armature shaft.

The pole assembly should not normally be disturbed. If, however, its removal is necessary, make a careful note of its position relative to the motor housing. The narrower pole piece is adjacent to the terminal locations. Also the position of the pole clamping member should be observed. When fitted correctly, it locates on both pole pieces but, if fitted incorrectly, pressure is applied to one pole piece only.

Bench-Testing

If the motor has been overheated, or if any part of the motor housing is damaged, a replacement motor unit must be fitted.

Armature:—

If the armature is damaged or if the windings are loose or badly discoloured, a replacement armature must be fitted.

The commutator must be cleaned with a fluffless cloth moistened in petrol, or, if necessary, polished with a strip of very fine glass paper.

The resistance of the armature winding should be checked with an ohmmeter. This resistance should be in accordance with that given in "Data".

Brushes:—

If the carbon is less than $\frac{1}{16}''$ (1·59 mm.) long, a new brush must be fitted. Check that the brushes bear firmly against the commutator.

Re-assembling

Re-assembling of the unit is the reverse of the dismantling procedure. However, the following points should be noted:—

Make sure the bearing recess in the motor is filled with Rocol Molypad molybdenised grease. Remove excessive grease from the face of the bearing boss.

Page O.X.s.2

HEATING AND WINDSCREEN WASHING EQUIPMENT

Check that the pole piece assembly does not rock and that the pole pieces are firmly located in the circular spigot. Ensure that the pole piece assembly and clamping member are the right way round.

Before replacing the motor unit on the cover, ensure that the armature coupling is pushed fully home and that the intermediate coupling is in place.

Performance Testing

Equipment required:—

D.C. supply of appropriate voltage.

D.C. voltmeter, first grade, moving coil 0–3 amp.

D.C. ammeter.

0–15 lb. sq. in. (0–1 kg. sq. cm.) pressure gauge.

Pushbutton with normally open contacts.

Two-jet nozzle.

On-off tap.

100 c.c. capacity measure.

4 ft. 6 in. (1·37 m.) length of plastic tubing.

Connect up the equipment as shown in Fig. 1. The water level in the container must be 4″ (101·6 mm.) above the base of the pump assembly. The pressure gauge and nozzle must be 18″ (45·72 cm.) above the water level.

Open the tap. Depress the button for approximately five seconds and check the voltmeter reading which should be the same as the supply voltage. On releasing the switch, close the tap to ensure that the plastic tubing remains charged with water. Empty the measuring cylinder.

Open the tap and operate the push switch for precisely ten seconds after which period release the switch and close the tap.

During the ten-second test, the current and pressure values should be in accordance with those given in Data and at least 35 c.c. of water should have been delivered.

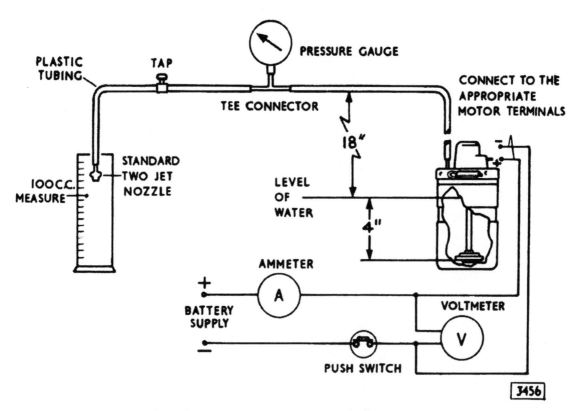

Fig. 1. *Performance testing the windscreen washing equipment.*

Page O.X.s.3

SECTION P

ELECTRICAL AND INSTRUMENTS

DISTRIBUTOR

From Engine No. 7E2459 a waterproof cover is incorporated in the distributor assembly, located between the distributor cap and body. This cover is detachable after removing the distributor cap and disconnecting the cable from the contact breaker spring post.

DATA

Ignition Distributor Type	22D6
8 to 1 Compression Ratio	41060A
9 to 1 Compression Ratio	41060A
Cam dwell angle	34° ± 3°
Contact breaker gap	0·014″–0·016″ (0·36—0·41 mm.)
Contact breaker spring tension (Measured at free contact)	18—24 ozs. (512—682 gms.)

IGNITION TIMING

8 to 1 Compression Ratio	9° BTDC
9 to 1 Compression Ratio	10° BTDC

IGNITION DISTRIBUTOR TEST DATA

Distributor Type	Lucas Service Number	Lucas Vacuum Unit Number	VACUUM TIMING ADVANCE TESTS The distributor must be run immediately below the speed at which the centrifugal advance begins to function to obviate the possibility of an incorrect reading being registered.		No advance in timing below-ins. of mercury	Lucas Advance Springs Number	CENTRIFUGAL TIMING ADVANCE TESTS Mount distributor in centrifugal advance test rig and set to spark at zero degrees at 100 r.p.m.				No advance in timing below-RPM
			Vacuum in inches of mercury and advance in degrees				Acelerate to-RPM and note advance in degrees		Decelerate to-RPM and note advance in degrees		
			Inches	Degrees			RPM	Degrees	RPM	Degrees	
22 D6	41060A	54415894	20 13 9 7½ 6	7—9 6—8½ 2½—5½ 0—3 0—½	4½	55415562	2,300	8½—10½	1800 1250 800 650 525	8½—10½ 6½—8½ 5—7 2—4 0—1½	300

Auto advance weights Lucas number 54413073.   One inch of mercury = 0·0345 kg/cm²

Page P.X.s.1

ELECTRICAL AND INSTRUMENTS

ROUTINE MAINTENANCE

EVERY 3,000 MILES (5,000 KM.)

Lubricate distributor and check contact points' gap.

FUSE UNITS

Fuse No.	CIRCUITS	Amps
1	Headlamps—Main Beam	35
2	Headlamps—Dip Beam	35
3	Horns	50
4	Spare	—
5	Side, Panel, Tail and Number Plate (not Germany) Lamps	35
6	Horn Relay, Washer, Radiator Fan Motor and Stop Lamps	35
7	Flashers, Heater, Wiper, Choke, Fuel, Water and Oil Gauges	35
8	Headlamp Flasher, Interior Lamps and Cigar Lighter	35
In line	Heated Backlight (when fitted)	15
In line	Radio, Optional Extras	5
In line	Traffic Hazard Warning System	35

THE ALTERNATOR

DESCRIPTION

The Lucas 11 AC alternator is a lightweight machine designed to give increased output at all engine speeds.

Basically the unit consists of a stationary output winding with built in rectification and a rotating field winding, energised from the battery through a pair of slip rings.

The stator consists of a 24 slot, 3 phase star connected winding on a ring shaped lamination pad housed between the slip ring end cover and the drive end bracket.

The rotor is of 8-pole construction and carries a field winding connected to two face type slip rings. It is supported by a ball bearing in the drive end bracket and a needle roller bearing in the slip ring end cover (see Fig. 1).

ELECTRICAL AND INSTRUMENTS

The brushgear for the field system is mounted on the slip ring end cover. Two carbon brushes, one positive and one negative, bear against a pair of concentric brass slip rings carried on a moulded disc attached to the end of the rotor. The positive brush is always associated with the inner slip ring. There are also six silicon diodes carried on the slip ring end cover, these being connected in a three phase bridge circuit to provide rectification of the generated alternating current output (see Fig. 2). The diodes are cooled by air flow through the alternator induced by a 6″ (15·24 cm.) ventilating fan at the drive end.

The alternator is matched to an output control unit, Model 4TR, which is described on Page P.X.s.8.

This unit controls the alternator field current and hence the alternator terminal voltage.

A cut-out is not included in the control unit as the diodes in the alternator prevent reverse currents from flying through the stator when the machine is stationary or is generating less than the battery voltage.

No separate current-limiting device is incorporated; the inherent self-regulating properties of the alternator effectively limit the output current to a safe value.

On later cars a Lucas 3AW warning light control unit is incorporated in the circuit.

The output control unit and the alternator field windings are isolated from the battery when the engine is stationary by a separate pair of contacts in the ignition switch.

On cars fitted with a steering column lock, the field windings are isolated by means of a relay replacing the ignition switch control.

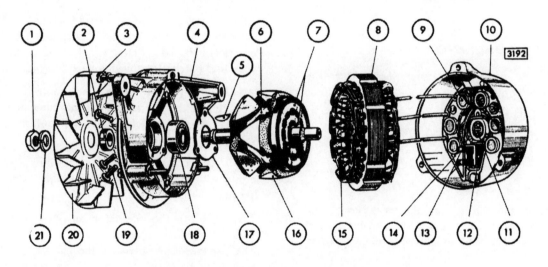

Fig. 1. *Exploded view of the Lucas 11 AC alternator.*

1. Shaft nut.
2. Bearing collar.
3. Through fixing bolts (3).
4. Drive end bracket.
5. Key.
6. Rotor (field) winding.
7. Slip rings.
8. Stator laminations.
9. Silicon diodes (6).
10. Slip ring end bracket.
11. Needle roller bearing.
12. Brush box moulding.
13. Brushes.
14. Diode heat sink.
15. Stator windings.
16. Rotor.
17. Bearing retaining plate.
18. Ball bearing.
19. Bearing retaining plate rivets.
20. Fan.
21. Spring washer.

ELECTRICAL AND INSTRUMENTS

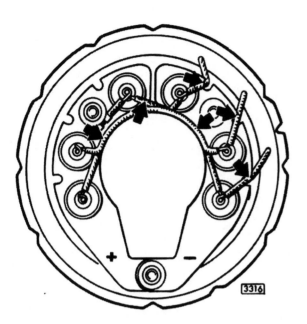

Fig. 2. *Showing the silicon diodes and connections in the slip ring end cover.*

ROUTINE MAINTENANCE

No routine maintenance is necessary with the alternator or control unit.

Occasionally wipe away any dirt or oil which may collect around the slip ring end cover.

REMOVAL

Disconnect the cables from the terminals on the slip ring end cover. Note the colour and location of the cables with Lucar termination for reference when refitting.

Note: Later cars having the 3AW warning light control unit to give an indication that the alternator is charging, also have a positive lock connector on the main alternator output cable making it impossible to connect the harness incorrectly. The cable to the 3AW control unit is connected to the fourth terminal on the slip ring.

Remove the drive belt by pushing the spring loaded jockey pulley inwards and lifting the belt over the alternator pulley.

Remove the two bolts securing the alternator to the mounting bracket and adjuster link. Withdraw the alternator.

REFITTING

Refitting is the reverse of the removal procedure.

When replacing the alternator belt, hold the spring loaded jockey pulley in towards the block and only release when the belt is sitting securely in the "vee" tracks.

SERVICE PRECAUTIONS

Important

4·2 "E" Type cars are equipped with transistors in the control box unit and diode rectifiers in the alternator.

The car electrical system must NOT be checked with an ohmmeter incorporating a hand driven generator until these components have been isolated.

REVERSED battery connections will damage the diode rectifiers.

Battery polarity must be checked before connections are made to ensure that the connections for the car battery are NEGATIVE earth. This is most important when using a slave battery to start the engine.

NEVER earth the brown/green cable if it is disconnected at the alternator. If this cable is earthed, with the ignition switched ON, the control unit and wiring may be damaged.

NEVER earth the alternator main output cable or terminal. Earthing at this point will damage the alternator or circuit.

NEVER run the alternator on open circuit with the field windings energised, that is with the main lead disconnected, otherwise the rectifier diodes are likely to be damaged due to peak inverse voltages.

SERVICING

Testing the Alternator in position

In the event of a fault developing in the charging circuit check by the following procedure to locate the cause of the trouble.

1. Disconnect the battery.

2. Lower the instrument panel and disconnect the brown and brown/white cables from the ammeter Connect the two cables to a good quality moving-coil ammeter registering at least 75 amperes.

Page P.X.s.4

ELECTRICAL AND INSTRUMENTS

3. Detach the terminal connectors from the base of the control unit and connect the black and brown/green cables together by means of a short length of cable with two Lucar terminals attached. This operation connects the alternator field winding across the battery terminals and by-passes the output control unit (Fig. 3).

4. Reconnect the battery earth lead. Switch on the ignition and start the engine. Slowly increase the engine speed until the alternator is running at approximately 4,000 r.p.m. (2,000 engine r.p.m.). Check the reading on the ammeter which should be approximately 40 amperes with the machine at ambient temperature.

   A low current reading will indicate either a faulty alternator or poor circuit wiring connections.

   If, after checking the latter, in particular the earth connections, a low reading persists on repeating the test refer to paragraph (5).

   In the case of a zero reading, switch on the ignition and check that the battery voltage is being applied to the rotor windings by connecting a voltmeter between the two cable ends normally attached to the alternator field terminals. No reading on this test indicates a fault in the field isolating contacts in the ignition switch or the wiring associated with this circuit. Check each item in turn and rectify as necessary.

   Note: There being no vibrating contact cut-out with the alternator, field isolation is by means of two extra contacts on the ignition switch. When a steering column lock is fitted, field isolation is by means of a relay.

5. If a low output has resulted from the test described in paragraph (4) and the circuit wiring is in order; measure the resistance of the rotor coil field by means of an ohmmeter connected between the field terminal blades with the external wiring disconnected.

   The resistance must approximate 3·8 ohms.

   When a ohmmeter is not available connect a 12 volt DC supply between the field terminals with an ammeter in series. The ammeter reading should be approximately 3·2 amperes Fig. 4.

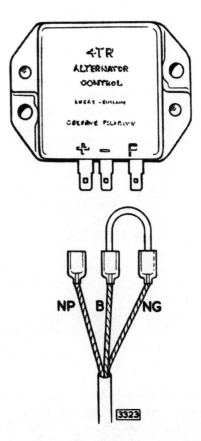

Fig. 3. *Detach the terminal connectors from the base of the control unit.*

A zero reading on the ammeter, or an infinity reading on the ohmmeter indicates an open circuit in the field system, that is, the brush gear slip rings or winding. Conversely, if the current reading is much above, or the ohmmeter is much below, the values given then it is an indication of a short circuit in the rotor winding in which case the rotor slip ring assembly must be changed.

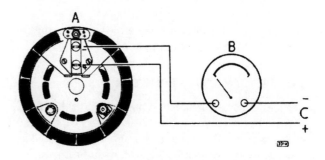

Fig. 4. *Testing the alternator with an ammeter.*
A—*Alternator.*   B—*Ammeter.*   C—*Battery.*

Page P.X.s.5

ELECTRICAL AND INSTRUMENTS

DISMANTLING THE ALTERNATOR (Fig. 1).

Disconnect the battery and remove the alternator as detailed on Page P.X.s.4.

Remove the shaft nut (1) and spring washer (21). Withdraw the pulley and fan (20).

Remove bolts (3) noting that the nuts are staked to the through bolts and that the staking must be removed before the nuts are unscrewed. If the threads of the nuts or bolts are damaged, new bolts must be fitted when reassembling.

Mark the drive end bracket (4), lamination pack (8) and slip ring end bracket (10) so that they may be reassembled in correct angular relation to each other. Care must be taken not to damage the lamination pack when marking.

Withdraw the drive end bracket (4) and rotor (16) from the stator (8). The drive end bracket and rotor need not be separated unless the bearing requires examination or the rotor is to replaced.

In the latter case the rotor should be removed from the drive end bracket by means of a hand press having first removed the shaft key (5) and bearing collar (2).

Remove the terminal nuts, washers and insulating pieces brush box screws and the 2 B.A., hexagon headed setscrew. Withdraw the stator and diode heat sink assemblies from the slip ring end cover.

Close up the retaining tongue at the root of each field terminal blade and withdraw the brush spring together with the terminal assemblies from the moulded brushbox.

REASSEMBLY

Reassembly of the alternator is the reverse of the dismantling procedure. Care must be taken to align the drive end bracket, lamination pack, slip ring and bracket correctly.

Tighten the three through bolts evenly to a maximum torque of 45 to 50 lb./ins. (0·518 to 0·576 kgm.). Restake the nuts after tightening.

Tighten the brush box fixing screws to a maximum torque of 10 lb./ins. (0·115 kgm.).

INSPECTION OF BRUSHGEAR

Measure brush length. A new brush is $\frac{5}{8}''$ (15·88 mm.) long; a fully worn brush is $\frac{5}{32}''$ (3·97 mm.) and must be replaced at, or approaching, this length. The new brush is supplied complete with brush spring and Lucar terminal blade and has merely to be pushed in until the tongue registers. To ensure that the terminal is properly retained, carefully lever up the retaining tongue with a fine screwdriver blade, so that the tongue makes an angle of 30° with the terminal blade.

The normal brush spring pressures are 4–5 oz. (113 to 142 gms.) with the spring compressed to $\frac{25}{32}''$ (19·84 mm.) in length and 7½ to 8½ oz. (212 to 242 gms.) with the spring compressed to $\frac{13}{32}''$ (10·31 mm.) in length. These pressures should be measured if the necessary equipment is available.

Check that the brushes move freely in their holders. If at all sluggish, clean the brush sides with a petrol moistened cloth or, if this fails to effect a cure, lightly polish the brush sides on a smooth file. Remove all traces of brush dust before re-housing the brushes in their holders.

INSPECTION OF SLIP RINGS

The surfaces of all slip rings should be smooth and uncontaminated by oil or other foreign matter. Clean the surfaces using a petrol moistened cloth, or if there is any evidence of burning, very fine glasspaper. On no account must emery cloth or similar abrasives be used. No attempt should be made to machine the slip rings, as any eccentricity in the machining may adversely affect the high-speed performance of the alternator. The small current carried by the rotor winding together with the unbroken surface of the slip rings mean that the likelihood of scored or pitted slip rings is almost negligible.

ROTOR

Test the rotor winding by connecting an ohmmeter or 12 volt D.C., supply between the slip rings (as described on page P.X.s.5) where this test was made with the brushgear in circuit. The readings of resistance or cuttent should be as given on page P.X.s.5.

Test for defective insulation between each of the slip rings and one of the rotor poles using a mains low-wattage test lamp for the purpose. If the lamp lights, the coil is earthing therefore a replacement rotor/slip ring assembly must be fitted.

No attempt should be made to machine the rotor poles or to true a distorted shaft.

STATOR

Unsolder the three stator cables from the heat sink assembly taking care not to overheat the diodes (see Fig. 1). By lettering these cables A, B and C, three pairs of cables AB, BC and AC—are available for

ELECTRICAL AND INSTRUMENTS

testing the stator windings. Measure the voltage drop across each of these pairs in turn while passing 20 amps between the cable ends. The voltage drop should be approximately 4·3 volts in each of the three measurements.

If any, or all, of the readings are other than these, a replacement stator must be fitted.

Test for defective insulation between stator coils and lamination pack with a mains test lamp. Connect the test probes between any one of the three cable ends and the lamination pack. If the lamp lights, the stator coils are earthing and a replacement stator must be fitted.

Before re-soldering the stator cable ends to the diode pins check the diodes.

DIODES

Each diode can be checked by connecting it in series with a 1·5 watt test bulb (Lucas No. 280) across a 12 volt D.C. supply and then reversing the connections.

Current should flow and the bulb light in one direction only. If the bulb lights up in both tests or does not light up in either then the diode is defective and the appropriate heat sink assembly must be replaced.

The above procedure is adequate for service purposes. Any accurate measurement of diode resistance requires factory equipment. Since the forward resistance of a diode varies with the voltage applied, no realistic readings can be obtained with battery-powered ohmmeters.

If a battery—ohmmeter is used, a good diode will yield "Infinity" on one direction and some indefinite, but much lower, reading in the other.

WARNING:

Ohmmeters of the type incorporating a hand-driven generator must never be used for checking diodes.

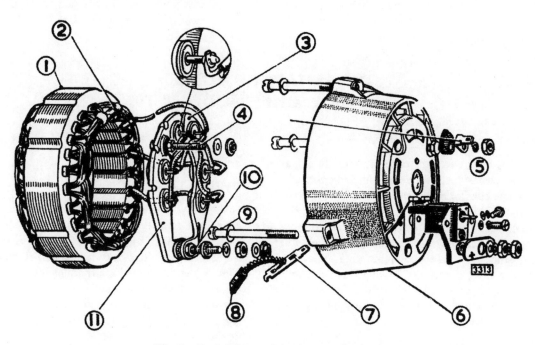

Fig. 5. *Exploded view of the slip ring end cover.*

1. Stator.
2. Star point.
3. Negative heat sink anode base diodes (black).
4. Warning light terminal 'AL'.
5. Field terminal (2).
6. Slip ring end cover.
7. Terminal blade retaining tongue.
8. Rotor slip ring brush (2).
9. "Through" bolts (3).
10. Output terminal (+).
11. Positive heat sink and cathode base diode (red).

Page P.X.s.7

ELECTRICAL AND INSTRUMENTS

ALTERNATOR DIODE HEAT SINK REPLACEMENT

The alternator heat sink assembly consists of two mutually insulated portions, one of positive and the other of negative polarity. The diodes are not individually replaceable but, for service purposes, are supplied already pressed into the appropriate heat sink portion. The positive carries three cathode base diodes marked black.

When soldering the interconnections, M grade 45–55 tin-lead solder should be used.

Great care must be taken to avoid overheating the diodes or bending the diode pins. The diode pins should be lightly gripped with a pair of suitable long-nosed pliers, acting as a thermal shunt and the operation of soldering carried out as quickly as possible.

After soldering to ensure adequate clearance of the rotor, the connections must be neatly arranged around the heat sinks and tacked down with "MMM" EC 1022 adhesive where indicated in Fig. 2. The stator connections must pass through the appropriate notches at the edge of the heat sink.

BEARINGS

Bearings which are worn to the extent that they allow excessive side movement of the rotor shaft must be renewed. The needle roller bearing in the slip ring end cover is supplied complete with the end cover.

To renew the drive end ball bearing following the withdrawal of the rotor shaft from the drive-end bracket, proceed as follows:—

(a) File away the roll-over on each of the three bearing retaining plate rivets and punch out the rivets.

(b) Press the bearing out of the bracket.

(c) Locate the bearing in the housing and press it home. Refit the bearing retaining plate using new rivets.

Note: Before fitting the replacement bearing see that it is clean and, if necessary, pack it with high-melting point grease such as Shell Alvania No. 3 or an equivalent lubricant.

ALTERNATOR OUTPUT CONTROL UNIT MODEL 4 TR.

GENERAL

Model 4 TR is an electronic control unit. In effect its action is similar to that of the vibrating contact type of voltage control unit but switching is achieved by transistors instead of vibrating contacts. A Zener diode provides the voltage reference in place of the voltage coil and tension spring system. No cut-out is required since the diodes incorporated in the alternator prevent reverse currents flowing. No current regulator is required as the inherent self-regulating properties of the alternator effectively limit the output current to a safe value.

The control unit and the alternator field windings are isolated from the battery, when the engine is stationary, by a special double-pole ignition switch.

On cars fitted with a steering column lock, the field windings are isolated by means of a relay replacing the ignition switch control.

Care must be taken at all times to ensure that the battery, alternator and control unit are correctly connected. Reversed connections will damage the semi conductor devices employed in the alternator and control unit.

OPERATION

When the ignition is switched on, the control unit is connected to the battery through the field isolating switch or relay. By virtue of the connection through R1 (see Fig. 6), the base circuit of the power transistor T2 is conducted so that, by normal transistor action, current also flows in the collector-emitter portion of T2 which thus acts as a closed switch to complete the field circuit and battery voltage is applied to the field winding.

As the alternator rotor speed increases, the rising voltage generated across the stator winding is applied to the potential divider consisting of R3, R2 and R4. According to the position of the tapping point on R2, a proportion of this potential is applied to the Zener diode (ZD). The latter is a device which opposes the passage of current through itself until a certain voltage is reached above which it conducts comparatively freely.

Page P.X.s.8

ELECTRICAL AND INSTRUMENTS

The Zener diode can thus be considered as a voltage-conscious switch which closes when the voltage across it reaches its "breakdown" voltage (about 10 volts) and, since this is a known proportion of the alternator output voltage as determined by the position of the tapping point on R2, the breakdown point therefore reflects the value of the output voltage.

Thus at "breakdown" voltage the Zener diode conducts and current flows in the base-emitter circuit of the driver transistor T1. Also, by transistor action, current will flow in the collector-emitter portion of T1 so that some of the current which previously passed through R1 and the base circuit of T2 is diverted through T1. Thus the base current of T2 is reduced and, as a result, so also is the alternator field excitation. Consequently, the alternator output voltage will tend to fall and this, in turn, will tend to reduce the base current in T1, allowing increased field current to flow in T2. By this means, the field current is continuously varied to keep the output voltage substantially constant at the value determined by the setting of R2.

To prevent overheating of T2, due to power dissipation, this transistor is operated only either in the fully-on or fully-off condition. This is achieved by the incorporation of the positive feed-back circuit consisting of R5 and C2. As the field current in transistor T2 starts to fall, the voltage at F rises and current flows through resistor R5 and capacitor C2 thus adding to the Zener diode current in the base circuit of transistor T1. This has the effect of increasing the current through T1 and decreasing, still further, the current through T2 so that the circuit quickly reaches the condition where T1 is fully-on and T2 fully-off. As C2 charges, the feed-back current falls to a degree at which the combination of Zener diode current and feed-back current in the base circuit of T1 is no longer sufficient to keep T1 fully-on. Current then begins to flow again in the base circuit of T2. The voltage at F now commences to fall, reducing the feed-back current eventually to zero. As T2 becomes yet more conductive and the voltage at F falls further, current in the feed-back circuit reverses in direction thus reducing, still further, the base current in T1.

This effect is cumulative and the circuit reverts to the condition where T1 is fully-off and T2 is fully-on.

The above condition is only momentary since C2 quickly charges to the opposite polarity when feed-back current is reduced and current again flows in

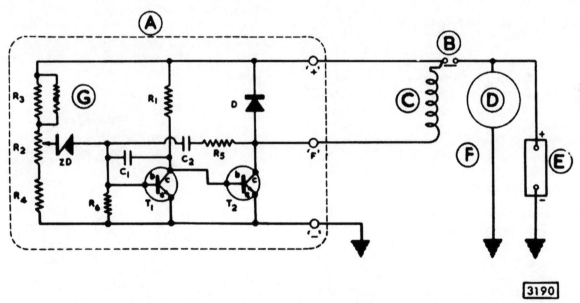

Fig. 6. *4TR Control Unit Circuit Diagram.*

A—Control unit.
B—Field isolating device.
C—Rotor field winding.
D—Alternator.
E—12-volt battery.
F—Stator winding (rectified) output.
G—Thermistor.

ELECTRICAL AND INSTRUMENTS

the base circuit of T1. The circuit thus oscillates, switching the voltage across the alternator field winding rapidly on and off.

Transistor T2 is protected from the high induced voltage surge, which results from the collapse of the field current, by the surge quench diode D connected across the field windings. This diode also provides a measure of field current smoothing since current continues to flow in the diode after the excitation voltage is removed from the field.

The elimination of radio interference is achieved by connecting condenser C1 between the base and collector terminals of T1 to provide negative feedback. At high temperatures, a small leakage current may flow through the Zener diode even though the latter is in the nominally non-conductive state. Resistor R6 provides a path for this leakage current which otherwise would flow through T1 base circuit and adversely affect the regulator action.

A thermistor is connected in parallel with resistor R3. The thermistor is a device whose resistance increases as the temperature falls and vice verse. Any alteration in its ohmic value will modify the voltage distribution across the potential divider and thus affect the voltage value at which the Zener diode begins to conduct, so matching the changes which take place in battery terminal voltage as the temperature rises.

CHECKING AND ADJUSTING THE CONTROL UNITS

Important:

Voltage checking and setting procedure may be carried out only if the alternator and associated wiring circuits have been tested and found satisfactory in conjunction with a well-charged battery, (i.e., charging current not exceeding 10 amperes).

VOLTAGE CHECKING

Run the alternator at charging speed for eight minutes. This operation applies when bench testing or testing on the car.

Leave the existing connections to the alternator and control unit undisturbed. Connect a high quality voltmeter between control unit terminals positive and negative. If available, use a voltmeter of the suppressed-zero type, reading 12 to 15 volts.

Switch on an electrical load of approximately 2 amperes (e.g., side and tail lighting).

Start the engine and run the alternator at 3,000 r.p.m. (1,500 engine r.p.m.).

The voltmeter should now show a reading of 13·9 to 14·3 volts at 68° to 78° F. (20° to 26° C.) ambient temperature. If not, but providing the reading obtained has risen to some degree above battery terminal voltage before finally reaching a steady value, the unit can be adjusted to control at the correct voltage (see Adjusting).

If, however, the voltmeter reading remains unchanged, at battery terminal voltage, or, conversely, increases in an uncontrolled manner, then the control unit is faulty and, as its component parts are not serviced individually a replacement unit must be fitted.

ADJUSTING

Stop the engine and withdraw the control unit mounting screws.

Invert the unit and chip away the sealing compound which conceals the potentiometer adjuster (see Fig. 7).

Check that the voltmeter is still firmly connected between terminals +ve and −ve. Start the engine and, while running the alternator at 3,000 r.p.m., turn the potentiometer adjuster slot (clockwise to increase the setting or anti-clockwise to decrease it) until the required setting is obtained.

Use care in making this adjustment as a small amount of adjuster movement causes an appreciable difference in the voltage reading.

Recheck the setting by first stopping the engine then again running the alternator at 3,000 r.p.m.

Remount the control unit and disconnect the voltmeter.

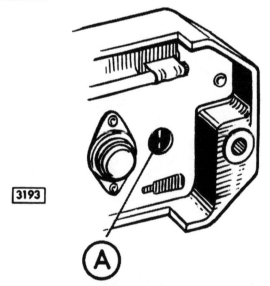

Fig. 7. *4 TR Alternator Control.*
A *Potentiometer adjuster.*

Page P.X.s.10

ELECTRICAL AND INSTRUMENTS

WARNING LIGHT CONTROL UNIT
Model 3AW

DESCRIPTION

The Model 3AW warning light unit fitted to later cars is a device connected to the centre point of one of the pairs of diodes in the alternator and operates in conjunction with the ignition warning light to give indication that the alternator is charging.

The unit is mounted on the bulkhead adjacent to the control box and is similar in appearance to the flasher unit but has different internal components consisting of an electrolytic (polarised) capacitor; a resistor and a silicone diode mounted on an insulated base with three "Lucar" terminals.

The unit is sealed, therefore servicing and adjustment is not possible. Faulty units must be replaced. Due to external similarity of the 3AW warning light unit and the flasher unit, a distinctive green label is attached to the aluminium case of the 3AW unit.

Checking Check by substitution after ensuring that the remainder of the charging circuit (including the drive belt) is functioning satisfactorily.

Warning. A faulty diode in the alternator or an intermittent or open-circuit in the alternator to battery circuit can cause excessive voltages to be applied to the warning light unit.

To prevent possible damage to a replacement unit, it is important to first check the voltage between the alternator "AL" terminal and earth. Run the engine at 1,500 r.p.m. when the voltage should be 7–7.5 volts measured on a good quality moving-coil voltmeter. If a higher voltage is registered, check that all charging circuit connections are clean and tight; then, if necessary, check the alternator rectifier diodes before fitting a replacement 3AW unit.

TRAFFIC HAZARD WARNING DEVICE (OPTIONAL EQUIPMENT)

Description

The system operates in conjunction with the four flashing (turn) indicator lamps fitted to the car. The operation of the dash panel switch will cause the four turn indicator lamps to flash simultaneously.

A red warning lamp is incorporated in the circuit to indicate that the hazard warning system is in operation.

A 35 amp. in-line fuse incorporated in the sub-panel circuit.

The flasher unit is located and is similar in appearance to the one used for the flashing turn indicators but has a differenti internal circuit. A correct replacement unit must be fitted in the even of failure.

The pilot lamp bulb is accessible after removing the bulb holder from the rear of the panel.

Failure of one or more of the bulbs due to an accident or other cause will not prevent the system operating on the remaining lamps.

Page P.X.s.11

ELECTRICAL AND INSTRUMENTS

THE STARTER MOTOR

DESCRIPTION

The purpose of the pre-engaged, or positive engagement, starting motor is to prevent premature pinion ejection.

Except on occasions of tooth to tooth abutment, for which special provision is made, the starter motor is connected to the battery only after the pinion has been meshed with the flywheel ring gear, through the medium of an electro-magnetically operated linkage mechanism.

After the engine has started, the current is automatically switched off before the pinion is retracted.

On reaching the out of mesh position, the spinning armature is brought rapidly to rest by a braking device. This device takes the form of a pair of moulded shoes driven by a cross peg in the armature shaft and spring loaded (and centrifuged) against a steel ring insert in the commutator end bracket. Thus, with the supply switched off and the armature subjected to a braking force, the possibility is minimised of damaged teeth resulting from attempts being made to re-engage a rotating pinion.

A bridge-shaped bracket is secured to the front end of the machine by the through bolts. This bracket carries the main battery input and solenoid winding

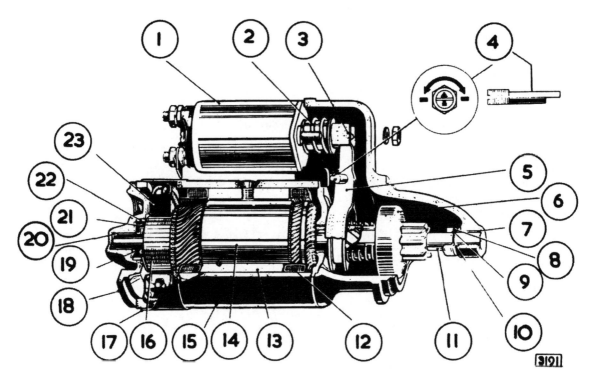

Fig. 8. *The Pre-engaged Starter Motor Model M45G.*

1. Actuating solenoid.
2. Return spring.
3. Clevis pin.
4. Eccentric pivot pin.
5. Engaging lever.
6. Roller clutch.
7. Porous bronze bush.
8. Thrust collar.
9. Jump ring.
10. Thrust washer.
11. Armature shaft extension.
12. Field coils.
13. Pole shoe.
14. Armature.
15. Yoke.
16. Commutator.
17. Band cover.
18. C.E. bracket.
19. Thrust washer.
20. Porous bronze bush.
21. Brake shoes and cross peg.
22. Brake ring.
23. Brushes.

Page P.X.s.12

ELECTRICAL AND INSTRUMENTS

terminals, short extension cables being connected between these and the corresponding solenoid terminals.

TOOTH TO TOOTH ABUTMENT

The electro-magnetically actuated linkage mechanism consists essentially of a pivoted engaging lever having two hardened steel pegs (or trunnion blocks) which locate with and control the drive through the medium of a groove in an operating bush. This bush is carried, together with the clutch and pinion assembly, on an internally splined outboard driving sleeve, the whole mechanism being housed in a cut-away flange mounting snout-shaped end bracket. This operating bush is spring loaded against a jump ring in the driving sleeve by an engagement spring located between the bush and the clutch outer cover. The system return or drive demeshing spring is located round the solenoid plunger.

On the occurrence of tooth to tooth abutment (between the ends of the starter pinion teeth and those of the flywheel ring gear), the pegs or trunnion blocks at the "lower" end of the engaging lever can move forward by causing the operating bush to compress the engagement spring, thus allowing the "upper" end of the lever to move sufficiently rearwards to close the starter switch contacts. The armature then rotates and the pinion slips into mesh with the flywheel ring gear under pressure of the compressed engagement spring.

THE "LOST MOTION" (SWITCH-OFF) DEVICE

As it is desirable that the starter switch contacts shall not close until the pinion has meshed with the flywheel ring gear therefore it is important that these same contacts should always re-open before the pinion has been retracted or can be opened in the event of a starter pinion remaining for some reason enmeshed with the flywheel ring gear. To ensure this, a measure of "lost motion" is designed into some part of the engagement mechanism, its effect being to allow the starter switch or solenoid contacts (which are always spring-loaded to the open position) to open before pinion retraction begins.

Several methods of obtaining "lost motion" have been adopted, but each depends upon the yielding of a weaker spring to the stronger system return (drive demeshing or dis-engagement) spring of the solenoid plunger.

This initial yielding results in the switch contacts being fully-opened within the first $\frac{1}{8}$" (3·18 mm.) of plunger return travel; this action being followed by normal drive retraction.

Solenoid model 10S has a weaker (lost motion) spring located inside the solenoid plunger. Here, enclosed at the outer end by a retaining cup, it forms a plunger within a plunger and it is spring loaded against the tip of the engaging lever inside the plunger clevis link.

THE ROLLER CLUTCH

Torque developed by the starting motor armature must be transmitted to the pinion and flywheel through an over-running or free-wheeling device which will prevent the armature from being rotated at an excessively high speed in the event of the engaged position being held after the engine has started. The roller clutch performs this function.

The operating principle of the roller clutch is the wedging of several plain cylindrical rollers between converging surfaces. The convergent form is obtained by matching cam tracks, to a perfectly circular bore. The rollers, of which there are three, are spring loaded and, according to the direction of drive, are either free or wedge-locked between the driving and driven members. The clutches are sealed in a rolled over steel outer cover and cannot be dismantled for subsequent reassembly.

THE STARTER SOLENOID

The starter solenoid is an electro-magnetic actuator mounted pick-a-back fashion on the yoke of the pre-engaged starter motor. It contains a soft iron plunger (linked to the engaging lever), the starter switch contacts and a coil consisting of a heavy gauge pull-in or series winding and a lighter-gauge hold-on or shunt winding.

Initially, both windings are energised in parallel when the starter device is operated but the pull-in winding is shorted out by the starter switch contacts at the instant of closure—its purpose having been effected.

Magnetically, the windings are mutually assisting.

Like the roller clutch assembly, the starter solenoid is sealed in a rolled-over steel outer case or body and cannot be dismantled for subsequent reassembly.

Page P.X.s.13

ELECTRICAL AND INSTRUMENTS

STARTER MOTOR PERFORMANCE DATA

Model	M45G Pre-engaged
Lock Torque	22·6 lb./ft. (3·13 kg./m.) with 465 amperes at 7·6 terminal volts
Torque at 1,000 r.p.m.	9·6 lb./ft. (1·33 kg./m.) with 240 amperes at 9·7 terminal volts
Light running current	70 amperes at 5,800 to 6,500 r.p.m.

SOLENOID SWITCH DATA

Model	10 S
Closing Coil Resistance (measured between terminal STA with copper link removed and Lucar terminal)	0·36 to 0·42 ohms
Hold on Coil Resistance (measured between Lucar terminal and solenoid outer case)	1·49 to 1·71 ohms

Page P.X.s.14

ELECTRICAL AND INSTRUMENTS

ROUTINE MAINTENANCE

EVERY 24,000 MILES (38,400 KM.)

Checking the Brushgear and Commutator

Remove the starter motor from the engine.

Release the screw and remove the metal band cover.

Check that the brushes move freely in the brush boxes by holding back the spring and pulling gently on the flexible connection. If a brush is inclined to stick, remove it from its holder and clean its sides with a petrol moistened cloth. Replace the brushes in their original position in order to retain "bedding". Brushes which will not "bed" properly or have worn to $\frac{5}{16}$" (7·94 mm.) in length must be renewed. See page P.S.s.17 for renewal procedure.

Check the tension of the brush springs with a spring balance. The correct tension should be 52 ozs. (1·47 kg.) on a new brush.

Replace each existing brush in turn with a new brush to enable the tension of the brush springs to be tested accurately.

Check that the commutator is clean and free from oil or dirt. If necessary clean with a petrol moistened cloth or, if this is ineffective, rotate the armature and polish the commutator with fine glass paper. DO NOT use emery cloth. Blow out all abrasive dust with a dry air blast.

A badly worn commutator can be reskimmed by first rough turning, followed by diamond finishing. DO NOT undercut the insulation. Commutators must not be skimmed below a diameter of $1\frac{17}{32}$" (38·89 mm.). Renew the armature if below this limit.

REMOVAL

DISCONNECT THE BATTERY EARTH LEAD.

Disconnect and remove the transmitter unit from the top of the oil filter.

Disconnect the battery cable and solenoid switch cable from the starter motor.

Remove the distributor clamping plate retaining screw and withdraw the distributor.

Remove the two setscrews and lock washers securing the motor to the housing, gently bend away the carburetter drain pipes and remove the starter motor through the chassis frame.

The two setscrews are accessible from beneath the car or through an access panel in the right-hand side of the gearbox tunnel. Remove the front carpet to expose the panel.

Refitting

Refitting is the reverse of the removal procedure.

Care must be taken when refitting the two setscrews, which have a fine thread, that they are not cross-threaded.

Insert the distributor and rotate the rotor until the drive dog engages correctly and secure with the clamping plate setscrew.

Note: If the clamping plate has been removed from the distributor or its position altered, the engine must be re-timed as detailed in Section B.

SERVICING

Testing in position

Check that the battery is fully charged and that the terminals are clean and tight. Recharge if necessary.

Switch on the lamps together with the ignition and operate the starter control. If the lights go dim and the starter does not crank the engine this indicates that the current is flowing through the starter motor windings but the armature is not rotating for some reason. The fault is due possibly to high resistance in the brush gear or an open circuit in the armature or field coils. Remove the starter motor for examination.

If the lights retain their full brilliance when the starter switch is operated check the starter motor and the solenoid unit for continuity.

If the supply voltage is found to be applied to the starter motor when the switch is operated the unit must be removed from the engine for examination.

Sluggish or slow action of the starter motor is usually due to a loose connection causing a high resistance in the motor circuit. Check as described above.

If the motor is heard to operate, but does not crank the engine, indication is given of damage to the drive.

Page P.X.s.15

ELECTRICAL AND INSTRUMENTS

BENCH TESTING

Remove the starter motor from the engine

Disconnect the battery. Disconnect and remove the starter motor from the engine (see page P.X.s.15 for the removal procedure).

Measuring the light running current

With the starter motor securely clamped in a vice and using a 12-volt battery, check the light running current and compare with the value given on page P.X.s.15. If there appears to be excessive sparking at the commutator, check that the brushes are clean and free to move in their boxes and that the spring pressure is correct.

Measuring lock torque and lock current

Carry out a torque test and compare with the values given on page P.X.s.15. If a constant voltage supply is used, it is important to adjust this to be 7·6 volts at the starter terminal when testing.

FAULT DIAGNOSIS

An indication of the nature of the fault, or faults, may be deduced from the results of the no-load and lock torque tests.

Symptom	Probable Fault
1. Speed, torque and current consumption correct.	Assume motor to be in normal operating condition.
2. Speed, torque and current consumption low.	High resistance in brush gear, e.g., faulty connections, dirty or burned commutator causing poor brush contact.
3. Speed and torque low, current consumption high.	Tight or worn bearings, bent shaft, insufficient end play, armature fouling a pole shoe, or cracked spigot on drive end bracket. Short circuited armature, earthed armature or field coils.
4. Speed and current consumption high, torque low.	Short circuited windings in field coils.
5. Armature does not rotate, high current consumption.	Open circuited armature, field coils or solenoid unit. If the commutator is badly burned, there may be poor contact between brushes and commutator.
6. Armature does not rotate, high current consumption.	Earthed field winding or short circuit solenoid unit. Armature physically prevented from rotating.
7. Excessive brush movement causing arcing at commutator.	Low brush spring tension or out-of-round commutator. "Thrown" or high segment on commutator.
8. Excessive arcing at the commutator.	Defective armature windings, sticking brushes or dirty commutator.
9. Excessive noise when engaged.	Pinion does not engage fully before solenoid main contacts are closed. Check pinion movement as detailed under Setting Pinion Movement.
10. Pinion engaged but starter motor not rotating.	Pinion movement excessive. Solenoid main contacts not closing. Check pinion movement as detailed under Setting Pinion Movement.

ELECTRICAL AND INSTRUMENTS

DISMANTLING

Disconnect the copper link between the lower solenoid terminal and the starting motor yoke.

Remove the two solenoid unit securing nuts. Detach the extension cables and withdraw the solenoid from the drive end bracket casting, carefully disengaging the solenoid plunger from the starter drive engagement lever.

Remove the cover band and lift the brushes from their holders.

Unscrew and withdraw the two through bolts from the commutator end bracket. The commutator end bracket and yoke can now be removed from the intermediate and drive end brackets.

Extract the rubber seal from the drive end bracket.

Slacken the nut securing the eccentric pin on which the starter drive engagement lever pivots. Unscrew and withdraw the pin.

Separate the drive end bracket from the armature and intermediate bracket assembly.

Remove the thrust washer from the end of the armature shaft extension using a mild steel tube of suitable bore. Prise the jump ring from its groove and slide the drive assembly and intermediate bracket from the shaft.

To dismantle the drive further prise off the jump ring retaining the operating bush and engagement spring.

BENCH INSPECTION

After dismantling the motor, examine individual items.

Replacement of brushes

The flexible connectors are soldered to terminal tags; two are connected to brush boxes and two are connected to free ends of the field coils. Unsolder these flexible connectors and solder the connectors of the new brush set in their place.

The brushes are pre-formed so that "bedding" to the commutator is unnecessary. Check that the new brushes can move freely in their boxes.

Commutator

A commutator in good condition will be burnished and free from pits or burned spots. Clean the commutator with a petrol moistened cloth. Should this be ineffective, spin the armature and polish the commutator with fine glass paper; remove all abrasive dust with a dry air blast. If the commutator is badly worn, mount the armature between centres in a lathe, rotate at high speed and take a light cut with a very sharp tool. Do not remove more metal than is necessary. Finally polish with very fine glass paper. The INSULATORS between the commutator segments MUST NOT BE UNDERCUT. Commutators must not be skimmed below a diameter of $1\frac{17}{32}''$ (38·89 mm.).

Armature
Lifted conductors

If the armature conductors are found to be lifted from the commutator risers, overspeeding is indicated. In this event, check that the clutch assembly is operating correctly.

Fouling of armature core against the pole faces

This indicates worn bearings or a distorted shaft. A damaged armature must in all cases be replaced and no attempt should be made to machine the armature core or to true a distorted armature shaft.

Insulation test

To check armature insulation, use a 110 volt a.c., test lamp. The test lamp must not light when connected between any commutator segment and the armature shaft.

If a short circuit is suspected, check the armature on a "growler". Overheating can cause blobs of solder to short circuit the commutator segments.

If the cause of an armature fault cannot be located or remedied, fit a replacement armature.

Field Coils
Continuity Test

Connect a 12-volt test lamp and battery between the terminal on the yoke and each individual brush (with the armature removed from the yoke). Ensure that both brushes and their flexible connectors are clear of the yoke. If the lamp does not light, an open circuit in the field coils is indicated.

Replace the defective coils.

Insulation test

Connect a 110-volt a.c., test lamp between the terminal post and a clean part of the yoke. The test lamp lighting indicates that the field coils are earthed to the yoke and must be replaced.

When carrying out this test, check also the insulated pair of brush boxes on the commutator end bracket.

Page P.X.s.17

ELECTRICAL AND INSTRUMENTS

Clean off all traces of brush deposit before testing. Connect the 110-volt test lamp between each insulated brush box and the bracket.

If the lamp lights this indicates faulty insulation and the end bracket must be replaced.

Replacing the field coils

Unscrew the four pole-shoe retaining screws, using a wheel operated screwdriver. Remove the insulation piece which is fitted to prevent the inter-coil connectors from connecting with the yoke.

Draw the pole-shoes and coils out of the yoke and lift off the coils. Fit the new field coils over the pole-shoes and place them in position inside the yoke. Ensure that the taping of the field coils is not trapped between the mating surfaces of the pole-shoes and the yoke.

Locate the pole-shoes and field coils by lightly tightening the retaining screws. Replace the insulation piece between the field coil connections and the yoke.

Finally, tighten the screws by means of the wheel operated screwdriver while the pole pieces are held in position by a pole shoe expander or a mandrel of suitable size.

Bearings and Bearing Replacement

The commutator and drive end brackets are each fitted with a porous bronze bush and the intermediate bracket is fitted with an indented bronze bearing.

Replace bearings which are worn to such an extent that they will allow excessive side play of the armature shaft.

The bushes in the intermediate and drive end brackets can be pressed out whilst that in the commutator bracket is best removed by inserting a $\frac{9}{16}''$ (14·29 mm.) tap squarely into the bearing and withdrawing the bush with the tap.

Before fitting a new porous bronze bearing bush, immerse it for 24 hours in clean engine oil (SAE 30 to 40). In cases of extreme urgency, this period may be shortened by heating the oil to 100° C. for 2 hours and then **allowing the oil to cool before removing the bush.** Fit new bushes by using a shouldered, highly polished mandrel approximately 0·0005″ (·013 mm.) greater in diameter than the shaft which is to fit in the bearing. **Porous bronze bushes must not be reamed out after fitting,** as the porosity of the bush will be impaired.

After fitting a new intermediate bearing bush, lubricate the bearing surface with Rocol "Molypad" molybdenised non-creep, or similar, oil.

Page P.X.s.18

CHECKING THE ROLLER CLUTCH DRIVE

A roller clutch drive assembly in good condition will:—

(i) Provide instantaneous take-up of the drive in the one direction.

(ii) Rotate easily and smoothly in the other.

(iii) Be free to move round or along the shaft splines without roughness or tendency to bind.

Similarly, the operating bush must be free to slide smoothly along the driving sleeve when the engagement spring is compressed. Trunnion blocks must pivot freely on the pegs of the engaging lever. All moving parts should be smeared liberally with Shell Retinax "A" grease or an equivalent alternative.

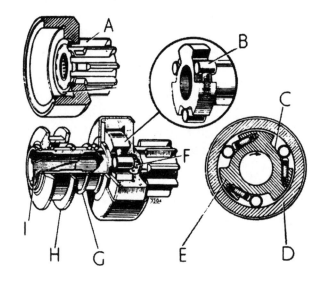

Fig. 9. *The roller clutch drive components.*

A—*Alternative contruction (pinion pressed and clear-ringed into driven member).*

B—*Spring loaded rollers.*

C—*Cam tracks.*

D—*Driven member (with pinion).*

E—*Driving member.*

F—*Bush.*

G—*Engagement spring.*

H—*Operating bush.*

I—*Driving sleeve.*

ELECTRICAL AND INSTRUMENTS

REASSEMBLY

After cleaning all parts, reassembly of the starting motor is a reversal of the dismantling procedure given on page P.X.s.15 but the following special points should be noted:—

(i) The following parts should be tightened to the maximum torques indicated:—

Nuts on solenoid copper terminals	20 lb./in. (0·23 kgm.)
Solenoid fixing bolts	4·5 lb./ft. (0·62 kgm.)
Starting motor through bolts	8·0 lb./ft. (0·83 kgm.)

(ii) When refitting the C.E. bracket see that the moulded brake shoes seat squarely and then turn them so that the ends of the cross peg in the armature shaft engage correctly with the slots in the shoes.

Setting Pinion Movement (Fig. 10)

Connect the solenoid Lucar terminal to a 6-volt supply. DO NOT use a 12-volt battery otherwise the armature will turn.

Connect the other side of the supply to the motor casing (this throws the drive assembly forward into the engage position).

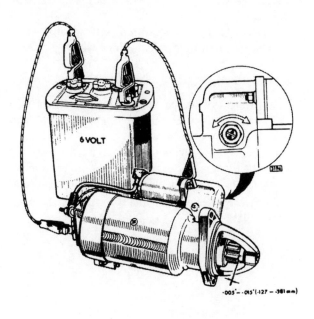

Fig. 10. *Setting pinion movement.*

Measure the distance between the pinion and the thrust washer on the armature shaft extension. Make this measurement with the pinion pressed lightly towards the armature.

For correct setting the dimension should be 0·005" to 0·015" (0·13 to 0·38 mm.).

Disconnect the battery.

Adjust the setting by slackening the eccentric pivot pin securing nut and turning the pin until the correct setting is obtained.

Note: The head of the arrow stamped on the end of the eccentric pivot pin should be set only between the ends of the arrows cast in the drive end bracket.

Turning the screw to the left (anti-clockwise) will increase the gap between the pinion and the thrust washer, turning to the right (clockwise) will decrease the gap.

Reconnect the battery and recheck the setting.

After setting tighten the securing nut to retain the pin position.

CHECKING OPENING AND CLOSING OF STARTER SWITCH CONTACTS

The following checks assume that pinion travel has been correctly set.

Remove the copper link connecting solenoid terminal STA with the starting motor terminal.

Connect, through a switch, a supply of 10 volts d.c., to the series winding, that is, connecting between the solenoid Lucar terminal and large terminal STA. DO NOT CLOSE THE SWITCH AT THIS STAGE.

Connect a separately energised test lamp circuit across the solenoid main terminals.

Insert a stop in the drive end bracket to restrict the pinion travel to that of the out of mesh clearance, normally a nominal ⅛" (3·17 mm.). An open-ended spanner or spanners of appropriate size and thickness can often be utilised for this purpose, its jaws embracing the armature shaft extension.

ELECTRICAL AND INSTRUMENTS

Energise the shunt winding with a 10-volt d.c., supply and then close the switch in the series winding circuit.

The solenoid contacts should close fully and remain closed, as indicated by the test lamp being switched on and emitting a steady light.

Switch off and remove the stop.

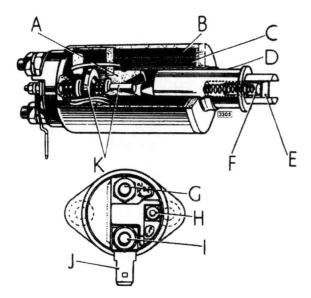

Switch on again and hold the pinion assembly in the fully engaged position.

Switch off and observe the test lamp.

The solenoid contacts should open, as indicated by the test lamp being switched off.

Fig. 11. *Checking the opening and closing of the starter switch contacts.*

A—*Core.*
B—*Shunt winding.*
C—*Series winding.*
D—*Plunger.*
E—*Clevis pin.*
F—*"Lost motion" device.*
G—*Starter terminal.*
H—*Solenoid terminal.*
I—*Battery terminal.*
J—*Accessories terminal.*
K—*Spindle and moving contact assembly.*

B—*Shunt winding.*

WINDSCREEN WIPER

Model	Chassis Number L.H. Drive
Open 2 Seater	1E.15980–U.S.A. Only
	1E.16010–Other than U.S.A.
Fixed Head Coupe	1E.34583–U.S.A. Only
	1E.34752–Other than U.S.A.
2+2	1E.77709

At the above chassis numbers, a Lucas DL3A windscreen wiper unit was introduced **on L.H. Drive cars only.**

It consists of a two-speed self-starting motor coupled by connecting rods to three wiper spindles.

Adjustment of the parked position is controlled by the location of the parking switch carrier plate mounted in the gear housing.

To adjust the parking position, if necessary, unscrew the three hexagon-headed drive screws sufficient to release the tension of the clamping plate, and rotate the switch carrier plate in the direction of the arrow on the plate.

Page P.X.s.20

ELECTRICAL AND INSTRUMENTS

Slight movement only should be necessary. Do not allow the blades to park below the glass lower edge.

Tighten the screws and recheck.

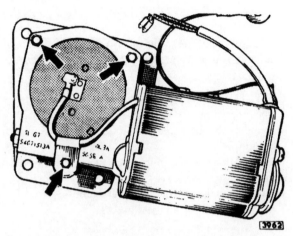

Fig. 1. The windscreen wiper motor parking adjustment. The arrows show the three screws to be released when adjusting the switch carrier plate.

WIPER MOTOR
Removal

Disconnect the battery.

Remove the spring clip retaining the throttle pedal link rod to the bellcrank lever and withdraw the rod.

Mark the location of the carrier bracket on the bulkhead, remove the two setscrews and detach the bracket.

Remove the plastic strap from the wiper motor and disconnect the cables from the snap connector block. Note the location of the cables for reference when refitting.

Lower the instrument panel after removing the two screws in the top corners and disconnect the ball joint from the centre windscreen wiper motor spindle bearing.

Remove the four setscrews retaining the motor to the bulkhead and withdraw the motor with attached link rod.

Note the sealing joint fitted between the motor bracket and the bulkhead.

Refitting

Refitting is the reverse of the removal procedure.

Note : It is essential when refitting to ensure that the length of the link rod is not altered. Any alteration in the length of this rod will place the windscreen wiper arms out of phase with each other.

When refitting the throttle bellcrank carrier bracket, care must be taken to ensure that the lever is central in its bearing. Adjustment is provided by means of the two slotted holes in the bracket.

WINDSCREEN WIPER SPINDLE HOUSINGS
Removal (Right or Left Hand Housing)

Disconnect the battery.

Withdraw the wiper arm from the spindle housing to be removed.

Unscrew the large nut securing the housing to the scuttle and remove the distance piece and rubber seal washer.

Lower the instrument panel after removing the two retaining screws in the top corners.

Remove the four nuts and washers retaining the screen rail facia assembly. Two are accessible from the centre aperture and one each at the outer edges below the screen rail.

Detach the two leads from the map light terminals.

Disconnect the demister ducts at the 'Y' pieces and withdraw the facia assembly.

Disconnect the ball joint from the spindle lever.

From inside the car remove the two nuts and washers securing the housing bracket to the base plate and withdraw the housing.

Remove the spring retainer and withdraw the pivot pin with attached outer link rods.

Complete the removal of the housing.

Page P.X.s.21

ELECTRICAL AND INSTRUMENTS

Removal (Central Housing)

Disconnect the battery.

Withdraw the wiper arm from the spindle.

Unscrew the large nut securing the housing to the scuttle and remove the distance piece and rubber seal washer.

Lower the instrument panel as detailed previously.

Disconnect the link ball joint from the spindle lever.

From inside the car, remove the two nuts and washers securing the housing bracket to the base plate and withdraw the housing.

Remove the spring retainer and withdraw the pivot pin with the attached outer link rods.

Complete the removal of the housing.

Model	Chassis Number	
	R.H. Drive	L.H. Drive
Open 2 Seater	1E.2037	1E.15980–U.S.A. Only
		1E.16010–Other than U.S.A.
Fixed Head Coupe	1E.21786	1E.34583–U.S.A. Only
		1E.34752–Other than U.S.A.
2+2	1E.51197	1E.77709

In subsequent text, the above mentioned are the commencing chassis numbers at which these items were introduced.

FAULT DIAGNOSIS

Poor performance can be electrical or mechanical in origin and not necessarily due to a faulty motor, for example:—

Low voltage at the motor due to poor connections or a discharged battery.

Excessive loading on the wiper blades.

Spindles binding in the housing.

THE INSTRUMENTS

ELECTRIC CLOCK
Description

The electric clock, fitted in the centre of the instrument panel, is a fully transistorised instrument powered by a mercury cell housed in a plastic holder attached to the back of the clock.

Frontal adjustment is provided by means of a small knurled knob for setting the hands and a slotted screw for time-keeping regulation.

To reset the hands, pull out the knurled knob, rotate and release.

To regulate the time-keeping, turn the slotted screw with a small screwdriver towards the positive (+) sign if gaining, and towards the minus (−) sign if losing.

Moving the indicator scale through one division will alter the time-keeping by five minutes per week.

The action of resetting the hands automatically restarts the movement.

The window of the clock is a plastic moulding, and should only be cleaned with a cloth or chamois leather slightly dampened with water. Oil, petrol or other fluids associated with cleaning, are harmful and must not be used.

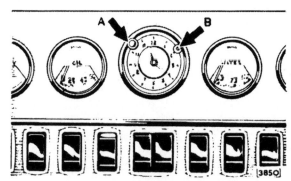

Fig. 13. Clock controls.
A — Handsetting.   B — Time regulator.

Page P.X.s.22

ELECTRICAL AND INSTRUMENTS

MAINTENANCE

The mercury cell life is in the region of 18 months, throughout which it ensures a steady and continuous voltage to the clock.

Renew the cell at this period to maintain perfect time-keeping.

Battery Replacement

Remove the instrument panel retaining screws and lower the panel.

Lever the battery out of the holder and discard.

Press the new battery into the holder.

Refit the panel.

Fig. 14. Renewing the electric clock battery.

Clock—Removal

Lower the instrument panel.

Withdraw the illumination bulb holder from the back of the clock.

Remove the two nuts and the clamp strap from the back of the clock.

Withdraw the clock, complete with the battery holder, from the instrument panel.

Refitting

Refitting is the reverse of the removal procedure.

THE REVOLUTION COUNTER (TACHOMETER)
Description

The revolution counter is an impulse tachometer instrument incorporating transistors and a printed circuit, the pulse lead (coloured WHITE) being wired in circuit with the S/W terminal on the ignition coil and the ignition switch.

Mechanical drive cables or an engine-driven generator are not required with this type of instrument.

The performance of this instrument is not affected by the distributor contact setting, by corrosion of the sparking plug points, or by differences in the gap settings.

Connection to the back of the instrument is by means of a locked plug and socket, the contacts being offset to prevent incorrect coupling.

Removal

Disconnect the battery.

Remove the screen rail facia assembly as detailed on Page P.X.s.00 to gain access to the instrument.

Remove the two knurled nuts, earth lead and instrument retaining pieces.

Withdraw the tachometer from the facia panel and remove the illumination bulb holders.

Disconnect the plug and socket as follows:—

Pinch together the prongs of the plastic retaining clip and withdraw from the plug and socket assembly (Fig. 15).

Detach the plug from the socket and complete the removal of the instrument.

IMPORTANT

Do not detach the green and white cables connected to the plug and the instrument.

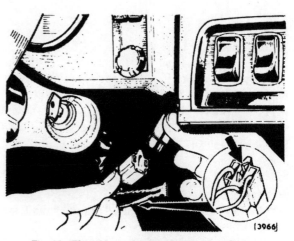

Fig. 15. The tachometer plug and socket assembly.
(Inset shows the clip in its fitted position).

Page P.X.s.23

ELECTRICAL AND INSTRUMENTS

Refitting

Refitting is the reverse of the removal procedure. Reconnect the plug and socket assembly and lock with the retaining clip.

THE INSTRUMENT PANEL

The instrument panel differs from that fitted to all previous cars in respect of the following items:—

(1) Rocker Switches—Replacing tumbler switches.
(2) Battery Indicator—Replacing Ammeter.
(3) Panel Light Dimming Resistance—Replacing resistance previously attached to the panel light switch.
(4) The combined Ignition/starter switch which is now mounted on a separate sub-panel. These switches were previously two separate items mounted in the instrument panel.
(5) The Cigar Lighter—Now located in the console below the instrument panel, was previously part of the instrument panel assembly.

THE SWITCHES

The rocker switches are mounted in a sub-panel which is attached to the instrument panel by four self-tapping screws.

Individual switches may be removed without detaching the sub-panel cluster as follows:—

Removal

Disconnect the battery.
Lower the instrument panel.
Remove the cables from the switch, noting location for reference when refitting.
Press in the two locking tabs located at the bottom and the top faces of the switch body and push the switch through the aperture.

Refitting

Press the switch into the panel aperture until the nylon locking tabs register.
Reconnect the cables as noted on removal.

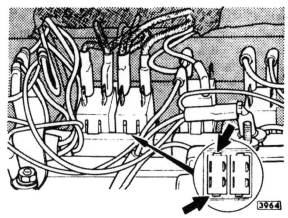

Fig. 16. Instrument panel rocker switch removal
(Inset shows arrowed the nylon locking tabs).

THE IGNITION/STARTER SWITCH

A Lucas 47SA combined ignition/starter switch replaces the separate switches previously used.

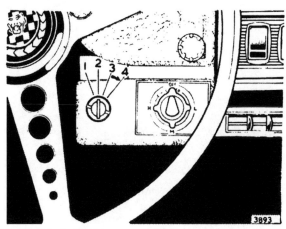

Fig. 17. The ignition/starter switch location when air-conditioning system is fitted.
1 — Auxiliaries.   2 — Ignition "OFF".
3 — Ignition "ON".   4 — Starter.

The switch is mounted on a bracket attached to the steering column (if Air-conditioning equipment is installed the bracket is attached to the evaporator unit).

In conjunction with the 47SA ignition/starter switch a Lucas 6RA relay is included in the alternator circuit. This functions as a field isolating relay, the relay coil being energised by operation of the ignition switch.

Removal

Remove the locking ring and withdraw the switch through the bracket with the brass locknut and wave washer.

Page P.X.s.24

ELECTRICAL AND INSTRUMENTS

Disconnect the cables and remove the switch. Note the location of the cables for reference when refitting.

The lock barrel can be withdrawn by inserting a thin rod through a hole in the body of the switch and depressing the plunger in the lock. Insert the key and turn to the 'OFF' position to gain access to the plunger.

Refitting

Refitting is the reverse of the removal procedure.

When refitting a new lock barrel, check that the number on the face of the barrel and the key is the same as that on the barrel removed. This will be identical to the door locks.

Insert the key in the lock and turn the switch to the 'OFF' position before inserting the barrel.

Battery Indicator

This instrument is a voltmeter with a specially calibrated dial which indicates the condition of the battery. It does not register the charging rate of the alternator.

The position of the needle with a charged battery will be within the area marked 'Normal'.

Removal

Disconnect the battery and lower the instrument panel.

Disconnect the cables, noting the location for reference when refitting.

Detach the illumination bulb holder.

Remove two nuts and clamp strap and withdraw the instrument forward through the panel.

Refitting

Refitting is the reverse of the removal procedure.

Check the condition of the battery by means of the panel shown below

RED (Off Charge)		NORMAL			RED (On Charge)
BATTERY CHARGE EXTREMELY LOW	BATTERY CHARGE LOW	WELL CHARGED BATTERY	CHARGING VOLTAGE LOW	CHARGING VOLTAGE SATISFACTORY	CHARGING VOLTAGE TOO HIGH
If with the ignition and electrical equipment e.g. headlamps etc., switched on, but with the engine not running the indicator settles in this section—your battery requires attention.		Ideally the indicator should settle in this section when the ignition and electrical equipment e.g. headlamps etc., are switched on and the engine is not running.	This condition may be indicated when the headlights and other equipment are in use.	The indicator should point to this section when the engine is running above idle.	If the indicator continues to point to this section after 10 minutes running either your voltage regulator requires adjustment or some other fault has developed.

IMPORTANT All readings on the indicator should be ignored when the engine is idling, since readings may vary at very slow engine speeds due solely to operation of the voltage regulator.

OFF CHARGE
This means more energy is being used from your battery than is being replaced by the alternator on your car. This condition is satisfactory provided it does not persist for long periods, when the engine is running above idle or at speed. If the indicator remains in the section, it may mean that you have a broken or slipping fan belt, a faulty alternator, a badly adjusted voltage regulator or some other fault.

ON CHARGE
This means your battery is having more energy put into it than is being taken out of it. In the ordinary way this condition predominates and your battery is continuously being recharged by the alternator whenever the engine is running above idle. If however the engine is continually running slowly as may be the case in traffic—or when, in winter, lights and cold starting make extra demands on the battery—you may find the rate of discharge exceeds the rate of charge—that is to say the battery is running down, as will be indicated on your Battery Condition Indicator and you may need an extra charge if "battery charge low or extremely low" is indicated by the instrument.

HEADLAMP

Sealed beam units are fitted to all cars with the exception of certain European Countries which retain the pre-focus bulb (see Bulb Data Chart).

The beam setting and unit replacement instructions differ from those stated on Page P.24 as follows:—

Beam Setting

If beam setting adjustment is required, prise off the headlamp rim (retained by spring clips). Switch on the headlamps and check that they are on Main beam.

LAMPS

The setting of the beams is controlled by two screws 'A' and 'B' on Fig. 18

The top screw 'A' is for vertical adjustment, i.e. to raise or lower the beam; turn the screw anti-clockwise to lower the beam or clockwise to raise the beam.

The side screw 'B' is for horizontal adjustment, i.e. to turn the beam to right or left. To move the beam to the right, turn the screw clockwise. To move the beam to the left, turn the screw anticlockwise.

Page P.X.s.25

ELECTRICAL AND INSTRUMENTS

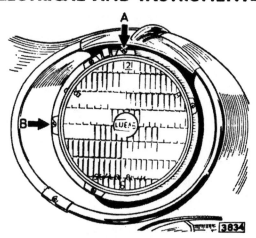

Fig. 18. Adjustment of the screw 'A' will alter the headlamp beams in the vertical plane; adjustment of the screw 'B' will alter the headlamp beams in the horizontal plane.

Sealed Beam Unit — Replacement

Prise off the headlamp rim (retained by spring clips).

Remove the three cross-headed screws and detach the retaining ring.

Note: Do not disturb the two beam setting screws.

Withdraw the sealed beam unit and unplug the adaptor.

Replace the sealed beam unit with one of the correct type (see 'Lamp Bulbs').

On cars fitted with bulb light units, proceed as directed above until the unit is removed. Release the bulb retaining clips and withdraw the bulb. Replace with a bulb of the correct type (see 'Lamp Bulbs').

When reassembling, note the groove in the bulb plate which must register with the raised portion on the bulb retainer.

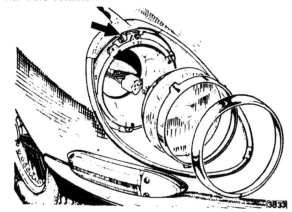

Fig. 19. Headlamp sealed beam unit removal. The arrow indicates one of the spring clips retaining the rim.

HORNS

DESCRIPTION

Lucas 9H horns are fitted replacing the WT618 Units previously fitted to early 4.2 'E' type cars.

The horns are now mounted on brackets attached to the sub-frame lower cross-member.

The horn circuit operates through a Lucas 6RA relay, the contacts C1 and C2 closing when the relay coil is energised by depressing the horn switch button located in the direction (turn) indicator switch lever.

Maintenance

In the event of the horns failing to sound or performance becoming uncertain, check before making adjustments that the fault is not due to external causes.

Check as follows and rectify as necessary:
(i) Battery condition.
(ii) Loose or broken connections in the horn circuit.
(iii) Loose fixing bolts. It is important to keep the horn mountings tight and to maintain rigid the mounting of any unit fitted near the horns.
(iv) Faulty relay. Check by substitution after verifying that current is available at terminal C2 (cable colour—brown/purple) and terminal W1 (cable colour—Green).
(v) Check that fuse No. 3 (50 amperes) and fuse No. 6 (35 amperes) have not blown.
   Note: Horns will not operate unless the ignition is switched on.

Adjustment

As the horns cannot conveniently be adjusted in position, remove and mount securely on a test fixture.

A small serrated adjusting screw located adjacent to the horn terminal is provided to take up wear of moving parts in the horn. Turning this screw does not alter the pitch of the horn note.

Connect a moving coil ammeter in series with the horn supply feed. The ammeter should be protected from overload by connecting on ON-OFF switch in parallel with its terminals.

Keep this switch ON except when taking readings, that is when the horn is sounding.

Turn the screw clockwise until the horn operates within the specified limits of 6.5-7.0 amperes.

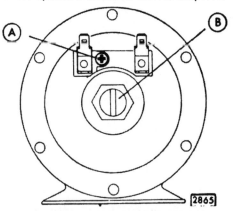

Fig. 20. The Lucas 9H horn.
A — Contact breaker adjustment screw.
B — Slotted centre core (Do not disturb).

Service Replacements

When fitting replacement horns it is essential that the following procedure be carried out:—

(i) Refit the lockwashers in their correct positions, one at each side of the mounting bracket centre fixing.

(ii) Ensure, after positioning the horn, that the $\frac{7}{16}''$ centre fixing bolt is secure but not over-tightened. Over-tightening of this bolt will damage the horn.

(iii) Ensure that, when a centre fixing bolt or washers other than the originals are used, the bolt is not screwed into the horn to a depth greater than $\frac{11}{16}''$ (17.5 mm).

Muted Horns (Holland only)

These horns are muted to comply with the Dutch Traffic Regulations and incorporate a rubber plug inserted in the trumpet.

Horn Relay—Checking

If the horn relay is suspected, check for fault by substitution or by the following method:—

(i) Check that fuses No. 3 and No. 6 have not blown. Replace if necessary.

(ii) Check with a test lamp that current is present at the relay terminal W1 (Green) and C2 (Brown/Purple). Switch on the ignition before checking terminal W1.

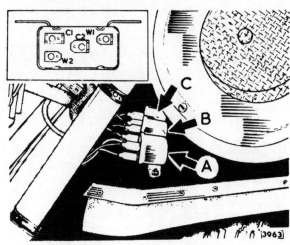

Fig. 21. Location of horn and alternator relays.
A — Horn relay.
B — Alternator/Ignition relay.
C — Air conditioning equipment relay (when fitted).
(Inset shows the connections).

(iii) Remove the cable from terminal W2 (Purple/Black) and earth the terminal to a clean part of the frame. The relay coils should now operate and close the contacts.
Reconnect cable.

ELECTRICAL AND INSTRUMENTS

(iv) Remove cable from terminal C2 (Brown/Purple). Check for continuity by means of an earthed test lamp. Check with the horn button depressed and the ignition 'ON'. Replace the relay if faulty.

THE ALTERNATOR

The alternator differs from that detailed on Page P.X.s.00 in respect of the following items only:—

(i) Inclusion of a Lucas 6RA relay in the alternator circuit due to the introduction of the Lucas 47SA Ignition switch.

(ii) Location on cars fitted with Air-conditioning Equipment. On cars so equipped the alternator is mounted centrally at the front of the engine. Belt adjustment and servicing details remain unaltered.

Alternator Relay—Checking

Check with test lamp that current is available at terminals C1 and W1. Switch on ignition before checking W1.

Check earth connection to terminal W2.

Switch on the ignition and check that the relay coil is energised and contacts C1-C2 have closed by means of an earthed test lamp connected to terminal C2.

The relay is mounted on the closing panel, adjacent to the battery, below the horn relay.

Refer to the wiring diagram when checking.

RADIATOR FANS

Radiator Fan Relays

A Lucas 6RA relay is included in the radiator fan/thermostat switch circuit to prevent overloading of the thermostat contacts. The relay is mounted on the front upper cross tube behind the radiator matrix.

When Air-conditioning Equipment is fitted, a second relay is included to over-ride the thermostat circuit when the car is stationary and the air-conditioning system is operational.

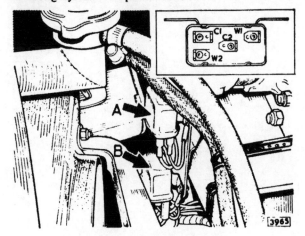

Fig. 22. The radiator fan relays.
"B" when air-conditioning is NOT fitted.
"A" is an over-riding relay when air-conditioning IS fitted.
(Inset shows the connections).

Page P.X.s.27

ELECTRICAL AND INSTRUMENTS

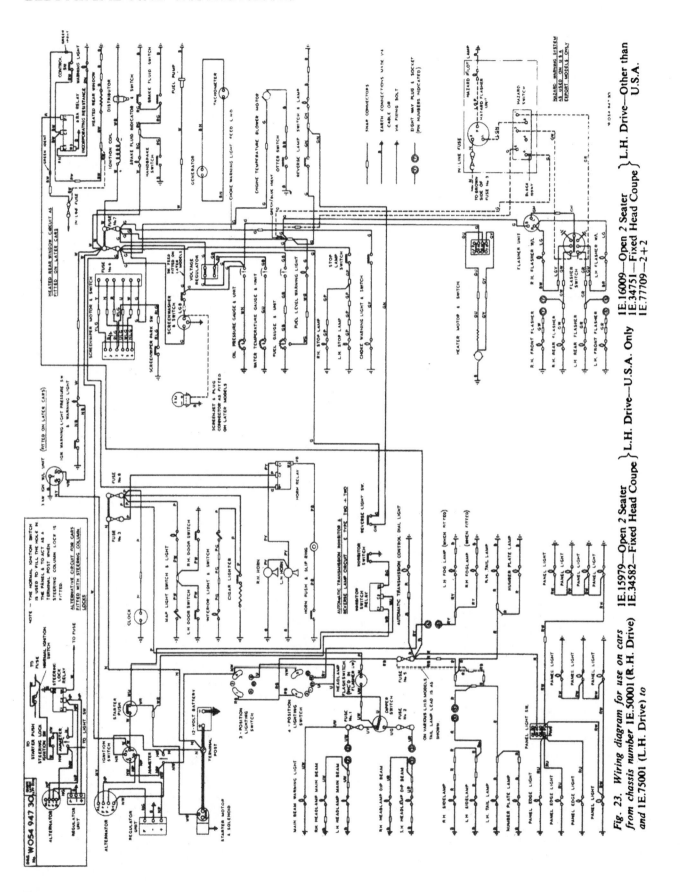

Fig. 23. Wiring diagram for use on cars from chassis number 1E.50001 (R.H. Drive) and 1E.75001 (L.H. Drive) to

1E.15979—Open 2 Seater } L.H. Drive—U.S.A. Only
1E.34582—Fixed Head Coupe

1E.16009—Open 2 Seater
1E.34751—Fixed Head Coupe } L.H. Drive—Other than U.S.A.
1E.77709—2+2

Page P.X.s.28

ELECTRICAL AND INSTRUMENTS

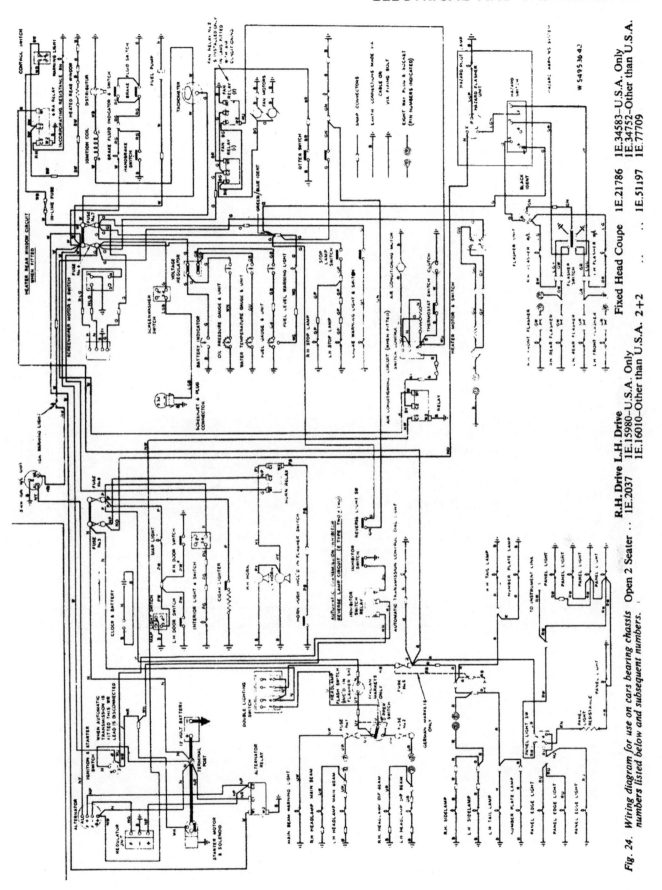

Fig. 24. Wiring diagram for use on cars bearing chassis numbers listed below and subsequent numbers.

	R.H. Drive	L.H. Drive	
Open 2 Seater	1E.2037	1E.15980—U.S.A. Only	1E.16010—Other than U.S.A.
Fixed Head Coupe	1E.21786	1E.34583—U.S.A. Only	1E.34752—Other than U.S.A.
2+2		1E.51197	1E.77709

Page P.X.s.29

ELECTRICAL AND INSTRUMENTS

CABLE COLOUR CODE

B	BLACK	**S**	SLATE
U	BLUE	**W**	WHITE
N	BROWN	**Y**	YELLOW
R	RED	**D**	DARK
P	PURPLE	**L**	LIGHT
G	GREEN	**M**	MEDIUM

When a cable has two colour code letters, the first denotes the main colour and the second denotes the tracer colour.

Page P.X.s.30

SUPPLEMENTARY INFORMATION

FOR

4·2 LITRE "E" TYPE AND 2+2 CARS

(SERIES 2)

This Supplement covers the variations between the 4·2 Litre (Series 2) "E" Type, 2+2 (Series 2) and the 3·8 Litre versions of the "E" Type. Insert the Supplement at the end of the 3·8 Litre "E" Type Service Manual, Publication No. E.123.

ISSUED BY

JAGUAR CARS LIMITED, COVENTRY, ENGLAND

Telephone — ALLESLEY 2121 (P.B.X.)
Code — BENTLEY'S SECOND
Telegraphic Address — "JAGUAR," COVENTRY. Telex. 31622

Publication No. E156/1

INDEX TO SECTIONS

SECTION TITLE	SECTION REFERENCE
GENERAL INFORMATION	AY
ENGINE	BY
CARBURETTERS AND FUEL SYSTEM	CY
COOLING SYSTEM	DY
CLUTCH	EY
GEARBOX	FY
AUTOMATIC TRANSMISSION	FFY
FRONT SUSPENSION	JY
REAR AXLE	HY
STEERING	IY
STEERING (POWER-ASSISTED)	IIY
BRAKES	LY
WHEELS AND TYRES	MY
BODY AND EXHAUST SYSTEM	NY
HEATING AND WINDSCREEN WASHER	OY
ELECTRICAL AND INSTRUMENTS	PY
EXHAUST EMISSION CONTROL	QY
AIR-CONDITIONING REFRIGERATION EQUIPMENT	RY

SECTION A

GENERAL INFORMATION

The Routine Maintenance Service periods have been increased to 3,000 mile (5,000 km.) stages as follows:—

1,000 MILES (1,600 KM.)
FREE SERVICE

1. Road test and check for oil, petrol, hydraulic fluid or coolant leaks.
2. Check torque loading of cylinder head nuts.
3. Check oil or fluid levels and top up as necessary:—
    (a) Brake reservoirs,
    (b) Clutch reservoir (if fitted),
    (c) Power steering reservoir (if fitted),
    (d) Top up carburetter hydraulic dampers and check carburation,
    (e) Battery,
    (f) Screen washer bottle,
    (g) Radiator header tank (add anti-freeze when necessary),
    (h) Manual gearbox,
    (i) Final drive unit.
4. Drain and refill
    (a) Engine sump,
    (b) Automatic transmission unit (if fitted).
5. Adjust front band on automatic transmission unit (if fitted).
6. Check driving belts for correct tension.
7. Clean and adjust contact-breaker points.
8. Check all brake pipe unions, petrol pipe unions, and hoses for leakage.
9. Check tightness of all front and rear suspension bolts and nuts.
10. Check tightness of nuts on all steering connections including column universal joints.
11. Check tightness of road wheel nuts and wheel alignment.
12. Check tyres for damage and adjust pressures.
13. Check operation of all lights and systems.
14. Check door locks and bonnet release control.
15. Lubricate all grease nipples (excluding wheel bearings).

Page AY.s.1

3,000 MILES (5,000 KM.)
CHECK SERVICE

Repeat these servicing items at the under-mentioned subsequent periods:—

      9,000 miles (15,000 Km.)
    15,000 miles (25,000 Km.)
    21,000 miles (35,000 Km.)
    27,000 miles (45,000 Km.)
    33,000 miles (55,000 Km.)
    39,000 miles (65,000 Km.)
    45,000 miles (75,000 Km.)
    51,000 miles (85,000 Km.)
    57,000 miles (95,000 Km.)
    63,000 miles (105,000 Km.)
    69,000 miles (115,000 Km.)

1. Check oil or fluid levels and top up as necessary:—
   (a) Engine sump,
   (b) Brake reservoirs,
   (c) Clutch reservoir (if fitted),
   (d) Power steering reservoir (if fitted),
   (e) Top up carburetter hydraulic dampers and check carburation,
   (f) Battery,
   (g) Screen washer bottle,
   (h) Radiator header tank (add anti-freeze when necessary),
   (i) Manual gearbox,
   (j) Final drive unit.
2. Check driving belts for correct tension.
3. Examine brake pads for wear and check operation of brake stop lights.
4. Examine tyres for damage and adjust pressures.
5. Check tightness of road wheel nuts.

6,000 MILES (10,000 KM.)
MINOR SERVICE

Repeat these servicing items at the under-mentioned subsequent periods:—

    18,000 miles (30,000 Km.)
    30,000 miles (50,000 Km.)
    42,000 miles (70,000 Km.)
    54,000 miles (90,000 Km.)
    66,000 miles (110,000 Km.)

1. Check oil or fluid levels and top up as necessary:—
   (a) Brake reservoirs,
   (b) Clutch reservoir (if fitted),
   (c) Power steering reservoir (if fitted),
   (d) Top up carburetter dampers,
   (e) Battery and check connections,
   (f) Screen washer bottle,
   (g) Radiator header tank (add anti-freeze when necessary),
   (h) Manual gearbox or automatic transmission unit,
   (i) Final drive unit.
2. Drain and refill:—
   (a) Engine sump. Fit new oil filter element and seal.
3. Check driving belts for correct tension.
4. Check brake pads for wear and advise wear-rate to owner.
5. Check tyres for damage and tread depth. If uneven wear evident, check wheel alignment. Adjust pressures.
6. Check tightness of road wheel nuts.
7. Check headlamp alignment and functioning of mandatory lights including stop lights.
8. Lubricate all grease nipples, excluding wheel bearings.

12,000 MILES (20,000 KM.)
MAJOR SERVICE

Repeat these servicing items at the under-mentioned subsequent periods:—
- 24,000 miles (40,000 Km.)
- 48,000 miles (80,000 Km.)
- 60,000 miles (100,000 Km.)

1. Check oil or fluid levels and top up as necessary:—
   - (a) Brake reservoirs,
   - (b) Clutch reservoir (if fitted),
   - (c) Power steering reservoir (if fitted),
   - (d) Top up carburetter hydraulic dampers,
   - (e) Battery and check connections,
   - (f) Screen washer bottle,
   - (g) Radiator header tank (add anti-freeze when necessary),
   - (h) Automatic transmission,
   - (i) Final drive unit.
2. Drain and refill:—
   - (a) Engine sump. Fit new oil filter element and seal,
   - (b) Manual gearbox. Clean overdrive filter (if fitted),
   - (c) Final drive unit (if 'Powr-Lok' differential fitted. Use only special limited slip oil).
3. Renew sparking plugs.
4. Renew air cleaner element and fuel line filter element.
5. Clean and adjust contact breaker points. Check operation of centrifugal advance mechanism. Lubricate distributor.
6. Check driving belts for wear and tension.
7. Adjust top timing chain if required.
8. Lubricate all grease nipples including front and rear wheel bearings.
9. Check all suspension and exhaust mountings for security.
10. Check all steering connections, ball joints etc., for security and wear.
11. Check brake pads for degree of wear and advise wear-rate to owner.
12. Check functioning of all mandatory lights including stop lights and alignment of headlamps.
13. Check tyres for damage and tread depth. If uneven wear evident, check wheel alignment. Adjust pressures.
14. Oil can lubrication of door locks, bonnet hinges and locks, boot hinges and lock, seat slides, fuel filler flap hinges, control linkages.
15. Detect and report any oil, petrol, water, hydraulic fluid leakage and damaged hoses or other damaged parts.

36,000 MILES (60,000 KM.)
MAJOR SERVICE

Repeat these servicing items at the under-mentioned subsequent period:—
- 72,000 miles (120,000 Km.)

1. Check oil or fluid levels and top up as necessary:—
   - (a) Clutch reservoir (if fitted),
   - (b) Power steering reservoir (if fitted),
   - (c) Top up carburetter hydraulic dampers,
   - (d) Battery and check connections,
   - (e) Screen washer bottle,
   - (f) Radiator header tank (add anti-freeze when necessary),
   - (g) Automatic transmission,
   - (h) Final drive unit.
2. Drain and refill:—
   - (a) Engine sump. Fit new oil filter element and seal,
   - (b) Manual gearbox. Clean overdrive filter (if fitted),
   - (c) Braking system. Retract wheel cylinder pistons to expell all old fluid,
   - (d) Final drive unit (if 'Powr-Lok' differential fitted. Use only special limited slip oil).
3. Renew sparking plugs.
4. Renew air cleaner element and fuel line filter element.
5. Clean and adjust contact breaker points. Check operation of centrifugal advance mechanism. Lubricate distributor.
6. Check driving belts for wear and tension.
7. Adjust top timing chain if required.
8. Lubricate all grease nipples including front and rear wheel bearings.
9. Check all suspension and exhaust mounting for security.
10. Check all steering connections, ball joints etc., for security and wear.
11. Check brake pads for degree of wear and advise wear-rate to owner.
12. Check functioning of all mandatory lights including stop lights and alignment of headlamps.
13. Check tyres for damage and tread depth. If uneven wear evident, check wheel alignment. Adjust pressures.
14. Oil can lubrication of door locks, bonnet hinges and lock, boot hinges, and lock, seat slides, fuel filler flap hinges, control linkages.
15. Detect and report any oil, petrol, water, hydraulic fluid leakage and damaged hoses or other damaged parts.

Page AY.s.3

RECOMMENDED LUBRICANTS

Component	MOBIL	CASTROL	SHELL	ESSO	B.P.	DUCKHAM	REGENT Caltex/Texaco
Engine	Mobil Super or Mobil Special	Castrol GTX	Shell Super Oil	Esso Extra Motor Oil 10W/30 Esso Extra Motor Oil 20W/40	Super Visco-Static 10W/40	Q20–50 or Q5500	Havoline 20W/40 or 10W/30
Upper cylinder lubrication	Mobil Upperlube	Castrollo	Shell U.C.L. or Donax U.	Esso U.C.L.	U.C.L.	Adcoid Liquid	Regent U.C.L.
Distributor oil can points Oil can lubrication	Mobiloil A	Castrol GTX	Shell Super Oil	Esso Extra Motor Oil 20W/40	Energol SAE 30	Q20–50	Havoline 30
Gearbox Final Drive Unit (not 'Powr-Lok')	Mobilube GX 90	Castrol Hypoy	Spirax 90 EP	Esso Gear Oil GP 90/140	Gear Oil SAE 90 EP	Hypoid 90	Multigear Lubricant EP.90
Front wheel bearings Rear wheel bearings Distributor cam Final drive half-shafts Steering tie-rods Wheel swivels Door hinges Steering housing	Mobil grease MP	Castrolease LM	Retinax A	Esso Multi-purpose Grease H	Energrease L.2	LB 10	Marfak All Purpose
Automatic transmission unit Power steering system	Mobilfluid 200	Castrol T.Q.	Shell Donax T6	Esso Automatic Transmission Fluid	Automatic Transmission Fluid Type A	Nolmatic	Texamatic Fluid

Page AY.s.4

SECTION B
ENGINE

DATA

Camshaft

    Permissible end float .. .. ·004″ to ·006″ (·10 to ·15 mm.)

Connecting Rod

    Big end—Diameter clearance .. .. ·0015″ to ·0033″ (·037 to ·083 mm.)

Crankshaft Main Bearings

    Journal diameter .. .. 2·750″ to 2·7505″ (69·85 to 69·86 mm.)

    Journal length

        —Front .. .. $1\frac{9}{16}″$ (39·06 mm.)

        —Centre .. .. $1\frac{3}{8}″$ + ·001″ (34·37 mm. + ·025 mm.)
                                                                                                                                       − ·0005″   − ·0125 mm.

        —Rear .. .. $1\frac{11}{16}″$ (42·86 mm.)

        —Intermediate .. .. $1\frac{7}{32}″$ ± ·002″ (30·96 mm. ± ·05 mm.)

Cylinder Block

    Bore size for fitting liners .. .. 3·761″ to 3·762″
                                                                                                              (94·03 to 94·05 mm.)

    Outside diameter of liner .. .. 3·765″ to 3·766″
                                                                                                            (94·13 to 94·15 mm.)

    Interference fit .. .. ·003″ to ·005″ (·08 to ·13 mm.)

    Overall length of liner .. .. 6·959″ to 6·979″ (17·39 to 17·45 cm.)

    Outside diameter of lead-in .. .. 3·758″ to 3·760″ (93·95 to 94·00 mm.)

    Size of bore honed after assembly—cylinder block—Nominal .. .. 92·07 mm. (3·625″)

Gudgeon Pin

    Length .. .. 3·00″ (75 mm.)

Page BY.s.1

ENGINE

Piston and Piston Rings

Gudgeon pin bore	·8571″ to ·8753″ (2·188 to 2·1883 mm.)
Piston rings—Width Compression	·0770″ to ·0780″ (1·97 to 2·00 mm.)
Oil Control	Self expanding (Maxiflex)
Piston rings—Thickness	·151″ to ·158″ (3·775 to 3·95 mm.)
Piston rings—Gap when fitted to cylinder bore Oil Control	·015″ to ·033″ (·38 to ·82 mm.)

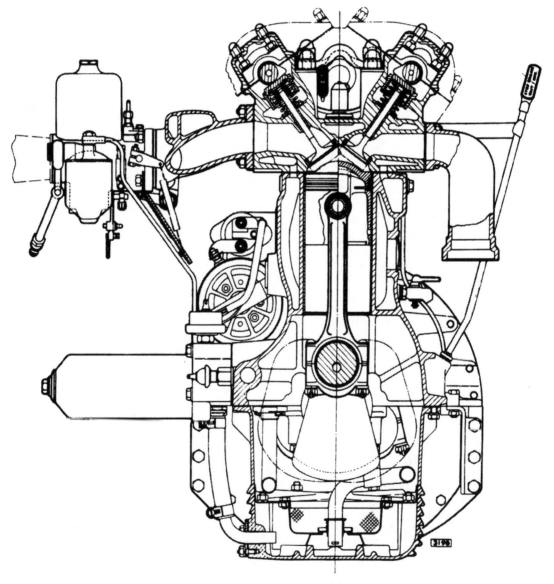

Fig. 1. *Cross sectional view of the engine.*

Page BY.s.2/3

ENGINE

ENGINE REMOVAL AND REFITTING

REMOVAL

Remove the bonnet.

Disconnect the battery.

Drain the cooling system and cylinder block; conserve the coolant if antifreeze is in use.

Slacken the clip on the breather pipe; unscrew the two wing nuts and withdraw the top of the air cleaner.

Disconnect the petrol feed pipe from under the centre carburetter.

Slacken the clamps and remove the water hoses from the cylinder head and radiator to the header tank. Slacken the two clamps and withdraw the water pump hose. Remove the heater hoses from the inlet manifold.

Disconnect the brake vacuum hose from the inlet manifold.

Pull off the two Lucar connectors from the fan thermostat control in the header tank.

Remove the two bolts securing the header tank mounting bracket to the front cross member. Remove the two nuts and two bolts securing the header tank straps to the radiator and fan cowl. Withdraw the header tank complete with mounting bracket and straps.

Disconnect the throttle linkage at the rear carburetter.

Disconnect:—

The two coil leads.

The water temperature transmitter.

The battery cable and solenoid switch cable from the starter.

The output cables from the alternator.

The engine earth strap from the left hand side member.

Withdraw the oil filter canister; catch the escaping oil in a drip tray.

Remove the crankshaft pulley, damper and drive belt. Mark the pulley and damper to facilitate re-fitting. Remove the ignition timing pointer from the sump.

Remove the revolution counter generator complete with cables.

Remove the four nuts and washers securing each exhaust downpipe from the manifold. Unclip the pipes at the silencers and withdraw the downpipes. Collect the sealing rings between the pipes and the manifold.

On standard transmission cars, proceed as follows:—

Remove the seats. Remove the knob and locknut from the gear lever. Remove two hexagon headed setscrews and two chromium-plated nuts and detach the radio/ash tray console panel from the gearbox tunnel. If a radio is fitted, disconnect the electrical cables from the control head to enable the panel to be completely removed.

On 2+2 cars, raise the central arm-rest; lift out the bottom panel; withdraw five self-tapping screws and remove the central arm-rest. Lift off the trimmed cover panel from the gearbox tunnel.

On all other cars, withdraw two pan-headed screws and two seat belt attachments before lifting off the trimmed cover. Withdraw the self-tapping screws and remove the gearbox cover.

Disconnect the reverse lamp cables from the switch on the gearbox top cover.

Disconnect the speedometer drive cable from the gearbox.

Remove the clutch slave operating cylinder from the clutch cover.

Disconnect the propeller shaft.

On automatic transmission cars, proceed as follows:—

Withdraw the transmission dipstick and unscrew the dipstick tube from the transmission oil pan.

Place the selector lever in L and, from underneath the car, unscrew the nut securing the selector cable adjustable ball joint to the transmission lever. Release the nut securing the outer cable clamp to the abutment bracket.

Remove the speedometer drive cable from the transmission extension housing.

Page BY.s.4

ENGINE

Disconnect the transmission oil cooler pipes from the right hand side of the radiator block and from the transmission unit. Withdraw the clips and remove the pipes.

Disconnect the kickdown cable at the rear of the cylinder head.

Remove the central arm-rest and lift off the trimmed cover panel from the gearbox tunnel. Withdraw the drive screws securing the cover plate on the transmission tunnel. Disconnect the propeller shaft.

For all models, proceed as follows:—

Remove the nuts securing the torsion bar reaction tie plate on each side and tap the bolts back flush with the face of the tie plate. With the aid of a helper, place a lever between the head of the bolt just released and the torsion bar. Exert pressure on the bolt head to release the tension on the upper bolt. Remove the nut and tap the upper bolt back flush with the face of the tie plate. Withdraw the bolts securing the tie plate on each side to the body underframe channels through the side members. Tap the tie plate off the four bolts.

Note: Failure to relieve the tension on the upper bolts when tapping them back to the tie plate will result in stripping the threads. If this occurs, new bolts must be fitted and the torsion bars re-set.

Support the engine by means of two individual lifting tackles using the hooks provided on the cylinder head. Insert a trolley jack under the transmission (or gearbox) and support the transmission.

Remove the self-locking nut and washer from the engine stabiliser.

Remove the bolts securing the rear engine mounting plate. Remove the bolts from the front engine mountings.

Raise the engine on the lifting tackles and, keeping the combined engine and transmission assembly level, move forwards ensuring that the water pump pulley clears the sub-frame top cross member. Carefully raise the front of the engine and withdraw forwards and upwards.

REFITTING

Refitting is the reverse of the removal procedure. After the unit is in place, it is important that the engine stabiliser is adjusted and that the clutch slave cylinder is mounted correctly.

On automatic transmission 2+2 cars, the kickdown cable must be adjusted and the manual linkage connected in accordance with the instructions given in Section FF.

ENGINE

THE CYLINDER BLOCK

OVERHAUL

Reboring is normally recommended when the bore wear exceeds ·006" (·15 mm). Reboring beyond the limit of ·030" (·76 mm) is not recommended and, when the bores will not clean out at ·030" (·76 mm), liners and standard size pistons should be fitted.

The worn liners must be pressed out from below utilising the stepped block illustrated.

PISTONS AND GUDGEON PINS

Piston Grades

Grade Identification Letter	To suit cylinder bore size
F	3·6250" to 3·6253" (92·075 to 92·0826 mm.)
G	3·6254" to 3·6257" (92·0852 to 92·0928 mm.)
H	3·6258" to 3·6261" (92·0953 to 92·1029 mm.)
J	3·6262" to 3·6265" (92·1055 to 92·1131 mm.)
K	3·6266" to 3·6269" (92·1156 to 92·1123 mm.)

Oversize Pistons

Oversize pistons are available in the following sizes: +·010" (·25 mm.) +·020" (·51 mm.) +·030" (·76 mm.).

There are no selective grades in oversize pistons as grading is necessary purely for factory production methods.

Tapered Periphery Rings

All engine units are fitted with tapered periphery piston rings and these must be fitted the correct way up.

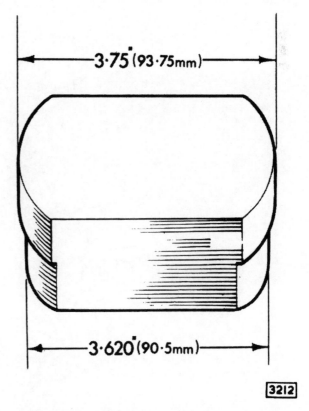

Fig. 2. *Stepped block for cylinder liner removal.*

Page BY.s.6

ENGINE

The narrowest part of the ring must be fitted uppermost; to assist in identifying the narrowest face a letter "T" or "Top" is marked on the side of the ring to be fitted uppermost.

The oil control ring consists of two steel rails with a spacer between. These rails are held together on assembly with an adhesive. The expander, which is fitted inside the oil control ring, should be assembled with the ends of the expander ring (internal ring) butted together. If the internal ring is fitted to the piston groove with the ends overlapping, the outer ring assembly cannot be seated properly.

Pistons
Skirt clearance (measured at bottom of skirt at 90 to gudgeon pin pin axis) ·0007" to ·0013" (·018 to ·03 mm.)

Ring gap—when fitted to bore
Top compression ·015" to ·020" (·38 to ·51 mm.)
Lower compression ·010" to ·015" (·254 to ·38 mm.)
Scraper ·015" to ·045" (·38 to 1·143 mm.)
Side clearance in groove ·001" to ·003" (·02 to ·07 mm.)

Gudgeon Pins
Grades (Red) ·8753" to ·8754" (22·23 to 22·24 mm.)
(Green) ·8752" to ·8753" (22·22 to 22·23 mm.)
Clearance in piston ·0001" to ·0003" (·0025 to ·0076 mm.)

Cargraph Treatment—Piston Rings
The chromium plated ring (top compression) is Cargraph treated on the outside diameter to assist in bedding in the chromium surface. This coating is coloured Red for identification purposes and should not be removed. Excess oil or grease may be removed with clean paraffin but rings should not be soaked in any degreasing agent.

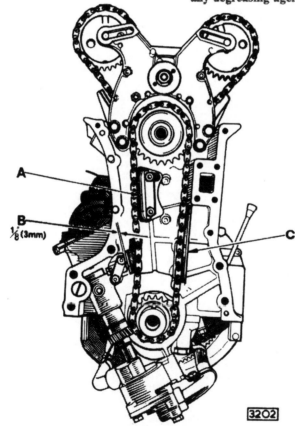

Fig. 3. *The timing gear arrangement.*

When fitting a new lower timing chain, set the intermediate damper (A) in light contact with the chain when there is a ⅛" (3 mm) gap between the rubber slipper and the tensioner body. In the case of a worn chain, the gap (B) may have to be increased to avoid fouling between the chain and the cylinder block. Set the lower damper (C) in light contact with the chain.

Page BY.s.7

ENGINE

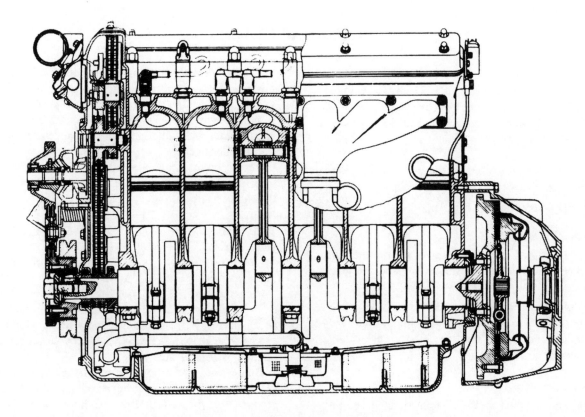

Fig. 4. *Longitudinal section of the engine.*

OIL SUMP

REMOVAL

Drain the sump; disconnect the oil return pipe and remove the crankshaft damper.

Remove the self-locking nut and washer from the top of the engine stabiliser. Screw down the lower flanged washer to the limit of the stud thread.

Sling the engine from the rear lifting loop and raise the engine approximately 1" (25·4 mm.).

Remove the sump securing screws, lower the front end and withdraw forward.

REFITTING

Refitting is the reverse of the removal procedure but care must be taken to ensure that the rear oil seal is positioned correctly. Adjust the engine stabiliser after refitting.

Check for oil leakage after refilling the sump and running the engine.

Page BY.s.8

ENGINE

VALVE GUIDES

Valve guides have circlips fitted to ensure positive location in the cylinder head. These valve guides are chamfered at the upper ends and have the outside diameter reduced at the lower end to provide a "lead-in" when fitting.

Oil seals are also fitted to the inlet valve guide—a second groove being machined in the guide above the circlip groove to seat the oil seal.

Checking Valve Guides

Examine the guides for evidence of wear in the bore. The clearance between the valve stem and the guide when new is ·001"–·004" (·025–·10 mm).

If it is found necessary to renew worn valve guides, they must be fitted in accordance with the following instructions and only genuine factory replacement parts used.

Valve Guide—Replacement

Heat the cylinder head by immersing in boiling water for 30 minutes. With a piloted drift, drive out the old valve guide from the combustion chamber end.

Note: If carbon deposits around the valve guide in the combustion chamber are quite heavy, they should be cleaned off thoroughly before attempting to drive out the old valve guide.

Valve guides when fitted during engine assembly are to the following dimensions and may be fitted in mixed form.
(1) ·501" to ·502" (12·70 to 12·725 mm.)
(2) ·503" to ·504" (12·776 to 12·801 mm.)

The valve guide (2) will be identified by the machining of one circular groove on the shank of the guide: valve guide (1) will **not** have the groove.

When removing worn guides, care must be taken to identify each individual guide to its particular bore in the cylinder head.

Replacement guides are available in the following sizes and will have identification grooves machined in the shank as noted:—

1st oversize (one groove)	·503" to ·504" (12·776 to 12·801 mm.)
2nd oversize (two grooves)	·506" to ·507" (12·852 to 12·877 mm.)
3rd oversize (three grooves)	·511" to ·512" (12·979 to 13·005 mm.)

Valve guides with one groove should only be fitted as replacements for those originally fitted without a groove: the bore in the cylinder head will not require reaming before fitting.

Guides with two grooves should be used as replacements for those with one groove and guides with three grooves for those with two. Cylinder head bores should be reamed to the following dimensions:—

Valve Guide	Ream to Size
2nd oversize (two grooves)	$.505"\ {}^{+.0005"}_{-.0002"}$ (12.83 mm. ${}^{+.012}_{-.005}$ mm.)
3rd oversize (three grooves)	$.510"\ {}^{+.0005"}_{-.0002"}$ (12.95 mm. ${}^{+.012}_{-.005}$ mm.)

Coat the valves with graphite grease and fit the circlips. Reheat the cylinder head. With a piloted drift, drive in the valve guide from the top until the circlip registers in the groove machined in the guide bore of the cylinder head. Visually check that the circlip has seated correctly.

Page BY.s.9

SECTION C

CARBURETTERS AND FUEL SYSTEM

CARBURETTERS

Removal

Drain the cooling system.

Disconnect the battery.

Slacken the hose clip securing the water hose from the inlet manifold to the header tank. Remove the hose.

Disconnect the two electrical connections from the thermostat fan control in the header tank.

Remove the throttle return springs.

Unclip hose connection to breather pipe.

Remove the two butterfly nuts at the carburetter trumpets and remove the air cleaner elbow.

Remove the carburetter trumpet from the carburetters having removed the six nuts and spring washers together with the three gaskets.

Disconnect the throttle linkage at the rear carburetter.

Remove the three banjo union bolts and six fibre washers from the float chambers.

Ensure that the three float chamber filters are not mislaid.

Disconnect the mixture control outer and inner cables.

Remove the suction pipe from the front carburetter.

Disconnect the brown/black cable from the oil pressure switch.

Slacken the clips and disconnect the heater pipes at the water manifold and below the inlet manifold.

On 2+2 cars fitted with automatic transmission, disconnect the kickdown cable at the rear of the cylinder head.

Fig. 1. *Refitting the mixture control rods with the jet levers against the stops.*

Page CY.s.1

CARBURETTERS AND FUEL SYSTEM

Remove the inlet manifold complete with the carburetters and linkage.

Remove the four nuts and spring washers, together with the return spring bracket from each carburetter. Remove all three carburetters together.

If necessary, remove the mixture control linkage from each carburetter by removing the split pins and withdrawing the clevis pins.

Refitting

Refitting is the reverse of the removal procedure except that new gaskets should be fitted to the inlet manifold, to either side of the heat insulating gasket and also to the carburetter trumpet flanges.

Adjust the kickdown cable as detailed on page FFY.s.24.

CARBURETTER TUNING

The method of tuning carburetters is identical with that given for 3·8 litre "E" Type cars, however, the idling speed on standard transmission cars should be 700 r.p.m. in order to eliminate any chatter from the constant mesh gears in the all-synchromesh gearbox.

On automatic transmission 2+2 cars, the idling speed should be 500 r.p.m.

The fuel feed line filter incorporates a renewable fibre filter element. This element should not be cleaned but must be renewed every 12,000 miles. When renewing, the two sealing washers should also be replaced.

If sediment build-up is excessive, the element should be renewed more frequently than stated above.

THE FUEL SYSTEM

THE PETROL PUMP

Description (Fig. 2)

The pump consists of three main assemblies, the main body casting (A); the diaphragm armature and magnet assembly (M) contained within the housing; and the contact breaker assembly housed within the end cap (T2). A non-return valve assembly (C) is affixed to the end cover moulding to assist in the circulation of air through the contact breaker chamber.

The main fuel inlet (B) provides access to an inlet air bottle (I) while access to the main pumping chamber (N) is provided by an inlet valve assembly. This assembly consists of a Melinex valve disc (F) permanently assembled within a pressed-steel cage, held in position by a valve cover (E1).

The outlet from the pumping chamber is provided by an identical valve assembly which operates in the reverse direction. Both inlet and outlet valve assemblies together with the filters are held in position by a clamp plate (H). The valve assemblies may be removed by detaching the clamp plate (H) after removing the self-tapping screws. A filter (E) is provided on the delivery side of the inlet valve assembly. The delivery chamber (O) is bounded by a flexible plastic spring loaded diaphragm (L) contained by the vented cover (P). Sealing of the diaphragm (L) is provided by the rubber sealing ring (L.2).

The magnetic unit consists of an iron coil housing, an iron core (Q), an iron armature (A1) provided with a central spindle (P1) which is permanently united with the diaphragm assembly (L1), a magnet coil (R) and a contact breaker assembly consisting of parts (P2), (U1), (U), (T1) and (V). Between the coil housing and the armature are located eleven spherically edged rollers (S). These rollers locate the armature (A1) centrally within the coil housing and permit freedom of movement in a longitudinal direction.

The contact breaker consists of a bakelite pedestal moulding (T) carrying two rockers (U) and (U1) which are both hinged to the moulding at one end by the rocker spindle (Z). These rockers are interconnected at their top ends by means of two small springs arranged to give a throw-over action. A trunnion (P2) is carried by the inner rocker and the armature spindle (P1) is screwed into this trunnion. The outer rocker (U) is fitted with two tungsten points which contact with corresponding tungsten points which form part of the

CARBURETTERS AND FUEL SYSTEM

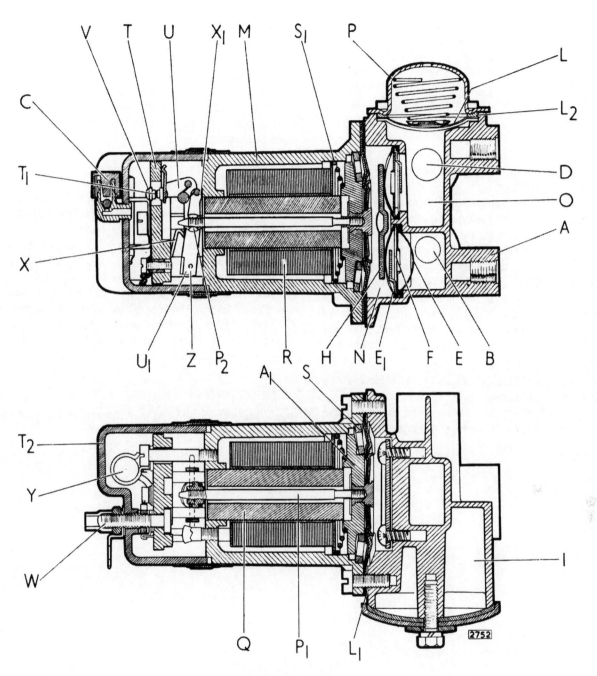

Fig. 2. *The Petrol Pump.*

WARNING: If at any time, it becomes necessary to blow through the fuel feed pipes the outlet pipes must be disconnected from the pumps. Failure to observe this procedure will cause the Melinex valves to be displaced or damaged.

Page CY.s.3

CARBURETTERS AND FUEL SYSTEM

spring blade (V) connected with one end of the coil. The other end of the coil is connected to a terminal (W) while a short length of flexible wire (X) connecting the outer rocker to one of the screws holding the pedestal moulding onto the coil housing provides an earth return to the body of the pump. It is important that the body of the pump be effectively earthed to the body of the vehicle by means of the earthing terminal provided on the flange of the coil housing.

OPERATION

When the pump is at rest the outer rocker (U) lies in the outer position and the tungsten points are in contact. Current passes from Lucar connector (W) through the coil and back to the blade (V), through the points and to earth, thus energising the coil and attracting the armature (A1). The armature, together with the diaphragm assembly, then retracts thereby sucking petrol through the inlet valve into the pumping chamber (N). When the armature has travelled nearly to the end of its stroke, the throw-over mechanism operates and the outer rocker moves rapidly backwards, thus separating the points and breaking the circuit.

The spring (S1) then reasserts itself forcing the armature and diaphragm away from the coil housing. This action forces petrol through the delivery valve at a rate determined by the requirements of the engine.

As the armature nears the end of its stroke the throw-over mechanism again operates, the tungsten points remake contact and the cycle of operations is repeated.

The spring blade (V) rests against the small projection moulding (T) and it should be set so that, when the points are in contact, it is deflected away from the moulding. The gap at the points should be approximately ·030" (·75 mm.) when the rocker (U) is manually deflected until it contacts the end face of the coil housing.

REMOVAL

Remove both inlet and outlet pipes from the side of the pump by withdrawing the banjo bolt and washers. Disconnect the electrical feed cable to the pump by unscrewing the knurled knob on the end of the pump. Remove the two self-locking nuts attaching the pump to the bracket and withdraw the two washers from each stud. The pump can now be withdrawn from the bracket leaving the two rubber grommets in position. The rubber grommets in the brackets should be examined for deterioration and replaced if necessary, otherwise excessive petrol pump noise may result.

REFITTING

Refitting is the reverse of the removal procedure.

CARBURETTERS AND FUEL SYSTEM

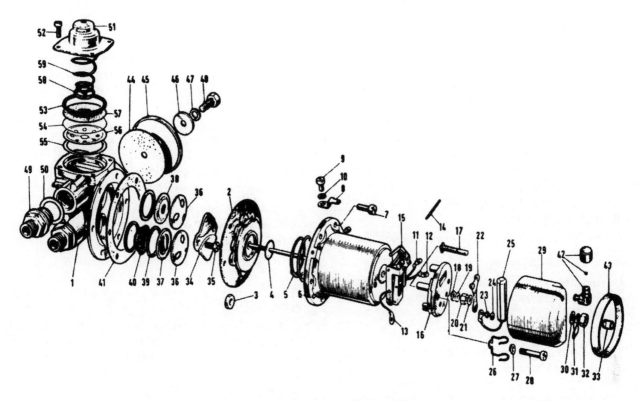

Fig. 3. *Exploded view of the petrol pump.*

1. Pump body.
2. Diaphragm and spindle assembly.
3. Roller—armature centralising.
4. Washer—impact.
5. Spring—armature.
6. Housing—coil.
7. Screw—securing housing—2 B.A.
8. Connector—earth.
9. Screw—4 B.A.
10. Spring washer.
11. Terminal tag.
12. Terminal tag.
13. Earth tag.
14. Rocker pivot pin.
15. Rocker mechanism.
16. Pedestal.
17. Terminal stud.
18. Spring washer.
19. Lead washer.
20. Terminal nut.
21. Washer.
22. Contact blade.
23. Washer.
24. Screw.
25. Condenser.
26. Clip.
27. Spring washer.
28. Screw.
29. End cover.
30. Shakeproof washer.
31. Lucas connector.
32. Nut.
33. Insulating sleeve.
34. Clamp plate.
35. Screw.
36. Valve cap.
37. Inlet valve.
38. Outlet valve.
39. Sealing washer.
40. Filter.
41. Gasket.
42. Vent valve.
43. Sealing band.
44. Joint.
45. Inlet air bottle cover.
46. Dished washer.
47. Spring washer.
48. Screw.
49. Outlet connection.
50. Fibre washer.
51. Cover.
52. Screw.
53. 'O' ring.
54. Diaphragm barrier.
55. Sealing washer.
56. Diaphragm plate.
57. Diaphragm.
58. Spring end cap.
59. Diaphragm spring.

Page CY.s.5

CARBURETTERS AND FUEL SYSTEM

DISMANTLING (Fig. 3).

Contact Breaker

Remove the insulated sleeve (33), terminal nut (32), and connector (31), together with its shakeproof washer (30). Remove the tape seal (if fitted) and take off the end-cover.

Unscrew the 5 B.A. screw (24) which holds the contact blade (22) to the pedestal (16) and remove the condenser (25) from its clip. This will allow the washer (23), terminal tag (11), and the contact blade to be removed.

Coil housing and diaphragm

Unscrew the coil housing securing screws (7), using a thick-bladed screwdriver to avoid damaging the screw heads.

Remove the earthing screw (9).

The coil housing (6) may now be removed from the body (1). Next remove the diaphragm and spindle assembly (2) by taking hold of the diaphragm and unscrewing it anti-clockwise until the armature spring (5) pushes the diaphragm away from the coil housing. It is advisable to hold the housing over the bench so that the 11 brass rollers (3) will not fall on the floor. The diaphragm and its spindle are serviced as a unit and should not be separated.

Pedestal and rocker

Remove the end-cover seal washer (21), unscrew the terminal nut (20), and remove the lead washer (19). This will have flattened on the terminal tag and thread and is best cut away with cutting pliers or a knife. Unscrew the two 2 B.A. screws (28), holding the pedestal to the coil housing, remove the earth terminal tag (13) together with the condenser clip (26). Tip the pedestal and withdraw the terminal stud (17) from the terminal tag (12). The pedestal (16) may now be removed with the rocker mechanism (15) attached.

Push out the hardened steel pin (14) which holds the rocker mechanism to the pedestal and separate the two.

Body and valves

Unscrew the two Phillips screws (35) securing the valve clamp plate (34), remove the valve caps (36), valves (37) and (38), sealing washers (39) and filter (40).

Note: Dismantling of the delivery flow-smoothing device should only be undertaken if the operation of it is faulty, and if the necessary equipment for pressure-testing after assembly is available. On this understanding proceed as follows:

Remove the four 4 B.A. screws (52) securing the delivery flow-smoothing device vented cover (51), remove the cover, the diaphragm spring (59), rubber 'O' ring (53), spring cap (58), diaphragm (57), barrier (54), diaphragm plate (56) and sealing washer (55).

Remove the single 2 B.A. screw (48), securing the inlet air bottle cover (45). Remove the cover and gasket (44).

Unscrew the inlet and outlet connections.

INSPECTION

If gum formation has occurred in the fuel used in the pump, the parts in contact with the fuel will have become coated with a substance similar to varnish. This has a strong stale smell and may attack the neoprene diaphragm. Brass and steel parts so affected can be cleaned by being boiled in a 20 per cent. solution of caustic soda, dipped in a strong nitric acid solution and finally washed in boiling water. Light alloy parts must be well soaked in methylated spirits and then cleaned.

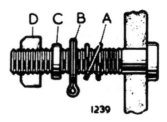

Fig. 4. *The terminal arrangement.*

A—*Double coil spring washer.*
B—*Cable tag.*
C—*Lead washer.*
D—*Countersunk nut.*

CARBURETTERS AND FUEL SYSTEM

Clean the pump and inspect for cracks, damaged joint faces and threads.

Examine the plastic valve assemblies for kinks or damage to the valve plates. They can best be checked by blowing and sucking with the mouth.

Check that the narrow tongue on the valve cage, which is bent over to retain the valve and to prevent it being forced out of position, has not been distorted but allows a valve lift of approximately $\frac{1}{16}$ in. (1·6 mm.).

Examine the delivery flow-smoothing device diaphragm, barrier, plate, spring, and spring cap for damage. If in doubt, renew the diaphragm.

Examine the inlet air bottle cover for damage. Examine the valve recesses in the body for damage and corrosion; if it is impossible to remove the corrosion, or if the recess is pitted, the body must be discarded.

Clean the filter with a brush and examine for fractures, renew if necessary.

Examine the coil lead tag for security and the lead insulation for damage.

Examine the contact breaker points for signs of burning and pitting; if this is evident, the rocker assembly and spring blade must be renewed.

Examine the pedestal for cracks or other damage, in particular to the narrow ridge on the edge of the rectangular hole on which the contact blade rests.

Examine the non-return vent valve in the end-cover for damage, ensure that the small ball valve is free to move.

Examine the diaphragm for signs of deterioration.

Renew the following parts: all fibre and cork washers, gaskets, and 'O' section sealing rings, rollers showing signs of wear on periphery, damaged bolts, and unions.

ASSEMBLY

Pedestal and rocker

Note: The steel pin which secures the rocker mechanism to the pedestal is specially hardened and must not be replaced by other than a genuine S.U. part.

Invert the pedestal and fit the rocker assembly to it by pushing the steel pin (14, Fig. 3) through the small holes in the rockers and pedestal struts. Then position the centre toggle so that, with the inner rocker spindle in tension against the rear of the contact point, the centre toggle spring is above the spindle on which the white rollers run. This positioning is important to

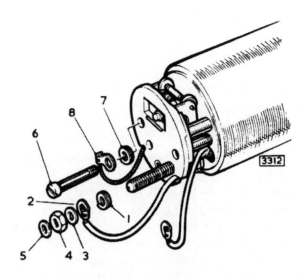

Fig. 5. *Attaching the pedestal to the coil housing.*

obtain the correct "throw over" action; it is also essential that the rockers are perfectly free to swing on the pivot pin and that the arms are not binding on the legs of the pedestal.

If necessary the rockers can be squared up with a pair of thin-nosed pliers.

Assemble the square-headed 2 B.A. terminal stud to the pedestal, the back of which is recessed to take the square head.

Assemble the 2 B.A. spring washer (1) (Fig. 5), and put the terminal stud through the 2 B.A. terminal tag (2), then fit the lead washer (3) and the coned nut (4) with its coned face to the lead washer. (This makes better contact than an ordinary flat washer and nut).

Tighten the 2 B.A. nut and finally add the end-cover seal washer (5).

Assemble the pedestal to the coil housing by fitting the two 2 B.A. pedestal screws (6), ensuring that the spring washer (7) on the left-hand screw (9 o'clock position) is between the pedestal and the earthing tag (8). When a condenser is fitted, its wire clip base is placed under the earthing tag and the spring washer is not required.

Tighten the screws, taking care to prevent the earthing tag (8) from turning, as this will strain or break the earthing flex. Do not tighten the screws or the pedestal will crack.

Do not fit the contact blade at this stage.

Page CY.s.7

CARBURETTERS AND FUEL SYSTEM

Diaphragm assembly

Place the armature spring into the coil housing with its larger diameter towards the coil (5, Fig. 3).

Before fitting the diaphragm make sure that the impact washer is fitted to the armature. (This is a small neoprene washer that fits in the armature recess). Do not use jointing compound or dope on the diaphragm.

Fit the diaphragm by inserting the spindle in the hole in the coil and screwing it into the threaded trunnion in the centre of the rocker assembly.

Screw in the diaphragm until the rocker will not "throw over"; this must not be confused with jamming the armature on the coil housing internal steps.

Fit the 11 brass centralizing rollers (3, Fig. 3) by turning back the diaphragm edge and dropping the rollers into the coil recess. The pump should be held in the left hand, rocker end downwards, to prevent the rollers from falling out.

Fit the contact blade and adjust the finger settings as described in "Contact gap setting", then carefully remove the contact blade.

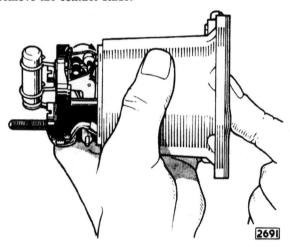

Fig. 6. *Setting the diaphragm.*

Holding the coil housing assembly in the left hand in an approximately horizontal position (see Fig. 6), push the diaphragm spindle in with the thumb of the right hand, pushing firmly but steadily. Unscrew the diaphragm, pressing and releasing with the thumb of the right hand until the rocker just "throws over". Now turn the diaphragm back (unscrew) to the nearest hole and again **4 holes** (two-thirds of a complete turn). The diaphragm is now correctly set.

Press the centre of the armature and fit the retaining fork at the back of the rocker assembly. This is done to prevent the rollers from falling out when the coil housing is placed on the bench prior to fitting the body, and is not intended to stretch the diaphragm before tightening the body screws.

Body components

The valve assemblies are retained internally in the body by a clamp plate secured with self-tapping screws (35, Fig. 3). The inlet valve recess in the body is deeper than the outlet recess to allow for the filter and extra washer. Another feature of these pumps is the incorporation of an air bottle on the inlet and a flow-smoothing device on the delivery side.

The inlet air bottle is a chamber in the body casting blanked off by a simple cover and joint washer held by a single screw. The delivery flow-smoothing device is formed by a perforated metal plate which is in contact with a plastic barrier backed by a rubber diaphragm, all held in position by a spring and end-cap retained by a vented cover. This assembly seals the delivery chamber in the body.

Screw in the inlet and outlet connections with their sealing rings. Assemble the outlet valve components into the outless recess in the following order, first a joint washer, then the valve, tongue side downwards, then the valve cap.

Assemble the inlet valve into the inlet recess as follows: first a joint washer, then the filter, dome side downwards, then another joint washer, followed by the valve assembly, tongue side upwards, then the valve cap.

Take care that both valve assemblies nest down into their respective recesses, place the clamp plate on top, and tighten down firmly to the body with the two screws.

Replace the inlet air bottle cover with its joint washer and tighten down the central screw.

Place the sealing washer in the bottom of the delivery flow-smoothing device recess, follow this with the perforated diaphragm plate, dome side downwards, then the plastic barrier, followed by the rubber diaphragm. Insert the "O" section sealing ring into the recess and ensure that it seats evenly. Place the diaphragm spring, large end towards the vented cover, into the cover, place the spring end-cap on the small end of the spring, pass the assembly tool through the cover, spring, and end cap and turn it through 90° so that tension may be applied to the spring during assembly. Finally fit the spring and cap assembly onto the diaphragm, tighten the four retaining screws, and

CARBURETTERS AND FUEL SYSTEM

release the assembly tool. The pump should be pressure-tested after disturbance of the flow-smoothing device.

Body attachment

Fit the joint washer to the body, aligning the screw holes.

Offer up the coil housing to the body, ensuring correct seating between them.

Line up the six securing screw holes, making sure that the cast lugs on the coil housing are at the bottom, insert the six 2 B.A. screws finger-tight. Fit the earthing screw with its Lucar connector.

Remove the roller retaining fork before tightening the body securing screws, making sure that the rollers retain their position; a displaced roller will cut the diaphragm. It is not necessary to stretch the diaphragm before tightening the securing screws.

Tighten the securing screws in sequence as they appear diametrically opposite each other.

Contact blade (Fig. 7)

Fit the contact blade and coil lead to the pedestal with the 5 B.A. washer and screw. The condenser should be fitted with the tag placed under the coil lead tag.

Adjust the contact blade so that the points are a little above the points on the rocker when closed, also that when the contact points make or break, one pair of points completely covers the other. As the contact blade is provided with a slot for the attachment screw, some degree of adjustment is possible.

Tighten the contact blade attachment screw when the correct setting is obtained.

Contact gap setting

Check that when the outer rocker is pressed onto the coil housing, the contact blade rests on the narrow rib or ridge which projects slightly above the main face of the pedestal. If it does not, slacken the contact blade attachment screw, swing the blade clear of the pedestal, and bend it downwards a sufficient amount so that when repositioned it rests against the rib lightly, over-tensioning of the blade will restrict the travel of the rocker mechanism.

Correct positioning gives a gap of $\cdot 035'' \pm \cdot 005''$ ($\cdot 9 \pm \cdot 13$ mm.) between the pedestal and tip of spring blade (Fig. 7).

Check the gap between rocker finger and coil housing with a feeler gauge, bending the stop finger, if necessary, to obtain a gap of $\cdot 070 \pm \cdot 005$ in. ($1 \cdot 8 \pm \cdot 13$ mm.).

End-cover

Tuck all spare cable into position so that it cannot foul the rocker mechanism. Ensure that the end-cover seal washer is in position on the terminal stud, fit the bakelite end-cover and lock washer, secure with the brass nut, fit the terminal tag or connector, and the insulated sleeve.

The pump is now ready for test.

After test, replace the rubber sealing band over the end cover gap and seal with adhesive tape.

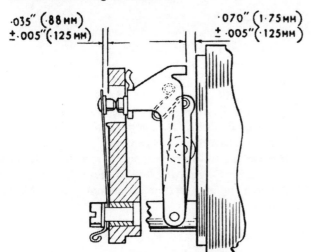

Fig. 7. *Rocker and contact clearances.*

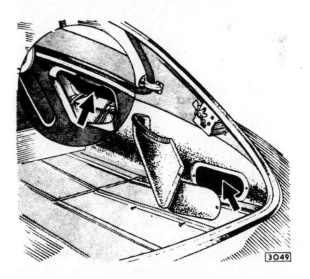

Fig. 8. *The location of the petrol pump. (Fixed head coupe). Inset shows location in open 2-seater model.*

SECTION D

COOLING SYSTEM

'E' Type series 2 have a sealed cooling system with a vertical flow radiator and an expansion tank.

The thermostat is retained in a revised housing.

The radiator top tank incorporates a plain (non-pressure) cap, the pressure cap being fitted to the expansion tank mounted on the bulkhead.

PRESSURE CAP RATING
With Standard Equipment .. .. 7 lb.
With Air-Conditioning System .. 13 lb.

Instructions for filling or checking the coolant level in the system differ from those stated for earlier cars fitted with the cross flow radiator, as follows:—

Filling Up
Remove the radiator and expansion tank filler caps.
Fill the radiator to the bottom of the filler neck.
Replace the filler cap and tighten down fully.
Top up the expansion tank to the half-way mark, refit the cap and tighten down fully.

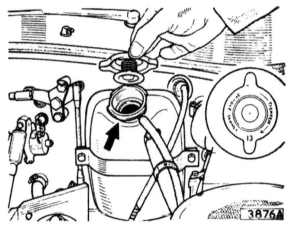

Fig. 1. The expansion tank and pressure cap. (Inset shows the pressure cap fitted to cars with Air-Conditioning Equipment).

Note: Care must be taken to ensure that the radiator and the expansion tank filler caps are not reversed.

Checking the Coolant Level
IMPORTANT: The coolant level must be checked at the expansion tank and NOT at the radiator top tank.

Check when the system is COLD.
Remove the pressure cap and top up to the half-way mark in the tank.
Replace the pressure cap and tighten down fully.

Refilling the Cooling System—Important

When refilling the cooling system following complete drainage, place the heater temperature control in the "Hot" position to allow the heater circuit to be filled with coolant. Re-check the level after running the engine for a short period.

THE RADIATOR
Removal
Release the filler cap, open the drain tap and drain the cooling system. Conserve the coolant if anti-freeze is in use.

Disconnect the multi-pin socket from the left-hand side of the bonnet.

Remove two bolts, self-locking nuts and washers securing the bonnet linkage to the sub-frame.

Withdraw two hexagon-headed pivot pins and washers securing the bonnet pivot to the sub-frame front lower cross tube, and remove the bonnet assembly.

Release the hose clips and disconnect the top and bottom hoses from the radiator.

Disconnect the oil cooler pipes (2 + 2 automatic transmission cars only) and blank off the unions to prevent loss of oil.

Remove six setscrews securing the cowl to the matrix side brackets. Disconnect the fan thermostat switch cables at the cable junction.

Release the radiator duct panel from the bottom of the matrix by removing two setscrews.

Remove the two bottom fixing nuts and rubber mounting washers.

Page DY.s.1

COOLING SYSTEM

Lift out the radiator matrix and collect the rubber washers fitted between the bottom tank and the mounting brackets.

NOTE : If air-conditioning equipment is fitted to the car, the condenser unit should be left in position after removal of the two setscrews securing the side support brackets to the matrix.

DO NOT DISTURB THE HOSE CONNECTIONS AT THE CONDENSER UNIT. IT IS DANGEROUS FOR AN UNQUALIFIED PERSON TO ATTEMPT TO DISCONNECT OR REMOVE ANY PART OF THE AIR-CONDITIONING SYSTEM.

Care must be taken when removing the radiator matrix that the fins of the condenser are not damaged.

Refitting
Refitting is the reverse of the removal procedure.

THERMOSTAT
The thermostat differs from that stated on Page D. 8. in respect of the mounting only.

Removal
Drain sufficient coolant from the system to allow the level to fall below the thermostat.

Disconnect the three hoses from the thermostat housing.

Remove three nuts and washers and detach the housing to gain access to the thermostat.

Refitting
Refitting is the reverse of the removal procedure. Renew all gaskets.

To avoid distortion of the flange faces do not over-tighten the nuts.

RADIATOR COWL
Removal
Disconnect the cables from the twin fan motors.

Remove six setscrews securing the cowl to the radiator and remove the cowl complete with fan motors and mounting brackets.

Refitting
Refitting is the reverse of the removal procedure.

FAN MOTORS
Remove the fan cowl as detailed above.

Remove three nuts and setscrews securing each fan mounting bracket to the cowl and detach the bracket assembly.

Remove four nuts and washers securing each motor and detach the motor units from the brackets.

WATER PUMP
The water pump and mounting remain basically the same as detailed in the 3.8 'E' Type Service Manual with the exception of the pump body and the impeller (See Fig. No. 3) which have been redesigned to give a higher flow rate of coolant.

It is important to note when fitting the impeller to the spindle that the dimension shown in Fig. 2 is obtained when measured with a feeler gauge.

Page DY.s.2

COOLING SYSTEM

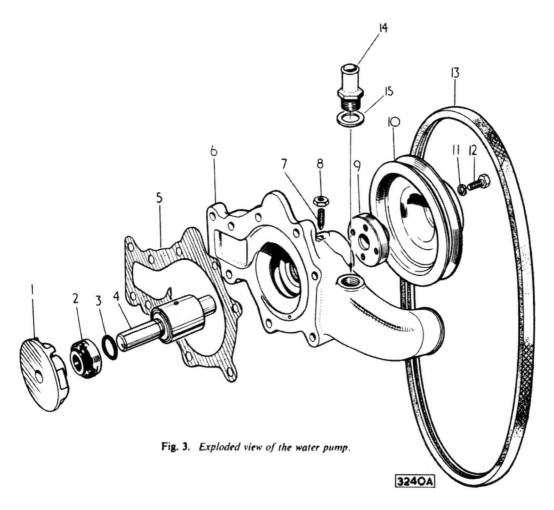

Fig. 3. *Exploded view of the water pump.*

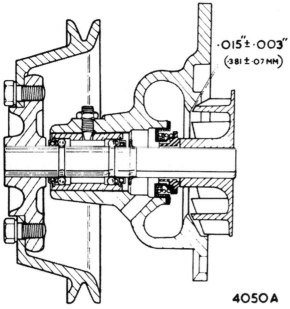

Fig. 2. *Sectioned view of water pump.*

1. Impeller.
2. Seal.
3. Thrower.
4. Spindle and bearing assembly.
5. Gasket.
6. Pump body.
7. Allen-headed lockscrew.
8. Locknut.
9. Pulley carrier.
10. Pulley.
11. Spring washer.
12. Setscrew.
13. Drive belt.
14. Adaptor for heater return pipe.
15. Copper washer.

Page DY.s.3

SECTION E

CLUTCH

DESCRIPTION

A Borg and Beck diaphragm spring clutch is fitted to all cars equipped with manual transmission.

The diaphragm spring is riveted inside the cover pressing with two fulcrum rings interposed between the shoulders of the rivets and the cover pressing. The diaphragm spring also pivots on these two fulcrum rings. Depressing the clutch pedal actuates the release bearing causing a corresponding deflection of the diaphragm spring thus pulling the pressure plate from the driven plate and freeing the clutch.

DATA

Make	Borg and Beck
Model	BB9/412G
Clutch Release Bearing	Graphite
Operation	Hydraulic
Hydraulic Fluid	Castrol/Girling Crimson Clutch/Brake Fluid

Page EY.s.1

CLUTCH

THE CLUTCH UNIT

1. Driven plate.
2. Pressure plate.
3. Rivet.
4. Centre sleeve.
5. Belleville washer.
6. Fulcrum ring.
7. Diaphragm spring.
8. Cover pressing.
9. Release plate.
10. Retainer.
11. Tab washer.
12. Setscrew.
13. Retainer.
14. Release bearing.

Fig. 1. *Exploded view of the diaphragm spring clutch.*

SERVICING

The Borg and Beck diaphragm spring clutch is serviced in the U.K. ONLY by fitting an exchange unit which is available from the Works, Spares Division, Coventry. Individual parts are available from the same source for the repair of this clutch in Overseas Markets where exchange units may not be readily available. IT IS ESSENTIAL when overhauling the diaphragm spring clutch, to rigidly observe the service instructions detailed below and particular attention is drawn to the necessary special tools required.

GENERAL INSTRUCTIONS

To enable the balance of the assembly to be preserved after dismantling, there are corresponding paint marks on the cover plate and driving plate. In addition, there are corresponding reference numbers stamped in the flanges of the cover and driving plate.

When reassembling ensure that the markings coincide, and that, when refitting the clutch to the flywheel, the letter "B" stamped adjacent to one of the dowel holes coincides with the "B" stamped on the edge of the flywheel.

The clutch is balanced in conjunction with the flywheel by means of loose balance pieces which are fitted under the appropriate securing bolt. Each balance piece must be refitted in its original position, the number stamped on the balance weight corresponding to the number stamped on the cover plate. There are three balance weights stamped 1, 2 and 3, the weight stamped 3 being the heaviest.

If it is necessary to fit a replacement unit, clutch units supplied as spares have no reference numbers and therefore must be balanced with the flywheel. The balance weight number should be stamped on the cover adjacent to the weight position. The letter 'B' should be stamped on the cover opposite the 'B' on the flywheel.

If the graphite release bearing ring is badly worn it should be replaced by a complete bearing assembly.

CLUTCH

CLUTCH REMOVAL

In order to remove the clutch, the engine and gearbox must first be removed (see Page B.Y.s.4).

Remove gearbox and clutch housing from engine.

Remove the bolts securing the clutch to the flywheel and withdraw the clutch assembly.

Retain any balance weight fitted.

DISMANTLING

Removing Release Plate

The centrally mounted release plate is held in position by a small centre sleeve which passes through the diaphragm spring and belleville washer into the release plate.

To free the plate, collapse the centre sleeve with a hammer and chisel. To avoid any possible damage whilst carrying out this operation, support the release plate in the locating boss of the special tool which should be held firmly in a vice.

Separating the Pressure Plate from Cover Pressing

Knock back the locking tabs and remove the three setscrews securing the pressure plate to the straps

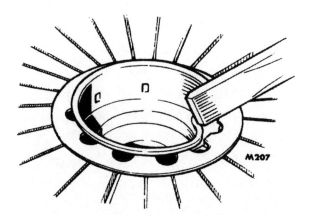

Fig. 3. *Collapsing the centre sleeve with a hammer and chisel.*

riveted to the cover pressing. These straps within the cover pressing must NOT be detached as this is an assembly reduced to its minimum as a spare part.

Dismantling the Cover Assembly

Remove the rivets securing the diaphragm spring and fulcrum rings by machining the shank of the rivets using a spot face cutter.

IT IS ESSENTIAL that the thickness of the cover is not reduced in excess of ·005″ (·127 mm.) at any point. The remaining portions of the rivets may be removed with a standard pin punch.

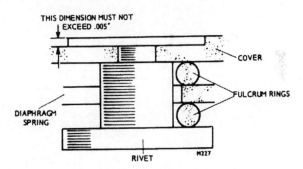

Fig. 4. *Do not reduce the thickness of the cover pressing in excess of ·005″ (·127 m.n.).*

REBUILDING

The Cover Assembly

Prior to rebuilding, check the cover pressing for distortion. Bolt the cover firmly to a flat surface plate and check that measurements taken at various points from the cover flange to the machined land inside the cover pressing do not vary by more than ·007″ (·2 mm.). If the measurement exceeds this figure the cover must be replaced.

Fig. 2. *Clutch and flywheel balancing.*

Page E.Y.s.3

CLUTCH

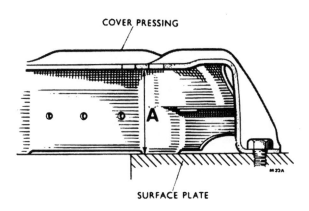

Fig. 5. *The measurement "A" must not vary by more than ·007" (·2 mm.).*

To achieve a satisfactory result when riveting the diaphragm spring into the cover pressing, a special tool must be fabricated to the specifications given in **Fig. 6**.

All parts except the spring can be made from mild steel. Position the fulcrum ring inside the cover pressing so that the location notches in the fulcrum ring engage a depression between two of the larger diameter holes in the cover pressing.

Place the diaphragm spring on the fulcrum ring inside the cover and line the long slots in the spring with the small holes in the cover pressing. Locate a further fulcrum ring on the diaphragm spring so that the location notches are diametrically opposite the location notches in the first ring. Fit new shouldered rivets, ensuring that the shouldered portion of each seats on the machined land inside the cover.

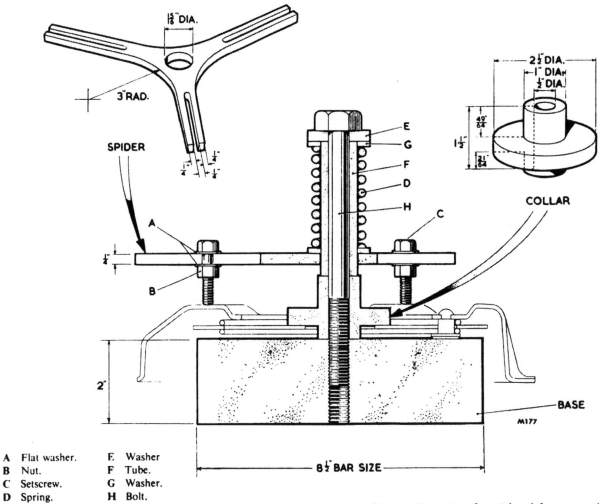

- A Flat washer.
- B Nut.
- C Setscrew.
- D Spring.
- E Washer.
- F Tube.
- G Washer.
- H Bolt.

Fig. 6. *Dimension of special tool for compressing the diaphragm spring when riveting the spring to cover pressing.*

CLUTCH

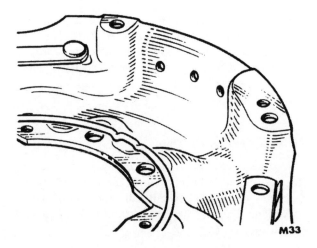

Fig. 7. *Assembly of cover pressing and fulcrum ring.*

Place the base of the special tool on to the rivet heads. Invert the clutch and base plate.

Fit the collar to the large bolt and fit the large bolt complete with spring, spider and collar into the tapped hole in the base. Position the three setscrews on the spider so that they contact the cover pressing. Tighten down the centre bolt until the diaphragm spring becomes flat and the cover pressing is held firmly by the setscrews.

Rivet securely with a hand punch.

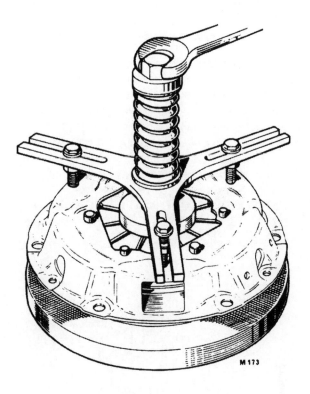

Fig. 9. *Tighten down the large nut so that the diaphragm spring is compressed flat.*

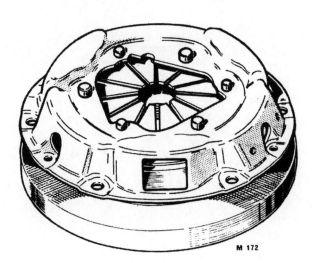

Fig. 8. *Clutch and base plate inverted.*

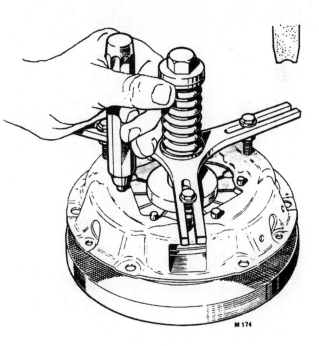

Fig. 10 *Riveting with a hand punch.*

Page E.Y.s.5

CLUTCH

Assembling the Pressure Plate to Cover Pressing

Before assembling the pressure plate to the cover pressing, examine the plate for any signs of wear. Should it have been damaged or have excessive scoring, it is strongly recommended that a new plate is fitted. If, however, renewal of the pressure plate is not possible, grinding of the original unit may be undertaken by a competent machinist, bearing in mind that incorrect grinding of the plate may seriously affect the operation of the clutch. IN NO CIRCUMSTANCES MUST THE PRESSURE PLATE BE GROUND TO A THICKNESS OF LESS THAN 1·070″ (27.178 mm.)

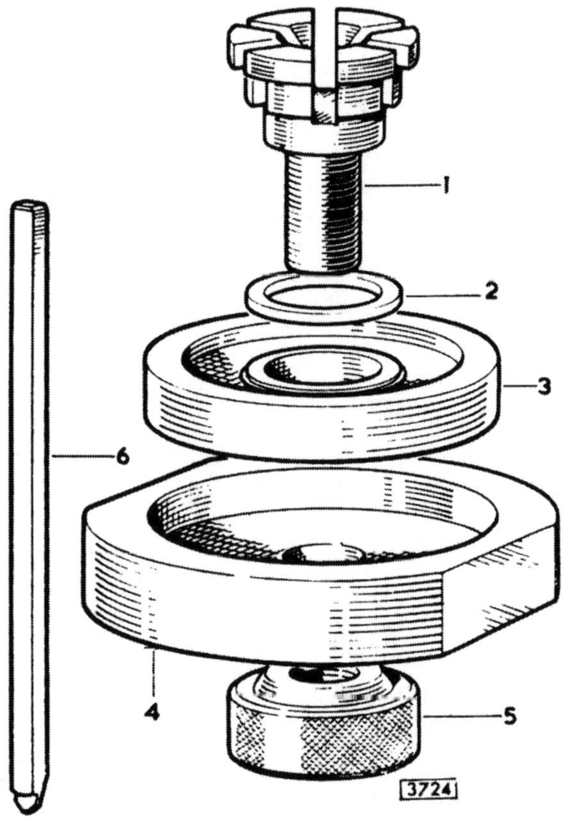

Position the pressure plate inside the cover assembly so that the lugs on the plate engage the slots in the cover pressing. Insert the three setscrews through the straps which are riveted to the cover pressing and lock with the tab washers.

Fitting a New Release Plate

A special tool (Part No. SSC.805) is available from Automotive Products Ltd., Service and Spares Division, Banbury, England, for completion of this operation. Ensure that all parts of the clutch and special tool are clean.

Grip the base of the tool in a vice and place the locating boss into the counterbore of the base plate. Place the release plate, face downwards, into the counterbore of the locating boss.

Apply a little high melting point grease to the tips of the diaphragm spring fingers and position the clutch, pressure plate friction face upwards, on to the release plate.

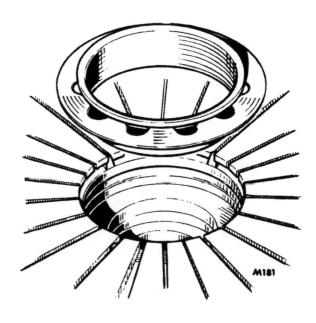

1. Staking guide.
2. Washer.
3. Locating base.
4. Base plate.
5. Knurled nut.
6. Punch.

Fig. 11. *Special Tool (SSC805).*

Fig. 12. *Fitting the sleeve and belleville washer.*

Page E.Y.s.6

CLUTCH

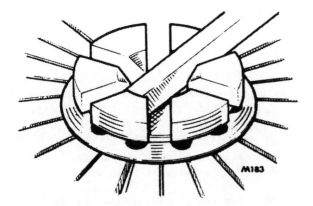

Fig. 13. *Staking the sleeve to the release plate.*

Place the belleville washer, concave surface towards the spring, on to the centre of the diaphragm spring and then push the centre sleeve through the spring into the release plate.

Drop the special washer into the sleeve and insert the staking guide into the centre of the assembly. Fit the knurled nut to the thread on the staking guide, tighten down until the whole assembly is solid. Using the special punch, stake the centre sleeve in six places into the groove in the release plate.

REFITTING

Place the driven plate on the flywheel, taking care that the larger part of the splined hub faces the gear-

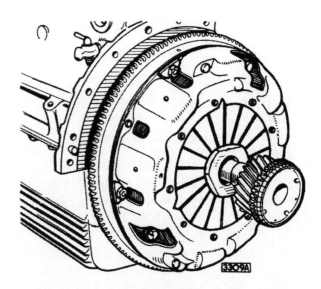

Fig. 14. *Centralising the driven plate on the flywheel by means of a dummy plate.*

box. Centralize the plate on the flywheel by means of the dummy shaft (a constant pinion shaft may be used for this purpose). Secure the cover assembly with the six setscrews and spring washers, tightening the screws a turn at a time by diagonal selection. Ensure that the "B" stamped adjacent to one of the dowel holes coincides with the "B" stamped on the periphery of the flywheel.

Page E.Y.s.7

CLUTCH

CONDITION OF CLUTCH FACINGS

The possibility of further use of the friction facings of the clutch is sometimes raised, because they have a polished appearance after considerable service. It is natural to assume that a rough surface will give higher frictional value against slipping, but this not correct. Since the introduction of non-metallic facings of the moulded asbestos type, in service a polished surface is a common experience, but it must not be confused with a glazed surface which is sometimes encountered due to the conditions discussed below.

The ideal smooth or polished condition will provide a normal contact, but a glazed surface may be due to a film or a condition introduced, which entirely alters the frictional value of the facings. These two conditions might be simply illustrated by the comparison between a polished wood and a varnished surface. In the former the contact is still made by the original material whereas, in the latter instance, a film of dried varnish is interposed between the contact surfaces.

The following notes are issued with a view to giving useful information on this subject:—

(a) After the clutch has been in use for some little time under perfect conditions (that is, with the clutch facings working on true and polished or ground surfaces of correct material, without the presence of oil, and with only that amount of slip which the clutch provides for under normal conditions) then the surface of the facings assumes a high polish, through which the grain of the material can be clearly seen. This polished facing is of mid-brown colour and is then in a perfect condition.

(b) Should oil in small quantities gain access to the clutch in such a manner as to come into contact with the facings, it will burn off due to the heat generated by slip which occurs under normal starting conditions. The burning off of the small amount of lubricant has the effect of gradually darkening the facings, but provided the polish on the facings remains such that the grain of the material can be clearly distinguished, it has very little effect on clutch performance.

(c) Should increased quantities of oil or grease obtain access to the facing, one or two conditions, or a combination of the two, may arise, depending upon the nature of oil, etc.

   (i) The oil may burn off and leave on the surface a carbon deposit which assumes a high glaze and causes slip. This is a very definite, though very thin deposit, and in general it hides the grain of the material.

   (ii) The oil may partially burn and leave a resinous deposit on the facings, which frequently produces a fierce clutch, and may also cause a "spinning" clutch due to tendency of the facings to adhere to the flywheel or pressure plate face.

   (iii) There may be a combination of (i) and (ii) conditions which is likely to produce a judder during clutch engagement.

(d) Still greater quantities of oil produces a black soaked appearance to the facings, and the effect may be slip, fierceness, or judder in engagement, etc., according to the conditions. If the conditions under (c) or (d) are experienced, the clutch driven plate should be replaced by one fitted with new facings, the cause of the presence of oil removed and the clutch and flywheel face thoroughly cleaned.

CLUTCH

FAULT FINDING

SYMPTOM	CAUSE	REMEDY
Drag or Spin	(a) Oil or grease on the driven plate facings.	Fit new facings or replace plate.
	(b) Misalignment between the engine and splined clutch shaft.	Check over and correct the alignment.
	(c) Air in clutch system.	"Bleed" system. Check all unions and pipes.
	(d) Bad external leak between the clutch master cylinder and the slave cylinder.	Renew pipe and unions.
	(e) Warped or damaged pressure plate or clutch cover.	Renew defective part.
	(f) Driven plate hub binding on splined shaft.	Clean up splines and lubricate with small quantity of high melting point grease.
	(g) Distorted driven plate due to the weight of the gearbox being allowed to hang on clutch plate during assembly.	Fit new driven plate assembly using a jack to take overhanging weight of the gearbox.
	(h) Broken facings of driven plate.	Fit new facings, or replace plate.
	(i) Dirt or foreign matter in the clutch.	Dismantle clutch from flywheel and clean the unit; see that all working parts are free.
		Caution: Never use petrol or paraffin for cleaning out clutch.
Fierceness or Snatch	(a) Oil or grease on driven plate facings.	Fit new facings and ensure isolation of clutch from possible ingress of oil or grease.
	(b) Misalignment.	Check over and correct alignment.
	(c) Worn out driven plate facings.	Fit new facings or replace plate.
Slip	(a) Oil or grease on driven plate facings.	Fit new facings and eliminate cause.
	(b) Seized piston in clutch slave cylinder.	Renew parts as necessary.
	(c) Master cylinder piston sticking.	Free off piston.
Judder	(a) Oil, grease or foreign matter on driven plate facings.	Fit new facings or driven plate.
	(b) Misalignment.	Check over and correct alignment.
	(c) Bent splined shaft or buckled driven plate.	Fit new shaft or driven plate assembly.

Page E.Y.s.9

CLUTCH

FAULT FINDING (continued)

SYMPTOM	CAUSE	REMEDY
Rattle	(a) Damaged driven plate. (b) Excessive backlash in transmission. (c) Wear in transmission bearings. (d) Bent or worn splined shaft. (e) Release bearing loose on throw out fork.	Fit new parts as necessary.
Tick or Knock	Hub splines worn due to misalignment.	Check and correct alignment then fit new driven plate.
Fracture of Driven Plate	(a) Misalignment distorts the plate and causes it to break or tear round the hub or at segment necks.	Check and correct alignment and fit new driven plate.
	(b) If the gearbox during assembly be allowed to hang with the shaft in the hub, the driven plate may be distorted, leading to drag, metal fatigue and breakage.	Fit new driven plate assembly and ensure satisfactory re-assembly.
Abnormal Facing Wear	Usually produced by over-loading and by excessive clutch slip when starting.	In the hands of the operator.

SECTION F
GEARBOX

DESCRIPTION

The gearbox is of the four speed type with baulk-ring synchromesh on all forward gears. With the exception of reverse, the detents for the gears are incorporated in the synchro assemblies, the three synchro balls engaging with grooves in the operating sleeve. The detent for reverse gear is a spring loaded ball which engages on a groove in the selector rod.

Two interlock balls and a pin located at the front of selector rods prevent the engagement of two gears at the same time.

The gears are pressure fed at approximately 5 lb. per sq. in. (0·35 kg/cm.^{2}) from a pump driven from the rear of the mainshaft.

DATA

Identification number	Open 2 seater and F.H. Coupe 2+2			KE 101 onwards KJS 101 onwards
	Ratios			
1st gear	2.933:1	3rd gear		1.389:1
2nd gear	1.905:1	4th gear		1.000:1
	Reverse		3.378:1	

1st gear—end float on mainshaft ·005" to ·007" (·13—·18 mm.)
2nd gear—end float on mainshaft ·005" to ·008" (·13—·20 mm.)
3rd gear—end float on mainshaft ·005" to ·008" (·13—·20 mm.)
Countershaft gear unit end float ·004" to ·006" (·10—·15 mm.)

RECOMMENDED LUBRICANTS

Mobilube GX 90	Castrol Hypoy	Spirax 90 E.P.	Esso Gear Oil GP 90/140	Gear Oil SAE 90 E.P.	Hypoid 90	Multigear Lubricant EP90

Page F Y.s.1

GEARBOX

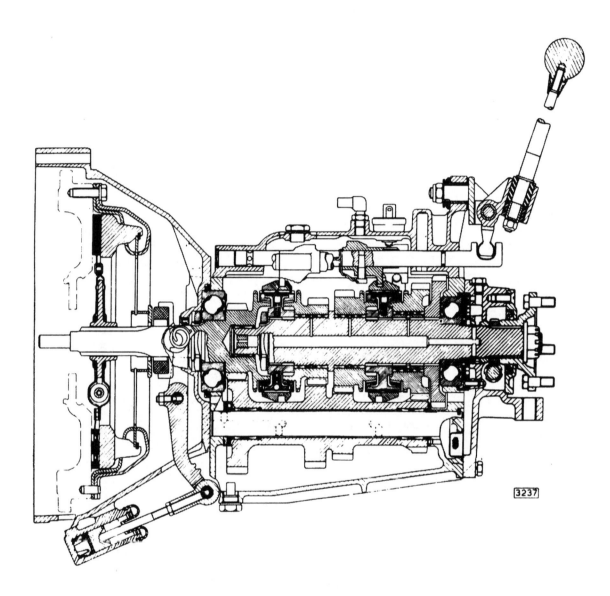

Fig. 1. *Longitudinal section of clutch and gearbox.*

GEARBOX

GEARBOX DISMANTLING

REMOVAL OF CLUTCH HOUSING

Detach the springs and remove the carbon thrust bearing.

Unscrew the two nuts and remove the clutch slave cylinder.

Remove the allen screw, push out the fulcrum pin and detach the clutch fork.

Tap back the locking tabs and break the locking wire and remove the eight setscrews.

Detach the clutch housing.

REMOVAL OF TOP COVER

Place the gear lever in neutral.

Remove the eight setscrews and two nuts and lift off the lid.

REMOVAL OF REAR EXTENSION

Engage first and reverse gears to lock the unit.

Remove the split pin and unscrew the flange nut.

Withdraw the flange.

Remove the four setscrews and detach the rear cover.

Remove the speedometer pinion and bush assembly after unscrewing the retaining bolt.

Withdraw the speedometer driving gear from the mainshaft.

Remove the seven setscrews and withdraw the extension.

Collect the distance piece and oil pump driving pin.

REMOVAL OF OIL PUMP

From the inside face of the rear extension break the staking and remove the three countersunk screws securing the oil pump gear housing. Withdraw the housing by entering two of the securing screws into the tapped holes in the housing; screw in the two screws evenly until the housing is free.

Mark the gears with marking ink so that they can be replaced the same way up in the housing.

REMOVAL OF COUNTERSHAFT

Remove the fibre plug from the front end of the countershaft.

Drive out the countershaft from the front of the casing.

Important:

Ensure that the rear washer (pegged to casing) drops down in a clockwise direction looking from the rear to avoid trapping the washer with the reverse gear when driving the mainshaft forward (see Fig. 2). This is effected by rocking the gearbox casing and moving the reverse lever backwards and forwards, or by pushing the washer down with a piece of wire bent at right angles.

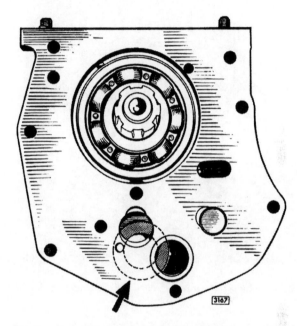

Fig. 2. Ensure that the rear washer (indicated by arrow) drops down in a clockwise direction.

REMOVAL OF CONSTANT PINION SHAFT

Rotate the constant pinion shaft until the cutaway portions of the driving gear are facing the top and bottom of the casing otherwise the gear will foul the countershaft.

With the aid of two levers ease the constant pinion shaft and front bearing assembly forward until it can be withdrawn (see Fig. 3).

DISMANTLING THE CONSTANT PINION SHAFT

Remove the roller bearing from inside the constant pinion shaft. On early cars, a spacer was also fitted along with the needle roller bearing.

Page F Y.s.3

GEARBOX

Tap back the tab washer and remove the large nut, tab washer and oil thrower.

Tap the shaft sharply against a metal plate to dislodge the bearing.

REMOVAL OF MAINSHAFT

Rotate the mainshaft until one of the cutaway portions in 3rd/Top synchro hub is in line with the countershaft (see Fig. 4), otherwise the hub will foul the constant gear or the countershaft.

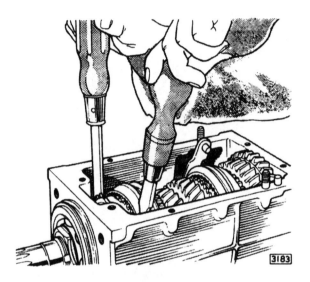

Fig. 3. *With the aid of two levers ease the constant pinion shaft forward.*

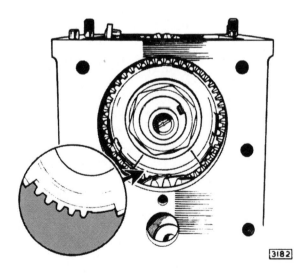

Fig. 4. *Rotate the mainshaft until one of the cutaway portions in 3rd/Top synchro hub is in line with the countershaft.*

Tap or press the mainshaft through the rear bearing ensuring that the reverse gear is kept tight against the first gear (see Fig. 5).

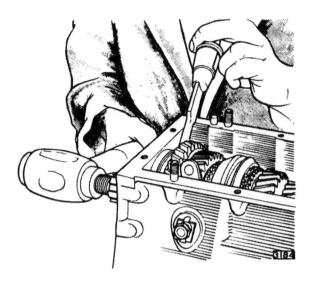

Fig. 5. *Tapping the mainshaft through the rear bearing.*

Remove the rear bearing from the casing and fit a hose clip to the mainshaft to prevent the reverse gear from sliding off (see Fig. 6).

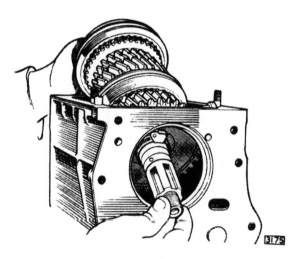

Fig. 6. *Removal of the mainshaft. Note the hose clip fitted to the mainshaft to retain the reverse gear.*

Slacken the reverse lever bolt until the lever can be moved to the rear.

Lift out the mainshaft forward and upward.

Lift out the countershaft gear unit and collect the needle bearings and retaining rings.

Withdraw the reverse idler shaft and lift out the gear.

Page FY.s.4

GEARBOX

DISMANTLING THE MAINSHAFT

Note: The needle rollers are graded on diameter and must be kept in sets for their respective positions.

Remove the hose clip.

Withdraw the reverse gear from the mainshaft.

Withdraw the 1st gear and collect the 120 needle rollers, spacer and sleeve.

Withdraw the 1st/2nd synchro assembly and collect the two loose synchro-rings.

Withdraw the 2nd speed gear and collect the 106 needle rollers leaving the spacer on the mainshaft.

Tap back the tab washer and remove the large nut retaining the 3rd/Top synchro assembly to the mainshaft.

Withdraw the 3rd/Top synchro assembly from the mainshaft and collect the two loose synchro-rings.

Withdraw the 3rd speed gear and collect the 106 needle rollers and spacer.

DISMANTLING THE SYNCHRO ASSEMBLY

Completely surround the synchro assembly with a cloth and push out the synchro hub from the operating sleeve. Collect the synchro balls and springs, and the thrust members, plungers and springs.

DISMANTLING TOP COVER

Unscrew the self-locking nut and remove the double coil spring, washer, flat washer and fibre washer securing the gear lever to the top cover.

Withdraw the gear lever and collect the remaining fibre washer.

Remove the locking wire and unscrew the selector rod retaining screws.

Withdraw the 3rd/Top selector rods and collect the selector, spacing tube and interlock ball. Note the loose interlock pin at the front end of the 1st/2nd selector rod.

Withdraw the reverse selector rod and collect the reverse fork, stop spring and detent plunger.

Withdraw the 1st/2nd selector rod and collect the fork and short spacer tube.

GEARBOX RE-ASSEMBLING

ASSEMBLING THE SYNCHRO ASSEMBLIES

The assembly procedure for the 1st/2nd and 3rd/Top synchro assemblies is the same.

Note: Although the 3rd/Top and 1st/2nd synchro hubs are similar in appearance they are not identical and to distinguish them a groove is machined on the edge of the 3rd/Top synchro hub (see Fig. 7).

Assemble the synchro hub to the operating sleeve with;

(i) The wide boss of the hub on the opposite side to the wide chamfer end of the sleeve (see Fig. 8).

(ii) The three ball and springs in line with the teeth having three detent grooves (see Fig. 10).

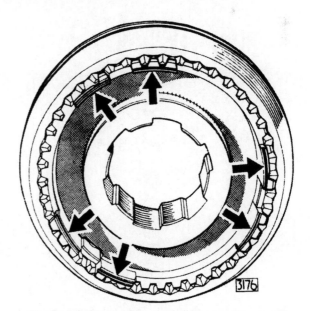

Fig. 7. *Identification grooves—3rd/Top, synchro assembly.*

Page FY.s.5

GEARBOX

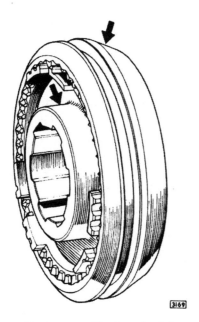

Fig. 8. *Assembly of synchro hub.*

Pack up the synchro hub so that holes for the ball and springs are exactly level with the top of the operating sleeve (see Fig. 11).

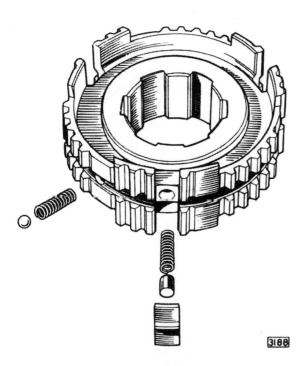

Fig. 9. *Showing the relative positions of the detent ball, plunger and thrust member.*

Fit the three springs, plungers and thrust members to their correct positions with grease; press down the thrust members as far as possible. Fit the three springs and balls to the remaining holes with grease.

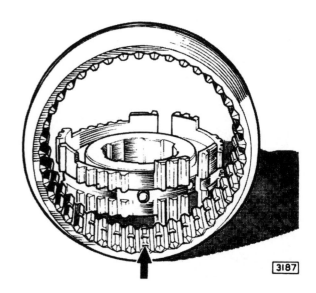

Fig. 10. *Fitting the synchro hub in the sleeve.*

Compress the springs with a large hose clip or a piston ring clamp as shown in Fig. 12 and carefully lift off the synchro assembly from the packing piece.

Depress the hub slightly and push down the thrust members with a screwdriver until they engage the neutral groove in the operating sleeve (see Fig. 13).

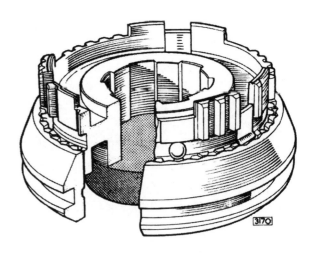

Fig. 11. *Fitting the springs, plungers and thrust members.*

Page FY.s.6

GEARBOX

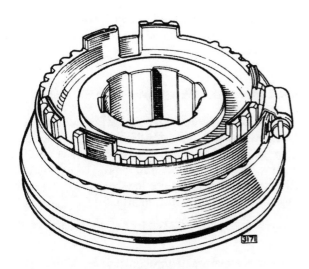

Fig. 12. *Compressing the springs.*

Finally tap the hub down until the balls can be heard and felt to engage the neutral groove (see Fig. 14).

Fig. 13. *Pushing down the thrust members.*

ASSEMBLING THE CLUSTER GEAR

Fit one retaining ring in the front end of the cluster gear. Locate the 29 needle roller bearings with grease and fit the inner thrust washer ensuring that the peg on the washer locates in a groove machined on the face of the cluster gear.

Fit a retaining ring, 29 needle roller bearings and a second retaining ring to the rear end of the cluster gear.

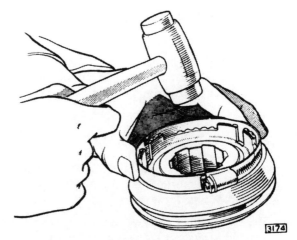

Fig. 14. *Tapping the hub into position.*

CHECKING THE CLUSTER GEAR END FLOAT

Fit the reverse idler gear, lever and idler shaft.

Fit the pegged rear washer to its boss on the casing with grease.

Locate the outer thrust washer to the front of the cluster gear with grease; lower the cluster gear into position carefully. Insert a dummy shaft and check the clearance between the rear thrust washer and the cluster gear. The clearance should be ·004″–·006″ (·10 mm.–·15 mm.) and is adjusted by means of the outer thrust washers. This is available in the following selective thicknesses:—

Part Number	Thickness
C.1862/3	·152″ (3·86 mm.)
C.1862	·156″ (3·96 mm.)
C.1862/1	·159″ (4·04 mm.)
C.1862/2	·162″ (4·11 mm.)
C.1862/4	·164″ (4·17 mm.)

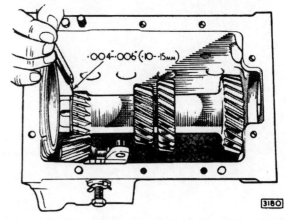

Fig. 15. *Checking the clearance between the rear thrust washer and the countershaft cluster gear.*

Page FY.s.7

GEARBOX

ASSEMBLING THE CONSTANT PINION SHAFT

Assembling is the reverse of the dismantling procedure but care must be taken to ensure that the bearing is seated squarely on the constant pinion shaft.

ASSEMBLING THE MAINSHAFT

The re-assembly of the mainshaft is the reverse of the dismantling instructions but the following instructions should be noted.

(i) The end float of the gears on the mainshaft is given in "Data" at the beginning of this section and if found to be excessive the end float can only be restored by the fitting of new parts.

(ii) The needle rollers which support the gears on the mainshaft are graded on diameter and rollers of one grade only must be used for an individual gear. The grades are identified by /1, /2, and /3 after the part number.

(iii) The "E" Type constant pinion, countershaft and 3rd speed gear have a groove machined around the periphery of the gear, see Fig. 16. This is to distinguish the "E" Type gears from those fitted to the same type of gearbox on other models which have different ratios.

Fig. 16. *Showing the groove which identifies the 'E' type gears.*

(iv) Fit a hose clip to prevent the reverse gear from sliding off when assembling the mainshaft to the casing.

ASSEMBLING THE GEARS TO THE CASING

Withdraw the dummy shaft from the cluster gear and, at the same time, substitute a thin rod keeping both the dummy shaft and the rod in contact until the dummy shaft is clear of the casing. The thin rod allows the cluster gear to be lowered sufficiently in the casing for insertion of the mainshaft.

Fit a new paper gasket to the front face of the casing.

Enter the mainshaft through the top of the casing and pass the rear of shaft through the bearing hole.

Enter the constant pinion shaft at the front of the casing with the cutaway portions of the tooth driving member at the top and bottom.

Tap the constant pinion shaft into position and enter the front end of the mainshaft into the spigot bearing of the constant pinion shaft.

Hold the constant pinion shaft in position and with a hollow drift tap the rear bearing into position.

Withdraw the thin rod from the front bore of the cluster gear approximately half way and lever the cluster gear upwards, rotating the mainshaft and constant pinion shaft gently until the cluster gear meshes. Carefully insert the countershaft from the rear and withdraw the rod. Fit the key locating the countershaft in the casing.

REFITTING REAR EXTENSION

Refit the gears to the oil pump the same way as removed, having previously coated the gears and the inside of the pump body with oil. Secure the pump housing with the three countersunk screws and retain by staking.

Fit a new paper gasket to the rear face of the casing.

Fit the distance piece and driving pin to the oil pump in the rear extension.

Offer up the rear extension and secure with the seven screws.

Fit the speedometer driving gear to the mainshaft.

Page FY.s.8

GEARBOX

Fit the speedometer driven gear and bush with the hole in the bush in line with the hole in the casing and secure with the retaining bolt.

Fit a new gasket to the rear cover face.

Fit a new oil seal to the rear cover with the lip facing forward.

Fit the rear cover to the extension noting that the setscrew holes are offset.

Fit the four bolts to the companion flange, slide on the flange and secure with flat washer with split pin.

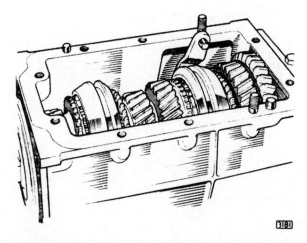

Fig. 17. *Re-assembled gearbox prior to refitting of top cover.*

RE-ASSEMBLING THE TOP COVER
(see Fig. 20).

Re-assembly of the top cover is the reverse of the dismantling instructions. When assembling the selector rods do not omit to fit the interlock balls and pin.

Renew the "O" rings on the selector rods.

To adjust the reverse plunger fit the plunger and spring.

Fit the ball and spring and start the screw and locknut; press the plunger inwards as far as possible and tighten the screw to lock the plunger.

Slowly slacken the screw until the plunger is released and the ball engages with the circular groove in the plunger. Hold the screw and tighten the locknut.

FITTING THE TOP COVER

Fit a new paper gasket.

Ensure that the gearbox and the top cover are in the neutral position.

Ensure that the reverse idler gear is out of mesh with the reverse gear on the mainshaft by pushing the lever rearwards.

Engage the selector forks with the grooves in the synchro assemblies.

Secure the top cover with the nuts and bolts noting that they are of different lengths.

REFITTING THE CLUTCH HOUSING

Refitting the clutch housing is the reverse of the removal procedure.

Fit a new oil seal to the clutch housing with the lip of the seal facing the gearbox. The oil seal has a metal flange and should be pressed in fully.

The two clutch housing securing bolts adjacent to the clutch fork trunnions are secured with locking wire; the remainder are secured with tab washers.

Note: After refitting the gearbox, run the car in top gear as soon as possible to attain the necessary mainshaft speed to prime the oil pump.

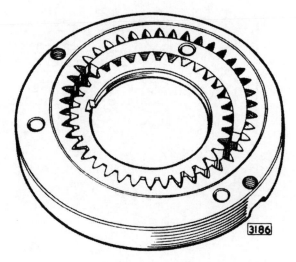

Fig. 18. *The oil pump.*

GEARBOX

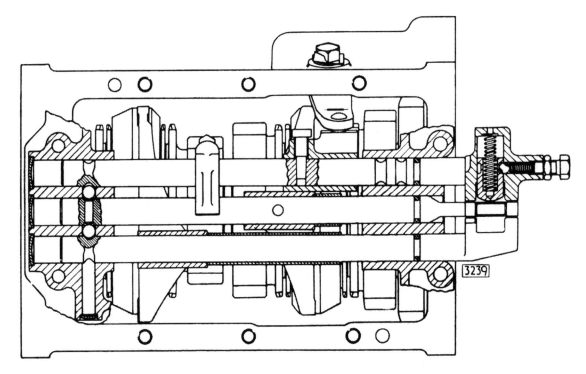

Fig. 19. *Plan view of gearbox showing selector arrangement.*

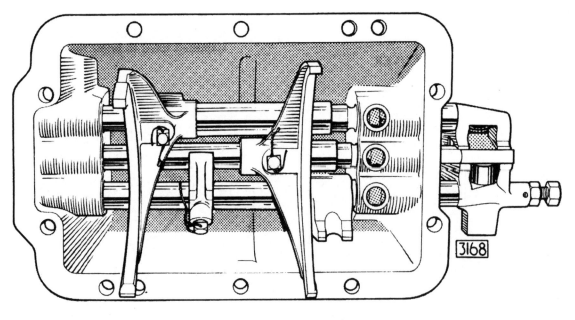

Fig. 20. *View of the underside of the top cover.*

SECTION FF
AUTOMATIC TRANSMISSION

GENERAL DATA

Maximum ratio of torque converter	2·00:1
1st Gear reduction	2·40:1
2nd Gear reduction	1·46:1
3rd Gear reduction	1·00:1
Reverse Gear reduction	2·00:1

AUTOMATIC SHIFT SPEEDS

2·88:1 Final Drive

Selector Position	Throttle Position	Upshifts 1 – 2	Upshifts 2 – 3	Downshifts 3 – 2	Downshifts 3 – 1	2 – 1
				M.P.H.		
	Minimum	7 – 9	12 – 15	8 – 14	—	4 – 8
D1	Full	38 – 44	66 – 71	23 – 37	—	—
	Kickdown	52 – 56	81 – 89	73 – 81	20 – 24	20 – 24
	Minimum	—	12 – 15	8 – 14	—	—
D2	Full	—	66 – 71	23 – 37	—	—
	Kickdown	—	81 – 89	73 – 81	—	—
L	Zero	—	—	60	—	12 – 20
				K.P.H.		
	Minimum	11 – 14	19 – 24	13 – 23	—	6 – 13
D1	Full	61 – 71	106 – 114	37 – 60	—	—
	Kickdown	83 – 90	130 – 143	118 – 130	32 – 39	32 – 39
	Minimum	—	19 – 24	13 – 23	—	—
D2	Full	—	106 – 114	37 – 60	—	—
	Kickdown	—	130 – 143	118 – 130	—	—
L	Zero	—	—	96	—	19 – 32

AUTOMATIC TRANSMISSION

AUTOMATIC SHIFT SPEEDS (Continued)

185 × 15 SP 41 HR Tyres – 3·31:1 Final Drive

Selector Position	Throttle Position	Upshifts 1–2	Upshifts 2–3	Downshifts 3–2	Downshifts 3–1	Downshifts 2–1
				M.P.H.		
D1	Minimum	6 – 8	11 – 13	7 – 13	—	3 – 7
	Full	33 – 40	58 – 62	19 – 33	—	—
	Kickdown	45 – 49	70 – 78	63 – 71	17 – 21	17 – 21
D2	Minimum	—	11 – 13	7 – 13	—	—
	Full	—	58 – 62	19 – 33	—	—
	Kickdown	—	70 – 78	63 – 71	—	—
L	Zero	—	—	60	—	10 – 18
				K.P.H.		
D1	Minimum	9 – 13	18 – 21	11 – 21	—	5 – 11
	Full	53 – 64	93 – 100	31 – 53	—	—
	Kickdown	73 – 80	113 – 126	101 – 114	28 – 34	28 – 34
D2	Minimum	—	18 – 21	11 – 21	—	—
	Full	—	93 – 100	31 – 53	—	—
	Kickdown	—	113 – 126	101 – 114	—	—
L	Zero	—	—	96	—	16 – 29

Note: Shift points are approximate and not absolute values. Reasonable deviations from the above values are permissible.

Page FFY.s.2

AUTOMATIC TRANSMISSION

TIGHTENING TORQUE FIGURES

	lb. ft.	kgm.
Front pump to transmission case bolts	17 – 22	2·35 – 3·04
Front servo to transmission case bolts	30 – 35	4·15 – 4·70
Rear servo to transmission case bolts	40 – 45	5·53 – 6·22
Centre support to transmission case bolts	20 – 25	2·76 – 3·46
Upper valve body to lower valve body bolts	4 – 6	0·55 – 0·83
Control valve body to transmission case bolts	8 – 10	1·11 – 1·38
Pressure regulator assembly to transmission case bolts	17 – 22	2·35 – 3·04
Extension assembly to transmission case bolts	28 – 33	3·87 – 4·56
Oil pan to transmission case bolts	10 – 13	1·38 – 1·80
Case assembly—gauge hole plug	10 – 15	1·38 – 2·07
Oil pan drain plug	25 – 30	3·46 – 4·15
Rear band adjusting screw lock nut	35 – 40	4·70 – 5·53
Front band adjusting screw lock nut	20 – 25	2·76 – 3·46
Detent lever attaching nut	35 – 40	4·70 – 5·53
Companion flange nut	90 – 120	12·44 – 16·58
Bearing retainer to extension housing bolts	28 – 33	3·87 – 4·56

	lb. in.	kgm.
Front pump cover attaching screws	25 – 35	0·29 – 0·40
Rear pump cover attaching screws ¼" (6·30 mm.)	50 – 60	0·58 – 0·69
Rear pump attaching screws Nos. 10–24	20 – 30	0·24 – 0·35
Governor inspection cover attaching screws	50 – 60	0·58 – 0·69
Governor valve body to counterweight screws	50 – 60	0·58 – 0·69
Governor valve body cover screws	20 – 30	0·24 – 0·35
Pressure regulator cover attaching screws	20 – 30	0·24 – 0·35
Control valve body screws	20 – 30	0·24 – 0·35
Control valve body plug	10 – 14	0·11 – 0·16
Control valve lower body plug	7 – 15	0·08 – 0·17

Page FFY.s.3

AUTOMATIC TRANSMISSION

SPECIAL SERVICE TOOLS

Service tools are not available from Borg-Warner Limited. Distributors and Dealers should obtain the following tools illustrated in this manual from Messrs. V. L. Churchill & Co. Ltd., London Road, Daventry, Northants.

Description

Mainshaft end play gauge (CB.W.33).

Rear clutch spring compressor (C.B.W. 37A used with W.G.37).

Hydraulic pressure test gauge equipment (C.B.W. 1A used with adaptor C.B.W.1A-5A).

Spring beam torque wrench (used in conjunction with the following adaptor) (C.B.W.547A-50).

Rear band adjusting adaptor (C.B.W.547A-50-2).

Torque screwdriver (used in conjunction with the following adaptor) (C.B.W.548).

Front band adjusting adaptor (C.B.W.548-2).

Front band setting gauge (C.B.W.34).

Circlip pliers (used with "J" points) (7066).

Bench cradle (C.W.G.35).

Rear clutch piston assembly sleeve (C.W.G.41).

Front clutch piston assembly sleeve (C.W.G.42).

Rear pump discharge tube remover (C.W.G.45).

AUTOMATIC TRANSMISSION

DESCRIPTION AND OPERATION

The Model 8 automatic transmission incorporates a fluid torque converter in place of the usual flywheel and clutch. The converter is coupled to a hydraulically operated planetary gearbox which provides three forward ratios and reverse. All forward ratios are automatically engaged in accordance with accelerator position and car speed.

Overriding control by the driver is available upon demand for engine braking by manual selection of "L".

TORQUE CONVERTER

The feature of using a hydraulic converter in conjunction with a three-speed automatic gearbox provides a means of obtaining a smooth application of engine power to the driving wheels and additional engine torque multiplication to the 1st and 2nd gears of the gearbox.

The converter also provides extreme low-speed flexibility when the gearbox is in 3rd gear and, due to the ability of multiplying engine torque, it provides good acceleration from very low road speed without having to resort to a down-shift in the gearbox.

Torque multiplication from the converter is infinitely variable between the ratios of 2:1 and 1:1. The speed range, during which the torque multiplication can be achieved, is also variable, depending upon the accelerator position.

The hydraulic torque converter for use in conjunction with the automatic gearbox has a mean fluid circuit diameter of 11" (27·9 cm.).

It is of the single-phase, three-element type, comprising an impeller connected to the engine crankshaft, a turbine connected to the input shaft of the gearbox, and a stator mounted on a sprag-type one-way clutch supported on a fixed hub projecting from the gearbox case.

THE GEAR SET

The planetary gear set consists of two sun gears, two sets of pinions, a pinion carrier, and a ring gear. Helical, involute tooth forms are used throughout. Power enters the gear set via the sun gears. In all forward gears power enters through the forward sun gear; in reverse, power enters through the reverse sun gear. Power leaves the gear set by the ring gear. The pinions are used to transmit power from the sun gears to the ring gear. In reverse a single set of pinions is used, which causes the ring gear to rotate in the opposite direction to the sun gear. In forward gears a double set of pinions is used to cause the ring gear to rotate in the same direction as the sun gear. The carrier locates the pinions in their correct positions relative to the sun gears and the ring gear (and also forms a reaction member for certain conditions). The various mechanical ratios of the gear set are obtained by the engagement of hydraulically operated multi-disc clutches and brake bands.

Page FFY.s.4

AUTOMATIC TRANSMISSION

CLUTCHES

Multi-disc clutches operated by hydraulic pistons connect the converter to the gear set. In all forward gears the front clutch connects the converter to the forward sun gear; for reverse the rear clutch connects the converter to the reverse sun gear.

BANDS

Brake bands, operated by hydraulic servos, hold elements of the gear set stationary to effect an output speed and a torque increase. In Lockup the rear band holds the planet carrier stationary and provides the 1st gear ratio of 2·40:1 and, in reverse, a ratio of 2·00:1. The front band holds the reverse sun gear stationary to provide the 2nd gear ratio of 1·46:1.

ONE-WAY CLUTCH

In D1, a one-way clutch is used in place of the rear band to prevent anti-clockwise rotation of the planet carrier, thus providing the 1st gear ratio of 2·40:1. This one-way clutch, allowing the gear set to freewheel in 1st gear, provides smooth ratio changes from 1st to 2nd, and vice-versa.

Selector Position	Ratio		Applied	Driving		Held
L	Lock-up	1st	Front Clutch Rear Band Sprag Clutch	Forward	Sun	Planet Carrier
D1	Drive One	1st	Front Clutch Sprag Clutch	Forward	Sun	Planet Carrier
L D1 D2	Lock-up Drive One Drive Two	2nd 2nd	Front Clutch Front Band	Forward	Sun	Reverse Sun
D1 D2	Drive One Drive Two	3rd	Front Clutch Rear Clutch	Forward Secondary	Sun Sun	
R	Reverse	Reverse	Rear Clutch Rear Band	Reverse	Sun	Planet Carrier

MECHANICAL POWER FLOW

First Gear (Lockup selected)

The front clutch is applied, connecting the converter to the forward sun gear. The rear band is applied, holding the planet carrier stationary, the gear set providing the reduction of 2·40:1. The reverse sun gear rotates freely in the opposite direction to the forward sun gear.

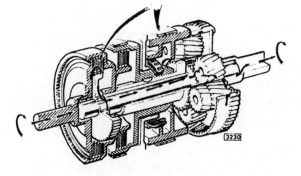

Fig. 1. *Mechanical power flow—1st gear (L) selected.*

AUTOMATIC TRANSMISSION

First Gear (Drive 1 selected)

The front clutch is applied, connecting the converter to the forward sun gear. The one-way clutch is in operation, preventing the planet carrier from rotating anti-clockwise; the gear set provides the reduction of 2·40:1. When the vehicle is coasting the one-way clutch over-runs and the gear set freewheels.

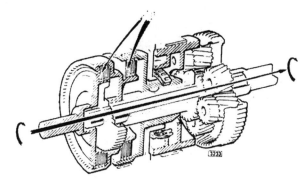

Fig. 4. *Mechanical power flow—3rd gear (D) selected.*

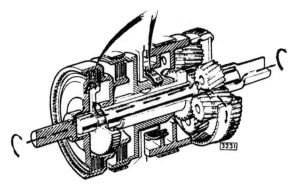

Fig. 2. *Mechanical power flow—1st gear (D) selected.*

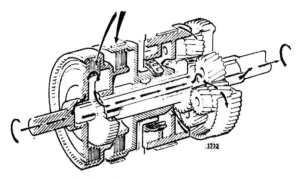

Fig. 3. *Mechanical power flow—2nd gear (L or D2) selected.*

Second Gear (Lockup or Drive 2 selected)

Again the front clutch is applied, connecting the converter to the forward sun gear. The front band is applied, holding the reverse sun gear stationary; the gear set provides the reduction of 1·46:1.

Third Gear

Again the front clutch is applied, connecting the converter to the forward sun gear. The rear clutch is applied, connecting the converter also to the reverse sun gear; thus both sun gears are locked together and the gear set rotates as a unit, providing a ratio of 1:1.

Neutral and Park

In neutral the front and rear clutches are off, and no power is transmitted from the converter to the gear set. The front and rear bands are also released. In "P" the Front Servo Apply and Release and Rear Servo circuits are pressurised while the engine is running, so that the rear band is applied.

Reverse Gear

The rear clutch is applied, connecting the converter to the reverse sun gear. The rear band is applied, holding the planet carrier stationary, the gear set providing the reduction of 2·00:1 in the reverse direction.

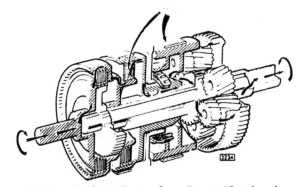

Fig. 5. *Mechanical power flow—Reverse (R) selected.*

THE HYDRAULIC SYSTEM

The hydraulic system contains a front and rear pump, both of the internal/external gear pattern, picking up fluid from the oil pan through a common strainer. Shift control is provided by a centrifugally operated hydraulic governor on the transmission output shaft. This governor works in conjunction with valves in the valve body assembly located in the base of the transmission. These valves regulate fluid pressure and direct it to appropriate transmission components.

The Front Pump

The front pump, driven by the converter impeller,

Page FFY.s.6

AUTOMATIC TRANSMISSION

is in operation whenever the engine is running. This pump, through the primary and secondary regulator valves supplies the hydraulic requirements of the transmission with the engine running when the vehicle is stationary, as well as at low vehicle speeds before the rear pump becomes effective.

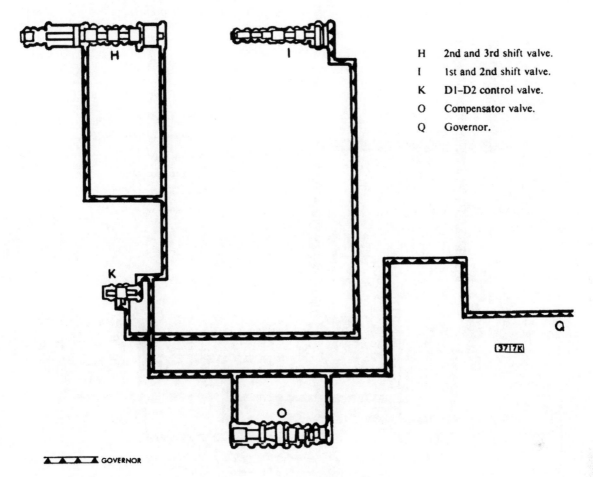

H   2nd and 3rd shift valve.
I   1st and 2nd shift valve.
K   D1-D2 control valve.
O   Compensator valve.
Q   Governor.

Fig. 6. *Governor circuit.*

The Rear Pump

The rear pump is driven by the output shaft of the transmission. It is fully effective at speeds above approximately 20 m.p.h. (32 k.p.h.) and then supplies most of the hydraulic requirements.

If, due to a dead engine, the front pump is inoperative, the rear pump, above approximately 20 m.p.h. (32 k.p.h.) can provide all hydraulic requirements, thus enabling the engine to be started through the transmission.

The Governor

The governor, revolving with the output shaft, is essentially a pressure regulating valve which reduces line pressure to a value which varies with output shaft speed. This variable pressure is utilised in the control system to effect up and down shifts through the 1-2 and 2-3 shift valves. Rotation of the governor at low speeds causes the governor weight and valve to be affected by centrifugal force. The outward force is opposed by an opposite and equal hydraulic force produced by pressure acting on the regulating area of the governor valve. The governor valve is a regulating valve and will attempt to maintain equilibrium. Governor pressure will rise in proportion to the increase in centrifugal force caused by higher output shaft speed.

As rotational speed increases the governor weight moves outward to rest on a stop in the governor body, and can move no further. When this occurs, a spring located between the counter weight and the valve

Page FFY.s.7

AUTOMATIC TRANSMISSION

becomes effective. The constant force of this spring then combines with the centrifugal force of the governor valve and the total force is opposed by governor pressure. This combination renders governor pressure less sensitive to output shaft speed variations.

It can be seen from the above, that the governor provides two distinct phases of regulation, the first of which is a fast rising pressure for accurate control of the low speed shift points.

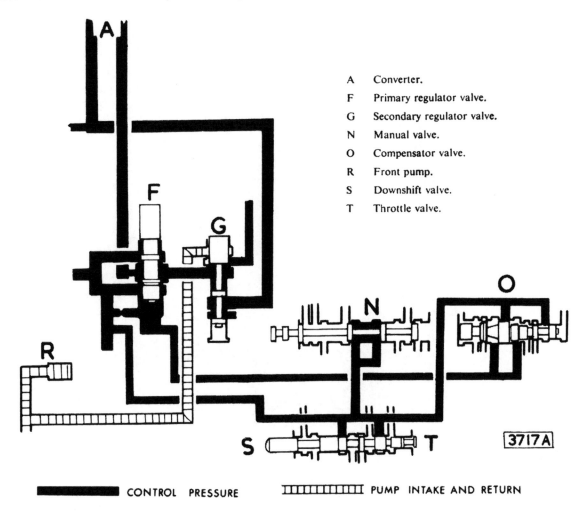

A   Converter.
F   Primary regulator valve.
G   Secondary regulator valve.
N   Manual valve.
O   Compensator valve.
R   Front pump.
S   Downshift valve.
T   Throttle valve.

■■■ CONTROL PRESSURE   ▥▥▥ PUMP INTAKE AND RETURN

Fig. 7. *Hydraulic circuit—neutral.*

THE CONTROL SYSTEM

Neutral—Engine Running (see Fig. 7)

When the selector is moved to the neutral position, the manual control valve is positioned so that control pressure cannot pass through the manual valve to the clutches or servos; therefore, the clutches and servos cannot apply. There is no transmission of power through the transmission in the neutral position.

The pressure regulation system, however, is functioning. With the engine running, the front pump is driven and fluid is picked up from the pan by the front pump inlet. Fluid, circulated by the front pump is directed to the control pressure regulator. The primary regulator valve will maintain correct control pressure by expelling the excess fluid to feed the secondary regulator valve. The secondary regulator valve maintains correct pressure for converter feed and lubrication, then forces the excess fluid back to the pump inlet.

Control pressure is directed to the manual control valve, where it is blocked by two lands on the valve. Control pressure is also directed to the throttle valve and the downshift valve and, with the valve closed

Page FFY.s.8

AUTOMATIC TRANSMISSION

(accelerator at idle position) it is blocked by lands on the valves. Control pressure to the compensator valve is regulated by that valve, and compensating pressure is directed to the primary regulator valve.

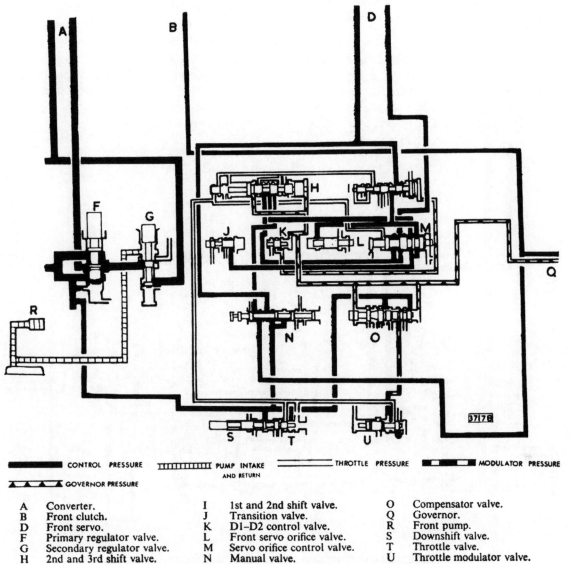

| | CONTROL PRESSURE | | PUMP INTAKE AND RETURN | | THROTTLE PRESSURE | | MODULATOR PRESSURE |
| | GOVERNOR PRESSURE | | | | | | |

A	Converter.	I	1st and 2nd shift valve.	O	Compensator valve.
B	Front clutch.	J	Transition valve.	Q	Governor.
D	Front servo.	K	D1–D2 control valve.	R	Front pump.
F	Primary regulator valve.	L	Front servo orifice valve.	S	Downshift valve.
G	Secondary regulator valve.	M	Servo orifice control valve.	T	Throttle valve.
H	2nd and 3rd shift valve.	N	Manual valve.	U	Throttle modulator valve.

Fig. 8. *Hydraulic circuit—1st gear (D1 range).*

First Gear, D1 Range (see Fig. 8)

When the selector lever is placed in the D1 position, with the car standing still, and the engine running, the manual control valve is moved to admit control pressure to apply the front clutch.

Control pressure is also directed to the governor, but with the car standing still, the control pressure is blocked at the governor valve.

Control pressure from the manual valve is directed through another passage to the apply side of the front servo and the 1–2 shift valve.

From the 1–2 shift valve pressure then passes to the servo orifice control valve and the front servo release valve where it is blocked.

Control pressure is then directed from the servo orifice control valve via the 2–3 shift valve and again through the control valve to the release side of the front servo.

Pressure is also present at the transition valve where it is blocked.

With pressure on both sides of the front servo piston, the servo is held in a released position. The

Page FFY.s.9

AUTOMATIC TRANSMISSION

one-way clutch takes the reaction torque on the rear drum, thus eliminating need for rear servo action.

The front pump supplies the pressure to operate the transmission and this pressure is controlled as it was in the neutral position.

When the accelerator is depressed and the car starts to move, centrifugal force, acting on the governor weight and valve, moves the valve to regulate governor pressure, which is directed to the 1–2 shift valve, 2–3 shift valve, and plug, and the compensator valve.

Movement of the accelerator also opens the throttle valve so that throttle pressure is directed to the modulator valve, orifice control valve, and the shift plug on the end of the 2–3 shift valve. Throttle pressure to the modulator valve is re-directed to the compensator valve to increase control pressure.

Throttle pressure to the shift plug on the 2–3 shift valve is reduced, and the reduced pressure is directed to the ends of the 1–2 shift valve and the 2–3 shift valve. This reduced pressure on the shift valves opposes governor pressure.

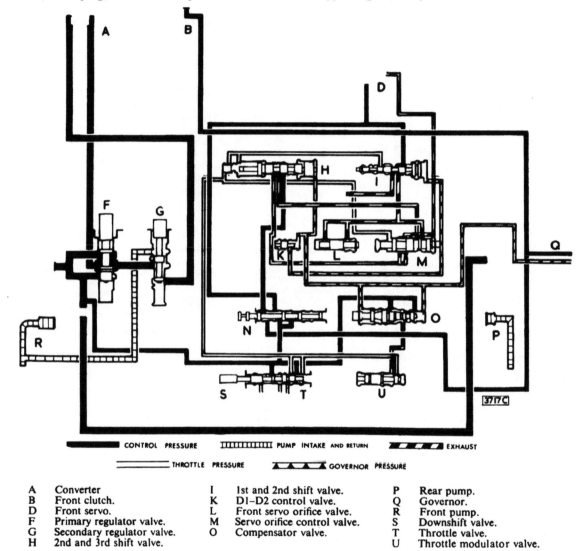

A	Converter	I	1st and 2nd shift valve.	P	Rear pump.
B	Front clutch.	K	D1–D2 control valve.	Q	Governor.
D	Front servo.	L	Front servo orifice valve.	R	Front pump.
F	Primary regulator valve.	M	Servo orifice control valve.	S	Downshift valve.
G	Secondary regulator valve.	O	Compensator valve.	T	Throttle valve.
H	2nd and 3rd shift valve.			U	Throttle modulator valve.

Fig. 9. *Hydraulic circuit—2nd gear (D1 range).*

Second Gear, D1 Range (Fig. 9)

As the car speed increases, the governor pressure builds up until it can overcome the opposite force of the 1–2 shift valve spring and reduced throttle pressure on the end of the valve and so moves the valve. When the 1–2 shift valve moves, control pressure at the valve is shut off and the front servo release pressure is

Page FFY.s.10

AUTOMATIC TRANSMISSION

exhausted, first slowly through a restricting orifice and then fast through the front servo release orifice valve. This leaves the front clutch and the front band applied.

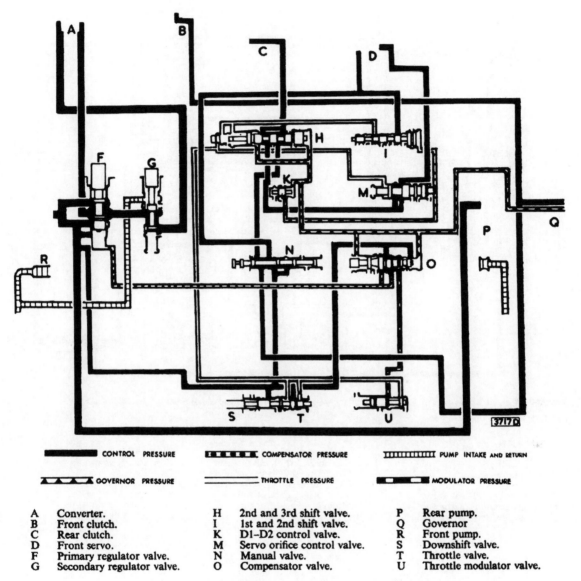

A	Converter.	H	2nd and 3rd shift valve.	P	Rear pump.
B	Front clutch.	I	1st and 2nd shift valve.	Q	Governor
C	Rear clutch.	K	D1–D2 control valve.	R	Front pump.
D	Front servo.	M	Servo orifice control valve.	S	Downshift valve.
F	Primary regulator valve.	N	Manual valve.	T	Throttle valve.
G	Secondary regulator valve.	O	Compensator valve.	U	Throttle modulator valve.

Fig. 10. *Hydraulic circuit—3rd gear (D1 or D2 range).*

Third Gear, D1 or D2 Range (Fig. 10)

As the car speed continues to increase, the governor pressure also increases until it overcomes the 2–3 shift valve spring and the reduced throttle pressure on the end of the 2–3 shift valve, thus causing the valve to move. When the valve moves, control pressure is admitted to the rear clutch and through the annulus of the servo orifice control valve to the release side of the front servo, thus applying the rear clutch and placing the front servo in the released position. This leaves the front clutch and the rear clutch applied.

As the governor pressure continues to increase, it acts against modulator pressure at the compensator valve to increase compensator pressure and decrease control pressure through the movement of the valve in the primary regulator.

Page FFY.s.11

AUTOMATIC TRANSMISSION

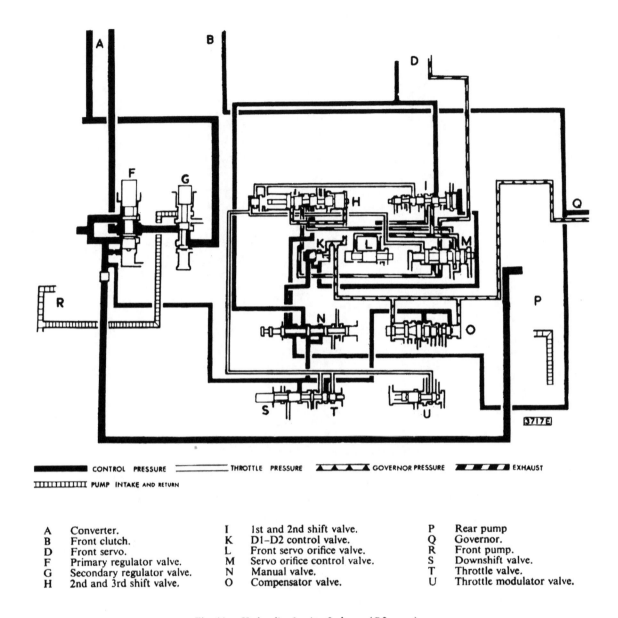

A	Converter.	I	1st and 2nd shift valve.	P	Rear pump
B	Front clutch.	K	D1-D2 control valve.	Q	Governor.
D	Front servo.	L	Front servo orifice valve.	R	Front pump.
F	Primary regulator valve.	M	Servo orifice control valve.	S	Downshift valve.
G	Secondary regulator valve.	N	Manual valve.	T	Throttle valve.
H	2nd and 3rd shift valve.	O	Compensator valve.	U	Throttle modulator valve.

Fig. 11. *Hydraulic circuit—2nd gear (D2 range).*

Second Gear, D2 Range (Fig. 11)

When the selector lever is placed in the D2 (drive) position, with the car standing still and the engine running, control pressure passes through the manual valve to the D1 and D2 control valve, overcomes any governor pressure acting on this valve and passes through the valve to the governor pressure area of the 1-2 shift valve, thus positioning it in the 2nd gear position.

Pressure is exhausted from the release side of the front servo, which results in the front clutch and front band being applied.

All upshifts from 2nd gear ratio direct will be similar to the description of 3rd gear D1 range.

Page FFY.s.12

AUTOMATIC TRANSMISSION

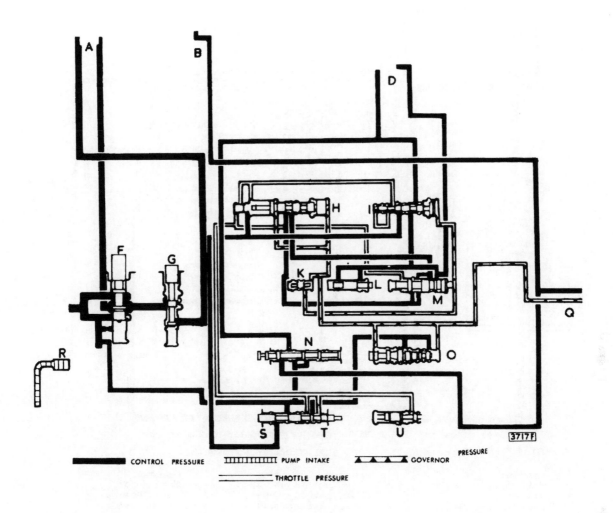

A	Converter.	I	1st and 2nd shift valve.	Q	Governor.
B	Front clutch.	K	D1-D2 control valve.	R	Front pump.
D	Front servo.	L	Front servo orifice valve.	S	Downshift valve.
F	Primary regulator valve.	M	Servo orifice control valve.	T	Throttle valve.
G	Secondary regulator valve.	N	Manual valve.	U	Throttle modulator valve.
H	2nd and 3rd shift valve.	O	Compensator valve.		

Fig. 12. *Hydraulic circuit—2-1 kickdown (D1 range).*

2-1 Kickdown, D1 Range (Fig. 12)

At car speeds up to approximately 20 m.p.h. (32 k.p.h.), after the transmission has shifted from 1st to 2nd or 3rd gear, the transmission can be downshifted to 1st gear by depressing the accelerator pedal beyond the wide open throttle position.

Movement of the accelerator to kickdown position causes the throttle cable to move the downshift valve to allow control pressure to pass through the downshift valve to another land on the 1-2 shift valve. The combination of control pressure and the 1-2 shift valve spring is sufficient to overcome governor pressure and return the valve to the 1st gear position. In this position, control pressure is admitted to the release side of the front servo. This places the front servo in the released position, leaving the front clutch applied and the one-way clutch holding the rear drum.

Page FFY.s.13

AUTOMATIC TRANSMISSION

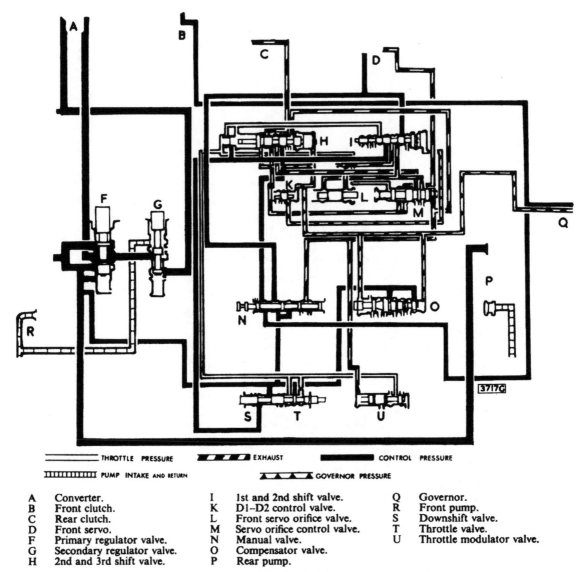

A	Converter.	I	1st and 2nd shift valve.	Q	Governor.
B	Front clutch.	K	D1–D2 control valve.	R	Front pump.
C	Rear clutch.	L	Front servo orifice valve.	S	Downshift valve.
D	Front servo.	M	Servo orifice control valve.	T	Throttle valve.
F	Primary regulator valve.	N	Manual valve.	U	Throttle modulator valve.
G	Secondary regulator valve.	O	Compensator valve.		
H	2nd and 3rd shift valve.	P	Rear pump.		

Fig. 13. *Hydraulic circuit—3–2 kickdown (D1 or D2 range).*

3–2 Kickdown, D1 or D2 Range (Fig. 13)

At car speeds between approximately 22 to 66 m.p.h. (35 to 106 k.p.h.) after the transmission has shifted to 3rd gear, the transmission can be downshifted from 3rd gear to 2nd gear by depressing the accelerator pedal beyond the wide open throttle position.

Movement of the accelerator causes the throttle cable to move the downshift valve to allow control pressure to pass through the downshift valve to the spring end of the 2–3 shift valve. The combination of control pressure at the end on the 2–3 shift valve and 2–3 shift valve springs is sufficient to overcome governor pressure to move the valve. When the valve is in 2nd gear position, control pressure to the rear clutch and through the servo orifice control valve to the release side of the front servo is shut off. The rear clutch circuit exhausts through the exhaust port of the manual control valve, whereas the front servo release circuit exhausts through the 1–2 shift valve, orifice and front servo release orifice valve. This leaves the front clutch and front band applied.

If the accelerator is left in the kickdown position, governor pressure will increase as the car speed increases until the governor pressure is greater than the combined pressures on the 2–3 shift valve, and the transmission will again upshift to 3rd gear.

At speeds above approximately 66 m.p.h. (106 k.p.h.) the governor pressure is so great that the combined pressures on the 2–3 shift valve cannot overcome the governor pressure; therefore, there is no kickdown.

Page FFY.s.14

AUTOMATIC TRANSMISSION

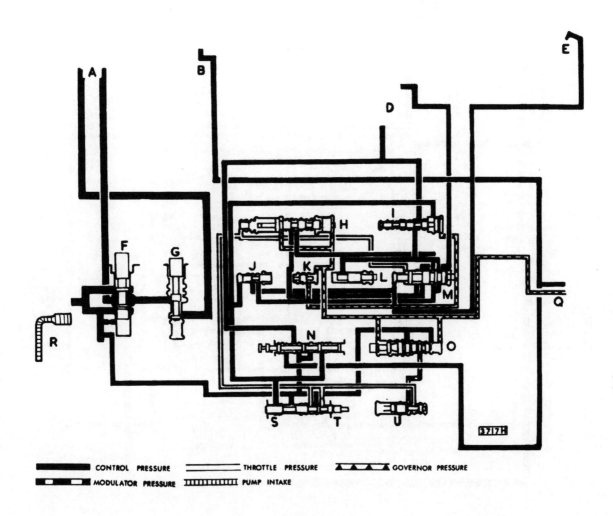

A	Converter.	I	1st and 2nd shift valve.	Q	Governor.
B	Front clutch.	J	Transition valve.	R	Front pump.
D	Front servo.	K	D1–D2 control valve.	S	Downshift valve.
E	Rear servo.	L	Front servo orifice valve.	T	Throttle valve.
F	Primary regulator valve.	M	Servo control valve.	U	Throttle modulator valve.
G	Secondary regulator valve.	N	Manual valve.		
H	2nd and 3rd shift valve.	O	Compensator valve.		

Fig. 14. *Hydraulic circuit—Lockup (1st gear).*

Lockup—First Gear (Fig. 14)

When the selector lever is placed in the Lockup position, the manual control valve is moved to admit through one port, control pressure to the governor feed and to apply the front clutch. Another port supplies both sides of the front servo which is held in the released position and also to the rear servo to apply the rear band through the servo orifice control and transition valves. A third port supplies pressure to move the transition valve and to an additional land on the 1–2 shift valve.

In this position, there is no automatic upshift to a higher gear ratio, since the combination of control pressure on the 1–2 shift valve and the 1–2 shift valve spring is greater than governor pressure acting against the valve, so that the valve cannot move. The combination of control pressure on the 2–3 shift valve and the 2–3 valve spring is also greater than the governor pressure acting against the valve so that the 2–3 shift valve cannot move.

AUTOMATIC TRANSMISSION

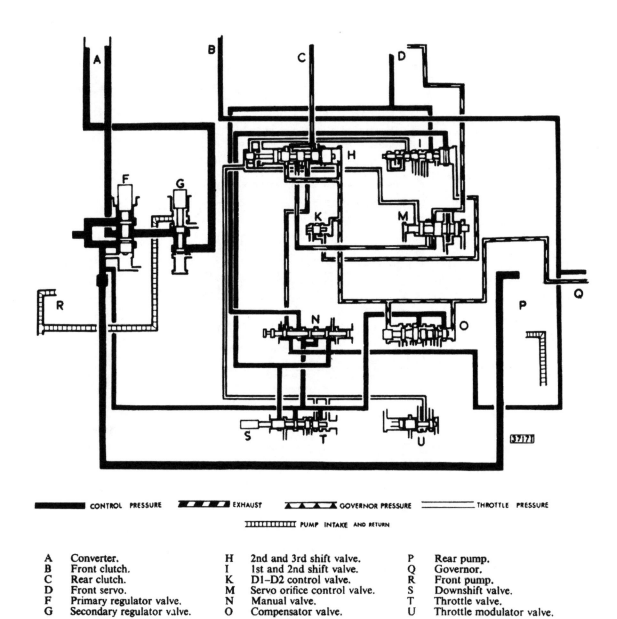

A	Converter.	H	2nd and 3rd shift valve.	P	Rear pump.
B	Front clutch.	I	1st and 2nd shift valve.	Q	Governor.
C	Rear clutch.	K	D1–D2 control valve.	R	Front pump.
D	Front servo.	M	Servo orifice control valve.	S	Downshift valve.
F	Primary regulator valve.	N	Manual valve.	T	Throttle valve.
G	Secondary regulator valve.	O	Compensator valve.	U	Throttle modulator valve.

Fig. 15. *Hydraulic circuit—Lockup (2nd gear).*

Lockup—Second Gear

In L the manual control valve opens to exhaust the rear clutch and front servo release circuit from the 2–3 shift valve. This causes a downshift from 3rd gear whenever L is selected at speed. In this condition, governor pressure will have moved the 1–2 shift valve; the result is that supply to the rear servo through the servo orifice control valve and transition valve is blocked and as front servo release pressure also exhausts through the 2–3 shift valve, the front band will be applied. This band, in conjunction with the front clutch, provides 2nd gear.

Page FFY.s.16

AUTOMATIC TRANSMISSION

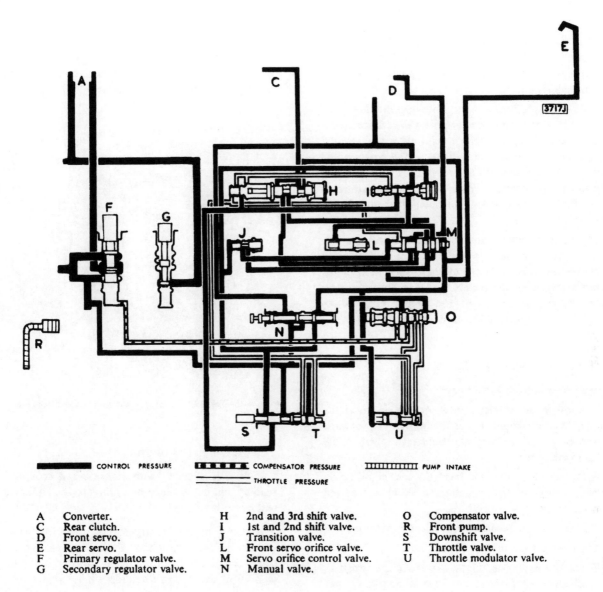

A	Converter.	H	2nd and 3rd shift valve.	O	Compensator valve.
C	Rear clutch.	I	1st and 2nd shift valve.	R	Front pump.
D	Front servo.	J	Transition valve.	S	Downshift valve.
E	Rear servo.	L	Front servo orifice valve.	T	Throttle valve.
F	Primary regulator valve.	M	Servo orifice control valve.	U	Throttle modulator valve.
G	Secondary regulator valve.	N	Manual valve.		

Fig. 16. *Hydraulic circuit—reverse gear.*

Reverse (Fig. 16)

When the selector lever is placed in the reverse position, the manual control valve moves to admit control pressure to the rear clutch, both sides of the front servo and the rear servo. This applies the rear clutch and the rear band.

Control pressure is also directed to the modulator valve to move the valve so when the throttle valve is opened by depressing the accelerator, the throttle pressure passes through the modulator valve to two lands on the compensator valve to reduce compensating pressure, thus increasing control pressure.

High control pressure is desired in reverse, since the reaction forces increase appreciably and higher pressure is required to hold the rear drum.

AUTOMATIC TRANSMISSION

MAINTENANCE

It is most IMPORTANT that the following maintenance instructions are closely followed and absolute cleanliness is maintained when topping-up or filling the transmission.

It is **vitally important** when checking the fluid level that no dirt or foreign matter enters the transmission, otherwise trouble will almost certainly arise. Before removing the transmission dipstick, the surrounding area must be cleaned off to prevent dirt from entering the dipstick aperture. When filling the transmission with fluid ensure that the fluid container and funnel are perfectly clean.

In countries where ambient temperatures are unusually high, dust and/or mud must not be allowed to decrease the effective areas of the stoneguards in the converter housing or the slots in the transmission case. Also any foreign matter on the oil pan must be removed as it would act as a temperature insulator.

EVERY 3,000 MILES (5,000 KM.)

Check Transmission Fluid Level

The transmission filler tube is located on the right-hand side of the engine under the bonnet just forward of the bulkhead. Check the fluid level every 3,000 miles (5,000 km.).

Before checking the fluid level, the car should be on level ground and the transmission should be at the normal operating temperature.

Set the handbrake firmly and select P position.

The engine should be at normal idle.

When the engine is running, remove the dipstick, wipe clean and replace in the filler tube in its correct position.

Withdraw immediately and check.

If necessary, add fluid to bring the level to the FULL mark on the dipstick. The difference between FULL and LOW marks on the stick represents approximately 1½ pints (2 U.S. pints or 0·75 litres).

Fig. 17. *Automatic transmission dipstick.*

Be careful not to overfill.

If fluid is checked with transmission cold, a false reading will be obtained and filling to the FULL mark will cause it to be overfilled.

If it is found necessary to add fluid frequently, it will be an indication that there is a leakage in the transmission and it should be investigated immediately to prevent damage to transmission.

Total fluid capacity (including cooler) 16 Imperial pints from dry (19 U.S. pints, 9 litres).

RECOMMENDED AUTOMATIC TRANSMISSION FLUIDS

Mobil	Castrol	Shell	Esso	B.P.	Duckham	Regent Caltex/Texaco
Mobilfluid 200	Castrol T.Q.	Shell Donax T.6	Esso Automatic Transmission Fluid	Automatic Transmission Fluid, Type A	Nolmatic	Teaxamatic Fluid

Page FFY.s.18

AUTOMATIC TRANSMISSION

If these recommended lubricants are not available, only a transmission fluid conforming to the following specification should be used:—

Automatic Transmission Fluid, Type "A" or Type "A" Suffix "A" (AQ-ATF)

ROAD TEST AND FAULT DIAGNOSIS

TESTING THE CAR

It is important to gain as much information as possible on the precise nature of any fault. In all cases the following road test procedure should be completely carried out, as there may be more than one fault.

Check that the starter will operate only with the selector in "P" and "N" and that the reverse light operates only in "R".

Apply the brakes and, with the engine at normal idling speed, select N-D, N-L, N-R. Transmission engagement should be felt in each position selected.

Check the engine stall speed (see converter diagnosis) with the transmission in "L" and "R". Check for slip or clutch break-away.

Note: Do not stall for longer than 10 seconds, or the transmission will overheat.

With the transmission at normal running temperature, select "D1". Release the brakes and accelerate with minimum throttle opening. Check for 1-2 and 2-3 shifts.

Note: At minimum throttle opening the shifts may be difficult to detect. Confirmation that the transmission is in 3rd gear may be obtained by selecting "L", when a 3-2 downshift will be felt.

At just over 30 m.p.h. (48 k.p.h.), select "N", switch off the ignition and let the car coast. At 30 m.p.h. (48 k.p.h.), switch on the ignition and select "L". The engine should start through the rear wheels, indicating that the rear oil pump of the transmission is operating.

Stop and restart, using full-throttle acceleration, i.e., accelerator at the detent. Check for 1-2 and 2-3 shifts according to the shift speed chart.

At 26 m.p.h. (42 k.p.h.), in 3rd gear, depress the accelerator to full-throttle position. The car should accelerate in 3rd gear and should not downshift to 2nd.

At 30 m.p.h. (48 k.p.h.), in 3rd gear, depress the accelerator to the kick-down position, i.e., through the detent. The transmission should downshift to 2nd gear.

At 18 m.p.h. (29 k.p.h.) in 3rd gear, depress the accelerator to the kick-down position. The transmission should downshift to 1st gear.

Stop and restart, using forced throttle acceleration (i.e., accelerator through the detent). Check for 1-2 and 2-3 shifts according to shift speed chart.

At 40 m.p.h. (64 k.p.h.) in 3rd gear, release the accelerator and select "L". Check for 3-2 downshift and engine braking. Check for inhibited 2-1 downshift and engine braking.

Stop, and with "L" still engaged, release the brakes and, using full throttle, accelerate to 20 m.p.h. (32 k.p.h.). Check for no slip or clutch break-away noise and no up-shifts.

Stop and select "R". Release the brakes and reverse, using full throttle if possible. Check for no slip or clutch break-away noise.

Stop on brakes facing downhill on gradient and select "P". Release the brakes and check that the parking pawl will hold the car. Re-apply brakes before disengaging the parking pawl. Repeat with car facing uphill.

Check that the selector is trapped by the gate in "Park" position.

At 30 m.p.h. (48 k.p.h.), in 3rd gear, D1, coast to a stop. Check roll out shifts for quality and speed in m.p.h. or k.p.h.

The front pump can be checked, with the selector in neutral, by revving the engine between idle and 2,000 r.p.m. A high pitched whine indicates a noisy front pump, a restricted front pump suction line, or a dirty oil screen.

At idle or slightly above idle speed in neutral, a gear whine indicates dragging front clutch plates. A tendency for the car to creep in neutral is a further

Page FFY.s.19

AUTOMATIC TRANSMISSION

indication of dragging front clutch plates. Check carefully, to avoid confusing this with front pump or engine noises.

PRESSURE TESTS

See "Throttle Cable Adjustment" section and ascertain correct adjustment of throttle cable and engine idle. The pressure gauge is used to check transmission pressures, which should correspond to values given below.

Note: Figures given in table are normal for transmission temperatures from 150° to 185°F. only (65·5°C. to 85°C.).

Selector Position	Control Pressure Idle r.p.m.	Control Pressure Stall r.p.m.
D2	50–60	150–185
D1	50–60	150–185
L	50–60	150–185
R	50–60	190–210
N	55–60	—

Recording stall speed and stall pressures at the time the coverter is being checked will reduce the overall stalling time, which should be kept to a minimum.

Pressures which have been recorded should be analysed as follows: Low pressure indicates leakage in the circuit tested. Low pressure in all selector positions would indicate leakage, faulty pump or incorrect pressure regulation. High pressures, in all selector positions, indicate faulty pressure regulation incorrect cable adjustment or stuck valves.

FAULT DIAGNOSIS

Converter

If the general vehicle performance is below standard, check the engine stall speed with the revolution indicator by applying maximum pressure on the foot brake pedal, selecting lock-up, and fully depressing the accelerator. If the engine stall speed is up to 300 r.p.m. below normal, the engine is not developing its full power.

Inability to start on steep gradients combined with poor acceleration from rest indicates that the converter stator one-way clutch is slipping. This condition permits the stator to rotate in an opposite direction to the turbine and torque multiplication cannot occur. Check the stall speed, and if it is more than 600 r.p.m. below normal the converter assembly must be renewed.

Below standard acceleration in 3rd gear above 30 m.p.h. (48 k.p.h.), combined with a substantially reduced maximum speed, indicates that the stator one-way clutch has locked in the engaged condition. The stator will not rotate with the turbine and impeller, therefore the fluid flywheel phase of the converter performance cannot occur. This condition will also be indicated by excessive overheating of the transmission, although the stall speed will remain normal. The converter assembly must be replaced.

Stall speed higher than normal indicates that the converter is not receiving its required fluid supply or that slip is occurring in the clutches of the automatic gearbox.

Note: When checking stall speeds ensure that the transmission is at normal operating temperature. Do not stall for longer than 10 seconds, or the transmission will overheat.

The torque converters are sealed by welding and serviced by replacement only.

The stoneguards in the converter housing must be unobstructed.

Stall Speed Test

This test provides a rapid check on the correct functioning of the converter as well as the gearbox.

The stall speed is the maximum speed at which the engine can drive the torque impeller while the turbine is held stationary. As the stall speed is dependent both on engine and torque converter characteristics, it will vary with the condition of the engine as well as with the condition of the transmission. It will be necessary, therefore, to determine the condition of the engine in order to correctly interpret a low stall speed.

To obtain the stall speed, allow the engine and the transmission to attain normal working temperature, set the handbrake, chock the wheels and apply the footbrake. Select "L" or "R" and fully depress the accelerator. Note the reading on the revolution indicator.

Note: To avoid overheating, the period of stall test must not exceed 10 seconds.

AUTOMATIC TRANSMISSION

R.P.M.	Condition Indicated
Under 1,000	Stator freewheel slip
1,600–1,700	Normal
Over 2,100	Slip in the transmission gearbox

Clutch and Band Checks

To determine if a clutch or band has failed, without removing a transmission, check as detailed below.

Refer to the chart on page FF.s.5, showing the clutches and bands applied in each gear position.

Apply the handbrake and start the engine.

Engage each gear ratio and determine if drive is obtained through the component to be checked. If a clutch or band functions in one selector position it is reasonable to assume that the element in question is normal and that trouble lies elsewhere. If the clutch or band is tried in two positions and no drive is obtained in either position, it can be assumed that the element is faulty.

Air Pressure Checks

Air pressure may be used to test various transmission components in the car on the bench. Care should be exercised when air pressure checks are being made to prevent oil blowing on the clothing or into the eyes.

Knowledge of various circuits should be acquired referring to Figs. 6 to 16. It is necessary to remove the valve body to complete these checks.

Apply air pressure to the front clutch passage. A definite thump will indicate engagement. A similar sound should be heard when the rear clutch circuit is tested.

If clutch engagement noise is indefinite it is almost certainly due to damaged piston rings.

Servo action may be watched as air is applied to apply circuits of each servo.

It can be assumed, that if air pressure checks indicate that clutches and servos are being applied normally with air pressure, then the trouble lies in the hydraulic system.

A. Front servo apply.
B. Front clutch.
C. Rear servo.
D. Rear clutch.
E. Governor feed.

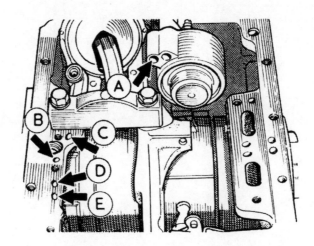

Fig. 18. *Showing pressure passages with valve body removed.*

Page FFY.s.21

AUTOMATIC TRANSMISSION

FAULT DIAGNOSIS

ENGAGEMENT	In Car	On Bench
Harsh	B, D, c, d	2, 4
Delayed	A, C, D, E, F, a, c, d	b
None	A, C, a, c, d	b, 9, 10, 11, 13
No forward	A, C, a, c, d	B, 1, 4, 7
No reverse	A, C, F, a, c, j, k, h	b, 2, 3, 6
Jumps in forward	C, D, E, F	4, 7, 8
Jumps in reverse	C, D, E	2
No neutral	C, c	2

UPSHIFTS	In Car	On Bench
No. 1–2	C, E, a, c, d, f, g, h, j	b, 5, 17
No. 2–3	C, a, c, d, f, g, h, k, l	b, 3, 17
Shift points too high	B, C, c, d, f, g, h, j, k, l	b
Shift points too low	B, c, f, g, h, l	B

UPSHIFT QUALITY	In Car	On Bench
1–2 slips or runs up	A, B, C, E, a, c, d, f, g, k	b, 1, 5
2–3 slips or runs up	C, a, c, d, f, g, h, k, l	b, 3, 5
1–2 harsh	B, C, E, c, d, f, g, h	1, 7, 8
2–3 harsh	B, C, E, s, d, f	4
1–2 Ties up or grabs	F, c	4, 7, 8
2–3 Ties up or grabs	E, F, C	4

DOWNSHIFTS	In Car	On Bench
No. 2–1	B, C, c, h, j	7
No. 3–2	B, c, h, k	4
Shift points too high	B, C, c, f, h, j, k, l	b
Shift points too low	B, C, c, f, h, j, k, l	b

DOWNSHIFT QUALITY	In Car	On Bench
2–1 Slides		7
3–2 Slides	B, C, E, a, c, d, f, g	b, 3, 5
2–1 Harsh		b, 1, 7
3–2 Harsh	B, E, c, d, f, g, 5	3, 4, 5

REVERSE	In Car	On Bench
Slips or chatters	A, B, F, d, c, g	b, 2, 3, 6

Page FFY.s.22

AUTOMATIC TRANSMISSION

FAULT DIAGNOSIS (continued)

	In Car	On Bench
LINE PRESSURE		
Low idle pressure	A, C, D, a, c, d	b, 11
High idle pressure	B, c, d, e, f, g	
Low stall pressure	A, B, a, c, d, f, g, h	b, 11
High stall pressure	B, c, d, f, g	
STALL SPEED		
Too low (200 r.p.m. or more)		13
Too high (200 r.p.m. or more)	A, B, C, F, a, c, d, f	b, 1, 3, 6, 7, 9, 13
OTHERS		
No push starts	A, C, E, F, c	12
Transmission overheats	E, F, e	1, 2, 3, 4, 5, 6, 13, 18
Poor acceleration		13
Noisy in neutral	m	2, 4
Noisy in park	m	14
Noisy in all gears	m	2, 4, 14, 16
Noisy during coast (30–20 m.p.h.)		16, 19
Park brake does not hold	C, 15	15

KEY TO THE FAULT DIAGNOSIS CHART

1. **Preliminary Checks in Car**
   A. Low fluid level.
   B. Throttle cable incorrectly assembled or adjusted.
   C. Manual linkage incorrectly assembled or adjusted.
   D. Engine idle speed.
   E. Front band adjustment.
   F. Rear band adjustment.

2. **Hydraulic Faults**
   (a) Oil tubes missing or broken.
   (b) Sealing rings missing or broken.
   (c) Valve body screws missing or not correctly tightened.
   (d) Primary valve sticking.
   (e) Secondary valve sticking.
   (f) Throttle valve sticking.
   (g) Compensator or modulator valve sticking.
   (h) Governor valve sticking leaking or incorrectly assembled.
   (i) Orifice control valve sticking.
   (j) 1–2 shift valve sticking.
   (k) 2–3 shift valve sticking.
   (l) 2–3 shift valve plunger sticking.
   (m) Regulator.

3. **Mechanical Faults**
   1. Front clutch slipping due to worn plates or faulty parts.
   2. Front clutch seized or plates distorted.
   3. Rear clutch slipping due to worn or faulty parts.
   4. Rear clutch seized or plates distorted.
   5. Front band slipping due to faulty servo, broken or worn band.
   6. Rear band slipping due to faulty servo, broken or worn band.
   7. One-way clutch slipping or incorrectly installed.
   8. One-way clutch seized.
   9. Broken input shaft.
   10. Front pump drive tangs on converter hub broken.
   11. Front pump worn.
   12. Rear pump worn or drive key broken.
   13. Converter blading and/or one-way clutch failed.
   14. Front pump.
   15. Parking linkage.
   16. Planetary assembly.
   17. Fluid distributor sleeve in output shaft.
   18. Oil cooler connections.
   19. Rear pump.

Page FFY.s.23

AUTOMATIC TRANSMISSION

SERVICE ADJUSTMENTS

THROTTLE/KICKDOWN CABLE ADJUSTMENT

The importance of correct throttle cable adjustment cannot be over-emphasised. The shift quality and correct shift positions are controlled by precise movement of the cable in relation to the carburetter throttle shaft movement.

Preliminary Testing

Test the car on a flat road.

With the selector in the D1 or D2 position and at a minimum throttle opening, the 2-3 upshift should occur at 1,100–1,200 r.p.m.

A "run-up" of 200–400 r.p.m. at the change point indicates LOW pressure.

At full throttle opening, a jerky 2-3 upshift or a sharp 2-1 downshift (in D1 when stopping the car) indicates HIGH pressure.

Install a pressure gauge, 0–200 lb./sq. in. (0–14 kg./sq. cm.) in the line pressure point at the left hand rear face of the transmission unit. Start the engine and allow to reach normal operating temperature.

Select D1 or D2, apply the handbrake firmly and increase the idling speed to exactly 1,250 r.p.m.

The pressure gauge reading should be 72·5 ± 2·5 lb./sq. in. (5·097 ± ·175 kg/cm. sq.).

Adjustment

If road and pressure tests indicate that the throttle/kickdown cable setting is incorrect, adjustment is made at the fork end (see Fig. 20).

Release the fork end locknut, remove the split pin and fork end clevis pin.

To LOWER the pressure, turn the fork end clockwise: to RAISE the pressure, turn anti-clockwise.

Note: One full turn will alter the setting by 9 lb./sq. in. (·63 kg./sq. cm.).

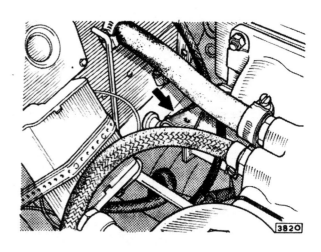

Fig. 20. *The kickdown cable adjustment.*

Slight adjustment only should be necessary; excessive adjustment will result in loss of "kickdown" or an increase in shift speeds.

Refit the fork end joint pin and split pin and tighten the locknut.

Restart the engine and check the pressure at 1,250 r.p.m.

Check that the carburetter butterfly valves are closed at idling speed after adjustment is completed.

If, after repeated attempts to stabilize the change points, the pressure still fluctuates, the throttle/kickdown inner cable may be binding or kinked and the cable should be replaced.

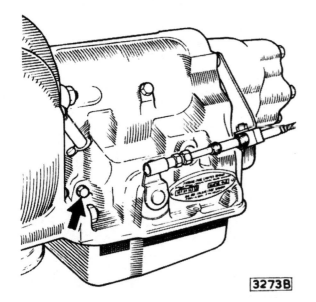

Fig. 19. *The transmission pressure take-off point.*

Page FFY.s.24

AUTOMATIC TRANSMISSION

Throttle/Kickdown Cable Renewal

Disconnect the cable at the fork end.

Remove the cable retaining clip after withdrawing the setscrew.

Lift the carpets and the underfelts from the gearbox tunnel on the left-hand side.

Remove six drive screws and detach the aperture cover plate now exposed.

Remove the Allen-headed screw and washer retaining the outer cable.

Withdraw the outer cable and locate the spring clip securing the inner cable to the control rod operating the kickdown cam in the transmission unit.

Spring the clip open with a small screw driver and withdraw the inner cable.

Refitting is the reverse of the removal procedure.

Adjust the length of the operating cable to $3\frac{6}{16}''$ (84·1 mm.) between the centre line of the clevis and the end of the outer cable.

Check that the carburetter butterfly valves are closed before commencing adjustments described under the previous heading.

MANUAL LINKAGE ADJUSTMENT

(See Fig. 23)

Remove the transmission tunnel finisher assembly and the carpet at the side of the transmission cover. Remove the rubberised felt and withdraw the setscrews securing the cover plate at the left-hand side of the transmission cover.

Loosen the linkage cable locknut and remove the cable from the transmission lever. Push the transmission lever fully forward to the Lockup detent. Place the gear selector lever in the Lockup position.

Adjust the cable end to fit freely on to the transmission lever. Temporarily re-attach the cable to the lever. Move the gear selector lever through the various positions checking that gating at positions L, D1, R and P does not interfere with the transmission lever setting at the detent positions. The transmission lever must locate the transmission detents positively. Once correct adjustment is established, be sure the linkage cable is secured to the transmission lever and the locknut is tightened.

REMOVAL OF OIL PAN

Prior to front band adjustment or a check of internal parts, the gearbox fluid must be drained and the oil pan removed. When this is done an inspection should be made. A few wear particles in the dregs of the fluid in the pan are normal. An excess of wear particles whether ferrous or non-ferrous, or pieces of band lining material, would indicate that further checking should be done. A new gasket should be used when refitting the pan and the 14 attaching screws torqued to 10–15 lb. ft. (1·38–2·07 kgm.). Always use fresh fluid when refilling.

FRONT BAND ADJUSTMENT

(See Fig. 21)

The front band should be adjusted after the first 1,000 miles (1,600 km.) of operation and at 21,000 mile (35,000 km.) intervals thereafter.

Drain the oil by removing the oil filler connection and remove the oil pan. Loosen the adjusting screw locknut on the servo, apply lever and check that the screws turn freely in the lever. Install a $\frac{1}{4}''$ (6·4 mm.) thick gauge block between the servo piston pin and the servo adjusting screw, then tighten the adjusting screw with a suitable torque wrench or adjusting tool until 10 lb. ins. (0·12 kgm.) is reached. Retighten the adjusting screw locknut to 20–25 lb. ft. (2·76–3·46 kgm.). Remove the $\frac{1}{4}''$ (6·3 mm.) spacer.

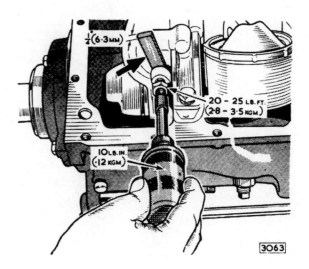

Fig. 21. *Front band adjustment.*

Page FFY.s.25

AUTOMATIC TRANSMISSION

REAR BAND ADJUSTMENT

The rear band adjustment at the first 1,000 miles (1,600 km.) and at 21,000 miles (35,000 km.) intervals thereafter is made externally. To make the adjustment, first loosen and back off the adjusting screw locknut three or four turns and then make sure that the adjusting screw works freely in the threads in the case. Turn the adjusting screw in with a torque wrench or special tool for this purpose to 10 lb. ft. (1·382 kgm.) torque reading. Back the adjusting screw off 1½ turns exactly, then retighten the locknut to 35-40 lb. ft. (4·84-5·53 kgm.). The adjusting screw is on the right-hand side of the casing and an access hole is provided in the transmission cowl.

GOVERNOR

The governor can be inspected without removal of the oil pan. Remove the inspection cover and gasket. This will expose the governor, but the output shaft may have to be turned to position the governor head at the opening. First check for freedom of the valve by pushing and pulling on the governor weight. If removal of the governor body is desired, take out the two screws which retain it, being careful that they are not dropped inside the extension housing After removal of the body, dismantle it completely and clean all parts. When reassembling the governor, torque the governor body plate screws to 20-30 lb. in. (0·24-0·36 kgm.). When replacing the governor body on to the transmission, torque the screws which retain it to 50-60 lb. in. (0·60-0·72 kgm.). Replace the governor inspection cover, using a new gasket and torque its retaining screws to 50-60 lb. in. (0·60-0·72 kgm.).

It should be noted that if any of the four governor screws mentioned above are loose, the governor will not function correctly.

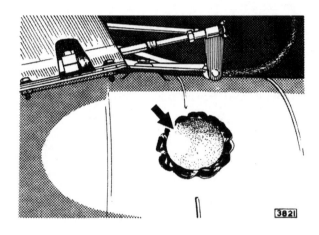

Fig. 22. *Rear band adjustment access point.*

Fig. 23. *Manual selector linkage adjustment.*

Page FFY.s.26

AUTOMATIC TRANSMISSION

TRANSMISSION UNIT
REMOVAL AND REFITTING

To remove the transmission unit, it is necessary to withdraw the engine and transmission as a complete unit from the car before separating the transmission.

Removal

Disconnect the battery.

Remove the bonnet.

Drain the cooling system and cylinder block. Conserve the coolant if antifreeze is in use.

Slacken the clip on the breather pipe; unscrew the two wing nuts and withdraw the top of the air cleaner.

Disconnect the petrol feed pipe under the centre carburetter.

Slacken the clamps and remove the water hoses from the cylinder head and radiator to the header tank.

Remove the transmission oil cooler pipes from the radiator block.

Remove the heater hoses from the inlet manifold.

Disconnect the brake vacuum pipe.

Pull off the two Lucar connectors from the fan control thermostat in the header tank.

Remove the two bolts securing the header tank mounting bracket to the front cross member. Remove two nuts and two bolts securing the header tank straps to the radiator and fan cowl. Remove the header tank complete with bracket and straps.

Disconnect the throttle linkage at the rear carburetter and the kickdown cable at the rear of the cylinder head.

Disconnect:—

The two coil leads.

The water temperature transmitter cable.

The battery cable and solenoid switch cable from the starter motor.

The oil pressure cable at the top of the oil filter body.

The main harness connector and the Lucar connector for the 3AW warning light control from the alternator.

The engine earth strap from the left-hand side member.

Withdraw the bolt securing the oil filter canister and remove the canister complete with filter. Catch the escaping oil in a drip pan.

Remove the crankshaft pulley; damper and drive belt. Remove the ignition timing pointer from the sump. Mark the pulley and damper to facilitate refitting.

Slacken the two clamps of the water pump hose and withdraw the hose.

Remove the revolution counter generator complete with cables.

Remove the four nuts and washers securing each exhaust downpipe to the manifold. Unclip the pipes at the silencers and withdraw the downpipes. Collect the sealing rings between the downpipes and the manifold.

Withdraw the transmission dipstick and unscrew the dipstick tube from the oil pan.

Place the selector lever in L and withdraw the nut securing the selector cable adjustable ball joint to the transmission lever. Release the nut securing the outer cable clamp to the abutment bracket.

Remove the two lower nuts securing the torsion bar reaction tie plate on each side and tap the bolts back flush with the face of the tie plate. With the aid of a helper, place a lever between the head of the bolt just released and the torsion bar. Exert pressure on the bolt head to relieve the tension on the upper bolt. Remove the nut and tap the upper bolt back flush with the face of the tie plate. Tap the tie plate off the four bolts.

Note: Failure to relieve the tension on the upper bolts when tapping them back against the face of the tie plate will result in stripping the threads. If this occurs, new bolts must be fitted and the torsion bars re-set.

Disconnect the speedometer cable from the rear extension of the transmission unit.

Support the engine by means of two individual lifting tackles using the hooks on the cylinder head. Insert a trolley jack under the transmission and support the unit.

Page FFY.s.27

AUTOMATIC TRANSMISSION

Remove the self-locking nut and stepped washer from the engine stabiliser.

Remove the bolts securing the rear mounting plate. Disconnect the propeller shaft at the front universal joint.

Remove the bolts from the front engine mountings.

Raise the engine on the lifting tackles and, keeping the unit level, move forwards ensuring that the converter housing clears the torsion bar anchor brackets and that the water pump pulley clears the sub-frame top cross member. Carefully raise the front of the engine and withdraw the complete unit forwards and upwards.

Refitting

Reverse the removal procedure to refit the transmission and engine. IT IS IMPORTANT that the engine stabiliser is adjusted properly and that the kickdown linkage is set correctly when refitting.

TRANSMISSION UNIT
Removal

Disconnect the kickdown linkage at the operating shaft. Drain the oil from the transmission unit. Remove the bolts securing the transmission to the converter housing and withdraw the unit.

TORQUE CONVERTER AND FLYWHEEL
Removal

Withdraw the cover from the front of the converter housing. Remove the starter motor and withdraw the setscrews securing the converter housing to the engine.

Remove the four setscrews, accessible through the starter motor mounting aperture, securing the torque converter to the flywheel. Rotate the engine to gain access to each setscrew in turn.

Remove the setscrews and locking plate securing the flywheel to the crankshaft and withdraw the flywheel.

TRANSMISSION
DISMANTLING AND ASSEMBLY

TRANSMISSION—DISMANTLING

Dismantling should not begin until the transmission exterior and work area have been thoroughly cleaned.

Place the transmission (bottom side up) on a suitable stand or holding fixture.

Remove the oil pan bolts, oil pan and gasket. Remove the oil screen retaining clip, lift off the oil screen from the regulator; lift and remove the screen from the rear pump suction tube. (See Fig. 24).

Use a screwdriver to prise the compensator tube from the valve body and regulator assemblies (Fig. 25).

The control pressure tube should be prised from the valve body, then removed from the regulator (Fig. 26).

Remove the rear pump suction tube by pulling and twisting it at the same time.

Fig. 24. *Removing the screen from the rear suction tube.*

Page FFY.s.28

AUTOMATIC TRANSMISSION

Loosen the front and rear servo adjusting screw locknuts and adjusting screws. This will aid in dismantling and later, in assembling, the transmission.

Fig. 25. *Removing the compensator tube.*

Fig. 27. *Removing the pressure spring retainer.*

Carefully remove the pressure regulator spring retainer. Maintain pressure on the retainer to prevent distortion of the retainer, and sudden release of the springs (Fig. 27).

Remove the three valve body attaching capscrews and lockwashers (Fig. 29).

Loosen the front servo to case capscrew and lockwasher approximately $\tfrac{5}{16}''$ (7·94 mm.) (Fig. 30).

Fig. 26. *Removing the line pressure tube.*

Fig. 28. *The regulator retaining screws.*

Remove springs and spring pilots, but do not remove the regulator valves at this time. The valves will be protected as long as they remain in the regulator body.

Remove the two regulator attaching capscrews and lockwashers, then lift the regulator assembly from the transmission case (Fig. 28).

Page FFY.s.29

AUTOMATIC TRANSMISSION

Place the manual selector lever in park or reverse position. Lift the valve body until the throttle control rod will clear the manual detent lever, then remove the hook from the throttle cam using the index finger or a screwdriver.

Remove the front servo apply and release tubes (Fig. 32).

Remove the front servo bolt and lift the servo from the transmission, catching the servo strut with the index and middle finger of the left hand (Fig. 33).

Fig. 29. *The valve body attaching screws.*

Fig. 31. *Lifting the valve body to clear the front servo.*

Lift the valve body and servo until the valve body will clear the linkage and slide it off the servo apply and release tubes (Fig. 31).

Remove the two rear servo attaching capscrews and lockwashers, then lift the rear servo assembly from the transmission (Fig. 34)

Fig. 30. *Slackening the front servo screws.*

Fig. 32. *Withdrawing the apply and release tubes.*

Page FFY.s.30

AUTOMATIC TRANSMISSION

Remove the rear band apply and anchor struts.

Remove the rear pump outlet tube, using special extractor tool Part No. CWG.45 (Fig. 35).

Fig. 33. *Removing the front servo.*

Check the end play at this time. Should the end play need correcting it will be done during assembly of the transmission (see Fig. 36). Place an indicator against the end of the input shaft. Prise between the front of the case and the front clutch to move clutch assemblies to their extreme rearward position. Set the indicator to "O". Prise between the planet carrier and the internal gear with a screwdriver to move the clutches to their extreme forward position. Read the end play on the indicator. The allowable limits are 0·008″–0·044″ (0·2–1·1 mm.). It is preferable to have approximately 0·020″ (0·5 mm.). Should correction be necessary, remove the output shaft, extension housing and companion flange as an assembly so that the selective washer can be changed.

Fig. 35. *Removing the rear pump outlet tube (Extractor Tool Part No. CWG45).*

Fig. 34. *Removing the rear servo.*

Fig. 36. *Checking end play.*

Page FFY.s.31

AUTOMATIC TRANSMISSION

Fig. 37. *Removing the selective thrust washer.*

Selective thrust washers are available in the following thicknesses:

0·061″–0·063″	0·074″–0·076″	0·092″–0·094″
(1·53–1·58 mm.)	(1·85–1·90 mm.)	(2·3–2·35 mm.)
0·067″–0·069″	0·081″–0·083″	0·105″–0·107″
(1·68–1·73 mm.)	(2·03–2·08 mm.)	(2·63–2·68 mm.)

Place the shift selector in park position to hold the output shaft, then remove the companion flange nut, lockwasher, flat washer and flange.

Remove the bearing retainer capscrews, the bearing retainer and the bearing retainer gasket.

Slide the speedometer drive gear off the output shaft.

Remove the governor inspection cover and gasket.

Remove the five extension housing capscrews and remove the output shaft and extension housing assembly.

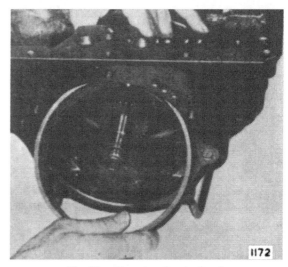

Fig. 39. *Removing the rear band.*

Remove the two hook type seal rings from the rear of the primary sun gear shaft.

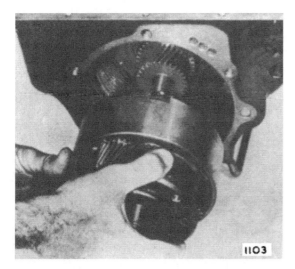

Fig. 38. *Removing the planet carrier.*

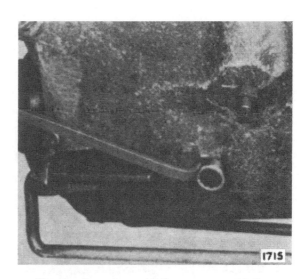

Fig. 40. *Removing one of the centre support bolts.*

Page FFY.s.32

AUTOMATIC TRANSMISSION

Remove the selective thrust washer from the rear of the planet carrier (Fig. 37).

Pull the planet carrier from the transmission (Fig. 38).

Fig. 41. *Removing the clutch assemblies.*

Pull the rear band through the rear opening of the transmission. Hold the two ends of the band together with the left hand while pulling rearward through the rear of the case with the right hand (Fig. 39).

Fig. 42. *Removing the front pump.*

Remove the two centre support bolts; one from each side of the case (Fig. 40).

Fig. 43. *Removing the attaching setscrew.*

Remove the centre support, push on the end of the input shaft to start the rearward movement of the centre support.

Remove the front and rear clutch assemblies, placing them in a suitable stand for dismantling (Fig. 41). (The planet carrier can be used as a stand for dismantling and assembling the clutches).

Remove the front band (up and out of the case).

Remove the front pump oil seal. Use a seal puller or punch.

Remove the four front pump attaching capscrews and lift off the front pump (Fig. 42).

Remove the front pump oil seal ring from the case.

Front Pump—Dismantling

Remove the stator support attaching screw and remove the stator support (Fig. 43). Mark the top of the internal and external gears with marking ink or a crayon. Lift the gears from the pump body.

Inspect the pump body, the internal and external tooth gears, and stator supports for scores, scratches and excessive wear.

Page FFY.s.33

AUTOMATIC TRANSMISSION

Minor scratches and scores can be removed with crocus cloth or jewellers' rouge. However, parts showing deep scratches, scores or excessive wear should be replaced. If excessive wear or scoring is observed, replace the complete pump assembly (since the gears and body are carefully matched when built, these parts should not be interchanged or individually replaced).

Front Pump—Assembling

Drive a new seal into the pump body until it bottoms.

Lubricate all pump parts with transmission fluid before assembly. Install the internal and external gears in the pump body with marks previously made in the upward position. Insert the stator support on the pump body and install the retaining screw. Torque the screw to 25–35 lb. in. (0·29–0·40 kgm.). Check the gears for free movement.

Manual Linkage—Dismantling

Pull the retainer clip from the forward end of the linkage rod (Fig. 44). Disconnect the rod from the manual valve detent lever. Release the detent ball and spring by rocking the manual valve lever to the extreme of its travel. The ball will be released with considerable force, but can be caught in a shop towel or even in the hands. Remove the manual lever locknut, the manual detent lever, and then pull the manual control lever from the transmission. Prise the manual lever oil seal from the transmission case with a screwdriver.

Manual Linkage—Assembling

Install a new manual lever oil seal. Assemble the manual control lever through the transmission case boss. Place the manual valve detent lever and locknut on the manual control lever shaft. Rock the manual valve lever to its extreme travel, then install the detent spring. Place the ball in position on the spring, then using the lubrication ball and spring (Fig. 45), rock the manual valve lever back over the ball and spring. Connect the linkage rod and insert the retainer spring clip.

Fig. 45. *Releasing the detent ball.*

Park Linkage—Dismantling

Pull the retainer clip from the rear of the parking brake linkage rod. Disconnect the linkage rod from the torsion lever. Remove the retainer spring from the torsion lever pin and slide the washer with the torsion lever off the pin. Tap the toggle lever rearward to loosen the pin retainer (Fig. 46), then pull the retainer using snap ring pliers (Fig. 47). The toggle lever pin and toggle lever can now be removed. A magnet may be used to pull the parking pawl anchor pin from the transmission case. The parking pawl is now free to be removed.

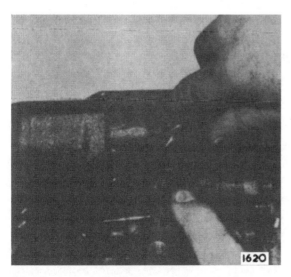

Fig. 44. *Removing the retainer clip from the linkage rod.*

Page FFY.s.34

AUTOMATIC TRANSMISSION

Fig. 46. *Tapping the toggle lever rearwards.*

Parking Linkage—Assembling

Assemble the parking pawl and shaft. Use a new toggle lever retainer to assemble the toggle lever and toggle pin. Assemble the torsion lever pin, then the washer, and then place the retainer spring on the torsion lever pin. Connect the linkage rod to the torsion lever and insert the spring clip.

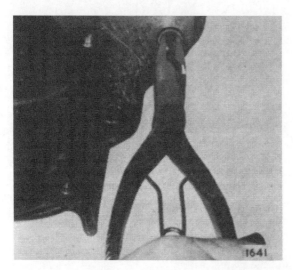

Fig. 47. *Removing the toggle lever pin retainer.*

Clutches—Dismantling

Place the clutch pack in a suitable stand. The planet carrier will work very well for this purpose.

Fig. 48. *Applying compressed air to the clutch feed hole.*

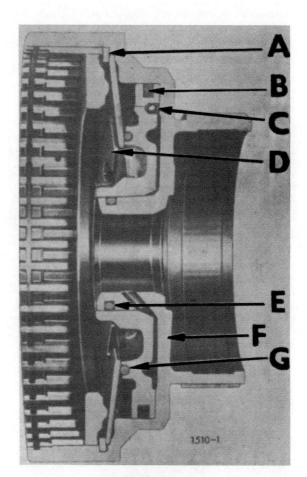

A. Clutch spring ring.
B. Sealing ring.
C. $\frac{3}{16}"$ steel ball.
D. Clutch spring.
E. Sealing ring.
F. Cylinder.
G. Piston.

Fig. 49. *Sectioned view of the clutch front drum.*

Page FFY.s.35

AUTOMATIC TRANSMISSION

Lift the complete front clutch assembly from the rear clutch and forward sun gear.

Remove the snap ring and lift the input shaft from the clutch cylinder. (The clutch hub thrust washer may stick to the input shaft).

Lift the clutch hub and thrust washer from the clutch assembly.

Lift the front clutch plates and the pressure plate from the assembly.

Remove the clutch return spring snap ring and then the return spring. It is not necessary to compress the spring to remove the snap ring.

Compressed air applied to the clutch feed hole in the clutch hub will force the piston from the clutch cylinder (Fig. 48).

Remove the rubber seal rings from the clutch hub and clutch piston.

Remove the two front clutch sealing rings from the forward sun gear shaft (Fig. 50).

Use the service tool to compress the clutch return spring, then remove the spring retainer snap ring. Release the spring, but do not permit the spring retainer to catch in the snap ring groove as the spring is being released (Fig. 51).

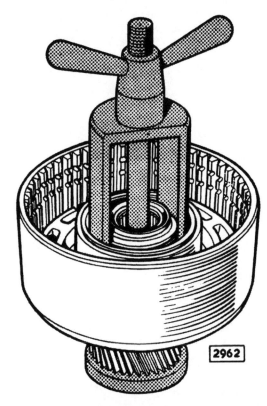

Fig. 51. *Dismantling the clutch using the special tool (Part No. CBW37A).*

Fig. 50. *Removing the two front clutch sealing rings.*

Remove the thrust washer and thrust plate from the shoulder of the rear clutch hub.

Lift the rear clutch assembly up and off the forward sun gear shaft.

Remove the rear clutch ring.

Remove the clutch pressure plate and the clutch plates.

Replace the forward sun gear shaft in the clutch hub, being careful not to break the cast iron sealing rings. The clutch piston can now be removed from the clutch cylinder by blowing compressed air through the rear clutch passage of the forward sun gear.

Remove the forward sun gear from the clutch cylinder and remove the two rear clutch sealing rings from their grooves in the shaft.

Remove the rubber seal rings from the clutch hub and the clutch piston.

Inspection of Clutches

Inspect all parts for burrs, scratches, cracks and wear. Check all the front clutch plates and the rear clutch friction plates for flatness. Check the rear

Page FFY.s.36

AUTOMATIC TRANSMISSION

clutch steel plates for proper cone. Lay plates on a flat surface when checking for flatness and cone. Cone should be 0·010″ to 0·020″ (0·25 to 0·5 mm.). Replace friction plates when wear has progressed so that the grooves are no longer visible. Replace all warped plates. Replace complete set of steel or friction plates in any clutch. Do not replace individual plates (Fig. 52).

Fig. 52. *Checking a clutch plate.*

Inspect the band surfaces of the drum for wear. If only slightly scored the drum may be refaced. Renew if excessive.

Inspect the clutch bushing and the needle bearing for wear and brinelling and for scores. The cast iron sealing rings are normally replaced. If the transmission is being rebuilt and has had little service, the rings may be re-used if they have not worn excessively and are not scratched or distorted.

Inspect the forward sun gear for broken or worn teeth. Inspect all journals and thrust surfaces for scores. Inspect all fluid passages for obstruction or leakage. Inspect the front clutch lubrication valve for freedom (Fig. 53).

Clutches—Assembling

Place the planet carrier on the assembly bench.

Place the forward sun gear in the carrier. Be sure the thrust washer is on the shaft (Fig. 54).

Fig. 54. *Placing the forward sun gear on the carrier.*

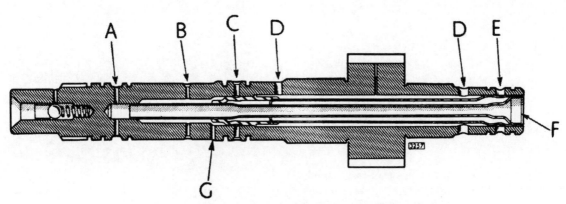

Fig. 53. *Longitudinal section of the forward sun gear showing oil ways.*
A, F—Front clutch: C, E—Rear clutch: B, D, G—lubrication.

Page FFY.s.37

AUTOMATIC TRANSMISSION

Assemble the rubber "O" ring in its groove on the rear clutch hub (Fig. 55).

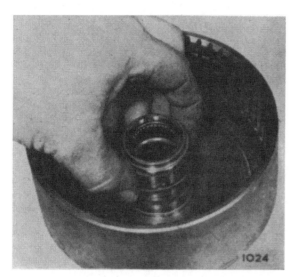

Fig. 55. *Fitting the "O" ring on the rear clutch hub.*

Assemble the square section rubber seal ring in its groove on the rear clutch piston (Fig. 56).

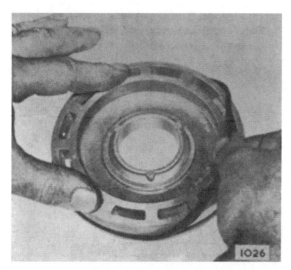

Fig. 56. *Fitting the rear clutch piston sealing ring.*

Assemble the clutch piston in the rear clutch cylinder using Tool Part No. CWG.41 to force it into position. Be sure to lubricate the seal rings so that they will assemble easier.

Place the rear clutch return spring and spring retainer in position on the clutch piston. The rear clutch spring fixture is then used to compress the spring, then the snap ring is assembled in its groove in the clutch.

Fig. 57. *Fitting rear clutch over primary sun gear ring.*

Install the rear clutch cast iron sealing rings in their grooves on the forward sun gear. Be sure that the rings are free in their grooves. Centre each ring in its groove, so that ends do not overlap edges of groove.

Fig. 58. *Fitting a rear clutch steel plate.*

Page FFY.s.38

AUTOMATIC TRANSMISSION

Place the rear clutch piston and cylinder assembly over the forward sun gear and gently slide it down over the sealing rings (Fig. 57).

Fig. 59. *Fitting a rear clutch friction plate.*

Install a rear clutch steel plate with its concave face up or forward facing in the transmission. Note that these plates are identified by missing teeth on the O.D. and are not interchangeable with front clutch steel plates (Fig. 58).

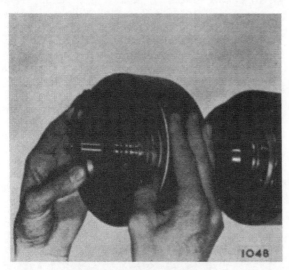

Fig. 60. *Fitting the snap ring.*

Install a rear clutch friction plate, then alternating with first a steel and then a friction plate, complete the clutch pack (Fig. 59).

Install the rear clutch pressure plate.

Install the rear clutch snap ring. This ring has one tanged end (Fig. 60).

Fig. 61. *Fitting the sealing rings.*

Install the front clutch cast iron sealing rings in their grooves on the forward sun gear. Centre each ring in its groove so that ends do not overlap edges of the groove (Fig. 61).

Fig. 62. *Fitting the front thrust plate.*

Install the front clutch cylinder thrust plate (Fig. 62). Be sure flats on the washer match flats on shaft.

Page FFY.s.39

AUTOMATIC TRANSMISSION

Install the front clutch cylinder thrust washer (Fig. 63).

Fig. 63. *Fitting the front clutch cylinder thrust washer.*

Assemble the front clutch hub "O" ring into its groove in the clutch hub.

Assemble the front clutch piston square section rubber sealing ring in the groove of the clutch piston.

Install the clutch piston into the clutch cylinder after thoroughly lubricating the parts. Press the piston into position using Tool Part No. WG.42.

Fig. 64. *Assembling the front clutch.*

Install the front clutch belleville spring and snap ring. This snap ring is thicker than the other two clutch snap rings and has two tanged ends instead of one.

Fig. 65. *Fitting the front pressure plate.*

Assemble the front clutch assembly over the forward sun gear shaft and into the rear clutch, being careful not to distort or break the cast iron sealing rings. Use a short oscillating movement to engage splines of the rear clutch friction plates (Fig. 64).

Install the front clutch pressure plate (Fig. 65).

Install the front clutch hub, followed by front clutch hub thrust washer (Fig. 66).

Fig. 66. *Fitting the front clutch hub thrust washer.*

Page FFY.s.40

AUTOMATIC TRANSMISSION

Install a front clutch friction plate over the splines of the hub (Fig. 67). Next, install a front clutch outer plate, meshing splines in the cylinder, alternating as above, complete assembly of plates (Fig. 68).

Centre Support

The centre support is serviced as an assembly. Therefore, there is no dismantling or assembly procedure.

Inspect the support for burrs or distortion, the race bearing surface for scores or scratches.

Fig. 67. *Fitting a front friction plate.*

Fig. 69. *Fitting the snap ring.*

Assemble the input shaft to the front clutch cylinder.

Assemble the snap ring that holds the input shaft in place (Fig. 69).

Fig. 68. *Fitting a front clutch outer plate.*

Fig. 70. *Placing the thrust washer in position.*

Pinion Carrier Assembly

The pinion carrier is serviced as an assembly. Therefore, there is no dismantling or assembly procedure.

Place the thrust washer on the input shaft and the clutch assemblies are complete (Fig. 70).

Page FFY.s.41

AUTOMATIC TRANSMISSION

Inspect the band surface and the inner and outer bushing for scores. Rotate pinions on their shafts to check for freedom of movement and for worn or broken teeth. Use a feeler gauge to check pinion end play. End play should be 0·010″ to 0·020″ (·23 to ·5 mm.). Inspect pinion shafts for tightness to the planet carrier.

Sprag Clutch

A sprag-type one-way clutch assembly is incorporated in the planet carrier assembly and is held in place by a snap ring.

When installing the sprag clutch, the flange side of the sprag cage is located down into the outer race of the planet carrier assembly with the copper tension springs towards the centre support.

After the planet carrier and sprag assembly are installed in the case, the planet carrier will freewheel when turned counterclockwise and lock when turned clockwise (from the rear).

Output Shaft

Remove the extension housing and bearing from the output shaft by lifting the housing and tapping the shaft with a heavy plastic hammer.

Remove the bearing spacer washer.

Slide the oil collector and tubes from the shaft.

Remove the four sealing rings.

Remove the governor snap ring, governor and governor drive ball from the output shaft.

Lift the rear pump from the shaft and remove the rear pump drive key.

The snap ring may be removed and the output shaft removed from the ring gear; however, this is not necessary unless replacing one of these parts.

Inspect the output shaft thrust surfaces and journals for scores and the internal gear for broken teeth. Check the ring grooves, splines and gear teeth for burrs, wear or damage. The output shaft is a two-piece assembly and is serviced separately. Inspect the distributor and sleeve mating surfaces for excessive wear and for burrs, scores or leakage.

Governor

Remove the governor body cover plate attaching screws and remove the plate (Fig. 71). Remove the governor body attaching screws, then remove the body from the counter weight. Slide the spring retainer from the stem of governor weight and remove the spring. Remove the valve and weight from the governor body.

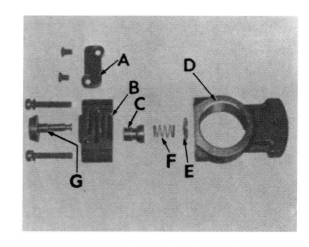

A. Governor body cover plate.
B. Governor body.
C. Valve.
D. Counter weight.
E. Spring retainer.
F. Spring.
G. Weight.

Fig. 71. *Exploded view of the governor.*

Inspect the governor weight, valve and bore for scores. Minor scores may be removed with crocus cloth. Replace the governor valve, weight or body if deeply scored. Check for free movement of the weight and valve in the bore. Inspect all fluid passages in the governor body and counterweight for obstruction. All fluid passages must be clean. Inspect the mating surfaces of the governor body and counterweight for burrs and distortion. Check governor spring retainer washer for burrs. The mating surfaces must be smooth and flat.

Re-install governor body cover plate, torqueing screws to 20–30 lb. in. (0.24 to 0.35 kgm.).

Install the governor valve in the bore of the body. Install the weight in the governor valve. Compress the spring and slide the retainer onto the stem of the weight and release the spring tension. Install the governor body on the counterweight.

Note: Make sure the fluid passages in the body and counterweight are aligned.

Torque the governor body attaching screws to 50–60 lb. in. (0·58 to 0·69 kgm.).

Page FFY.s.42

AUTOMATIC TRANSMISSION

Rear Pump

Withdraw the five ¼" (6·4 mm.) screws, also the No. 10 U.N.C. screw and remove the cover. Mark the top face of the gears with marking ink or a crayon to assure correct re-installation of gears upon assembly (Fig. 72). Remove the drive and driven gears from the pump body.

Check the pump for free movement of the gears.

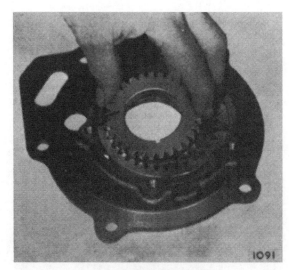

Fig. 73. *Replacing the gears.*

Output Shaft and Rear Pump—Assembling

Install the rear pump drive key in the output shaft.

Install rear pump assembly over the shaft.

Install the governor drive ball into the recess in the output shaft, using a spot of petrolatum to hold in place.

Install governor assembly, with plate on the governor body down (facing pump assembly). Install snap ring to lock governor in place (Fig. 74).

Fig. 72. *Marking the top face of the gears.*

Inspect the gear pockets and crescent of the pump body for scores or pitting. Inspect the bushing and drive and driven gear bearing surfaces for scores. Check all fluid passages for obstructions and clean if necessary. Inspect the mating surfaces, gear teeth, pump body and cover for burrs. If any pump parts are defective beyond minor burrs or scores, which cannot be removed with a crocus cloth, replace complete pump as a unit.

Lubricate parts with transmission fluid and replace both gears with the marks facing upward. Install the pump cover, attaching screws and lock-washers. Tighten the ¼" (6·4 mm.) screws to 50–60 lb. in. (0·58 to 0·69 kgm.) torque and the number 10 screw to 20–30 lb. in. (0·24 to 0·35 kgm.) torque (Fig. 73).

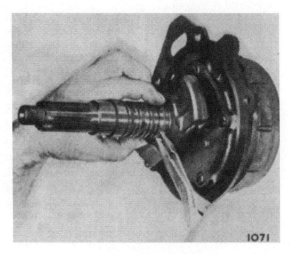

Fig. 74. *Fitting the snap ring.*

Page FFY.s.43

AUTOMATIC TRANSMISSION

Install the four output shaft sealing rings, making sure they are free in their grooves (Fig. 75).

Install oil collector sleeve and tube assembly. Compress each ring with the fingers and carefully slide the sleeve over them (Fig. 76).

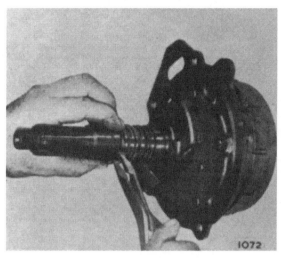

Fig. 75. *Fitting the output shaft sealing ring.*

Fig. 76. *Installation of the oil collector sleeve and tube.*

Assemble the bearing spacer washer against the shoulder on the output shaft (Fig. 77).

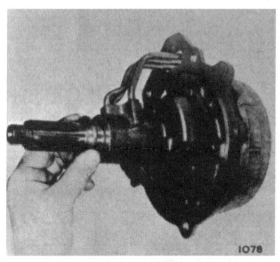

Fig. 77. *Fitting the bearing spacer washer.*

AUTOMATIC TRANSMISSION

Fig. 78. *Exploded view of the front servo.*

Front Servo—Dismantling

Use a small screwdriver to remove the snap ring.

Pull the sleeve and piston from the servo body.

Remove the piston from the servo sleeve.

Remove all sealing rings.

If the servo lever needs attention, it may be removed by first driving the roll pin from the servo and then removing the pivot pin and lever. Use a ⅛" (3.1 mm.) drift punch to remove the roll pin.

Inspect the servo parts for cracks, scratches and wear. Check the adjusting screw for freedom in the lever. Check the lever for freedom of movement.

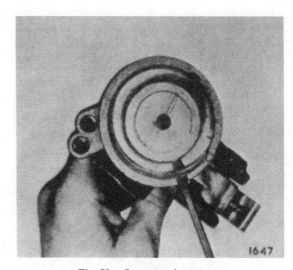

Fig. 79. *Removing the snap ring.*

Page FFY.s.45

AUTOMATIC TRANSMISSION

Front Servo—Assembling

Assemble the servo lever, pivot pin and the roll pin.

Assemble the sealing rings on the sleeve and piston.

Assemble the piston to the sleeve, place the spring in the piston, and assemble the sleeve, piston and spring into the housing.

Replace the snap ring.

Remove the lever and shaft.

Depress the spring retainer while removing the snap ring.

Remove the servo release spring, piston and rubber "O" ring.

Inspect the servo body for cracks, burrs and obstructed passages and the piston bore and stem for scores. Inspect the actuating lever and shaft for wear and brinnelling.

Fig. 80. *Assembling the front servo.*

Rear Servo—Dismantling

Remove the actuating lever roll pin with a ⅛" (3·1 mm.) drift punch.

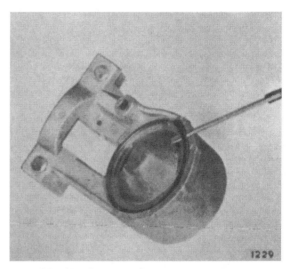

Fig. 82. *Removing the rear servo snap ring.*

Fig. 81. *Removing the rear servo roll pin.*

Fig. 83. *Replacing the roll pin.*

Page FFY.s.46

AUTOMATIC TRANSMISSION

Rear Servo—Assembling

Lubricate all parts of the servo with transmission fluid before starting assembly.

Install a new "O" ring and then install piston in the servo body.

Install the release spring, retainer and snap ring.

Replace the servo lever, shaft and roll pin.

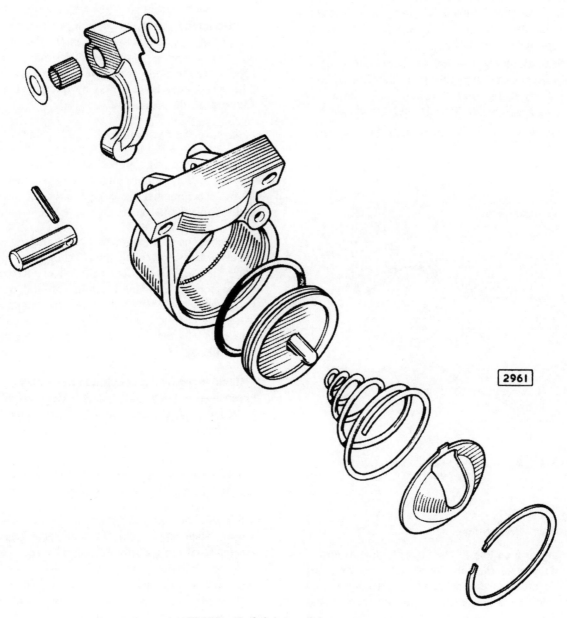

Fig. 84. *Exploded view of the rear servo.*

Page FFY.s.47

AUTOMATIC TRANSMISSION

Pressure Regulator

Remove the valves from the regulator body. Remove the regulator body cover attaching screws and remove the cover. Remove the separator plate from the regulator body.

Wash all parts thoroughly in cleaning solvent and dry with compressed air. Inspect the regulator body and cover mating surfaces for burrs. Check all fluid passages for obstructions. Inspect the control pressure and converter pressure valves and bores for burrs and scores. Remove all burrs carefully with crocus cloth. Check free movement of the valves in their respective bores. The valves should fall freely into the bores when both the valve and bore are dry. Inspect the valve springs for distortion.

When assembling, be careful to avoid damaging the parts. Replace the separator plate and then the cover on the regulator body. Install and torque the attaching screws to 20–30 lb. in. (0·24–0·35 kgm.).

Insert the valves in the pressure regulator body.

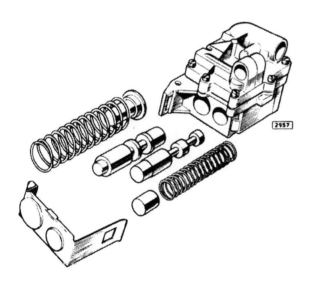

Fig. 85. *Regulator assembly. Valves, springs and retainer shown exploded.*

Valve Body—Dismantling

During dismantling of the control valve assembly, avoid damage to the valve parts and keep the parts clean. Place the valve parts and the assembly on a clean surface while performing the dismantling operation.

Remove the manual valve from the upper valve body.

Remove the four cap screws that retain the valve bodies.

Remove the cover and separator plates from the valve bodies. The body plate is attached to the lower valve body by a cheese head screw and to the upper valve body by a cheese head and a flat head screw. The separator plate and the lower valve body cover are held together by two cheese head screws.

Remove the front upper valve body plate retained by two screws. Remove the compensator valve plug, sleeve, springs and valve. Remove the modulator valve and spring assembly. The outer spring is retained to the modulator valve by a stamped retainer. The spring may be removed by tilting and pressing outward on the retainer.

Remove the downshift valve and spring.

Remove the rear upper valve body plate and throttle return spring retained by three screws to the body. Then remove the compensator cut back valve and the throttle valve.

Remove the four screws that retain the end body to the lower body. Remove the 2–3 shift valve inner and outer springs and the 2–3 shift valve. Remove the orifice control valve and spring and the transition valve spring and valve. Remove the orifice control valve plug and the 2–3 shift valve plug from end body. The end body plate should be removed for cleaning the end body.

Remove the four cheese head screws that retain the lower valve body side plate. Remove the 2–3 governor plug, the D1 and D2 control valve spring and valve.

The rear pump check valve, spring and sleeve generally should not be removed. The sleeve may be removed with snap ring pliers, if necessary.

Remove the end plate from the lower valve body cover. Then remove the 1–2 shift valve and spring and the front servo release orifice valve and spring.

Note: When removing all plates, be sure to hold the plates until screws are removed and release slowly as they are spring loaded.

Page FFY.s.48

AUTOMATIC TRANSMISSION

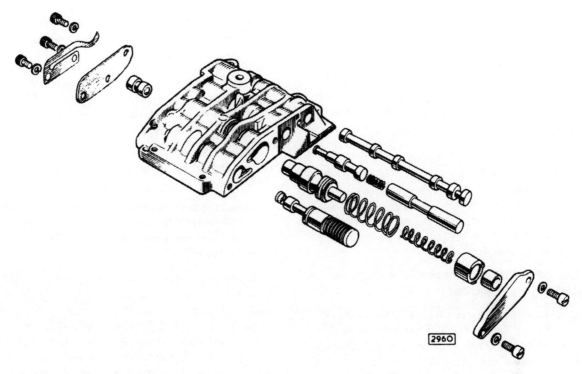

Fig. 86. *Upper valve body exploded - Manual valve, downshift valve, compensator valve and throttle modulator valve exploded.*

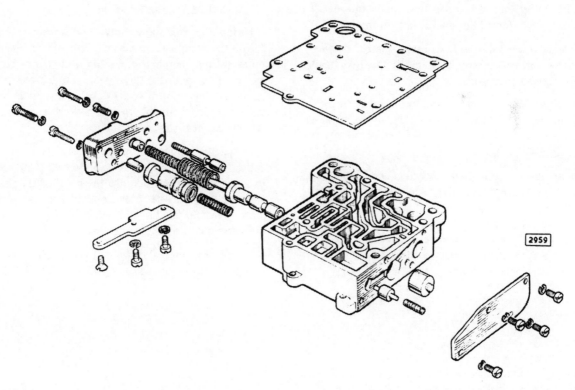

Fig. 87. *Lower valve body - transition valve, 2-3 shift valve and servo orifice control valve (from right to left on left of body 2nd and 3rd governor plug and D1, D2, control valve exploded on right of body.*

Page FFY.s.49

AUTOMATIC TRANSMISSION

Inspection

Clean all parts thoroughly in a cleaning solvent, then dry them with compressed air. Inspect all fluid passages for obstructions. Inspect the check valve for free movement. Inspect all mating surfaces for burrs and distortion. Inspect all plugs and valves for burrs and scores.

Note: Crocus cloth can be used to polish the valves and plugs if care is taken to avoid rounding the sharp edges.

Valve Body—Assembling

When assembling the control valve bodies, always use the following procedure:

Install the valve body plate on the upper valve body (retained by one cheese head and one flat head screw). Do not tighten the screws. If the rear pump check valve sleeve, valve and spring were removed from the lower valve body, install them, carefully staking the sleeve in the bore with the smooth end against the valve.

Place the upper body on the lower body and install the cheese head screw, but do not tighten the screw.

Place the lower valve body separator plate and cover on the lower valve body and install the two head screws, leaving them loose.

Install the four cap screws and lockwashers; torque the four screws to 72 lb. in. (·84 kgm.), then tighten the cheese head screws and flat head screw to 20–30 lb. in. (0·23–0·35 kgm.).

Try all valves dry in their respective bores, rotating them to make sure that they are free before final assembly in the valve body. If any sticking or binding occurs, the valve bodies will have to be separated and each surface lapped on crocus cloth, using a surface plate or a glass plate, to ensure against low or high spots or a warped condition.

Note: Lubricate all valves and plugs with automatic transmission fluid before final assembly in their respective bores.

Install the 1–2 shift valve spring and valve in the lower valve body cover. Install the front servo release orifice valve spring and valve and the cover end plate with two cheese head screws.

Install the range control valve and spring, the governor plug, and then install the side plate with four cheese head screws.

Install the orifice control valve spring and valve, the 2–3 shift valve, the 2–3 shift valve inner and outer springs, the transition valve, and spring in the lower valve body.

Replace the end body plate using one flat head and two cheese head screws and torque to 20–30 lb. in. (0·23–0·35 kgm.). Install the orifice control valve plug and the 2–3 shift valve plug in the lower valve body. Install the end body to the lower valve body, guiding the 2–3 shift valve inner spring into the 2–3 shift valve plug. Three long and one short special cheese head screws are used to retain the end body.

Note: Make sure the inner spring is piloted on the 2–3 shift valve plug.

Install the modulator valve and spring assembly. Install the compensator valve, compensator inner and outer springs, compensator plug and sleeve (be sure end of sleeve with the three protrusions is toward the plate and the smooth end to the spring in the upper valve body). Assemble the plate which is retained by two cheese headed screws.

Install the compensator cut-back valve in the rear end of the upper body. Install the rear plate so that the edge of the plate fits into the band of the throttle valve and install one screw to hold the rear plate in place. Install the throttle return spring and install the two remaining cheese headed screws.

Install the manual valve. Torque on all cheese headed screws should be 20–30 lb. in. (0·23 to 0·35 kgm.)

Page FFY.s.50

AUTOMATIC TRANSMISSION

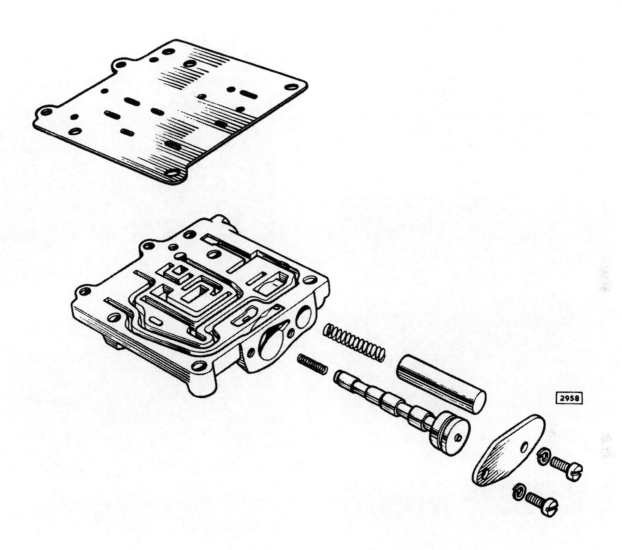

Fig. 88. *Lower valve body cover—front servo release orifice valve and 1-2 shift valve exploded.*

AUTOMATIC TRANSMISSION

TRANSMISSION ASSEMBLING

Lubricate all parts as they are assembled, with the same fluid used for filling the transmission. Petrolatum can be used sparingly to hold gaskets or thrust washers in position during assembly.

Wash the transmission case and dry with compressed air.

Install a new front pump to case gasket, then install the front pump. Torque the four attaching cap screws to 17–22 lb. ft. (2·35 to 3·04 kgm.).

Install the front band through the bottom of the case, positioning the band so that the anchor end is aligned with the anchor in the case.

Fig. 89. *Fitting a new front pump gasket.*

Fig. 90. *Installing the front pump.*

Install the front clutch, rear clutch and forward sun gear assembly in the case. Handle the clutch assemblies in a manner that will prevent the clutches being pulled apart.

Fig. 91 *Installing the front band.*

Fig. 92. *Installing the front clutch.*

Install the centre support in the transmission case with the three positioning holes aligned with the holes in the case.

Install the centre support cap screws with the rolled edge of each lockwasher towards the case. Torque to 20–25 lb. ft. (2·76 to 3·46 kgm.).

Page FFY.s.52

AUTOMATIC TRANSMISSION

Install the rear band through the rear of the case. Be sure that the end with the depression or dimple is placed toward the adjusting screw.

Fig. 93. *Installing the centre support.*

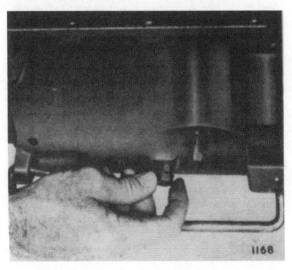

Fig. 94. *Fitting the centre support cap screws.*

Use petrolatum sparingly to hold the forward sun gear thrust plate and needle bearing in the planet carrier, while the carrier is assembled over the sun gear.

Install the hook type seal rings on the rear of the forward sun gear. Check the rings for free movement in their grooves.

Choose a selective washer to give the correct end play (end play determined during dismantling is used to determine the need for a different thrust washer).

Fig. 95. *Fitting the rear band.*

Fig. 96. *Fitting the thrust plate and needle bearing.*

Install washer on the rear of the planet carrier.

Use petrolatum to hold the rear pump to case gasket to rear of the case.

Install the ring gear and output shaft assembly. Align the three oil tubes as the assembly is fitted and tap them in position.

Page FFY.s.53

AUTOMATIC TRANSMISSION

Place the rear pump to extension housing gasket in position, then assemble the extension housing. Torque the five extension housing cap screws to 28–33 lb. ft. (3·87 to 4·56 kgm.).

Install the companion flange, flat washer, lockwasher and nut. Torque the nut to 90–120 lb. ft. (12·44–16·58 kgm.).

Fig. 97. *Assembling the carrier over the sun gear.*

Fig. 99. *Fitting the washer on the rear of planet carrier.*

Fig. 98. *Fitting the sealing rings.*

Fig. 100. *Tapping the output shaft assembly into position.*

Install the bearing snap ring, and then tap the ball bearing into position in the extension housing and on the output shaft (be sure spacer washer is on shaft ahead of bearing).

Slide the speedometer drive gear on the output shaft.

Install rear seal in bearing retainer. Assemble the bearing retainer in its gasket.

Front Servo Installation

Rotate the front band into position so that the anchor end is positioned over the anchor pin in the case.

Position the servo strut with the slotted end aligned with the servo actuating lever, and hold it in position with the middle and index fingers of the left hand.

AUTOMATIC TRANSMISSION

Engage the end of the band with the small end of the strut then position the servo over the dowel pin.

Install the attaching cap screw but do not screw it in more than two or three threads at this time.

Fig. 101. *Installing the front servo.*

Fig. 102. *Engaging the servo anchor strut.*

Rear Servo Installation

Position the servo anchor strut over the adjusting screw, then rotate the rear band to engage this strut. Place the servo actuating lever strut with the notched end to the band and lift the other end with index finger or screwdriver, while locking the servo lever over the strut.

Install the long pointed bolt in the forward servo hole so that it will engage the centre support.

The other shorter bolt is used in the rear position.

Torque the bolts to 40–50 lb. ft. (5·53–6·91 kgm.).

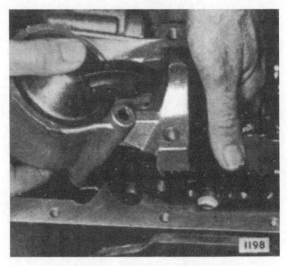

Fig. 103. *Fitting the rear servo.*

Valve Body Installation

Place the manual selector in park or reverse position. Carefully align the valve body with the servo tubes and gently slide the valve body further onto the tubes.

The front servo must be pulled up off the dowel to allow easy assembly. Be careful at this point—the servo apply strut may become disengaged from the servo. Before seating the valve body on the case, install the nipple end of the throttle cable into the throttle cam.

Page FFY.s.55

AUTOMATIC TRANSMISSION

Next, align the manual valve with the inside lever pin and the valve body will then drop into position. Torque the three valve body attaching cap screws to 8–10 lb. in. (0·09–0·12 kgm.).

Replace the control pressure tube, by first assembling the long straight end into the regulator, then rocking the tube downward into the control valve body. If too much resistance is encountered, it will help to loosen the control body attaching cap screws until the tube can be assembled.

Fig. 104. *Fitting the servo tubes.*

Fig. 106. *The valve body in position.*

Fig. 105. *Positioning the valve body.*

Torque the front servo attaching cap screw to 30–35 lb. ft. (4·15–4·84 kgm.) and adjust the front servo.

Fig. 107. *Replacing the control pressure tube.*

Page FFY.s.56

AUTOMATIC TRANSMISSION

Pressure Regulator Installation

Assemble the regulator, with the valves in position in their bores, to the case with the attaching cap screws.

Torque cap screws to 17–22 lb. ft. (2·35–3·04 kgm.). Install both springs and guides, then install the spring retainer.

Install the front servo apply and release tubes in the servo.

Install the rear pump inlet and outlet tubes, using new "O" rings.

Replace the compensator tube by aligning one end with the pressure regulator and the other end with the control valve body and then tap it into position.

Assemble the long end of the lubrication tube into the rear pump, then rock the other end into position and tap it into the pressure regulator assembly.

Fig. 108. *The pressure regulator installed.*

Fig. 110. *Fitting the apply and release tubes.*

Fig. 109. *Fitting the pressure regulator springs.*

Fig. 111. *Fitting the lubrication tube.*

Page FFY.s.57

AUTOMATIC TRANSMISSION

Replace the front band lubrication tube. Be sure the tube is aligned so that the open end will direct oil onto the front drum surface at the front band gap. Tube should point at approximately the centre of the gap.

Assemble the oil screen assembly onto the rear pump inlet tube and then rock into position over the front pump inlet on the pressure regulator assembly. Hook the screen retainer under the lubrication tube, lay across screen, and snap onto compensator tube.

Install the oil pan gasket, the oil pan and torque the 14 cap screws to 10–20 lb. ft. (1·38–2·76 kgm.).

Adjust the rear band.

Fig. 113. *Fitting the rear pump inlet tube.*

Fig. 112. *Replacing the compensator tube.*

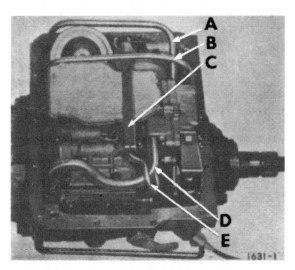

Fig. 114. *View of the Model 8 transmission inverted.*

CONVERTER AND CONVERTER HOUSING

When installing the converter housing, the maximum allowable runout should not exceed 0·010″ (0·25 mm.) for bore or face indicator readings relative to crankshaft centre line; however, it is preferable to have less than 0·006″ (0·015 mm.) reading for both.

When installing the transmission to the converter housing and converter assembly, be certain that the converter lugs are properly aligned with the front pump drive gear, so that the parts will not be damaged by forcing impeller hub drive tangs against the pump drive gear lugs.

AUTOMATIC TRANSMISSION

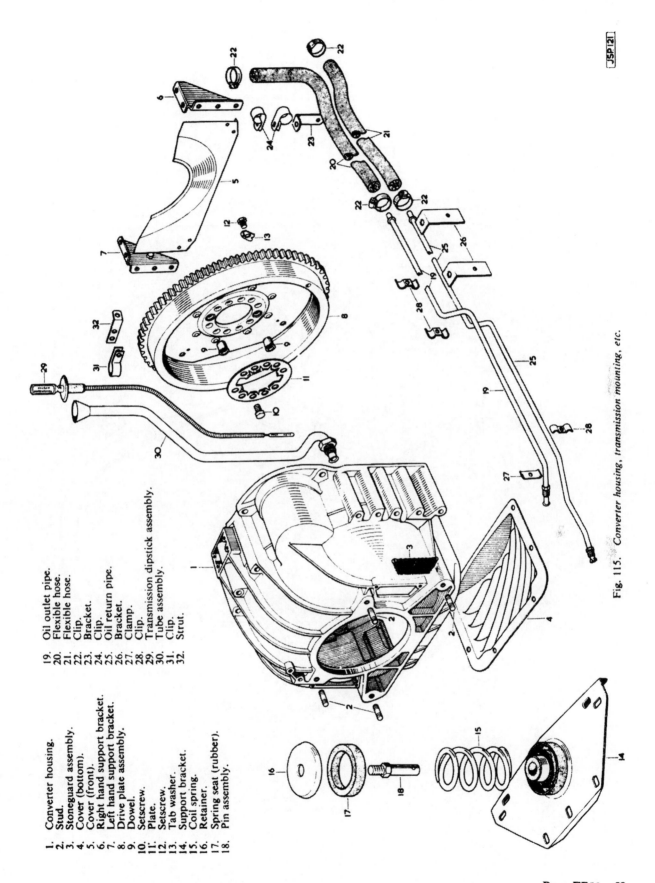

Fig. 115. *Converter housing, transmission mounting, etc.*

1. Converter housing.
2. Stud.
3. Stoneguard assembly.
4. Cover (bottom).
5. Cover (front).
6. Right hand support bracket.
7. Left hand support bracket.
8. Drive plate assembly.
9. Dowel.
10. Setscrew.
11. Plate.
12. Setscrew.
13. Tab washer.
14. Support bracket.
15. Coil spring.
16. Retainer.
17. Spring seat (rubber).
18. Pin assembly.
19. Oil outlet pipe.
20. Flexible hose.
21. Flexible hose.
22. Clip.
23. Bracket.
24. Clip.
25. Oil return pipe.
26. Bracket.
27. Clamp.
28. Clip.
29. Transmission dipstick assembly.
30. Tube assembly.
31. Clip.
32. Strut.

Page FFY.s.59

AUTOMATIC TRANSMISSION

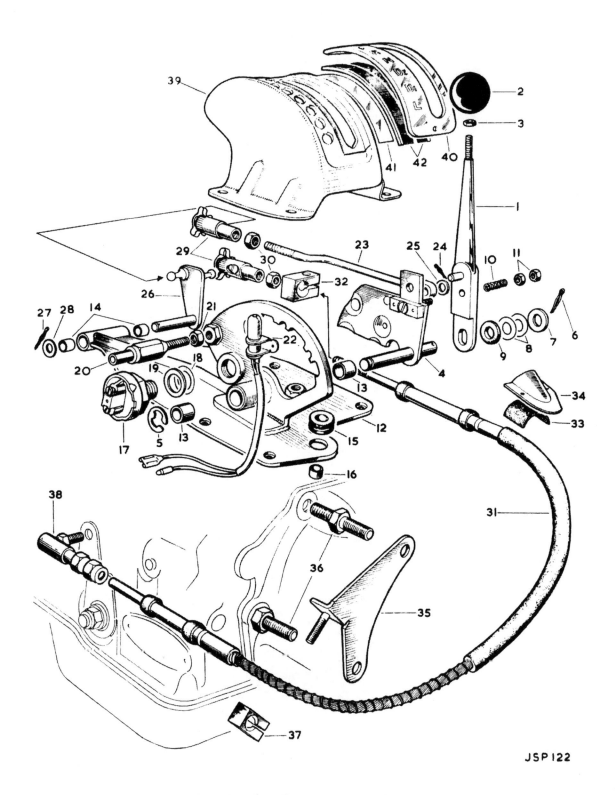

Fig. 116. *The transmission controls.*

Page FFY.s.60

AUTOMATIC TRANSMISSION

1. Selector lever assembly.
2. Knob.
3. Nut.
4. Cam plate assembly.
5. Circlip.
6. Split pin.
7. Washer.
8. Shim.
9. Washer (rubber).
10. Spring.
11. Nut.
12. Mounting plate and selector gate assembly.
13. Bush.
14. Bush.
15. Grommet.
16. Distance tube.
17. Reverse lamp switch.
18. Shim.
19. Shim.
20. Starter cut-out switch.
21. Nut.
22. Lamp assembly.
23. Operating rod assembly.
24. Split pin.
25. Washer.
26. Transfer lever assembly.
27. Split pin.
28. Washer.
29. Ball joint.
30. Nut.
31. Gear control cable assembly.
32. Clamp.
33. Pad.
34. Plate.
35. Abutment bracket.
36. Stud.
37. Clamp.
38. Adjustable ball joint.
39. Cover assembly.
40. Indicator plate.
41. Light filter.
42. Seal.

Page FF Y.s.61

AUTOMATIC TRANSMISSION

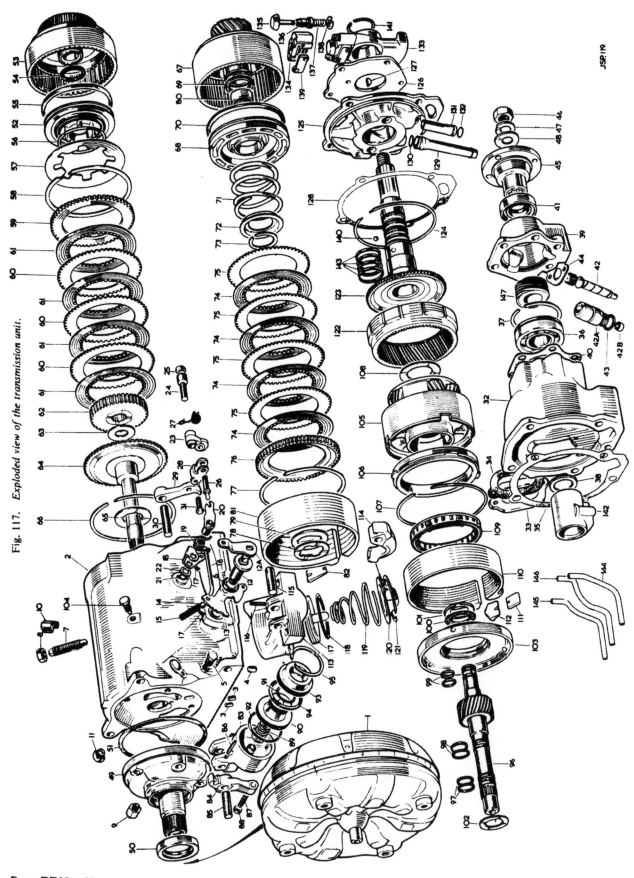

Fig. 117. Exploded view of the transmission unit.

Page FFY.s.62

AUTOMATIC TRANSMISSION

1. Converter assembly.
2. Transmission case assembly.
3. Plug.
4. Dowel.
5. Plug.
6. Oil seal.
7. Screw.
8. Nut.
9. Union.
10. Union.
11. Breather assembly.
12. Manual control shaft assembly.
12A. Selector lever.
13. Lever assembly.
14. ⅜″ ball.
15. Spring.
16. Link.
17. Clip.
18. Torsion lever.
19. Spring.
20. Forked lever.
21. Clip.
22. Washer.
23. Toggle lever.
24. Toggle pin.
25. Plug.
26. Ball pin.
27. Spring.
28. Link.
29. Pawl.
30. Pivot pin.
31. Pin.
32. Extension case assembly.
33. Cover plate.
34. Gasket.
35. Gasket.
36. Bearing.
37. Snap ring.
38. Spacing washer.
39. Speedometer housing.
40. Gasket.
41. Oil seal assembly.
42. Speedometer driven gear.
42A. Bearing.
42B. Oil seal.
43. "O" ring.
44. Plate.
45. Flange.
46. Nut.
47. Lockwasher.
48. Washer.
49. Front pump assembly.
50. Oil seal assembly.
51. Sealing ring.
52. Piston assembly.
53. Cylinder.
54. Sealing ring (inner).
55. Sealing ring (outer).
56. Split ring.
57. Spring.
58. Snap ring.
59. Pressure plate.
60. Clutch plate (drive).
61. Clutch plate (friction).
62. Hub.
63. Thrust washer (fibre).
64. Input shaft assembly.
65. Thrust washer.
66. Snap ring.
67. Front drum assembly.
68. Piston assembly.
69. Sealing ring (inner).
70. Sealing ring (outer).
71. Spring.
72. Seat.
73. Snap ring.
74. Clutch plate (friction).
75. Clutch plate (drive).
76. Pressure plate.
77. Snap ring.
78. Thrust washer (bronze).
79. Thrust washer (steel).
80. Needle bearing.
81. Brake band.
82. Strut (servo).
83. Body.
84. Lever.
85. Pivot pin.
86. Roll pin.
87. Screw.
88. Nut.
89. Return spring.
90. Piston assembly.
91. "O" ring (small).
92. "O" ring (large).
93. Piston sleeve.
94. Sealing ring.
95. Snap ring.
96. Forward sun gear assembly.
97. Sealing ring.
98. Sealing ring.
99. Sealing ring.
100. Thrust bearing.
101. Race.
102. Thrust washer (bronze).
103. Centre support assembly.
104. Screw.
105. Planetary gears and rear drum assembly.
106. Outer race.
107. Snap ring.
108. Thrust washer.
109. One way clutch assembly.
110. Brake band for rear drum.
111. Strut (servo).
112. Anchor strut.
113. Body assembly.
114. Lever.
115. Shaft.
116. Roll pin.
117. Piston.
118. "O" ring.
119. Return spring.
120. Plate.
121. Snap ring.
122. Ring gear.
123. Mainshaft assembly.
124. Snap ring.
125. Rear pump assembly.
126. Plate.
127. Key.
128. Gasket.
129. Oil inlet tube.
130. "O" ring.
131. Oil outlet tube.
132. "O" ring.
133. Governor assembly.
134. Governor body.
135. Governor weight.
136. Governor valve.
137. Spring.
138. Retainer.
139. Cover plate.
140. ¼″ ball.
141. Snap ring.
142. Oil collector sleeve.
143. Piston ring.
144. Oil collector tube (front).
145. Oil collector tube (intermediate).
146. Oil collector tube (rear).
147. Speedometer drive gear.

Page FFY.s.63

SECTION H
REAR AXLE

The rear axle (final drive unit) remains identical to that stated in the 3·8 "E" Type Service Manual with the exception of the following details:—

Axle Ratios

U.S.A., CANADA (Manual transmission only)	3·54:1
All other countries (Manual transmission only)	3·07:1
Automatic Transmission (2+2 only, U.S.A. and CANADA)	3·31:1
Automatic Transmission (2+2 only, all other countries)	2·88:1

HALF SHAFT UNIVERSAL JOINTS

Grease nipples (see the 3·8 "E" Type Service Manual—Page H8—Early cars) were reintroduced from the commencement of production of Series 2 cars.

Access to the nipples of the outer joints is gained by removing the plastic sealing plugs from the joint covers. The universal joints should be greased every 6,000 miles (10.000 km.).

SECTION I

STEERING

GENERAL DESCRIPTION

The upper and lower steering columns and the mountings are of the collapsible type designed to comply with the U.S.A. Federal Safety Regulations.

The collapse points are retained by nylon plugs which will shear on impact, allowing the steering wheel and columns (upper and lower) to move forward.

NO ATTEMPT must be made to repair the units if damaged due to accident.

NEW replacement items MUST be fitted.

UPPER STEERING COLUMN

Description

The upper steering column (inner) is composed of two separate sliding shafts retained to a fixed length by nylon plugs, the outer column being pierced in a lattice form.

The inner shaft assembly is supported in the outer column by two pre-lubricated taper roller bearings.

A gaiter covers the pierced portion of the outer column to seal against the ingress of dirt.

Removal

Disconnect the battery.

Withdraw the self-tapping screws and remove the under-scuttle casing above the steering column.

Disconnect the cables contained in the direction indicator switch harness.

Note the location of the connections for reference when refitting.

Withdraw the ignition key, remove the ring nut and detach the ignition switch from the mounting bracket on the steering column.

Note : If the car is fitted with air-conditioning equipment the switch will be mounted on a bracket attached to the evaporator unit and need not be removed.

Release three grub screws in the steering wheel hub and remove the steering wheel motif.

Remove the locknut, hexagon nut and flat washer and withdraw the steering wheel from the splines on the inner column.

Disconnect the cables from the steering column lock (if fitted). Note the location of the cables for reference when refitting.

Remove the nut, lockwasher and pinch bolt securing the upper universal joint to the lower steering column.

Remove two nuts and lockwashers securing the upper column lower mounting bracket to the underside of the scuttle.

Remove two bolts, nuts, lockwashers and distance pieces securing the upper mounting bracket to the support bracket on the body.

Withdraw the upper column from the splines on the lower column.

Note : If the steering column has not been damaged by impact, i.e., if the nylon plugs in the inner column or the top mounting bracket have not sheared, excessive force must NOT be used to separate the upper universal joint from the lower column.

Refitting

Refitting is the reverse of the removal procedure.

Set the road wheels in the straight ahead position and check that the bolt holes in the lugs of the upper column universal joint register correctly with the groove machined in the lower column splines. Tighten the pinch bolt to a torque of 16-18 lb. ft. (2.2-2.5 kgm).

IMPORTANT

Excessive force as noted under 'Removal' must not be used when reassembling the universal joint to the column.

UNDER NO CIRCUMSTANCES should a mallet or similar tool be used when engaging the splines in the joint and column.

If the splines will not engage freely, inspect for damage or burrs and remove with a fine file.

NO ATTEMPT must be made to repair any nylon plugs which have sheared due to impact.

Dismantling

Dismantling is confined to removing the steering column adjuster locknut, the splined shaft and the direction indicator switch as detailed on **page I.8**.

LOWER STEERING COLUMN

Description

The lower steering column comprises two sliding shafts retained to a fixed length by nylon plugs.

Page IY.s.1

STEERING

Removal

Remove the upper steering column as detailed previously.

Remove the nut, lockwasher and bolt securing the column to the lower universal joint and withdraw the column rearwards through the grommet.

Note : If the steering column has not been damaged by impact, i.e., if the nylon plugs in the column have not sheared, excessive force must NOT be used to separate the column and the lower universal joint.

Refitting

Refitting is the reverse of the removal procedure.

Check that the bolt holes in the universal joint register correctly with the groove machined in the column splines. Tighten the pinch bolt to a torque of 16 - 18 lb. ft. (2.2 - 2.5 kgm).

IMPORTANT

Excessive force as noted under 'Removal' must not be used when reassembling the universal joint to the column. **UNDER NO CIRCUMSTANCES** should a mallet or similar tool be used when engaging the splines in the joint or the column.

If the splines will not engage freely, inspect for damage or burrs and carefully remove with a fine file. NO ATTEMPT must be made to repair any nylon plugs which have sheared due to impact.

STEERING HOUSING

The rack and pinion assembly is identical to that shown in Fig. 3—page I5 of the 3·8 "E" Type Service Manual with the exception of the rack pre-load components.

The belleville washer and disc are replaced by a spring and retaining cap.

The method of adjustment and the rack and float figure remain as stated on page I.7.

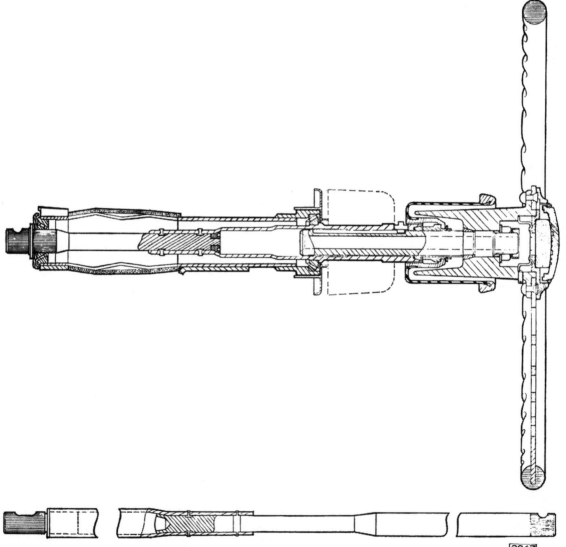

Fig. 1. Sectioned view of the upper and lower steering columns showing the nylon plugs.

Page IY.s.2

SECTION II
STEERING
(Power assisted)

DESCRIPTION

The power-assisted steering system, available as optional equipment, consists of three main components; the rack and pinion steering unit, the pump, and the reservoir interconnected by flexible hoses.

The pump is mounted on the left-hand side of the engine, the reservoir being attached to brackets secured to the left-hand sub-frame.

A shield protects the reservoir on left-hand drive cars from the exhaust manifold heat.

The upper and lower columns remain the same in detail as stated in Section I, page IY.s.1.

DATA

Steering Gear

Make	Adwest Engineering Co. Ltd.
Type	Rack and Pinion
Number of turns—lock to lock	
Turning circle	

Oil Pump

Make	Hobourn Eaton
Location	Left-front of engine
Operating Pressure	1,000 lb./sq. in. (70·3 kg./cm.^{2})
Front Wheel Alignment	$\frac{1}{16}-\frac{1}{8}''$ (1·6–3·2 mm.) toe-in

Page IIY.s.1

STEERING (POWER-ASSISTED)

OPERATION

STEERING GEAR

Oil is supplied from the reservoir via the output side of the pump to the steering unit (pressure hose) and is then returned from the steering unit to the reservoir (return hose).

A continuous flow of oil is pumped through the system whilst the engine is running but pressure builds up only when the steering wheel is turned.

The steering gear is basically a normal rack and pinion manual steering with a torsion bar controlled rotary value embodied in the input shaft and a hydraulic cylinder.

The piston in the hydraulic cylinder is connected to the rack.

Steering lock stops are incorporated in the gear unit.

THE VALVE UNIT

This is a rotary type control valve. The valve rotor, which is also the input shaft to the steering gear, has three grooves machined in it.

These grooves lie between three grooves in the valve sleeve when no load is applied to the steering wheel, the rotor being centred in the sleeve by a torsion bar.

When steering effort is applied at the wheel, this is transmitted via the torsion bar to the rotor. The torsion bar is, however, slender and the manual effort causes it to twist, thus allowing the rotor to rotate in the sleeve.

The relative movement of the grooves in the rotor to the grooves in the sleeve allows hydraulic pressure from the pump to operate on either side of the piston thus assisting the movement of the rack.

THE PRESSURE PUMP

The pressure pump is a roller type, belt driven unit. The relief valve is set to operate between 950 and 1,000 lb./sq. in. (66·8–70·3 kg./cm.^{2}). The flow control valve is set at 2·2 Imp. galls. per min. (10 litres/min.) (21 U.S. pints per min.).

No servicing or adjustment is possible with the pump. Replacement units can be obtained on an exchange basis from:—

THE SPARES DIVISION,

JAGUAR CARS LTD.,

COVENTRY,

ENGLAND.

STEERING (POWER-ASSISTED)

SERVICING

Checking The Reservoir Oil Level

The oil reservoir is mounted on the right-hand side of the engine. It is important that absolute cleanliness is observed when replenishing with oil as any foreign matter that enters may affect the hydraulic system.

Remove the filler cap, check the oil level and top up if necessary with the recommended grade of fluid. The correct level of the oil is just above the filter element.

Rack and Pinion Housing

The rack and pinion housing is attached to the front suspension crossbeam.

A grease nipple, located in the rack adjuster pad for the lubrication of the rack and pinion assembly, is accessible from underneath the front of the car from the driver's side.

Lubricate sparingly with the recommended grade of lubricant. Do not over-lubricate the housing to the extent where the bellows at the end of the housing become distended. Over-lubricating may also block the air transfer pipe.

Check that the clips at the ends of the bellows are fully tightened, otherwise the grease will escape from the housing.

Steering Tie-Rods

Lubricate the ball joints of the two steering tie-rods with the recommended lubricant. When carrying out this operation, examine the rubber seals at the bottom of the ball housing to see if they have become displaced or split. In this event they should be repositioned or replaced as any dirt or water that enters the joint will cause premature wear.

Do not over-lubricate the ball joints to the extent where grease escapes from the rubber seal.

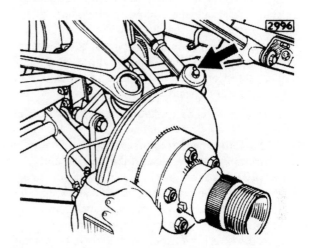

Fig. 1.

Front Wheel Alignment

Check the front wheel alignment as detailed on page IIY.s.9. if uneven wear is evident on the tyres.

CHECKING AND ADJUSTMENT ON CAR

The following adjustments can be carried out on the car; all others which may develop require the removal of the unit from the chassis.

RACK RATTLE

This is usually apparent when travelling on rough surfaces.

Adjust as follows:—

(1) Release the locknut retaining the rack pad adjusting screw.
(2) Screw the rack adjusting screw until a firm resistance is felt, and back off $\frac{1}{16}$th of a turn ($22\frac{1}{2}°$) maximum.

Firmly grip the ball pin arm protruding from the pinion end of the steering gear and by moving it towards the rack back-up pad, a spring resistance should be felt.

The total amount of play at the rack pad should not exceed ·010″ (·254 mm.). Check by removing the grease nipple and inserting a dial indicator through the rack pad and rack adjusting screw until the stem contacts the back of the rack. By pulling the rack against the spring the total amount of end play can be measured.

If the spring resistance is negligible, remove the

Page IIY.s.3

STEERING (POWER-ASSISTED)

rack pad screw and check that the spring is not broken.

The clearance should be the minimum that will allow smooth operation of the steering unit with no binding at any point throughout the full travel.

STEERING VEERING TO RIGHT OR LEFT

If the car steers to the right or left when being driven in the straight ahead position, or if unequal efforts are required to turn the steering to the right or left, carry out the following preliminary tests before proceeding further:—

Check the tyre pressures and tyre wear and change the front tyres from one side to the other. If the pull changes direction, then the trouble lies with one or both of the front tyres.

If the pull remains unchanged, check the steering geometry.

If no improvement is apparent, the fault must be in the trimming of the valve in the steering unit.

Fit a 100 lb. per sq. in. (7 kg./cm.²) pressure gauge in the return line, start engine and allow to idle. Note the pressure gauge reading which should be 40 lb. per sq. in. (2·8 kg./cm.²) approximately.

Turn the steering to the left and right by a small equal amount. The pressure should increase by an equal amount irrespective of the direction the steering is turned.

If the pressure is not balanced as indicated by a slight fall in pressure on one side before rising, the valve and the pinion assembly must be replaced.

If, on starting the engine, the steering kicks to one side, replace valve and pinion assembly as detailed on page IIY.s.6, under "Dismantling and Reassembling".

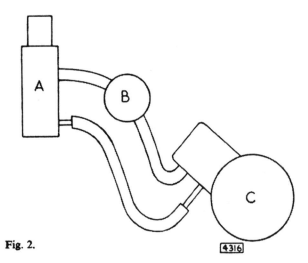

Fig. 2.

The pinion assembly can be removed without detaching the rack housing assembly from the car, if necessary.

Back off the rack adjuster pad fully before removing the pinion housing and readjust to give ·010" (·254 mm.) end play as detailed under "Rack Rattle" after refitting.

Note the position of the pinch bolt slot in the input shaft before removing, and ensure that the slot is in the same position after refitting. Allow for the spiral in the pinion when reassembling.

BALL PIN KNOCK

Ball pin knock, evident when turning to left or right, is due to wear in the inner ball assembly.

This will only be apparent after long periods of service, and on no account must any adjustment be attempted to reduce wear which may have developed.

A new ball pin/track rod assembly must, in ALL cases, be fitted.

The new assembly will be supplied less the outer ball pin and bellows which must be ordered separately if required.

The ball pin/track rod assembly can be removed with the rack *in situ* as follows:—

Disconnect the track rod on the side to be removed, from the steering arm.

Remove the bellows retaining clip from the rack housing. Fold the bellows back until the inner ball joint is exposed.

Knock back the ears of the tab washer which locks the inner ball joint assembly to the rack shaft, remove the ball joint and track rod as a unit and collect the spring.

Check the length of the track rod between the inner and outer ball pin centres.

Release the outer ball joint locknut, remove the joint and nut and withdraw the bellows after releasing the clip.

Check the outer ball joint and replace if necessary. Re-assemble to the track rod and adjust the length between the ball pin centres to the figure as noted on removal. This should be 8·75" (22·2 cm.). IT IS IMPORTANT that both track rods are of equal length.

Refit the inner ball joint and spring to the rack shaft and tighten fully. Secure with new tabwasher.

Apply a generous coating of the recommended grade of grease to the inner ball housing and refit the bellows and tighten the clips.

Reconnect the track rod to the steering arm and check the front wheel alignment as detailed on page IIY.s.9.

STEERING (POWER-ASSISTED)

CHECKING THE HYDRAULIC SYSTEM

A number of faults in the steering system can be caused by inefficiencies in the hydraulic circuit, see page IIY.s.10 for "Fault Finding" chart. The following checks can be carried out without removing any components from the car. Before starting any of this work the fluid should be checked for correct level and for lack of froth.

Pump Blow Off Pressures

Fit a pressure gauge into the return line, start the engine and run at idling speed.

Turn the steering to full lock and continue to increase the steering wheel effort until the pressure ceases to increase. The peak pressure should lie between 950 and 1,000 lb./sq. in. (66·8–70·3 kg./cm.^{2}) and should not increase with engine r.p.m.

If however, the pressure is below 950 lb./sq. in. (66·8 kg./cm.^{2}) at tickover but rises to the correct figure with increased engine speed, then the trouble is caused either by a faulty relief valve in the pump or by excessive internal leakage in the steering gear.

Fit a pressure gauge into the pressure line with an "ON–OFF" tap in series with the gauge and the steering unit.

Start the engine, open the tap and turn the steering to full lock. Check the pressure reading on the gauge. This should read 1,000 lb./sq. in. (70·3 kg./cm.^{2}).

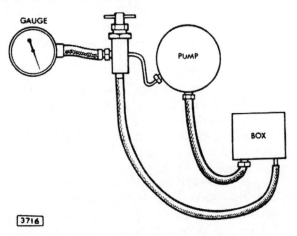

Fig. 3.

If the pressure does not rise to this figure **close the tap for a maximum of 5 seconds** and note the gauge reading. This should be 1,000 lb./sq. in. (70·3 kg./cm.^{2}) —relief valve pressure.

If this reading is obtained, the leaks are confined to the steering unit which should be removed and overhauled. If the reading is not obtained the fault lies in the pump.

Faulty pumps cannot be serviced. New replacement units can be obtained on an exchange basis from:—

> The Spares Division,
> Jaguar Cars Ltd.,
> Coventry,
> England.

STEERING GEAR

Removal

Remove the bonnet as detailed in Section N—page NY.s.3.

Remove the radiator as detailed in Section C—page DY.s.1.

Turn the steering wheel until the Allen screw in the lower column universal joint is accessible, insert an Allen key and remove the screw.

Disconnect the hoses from the steering unit and catch the oil which will drain away. Blank off the connections and unions to prevent the ingress of dirt.

Remove the nuts and washers and disconnect the track rod ball joints from the steering levers using a suitable extractor.

Remove the bolts, nuts and washers securing the steering gear unit to the frame assembly as detailed under "Steering Housing—Removal", on page I.6 and withdraw the unit.

Note: If the steering column has not been damaged by impact, i.e. if the nylon plugs in the column have not sheared, excessive force must NOT be used to separate the pinion shaft and the lower universal joint.

Page IIY.s.5

STEERING (POWER-ASSISTED)

Dismantling

Thoroughly clean the outside of the unit before attempting to dismantle.

Remove the sub-assemblies or components as follows:—

(a) Remove the external pipes.
(b) Remove the wire clips retaining the bellows to the steering unit and fold back the bellows to expose the ball joints.
(c) Knock back the tab washer securing the inner ball pin to the rack shaft.
   Remove the inner ball pin and track rod as a unit. DO NOT dismantle the ball pin assembly. Collect the thrust spring and spacer.
(d) Release the locknut retaining the rack adjusting pad screw, remove screw, spring and pad.
(e) Mark the location of the pinion housing in relation to the rack housing. Remove three nylon locking nuts retaining valve body assembly to the rack housing. Withdraw the assembly and discard the joint. Note the location of the pinch bolt slot before withdrawing the housing.
(f) The unit is now separated into its two major components, that is, valve and pinion assembly and the rack and tube assembly.
   Depending on the fault, either of these or both can be dismantled and the faulty component replaced.

Do not disturb the outer ball joints unless these are to be removed for replacement. If the ball joints are removed for any purpose, check the total length of the tie rods before releasing the locknut.

Tie rods must be re-assembled to an equal length of 8·75″ (22·2 cm.).

EXAMINATION OF COMPONENTS

Valve and Pinion/Housing Assembly

The valve and pinion/housing assembly will be available for Service Replacement purposes as a complete unit only, with the exception of the top seal and the associated back-up seal, the housing gasket and the pipe union seats.

With the assembly removed from the rack tube, carry out the following checks:—

With a soft mallet drift out the valve and pinion assembly from the housing.

Examine the teflon rings. These should be a loose fit in their grooves and the outer diameter should be free from cuts, scratches or similar blemishes.

Ensure that there is no relative movement at the trim pin between the valve sleeve and the shaft.

Check that there is no wear in the torsion bar assembly pins by ensuring that there is no free movement between the input and output pinion shafts.

Examine the housing bore for signs of wear, particularly on the rubbing surfaces of the teflon rings.

Examine the needle roller bearings for damage or wear.

If, during the above checks, any fault is found, the complete assembly must be replaced as a unit.

Replacing the Top Seal

Drift out the shaft assembly as detailed above.

Remove the circlip and extract the top and back-up seals.

Replace with the new seals contained in the seal kit and refit the circlip.

Refit the shaft assembly and reassemble the housing to the rack tube with a new gasket.

Renewing the Pipe Union Seats

If worn or damaged, the pipe union seats can be renewed by tapping a suitable thread in the internal bore of the seat and inserting a setscrew with an attached nut and plain washer.

Tighten down the nut against the housing base and withdraw the seat.

Fit new seat by inserting in the housing and tapping home square with a soft drift.

Rack and Rack Housing

The following items will be available as replacement parts:—

1. Rack
2. Rack Housing
3. "Clevite" Bearing
4. Seal (contained in Seal Kit).

Replacement rack housings will be complete with end cap, seals, "Clevite" bearing and needle bearing.

1. Remove the valve and pinion housing as detailed previously.

Page IIY.s.6

STEERING (POWER-ASSISTED)

Mark the location of the end cap in relation to the rack tubing.

Unscrew the ring nut retaining the end cap to the rack tube and withdraw the cap.

Remove the union from the rack tube and push out the centre seal housing from the pinion housing end of the tube.

Check the condition of the piston and ring and if worn or scratched renew complete.

Remove the outer circlip and withdraw the piston. Note any shims which may be fitted between the inner circlip and the piston.

Remove the "O" ring from the shaft and replace with the new part contained in the Seal Kit. Refit the piston and secure with circlips.

Check that end float between piston and circlip does not exceed ·010". If this condition exists reduce end float by adding a shim.

IMPORTANT: Check that the piston rotates freely between the circlips on completion.

Remove the "O" rings and seal from the centre seal housing and discard. Replace with new parts contained in the Seal Kit.

Insert the housing into the tube with the lips of the seal facing towards the centre.

Line up the hole in the housing with the hole in the tube and secure with the union.

Insert the rack in the rack housing. Extreme care must be taken to ensure that the oil seal in the housing is not damaged by the rack teeth.

Remove the "O" ring and oil seal from the end cap. Check the condition of the "Clevite" bearing and replace if worn or damaged with a new bearing. A mandrell machined to the internal diameter of the bearing should be used when refitting to prevent the bearing collapsing.

Renew the oil seal and "O" ring and refit the end cap. Line up the location marks made during removal and secure with the ring nut. Care must be taken to ensure that the end cap does not turn when tightening the ring nut. Any movement will place the mounting brackets out of phase with each other.

Refit the rack adjuster pad, spring, adjuster screw and locknut, but do not attempt to carry out any rack adjustment at this stage.

Final Assembly

Place a new seal joint over the three studs in the rack housing.

Refit the pinion housing, noting the position as marked on removal.

Check that the pinch bolt slot is in the same relative location as noted on removal when engaging the pinion with the rack teeth.

Allow for the spiral in the pinion assembly when reassembling. Fit the self-locking nuts and tighten down evenly.

Refit the inner ball joints and track rods as an assembly as detailed on page IIY.s.4 under "Ball Pin Knock". If the inner ball joint is to be replaced due to wear, a new unit complete with track rod must be obtained. **ADJUSTMENT OF THE JOINT IS NOT PERMISSIBLE.**

Refit the outer ball joints if removed; adjust the length of each track rod to 8·75"(22·2 cm.) between the ball joint centres. IT IS IMPORTANT that the track rods are of equal length.

Adjust the rack back-up pad as detailed on page IIY.s.3.

Refit the grease nipple and the external pipes.

Coat both rack ball housings with 2 oz. (56·7 grammes) of the recommended grade of grease, fit the bellows and secure with the clips to the track rod and steering housing.

Apply a grease gun to the nipple in the back-up adjuster pad and inject 1 oz. (28·35 grammes) of the recommended lubricant.

Do not lubricate the housing to the extent where the bellows become distended. Over-lubrication may also block the air transfer pipe.

Refitting

Refitting is the reverse of the removal procedure.

Reconnect the high and low pressure hoses, care being taken to ensure that the connections are perfectly clean.

Refit the lower and upper steering columns as detailed on page IIY.s.8.

Refill the reservoir to the full mark of the dipstick with the recommended grade of Automatic Transmission Fluid.

Bleed the system as follows:—

(a) With the engine running, turn the steering from lock to lock a few times to expel any air which may be present, indicated when all lumpiness has disappeared.

(b) Check the fluid level in the reservoir and top up if necessary with the recommended grade of fluid. The correct level is just above the filter element.

STEERING (POWER-ASSISTED)

STEERING COLUMN

The upper and lower steering columns and mountings are of the collapsible type designed to comply with U.S.A. Federal Safety Regulations.

The collapse points are retained by nylon plugs which will shear on impact, allowing the steering wheel and columns (upper and lower) to move forward.

NO ATTEMPT must be made to repair the units if damaged due to accident. NEW replacement items MUST be fitted.

UPPER STEERING COLUMN
Description

The upper steering column (inner) is composed of two separate sliding shafts retained to a fixed length by nylon plugs, the outer column being pierced in a lattice form.

The inner shaft assembly is supported in the outer column by two pre-lubricated taper roller bearings.

A gaiter covers the pierced portion of the outer column to seal against the ingress of dirt.

Removal

Disconnect the battery.

Withdraw the self-tapping screws and remove the under-scuttle casing above the steering column.

Disconnect the cables contained in the direction indicator switch harness.

Note the location of the connections for reference when refitting.

Withdraw the ignition key, remove the ring nut and detach the ignition lock from the mounting bracket on the steering column.

Note: If the car is fitted with air-conditioning equipment, the switch will be mounted on a bracket attached to the evaporator unit and need not be removed.

If the car is fitted with a steering column lock, removal of the lock will not be possible, and it will be necessary to disconnect the attached cables at the snap connectors.

Release three grub screws in the steering wheel hub and remove the steering wheel motif.

Remove the locknut, hexagon nut and flat washer, and withdraw the steering wheel from the splines on the inner column.

Remove the nut, lockwasher and pinch securing bolt the upper universal joint to the lower steering column.

Remove two nuts and lockwashers securing the upper column lower mounting bracket to the underside of the scuttle.

Remove two bolts, nuts, lockwashers and distance pieces securing the upper mounting bracket to the support bracket on the body.

Withdraw the upper column from the splines on the lower column.

Note: If the steering column has not been damaged by impact, i.e. if the nylon plugs in the inner column or the top mounting bracket have not sheared, excessive force must NOT be used to separate the upper universal joint from the lower column.

Refitting

Refitting is the reverse of the removal procedure.

Set the road wheels in the straight ahead position and check that the bolt holes in the lugs of the upper column universal joint register correctly with the groove machined in the lower column splines. Tighten the pinch bolt to a torque of 16–18 lb. ft. (2·2–2·5 kgm.).

IMPORTANT

Excessive force as noted under "Removal" must not be used when reassembling the universal joint to the column.

UNDER NO CIRCUMSTANCES should a mallet or similar tool be used when engaging the splines in the joint and column.

If the splines will not engage freely, inspect for damage or burrs and remove with a fine file.

NO ATTEMPT must be made to repair any nylon plugs which have sheared due to impact.

Dismantling

Dismantling is confined to removing the steering column adjuster locknut, the splined shaft and the direction indicator switch as detailed on page IIY.s.8.

LOWER STEERING COLUMN
Description

The lower steering column comprises two sliding shafts retained to a fixed length by nylon plugs.

Removal

Remove the upper steering column as detailed

Page IIY.s.8

STEERING (POWER-ASSISTED)

previously.

Remove the nut, lockwasher and bolt securing the column to the lower universal joint and withdraw the column rearwards through the grommet.

Note: If the steering column has not been damaged by impact, i.e. if the nylon plugs in the column have not sheared, excessive force must NOT be used to separate the column and the lower universal joint.

Refitting

Refitting is the reverse of the removal procedure.

Check that the bolt holes in the universal joint register correctly with the groove machined in the column splines. Tighten the pinch bolt to a torque of 16–18 lb. ft. (2·2–2·5 kgm.).

IMPORTANT

Excessive force as noted under "Removal" must not be used when reassembling the universal joint to the column. UNDER NO CIRCUMSTANCES should a mallet or similar tool be used when engaging the splines in the joint or the column.

If the splines will not engage freely, inspect for damage or burrs and carefully remove with a fine file. NO ATTEMPT must be made to repair any nylon plugs which have sheared due to impact.

FRONT WHEEL ALIGNMENT

It is ESSENTIAL that the following instructions are carried out when checking the front wheel alignment, otherwise steering irregularities may result.

Important

Inflate all tyres to the recommended pressures.

Each wheel must be adjusted individually by means of the tie-rods to give half the total toe-in of $\frac{1}{16}''$ to $\frac{1}{8}''$ (1·6 to 3·2 mm.).

Procedure

Set the front wheels in the straight ahead position.

Centralise the steering by removing the grease nipple from the rack adjuster pad and inserting the centralising tool (Jaguar Part No. 12297).

Check the alignment of the front wheels by using light beam equipment or an approved tract setting gauge.

If adjustment is required, slacken the locknuts at the outer end of each tie-rod, release the outer clips securing the rack housing bellows to avoid distortion after turning the tie-rods.

Turn the tie-rods by an equal amount in the necessary direction until the alignment is correct. Tighten the locknut and re-check.

Ensure that the bellows are not twisted and tighten the clips.

REMOVE THE CENTRALISING TOOL and refit the grease nipple to the rack adjuster pad.

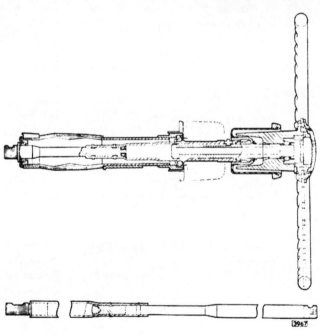

Fig. 4.

FAULT FINDING CHART

FAULT	POSSIBLE CAUSE	REMEDY
External oil leaks from Rack and Pinion unit.	Damage or wear to seals or incorrect tightening of unions or bolts.	It is most important that source of the leak is traced before any attempt is made to rectify. Once the leak is located, tighten the unions or bolts or replace the seals as necessary.
Leak at reservoir.	Cover gasket damaged. Hose connection loose.	Renew gasket. Tighten hoses.
Leak at pump shaft.	Worn or damaged seals on shaft.	Replace pump.
Steering veering to right or left.	Unbalanced tyre pressures or faulty tyres. Steering gear out of trim.	Check as detailed on Page IIY.s.4.
Heavy steering when driving.	Low tyre pressures. Tightness or stiffness in column and/or steering and suspension joints.	Inflate tyres. Grease or replace components.
Heavy steering when parking.	Loose pump belt (nearly always accompanied by a squealing noise). Insufficient pressure from pump due to defective pump valve or restricted hoses. Insufficient pressure due to high leaks in steering gear.	Check pump belt, replace if necessary. Remove restriction or check pump pressure as detailed on Page IIY.s.5. Replace pump if faulty. Confirm high internal leaks. If proven, remove the rack unit from the car and replace seals.
Steering effort too light.	Worn torsion dowel pins or torsion bar broken.	Remove valve housing and fit new unit.
Poor straight running.	Incorrect tyre pressures. Incorrect toe-in.	Inflate. Check and reset as necessary.
Noise from pump.	Belt loose, indicated by squealing during parking manoeurvres. Other pump noises are due to wear or damage.	Check belt and replace as necessary. Replace pump.
Rattle when travelling on rough roads.	Wear between rack and pinion assembly. Wear at ball joints at the ends of the rack. Wear in the rack housing bush.	Adjust rack pad adjuster screw as detailed on Page IIY.s.3. Renew ball joint and track rod as an assembly. Remove the rack and renew the bush.

SECTION J

FRONT SUSPENSION

The front suspension remains basically the same as that detailed in Section J—3·8 "E" Type Service Manual with the exception of the removal of the wheel hubs when steel spoked wheels are fitted and the torsion bar settings.

The routine maintenance periods are increased from those stated previously.

Wheel Swivels

Lubricate the nipples (four per car) fitted to the top and bottom of the wheel swivels.

A bleed hole is provided in each ball joint; The hole being covered by a nylon washer which lifts under pressure and indicates when sufficient lubricant has been applied.

The nipples are accessible from underneath the car.

Wheel Bearings

Removal of the wheel bearings will expose a grease nipple in the wheel hubs.

Lubricate sparingly with the recommended grade of lubricant. If excess grease is pumped into the bearing the grease will exude into the bore of the hubs (Wire wheels) or through a small hole drilled centrally in the hub end cap (Pressed steel wheels).

TORSION BAR SETTING

Check that the car is full of petrol, oil and water and that the tyre pressures are correct.

Place the car on a perfectly level surface, with the wheels in the straight ahead position.

Check the measurement from the centre line of the inner fulcrum of each lower wishbone assembly to the ground. This should be 9" (22·86 cm.) minimum.

If any adjustment is required this should be carried out in accordance with the instructions given in Section J—Page J.15 of the 3·8 "E" Type Service Manual.

The correct dimensions for the hole centres for setting links are as follows:—

Without air-conditioning equipment

4·2 "E" Type—Open Sports	17$\frac{11}{16}$" (45·16 cm.) L.H.D.
4·2 "E" Type—F.H.C.	17$\frac{11}{16}$" (45·16 cm.) L.H.D.
4·2 "E" Type—Open Sports	17$\frac{15}{16}$" (45·6 cm.) R.H.D.
4·2 "E" Type—F.H.C.	17$\frac{15}{16}$" (45·6 cm.) R.H.D.
4·2 "E" Type—2+2	18$\frac{1}{8}$" (46·06 cm.) L.H.D./R.H.D.

With air-conditioning equipment

4·2 "E" Type—Open Sports	17$\frac{15}{16}$" (45·6 cm.) L.H.D./R.H.D.
4·2 "E" Type—F.H.C.	17$\frac{15}{16}$" (45·6 cm.) L.H.D./R.H.D.
4·2 "E" Type—2+2	18$\frac{5}{16}$" (46·5 cm.) L.H.D./R.H.D.

This method ensures that the headlamps centres are maintained at the correct height from the ground level necessary to conform to U.S.A. Federal Regulations which is 24" (60·9 cm.).

FRONT HUBS (Pressed Steel Wheels)

If the car is fitted with pressed steel wheels, available as optional equipment, the hub nut and split pin retaining the hub to the stub axle is accessible after prising off the end cap.

Further removal, dismantling, bearing end float and refitting details remain as stated on **Page J.13** of the 3·8 "E" Service Manual.

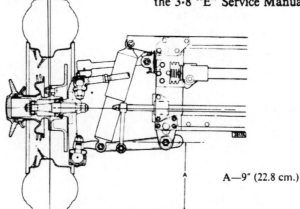

Fig. 1. *The dimension for checking the front suspension riding height.*

Page JY.s.1

SECTION L
BRAKES

DATA

Caliper type	Girling bridge type with quick change pads
Brake disc diameter—front	11″ (27·9 cm.)
—rear	10″ (25·4 cm.)
Master cylinder bore diameter	⅞″ (22·23 mm.)
Master cylinder stroke	1·30″ (3·3 cm.)
Servo unit type	Lockheed Dual—line
Main friction pad material	Mintex M.59
Handbrake friction pad material	Mintex M.34

Key to Figs. 1

1. Fluid at feed pressure
2. Fluid at master cylinder delivery pressure
3. Fluid at system delivery pressure
4. Vacuum
5. Air at atmospheric pressure
A Slave cylinder primary chamber
B Outlet port—front brakes
C Inlet port for secondary piston
D Outlet port—rear brakes
E Vacuum
F Air pressure
G Diaphragm
H Filter
I Air control spool
J To rear brakes
K To front brakes
L Dual slave cylinder
M Servo unit
N Master cylinder
O Brake reservoirs
P To manifold
Q To reservac
R Reaction valve
S Atmospheric pressure

Page LY.s.1/2

BRAKES

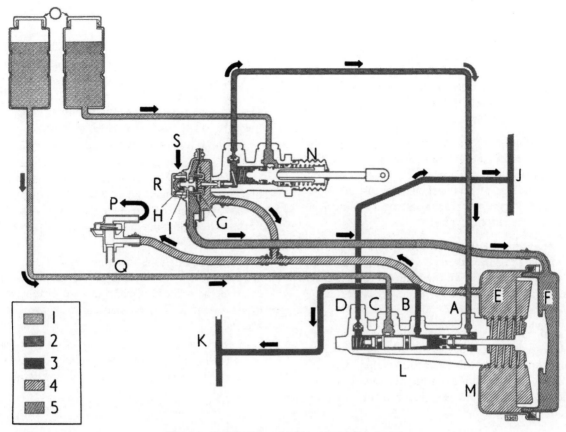

Fig. 1. *Dual-line servo braking system (Later cars)*

DESCRIPTION

The dual-line servo braking system consists of an integral vacuum booster with tandem slave cylinder, a master cylinder combined with a booster reaction valve and two fluid reservoirs.

The master cylinder is of conventional design consisting of a single cast iron cylinder housing a steel, black oxided piston sealed by a single hydraulic cup. This piston is deeply skirted to engage the operating push rod. The smaller intermediate piston, housed in its own bore in the nose of the master cylinder, is actuated by hydraulic pressure generated within the main chamber.

Mounted on the end of the master cylinder, the reaction valve consists of a pair of flow control valves which sequence the flow of air to the booster. Both control valves are operated by the intermediate piston in the master cylinder. A flat plate, interposed between the two master cylinder pistons, enables the intermediate piston to function mechanically in the event of an hydraulic failure.

The booster portion of the integral booster and slave cylinder assembly consists of a pressed steel tank which houses a moulded phenolic resin piston and a rubber rolling diaphragm. A push rod, secured to the piston, extends through the forward face of the tank into the slave cylinder. This push rod provides the principal motive force for the tandem pistons.

On the forward face of the boost tank is mounted the tandem slave cylinder which consists of a single cast iron cylinder housing two pistons in tandem, each piston having its own inlet and outlet port. Either piston will, in the event of a failure, operate independently.

OPERATION (Figs. 1)

When the system is at rest, both sides of the boost system are continuously exhausted by the engine manifold depression.

As the brake pedal is depressed, the master cylinder

Page LY.s.3

BRAKES

piston moves along the cylinder building up pressure and forcing fluid out to the primary chamber of the slave cylinder (A). Simultaneously, the intermediate piston, in the end of the master cylinder, closes the diaphragm valve (G) in the reaction valve and, in so doing, isolates the vacuum (E) from the air pressure side (F) of the boost system.

Further progress of the intermediate piston along its bore will crack the air control spool (I) in the reaction valve thus admitting air at atmospheric pressure to the rear of the boost cylinder piston. The air enters the system through a small cylindrical filter (H) on the reaction valve.

The pressure imbalance, created by the admission of air to the pressure side of the boost system, will push the boost piston down the cylinder transmitting a linear force, through the push rod, to the primary piston of the slave cylinder.

Forward motion of the primary piston, supplemented by the output of the master cylinder, transmits hydraulic pressure to the secondary piston (C) and fluid under pressure flows simultaneously from the two output ports (B and D), to the front and rear brakes.

SAFETY FACTORS

In the event of a fluid line failure in the pipe linking the master cylinder to the slave cylinder or the pipe linking the master cylinder to the fluid supply tanks, the reaction valve will be actuated mechanically by the master cylinder piston providing the booster pressure to the front and rear brakes.

A failure in the fluid line coupling the slave cylinder to the front brakes will result in the slave cylinder secondary piston travelling to its fullest extent, down the bore. This has the effect of isolating the front brake line from the rest of the system and permitting normal fluid pressure to build up in the rear brake line.

If a fault exists in the rear brake line, the slave cylinder piston will travel along the bore until it contacts the other piston and the two pistons will then travel along the bore together to apply the front brakes.

In the case of leaks in either the air or vacuum pipes both front and rear brakes may still be applied by the displacement of fluid at master cylinder pressure.

REMOTE SERVO AND SLAVE CYLINDER

Removal

Remove the trim on the floor recess panel on the left-hand side of the car. This will disclose the three nuts securing the remote servo to the bulkhead. Withdraw the three nuts.

Drain the fluid from the system as detailed on page L.10.

Disconnect the four brake pipe unions and the two flexible hoses.

Remove the battery and carrier bracket for the battery tray.

Withdraw the bolt securing the slave cylinder to the mounting bracket on the outer side member. Remove the servo together with the slave cylinder.

Refitting

Refitting the servo is the reverse of the removal procedure. Bleed the system after replenishing with fresh fluid.

MASTER CYLINDER AND REACTION VALVE

Removal

Drain the fluid from the system. Disconnect the two hydraulic pipes from the master cylinder. Disconnect the vacuum hose from the reaction valve.

Remove the clevis pin, which is retained by a split pin, securing the brake pedal to the master cylinder push rod from inside the car. In the case of right-hand drive cars, remove the top of the air cleaner and reaction valve prior to removing the two nuts securing the master cylinder to the mounting.

On left-hand drive cars the master cylinder and reaction valve can be removed as a complete unit.

Refitting

Refitting is the reverse of the removal procedure. Bleed the system after replenishing with fresh fluid.

Page LY.s.4

BRAKES

SERVICING THE UNIT

General

Prior to dismantling either the remote servo or the master cylinder reaction valve assembly, it is advisable to obtain repair kits containing all the necessary rubber parts required during overhaul. Three separate repair kits are available as follows:—

(a) Remote servo repair kit.

(b) Reaction valve repair kit.

(c) Master cylinder repair kit.

When either of the units have been dismantled the component parts should be washed in denatured alcohol (industrial methylated spirits). Parts that have been washed should be thoroughly dried using a clean lint-free cloth or pressure line and then laid out on clean paper to prevent dirt being assembled into the servo or master cylinder and reaction valve assembly.

Examine all metal parts for damage, with particular reference to those listed below and make renewals where necessary:—

(a) the reaction valve piston and bore.

(b) the master cylinder piston and bore.

(c) the servo slave cylinder pistons and bore.

(d) the servo push rod stem.

If any of the vacuum hose connections have become loose in service these must be rectified prior to reassembly.

The vacuum non-return valve is a sealed unit and, if faulty, it must be replaced by a new assembly.

THE REMOTE SERVO (Fig. 2)

Dismantling

Support the servo slave cylinder in the jaws of a vice, shell uppermost, with specially formed wooden blocks placed either side of the cylinder and against the jaws of the vice.

Fit the cover removal tool (Churchill Tool No. J.31) to the end cover and secure it by fitting the three nuts.

Turn the end cover in an anti-clockwise direction until the indents in the servo shell line-up with the small radii around the periphery of the end cover. At this stage the end cover may be removed from the servo.

Remove the diaphragm (11) from its groove in the diaphragm support (10) and, with the servo removed from the jaws of the vice, apply a gentle pressure to the diaphragm support and shake out the key (12).

The diaphragm support (10) and diaphragm support return spring (8) can then be removed.

Bend down the tabs on the locking plate (16) and remove the locking plate, abutment plate (17) and servo shell (14) from the slave cylinder by unscrewing and removing three screws (15).

Extract the seal (19) and bearing (18) from the mouth of the slave cylinder bore which will permit the removal of the push rod (9) together with the slave cylinder piston assembly.

The push rod may be separated from the piston by sliding back the spring steel clip (6) around the piston and removing the pin (5). It is not necessary to remove the cup (21) from the piston as a new piston together with a cup are contained in the repair kit.

Unscrew and remove the fluid inlet connection (3) and extract the piston stop pin (30) from the base of the inlet fluid port. To facilitate this operation, apply gentle pressure to the secondary piston (4).

Tap the open end of the slave cylinder body with a hide or rubber hammer to remove the secondary piston together with the piston return spring (28) from the bore.

The rubber seal (25) located in the groove adjacent to the heel of the piston may be removed but it is advisable to first remove the spring retainer (26) from the piston head extension before attempting to remove the seal (25) and piston washer (24). Removal of the plastic spring retainer (26) is sometimes difficult but, as a new one is provided in the repair kit, this part can be replaced if damaged.

To remove the trap valve assembly, unscrew and remove the adaptor (1) from the fluid outlet port. If it is necessary to remove the shim-like clip from the body of the trap valve (29) ensure that this part is not distorted in any way.

Page LY.s.5

BRAKES

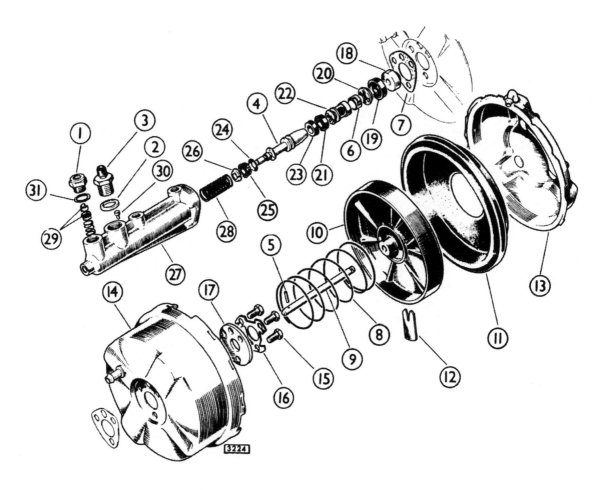

Fig. 2. *Exploded view of the remote servo.*

1. Outlet connection.
2. Gasket.
3. Inlet connection.
4. Piston.
5. Pin.
6. Retaining clip.
7. Gasket.
8. Spring.
9. Push rod.
10. Diaphragm support.
11. Diaphragm.
12. Key.
13. Cover.
14. Vacuum cylinder shell.
15. Screw.
16. Locking plate.
17. Abutment plate.
18. Bearing.
19. Seal.
20. Spacer.
21. Cup.
22. Piston.
23. Cup.
24. Piston washer.
25. Seal.
26. Retainer.
27. Slave cylinder body.
28. Spring.
29. Trap valve.
30. Stop pin.
31. Gasket.

Page LY.s.6

BRAKES

Assembling

Assemble the trap valve (29) complete with spring and clip into the outlet port and secure it by fitting the fluid outlet adaptor (1) together with the copper gasket (31).

Prior to further assembly, lightly coat the four rubber seals to be replaced in the slave cylinder bore with Lockheed Disc Brake Lubricant.

Locate the piston washer (24) over the piston head extension, convex face towards the piston flange and, using the fingers only, assemble the two rubber seal (23 and 25) onto the piston so that their concave faces oppose each other.

Press the spring retainer (26) onto the piston head extension with both seals in position.

Fit the piston return spring (28) to the secondary piston complete and assemble into the slave cylinder bore, spring leading.

Press the piston assembly down the cylinder bore, using a short length of brass bar, until the drilled piston flange passes the piston stop pin hole.

Insert the piston stop pin (30) into the fluid inlet port and secure it by fitting the inlet adaptor (3) complete with the copper gasket (2). Place the push rod (9) in the primary piston and, with the aid of a small screwdriver, compress the small spring within the piston to enable the pin (5) to be inserted. Prior to fitting the pin retainer (6), it is important to establish that the small coil spring is loaded between the heel of the piston and the pin. Ensure that the pin does not pass through the coils of the spring.

Fit the spring retainer by sliding it into position along the piston ensuring that no corners are left standing proud after assembly.

Using fingers only, fit a new cup (21) into the groove on the piston so that its lip (concave face) faces towards the piston head and assemble the piston into the slave cylinder bore.

Insert the spacer (20), gland seal (19) and plastic bearing (18) into the slave cylinder counterbore leaving the bearing projecting slightly from the mouth of the bore.

Place the gasket (7) in position on the end face of the slave cylinder, using the plastic bearing as a location spigot and fit the vacuum shell (14), abutment plate (17) and locking plate (16).

Insert the three securing screws (15) and tighten down to a torque of 150/170 lb./ins. (1·7–1·9 kgm.).

Bend the tabs on the locking plate against the flats on the three screws.

Locate the diaphragm support return spring (8) centrally inside the vacuum shell, fit the diaphragm support (10) to the push rod and secure it by dropping the key (12) into the slot provided in the diaphragm support.

Stretch the rubber diaphragm (11) into position on the diaphragm support ensuring that the bead around its inside diameter fits snugly into the groove in the diaphragm support.

If the surface of the rubber diaphragm appears wavy or crinkled this indicates that it is not correctly seated. To ease assembly, smear the outside edges of the diaphragm liberally with Lockheed disc brake lubricant.

Fit the end cover using Churchill Tool No. J.31.

Note: As it is possible to fit the end cover in three different positions, ensure that the end cover hose connections line up with the slave cylinder inlet and outlet ports when assembly is complete.

Page LY.s.7

BRAKES

1. Diaphragm.
2. Screw.
3. Shakeproof washer.
4. Gasket.
5. Bolt.
6. Outlet adaptor.
6A. Copper gasket.
7. Trap valve body.
8. Washer.
9. Banjo.
10. Copper gasket.
11. Body.
12. Bearing.
13. Secondary cup.
14. Seal.
15. Piston.

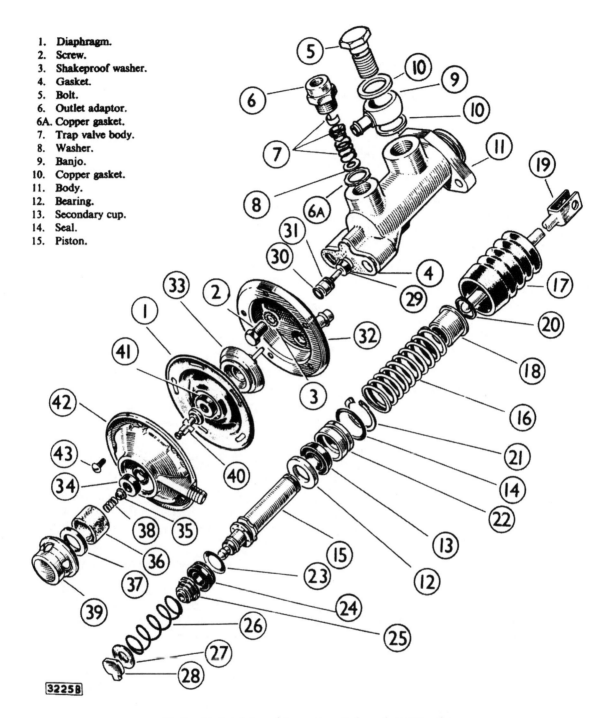

Fig. 3. *Exploded view of the master cylinder and reaction valve.*

16. Return spring.
17. Rubber boot.
18. Spring retainer.
19. Push rod.
20. Spirolox circlip.
21. Circlip.
22. Bearing.
23. Piston washer.
24. Main cup.
25. Retainer.
26. Spring.
27. Retainer.
28. Lever.
29. Seal.
30. Seal.
31. Piston.
32. Valve housing.
33. Diaphragm support.
34. Valve rubber.
35. Valve cap.
36. Filter.
37. Sorbo washer.
38. Spring.
39. Filter cover.
40. Valve stem.
41. Valve rubber.
42. Valve cover.
43. Screw.

Page LY.s.8

BRAKES

MASTER CYLINDER AND REACTION VALVE

Dismantling (Fig. 3)

Unscrew and remove the fluid outlet adaptor (6) and extract the trap valve assembly (7) from the outlet port.

Remove the rubber boot (17) from the mouth of the cylinder bore, compress the piston return spring (16) and unwind the spirolox circlip (20) from the heel of the piston. The spring retainer (18) and piston return spring (16) can at this stage be removed.

Press the piston (15) down the bore and, with the aid of special circlip pliers (Tool number 7066) extract the circlip (21) from the mouth of the cylinder bore. Care should be taken during this operation not to damage the finely machined cylinder piston.

The piston assembly, complete with nylon bearings and rubber seals, can be withdrawn from the cylinder bore.

Remove the plastic bearing (22), complete with "O" ring (14), secondary cup (13) and rectangular section plastic bearing (12) from the piston by sliding the assembly along the finely machined portion.

Due to the plastic spring retainer (25) being an interference fit onto the piston head extension, this part is likely to become damaged during dismantling. In view of this a new spring retainer is contained in the appropriate repair kit. To remove the spring retainer, hold the piston on a bench, piston head downwards, applying a downwards force to the back face of the spring retainer with a slim open-ended spanner. The piston return spring (26), pressed steel retainer (27) and lever (28) may, at this stage, be withdrawn from the cylinder bore.

Remove the filter cover (39) and collect the filter (36) sorbo washer (37) and spring (38).

Unscrew and remove the five screws securing the valve cover (42), remove the valve cover assembly from the valve housing (32) which can be dismantled further by prising off the snap-on clip securing the valve rubber (34).

The valve stem (40) complete with the other valve rubber (41) can now be withdrawn from the valve housing and the valve rubber removed from the valve stem flange. The reaction valve diaphragm (1) can now be separated from the diaphragm support (33) and, by unscrewing the two hexagon-headed screws (2), the valve housing can be separated from the master cylinder body.

Removal of the valve piston (31) assembly can be effected by inserting a small blunt instrument into the master cylinder fluid outlet port and easing the valve piston assembly along its bore until it can be removed by hand.

Important: No attempt should be made to withdraw the valve piston assembly along its bore by using pliers.

Assembling

Prior to assembly liberally coat all rubber seals and plastic bearings, with the exception of the two valve rubbers, with Lockheed disc brake lubricant.

Holding the master cylinder body at an angle of approximately 25° to the horizontal, insert the lever (28), tab foremost, into the cylinder bore ensuring that, when it reaches the bottom of the bore, the tab on the lever drops into the recessed portion provided.

Place the piston washer (23) on the piston head, convex face towards the piston flange, together with a new main cup (24) and press the plastic spring retainer (25) onto the piston head extension.

Drop the pressed steel spring retainer (27) into the bottom of the bore following up with the piston return spring (26). When these two parts have been assembled it is advisable to recheck the position of the lever.

Press the piston assembly into the cylinder bore and locate the rectangular section plastic bearing (12), secondary cup (13) and bearing (22) together with seal (14) into the mouth of the cylinder bore.

Press the assembly down the bore to its fullest extent and with the aid of the special circlip pliers (Tool number 7006 with "K" points) fit the circlip to retain the internal parts.

Locate the other piston return spring (16) over the heel of the piston together with the pressed steel spring retainer (18), slide the spring retainer down the finely machined portion of the piston against the load of the spring and fit the spirolox circlip (20) into the groove ground around the heel of the piston.

Using the fingers only, stretch a new valve seal (29) and "O" ring into position on the valve piston and insert the assembly into the valve box.

Page LY.s.9

BRAKES

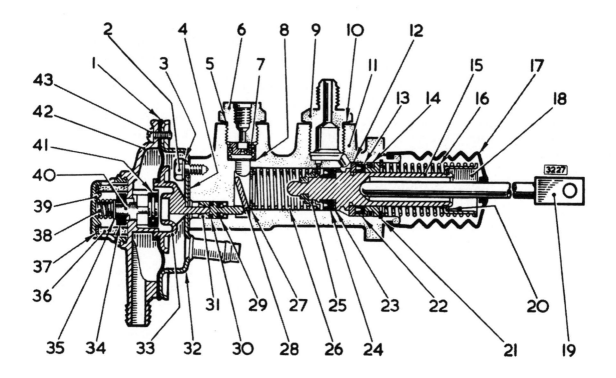

Fig. 4. *Sectioned view of the master cylinder and reaction valve.*

1. Diaphragm.
2. Screw.
3. Shakeproof washer.
4. Gasket.
5. Gasket.
6. Outlet adaptor.
7. Trap valve body.
8. Washer.
9. Inlet adaptor.
10. Copper gasket.
11. Body.
12. Bearing.
13. Secondary cup.
14. Seal.
15. Piston.
16. Return spring.
17. Rubber boot.
18. Spring retainer.
19. Push rod.
20. Spirolox circlip.
21. Circlip.
22. Bearing.
23. Piston washer.
24. Main cup.
25. Retainer.
26. Spring.
27. Retainer.
28. Lever.
29. Seal.
30. Seal.
31. Piston.
32. Valve housing.
33. Diaphragm support.
34. Valve rubber.
35. Valve cap.
36. Filter.
37. Sorbo washer.
38. Spring.
39. Filter cover.
40. Valve stem.
41. Valve rubber.
42. Valve cover.
43. Screw.

Page LY.s.10

BRAKES

Secure the valve housing to the master cylinder body by fitting the two hexagon headed screws (2) complete with spring washers and tighten each screw to a torque of 160/180 lb./ins. (1·8–2 kgm.). A new gasket should be fitted between the valve housing and the master cylinder body.

Stretch the reaction valve diaphragm onto the diaphragm support through the hole in the valve housing so that it engages the depression in the valve piston.

Using the fingers only, stretch the valve rubber, which is formed with the groove around its inside diameter, onto the valve stem flange, insert the valve stem through the hole in the valve cover and secure it by placing the other valve rubber over the valve stem and fitting the snap-on clip.

The valve cover assembly can now be placed into position on the valve housing ensuring that all the holes line up and that the hose connections are in line with each other at the bottom of the unit. Secure the valve cover assembly by fitting the five self-tapping screws.

Hold the master cylinder in an upright position (valve uppermost) and place the air filter together with the rubber washer in position upon the valve cover with the small spring on the snap-on valve stem clip.

Carefully locate the air filter cover over the air filter and press it firmly home.

If the trap valve assembly has been dismantled; insert the small clip into the trap valve body ensuring that it does not become distorted and locate the spring on the reduced diameter of the trap valve body.

Assemble the trap valve complete (spring innermost) into the master cylinder fluid outlet port.

Place a copper gasket under the head of the fluid outlet adaptor and screw the adaptor into the fluid outlet port. If the fluid inlet adaptor has been removed, this must be replaced in the same manner using a copper gasket under the head.

The master cylinder push rod and convoluted rubber boot can best be fitted during the installation of the assembly.

FRICTION PADS
Renewal

Friction pads should be renewed if it is found, on visual examination through the caliper apertures, that they have worn down to an approximate thickness of $\frac{1}{8}''$ (3·2 mm.).

Withdraw the hairpin clips and extract the pad retaining pins. On front brakes, remove the anti-chatter clips from around the retaining pins and pad backing plates. Withdraw the pads.

To enable new pads to be inserted it will be necessary to lever the pistons back down the cylinder bores. It is advisable to half empty the brake fluid reservoirs otherwise forcing the pistons back will eject fluid from the reservoirs with possible resultant paint damage.

Insert new pads. Line up the holes in the backing plates and caliper bodies. Fit the retaining pins and hairpin clips: fit the anti-chatter clips to front pads. Ensure that the pads are free to move on the pins to allow for brake application and automatic adjustment.

Top up the reservoirs to the correct level and apply the brake several times until the pedal feels "solid".

FRONT CALIPERS
Removal

Jack up the car and remove the front wheel(s). Disconnect the caliper fluid feed pipe from the union and seal the pipe and union.

Remove the locking wire, withdraw the mounting bolts and lockwashers and detach the caliper.

Locate the caliper in position and secure with the mounting bolts and lockwashers. Lockwire the bolts after fully tightening.

Reconnect the caliper feed pipe to the union and bleed the braking system as detailed on page L10.

REAR CALIPERS
Removal

The rear suspension unit must be removed in order to withdraw the rear calipers.

Proceed as described in Section K "Rear Suspension" and support the suspension unit under its centre.

Disconnect the handbrake compensator linkage from the handbrake operating levers. Discard the split pins and withdraw the clevis pins.

Lift the locking tabs and remove the pivot bolts together with the retraction plate.

Remove the friction pad carriers from the caliper bridges by moving them rearwards round the discs and withdrawing from the rear of the rear suspension assembly.

Remove the hydraulic feed pipe at the caliper and plug the hole to prevent the entry of dirt.

Remove the friction pads from the caliper as described previously.

Remove the front hydraulic damper and road spring unit (as described in Section K "Rear Suspension")

Page LY.s.11

BRAKES

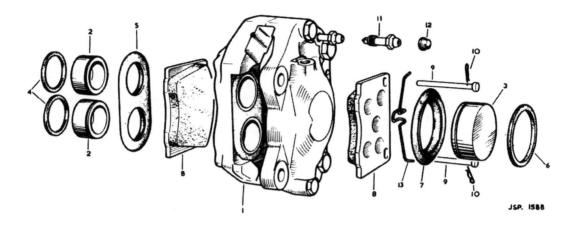

Fig. 5. *Exploded view of front caliper.*

1. Caliper body
2. Outer piston
3. Inner piston
4. Seal
5. Dust seal
6. Seal
7. Dust seal
8. Friction pad
9. Retaining pin
10. Clip
11. Brake bleed nipple
12. Dust cap
13. Anti-chatter clip

and remove the four self-locking nuts from the halfshaft inner universal joint.

Withdraw the joint from the bolts and allow the hub carrier to move outwards—support the carrier in this position.

The caliper can now be removed from the aperture at the front of the cross-member.

Refitting

Refitting is the reverse of the removal procedure.

Fit the fluid supply pipe and the bridge pipe. Bleed the braking system.

THE FRONT BRAKE DISCS
Removal

Jack up the car and remove the road wheel. Disconnect the flexible hydraulic pipe from the frame connection and plug the connector to prevent ingress of dirt and loss of fluid.

Discard the locking wire and remove the two caliper mounting bolts. Remove the caliper.

Remove the hub (as described in Section J "Front Suspension").

THE REAR BRAKE DISCS
Removal

Remove the rear suspension unit (as described in Section K "Rear Suspension").

Invert the suspension and remove the two hydraulic damper and road spring units (as described in Section K "Rear Suspension").

Remove the four steel type self-locking nuts securing the halfshaft inner universal joint and brake disc to the axle output shaft flange.

Withdraw the halfshaft from the bolts noting the number of camber shims between the universal joint and the brake disc.

Knock back the tabs and unscrew the two pivot bolts securing the hand brake pad carriers to the caliper. Remove the pivot bolts and the retraction plate.

Withdraw the handbrake pad carriers from the aperture at the rear of the cross members.

Knock back the tabs at the caliper mounting bolts.

Remove the keeper plate on the caliper and using a hooked implement, withdraw both brake pads.

Disconnect the brake fluid feed pipe at the caliper.

Unscrew the mounting bolts through the access holes in the brake disc. Remove the bolts.

Withdraw the caliper through the aperture at the front of the cross member.

Tap the halfshaft universal joint and brake disc securing bolts back as far as possible.

Lift the lower wishbone, hub carrier and halfshaft assembly upwards until the brake disc can be withdrawn from the mounting bolts.

Refitting

Refitting the brake discs is the reverse of the removal procedure. The securing bolts must be knocked back against the drive shaft flange when the new disc has been fitted.

Refit the rear suspension (as described in Section K "Rear Suspension").

Bleed the brakes.

Page LY.s.12

BRAKES

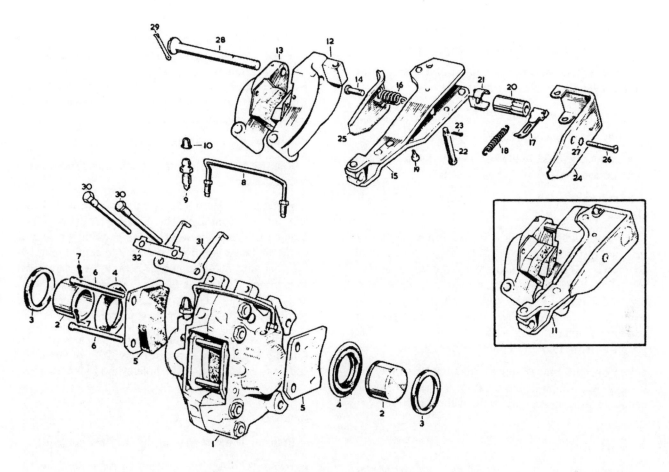

Fig. 5. *Exploded view of the rear brake caliper*

1 Rear caliper assembly (R.H.)
2 Piston
3 Seal
4 Dust seal
5 Friction pad
6 Pin
7 Clip
8 Bridge pipe
9 Bleed screw
10 Dust cap
11 Handbrake mechanism assembly
12 Pad carrier assembly (R.H. outer)
13 Pad carrier assembly (R.H. inner)
14 Anchor pin
15 Operating lever
16 Return spring
17 Pawl assembly
18 Tension spring
19 Anchor pin
20 Adjusting nut
21 Friction spring
22 Hinge pin
23 Split pin
24 Protection cover
25 Protection cover
26 Bolt
27 Washer
28 Bolt
29 Split pin
30 Bolt
31 Retraction plate
32 Tab washer

JSP159A

Page LY.s.13

BRAKES

THE BRAKE/CLUTCH PEDAL BOX ASSEMBLY

Removal (L.H. Drive)

Remove the servo vacuum pipe and clips.

Drain the brake and clutch fluid reservoirs.

Remove fluid inlet pipes from the clutch and brake master cylinders. Plug the holes.

Remove the brake fluid warning light wires.

Remove the brake and clutch reservoirs.

Remove the fluid outlet pipes from the brake and clutch master cylinders. Plug all holes.

Remove the brake and clutch pedal pads from inside the car.

Remove the dash casing in accordance with the instructions contained in Section N (Body and Exhaust). The nuts securing the pedal box assembly to the bulkhead are now exposed and can be removed together with two distance pieces and the brake pedal stop plate. Note that there are six self-locking nuts and one plain nut with a shakeproof washer. The plain nut is located on the bottom centre stud.

Remove the brake/clutch pedal box assembly by turning it through approximately 90° to allow the pedals to pass through the hole in the bulkhead.

Removal (R.H. Drive)

Remove the air cleaner elbow and the carburetter trumpets.

Remove the servo vacuum pipe and clips.

Drain the brake and clutch fluid reservoirs.

Remove the fluid inlet pipes from the clutch and brake master cylinders. Plug the holes.

Slacken the rear carburetter float chamber banjo nut and bend the petrol feed pipe towards the float chamber.

Remove the brake fluid warning light wires.

Remove the brake and clutch reservoirs.

Remove the fluid outlet pipes from the brake and clutch master cylinders. Plug all holes.

Remove the five screws securing the reaction valve assembly to the valve housing and withdraw the complete assembly. The valve housing can be removed by unscrewing the two setscrews, together with the shakeproof washers, which secure the housing to the body of the master cylinder.

Remove the throttle bell crank bracket.

Remove the brake and clutch pedal pads from inside the car.

Remove the dash casing in accordance with the instructions contained in Section N (Body and Exhaust). The nuts securing the pedal box assembly to the bulkhead are now exposed and can be removed together with two distance pieces and the brake pedal stop plate. Note that there are six self-locking nuts and one plain nut with a shakeproof washer. The plain nut is located on the bottom centre stud.

Remove the brake/clutch pedal box assembly by turning it through approximately 90° to allow the pedals to pass through the hole in the bulkhead.

Refitting

Refitting is the reverse of the removal procedure.

When refitting the securing nuts inside the car ensure that the plain nut and the shakeproof washer are fitted on the short stud in the bottom centre position.

Ensure that the brake fluid warning light wires are fitted with one feed wire (red and green) and one earth wire (black) to each reservoir cap.

When tightening the banjo union nut ensure that the petrol feed pipe is clear of the rear float chamber.

Bleed the brake and clutch hydraulic systems.

Page LY.s.14

SECTION M

WHEELS AND TYRES

DESCRIPTION

Pressed spoke or wire spoke wheels, the latter in spray painted form, are available as alternative standard equipment.

Chromium plated wire spoke wheels are fitted as Special Equipment only.

Dunlop 185VR15 SP. SPORT tubed tyres are fitted as standard equipment, whitewall tyres being available to special order only.

Whitewall tyres only will be fitted to cars sold in U.S.A. or CANADA.

These tyres will have the "MAXIMUM LOADING" and "MAXIMUM PRESSURE" information moulded on the wall of the tyre, necessary to conform to U.S.A. FEDERAL REGULATIONS.

This information, together with Seating Capacity, Seating Distribution and Recommended Tyre Size data is also quoted on a panel attached to the inside of the glove box lid.

Type—standard equipment	Pressed spoke
	Wire spoke—spray painted
—special equipment	Wire spoke—chromium plated
Fixing—Pressed spoke	Five studs and nuts
—Wire spoke	Centre lock, knock-on hub cap

Rim Section—Pressed spoke 6 JK
 —Wire spoke 5 K

Tyres
Make Dunlop
Type 185VR15 SP. SPORT

IMPORTANT

It is particularly important that tyres of different makes or types, or, those having different tread patterns, should not be mixed on individual cars as this may adversely affect the handling and steering characteristics.

A car should not, of course, be driven on bald tyres, or on tyres which have only part of the tread showing.

Driving with badly worn tyres on wet roads also greatly increases the risk of "aquaplaning" with consequent loss of steering and braking.

The importance of having tyres that are in good condition of the correct type cannot be overstressed. The Dunlop 185VR15 SP. SPORT tyres fitted as original equipment are specially produced to suit the performance of the car, and a change of make or tyre should not be made unless an assurance is given by the tyre manufacturer concerned that the alternative type is suitable for the car under maximum performance conditions.

INFLATION PRESSURES

PRESSURES SHOULD BE CHECKED WHEN THE TYRES ARE COLD, SUCH AS STANDING OVERNIGHT AND NOT WHEN THEY HAVE ATTAINED THEIR NORMAL RUNNING TEMPERATURES.

Dunlop SP Sport 185VR15	Front	Rear
For speeds up to 125 m.p.h. (200 k.p.h.)	32 lb/sq. in. 2.25 kg/cm²	32 lb/sq. in. 2.25 kg/cm²
For speeds up to maximum	40 lb/sq. in. (2.81 kg/cm²)	40 lb/sq. in. (2.81 kg/cm²)

Tyres for Winter Use
(When snow conditions make the use of special tyres necessary).

Dunlop Weathermaster SP. 44.185x15
(For use on rear wheels to replace SP. Sports tyres)
Maximum permitted speed 100 m.p.h. 32 lb/sq. in.
(160 k.p.h.) (2.25 kg/cm²)

TYRES

General Information

The Dunlop tyres specified have been specially designed for the high speeds possible with this car.

When replacing worn or damaged tyres and tubes it is essential that tyres with exactly the same characteristics are fitted.

Due to the high speed performance capabilities of the car, it is important that no attempt is made to

WHEELS AND TYRES

repair damaged or punctured tyres.

All tyres which are suspect in any way should be submitted to the tyre manufacturer for their examination and report. The importance of maintaining all tyres in perfect condition cannot be too highly stressed.

IMPORTANT

The use of standard inner tubes with Weathermaster tyres is NOT permissible.

Special tubes are available and are identified by the size and lettering "Weathermaster Only".

These special tubes should NOT be used with standard tyres.

WIRE SPOKE WHEELS
Description

Dunlop cross-spoked wheels are fitted as optional equipment.

Cross spoking refers to the spoke pattern, where the spokes radiate from the well of the wheel rim to the nose or outer edge of the hub shell, and from the rim to the flanged or inner end of the shell.

Dismantling, reassembling, and adjustment details remain the same as that detailed in the 3.8 'E' Type Service Manual Section M.

Warning

Chromium plated wire wheels are protected by a clear lacquer which, under normal circumstances, should never be removed. Should removal become necessary, due to dismantling the wheel, however, the best results can be obtained by using British Domolac L10–12 Cellulose Thinners.

UNDER NO CIRCUMSTANCES SHOULD A WIRE SPOKED WHEEL BE FITTED TO THE CAR IN AN UN-LACQUERED CONDITION.

To re-lacquer, the wheels should be treated with "NECOL" which is an I.C.I. clear cellulose air-drying lacquer. This will obviate rust stains originating at the unprotected threaded portion of the spokes.

Page MY.s.2

SECTION N

BODY AND EXHAUST

SIDE FACIA PANEL
Removal

Disconnect the battery.

Remove the screen rail facia.

Withdraw all warning light and panel illumination bulb holders. Note the location for reference when refitting.

Remove the chrome ring nut and withdraw the dipper switch from the panel.

Disconnect the speedometer drive cable from the instrument head.

Withdraw the plastic retaining clip and seperate the plug and socket connection attached to the tachometer cables.

Disconnect the brake fluid warning light cables.

Disconnect the control cable from the heater air outlet ducts after releasing the locking screw securing the inner cable to the air duct operating spindle.

Remove the locknut securing the outer cable to the air duct bracket. Withdraw the cable and collect the loose adaptor.

Remove the two thumbscrews and lower the centre instrument panel.

Release two setscrews securing the two heater inner control cables to the control levers and withdraw the cables.

Withdraw two slotted screws and one setscrew, nut, and washer securing the side facia panel to the centre panel support bracket and two nuts and washers securing the panel to the support bracket at the base of the screen pillar and remove the panel.

Refitting

Refitting is the reverse of the removal procedure.

Reconect the heater and air outlet control cables ensuring that the full movement of the levers is maintained.

Reconnect the warning and panel light illumination bulb holders as noted on removal.

GLOVE BOX
Removal

Disconnect the battery.

Remove the screen rail facia.

Withdraw two screws and nuts and detach the grab handle from the mounting brackets.

Withdraw the choke warning light bulb holder from the socket at the rear of the glove box panel.

Release the locking screw and disconnect the choke operating cable from the lever (cars equipped with Exhaust Emission Control only).

Lower the centre instrument panel.

Remove the securing setscrews, nuts and washers as detailed for the side facia panel and detach the glove box.

Refitting

Refitting is the reverse of the removal procedure.

Check when reconnecting the choke control cable that the full movement of the lever is maintained when the choke is operated.

SCREEN RAIL FACIA
Removal

Disconnect the battery and lower the centre instrument panel.

Remove two nuts, lock and plain washers securing the facia to the centre panel supports.

Remove two drive screws securing the facia to the demister panel (2+2 cars only).

Remove two nuts, lock and plain washers securing the facia to the body side panels at the base of the screen pillars.

Detach the flexible demister conduit pipes from the demister nozzles.

Disconnect the two cables from the map light.

Remove the facia complete with demister nozzles.

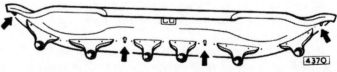

Fig. 1. *Screen rail facia.*

Page NY.s.1

BODY AND EXHAUST SYSTEM

Refitting
Refitting is the reverse of the removal procedure.

Utilizing the elongated holes in the mounting brackets adjust the forward edge of the facia to the screen rail.

DEMISTER PANEL (2+2 cars only)
Removal
Remove the screen rail facia as detailed previously.

Remove four setscrews, nuts and lockwashers securing the panel support brackets to the instrument panel support brackets.

Detach the conduits from the four demister nozzles.

Remove the panel complete with the demister nozzles.

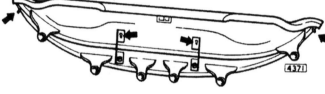

Fig. 2. *Demister panel (2+2 cars).*

Refitting
Refitting is the reverse of the removal procedure.

Utilising the elongated holes in the centre brackets adjust the forward edge of the panel to the screen.

FRONT BUMPER
Removal
The front bumper is comprised of three components, right and left hand outer sections and a centre section.

Sections are detachable after removing the bumper as a complete assembly.

Remove the four setscrews, lock and plain washers securing the outer sections to the bonnet. The setscrews are accessible through cut-out portions in the underside of the sections.

Dismantle the bumper after removing the two setscrews, lock and plain washers securing the outer sections to the centre and the two setscrews securing the over-riders.

Remove the motif, if required, after withdrawing two drive screws, spring clip, and the backing plate.

Refitting
Reassembly and refitting is the reverse of the removal procedure.

Renew the beading between the bumper sections and the over-riders if worn or damaged.

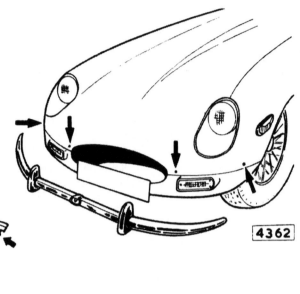

Fig. 3. *Front bumper removal The arrows indicate the mounting points.*

REAR BUMPER
Removal
The rear bumper is comprised of three components, right and left hand and centre sections.

Sections are detachable after removal of the bumper as an assembly.

Remove the two setscrews, lock and plain washers and two nuts securing the bumper outer sections to the body; the setscrews are located within the wheel arch.

Release the two setscrews located above the rear lamps. Withdraw the two setscrews securing the reverse lamp carrier brackets and remove the bumper assembly.

Dismantle the bumper after removing the two setscrews, lock and plain washers securing the outer sections to the centre and the two setscrews securing the over-riders.

Page NY.s.2

BODY AND EXHAUST SYSTEM

Refitting

Refitting is the reverse of the removal procedure.
Renew the beading between the bumper sections and the over-riders if worn or damaged.

Fig. 4 *Rear bumper removal. The arrows indicate the mounting points.*

BONNET

The removal and refitting procedure remains identical to that stated in the 3·8 "E" Type Service Manual—Page N8—with the addition of the following:—

Withdraw the split pin, washer and clevis pin and detach the front number plate tie-rod fork end from the chassis front cross tube.

Fig. 5. *Bonnet hinge mounting points.*

WINDOW REGULATOR

The window regulator remains the same basically as that detailed on Page N21 of the 3·8 "E" Type Service Manual with the exception of the method of securing the regulator handle.

This is now secured to the control unit by a central fixing screw and not by a pin. Removal of this screw will permit the handle to be withdrawn.

Fig. 6. *Location of the screws and nuts securing the window regulator to the door panel.*

DOOR LOCK MECHANISM

Door locks fitted to the "E" Type Series 2 cars incorporate an anti-burst feature, while the remote control units have recessed handles.

The removal, refitting and adjusting details remain the same as those quoted in the "E" Type Service Manual with the exception of the following details:—

The recessed handle is secured to the remote control unit by a central fixing screw and not by a pin.

Removal of the screw will permit the handle to be withdrawn.

The remote control unit is not supplied in the locked position as was the previous model and no provision is made for the insertion of a pin when refitting (see Page N23 under "Locating the Remote Control Unit").

Any adjustment necessary when reconnecting the link can be made by utilising the elongated holes in the control unit mounting plate.

Page NY.s.3

BODY AND EXHAUST SYSTEM

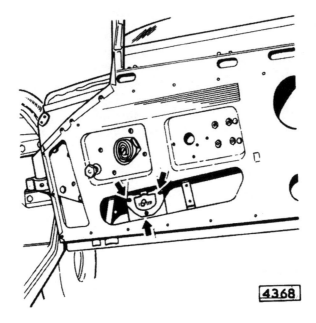

Fig. 7. *Location of the screws securing the door lock control.*

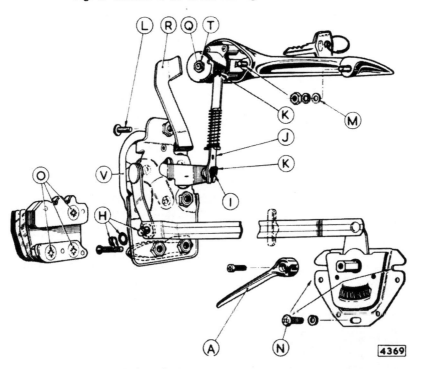

Fig. 8. *Exploded view of the door lock mechanism.*

- A   Interior handle.
- H   Lever pin, wave washer and clip.
- I   Spring clip.
- J   Adjustable link.
- K   Dowel.
- L   Latch fixing screw.
- M   Exterior handle fixings.
- N   Remote control fixing screw.
- O   Striker fixing screws.
- Q   Striker.
- R   Striker lever.
- T   Locknut.

Page NY.s.4

SECTION O

HEATING AND WINDSCREEN WASHING EQUIPMENT

HEATER

The heater unit remains the same as that stated in the 3·8 "E" Type Service Manual with the exception of the following items:—
1. Heater air controls.
2. Heater temperature controls.
3. Heater fan switch.
4. Air distributor controls.

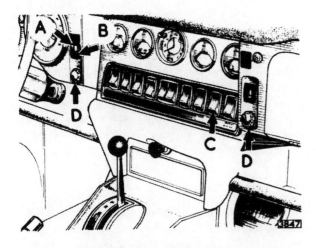

Fig. 1. *Heating and ventilating controls.*

    A Heater air controls.
    B Heater temperature controls.
    C Heater fan switch.
    D Heater outlet controls.

The outlets situated under the duct behind the facia panel, are fitted with finger operated direction controls.

Fully rotating the right-hand knob clockwise and the left-hand knob anti-clockwise will cut off all air to the interior of the car and direct the supply to the ducts at the base of the windscreen. Reverse rotation of the knobs will re-direct air progressively from the windscreen to the car interior.

VENT CONTROL CABLES
Removal

Withdraw the parcel tray on each side of the dash by removing four drive screws and four thumb screws.

Release the locknuts securing the outer cables to the vent bracket. Disconnect the cables and collect the loose adaptor. Unscrew the cable from the centre finisher and withdraw the assemblies. A thin spanner will be required to remove the outer casing from the finisher.

Refitting
Reverse the removal procedure to refit the cables.

AIR/TEMPERATURE CONTROLS
Removal

Remove the screen rail facia as detailed in Section N Page NY.s.1.

Withdraw two small screws and detach the knobs from the levers.

Release the locking screws and disconnect the control cables.

Remove the self locking nut and withdraw the lever pivot pin. Note the plain washer located between the levers.

Refitting

Refitting is the reverse of the removal procedure. Check that full movement of the levers is maintained when connecting the cables.

WINDSCREEN WASHER
Description

The Lucas 5SJ windscreen washer replaces the unit detailed in the 3·8 Service Manual.

The Lucas 5SJ screen jet is an electrically operated unit comprising a small permanent—magnet motor driving a centrifugal pump through a 3-piece Oldham type coupling. The water container is moulded in high density polythene.

DATA

Minimum water delivery pressure	4·5 lb./sq. in. (0·32 kg./sq. cm.).
Minimum water delivery per second	3·5 c.c.
Container capacity	2¼ pints (1·1 litres).
Usable quantity of water	2 pints (1 litre).
Diameter of nozzle orifice	0·25″–0·28″ (6·3–7 mm.).
Nominal voltage of unit	12
Maximum current consumption	2 amps.
Resistance between commutator segments	2·8–3·1 ohms.

Page OY.s.1

HEATING AND WINDSCREEN WASHING EQUIPMENT

Filling Up

The correct water level is up to the bottom of the container neck. Do not overfill or unnecessary splashing may result. Always replace the filler cover correctly after filling up. It is not possible to empty the container with the pump. Refilling is necessary when the water level has fallen below the level of the pump.

Keep the pump filter clean and the container free from sediment.

Cold Weather

The water container can be given a safe degree of protection down to —28°F. (—33°C.) by the use of proprietary antifreeze solutions such as marketed by Trico or Holts. Instructions regarding the use of the solvent will be found on the container.

Denatured alcohol (methylated spirits) must NOT be used. The use of this chemical will discolour the paintwork.

SERVICING

Testing in Position

(a) Testing with a voltmeter:-

Connect a suitable direct current voltmeter to the motor terminals observing the polarity as indicated on the moulding housing. Operate the switch. If a low or zero voltage is indicated, the No. 6 fuse, switch and external connections should be checked and corrected as necessary.

If the voltmeter gives a reverse reading, the connections to the motor must be transposed.

If supply voltage is registered at the motor terminals but the unit fails to function, an open-circuit winding or faulty brush gear can be suspected. Dismantle the motor as described under the heading "Dismantling".

(b) Checking the external nozzles and tubes:—

If the motor operates but little or no water is delivered to the screen, the external tubes and nozzles may be blocked.

Remove the external plastic tube from the short connector on the container and, after checking that the connector tube is clear, operate the washer switch. If a jet of water is ejected, check the external tubes and nozzles for damage or blockage.

If no water is ejected, proceed as detailed under "Dismantling".

(c) Testing with an ammeter:—

Connect a suitable direct current ammeter in series with the motor and operate the switch. If the motor does not operate but the current reading exceeds that given in "Data", remove the motor and check that the pump impeller shaft turns freely.

If the shaft is difficult to turn, the water pump unit must be replaced. If the shaft turns freely, the fault lies in the motor which must be dismantled and its component parts inspected.

Dismantling

Disconnect the external tube and the electrical connections and remove the cover from the container. Remove the self-tapping screw which secures the motor to the cover and pull away the motor unit. Take care not to lose the loose intermediate coupling which connects the armature coupling to the pump spindle coupling.

Remove the armature coupling from the armature shaft as follows:—

Hold the armature shaft firmly with a pair of snipe-nosed pliers and, using a second pair of pliers, draw off the armature coupling.

Remove the two self-tapping screws from the bearing plate. The bearing plate and rubber gasket can now be removed. Remove the two terminal screws. The terminal nuts and brushes can now be removed and the armature withdrawn. Take care not to lose the bearing washer which fits loosely on the armature shaft.

The pole assembly should not normally be disturbed. If, however, its removal is necessary, make a careful note of its position relative to the motor housing. The narrower pole piece is adjacent to the terminal locations. Also the position of the pole clamping member should be observed. When fitted correctly, it locates on both pole pieces but, if fitted incorrectly, pressure is applied to one pole piece only.

Bench-Testing

If the motor has been overheated, or if any part of the motor housing is damaged, a replacement motor unit must be fitted.

Page OY.s.2

HEATING AND WINDSCREEN WASHING EQUIPMENT

Armature:—

If the armature is damaged or if the windings are loose or badly discoloured, a replacement armature must be fitted.

The commutator must be cleaned with a fluffless cloth moistened in petrol, or, if necessary, polished with a strip of very fine glass paper.

The resistance of the armature winding should be checked with an ohmmeter. This resistance should be in accordance with that given in "Data".

Brushes:—

If the carbon is less than $\frac{1}{16}''$ (1·59 mm.) long, a new brush must be fitted. Check that the brushes bear firmly against the commutator.

Re-assembling

Re-assembling of the unit is the reverse of the dismantling procedure. However, the following points should be noted:—

Make sure the bearing recess in the motor is filled with Rocol Molypad molybdenised grease. Remove excessive grease from the face of the bearing boss.

Check that the pole piece assembly does not rock and that the pole pieces are firmly located in the circular spigot. Ensure that the pole piece assembly and clamping member are the right way round.

Before replacing the motor unit on the cover, ensure that the armature coupling is pushed fully home and that the intermediate coupling is in place.

Performance Testing

Equipment required:—

    D.C. supply of appropriate voltage.

    D.C. voltmeter, first grade, moving coil 0–3 amp.

    D.C. ammeter.

    0–15 lb. sq. in. (0–1 kg. sq. cm.) pressure gauge.

    Pushbutton with normally open contacts.

    Two-jet nozzle.

    On-off tap.

    100 c.c. capacity measure.

    4 ft. 6 in. (1·37 m.) length of plastic tubing.

Connect up the equipment as shown in Fig. 1. The water level in the container must be 4" (101·6 mm.) above the base of the pump assembly. The pressure gauge and nozzle must be 18" (45·72 cm.) above the water level.

Open the tap. Depress the button for approximately five seconds and check the voltmeter reading which should be the same as the supply voltage. On releasing the switch, close the tap to ensure that the plastic tubing remains charged with water. Empty the measuring cylinder.

Open the tap and operate the push switch for precisely ten seconds after which period release the switch and close the tap.

During the ten-second test, the current and pressure values should be in accordance with those given in Data and at least 35 c.c. of water should have been delivered.

HEATING AND WINDSCREEN WASHING EQUIPMENT

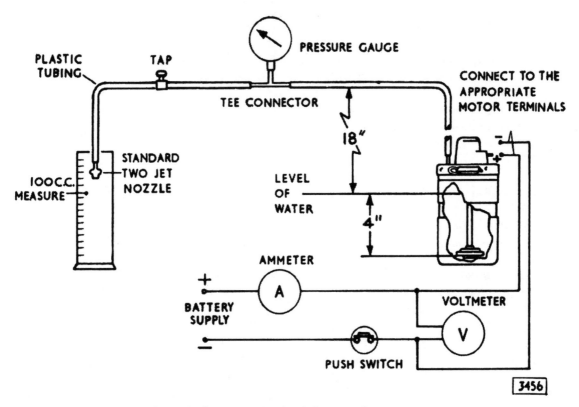

Fig. 1. *Performance testing the windscreen washing equipment.*

SECTION P
ELECTRICAL AND INSTRUMENTS
BATTERY
LUCAS CA11/7

Description
The Lucas Model C.A. battery, as listed above, is a new type fitted with an air lock device (Aqualok) which simplifies the topping up procedure. This device consists of a one-piece vent cover and six sliding tubes, perforated to act as air valves and fitted one to each cell aperture.

The tubes are automatically raised when the vent cover is removed and depressed when the cover is refitted.

Air or added water is admitted to the cell venting chamber (via the tube perforations) only while the tubes are in the depressed position.

No provision is provided in the battery top cover for the insertion of the prong of a heavy discharge tester.

DATA

Battery type	CA11/7
Voltage	12
Number of plates per cell	11
Capacity at 10-hour rate	53
Capacity at 20-hour rate	60

Maintenance
Wipe away any foreign matter or moisture from the top of the battery and ensure that the connections and fixings are clean and tight.

Check the electrolyte level weekly. In extreme cold conditions the battery should be topped up immediately prior to driving the car so that the electrolyte mixing can occur to prevent freezing of the added water.

Topping up the battery should be carried out with the car on a reasonably level surface.

Remove the vent cover. If the acid level is below the bottom of the tubes pour distilled water into the trough until all the tubes are filled.

Replace the vent cover. The electrolyte level is now correct.

DO NOT, under any circumstances, top up the battery by using the normal type of battery filler which incorporates a plunger valve in the filler neck.

The use of this type of filler will depress the sliding tubes and allow the battery to be overfilled.

Important: The vent cover must be kept in position at all times except when topping up.

Distilled water should always be used for topping-up. In an emergency, however, clean soft rain water collected in an earthernware container may be used.

Note: Never use a naked light when examining a battery as the mixture of oxygen and hydrogen given off by the battery when on charge and to a lesser extent when standing idle, can be dangerously explosive.

Clean off any corrosion from the battery cable terminals and coat with vaseline before re-connecting.

Removal
Unscrew the two wing nuts retaining the battery strap; remove the fixing rods and strap. Disconnect terminals and lift out the battery from the tray.

WARNING: Rubber sealing plugs are not incorporated in the manifold filler cover.
When removing the battery it is ESSENTIAL that extreme care is taken to ensure that it is NOT tipped to any degree.
Failure to ensure this will result in acid spillage which may cause severe acid burning to the operator and to the car.

Refitting
Refitting is the reverse of the removal procedure.

Page PY.s.1

ELECTRICAL AND INSTRUMENTS

Before refitting the cable connectors, clean the terminals and coat with petroleum jelly.

Persistent low state of charge

First consider the conditions under which the battery is used. If the battery is subjected to long periods of discharge without suitable opportunities for recharging, a low state of charge can be expected. A fault in the alternator or control unit, or neglect of the battery during a period of low or zero mileage may also be responsible for the trouble.

Manifold Vent Cover

See that the ventilating holes in the cover are clear.

Level of Electrolyte

The surface of the electrolyte should be just level with the tops of the separator guards. If necessary, top up with distilled water as detailed on page PY.s.1. Any loss of acid from spilling or spraying (as opposed to the normal loss of water by evaporation) should be made good by dilute acid of the same specific gravity as that already in the cell.

Cleanliness

See that the top of the battery is free from dirt or moisture which might provide a discharge path. Ensure that the battery connections are clean and tight.

Hydrometer Tests

Measure the specific gravity of the acid in each cell in turn with a hydrometer. To avoid misleading readings, do not take hydrometer reading immediately after topping-up.

The reading given by each cell should be approximately the same.

If one cell differs appreciably from the others, an internal fault in the cell is indicated.

The appearance of the electrolyte drawn into the hydrometer when taking a reading gives useful indication of the state of the plates. If the electrolyte is very dirty, or contains small particles in suspension, it is possible that the plates are in a bad condition.

The specific gravity of the electrolyte varies with the temperature, therefore, for convenience in comparing specific gravities, this is always corrected to 60°F (16°C) which is adopted as a reference temperature.

The method of correction is as follows:—

For every 5°F (2.8°C) below 60°F (16°C) deduct 0.002 from the observed reading to obtain the true specific gravity at 60°F (16°C).

For every 5°F (2.8°C) above 60°F (16°C) add 0.002 to the observed reading to obtain the true specific gravity at 60°F (16°C).

The temperature must be that indicated by a thermometer actually immersed in the electrolyte and not in the air temperature.

Compare the specific gravity of the electrolyte with the values given in the table and so ascertain the state of charge of the battery.

If the battery is in a discharged state, it should be recharged, either on the vehicle by a period of daytime running or on the bench from an external supply, as described under "Recharging from an External Supply".

All Service procedure concerning the following items remains as detailed on pages P8–P9 of the 3.8 'E' Type Service Manual.

(1) Recharging from an external supply

(2) Preparing new unfilled, uncharged batteries for Service

(3) Preparing new "Dry-charged" batteries for Service.

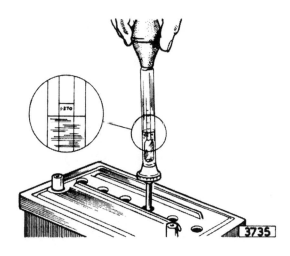

Fig. 1. *Method of topping up the Lucas C.A. battery.*

Page PY.s.2

ELECTRICAL AND INSTRUMENTS

DISTRIBUTOR
(NOT U.S.A. AND CANADA)

A waterproof cover is incorporated in the distributor assembly, located between the distributor cap and body. This cover is detachable after removing the distributor cap and disconnecting the cable from the contact breaker spring post.

DATA

Ignition Distributor Type	22D6
8 to 1 Compression Ratio	41060A
9 to 1 Compression Ratio	41060A
Cam dwell angle	34°±3°
Contact breaker gap	0·014″–0·016″ (0·36—0·41 mm.)
Contact breaker spring tension (Measured at free contact)	18—24 ozs. (512—682 gms.)

IGNITION TIMING

8 to 1 Compression Ratio	9° BTDC
9 to 1 Compression Ratio	10° BTDC

IGNITION DISTRIBUTOR TEST DATA

Distributor Type	Lucas Service Number	Lucas Vacuum Unit Number	VACUUM TIMING ADVANCE TESTS The distributor must be run immediately below the speed at which the centrifugal advance begins to function to obviate the possibility of an incorrect reading being registered.		No advance in timing below-ins. of mercury	CENTRIFUGAL TIMING ADVANCE TESTS Mount distributor in centrifugal advance test rig and set to spark at zero degrees at 100 r.p.m.					
			Vacuum in inches of mercury and advance in degrees			Lucas Advance Springs Number	Acelerate to-RPM and note advance in degrees		Decelerate to-RPM and note advance in degrees		No advance in timing below-RPM
			Inches	Degrees			RPM	Degrees	RPM	Degrees	
22 D6	41060A	54415894	20 13 9 7½ 6	7—9 6—8½ 2½—5½ 0—3 0—½	½ 4½	55415562	2,300	8½—10½	1800 1250 800 650 525	8½—10½ 6½—8½ 5—7 2—4 0—1½	300

Auto advance weights Lucas number 54413073.   One inch of mercury = 0·0345 kg/cm²

Page PY.s.3

ELECTRICAL AND INSTRUMENTS

FUSE UNITS

Fuse No.	CIRCUITS	Amps
1	Headlamps—Main Beam	35
2	Headlamps—Dip Beam	35
3	Horns	50
4	Traffic Hazard Warning System	35
5	Side, Panel, Tail and Number Plate (not Germany) Lamps	35
6	Horn Relay, Washer, Radiator Fan Motor and Stop Lamps	35
7	Flashers, Heater, Wiper, Choke, Fuel, Water and Oil Gauges	35
8	Headlamp Flasher, Interior Lamps and Cigar Lighter	35
In line	Heated Backlight (when fitted)	15
In line	Radio, Optional Extras	5

THE ALTERNATOR

MODEL 11AC (43 AMP)

DESCRIPTION

The Lucas 11 AC alternator is a lightweight machine designed to give increased output at all engine speeds.

Basically the unit consists of a stationary output winding with built in rectification and a rotating field winding, energised from the battery through a pair of slip rings.

The stator consists of a 24 slot, 3 phase star connected winding on a ring shaped lamination pad housed between the slip ring end cover and the drive end bracket.

The rotor is of 8-pole construction and carries a field winding connected to two face type slip rings. It is supported by a ball bearing in the drive end bracket and a needle roller bearing in the slip ring end cover (see Fig. 2).

Page PY.s.4

ELECTRICAL AND INSTRUMENTS

The brushgear for the field system is mounted on the slip ring end cover. Two carbon brushes, one positive and one negative, bear against a pair of concentric brass slip rings carried on a moulded disc attached to the end of the rotor. The positive brush is always associated with the inner slip ring. There are also six silicon diodes carried on the slip ring end cover, these being connected in a three phase bridge circuit to provide rectification of the generated alternating current output (see Fig. 2). The diodes are cooled by air flow through the alternator induced by a 6" (15·24 cm.) ventilating fan at the drive end.

The alternator is matched to an output control unit, Model 4TR, which is described on page PY.s.11.

This unit controls the alternator field current and hence the alternator terminal voltage.

A cut-out is not included in the control unit as the diodes in the alternator prevent reverse currents from flowing through the stator when the machine is stationary or is generating less than the battery voltage.

No separate current-limiting device is incorporated; the inherent self-regulating properties of the alternator effectively limit the output current to a safe value.

A Lucas 3AW warning light control unit is incorporated in the circuit.

The output control unit and the alternator field windings are isolated from the battery when the engine is stationary by a Lucas 6RA relay incorporated in the circuit.

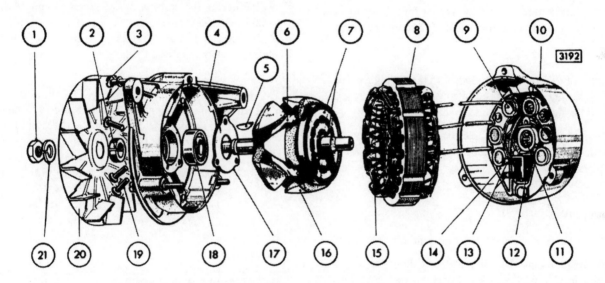

Fig. 2. *Exploded view of the Lucas 11 AC alternator.*

1. Shaft nut.
2. Bearing collar.
3. Through fixing bolts (3).
4. Drive end bracket.
5. Key.
6. Rotor (field) winding.
7. Slip rings.
8. Stator laminations.
9. Silicon diodes (6).
10. Slip ring end bracket.
11. Needle roller bearing.
12. Brush box moulding.
13. Brushes.
14. Diode heat sink.
15. Stator windings.
16. Rotor.
17. Bearing retaining plate.
18. Ball bearing.
19. Bearing retaining plate rivets.
20. Fan.
21. Spring washer.

Page PY.s.5

ELECTRICAL AND INSTRUMENTS

PERFORMANCE DATA

Nominal voltage	12 volts
Nominal d.c. output (hot) in amperes	43 amperes
Stator phases	3
Phase connections	Star
Resistance/phase at 68°F (20°C)±5%	0.107 ohms
Resistance of rotor winding in ohms at 68°F (20°C)	3.8±5%

REMOVAL

Disconnect the cables from the terminals on the slip ring end cover. Note the colour and location of the cables with Lucar termination for reference when refitting.

Remove the drive belt by pushing the spring loaded jockey pulley inwards and lifting the belt over the alternator pulley.

Remove the two bolts securing the alternator to the mounting bracket and adjuster link. Withdraw the alternator.

REFITTING

Refitting is the reverse of the removal procedure.

When replacing the alternator belt, hold the spring loaded jockey pulley in towards the block and only release when the belt is sitting securely in the "vee" tracks.

SERVICE PRECAUTIONS

Important

4.2 "E" Type cars are equipped with transistors in the control box unit and diode rectifiers in the alternator.

The car electrical system must NOT be checked with an ohmmeter incorporating a hand driven generator until these components have been isolated.

REVERSED battery connections will damage the diode rectifiers.

Battery polarity must be checked before connections are made to ensure that the connections for the car battery are NEGATIVE earth. This is most important when using a slave battery to start the engine.

NEVER earth the brown/green cable if it is disconnected at the alternator. If this cable is earthed, with the ignition switched ON, the control unit and wiring may be damaged.

NEVER earth the alternator main output cable or terminal. Earthing at this point will damage the alternator or circuit.

NEVER run the alternator on open circuit with the field windings energised, that is with the main lead disconnected, otherwise the rectifier diodes are likely to be damaged due to peak inverse voltages.

WARNING: When using electric welding equipment for car accident repair it is advisable to carry out the following precautions.
1. Disconnect the battery.
2. Disconnect the main output and AL cables at the alternator.
3. Disconnect all cables at the 4TR control unit.

SERVICING

Testing the Alternator in position

In the event of a fault developing in the charging circuit check by the following procedure to locate the cause of the trouble.

1. Disconnect the battery.

2. Disconnect the cable (brown) from the alternator output terminal and connect a good quality moving-coil ammeter between the disconnected cable and the output terminal.

Page PY.s.6

ELECTRICAL AND INSTRUMENTS

3. Detach the terminal connector block from the base of the control unit and connect the black and brown/green cables together by means of a short length of cable with two Lucar terminals attached. This operation connects the alternator field winding across the battery terminals and by-passes the output control unit (Fig. 3).

4. Reconnect the battery earth lead. Switch on the ignition and start the engine. Slowly increase the engine speed until the alternator is running at approximately 4,000 r.p.m. (2,000 engine r.p.m.). Check the reading on the ammeter which should be approximately 40 amperes with the machine at ambient temperature.

A low current reading will indicate either a faulty alternator or poor circuit wiring connections.

If, after checking the latter, in particular the earth connections, a low reading persists on repeating the test refer to paragraph (5).

In the case of a zero reading, switch on the ignition and check that the battery voltage is being applied to the rotor windings by connecting a voltmeter between the two cable ends normally attached to the alternator field terminals. No reading on this test indicates a fault in the field isolating relay or the wiring associated with this circuit. Check each item in turn and rectify as necessary.

5. If a low output has resulted from the test described in paragraph (4) and the circuit wiring is in order; measure the resistance of the rotor coil field by means of an ohmmeter connected between the field terminal blades with the external wiring disconnected.

The resistance must approximate 3·8 ohms.

When a ohmmeter is not available connect a 12 volt DC supply between the field terminals with an ammeter in series. The ammeter reading should be approximately 3·2 amperes Fig. 4.

A zero reading on the ammeter, or an infinity reading on the ohmmeter indicates an open circuit in the field system, that is, the brush gear slip rings or winding. Conversely, if the current reading is much above, or the ohmmeter is much below, the values given then it is an indication of a short circuit in the rotor winding in which case the rotor slip ring assembly must be changed.

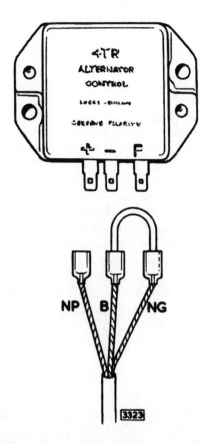

Fig. 3. *Detach the terminal connectors from the base of the control unit.*

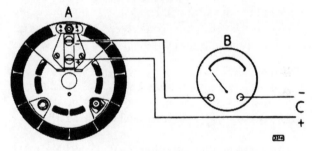

Fig. 4. *Testing the alternator with an ammeter.*
A—*Alternator.* B—*Ammeter.* C—*Battery.*

DISMANTLING THE ALTERNATOR (Fig. 2).

Disconnect the battery and remove the alternator as detailed on page PY.s.6.

Remove the shaft nut (1) and spring washer (21). Withdraw the pulley and fan (20).

Remove bolts (3) noting that the nuts are staked to the through bolts and that the staking must be removed before the nuts are unscrewed. If the threads of the nuts or bolts are damaged, new bolts must be fitted when reassembling.

Page PY.s.7

ELECTRICAL AND INSTRUMENTS

Mark the drive end bracket (4), lamination pack (8) and slip ring end bracket (10) so that they may be reassembled in correct angular relation to each other. Care must be taken not to damage the lamination pack when marking.

Withdraw the drive end bracket (4) and rotor (16) from the stator (8). The drive end bracket and rotor need not be separated unless the bearing requires examination or the rotor is to replaced.

In the latter case the rotor should be removed from the drive end bracket by means of a hand press having first removed the shaft key (5) and bearing collar (2).

Remove the terminal nuts, washers and insulating pieces brush box screws and the 2 B.A., hexagon headed setscrew. Withdraw the stator and diode heat sink assemblies from the slip ring end cover.

Close up the retaining tongue at the root of each field terminal blade and withdraw the brush spring together with the terminal assemblies from the moulded brushbox.

REASSEMBLY

Reassembly of the alternator is the reverse of the dismantling procedure. Care must be taken to align the drive end bracket, lamination pack, slip ring and bracket correctly.

Tighten the three through bolts evenly to a maximum torque of 45 to 50 lb./ins. (0·518 to 0·576 kgm.). Restake the nuts after tightening.

Tighten the brush box fixing screws to a maximum torque of 10 lb./ins. (0·115 kgm.).

IMPORTANT

It is important to ensure that a .045″ (1.28 mm) gap exists between the non-pivotal end of the heat sinks (see Fig. 15) when reassembling the alternator.

INSPECTION OF BRUSHGEAR (EARLY MODELS)

Measure brush length. A new brush is $\frac{5}{8}$″ (15·88 mm.) long; a fully worn brush is $\frac{5}{32}$″ (3·97 mm.) and must be replaced at, or approaching, this length. The new brush is supplied complete with brush spring and Lucar terminal blade and has merely to be pushed in until the tongue registers. To ensure that the terminal is properly retained, carefully lever up the retaining tongue with a fine screwdriver blade, so that the tongue makes an angle of 30° with the terminal blade.

The normal brush spring pressures are 4-5 oz. (113 to 142 gms.) with the spring compressed to $\frac{25}{32}$″ (19·84 mm.) in length and 7½ to 8½ oz. (212 to 242 gms.) with the spring compressed to $\frac{13}{32}$″ (10·31 mm.) in length. These pressures should be measured if the necessary equipment is available.

Check that the brushes move freely in their holders. If at all sluggish, clean the brush sides with a petrol moistened cloth or, if this fails to effect a cure, lightly polish the brush sides on a smooth file. Remove all traces of brush dust before re-housing the brushes in their holders.

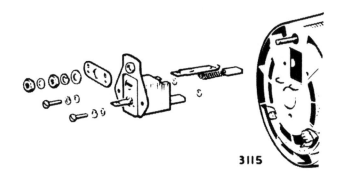

Fig. 5. *Showing the brush removal (early cars).*

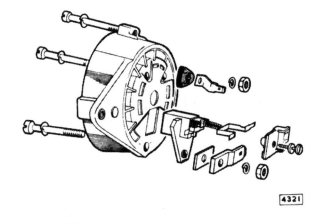

Fig. 6. *Showing the brush removal (later cars).*

INSPECTION OF BRUSHGEAR (LATER MODELS)

Later model alternators will have side entry cables. The characteristics of the alternator remain the same as the previous model (side entry cables) with the exception of the method of inspection and the fixing of the brush gear as detailed below:—

The brush length when new is $\frac{5}{8}$″ (15.9 mm.). The serviceability of the brushes may be gauged by measuring the amount by which they protrude beyond the brushbox moulding when in the free position. For a brush to remain serviceable this should exceed 0.2″ (5 mm.).

Page PY.s.8

ELECTRICAL AND INSTRUMENTS

Renew the brush assemblies if the brushes are worn to or below this length.

The new brush is supplied complete with brush spring and 'Lucar' terminal blade and is retained in position by a plate secured with a single fixing screw.

Check the brush spring pressure by using a push type spring gauge. Push each spring back against its spring until the brush face is flush with the housing.

The gauge should then register 8–16 oz. (227–454 grammes). Replace a brush assembly which gives a reading appreciably outside these limits where this is not due to the brush movement being impeded for any reason.

INSPECTION OF SLIP RINGS

The surfaces of all slip rings should be smooth and uncontaminated by oil or other foreign matter. Clean the surfaces using a petrol moistened cloth, or if there is any evidence of burning, very fine glasspaper. On no account must emery cloth or similar abrasives be used. No attempt should be made to machine the slip rings, as any eccentricity in the machining may adversely affect the high-speed performance of the alternator. The small current carried by the rotor winding together with the unbroken surface of the slip rings mean that the likelihood of scored or pitted slip rings is almost negligible.

ROTOR

Test the rotor winding by connecting an ohmmeter (Fig. 7) or 12 volt D.C. (Fig. 8) supply between the slip rings. The readings of resistance or current should be as given on page PY.s.6.

Test for defective insulation between each of the slip rings and one of the rotor poles using a mains low-wattage test lamp for the purpose. If the lamp lights, the coil is earthing therefore a replacement rotor/slip ring assembly must be fitted.

No attempt should be made to machine the rotor poles or to true a distorted shaft.

STATOR

Unsolder the three stator cables from the heat sink assembly, taking care not to overheat the diodes—(see 4h page 6). Check the continuity of the stator windings by first connecting any two of the three stator cables in series with a test lamp of not less than 36 watts and a 12-volt battery as shown in Fig. 10. Repeat the test, replacing one of the two cables by the third cable. Failure of the test lamp to light on either occasion means that part of the stator winding is open-circuit and a replacement stator must be fitted.

Test for defective insulation between stator coils and lamination pack with a mains test lamp. Connect the test probes between any one of the three cable ends and the lamination pack. If the lamp lights, the stator coils are earthing and a replacement stator must be fitted.

Before re-soldering the stator cable ends to the diode pins check the diodes.

DIODES

Each diode can be checked by connecting it in series with a 1·5 watt test bulb (Lucas No. 280)

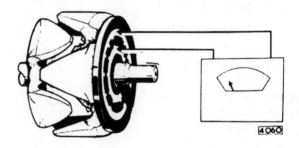

Fig. 7. *Measuring the rotor winding resistance with an ohmmeter.*

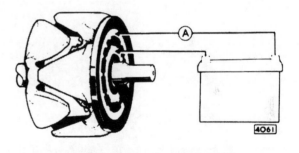

Fig. 8. *Measuring the rotor winding resistance with an ammeter and battery.*

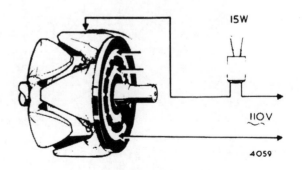

Fig. 9. *Insulation test of rotor winding.*

Page PY.s.9

ELECTRICAL AND INSTRUMENTS

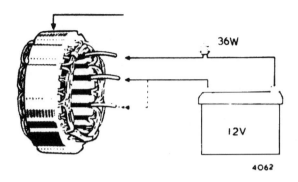

Fig. 10. *Stator winding continuity test.*

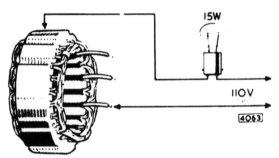

Fig. 11. *Stator winding insulation test.*

across a 12 volt D.C. supply and then reversing the connections.

Current should flow and the bulb light in one direction only. If the bulb lights up in both tests or does not light up in either then the diode is defective and the appropriate heat sink assembly must be replaced.

The above procedure is adequate for service purposes. Any accurate measurement of diode resistance requires factory equipment. Since the forward resistance of a diode varies with the voltage applied, no realistic readings can be obtained with battery-powered ohmmeters.

If a battery—ohmmeter is used, a good diode will yield "Infinity" on one direction and some indefinite, but much lower, reading in the other.

WARNING:

Ohmmeters of the type incorporating a hand-driven generator must never be used for checking diodes.

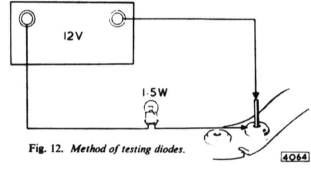

Fig. 12. *Method of testing diodes.*

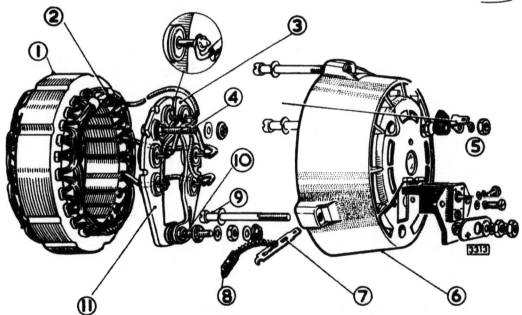

Fig. 13. *Exploded view of the slip ring end cover.*

1. Stator.
2. Star point.
3. Negative heat sink anode base diodes (black).
4. Warning light terminal 'AL'.
5. Field terminal (2).
6. Slip ring end cover.
7. Terminal blade retaining tongue.
8. Rotor slip ring brush (2).
9. "Through" bolts (3).
10. Output terminal (+).
11. Positive heat sink and cathode base diode (red).

Page PY.s.10

ALTERNATOR DIODE HEAT SINK REPLACEMENT

The alternator heat sink assembly consists of two mutually insulated portions, one of positive and the other of negative polarity. The diodes are not individually replaceable but, for service purposes, are supplied already pressed into the appropriate heat sink portion. The positive carries three cathode base diodes marked black.

When soldering the interconnections, M grade 45–55 tin-lead solder should be used.

Great care must be taken to avoid overheating the diodes or bending the diode pins. The diode pins should be lightly gripped with a pair of suitable long-nosed pliers, acting as a thermal shunt and the operation of soldering carried out as quickly as possible.

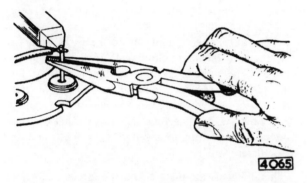

Fig. 14. *Use of thermal shunt when soldering diode connections*

After soldering to ensure adequate clearance of the rotor, the connections must be neatly arranged around the heat sinks and tacked down with "MMM" EC 1022 adhesive where indicated in Fig. 15. The stator connections must pass through the appropriate notches at the edge of the heat sink.

ELECTRICAL AND INSTRUMENTS

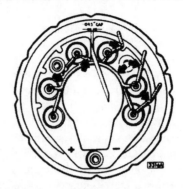

Fig. 15. *Showing the silicon diodes and connection in the slip ring end cover.*
The feeler gauge inserted between the diode carriers.

BEARINGS

Bearings which are worn to the extent that they allow excessive side movement of the rotor shaft must be renewed. The needle roller bearing in the slip ring end cover is supplied complete with the end cover.

To renew the drive end ball bearing following the withdrawal of the rotor shaft from the drive-end bracket, proceed as follows:—

(a) File away the roll-over on each of the three bearing retaining plate rivets and punch out the rivets.

(b) Press the bearing out of the bracket.

(c) Locate the bearing in the housing and press it home. Refit the bearing retaining plate using new rivets.

Note: Before fitting the replacement bearing see that it is clean and, if necessary, pack it with high-melting point grease such as Shell Alvania No. 3 or an equivalent lubricant.

ALTERNATOR OUTPUT CONTROL UNIT MODEL 4 TR.

GENERAL

Model 4 TR is an electronic control unit. In effect its action is similar to that of the vibrating contact type of voltage control unit but switching is achieved by transistors instead of vibrating contacts. A Zener diode provides the voltage reference in place of the voltage coil and tension spring system. No cut-out is required since the diodes incorporated in the alternator prevent reverse currents flowing. No current regulator is required as the inherent self-regulating properties of the alternator effectively limit the output current to a safe value.

The control unit and the alternator field windings are isolated from the battery, when the engine is stationary, by a special double-pole ignition switch.

On cars fitted with a steering column lock, the field windings are isolated by means of a relay replacing the ignition switch control.

Care must be taken at all times to ensure that the battery, alternator and control unit are correctly connected. Reversed connections will damage the semi conductor devices employed in the alternator and control unit.

OPERATION

When the ignition is switched on, the control unit is connected to the battery through the field isolating switch or relay. By virtue of the connection through

Page PY.s.11

ELECTRICAL AND INSTRUMENTS

R1 (see Fig. 16), the base circuit of the power transistor T2 is conducted so that, by normal transistor action, current also flows in the collector-emitter portion of T2 which thus acts as a closed switch to complete the field circuit and battery voltage is applied to the field winding.

As the alternator rotor speed increases, the rising voltage generated across the stator winding is applied to the potential divider consisting of R3, R2 and R4. According to the position of the tapping point on R2, a proportion of this potential is applied to the Zener diode (ZD). The latter is a device which opposes the passage of current through itself until a certain voltage is reached above which it conducts comparatively freely.

The Zener diode can thus be considered as a voltage-conscious switch which closes when the voltage across it reaches its "breakdown" voltage (about 10 volts) and, since this is a known proportion of the alternator output voltage as determined by the position of the tapping point on R2, the breakdown point therefore reflects the value of the output voltage.

Thus at "breakdown" voltage the Zener diode conducts and current flows in the base-emitter circuit of the driver transistor T1. Also, by transistor action, current will flow in the collector-emitter portion of T1 so that some of the current which previously passed through R1 and the base circuit of T2 is diverted through T1. Thus the base current of T2 is reduced and, as a result, so also is the alternator field excitation. Consequently, the alternator output voltage will tend to fall and this, in turn, will tend to reduce the base current in T1, allowing increased field current to flow in T2. By this means, the field current is continuously varied to keep the output voltage substantially constant at the value determined by the setting of R2.

To prevent overheating of T2, due to power dissipation, this transistor is operated only either in the fully-on or fully-off condition. This is achieved by the incorporation of the positive feed-back circuit consisting of R5 and C2. As the field current in transistor T2 starts to fall, the voltage at F rises and current flows through resistor R5 and capacitor C2 thus adding to the Zener diode current in the base circuit of transistor T1. This has the effect of increasing the current through T1 and decreasing, still further, the current through T2 so that the circuit quickly reaches the condition where T1 is fully-on and T2 fully-off. As C2 charges, the feed-back current falls to a degree at which the combination of Zener diode current and feed-back current in the base circuit of T1 is no longer sufficient to keep T1 fully-on. Current then begins to flow again in the base circuit of T2. The voltage at F now commences to fall, reducing the feed-back current eventually to zero. As T2 becomes yet more conductive and the voltage at F falls further, current in the feed-back circuit reverses in direction thus reducing, still further, the base current in T1.

This effect is cumulative and the circuit reverts to the condition where T1 is fully-off and T2 is fully-on.

The above condition is only momentary since C2 quickly charges to the opposite polarity when feed-back current is reduced and current again flows in

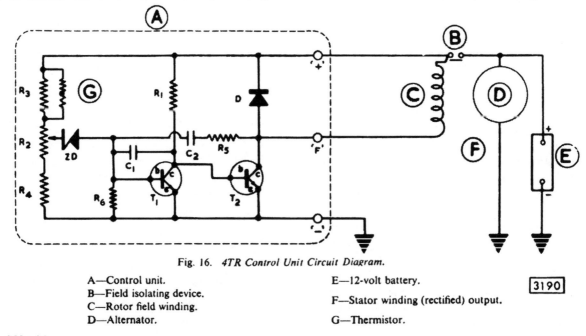

Fig. 16. *4TR Control Unit Circuit Diagram.*

A—Control unit.
B—Field isolating device.
C—Rotor field winding.
D—Alternator.
E—12-volt battery.
F—Stator winding (rectified) output.
G—Thermistor.

Page PY.s.12

ELECTRICAL AND INSTRUMENTS

the base circuit of T1. The circuit thus oscillates, switching the voltage across the alternator field winding rapidly on and off.

Transistor T2 is protected from the high induced voltage surge, which results from the collapse of the field current, by the surge quench diode D connected across the field windings. This diode also provides a measure of field current smoothing since current continues to flow in the diode after the excitation voltage is removed from the field.

The elimination of radio interference is achieved by connecting condenser C1 between the base and collector terminals of T1 to provide negative feedback. At high temperatures, a small leakage current may flow through the Zener diode even though the latter is in the nominally non-conductive state. Resistor R6 provides a path for this leakage current which otherwise would flow through T1 base circuit and adversely affect the regulator action.

A thermistor is connected in parallel with resistor R3. The thermistor is a device whose resistance increases as the temperature falls and vice verse. Any alteration in its ohmic value will modify the voltage distribution across the potential divider and thus affect the voltage value at which the Zener diode begins to conduct, so matching the changes which take place in battery terminal voltage as the temperature rises.

CHECKING AND ADJUSTING THE CONTROL UNITS

Important:

Voltage checking and setting procedure may be carried out only if the alternator and associated wiring circuits have been tested and found satisfactory in conjunction with a well-charged battery, (i.e., charging current not exceeding 10 amperes).

VOLTAGE CHECKING

Run the alternator at charging speed for eight minutes. This operation applies when bench testing or testing on the car.

Leave the existing connections to the alternator and control unit undisturbed. Connect a high quality voltmeter between control unit terminals positive and negative. If available, use a voltmeter of the suppressed-zero type, reading 12 to 15 volts.

Switch on an electrical load of approximately 2 amperes (e.g., side and tail lighting).

Start the engine and run the alternator at 3,000 r.p.m. (1,500 engine r.p.m.).

The voltmeter should now show a reading of 13·9 to 14·3 volts at 68° to 78° F. (20° to 26° C.) ambient temperature. If not, but providing the reading obtained has risen to some degree above battery terminal voltage before finally reaching a steady value, the unit can be adjusted to control at the correct voltage (see Adjusting).

If, however, the voltmeter reading remains unchanged, at battery terminal voltage, or, conversely, increases in an uncontrolled manner, then the control unit is faulty and, as its component parts are not serviced individually a replacement unit must be fitted.

ADJUSTING

Stop the engine and withdraw the control unit mounting screws.

Invert the unit and chip away the sealing compound which conceals the potentiometer adjuster (see Fig. 7).

Check that the voltmeter is still firmly connected between terminals +ve and —ve. Start the engine and, while running the alternator at 3,000 r.p.m., turn the potentiometer adjuster slot (clockwise to increase the setting or anti-clockwise to decrease it) until the required setting is obtained.

Use care in making this adjustment as a small amount of adjuster movement causes an appreciable difference in the voltage reading.

Recheck the setting by first stopping the engine then again running the alternator at 3,000 r.p.m.

Remount the control unit and disconnect the voltmeter.

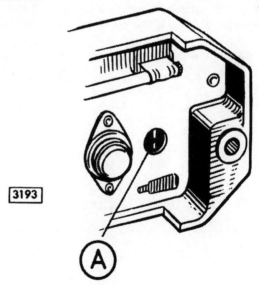

Fig. 17 *4 TR Alternator Control.*
A *Potentiometer adjuster.*

Page PY.s.13

ELECTRICAL AND INSTRUMENTS

WARNING LIGHT CONTROL UNIT
Model 3AW

DESCRIPTION

The Model 3AW warning light unit is a device connected to the centre point of one of the pairs of diodes in the alternator and operates in conjunction with the ignition warning light to give indication that the alternator is charging.

The unit is mounted on the bulkhead adjacent to the control box and is similar in appearance to the flasher unit but has different internal components consisting of an electrolytic (polarised) capacitor; a resistor and a silicone diode mounted on an insulated base with three "Lucar" terminals.

The unit is sealed, therefore servicing and adjustment is not possible. Faulty units must be replaced. Due to external similarity of the 3AW warning light unit and the flasher unit, a distinctive green label is attached to the aluminium case of the 3AW unit.

Checking Check by substitution after ensuring that the remainder of the charging circuit (including the drive belt) is functioning satisfactorily.

Warning. A faulty diode in the alternator or an intermittent or open-circuit in the alternator to battery circuit can cause excessive voltages to be applied to the warning light unit.

To prevent possible damage to a replacement unit, it is important to first check the voltage between the alternator "AL" terminal and earth. Run the engine at 1,500 r.p.m. when the voltage should be 7–7.5 volts measured on a good quality moving-coil voltmeter. If a higher voltage is registered, check that all charging circuit connections are clean and tight; then, if necessary, check the alternator rectifier diodes before fitting a replacement 3AW unit.

TRAFFIC HAZARD WARNING DEVICE (OPTIONAL EQUIPMENT)

Description

The system operates in conjunction with the four flashing (turn) indicator lamps fitted to the car. The operation of the dash panel switch will cause the four turn indicator lamps to flash simultaneously.

A red warning lamp is incorporated in the circuit to indicate that the hazard warning system is in operation.

A 35 amp. in-line fuse incorporated in the sub-panel circuit.

The flasher unit is located and is similar in appearance to the one used for the flashing turn indicators but has a differenti internal circuit. A correct replacement unit must be fitted in the even of failure.

The pilot lamp bulb is accessible after removing the bulb holder from the rear of the panel.

Failure of one or more of the bulbs due to an accident or other cause will not prevent the system operating on the remaining lamps.

Page PY.s.14

ELECTRICAL AND INSTRUMENTS

THE STARTER MOTOR

DESCRIPTION

The purpose of the pre-engaged, or positive engagement, starting motor is to prevent premature pinion ejection.

Except on occasions of tooth to tooth abutment, for which special provision is made, the starter motor is connected to the battery only after the pinion has been meshed with the flywheel ring gear, through the medium of an electro-magnetically operated linkage mechanism.

After the engine has started, the current is automatically switched off before the pinion is retracted.

On reaching the out of mesh position, the spinning armature is brought rapidly to rest by a braking device. This device takes the form of a pair of moulded shoes driven by a cross peg in the armature shaft and spring loaded (and centrifuged) against a steel ring insert in the commutator end bracket. Thus, with the supply switched off and the armature subjected to a braking force, the possibility is minimised of damaged teeth resulting from attempts being made to re-engage a rotating pinion.

A bridge-shaped bracket is secured to the front end of the machine by the through bolts. This bracket carries the main battery input and solenoid winding

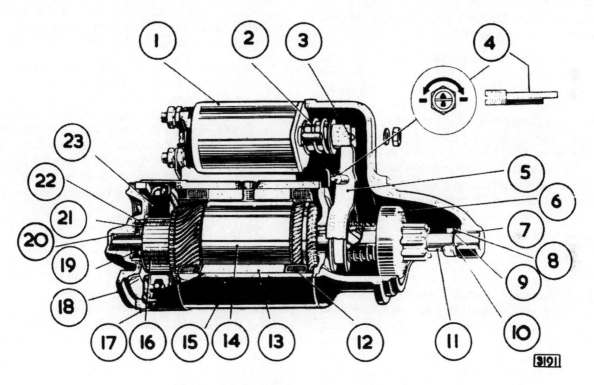

Fig. 18. *The Pre-engaged Starter Motor Model M45G.*

1. Actuating solenoid.
2. Return spring.
3. Clevis pin.
4. Eccentric pivot pin.
5. Engaging lever.
6. Roller clutch.
7. Porous bronze bush.
8. Thrust collar.
9. Jump ring.
10. Thrust washer.
11. Armature shaft extension.
12. Field coils.
13. Pole shoe.
14. Armature.
15. Yoke.
16. Commutator.
17. Band cover.
18. C.E. bracket.
19. Thrust washer.
20. Porous bronze bush.
21. Brake shoes and cross peg.
22. Brake ring.
23. Brushes.

Page PY.s.15

ELECTRICAL AND INSTRUMENTS

terminals, short extension cables being connected between these and the corresponding solenoid terminals.

TOOTH TO TOOTH ABUTMENT

The electro-magnetically actuated linkage mechanism consists essentially of a pivoted engaging lever having two hardened steel pegs (or trunnion blocks) which locate with and control the drive through the medium of a groove in an operating bush. This bush is carried, together with the clutch and pinion assembly, on an internally splined outboard driving sleeve, the whole mechanism being housed in a cut-away flange mounting snout-shaped end bracket. This operating bush is spring loaded against a jump ring in the driving sleeve by an engagement spring located between the bush and the clutch outer cover. The system return or drive demeshing spring is located round the solenoid plunger.

On the occurrence of tooth to tooth abutment (between the ends of the starter pinion teeth and those of the flywheel ring gear), the pegs or trunnion blocks at the "lower" end of the engaging lever can move forward by causing the operating bush to compress the engagement spring, thus allowing the "upper" end of the lever to move sufficiently rearwards to close the starter switch contacts. The armature then rotates and the pinion slips into mesh with the flywheel ring gear under pressure of the compressed engagement spring.

THE "LOST MOTION" (SWITCH-OFF) DEVICE

As it is desirable that the starter switch contacts shall not close until the pinion has meshed with the flywheel ring gear therefore it is important that these same contacts should always re-open before the pinion has been retracted or can be opened in the event of a starter pinion remaining for some reason enmeshed with the flywheel ring gear. To ensure this, a measure of "lost motion" is designed into some part of the engagement mechanism, its effect being to allow the starter switch or solenoid contacts (which are always spring-loaded to the open position) to open before pinion retraction begins.

Several methods of obtaining "lost motion" have been adopted, but each depends upon the yielding of a weaker spring to the stronger system return (drive demeshing or dis-engagement) spring of the solenoid plunger.

This initial yielding results in the switch contacts being fully-opened within the first $\frac{1}{8}''$ (3·18 mm.) of plunger return travel; this action being followed by normal drive retraction.

Solenoid model 10S has a weaker (lost motion) spring located inside the solenoid plunger. Here, enclosed at the outer end by a retaining cup, it forms a plunger within a plunger and it is spring loaded against the tip of the engaging lever inside the plunger clevis link.

THE ROLLER CLUTCH

Torque developed by the starting motor armature must be transmitted to the pinion and flywheel through an over-running or free-wheeling device which will prevent the armature from being rotated at an excessively high speed in the event of the engaged position being held after the engine has started. The roller clutch performs this function.

The operating principle of the roller clutch is the wedging of several plain cylindrical rollers between converging surfaces. The convergent form is obtained by matching cam tracks, to a perfectly circular bore. The rollers, of which there are three, are spring loaded and, according to the direction of drive, are either free or wedge-locked between the driving and driven members. The clutches are sealed in a rolled over steel outer cover and cannot be dismantled for subsequent reassembly.

THE STARTER SOLENOID

The starter solenoid is an electro-magnetic actuator mounted pick-a-back fashion on the yoke of the pre-engaged starter motor. It contains a soft iron plunger (linked to the engaging lever), the starter switch contacts and a coil consisting of a heavy gauge pull-in or series winding and a lighter-gauge hold-on or shunt winding.

Initially, both windings are energised in parallel when the starter device is operated but the pull-in winding is shorted out by the starter switch contacts at the instant of closure—its purpose having been effected.

Magnetically, the windings are mutually assisting.

Like the roller clutch assembly, the starter solenoid is sealed in a rolled-over steel outer case or body and cannot be dismantled for subsequent reassembly.

Page PY.s.16

ELECTRICAL AND INSTRUMENTS

STARTER MOTOR PERFORMANCE DATA

Model	M45G Pre-engaged
Lock Torque	22·6 lb./ft. (3·13 kg./m.) with 465 amperes at 7·6 terminal volts
Torque at 1,000 r.p.m.	9·6 lb./ft. (1·33 kg./m.) with 240 amperes at 9·7 terminal volts
Light running current	70 amperes at 5,800 to 6,500 r.p.m.

SOLENOID SWITCH DATA

Model	10 S
Closing Coil Resistance (measured between terminal STA with copper link removed and Lucar terminal)	0·36 to 0·42 ohms
Hold on Coil Resistance (measured between Lucar terminal and solenoid outer case)	1·49 to 1·71 ohms

Page PY.s.17

ELECTRICAL AND INSTRUMENTS

REMOVAL

DISCONNECT THE BATTERY EARTH LEAD.

Disconnect and remove the transmitter unit from the top of the oil filter.

Disconnect the battery cable and solenoid switch cable from the starter motor.

Remove the distributor clamping plate retaining screw and withdraw the distributor.

Remove the two setscrews and lock washers securing the motor to the housing, gently bend away the carburetter drain pipes and remove the starter motor through the chassis frame.

The two setscrews are accessible from beneath the car or through an access panel in the right-hand side of the gearbox tunnel. Remove the front carpet to expose the panel.

Refitting

Refitting is the reverse of the removal procedure.

Care must be taken when refitting the two setscrews, which have a fine thread, that they are not cross-threaded.

Insert the distributor and rotate the rotor until the drive dog engages correctly and secure with the clamping plate setscrew.

Note: If the clamping plate has been removed from the distributor or its position altered, the engine must be re-timed as detailed in Section B.

SERVICING

Checking the Brushgear and Commutator

Remove the starter motor from the engine.

Release the screw and remove the metal band cover.

Check that the brushes move freely in the brush boxes by holding back the spring and pulling gently on the flexible connection. If a brush is inclined to stick, remove it from its holder and clean its sides with a petrol moistened cloth. Replace the brushes in their original position in order to retain "bedding". Brushes which will not "bed" properly or have worn to $\frac{5}{16}$" (7.94 mm.) in length must be renewed. See page PY.s.20 for renewal procedure.

Check the tension of the brush springs with a spring balance. The correct tension should be 52 ozs. (1.47 kg.) on a new brush.

Replace each existing brush in turn with a new brush to enable the tension of the brush springs to be tested accurately.

Check that the commutator is clean and free from oil or dirt. If necessary clean with a petrol moistened cloth or, if this is ineffective, rotate the armature and polish the commutator with fine glass paper. DO NOT use emery cloth. Blow out all abrasive dust with a dry air blast.

A badly worn commutator can be reskimmed by first rough turning, followed by diamond finishing. DO NOT undercut the insulation. Commutators must not be skimmed below a diameter of $1\frac{17}{32}$" (38.89 mm.). Renew the armature if below this limit.

SERVICING

Testing in position

Check that the battery is fully charged and that the terminals are clean and tight. Recharge if necessary.

Switch on the lamps together with the ignition and operate the starter control. If the lights go dim and the starter does not crank the engine this indicates that the current is flowing through the starter motor windings but the armature is not rotating for some reason. The fault is due possibly to high resistance in the brush gear or an open circuit in the armature or field coils. Remove the starter motor for examination.

If the lights retain their full brilliance when the starter switch is operated check the starter motor and the solenoid unit for continuity.

If the supply voltage is found to be applied to the starter motor when the switch is operated the unit must be removed from the engine for examination.

Sluggish or slow action of the starter motor is usually due to a loose connection causing a high resistance in the motor circuit. Check as described above.

If the motor is heard to operate, but does not crank the engine, indication is given of damage to the drive.

Page PY.s.18

ELECTRICAL AND INSTRUMENTS

BENCH TESTING

Remove the starter motor from the engine

Disconnect the battery. Disconnect and remove the starter motor from the engine (see page PY.s.18 for the removal procedure).

Measuring the light running current

With the starter motor securely clamped in a vice and using a 12-volt battery, check the light running current and compare with the value given on page P.X.s.15. If there appears to be excessive sparking at the commutator, check that the brushes are clean and free to move in their boxes and that the spring pressure is correct.

Measuring lock torque and lock current

Carry out a torque test and compare with the values given on page PY.s.17. If a constant voltage supply is used, it is important to adjust this to be 7.6 volts at the starter terminal when testing.

FAULT DIAGNOSIS

An indication of the nature of the fault, or faults, may be deduced from the results of the no-load and lock torque tests.

Symptom	Probable Fault
1. Speed, torque and current consumption correct.	Assume motor to be in normal operating condition.
2. Speed, torque and current consumption low.	High resistance in brush gear, e.g., faulty connections, dirty or burned commutator causing poor brush contact.
3. Speed and torque low, current consumption high.	Tight or worn bearings, bent shaft, insufficient end play, armature fouling a pole shoe, or cracked spigot on drive end bracket. Short circuited armature, earthed armature or field coils.
4. Speed and current consumption high, torque low.	Short circuited windings in field coils.
5. Armature does not rotate, high current consumption.	Open circuited armature, field coils or solenoid unit. If the commutator is badly burned, there may be poor contact between brushes and commutator.
6. Armature does not rotate, high current consumption.	Earthed field winding or short circuit solenoid unit. Armature physically prevented from rotating.
7. Excessive brush movement causing arcing at commutator.	Low brush spring tension or out-of-round commutator. "Thrown" or high segment on commutator.
8. Excessive arcing at the commutator.	Defective armature windings, sticking brushes or dirty commutator.
9. Excessive noise when engaged.	Pinion does not engage fully before solenoid main contacts are closed. Check pinion movement as detailed under Setting Pinion Movement.
10. Pinion engaged but starter motor not rotating.	Pinion movement excessive. Solenoid main contacts not closing. Check pinion movement as detailed under Setting Pinion Movement.

Page PY.s.19

ELECTRICAL AND INSTRUMENTS

DISMANTLING

Disconnect the copper link between the lower solenoid terminal and the starting motor yoke.

Remove the two solenoid unit securing nuts. Detach the extension cables and withdraw the solenoid from the drive end bracket casting, carefully disengaging the solenoid plunger from the starter drive engagement lever.

Remove the cover band and lift the brushes from their holders.

Unscrew and withdraw the two through bolts from the commutator end bracket. The commutator end bracket and yoke can now be removed from the intermediate and drive end brackets.

Extract the rubber seal from the drive end bracket.

Slacken the nut securing the eccentric pin on which the starter drive engagement lever pivots. Unscrew and withdraw the pin.

Separate the drive end bracket from the armature and intermediate bracket assembly.

Remove the thrust washer from the end of the armature shaft extension using a mild steel tube of suitable bore. Prise the jump ring from its groove and slide the drive assembly and intermediate bracket from the shaft.

To dismantle the drive further prise off the jump ring retaining the operating bush and engagement spring.

BENCH INSPECTION

After dismantling the motor, examine individual items.

Replacement of brushes

The flexible connectors are soldered to terminal tags; two are connected to brush boxes and two are connected to free ends of the field coils. Unsolder these flexible connectors and solder the connectors of the new brush set in their place.

The brushes are pre-formed so that "bedding" to the commutator is unnecessary. Check that the new brushes can move freely in their boxes.

Commutator

A commutator in good condition will be burnished and free from pits or burned spots. Clean the commutator with a petrol moistened cloth. Should this be ineffective, spin the armature and polish the commutator with fine glass paper; remove all abrasive dust with a dry air blast. If the commutator is badly worn, mount the armature between centres in a lathe, rotate at high speed and take a light cut with a very sharp tool. Do not remove more metal than is necessary. Finally polish with very fine glass paper. The INSULATORS between the commutator segments MUST NOT BE UNDERCUT: Commutators must not be skimmed below a diameter of $1\frac{17}{32}''$ (38·89 mm.).

Armature
Lifted conductors

If the armature conductors are found to be lifted from the commutator risers, overspeeding is indicated. In this event, check that the clutch assembly is operating correctly.

Fouling of armature core against the pole faces

This indicates worn bearings or a distorted shaft. A damaged armature must in all cases be replaced and no attempt should be made to machine the armature core or to true a distorted armature shaft.

Insulation test

To check armature insulation, use a 110 volt a.c., test lamp. The test lamp must not light when connected between any commutator segment and the armature shaft.

If a short circuit is suspected, check the armature on a "growler". Overheating can cause blobs of solder to short circuit the commutator segments.

If the cause of an armature fault cannot be located or remedied, fit a replacement armature.

Field Coils
Continuity Test

Connect a 12-volt test lamp and battery between the terminal on the yoke and each individual brush (with the armature removed from the yoke). Ensure that both brushes and their flexible connectors are clear of the yoke. If the lamp does not light, an open circuit in the field coils is indicated.

Replace the defective coils.

Insulation test

Connect a 110-volt a.c., test lamp between the terminal post and a clean part of the yoke. The test lamp lighting indicates that the field coils are earthed to the yoke and must be replaced.

When carrying out this test, check also the insulated pair of brush boxes on the commutator end bracket.

Page PY.s.20

ELECTRICAL AND INSTRUMENTS

Clean off all traces of brush deposit before testing. Connect the 110-volt test lamp between each insulated brush box and the bracket.

If the lamp lights this indicates faulty insulation and the end bracket must be replaced.

Replacing the field coils

Unscrew the four pole-shoe retaining screws, using a wheel operated screwdriver. Remove the insulation piece which is fitted to prevent the inter-coil connectors from connecting with the yoke.

Draw the pole-shoes and coils out of the yoke and lift off the coils. Fit the new field coils over the pole-shoes and place them in position inside the yoke. Ensure that the taping of the field coils is not trapped between the mating surfaces of the pole-shoes and the yoke.

Locate the pole-shoes and field coils by lightly tightening the retaining screws. Replace the insulation piece between the field coil connections and the yoke.

Finally, tighten the screws by means of the wheel operated screwdriver while the pole pieces are held in position by a pole shoe expander or a mandrel of suitable size.

Bearings and Bearing Replacement

The commutator and drive end brackets are each fitted with a porous bronze bush and the intermediate bracket is fitted with an indented bronze bearing.

Replace bearings which are worn to such an extent that they will allow excessive side play of the armature shaft.

The bushes in the intermediate and drive end brackets can be pressed out whilst that in the commutator bracket is best removed by inserting a $\frac{9}{16}''$ (14·29 mm.) tap squarely into the bearing and withdrawing the bush with the tap.

Before fitting a new porous bronze bearing bush, immerse it for 24 hours in clean engine oil (SAE 30 to 40). In cases of extreme urgency, this period may be shortened by heating the oil to 100° C. for 2 hours and then **allowing the oil to cool before removing the bush.** Fit new bushes by using a shouldered, highly polished mandrel approximately 0·0005" (·013 mm.) greater in diameter than the shaft which is to fit in the bearing. **Porous bronze bushes must not be reamed out after fitting,** as the porosity of the bush will be impaired.

After fitting a new intermediate bearing bush, lubricate the bearing surface with Rocol "Molypad" molybdenised non-creep, or similar, oil.

CHECKING THE ROLLER CLUTCH DRIVE

A roller clutch drive assembly in good condition will:—

(i) Provide instantaneous take-up of the drive in the one direction.

(ii) Rotate easily and smoothly in the other.

(iii) Be free to move round or along the shaft splines without roughness or tendency to bind.

Similarly, the operating bush must be free to slide smoothly along the driving sleeve when the engagement spring is compressed. Trunnion blocks must pivot freely on the pegs of the engaging lever. All moving parts should be smeared liberally with Shell Retinax "A" grease or an equivalent alternative.

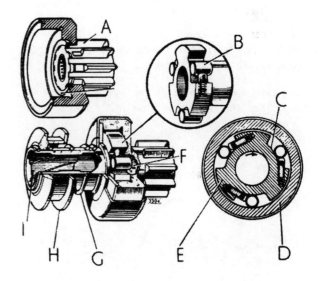

Fig. 19. *The roller clutch drive components.*

A—*Alternative contruction (pinion pressed and clear-ringed into driven member).*

B—*Spring loaded rollers.*

C—*Cam tracks.*

D—*Driven member (with pinion).*

E—*Driving member.*

F—*Bush.*

G—*Engagement spring.*

H—*Operating bush.*

I—*Driving sleeve.*

Page PY.s.21

ELECTRICAL AND INSTRUMENTS

REASSEMBLY

After cleaning all parts, reassembly of the starting motor is a reversal of the dismantling procedure given on page PY.s.20 but the following special points should be noted:—

(i) The following parts should be tightened to the maximum torques indicated:—

Nuts on solenoid copper terminals	20 lb./in. (0·23 kgm.)
Solenoid fixing bolts	4·5 lb./ft. (0·62 kgm.)
Starting motor through bolts	8·0 lb./ft. (0·83 kgm.)

(ii) When refitting the C.E. bracket see that the moulded brake shoes seat squarely and then turn them so that the ends of the cross peg in the armature shaft engage correctly with the slots in the shoes.

Setting Pinion Movement (Fig. 10)

Connect the solenoid Lucar terminal to a 6-volt supply. DO NOT use a 12-volt battery otherwise the armature will turn.

Connect the other side of the supply to the motor casing (this throws the drive assembly forward into the engage position).

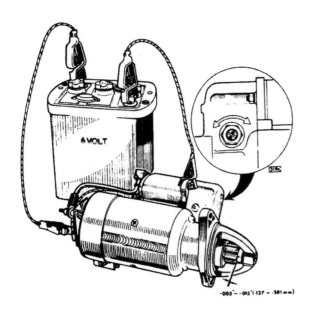

Fig. 20. *Setting pinion movement.*

Measure the distance between the pinion and the thrust washer on the armature shaft extension. Make this measurement with the pinion pressed lightly towards the armature.

For correct setting the dimension should be 0·005" to 0·015" (0·13 to 0·38 mm.).

Disconnect the battery.

Adjust the setting by slackening the eccentric pivot pin securing nut and turning the pin until the correct setting is obtained.

Note: The head of the arrow stamped on the end of the eccentric pivot pin should be set only between the ends of the arrows cast in the drive end bracket.

Turning the screw to the left (anti-clockwise) will increase the gap between the pinion and the thrust washer, turning to the right (clockwise) will decrease the gap.

Reconnect the battery and recheck the setting.

After setting tighten the securing nut to retain the pin position.

CHECKING OPENING AND CLOSING OF STARTER SWITCH CONTACTS

The following checks assume that pinion travel has been correctly set.

Remove the copper link connecting solenoid terminal STA with the starting motor terminal.

Connect, through a switch, a supply of 10 volts d.c., to the series winding, that is, connecting between the solenoid Lucar terminal and large terminal STA. DO NOT CLOSE THE SWITCH AT THIS STAGE.

Connect a separately energised test lamp circuit across the solenoid main terminals.

Insert a stop in the drive end bracket to restrict the pinion travel to that of the out of mesh clearance, normally a nominal $\frac{1}{8}$" (3·17 mm.). An open-ended spanner or spanners of appropriate size and thickness can often be utilised for this purpose, its jaws embracing the armature shaft extension.

Page PY.s.22

Energise the shunt winding with a 10-volt d.c., supply and then close the switch in the series winding circuit.

The solenoid contacts should close fully and remain closed, as indicated by the test lamp being switched on and emitting a steady light.

Switch off and remove the stop.

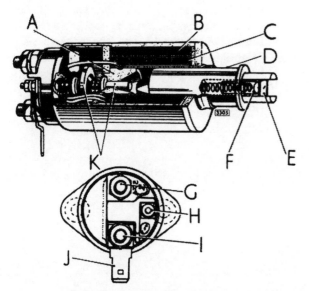

Fig. 11. *Checking the opening and closing of the starter switch contacts.*

A—*Core.*
B—*Shunt winding.*
C—*Series winding.*
D—*Plunger.*
E—*Clevis pin.*
F—*"Lost motion" device.*
G—*Starter terminal.*
H—*Solenoid terminal.*
I—*Battery terminal.*
J—*Accessories terminal.*
K—*Spindle and moving contact assembly.*

Switch on again and hold the pinion assembly in the fully engaged position.

Switch off and observe the test lamp.

The solenoid contacts should open, as indicated by the test lamp being switched off.

ELECTRICAL AND INSTRUMENTS

WINDSCREEN WIPER

(LUCAS MODEL 15W)

DESCRIPTION

The windscreen wiper assembly consists of a two-speed motor coupled by connecting rods to three wiper arm spindle bearings (Open Sports and F.H.C.) or two spindles (2+2).

Windshield wiper motor model 15W is designed to operate a link-type wiper installation. The motor is self-switching to the OFF (or park) position. A two-pole permanent magnet field is provided by two ceramic magnets which form part of the yoke assembly. Inside the motor gearbox a worm gear on the armature shaft drives a shaft-and-gear assembly comprising a moulded gearwheel assembled to a location-plate-and-shaft. Power from the motor is transmitted through the gearwheel, location-plate-and-shaft to, finally, a rotary link which serves as a coupling between the motor and the links which operate the wiper arm spindles.

Associated with the terminal assembly is a two-stage plunger operated limit switch. The plunger is actuated by a cam on the underside of the moulded gearwheel inside the gearbox. When the manually-operated control switch is moved to OFF (or park) the motor continues to operate under the automatic control of the limit switch. As the wiper blades near the parked position the first-stage contacts open and the motor is switched off but continues to rotate under its own momentum. The second-stage contacts, to which are connected the positive and negative brushes, then close and regenerative braking of the armature takes place to maintain consistent parking of the blades.

Two-speed operation is provided by a third (stepped) brush incorporated in the brushgear assembly. When the main control switch is moved to the high speed position, the positive feed to the normal brush is transferred to the third brush, and a higher-than-normal wiping speed is obtained. (The higher speed should not be used in heavy snow or on a partially wet windshield).

The blades and arms fitted to the two models are of different lengths and are individually cranked for Right-hand and Left-hand drive cars.

Note: The wiper blades are manufactured with special anti-smear properties. Renew only with genuine Jaguar replacement parts.

Page PY.s.23

ELECTRICAL AND INSTRUMENTS

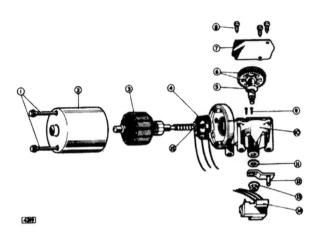

4.2 'E' Type RHD   10½" (R.H. crank)   12"
           LHD   10½" (L.H. crank)   12"

SERVICING

Note: Since the motor is of permanent magnet design, the direction of rotation of the armature depends on the polarity of the supply of its terminals. If it is necessary to run the motor while it is removed from the vehicle, the negative supply cable must be connected to motor terminal number 1 and the positive supply cable to terminal number 5 for normal speed or terminal number 3 for high speed (see Fig. 23).

Fig. 22. *Exploded view of the 15W windscreen wiper motor.*

1—*Yoke fixing bolts.*
2—*Yoke assembly comprising two permanent-magnet poles and retaining clips and armature bearing bush.*
3—*Armature.*
4—*Brushgear, comprising insulating plate and brushboxes, brushes, springs and fixing bolts.*
5—*'Dished' washer.*
6—*Shaft and gear.*
7—*Gearbox cover.*
8—*Cover fixing screws.*
9—*Limit switch fixing screws.*
10—*Gearbox.*
11—*Flat washer.*
12—*Rotary link.*
13—*Link fixing nut.*
14—*Limit switch assembly.*
15—*Nylon thrust cap.*

MAINTENANCE

All bearings are adequately lubricated during manufacture and require no maintenance.

Oil, tar spots or similar deposits should be removed from the windshield with methylated spirits (denatured alcohol). Silicone or wax polishes must not be used for this purpose.

Efficient wiping is dependent upon keeping wiper blades in good condition. Worn or perished blades are easily removed for replacement.

DATA

MOTOR

(i) Typical light running current (i.e. with the rotary link disconnected from the transmission) after 60 seconds from cold: 1.5 amp. (normal speed)   2.0 amp. (high speed)

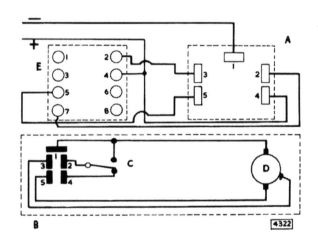

Fig. 23. *15W Wiper wiring diagram.*

A—*Moulded terminal connector on cable harness.*
B—*Terminal connector on wiper motor.*
C—*Limit switch.*
D—*Armature.*
E—*152 SA switch.*
Switch internal connections.

OFF   NORMAL SPEED   HIGH SPEED
(5–7)   (4–5)   (2–4)

Page PY.s.24

ELECTRICAL AND INSTRUMENTS

Systematic Check of Faulty Wiping Equipment

Unsatisfactory operation (if the supply voltage to the motor is adequate) may be caused by a fault that is mechanical or electrical in origin. Before resorting to dismantling, consideration should be given to the nature of the fault.

The symptoms and remedial procedure associated with the more common causes of wiper failure (or poor performance) are described in (i) and (ii) below.

(i) **Frictional Wiper Blades**

Excessive friction between apparently satisfactory wiper blades and the windshield may result in a marked reduction in wiping speed when the blades are operating on a windshield that is only partially wet. A further symptom is that the blades become noisy at each end of the wiping arc. When possible, the blades should be temporarily replaced with a pair known to be in good condition. If this rectifies the fault, fit new blades.

(ii) **Low Wiping Speed or Irregular Movement of the Blades**

To determine whether a low wiping speed is due to excessive mechanical loading or to poor motor performance, the rotary link must first be disconnected from the transmission linkage and the light running current and speed of the motor can then be checked under no load conditions.

Measuring Light Running Current and Speed

Connect a first-grade moving coil ammeter in series with the motor supply cable and measure the current consumption. Also check the operating speed by timing the speed of rotation of the rotary link or moulded gearwheel. The current consumption and speed are given in Data.

If the motor does not run, or current consumption and speed are not as stated, an internal fault in the motor is indicated and a replacement unit should be fitted or the motor removed for detailed examination.

If current consumption and speed are correct, check for proper functioning of the transmission linkage and wiper arm spindles.

Removal

4.2 litre Open Sports and F.H.C.

Disconnect the battery.

Remove the two thumbscrews and lower the instrument panel.

Remove two drive screws and detach the hazard warning unit carrier plate. Disconnect the drive link from the ball joint on the centre wheelbox.

Remove the P.V.C. cable strap from the body of the windscreen wiper motor and disconnect the cables by withdrawing the cable plug from the socket.

Mark the position of the throttle fulcrum lever bracket relative to the bulkhead, remove two setscrews, and detach the bracket. It is not necessary to disconnect the throttle control pads.

Remove four setscrews securing the motor mounting plate to the bulkhead and withdraw the motor with the attached drive link.

4.2 litre 2+2

Disconnect the battery.

Remove the top facia panel as detailed in Body and Exhaust System—Section N, page NY.s.1.

Disconnect the motor drive link from the ball joint on the L/H wheel box.

Remove the cable strap and plug as detailed above.

Remove the setscrews securing the mounting plate to the bulkhead and withdraw the motor.

Dismantling

Remove the gearbox cover.

The rotary link may be fitted to the gearwheel shaft in one of two positions (180° apart) depending on the parking requirement of the windshield installation. To ensure that the original parking position is maintained, the position of the rotary link in relation to the zero mark on the gearwheel location plate must be noted before removing the link.

Important: The moulded gearwheel inside the gearbox must be prevented from moving while the rotary link fixing nut is slackened (or, on reassembly, tightened). This is most easily achieved by securing the rotary link in a vice while the nut is turned.

Remove the fixing nut and withdraw the rotary link and flat washer.

Remove the shaft-and-gear from the gearbox, taking care not to lose the dished washer fitted beneath the gearwheel. It is not normally necessary to dismantle the shaft-and-gear assembly since this is serviced only in an assembled condition. However, should it become necessary to assemble the moulded gearwheel

Page PY.s.25

ELECTRICAL AND INSTRUMENTS

to the location-plate-and-shaft, it is essential to fit the gear wheel in the correct one of the two alternative positions to maintain the original parking position of the wiper blades. The gearwheel is correctly fitted to the location-plate-and-shaft when the 'zero' mark on the location plate is positioned furthest away from the gearwheel cam.

Unscrew and remove the two fixing bolts from the motor yoke and carefully remove the yoke assembly and armature. While removed, the yoke must be kept well clear of swarf, etc., which may otherwise be attracted to the pole pieces.

Undo the two sets of fixing screws and remove from the gearbox the brushgear and the terminal and switch unit assemblies, linked together by the connecting cables.

Bench Inspection

After dismantling, examine individual items.

(i) **Brush replacement**

The original specified length of the brushes is sufficient to last the life of the motor. If, due to accidental damage to the brushes, or faulty commutator action, it becomes necessary to renew the brushes, the complete brushgear service-assembly must be fitted. The brushgear assembly must be renewed if the main (diametrically-opposed) brushes are worn to $\frac{3}{16}$ in. (4.8 mm.), or if the narrow section of the third brush is worn to the full width of the brush. Check that the brushes move freely in the boxes.

(ii) **Check Brush Springs**

The design of the brushgear does not allow for easy removal of the brush springs. This is due to the fact that, similar to the brushes, the springs are expected to last the life of the motor and should not normally require renewing. In the unlikely event of the spring pressure failing to meet the specified requirements, the complete brushgear service-assembly must be renewed in a similar manner to that necessary for servicing the brushes.

Note the location of the cables before unsoldering for reference when refitting.

To check the spring pressure on the end face of the brush with a push-type spring gauge push until the bottom of the brush is level with the bottom of the slot in the brush box, when the spring pressure reading should be 5–7 oz.f. (140–200 gf).

(iii) **Testing and Servicing the Armature**

Use armature testing equipment to check the armature windings for open and short circuits. Test the insulation by using a mains test lamp (Fig. 24). Lighting of the lamp indicates faulty insulation.

If the commutator is worn, it can be lightly skimmed while the armature is mounted in a lathe.

Afterwards, clear the inter-segment spaces of copper swarf.

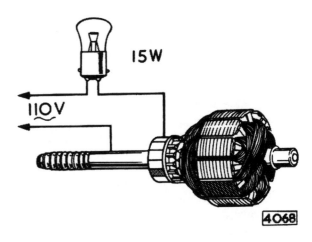

Fig. 24. *Armature insulation test.*

(iv) **Inspection of Moulded Gear**

Examine the gearwheel, especially the teeth, for signs of wear or damage. If the gearwheel needs renewing, a shaft-and-gear service replacement assembly will have to be fitted.

Re-assembly

Reassembly is the reverse of the dismantling procedure.

A liberal quantity of Ragosine Listate grease is necessary for lubrication of the gearwheel teeth, the gearwheel cam and the worm gear on the armature shaft. The total quantity of grease must not be less than 15 cc.

Apply Shell Turbo 41 oil to the bearing bushes, armature shaft bearing surfaces (sparingly), gearwheel shaft, and the felt-oiler washer in the yoke bearing (thoroughly soak).

Reassembly of Yoke

Before refitting the armature to the yoke, inspect the inside of the yoke and ensure that the thrust disc and the felt-oiler washer are in place in the yoke bearing.

The correct method of assembly is with the thrust disc flat against the end face of the bearing, followed by the felt-oiler washer which must have a hole in the centre to allow the captive ball bearing in the end of the armature shaft to contact the thrust disc.

If the felt-oiler is renewed, check that the replacement is provided with the necessary hole and, if not, make a $\frac{1}{8}''$ (3 mm.) diameter hole in the centre of the felt. (A felt-oiler without a hole could result in the armature end-float becoming excessive in service due to the ball bearing wearing away the felt after the end-float adjustment has been made). Soak the felt-oiler in Shell Turbo 41 oil.

The yoke fixing bolts should be tightened to a torque of 12–16lb in. (0.138–0.184 kg.m.). If a service replacement armature is being fitted, it is advisable to first slacken the armature end-float thrust screw before tightening the yoke fixing bolts. Afterwards, reset the thrust screw.

Armature End-Float Adjustment

Armature end-float is 0.002–0.008" (0.05–0.2 mm.).

To obtain a satisfactory end-float adjustment with the motor and gearbox completely assembled, position the unit with the thrust screw uppermost, tighten the thrust screw until abutment takes place and then slacken it off one quarter turn and secure it in this position by tightening the locknut.

WINDSCREEN WIPER SPINDLE HOUSINGS
2+2 Models
Removal

Disconnect the battery. Remove both windscreen wiper arms.

Lower the instrument panel and remove the screen rail facia assembly. Disconnect the motor link rod from the ball joint on the left-hand spindle lever.

Unscrew the large nuts securing the housings to the scuttle and remove the distance pieces and rubber seal washers. Withdraw the twin spindle and carrier plate assembly through the scuttle panel and complete the removal through the left-hand aperture in the bulkhead inner panel.

Housings are replaceable only as a complete assembly comprising both housings and carrier plate.

Open Sports and F.H.C. Models
Removal (Right or Left Hand Housing)

Disconnect the battery.

ELECTRICAL AND INSTRUMENTS

Withdraw the wiper arm from the spindle housing to be removed. Unscrew the large nut securing the housing to the scuttle and remove the distance piece and rubber seal washer.

Lower the instrument panel after removing the two retaining screws in the top corners. Remove the four nuts and washers retaining the screen rail facia assembly. Two are accessible from the centre aperture and one each at the outer edges below the screen rail.

Disconnect the ball joint from the spindle lever.

From inside the car remove two nuts and washers securing the housing bracket to the base plate and withdraw the housing.

Removal (Central Housing)

Disconnect the battery. Remove all wiper arms from the spindles.

Lower the instrument panel, remove the screen rail facia and outer housings as detailed previously.

Remove the large nut, distance piece and rubber seal washer from the central housing. Withdraw the housing and carrier plate assembly through the scuttle panel and complete the removal through the left-hand aperture in the bulkhead inner panel.

WIPER MOTOR LINKAGE SETTING (2+2 cars)

It is essential that the wiper motor primary linkage (motor to wheelbox) is adjusted as detailed below if the link length is altered from the original dimension or, if a new link is fitted. Failure to carry out these instructions will result in a knock when the wiper arms are operating.

R.H.D. Cars

Disconnect the primary link (motor to wheelbox) from the ball joint on the wheelbox spindle.

Manually operate the non-adjustable link, connecting the two wiper spindles, to the extreme left position.

Adjust the ball joint socket on the link until it fits on the ball whilst maintaining the position of the lever to the left.

Disconnect the socket and turn to the **right (clockwise) four complete** turns, i.e. shorten the length of the rod between the socket and the motor.

Refit the socket on the ball and secure with the locknut.

L.H.D. Cars

Proceed as for R.H.D. cars, but for the following exceptions.

Manually operate the non-adjustable link to the extreme right and after adjustment turn the ball socket four complete turns to the left (anti-clockwise) to increase the length of the rod.

Page PY.s.27

ELECTRICAL AND INSTRUMENTS

THE INSTRUMENTS

ELECTRIC CLOCK
Description

The electric clock, fitted in the centre of the instrument panel, is a fully transistorised instrument powered by a mercury cell housed in a plastic holder attached to the back of the clock.

Frontal adjustment is provided by means of a small knurled knob for setting the hands and a slotted screw for time-keeping regulation.

To reset the hands, pull out the knurled knob, rotate and release.

To regulate the time-keeping, turn the slotted screw with a small screwdriver towards the positive (+) sign if gaining, and towards the minus (−) sign if losing.

Moving the indicator scale through one division will alter the time-keeping by five minutes per week.

The action of resetting the hands automatically restarts the movement.

The window of the clock is a plastic moulding, and should only be cleaned with a cloth or chamois leather slightly dampened with water. Oil, petrol or other fluids associated with cleaning, are harmful and must not be used.

Battery Replacement

Remove the instrument panel retaining screws and lower the panel.

Lever the battery out of the holder and discard.

Press the new battery into the holder.

Refit the panel.

Fig. 26. *Renewing the electric clock battery.*

Clock—Removal

Lower the instrument panel.

Withdraw the illumination bulb holder from the back of the clock.

Remove the two nuts and the clamp strap from the back of the clock.

Withdraw the clock, complete with the battery holder, from the instrument panel.

Refitting

Refitting is the reverse of the removal procedure.

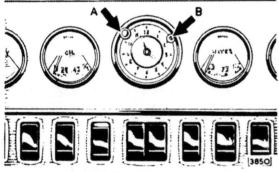

Fig. 25. *Clock controls.*
A — Handsetting.   B — Time regulator.

MAINTENANCE

The mercury cell life is in the region of 18 months, throughout which it ensures a steady and continuous voltage to the clock.

Renew the cell at this period to maintain perfect time-keeping.

THE REVOLUTION COUNTER (TACHOMETER)
Description

The revolution counter is an impulse tachometer instrument incorporating transistors and a printed

Page PY.s.28

ELECTRICAL AND INSTRUMENTS

circuit, the pulse lead (coloured WHITE) being wired in circuit with the S/W terminal on the ignition coil and the ignition switch.

Mechanical drive cables or an engine-driven generator are not required with this type of instrument.

The performance of this instrument is not affected by the distributor contact setting, by corrosion of the sparking plug points, or by differences in the gap settings.

Connection to the back of the instrument is by means of a locked plug and socket, the contacts being offset to prevent incorrect coupling.

Removal

Disconnect the battery.

Remove the screen rail facia assembly as detailed on Page NY.s.1 to gain access to the instrument.

Remove the two knurled nuts, earth lead and instrument retaining pieces.

Withdraw the tachometer from the facia panel and remove the illumination bulb holders.

Disconnect the plug and socket as follows:—

Pinch together the prongs of the plastic retaining clip and withdraw from the plug and socket assembly (Fig. 27).

Detach the plug from the socket and complete the removal of the instrument.

IMPORTANT

Do not detach the green and white cables connected to the plug and the instrument.

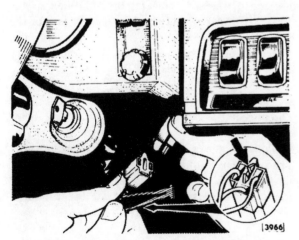

Fig. 27. *The tachometer plug and socket assembly.*
(Inset shows the clip in its fitted position).

Refitting

Refitting is the reverse of the removal procedure. Reconnect the plug and socket assembly and lock with the retaining clip.

THE INSTRUMENT PANEL

The instrument panel differs from that fitted to all previous cars in respect of the following items:—

(1) Rocker Switches—Replacing tumbler switches.
(2) Battery Indicator—Replacing Ammeter.
(3) Panel Light Dimming Resistance—Replacing resistance previously attached to the panel light switch.
(4) The combined Ignition/starter switch which is now mounted on a separate sub-panel. These switches were previously two separate items mounted in the instrument panel.
(5) The Cigar Lighter—Now located in the console below the instrument panel, was previously part of the instrument panel assembly.

THE SWITCHES

The rocker switches are mounted in a sub-panel which is attached to the instrument panel by four self-tapping screws.

Individual switches may be removed without detaching the sub-panel cluster as follows:—

Removal

Disconnect the battery.

Lower the instrument panel.

Remove the cables from the switch, noting location for reference when refitting.

Press in the two locking tabs located at the bottom and the top faces of the switch body and push the switch through the aperture.

Refitting

Press the switch into the panel aperture until the nylon locking tabs register.

Reconnect the cables as noted on removal.

Page PY.s.29

ELECTRICAL AND INSTRUMENTS

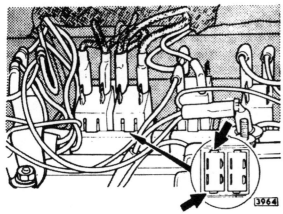

Fig. 28. *Instrument panel rocker switch removal (Inset shows arrowed the nylon locking tabs).*

THE IGNITION/STARTER SWITCH

A Lucas 47SA combined ignition/starter switch replaces the separate switches previously used.

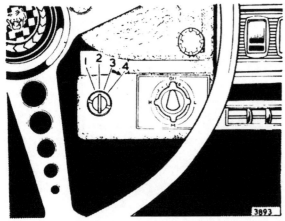

Fig. 29. *The ignition/starter switch location when air-conditioning system is fitted.*
1 — Auxiliaries. 2 — Ignition "OFF".
3 — Ignition "ON". 4 — Starter.

The switch is mounted on a bracket attached to the steering column (if Air-conditioning equipment is installed the bracket is attached to the evaporator unit).

In conjunction with the 47SA ignition/starter switch a Lucas 6RA relay is included in the alternator circuit. This functions as a field isolating relay, the relay coil being energised by operation of the ignition switch.

Removal

Remove the locking ring and withdraw the switch through the bracket with the brass locknut and wave washer.

Disconnect the cables and remove the switch. Note the location of the cables for reference when refitting.

The lock barrel can be withdrawn by inserting a thin rod through a hole in the body of the switch and depressing the plunger in the lock. Insert the key and turn to the 'OFF' position to gain access to the plunger.

Refitting

Refitting is the reverse of the removal procedure.

When refitting a new lock barrel, check that the number on the face of the barrel and the key is the same as that on the barrel removed. This will be identical to the door locks.

Insert the key in the lock and turn the switch to the 'OFF' position before inserting the barrel.

PANEL LIGHT DIMMING RESISTOR

The resistor unit is comprised of a wire resistance attached to two 'Lucar' cable contact blades mounted on an insulating carrier plate.

The plate is secured to two studs on the back face of the instrument panel by means of distances peices, nuts and lockwashers.

Excessive force should not be used when connecting the cable terminals to the blades.

This may force the blade retaining tongues through the insulating plate and allow the resistor to make contact with the metal face of the panel, resulting in a short circuit in the side light feed line.

BATTERY INDICATOR

This instrument is a voltmeter with a specially calibrated dial which indicates the condition of the battery. It does not register the charging rate of the alternator.

The position of the needle with a charged battery will be within the area marked 'Normal'.

Removal

Disconnect the battery and lower the instrument panel.

Disconnect the cables, noting the location for reference when refitting.

Detach the illumination bulb holder.

Remove two nuts and clamp strap and withdraw the instrument forward through the panel.

Refitting

Refitting is the reverse of the removal procedure.

Page PY.s.30

ELECTRICAL AND INSTRUMENTS

Check the condition of the battery by means of the panel shown below

RED (Off Charge)		NORMAL			RED (On Charge)
BATTERY CHARGE EXTREMELY LOW	BATTERY CHARGE LOW	WELL CHARGED BATTERY	CHARGING VOLTAGE LOW	CHARGING VOLTAGE SATISFACTORY	CHARGING VOLTAGE TOO HIGH
If with the ignition and electrical equipment e.g. headlamps etc., switched on, but with the engine not running the indicator settles in this section—your battery requires attention.		Ideally the indicator should settle in this section when the ignition and electrical equipment e.g. headlamps etc., are switched on and the engine is not running.	This condition may be indicated when the headlights and other equipment are in use.	The indicator should point to this section when the engine is running above idle.	If the indicator continues to point to this section after 10 minutes running either your voltage regulator requires adjustment or some other fault has developed.

IMPORTANT All readings on the indicator should be ignored when the engine is idling, since readings may vary at very slow engine speeds due solely to operation of the voltage regulator.

OFF CHARGE
This means more energy is being used from your battery than is being replaced by the alternator on your car. This condition is satisfactory provided it does not persist for long periods, when the engine is running above idle or at speed. If the indicator remains in the section, it may mean that you have a broken or slipping fan belt, a faulty alternator, a badly adjusted voltage regulator or some other fault.

ON CHARGE
This means your battery is having more energy put into it than is being taken out of it. In the ordinary way this condition predominates and your battery is continuously being recharged by the alternator whenever the engine is running above idle. If however the engine is continually running slowly as may be the case in traffic—or when, in winter, lights and cold starting make extra demands on the battery—you may find the rate of discharge exceeds the rate of charge—that is to say the battery is running down, as will be indicated on your Battery Condition Indicator and you may need an extra charge if "battery charge low or extremely low" is indicated by the instrument.

LAMPS

HEADLAMP

Sealed beam units are fitted to all cars with the exception of certain European Countries which retain the pre-focus bulb (see Bulb Data Chart).

The beam setting and unit replacement instructions differ from those stated on Page P.24 as follows:—

Beam Setting

If beam setting adjustment is required, prise off the headlamp rim (retained by spring clips). Switch on the headlamps and check that they are on Main beam.

The setting of the beams is controlled by two screws 'A' and 'B' on Fig. 30.

The top screw 'A' is for vertical adjustment, i.e. to raise or lower the beam; turn the screw anti-clockwise to lower the beam or clockwise to raise the beam.

The side screw 'B' is for horizontal adjustment, i.e. to turn the beam to right or left. To move the beam to the right, turn the screw clockwise. To move the beam to the left, turn the screw anticlockwise.

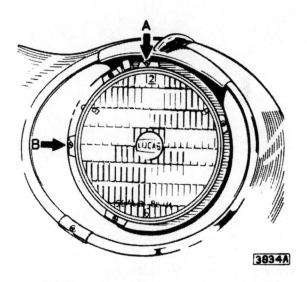

Fig. 30. *Adjustment of the screw 'A' will alter the headlamp beams in the vertical plane; adjustment of the screw 'B' will alter the headlamp beams in the horizontal plane.*

ELECTRICAL AND INSTRUMENTS

Sealed Beam Unit — Replacement

Prise off the headlamp rim (retained by spring clips).

Remove the three cross-headed screws and detach the retaining ring.

Note: Do not disturb the two beam setting screws.

Withdraw the sealed beam unit and unplug the adaptor.

Replace the sealed beam unit with one of the correct type (see 'Lamp Bulbs').

On cars fitted with bulb light units, proceed as directed above until the unit is removed. Release the bulb retaining clips and withdraw the bulb. Replace with a bulb of the correct type (see 'Lamp Bulbs').

When reassembling, note the groove in the bulb plate which must register with the raised portion on the bulb retainer.

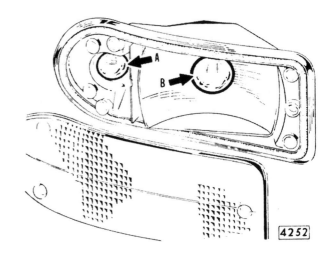

Fig. 32.   A—Side lamp bulb.
          B—Flasher bulb.

FRONT FLASHER BULB—REPLACEMENT

Proceed as detailed for 'Side Lamp Bulb'.

REAR/BRAKE LIGHT BULB—REPLACEMENT

Remove the four screws and detach the glass. The rear/braking light bulb is the lower one of the two exposed and is removed by pressing inwards and rotating anti-clockwise. When refitting a replacement bulb note that the pins are offset.

Fig. 31. Headlamp sealed beam unit removal. The arrow indicates one of the spring clips retaining the rim.

SIDE LAMP BULB—REPLACEMENT

Remove three screws and detach the lens. Remove the bulb by pressing inwards and rotating outwards. Check the condition of the lens seal when refitting.

On cars for certain European countries the side lamp bulb is mounted in the headlight unit and is accessible after removing the light unit as detailed under 'Headlamps'.

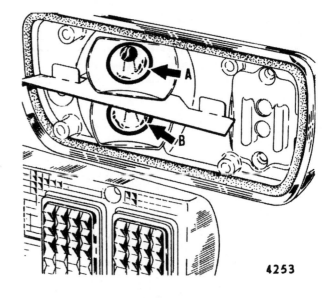

Fig. 33.   A—Rear flasher bulb.
          B—Rear/Brake bulb.

Page PY.s.32

ELECTRICAL AND INSTRUMENTS

REAR FLASHER BULB—REPLACEMENT
Proceed as detailed for 'Rear/Brake Light Bulb—Replacement.' The flasher bulb is the top one of the two exposed.

INTERIOR LIGHT BULB—REPLACEMENT (2+2).
Release the spring side clip and withdraw the retaining tongue on the glass cover from the slot in the lamp base. Remove the faulty bulb and replace with one of the correct value.

When refitting ensure that the retaining tongue is inserted in the slot in the base before locking into position.

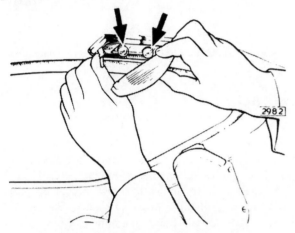

Fig. 34. *Interior lamp bulb removal (2+2).*

INTERIOR/LUGGAGE LIGHT BULB—REPLACEMENT (Open 2 seater)
Proceed as detailed in the 3.8 'E' type Service Manual—page P.26.

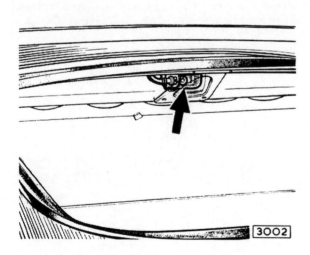

Fig. 35. *Interior lamp bulb removal (open 2 seater).*

NUMBER PLATE LAMP BULB—REPLACEMENT
Remove two screws and detach the glass and rim. Replace the faulty bulb with one of the correct value. Check the condition of the seal before refitting.

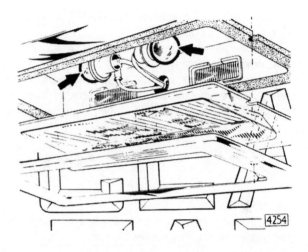

Fig. 36. *Number plate lamp bulb removal.*

REVERSE LAMP BULB—REPLACEMENT
Remove the two retaining screws and detach the lamp glass. Lift the upper contact and withdraw the bulb. Check the condition of the lens seal before refitting.

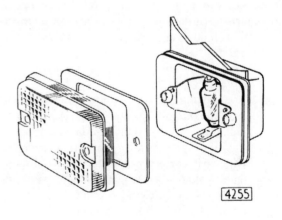

Fig. 37. *Reverse lamp bulb removal.*

Page PY.s.33

ELECTRICAL AND INSTRUMENTS

TRAFFIC HAZARD INDICATOR BULB—REPLACEMENT

Remove the chrome bezel and unscrew the bulb from the holder.

HEATED BACK LIGHT INDICATOR BULB—REPLACEMENT

Proceed as detailed under 'Traffic Hazard Indicator Bulb'.

AUTOMATIC TRANSMISSION INDICATOR BULB—REPLACEMENT

Remove the drive screws, detach the arm rest and transmission unit cover.

Unscrew the gear control knob. Withdraw two screws and remove the gear indicator cover. Replace the bulb with one of the same value (24 volts).

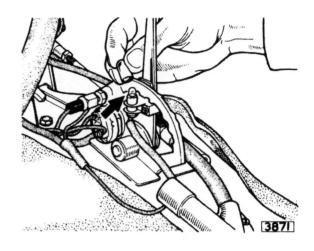

Fig. 38. *Automatic transmission indicator bulb removal.*

HORNS—MODEL 9H

DESCRIPTION

The horns are mounted on brackets attached to the sub-frame lower cross-member.

The horns are now mounted on brackets attached to the sub-frame lower cross-member.

The horn circuit operates through a Lucas 6RA relay, the contacts C1 and C2 closing when the relay coil is energised by depressing the horn switch button located in the direction (turn) indicator switch lever.

Maintenance

In the event of the horns failing to sound or performance becoming uncertain, check before making adjustments that the fault is not due to external causes.

Check as follows and rectify as necessary:
(i) Battery condition.
(ii) Loose or broken connections in the horn circuit.
(iii) Loose fixing bolts. It is important to keep the horn mountings tight and to maintain rigid the mounting of any unit fitted near the horns.
(iv) Faulty relay. Check by substitution after verifying that current is available at terminal C2 (cable colour—brown/purple) and terminal W1 (cable colour—Green).
(v) Check that fuse No. 3 (50 amperes) and fuse No. 6 (35 amperes) have not blown.
Note: Horns will not operate unless the ignition is switched on.

Adjustment

As the horns cannot conveniently be adjusted in position, remove and mount securely on a test fixture.

A small serrated adjusting screw located adjacent to the horn terminal is provided to take up wear of moving parts in the horn. Turning this screw does not alter the pitch of the horn note.

Connect a moving coil ammeter in series with the horn supply feed. The ammeter should be protected from overload by connecting on ON-OFF switch in parallel with its terminals.

Keep this switch ON except when taking readings, that is when the horn is sounding.

Turn the screw clockwise until the horn operates within the specified limits of 6.5-7.0 amperes.

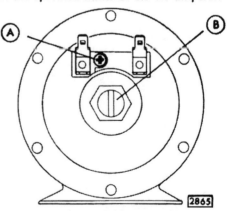

Fig. 39. *The Lucas 9H horn.*
A — *Contact breaker adjustment screw.*
B — *Slotted centre core (Do not disturb).*

Page PY.s.34

ELECTRICAL AND INSTRUMENTS

Service Replacements

When fitting replacement horns it is essential that the following procedure be carried out:—

(i) Refit the lockwashers in their correct positions, one at each side of the mounting bracket centre fixing.

(ii) Ensure, after positioning the horn, that the $\frac{5}{16}"$ centre fixing bolt is secure but not over-tightened. Over-tightening of this bolt will damage the horn.

(iii) Ensure that, when a centre fixing bolt or washers other than the originals are used, the bolt is not screwed into the horn to a depth greater than $\frac{11}{16}"$ (17.5 mm).

Muted Horns (Holland only)

These horns are muted to comply with the Dutch Traffic Regulations and incorporate a rubber plug inserted in the trumpet.

Horn Relay—Checking

If the horn relay is suspected, check for fault by substitution or by the following method:—

(i) Check that fuses No. 3 and No. 6 have not blown. Replace if necessary.

(ii) Check with a test lamp that current is present at the relay terminal W1 (Green) and C2 (Brown/Purple). Switch on the ignition before checking terminal W1.

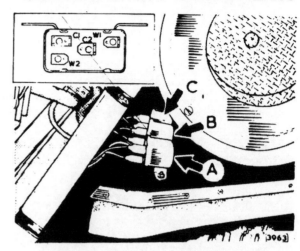

Fig. 40. *Location of horn and alternator relays.*
A — *Horn relay.*
B — *Alternator/Ignition relay.*
C — *Air conditioning equipment relay (when fitted).*
(Inset shows the connections).

(iii) Remove the cable from terminal W2 (Purple/Black) and earth the terminal to a clean part of the frame. The relay coils should now operate and close the contacts.

Reconnect cable.

(iv) Remove cable from terminal C2 (Brown/Purple). Check for continuity by means of an earthed test lamp. Check with the horn button depressed and the ignition 'ON'. Replace the relay if faulty.

RADIATOR COOLING FANS

Twin electrically motor driven cooling fans are fitted, automatic operation being controlled by a thermostat switch mounted in the radiator header tank.

A Lucas 6RA relay is incorporated in the circuit to prevent over-loading the thermostat switch contacts.

When air-conditioning is fitted a second relay is also included to over-ride the thermostat switch when the car is stationary and the air-conditioning system is working.

THERMOSTAT
Checking

Check by substitution or by the following test procedure:—

Drain off sufficient water from the radiator and remove the switch from the header tank.

Wire the switch in series with a 12 volt battery and a 1.5 watt bulb and suspend in water with a thermometer.

Heat the water and note the temperature at which the contacts close and the bulb lights up. Cool the water and note the temperature at which the contacts open. Replace if faulty.

Renew the joint between the tank and the thermostat switch if damaged.

TEST DATA

Closing temperature	..75°C±2°C
Opening differential	..3°–5°

FAN MOTOR
Checking

Disconnect the cables and check the fan motors by connecting to a 12 volt battery.

Remove for inspection if faulty.

Page PY.s.35

ELECTRICAL AND INSTRUMENTS

Inspection

Withdraw the two through bolts, and detach the end cover. Expand the retaining spring and lift off the two brush carriers as an assembly.

Note: The brushes are loose in the carriers and care must be taken that they are not misplaced when removed.

Examine the commutator and clean with a petrol moistened cloth or fine glass paper if dirty or scored.

Check that the current is present at terminal C2 (Green) with the ignition 'ON'.

Earth the terminal W1 (Black/red), switch on the ignition and check by means of an earthed test lamp that current is available at terminal C1 (Black/green).

If air-conditioning system is fitted, check the over-riding relay as detailed under 'Horn relay'. Ignore reference to horn button.

Refer to the wiring diagram when checking.

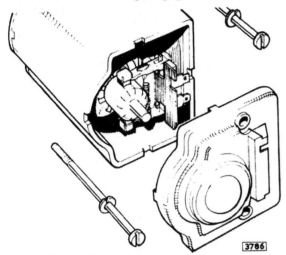

Fig. 41. *Exploded view of the fan motor.*

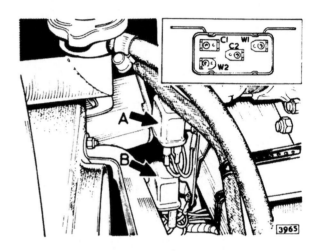

Fig. 42. *The radiator fan relays.*
A—Fan relay.
B—Over-riding relay.

FAN RELAY

Checking

Check that fuse No. 6 has not blown. Replace if necessary.

STEERING COLUMN LOCKS

Description

Steering column locks, if fitted, replace the normal Starter/Ignition Switch.

The lock, mounted on a bracket attached to the steering column, has five operative positions as follows:—

(1) Lock, (2) Park, (3) Accessories, (4) Ignition, (5) Start.

During the assembly of the switch/lock to the column, the hexagon portions of the lock securing bolts which are of the necked type, are sheared when fully tightened and cannot subsequently be removed, thus preventing unauthorised removal of the lock.

IMPORTANT

The steering column lock is brought into action when the key is turned to the 'LOCK' position and then removed.

IMMEDIATELY THIS IS DONE IT BECOMES IMPOSSIBLE TO STEER THE CAR.

It is, therefore, important to remember that if the ignition is switched off whilst the car is in motion the key should not be turned past the 'PARK' position. The ignition key should NEVER be removed from the lock whilst the car is moving.

Page PY.s.36

ELECTRICAL AND INSTRUMENTS

OPERATION

(1) Lock
This is a locked stop position. The key can be removed leaving the steering locked by engagement of the lock bolt with the register in the inner steering column.

(2) Park
This is the normal stop position. The key can be removed leaving the car capable of being steered with the ignition "OFF".

(3) Accessories
This position will allow the operation of accessories such as Radio and Electric Window Lift control (when either is fitted) with the ignition 'OFF'.

The key cannot be removed.

(4) ON
This is the normal starting position. On release, the key will automatically return to the ignition 'ON' position.

IMPORTANT

Re-engagement of the starter (cranking) motor will not be possible until the key is returned to the 'Park' position. This is a safety device introduced to prevent damage to the starter drive through accidental engagement when the engine is running.

SERVICING
No servicing is possible with the exception of the switch carrier contact plate which can be replaced if faulty as follows:—
(1) Disconnect the cables at the connectors.
(2) Remove two hexagon headed screws and plain washers and withdraw the contact plate with attached harness.
(3) Refit by reversing the removal procedure.

Note: The contact plate is indexed and cannot be fitted incorrectly.

Two Lucas 6RA relays are incorporated in the circuits controlled by the ignition switch to prevent overloading of the switch contacts.

Both units are located under the screen rail facia.

Operation of the individual relays should be checked when testing for a fault in the ignition/starter switch circuits.

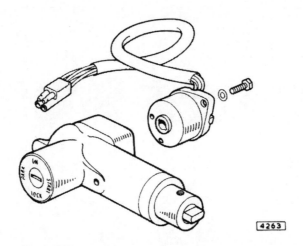

Fig. 43. *The steering column lock dismantled.*
(*The lock is shown removed from the column for clarity*).

Page PY.s.37

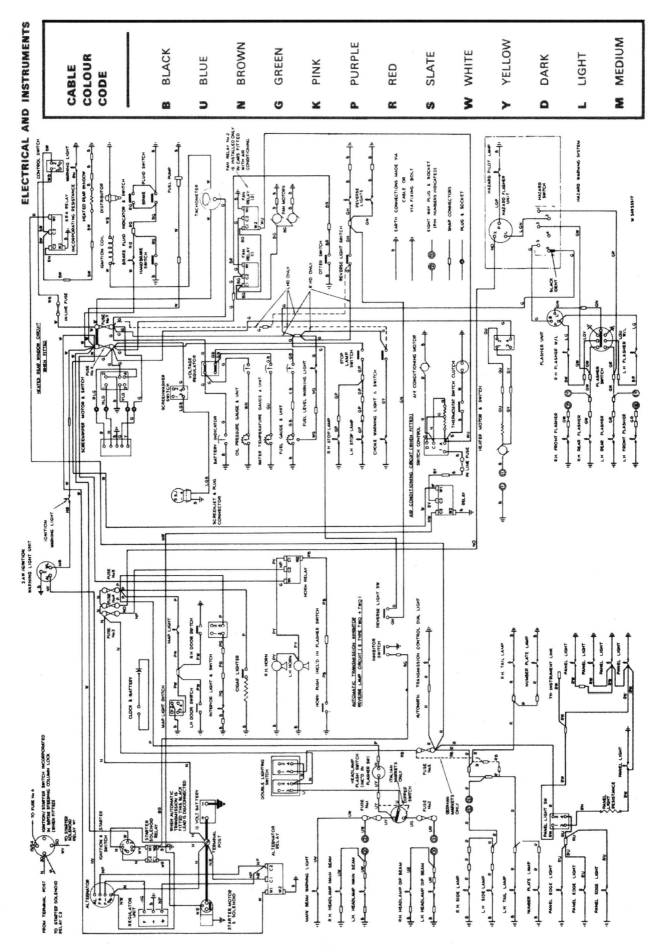

SECTION Q

EXHAUST EMISSION CONTROL

To meet U.S.A. Federal and Canadian engine emission requirements, the Jaguar 4·2 Litre engine has been modified in relation to carburation, induction system and ignition as follows.

The Duplex Manifolding System employs the metering system of the carburetters to feed fuel mixture to the combustion chambers through two manifolds. Two throttles are employed—the primary being in its normal position in the carburetter and the secondary located in the water-jacketed secondary throttle housing. The linkage between the two throttles is so arranged that on part throttle opening (employed during most city driving and cruising) the secondary throttle remains closed, routing the mixture through the primary mixture pipe and returning it to the inlet manifold downstream of the secondary throttles.

This circuit ensures complete homogeneity of fuel mixture resulting in a constant air/fuel ratio being distributed to the cylinders so that an overall leaner mixture can be employed. It also avoids the deposition of wet fuel in the inlet manifold which is a major cause of engine emissions.

After approximately 25° of primary throttle opening, the secondary throttle comes into operation until, at full throttle, both butterflies are fully opened and the mixture passes straight through both manifolds so that maximum power is maintained.

A modified distributor with redesigned power curve is fitted in conjunction with this system, the ignition timing being 5° B.T.D.C. static (10° B.T.D.C. at 1,000 r.p.m.).

ROUTINE MAINTENANCE

Owing to the critical emission limits which must be maintained throughout the lifetime of the engine, it is imperative that the following routine maintenance instructions are carried out conscientiously at the recommended mileage intervals.

Failure to comply with these recommendations may result in engine emissions falling outside U.S.A. Federal or Canadian limitations.

1,000 MILES FREE SERVICE

After the car has completed 1,000 miles from delivery, the following checks relating to Engine Emission Control should be carried out together with other Free Service details as shown in the Service Maintenance Voucher Booklet.

Engine Oil
Change the engine oil.

Distributor Contact Breaker Gap
Adjust contact breaker points to ·014"–·016" gap as detailed in the Service Manual. Verify correct dwell angle—see Diagnosis Chart.

Distributor Lubrication
Lubricate the distributor as detailed in the Service Manual.

Sparking Plugs
Clean the sparking plugs and adjust the gap to ·025".

Engine Idle Speed
Allow the engine to warm up to normal operating temperature. **Adjust the idle speed by turning each adjuster screw an equal amount** to give slow running speed of 750 r.p.m. on standard transmission cars; 650 r.p.m. on cars with automatic transmission with the selector lever in the neutral position.

Check the synchronisation of the carburetters by using a balance meter.

Fast Idle Speed
Ensure that the choke control cam on the rear carburetter is in the "fully off" position.

Release the locknut and turn the fast idle abutment screw until the gap between the cam and the screw is ·067". Tighten the locknut and re-check the gap.

Page QY.s.1

EXHAUST EMISSION CONTROL

Carburetter Hydraulic Piston Damper

Unscrew and withdraw the hydraulic piston damper from the piston cover.

Top up the hollow guide rod of the piston with Zenith Lube Pack, or, if this is not available, use SAE 20 engine oil, to within ¼" of the top of the rod. Replace the damper securely.

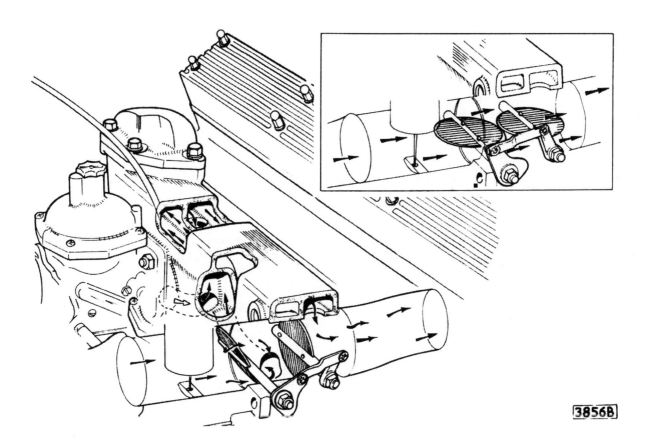

Fig. 1. *Schematic layout of the carburetters and linkage showing the direction of the gas flow.*

Page QY.s.2

EXHAUST EMISSION CONTROL

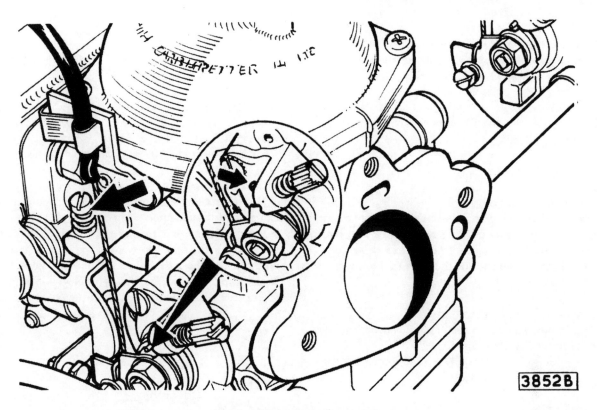

Fig. 2. *Slow running adjustment screw.*

Ignition Timing

As the ignition timing of an emission controlled engine is critical, the greatest possible care must be taken to ensure that the correct figure is obtained.

Adjust the distributor vernier scale to the central position. With the engine running at exactly 1,000 r.p.m. check the timing with a stroboscope and adjust by means of the vernier control on the distributor to 10° B.T.D.C.

The timing scale is located on the rim of the crankshaft damper.

To check the distributor advance characteristics at higher r.p.m., refer to Distributor Test Data on page QY.s.5.

Cylinder Head Nuts

Check the torque of the cylinder head nuts (Torque 696 lbs. ins.) tightening in sequence as detailed in the Service Manual.

Inlet Manifolds

Check the tightness of all inlet manifold system securing nuts.

Tighten the carburetter mounting nuts.

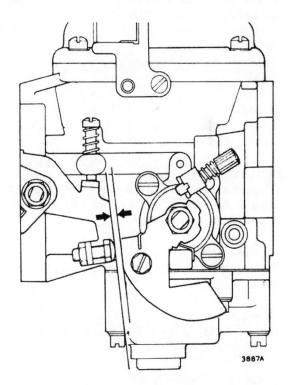

Fig. 3. *Fast idle adjustment. Gap between screw and cam should be ·067".*

Page QY.s.3

EXHAUST EMISSION CONTROL

It is important to ensure that no air leaks exist at any of these joints otherwise the mixture will be weakened to a point where idling will be effected.

Idle "Trim" Screw
Screw in the idle "trim" screw to obtain the optimum quality of idling.

EVERY 3,000 MILES
Engine Oil
Change the engine oil.

Low speed city driving in hot dusty territory or in very cold weather may produce conditions conducive to oil dilution and sludge formation. In these conditions the engine oil and the filter should be changed every 1,000 miles.

Distributor Contact Breaker Gap
Clean points and adjust the contact breaker gap as detailed in the Service Manual.

Adjust the gap to ·014″–·016″. Verify correct dwell angle—see Diagnosis Chart.

Ignition Timing
Check ignition timing as detailed in the 1,000 miles Free Service.

Distributor Lubrication
Lubricate the distributor as detailed in the Service Manual.

Sparking Plugs
Clean and adjust and test the spark plugs. Check on an oscilloscope and renew any doubtful plugs. Set the gap between the side wire and the centre electrode to ·025″.
NOTE: Misfiring of a plug will cause incomplete combustion of the mixture and raise the engine emission levels above the specified limits.

Engine Idle Speed
Adjust the engine idle speed as detailed in the 1,000 miles Free Service.

Fast Idle Speed
Adjust the fast idle speed as detailed in the 1,000 miles Free Service.

Carburetter Hydraulic Damper
Top up the hydraulic piston damper as detailed in the 1,000 miles Free Service.

EVERY 12,000 MILES
Fit Emission Pack Part No. 11549 (coloured Yellow) to the carburetters. See page QY.s.11 for details.

Air Filter
Renew the air filter element as detailed in the Service Manual. If the car is operating in dusty territory inspect at 6,000 miles and renew if necessary.

Crankcase Breather
Disconnect the breather pipe from the front of the engine and the air filter. Remove the pipe. Remove the nuts securing the breather and withdraw the flame trap. Wash the flame trap and pipe in gasoline and refit. Renew the gaskets located on each side of the flame trap. Examine all hoses, renew if necessary. Check that all clamps are tight allowing no air leakage.

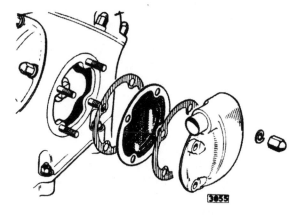

Fig. 4. *The crankcase breather.*

Inlet Manifolds
Check tightness of all secondary throttle housing and inlet manifold securing nuts. Check tightness of nuts securing primary mixture pipe to secondary throttle housing.

It is important to ensure that no leaks exist at any of these joints.

Exhaust System
Check the exhaust system for leaks. Renew any parts showing signs of deterioration.

Spark Plugs
Renew the spark plugs with the recommended grade.

Cylinder Head
Check the torque of the cylinder head securing nuts (696 lb. ins.) and check the cylinder head gasket for leaks.

EXHAUST EMISSION CONTROL

Fuel Line Filter
At the recommended interval, or more frequently if sediment build-up is evident, slacken the locknut, swing the retaining clip to one side and remove the glass bowl, sealing washers and filter.

Wash the glass bowl in gasoline. Fit a new filter element with new sealing washers and re-assemble.

Distributor Contact Breaker Gap
Clean points and adjust contact breaker gap as detailed in the Service Manual. Adjust the gap to ·014″–·016″. Verify correct dwell angle—see Diagnosis Chart.

Ignition Timing
Check ignition timing as detailed in the 1,000 mile Free Service.

EVERY 24,000 MILES
Carburetters
Remove lead seal and fit Red Emission Pack Part No. 11791 to carburetters. See pages QY.s.12 to QY.s.13. Fit new lead seal after completion.

Valve Clearances
Check the valve clearances as detailed in the Service Manual. Clearances (cold)—inlet ·004″; exhaust ·006″

Valve Timing
Check valve timing as detailed in the Service Manual.

Contact Breaker Points
Renew contact breaker points as detailed in the Service Manual. Adjust points gap to ·014″–·016″. Check ignition timing as detailed in the 1,000 Miles Free Service. Verify correct dwell angle—see Diagnosis Chart.

Compression Pressures
Compression pressures must be checked with all spark plugs removed, carburetter throttles wide open and the engine at normal running temperature.

Disconnect the black/white low tension lead from the coil before operating the ignition/starter switch to check pressures. All cylinders should be even and approximately 150 p.s.i.

If one or more cylinders show low compression, a full investigation into engine condition must be made on an Electronic Engine Tester such as a Sun 1020. See diagnosis chart.

DISTRIBUTOR TEST DATA

CENTRIFUGAL TIMING ADVANCE
With a stroboscopic timing light, check the advance characteristics of the distributor at the following r.p.m.

R.P.M.	DEGREES
1200	13—17
1600	22—26
2900	29—33
4400	37—41

THE STROMBERG 175 CD2SE EMISSION CARBURETTER

DESCRIPTION
The STROMBERGE 175 CD2SE carburetter is a development of the constant depression carburetter which operates on the principle of varying the effective areas of choke and jet orifice in accordance with the degree of throttle opening, engine speed and engine load. A number of special features have been introduced to meet the needs of engine emission control.

Fuel passes into the float chamber via a needle valve where flow is controlled by the needle valve and twin floats mounted on a common arm. Fuel in the jet orifice is controlled at the same level as that in the float chamber by means of cross drillings in the jet assembly.

Clearance around the piston in its vertical bore permits air to "leak" into the mixing chamber and thus lower the depression. A drilling is taken from the atmospherically vented region beneath the diaphragm to meet a further drilling that breaks into the mixing chamber downstream of the piston. An adjusting screw with a conical tip is inserted into the drilling and is adjusted by the manufacturer to bring each carburetter to a common "leak" datum and sealed with a plug **which must not be disturbed in any circumstance.**

EXHAUST EMISSION CONTROL

COLD STARTING

Pulling out the choke control on the instrument panel operates a lever at the side of each carburetter which rotates a disc in the starting device in which a series of holes of different diameters are drilled. In the fully rich position all these holes will be in communication with the starting circuit and will provide the richest mixture. Gasoline is drawn from the float-chamber via a vertical drilling adjacent to the central jet, through the starting device and into the throttle body between the piston and the throttle disc. Simultaneously the cam on the choke lever will open the throttle disc beyond the normal idle position to ensure a faster idle speed and prevent stalling.

As the choke is gradually pushed to the "OFF" position, fewer and smaller holes will limit the gasoline feed from the float chamber thereby progressively weakening the mixture to a point where the choke is fully home and the mixture strength is governed by the Factory setting of the main jet and the idling speed determined by the setting of the throttle stop screw.

NOTE: DO NOT DEPRESS THE ACCELERATOR PEDAL WHEN STARTING FROM COLD.

A control in each carburetter enables the choke to be varied for summer and winter operation, and takes the form of a spring loaded plunger operating against the cam. To check the setting, note the position of the stop cross-pin. If lying in the horizontal slot in the casting the choke is set for winter operation. To adjust for summer running depress the spring loaded pin and turn through 90°. Release and check that the cross-pin is at right angles to the slot.

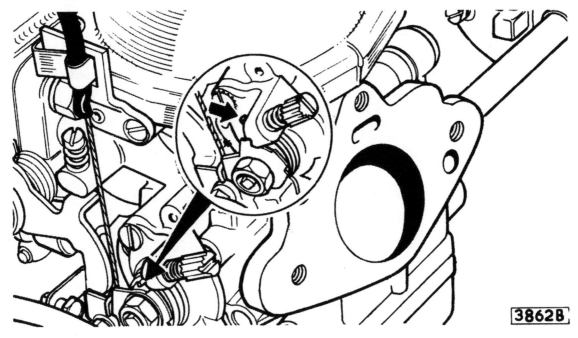

Fig. 5. *The choke limiting spindle in the Winter setting (inset shows the Summer setting).*

IDLING

There is no separate circuit for idling. The fuel is provided by the jet orifice (the amount being controlled by jet/needle relationship established during manufacture) and the speed of idle by adjustment of the throttle stop screw which limits the closure of the throttle when the accelerator pedal is released.

To cater for variations in engine "stiffness" when manufactured, an idle "trim screw" is provided. Engine stiffness dictates idling air consumption and a new and very stiff engine will require more air than one which has become "free". The trim screw may be adjusted to provide a slightly leaner mixture for any engine found to be extremely stiff on production engine test. When fully seated, the maximum enrichment is achieved and emission figures will be within requirements, the engine having freed to a value equivalent to the datum at which the original trim screw setting was carried out.

Finger pressure only should be used when tightening the trim screw, care being taken not to over-tighten.

Page QY.s.6

EXHAUST EMISSION CONTROL

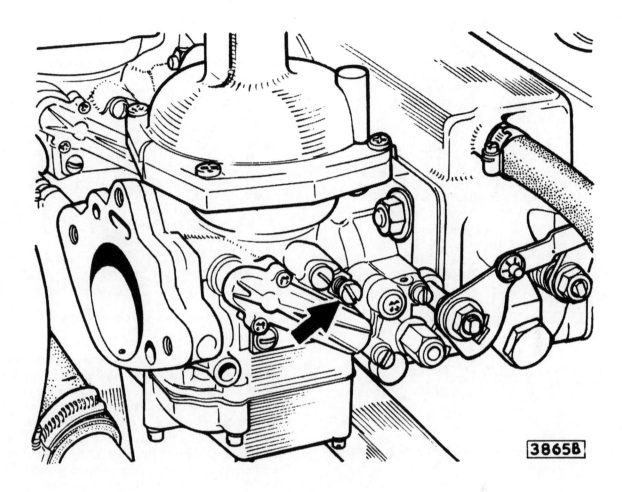

Fig. 6. *The idle trim screw.*

JET/NEEDLE RELATIONSHIP

The jet/needle relationship not only governs the correct idle mixture but also the correct mixture strength throughout the range. During development, it was found desirable to have the needle central in the jet. This not being practicable to achieve the needle has been biased permanently on one side of the jet to rub lightly against the jet orifice.

The needle profile has been evolved to compensate for the known air leak (consistency being obtained by manufacturer's setting of the "leak adjuster screw") and therefore a constant fuel/air ratio is maintained. All carburetters therefore, produce a consistent flow from the given needle profile.

To meet emission control requirements, carburetters must be kept within very narrow "flow bands". Exhaustive testing on Jaguar engines decided the optimum jet position in the orifice and, therefore, all carburetters have the jets pressed into position to a predetermined depth thereby eliminating any possible maladjustment in service. Every unit is flow tested by the carburetter manufacturer ensuring that all carburetters are supplied within the desired limits.

On throttle opening, the piston rises withdrawing the tapered jet metering needle, held in its base, from the jet orifice so that the fuel flow is increased proportionate to the greater air flow.

The metering needle is variable along its length and has been machined to very close limits.

Page QY.s.7

EXHAUST EMISSION CONTROL

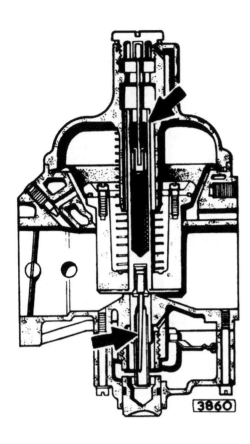

Fig. 7. *Cut-away view showing the Jet/Needle relationship.*

As the needle profile has been developed from exhaustive testing, it is vitally important, to maintain correct results in regard to engine emission control, that only the recommended needle is used.

For correct identification the needle is stamped with the figure B1E on the shank.

Variations in mixture strength caused by heat transfer to the carburetter castings are significant in the context of the extraordinary precision demanded by emission requirements. A temperature compensator is incorporated to cater for this condition.

An air flow channel permits air passing through the carburetter to by-pass the bridge section. A bimetallic blade regulated the movement of a tapered plug which adjusts the quantity of air by-passed to the mixing chamber. Two screws attach the temperature compensator assembly to the body and two seals are provided to ensure that no leakage can occur at the joint with the body.

The assmbly is preset, and unless necessary due to the tapered plug sticking, should not be readjusted in service. If malfunctioning of the compensator is suspected and the tapered plug moves freely when tested carefully by hand with the engine hot or cold, the compensator assembly must be changed for a new unit.

Fig. 8. *Cut-away view of the temperature compensator.*

THROTTLE BY-PASS

During periods of engine over-run, high emissions will occur if the fuel/air mixture in the combustion chambers is not of sufficient strength, when diluted by exhaust gas, to support combustion. To overcome this problem, a device is fitted to the carburetters which consists of a by-pass formed in the carburetter around the primary throttle under the control of a vacuum operated valve. The vacuum signal to the valve is via an internal drilling in each carburetter. The flow of this circuit is determined by the size of the ports, the valve always lifting to full travel. As the throttle remains on its stop, the primary induction circuit only is in use ensuring that even mixture is fed through the primary system to all six combustion chambers. This valve is pre-set and provided it is free from air leaks, requires no adjustment. It is possible however, that small particles of foreign matter may lodge under the valve seating causing leakage and consequent high idling speed. In these circumstances the valve cover should be removed and the valve and seating cleaned.

Manifold depression acting on the valve diaphragm will cause the valve to open when the value is reached that will overcome the valve spring tension.

This allows fuel to feed from the mixing chamber to the downstream side of the primary throttle enriching the gases in the combustion chamber to a combustible level.

Page QY.s.8

EXHAUST EMISSION CONTROL

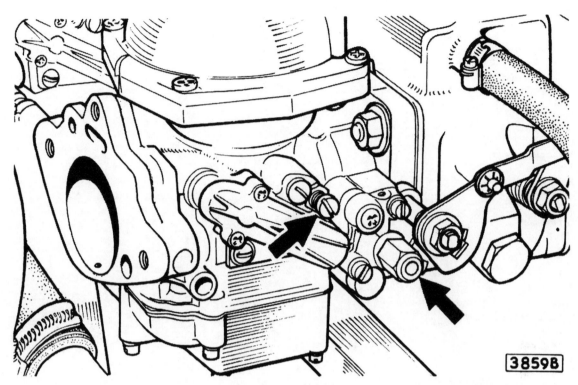

Fig. 9. *The throttle by-pass valve.*

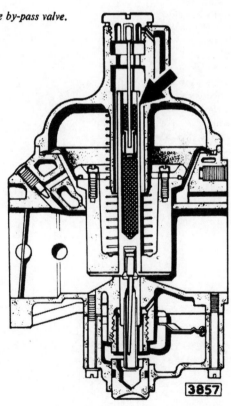

C. *The hydraulic damper. Sectioned view showing the oil level in the piston guide.*

HYDRAULIC DAMPER

At any point in the throttle range, a temporary enrichment is required when the throttle is suddenly opened. A hydraulic damper is arranged inside the hollow guide rod of the piston to provide this.

The guide rod itself is filled with "Zenith Lube Pack" or SAE 20 engine oil to within $\frac{1}{4}''$ of the end of the rod.

When the throttle is suddenly opened the immediate upward motion of the piston is resisted by the damper. For this brief period a temporary increase in the depression over the jet orifice is achieved and the mixture is enriched. Downward movement of the piston is assisted by a coil spring.

RECOMMENDED SERVICE PROGRAMME

It is necessary to maintain the carburetters at peak efficiency to control engine emission, therefore the following service recommendations should be carried out.

Page QY.s.9

EXHAUST EMISSION CONTROL

1. Hydraulic damper.
2. "O" ring.
3. Cover.
4. Diaphragm securing ring.
5. Piston return spring.
6. Needle securing screw.
7. Butterfly.
8. Bush.
9. Pick-up lever.
10. Floating lever.
11. Washer.
12. Shakeproof washer.
13. Nut.
14. Diaphragm.
15. Idle trim screw.
16. Gasket.
17. By-pass valve.
18. Gasket.
19. Spring.
20. Cover.
21. Seal.
22. Seal.
23. Gasket.
24. Temperature compensator housing.
25. Tapered plug.
26. Bi-metallic blade.
27. Plastic cover.
28. Jet assembly.
29. Float assembly.
30. Float chamber.
31. Pivot pin.
32. "O" ring.
33. Needle valve.
34. Special washer.
35. Choke assembly.
36. Needle.
37. Spring.
38. Throttle stop screw.
39. Throttle spindle assembly.
40. Piston.
41. Diaphragm.
Inset—Lead seal.

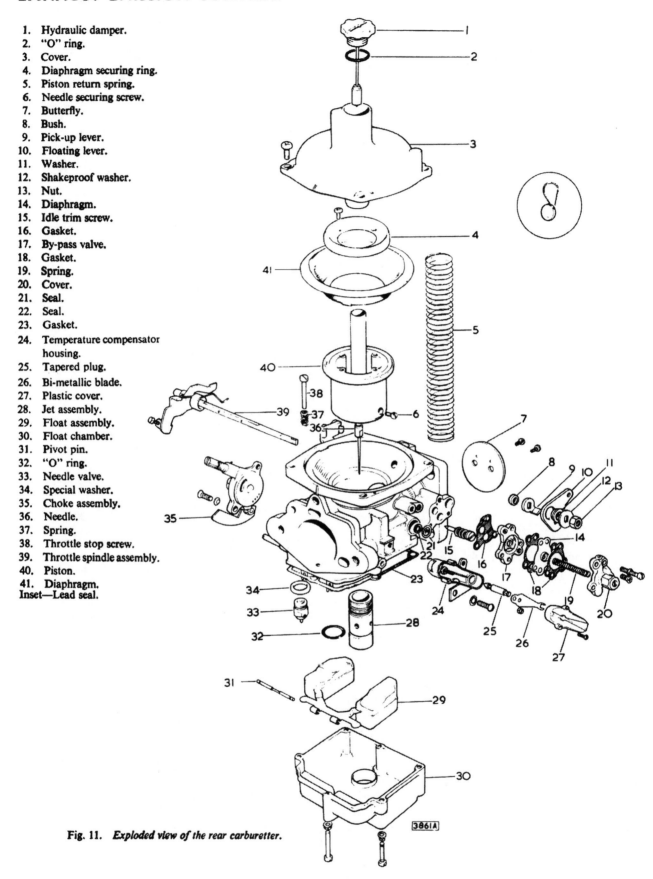

Fig. 11. *Exploded view of the rear carburetter.*

Page QY.s.10

EXHAUST EMISSION CONTROL

12,000 Miles Service

For this service period, one Yellow Emission Pack (Part No. 11549) is required for the two carburetters. This pack contains 2 float-chamber gaskets, 2 "O" rings for the float-chamber plugs and 2 needle valve washers.

Additionally, 4 manifold/carburetter gaskets and 2 spacers will be required.

Remove the carburetters as follows:

Remove three setscrews securing the air trumpet to each carburetter and withdraw the engine breather pipe from the air box. Lift off the air cleaner. Thin gaskets are fitted between each trumpet and carburetter and these should be carefully placed to one side for use when refitting.

Remove the gasoline pipes from the float-chambers.

Remove the nut, lock washer and plain washer, securing the linkage from the primary to the secondary throttles, at the primary spindles of the rear carburetter.

Fig. 12. *The choke cable connections.*

Detach the link. Replace the washers and nuts on the spindle for safe-keeping.

Repeat the operation for the front carburetter.

Release the bolts securing the inner cables to the choke levers and withdraw the outer casings from the clips at the sides of the carburetters.

Remove four nuts, spring washers and plain washers securing each carburetter to the studs on the primary inlet manifold. Disconnect the central link between the throttle slave shaft and the carburetter spindles. Withdraw both carburetters as an assembly.

Separate the units by slackening the clamps on the throttle spindles.

It is important to dismantle and assemble each carburetter individually to avoid the possibility of similar parts being interchanged between carburetters.

Page QY.s.11

EXHAUST EMISSION CONTROL

Unscrew the float-chamber fixing screws and withdraw the float-chambers vertically away from the body to clear the float mechanism. Remove the float-chamber gaskets.

Unclip the float pivot pin.

Note the fitted position of the float assembly. The flat portion of the float must be uppermost when refitted, with the carburetter in an inverted position.

Unscrew the hexagon-bodied needle valve from the float-chamber body.

Remove the "O" ring from the centre plug and wash all metal parts in cleansing solvent.

Re-assemble the carburetters as follows:

Refit the needle valve with the new washer to the float-chamber body and screw home tightly. Replace the float assembly, after inspecting for distortion or damage.

To ensure correct float level, measure the float height as follows.

Invert the carburetter so that the float tag closes the needle valve. Measure from the face of the carburetter body (with the gasket removed) to the top of each float.

The correct height should be 16·5 ± ·5 mm.

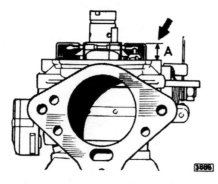

Fig. 13. Checking the float height. Dimension A should be 16·5 ± ·5 mm.

Fit a new "O" ring to the centre plug.

With the new gasket in position refit the float chamber and tighten the securing screws from the centre outwards.

Refit the carburetters to the manifold with the new gaskets and spacers. Reconnect the carburetter linkage. Reconnect the gasoline pipes and top up the hydraulic piston damper of each carburetter with "Zenith Lube Pack" or SAE 20 engine oil to within ¼" of the top of the centre rod.

Leaving the clamping bolts on the throttle spindles loose, unscrew the throttle stop screws to permit the primary throttles in each carburetter to close completely. Screw in the stop screws to the point where the ends of the screws are just contacting the casting. Rotate each screw 1½ turns to open the throttles an equal amount and to provide a basis from which the final idling speed can be set.

Ensure that the fast idle screw is clear of the choke cam otherwise incorrect synchronising will result.

Check that both choke cams are in contact with the stops. With the instrument panel choke control pushed home fully, reconnect the choke cables to the cams.

Check that both cams operate simultaneously.

Start the engine and warm up to the normal operating temperature.

Check the synchronising of the throttles with a balance meter, and tighten the clamping bolts on the throttle spindles. Set the throttle stop screws to give the correct idling speed as stated in "Routine Maintenance". Turn each screw by an equal amount.

Adjust the fast idle as detailed on page QY.s.6.

If care is exercised in setting each throttle opening to the same extent, no difficulty should be encountered in obtaining satisfactory idling and smooth acceleration.

Refit the vacuum pipe to the by-pass valve.

Refit the air cleaner and the engine breather pipe.

NOTE: The idling quality and acceleration depend to a large extent upon general engine condition and it is therefore essential to check the engine on an Electronic Engine Tester such as the Sun 1020 or other make of similar capacity. See Diagnosis Chart for test procedure.

24,000 Miles Service

This is a more comprehensive service for which one **RED Emission Pack**, Part No. 11791 will be required for each carburetter. This Pack contains 2 float-chamber gaskets, 2 "O" rings, 2 needle valves and washers, 4 throttle spindle seals, 4 temperature compensator seals, 2 by-pass body gaskets, 2 piston diaphragms, and 2 hydraulic damper seals.

Additionally, the following will also be required 4 secondary throttle housing/carburetter gaskets and spacers, 1 secondary throttle housing/manifold gasket. 1 primary mixture pipe/secondary throttle housing gasket.

IMPORTANT: Dismantle and assemble each carburetter individually to avoid the possibility of interchanging similar parts between carburetters.

Remove the carburetters as detailed under the 12,000 miles service.

EXHAUST EMISSION CONTROL

Drain sufficient water from the cooling system to allow the level of coolant to fall below the throttle housing.

Disconnect the water pipes from the housing.

Remove four nuts and washers securing the primary mixture pipe to the secondary throttle housing.

Release the four nuts securing the secondary throttle housing to the inlet manifold; these are located behind the manifold.

Disconnect the clamping bolt securing the front throttle slave shaft to the rear throttle slave shaft.

If automatic transmission is fitted, disconnect the link between the automatic transmission throttle control shaft and the front throttle slave shaft.

Withdraw the secondary throttle housing together with the front throttle slave shaft.

For each carburetter, carry out the instructions detailed under 12,000 miles Service.

In addition to this service however, fit the new needle valve assemblies with new washers.

Remove the damper assembly from the top cover and break the wire seal.

Unscrew the four cover fixing screws and carefully lift off the cover. Remove the piston return spring and lift out the piston assembly.

Drain off the oil from the damper reservoir (centre of guide rod). Slacken the metering needle clamping screw and withdraw the needle from the piston.

Place the needle carefully to one side to avoid damage.

Remove the four screws attaching the diaphragm retaining ring to the top of the piston.

Lift off the ring and diaphragm.

Fit the new diaphragm into the top of the piston ensuring that the locating tag is recessed into the aperture provided. Secure in position with the retaining ring and the four screws.

Check the spring action of the needle in its housing at the top of the shank. Fit the needle into the base of the piston lining up the flat portion with the locking screw. Using a straight edge placed lightly against the small shoulder on the needle, (not the casing) press the assembly into the piston until the straight edge aligns the shoulder of the needle with the flat surface of the piston. Lightly tighten the locking screw, taking care not to collapse the needle housing. Shoulder alignment of the needle is critical and great care must be taken during this operation. Correctly fitted, the needle will be baised towards the throttle and the shoulder of the needle will be exactly flush with the piston face.

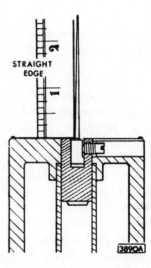

Fig. 14. *Positioning the needle in the piston.*

Carefully enter the piston and diaphragm assembly into the main body, guiding the needle into the jet with a finger in the air intake. Locate the outer tag of the diaphragm into the recess at the top of the body. Check the assembly by looking down the piston to ascertain that the two depression transfer holes are towards and in line with the throttle spindle and that the needle is biased towards the throttle.

Replace the piston spring, hold the piston against the spring with a finger through the air intake and fit the cover. This method will prevent the piston from moving out of position. The cover must be fitted with the damper ventilation boss towards the air intake.

Replace the four cover screws and tighten down evenly.

Check the movement of the piston; freedom of movement over the full travel is essential and when released from the uppermost position, the piston should fall with a sharp click onto the bridge of the carburetter.

Page QY.s.13

EXHAUST EMISSION CONTROL

Top up the piston damper, fit new seal and refit the damper assembly.

Remove the two screws securing the temperature compensator unit to the body and withdraw the assembly. Take out the inner seal from the carburetter body and remove the outer seal from the valve. Renew both seals and refit the assembly to the carburetter body.

Tighten the two screws evenly.

Remove the two screws and detach the compensator cover. Check for free movement of the valve by lifting the plug from its seat. On releasing, the valve should return freely.

Do not strain the bi-metal strip, or attempt to alter the adjustment. It is permissible to ensure that there is consistent radial clearance around the valve to allow for thermal expansion. If the valve is sticking, remove the securing nut and screw. Lightly clean the bore and the plug with a gasoline moistened cloth. Refit the bi-metal strip and re-tension by tightening the nut until the valve is just seated. DO NOT TIGHTEN BEYOND THIS POINT.

Unscrew the three setscrews securing the by-pass valve body and lift the assembly from its seat. Prise out the throttle spindle seal and renew before replacing the by-pass body with a new gasket. Renew the throttle spindle seal on the other side of the carburetter spindle. Repeat this operation on the rear carburetter.

Refit the secondary throttle housing and mixture pipe with new gaskets. Refit the water pipes to the secondary throttle housing and top up the coolant to the correct level.

Refit the carburetters to the secondary throttle housing with new gaskets and spacers. Fit new wire and seal to the dashpot. Re-connect the throttle spindles controls and set the carburetters as detailed in the 12,000 Miles Service.

Check the engine performance on an Electronic Engine Tester such as the Sun 1020 or other make of similar capacity. See Diagnosis Chart for test procedure.

CARBURETTER FAULT FINDING

Service complaints can generally be broken down into three main categories:—
(a) Erratic or poor idling.
(b) Hesitation or flat spot.
(c) Heavy fuel consumption.

Under these headings, possible causes are put forward together with suggestions.

Erratic or Poor Idling

(1) Incorrect fuel level caused by maladjustment of the floats and/or worn or dirty needle valves. Check the float level. Wash the needle valve in clean gasoline, replace the valve if worn.
(2) Piston sticking. Check for free movement of the spring loaded metering valve needle. Clean the piston rod and guide. Lubricate the rod and guide with a few drops of light oil.
(3) Metering needle incorrectly fitted. Check that the shoulder of the needle is flush with the face of the piston and that the needle is biased towards the throttle. Check the needle identification and ascertain that the correct needle is fitted. Check that the needle housing has not been distorted by over-tightening of the securing screw.
(4) Partially or fully obstructed diaphragm ventilation holes. Check that the air cleaner element and casing are correctly fitted and that the air trumpet/carburetter gaskets are not causing obstruction.
(5) Diaphragm incorrectly fitted or damaged. Check the location with the depression chamber cover removed. The two depression holes at the base of the piston should be in line with and towards the throttle spindle. Replace the diaphragm if damaged. When replacing the depression chamber cover, the damper ventilation boss must be towards the air intake.
(6) Throttles not synchronised. Reset correctly using a balance meter.
(7) Temperature compensator not working properly. With the engine and the carburetters cold, remove the cover from the temperature compensator assembly. The tapered valve should be seated in this instance. Check the operation by carefully lifting of the valve off its seat. When released, the valve should return freely. If damage prevents the mechanical operation functioning correctly, renew the compensator unit.
(8) If high mileage has been covered, inspect the throttle spindles and end seals for wear. Check the spindles for fractures: renew if suspect.

Hesitation or Flat Spot

Possible causes are as enumerated for "Erratic or Poor Idling" but with the addition of the following:
(1) Damper inoperative. Check the oil level and top up with light engine oil.
(2) Piston return spring omitted.

Page QY.s.14

EXHAUST EMISSION CONTROL

Heavy Fuel Consumption

Points covered under the two previous headings may contribute to heavy fuel consumption. Additionally, check that there is no fuel leakage from the float-chambers or centre plug "O" ring. Replace as necessary.

SPECIAL PARTS

In an endeavour to maintain engine emission within the legislated limits, the following parts must not be changed in service.

ITEMS THAT MUST NOT BE CHANGED

(a) The jet assembly.
(b) The piston.
(c) The depression chamber cover.
(d) The **position** of the metering needle.

If any of the above items require changing, the sub-assemblies or the complete carburetter(s) must be renewed. In the case of the metering needle it is quite permissible to replace the needle providing the procedure outlined previously is carefully followed.

EQUIPMENT

The recommended equipment for servicing should include at least the following:

Ignition Analyser oscilloscope	Cam Angle Dwell Meter
Ohmeter	Ignition Timing Light
Voltmeter	Engine Exhaust Combustion Analyser
Tachometer	
Vacuum Gauge	Cylinder Leak Tester
Carburetter Balance Meter	Distributor Advance Tester

It is important that test equipment has regular maintenance and calibration.

The following equipment covers most of the requirements for engine testing and tuning of vehicles fitted with exhaust emission control devices.

Equipment made by other suppliers may also be adequate.

Equipment	Type/Model	Manufacturer
Oscilloscope engine tuning set and exhaust gas analyser	1020 or 720	Sun Electric Corp

Page QY.s.15

DIAGNOSIS CHART

ENGINE SPEED	TEST	COMPONENT CONDITION	READ/OBSERVE	CORRECT READINGS	CHECK SEQUENCE—FAULT LOCATION
START (CRANKING)	Cranking voltage	Battery; Starting system	Voltmeter	10·2 volts minimum at the battery	Battery—Starter Motor—Connection/Cables—Alternator
	Cranking Coil output	Coil; Ign. Primary circuit	Scope Trace	16 kv. max. 14 kv. min.	Ignition coil—Battery—Condenser—Resistance in distributor wiring and/or points—Ignition switch
	Cranking vacuum	Engine	Vacuum Gauge	16-18 in. hg. even pulse	Hoses and connections—Valve tappet clearance—Servo—Inlet manifold system leaks—Valves or seats—Piston rings
IDLING	Idle Speed	Carburetter idle setting	Tachometer	750 r.p.m. (Standard trans.) 650 r.p.m. (Automatic trans.)	Carburetter adjustment—Hoses and connections—Servo—Carburetter mechanical condition—Engine condition
	Dwell Initial Timing	Distributor/drive; Points Spark Timing Setting	Dwell Meter; Scope Timing light	33°–37° 5° B.T.D.C. (Static) 10° B.T.D.C. (1,000 r.p.m.)	Breaker points—Distributor and drive mechanical condition Distributor adjustment
	Fuel Mixture	Carburetter float level and needle position	Exhaust Gas Analyser	2–4% c.o. at idle when hot	Carburetter float level and needle position—Hoses and connections—Manifold system leaks—Servo—Spark plugs—Ignition timing
	Manifold Vacuum	Engine Idle Efficiency	Vacuum Gauge	16 to 18 in. hg.	Hoses and connections—Inlet manifold system leaks—Valves or seats—Piston rings
CRUISE (1,000 r.p.m.)	Dwell Variation Coil Polarity	Distributor mechanical Ignition circuit polarity	Dwell Meter Scope Trace	Variation of 3° maximum Pattern inverted	Distributor and drive mechanical condition Ignition circuit connections—Ignition coil—Battery or charging system polarity reversed
	Cam Lobe Accuracy Secondary Circuit	Distributor Cam Plugs; Leads; Cap; Rotor	Scope Trace Scope Trace	2° max. Variation Standard pattern	Distributor cam defective Spark plugs and leads—Breaker points—Distributor cap towers—Hoses and connections—Servo—Coil
	Coil and Condenser Condition Breaker Point Condition Spark Plug Firing Voltage	Coil Windings; Condenser Points Closing/Opening/Bounce Fuel Mixture; Compression; Plug/Rotor Gaps	Scope Trace Scope Trace Scope Trace	Lack of oscillation in Intermediate Section Unusual Dwell Section Abnormal Spark Plug Firing Voltage at Rotor Caps—4 kv. max.	Ignition coil—Condenser Breaker points—Condenser Spark plugs and leads—Breaker points—Ignition timing—Distributor cap and rotor—Carburetter float level or needle position—Hoses and connections—Servo
	Engine/Cylinder Balance/Power Drop	Cylinder compression	Tachometer/Cylinder Leak Tester (150 r.p.m. scale)	Max. variation/cylinder 40 r.p.m. Complete cut 170 r.p.m.	Valve tappet clearance—Valves and seats—Piston rings
ACCELERATE	Spark Plugs Under Load	Spark Plugs	Scope Trace	Abnormal Scope Firing Lines under Load 10 kv./plug maximum	Spark plugs and leads—Carburetter float level or needle position—Hoses and connections—Servo
TURNPIKE (2,500 r.p.m.)	Timing Advance	Distributor Mech.	Timing light advance Meter	5° B.T.D.C. (Static) 10° B.T.D.C. (1,000 r.p.m.)	Distributor mechanical condition, centrifugal weights and springs
	Maximum Coil Output	Coil; Condenser; Ign. Primary	Scope Trace	Standard pattern minimum reserve ⅓ more than requirement	Ignition coil/condenser—H.T. circuit insulation—Charging circuit—Battery
	Secondary Circuit Insulation Charging Voltage Exhaust Restriction	H.T. Cables, Cap, Rotor Alternator—Alternator control unit Exhaust system	Scope Trace Voltmeter Vacuum Gauge	Standard pattern 14·5 volts steady reading No variation in reading at constant speed for 10 sec.	H.T. leads—Distributor cap and rotor—Coil tower Alternator—Alternator control unit Exhaust system

SECTION R

AIR-CONDITIONING REFRIGERATION EQUIPMENT

DESCRIPTION

Air conditioning equipment is available as an optional extra, and is fitted in addition to the car heating and ventilating system standard on all cars.

The air-conditioning equipment is comprised of the following components:—

A compressor, magnetic clutch, condenser unit, a receiver drier, and evaporator unit, and expansion valve, a thermostatic control and interconnecting lines.

The expansion valve and thermostatic control are contained in the evaporator case.

The refrigerant used as to Specification R.12 (Refrigerant 12) which is a halogenated hydrocarbon (dichlorodifluoromethane).

A basic knowledge of refrigeration systems and the use of the special tools required is necessary before any Service operations can be attempted. It is, therefore, ESSENTIAL that only qualified Refrigeration Service Engineers should carry out any repair work necessary.

IT IS DANGEROUS FOR ANY UNQUALIFIED PERSON TO ATTEMPT TO DISCONNECT OR REMOVE ANY PART OF THE AIR-CONDITIONING SYSTEM.

If, during repair work on the car, it becomes necessary to remove any part of the air-conditioning system, DO NOT DISCONNECT THE HOSE CONNECTIONS until the system has been "pumped down", that is, until all the refrigerant has been removed.

WARNING:

EXTREME CARE SHOULD BE EXERCISED IN HANDLING THE REFRIGERANT. LIQUID REFRIGERANT AT ATMOSPHERIC PRESSURE BOILS AT $-20°F$ ($-29°C$). SERIOUS INJURY MAY OCCUR IF ALLOWED TO CONTACT THE EYES. DO NOT SMOKE WHILST CHARGING THE SYSTEM.

For operating details refer to page RY.s.13 under "The Electrical System".

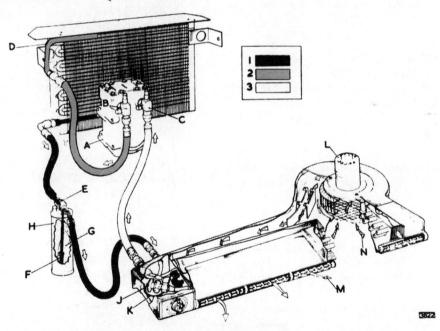

Fig. 1. *The schematic refrigeration circuit*

A. Compressor.	D. Condenser.	G. Dessicant.	K. Capillary tube.
B. Discharge valve.	E. Sight glass.	H. Receiver drier assembly.	L. Blower assembly.
C. Suction valve.	F. Cotton bobbin.	J. Expansion valve.	M. Air flow (outlet).

1. HOT VAPOUR. 2. LIQUID. 3. COLD VAPOUR.

N. Air flow (inlet).

AIR-CONDITIONING REFRIGERATION EQUIPMENT

PERIODICALLY
Compressor Drive Belt—Adjusting

Periodically check the compressor drive belt and adjust to the correct tension by means of the adjuster pulley.

Fig. 2. *The compressor drive belt adjustment point.*

Every 12 Months
Compressor Oil Level—Checking

A manifold gauge set must be available before this maintenance operation can be carried out.

At every 12 monthly period check the oil level in the compressor as follows:—

Operate the system until the desired interior body temperature is obtained and the crankcase is warm.

Stop the engine and connect the manifold gauge set.

Start the engine and note the gauge readings.

Slowly forward seat (turn clockwise) the suction service valve until the compressor gauge reads 2 lb. per sq. in. (0·1406 kg./cm.²).

Stop the engine at this point and quickly **fully** close the suction valve.

Forward seat the discharge service valve.

Note: It is important that the suction service valve is closed slowly when pumping the system down, otherwise an abnormal amount of oil may leave the compressor due to the sudden pressure of reduction on the refrigerant oil in the crankshaft case.

Unscrew the oil level plug slowly and bleed off the remaining pressure in the crankcase until the gauge reads zero. Complete the removal of the plug. The oil level plug is located in the **rear face of the crankcase**. Two plugs are fitted but one only need be removed.

Insert the dipstick, made from a piece of wire suitably bent, through the plug-hole until it contacts the bottom of the crankcase. Withdraw the dipstick and measure the wetted portion.

The oil depth obtained should be approximate to the figure shown in the Oil Level chart below.

The oil should not be allowed to fall below the "Minimum" level shown in the chart.

If oil is added, do not exceed the figure shown in the chart.

It is essential that only oil of the correct specification and grade is used.

The oil level should also be checked after placing a new compressor in operation, charging or repairing a compressor, or after adding refrigerant to the system.

OIL LEVEL

Factory charge of 11 fluid ounces $1\frac{5}{16}''$ (33·3 mm.)
Minimum height .. .. .. $\frac{7}{8}''$ (22·2 mm.)
Maximum height .. .. .. $1\frac{3}{8}''$ (34·9 mm.)

RECOMMENDED LUBRICANTS

SUN OIL CO. .. "3 G "3 G dual-inhibited" Oil
TEXACO .. "Cappela B dual-inhibited" Oil
SHELL .. .. Clavus 53
B.P. .. .. Energol LPT.100

CHARGING AND SERVICING A REFRIGERATION SYSTEM

Charging an air-conditioning system will not be necessary unless leaks develop in the system resulting in loss of the refrigerant, or in the event of any components being disconnected or removed.

This service can only be performed by a fully qualified Refrigeration Service Engineer who will have the necessary equipment.

IMPORTANT: The air-conditioning equipment is manufactured for use only with Refrigerant 12 (dichlorodifluoromethane) and **extreme care** must be taken **never** to use methylchloride refrigerants.

Page RY.s.2

AIR-CONDITIONING REFRIGERATION EQUIPMENT

The chemical re-action between methylchloride and the aluminium parts of the compressor will result in the formation of products which burn spontaneously on exposure to air, or decompose with violence in the presence of moisture.

To ensure efficient operation of a refrigeration unit, all air and non-condensable gases must be completely evacuated from the system before inserting the refrigerant charge.

A sufficient quantity of refrigerant should be obtained before commencing operations, and should be available from any qualified Refrigerant Service Suppliers under the following trade names:—

FREON 12
ARCTON 12
ISCEON 12

or any refrigerant to specification R.12. The refrigerant is available in 1, 2, 10, 25, or 145 lb. (0·454, 0·907, 4·536, 11·34 or 65·7 kgm.) containers. The lower weight being in canisters, the higher weights being in steel cylinders.

1 lb. 9 oz. (0·862 kgm.) will be the approximate weight of refrigerant 12 required to complete the charging operation.

SERVICE DIAGNOSIS

The following Service Diagnosis chart is included to assist the Service Engineer in fault finding:—

SERVICE DIAGNOSIS CHART

SYMPTOM	Low refrigerant charge (a)	Very light heat load (b)	Partial restriction in expansion valve (c)	Low capillary charge in expansion valve (d)	Defective expansion valve (e)	Evaporator coil blocked with ice (f)	Excessive refrigerant charge (g)	Very high heat load (h)	Loose capillary tube connection at evaporator coil outlet (i)	Restriction in high pressure side (j)	Insufficient air over condensers (k)	Unusually hot running engine (l)	Air or non-condensible gases in system (m)	Defective compressor head gasket (n)	Defective compressor plate gasket (o)	Defective compressor plate (p)
Unusually low reading of compound gauge	X	X	X	X	X	X										
Unusually high reading of compound gauge					X			X	X	X				X	X	X
Unusually low reading of high pressure gauge	X	X	X	X	X	X								X	X	X
Very low reading on high pressure gauge coupled with very high reading on compound gauge														X	X	X
Unusually high reading of high pressure gauge							X	X		X	X	X	X			

Page RY.s.3

AIR-CONDITIONING REFRIGERATION EQUIPMENT

Preliminary Check

Carry out preliminary checks as detailed below, before proceeding with any further tests:—

(1) Check that the fan is operating and that the blades are not fouling.
(2) Check that the compressor clutch is engaging or releasing satisfactorily when the air-conditioning system is switched on.
(3) Check that air is not present in the system by observing the sighting glass attached to the receiver/drier unit.
   Run the engine at fast idle speed (1,000 r.p.m.) and check the sighting glass.
   Repeat at 1,800 r.p.m.
   Gradually increase the speed of the engine to the high range and check the sighting glass at intervals.
   Bubbles in the sighting glass will indicate that air is present in the system.
(4) Check for frosting on the compressor valves.
(5) Check by feel for varying temperatures in the various pipe lines indicating blockage in the line system.
(6) Place a thermometer in the air duct, run the car on the road and note the drop in temperature with the system on or off or, if available, place a cooling fan in front of the condenser.
   Check that the condenser is clear of mud, road dirt or flies, preventing the free passage of air over the condenser unit.

If the fault persists after carrying out these tests, the services of a qualified Refrigerant Engineer should be obtained.

THE COMPONENTS

The following pages contain the General Description, Removal and Refitting details for the components used in the system.

THE COMPRESSOR UNIT

Description

The compressor unit used in the JAGUAR Air-conditioning system is the "TECUMSEH" H.G.500.

The compressor is a completely sealed unit with the exception of the suction and discharge ports to which are attached the service valves.

The cold low pressure refrigerant is pulled into the suction service valve, indicated by the word "SUCTION" stamped on the cylinder head, through the valve plate and into the suction chamber.

The compression stroke of the piston closes the valve and forces the compressed vapour into the discharge chamber.

The vapour is them pumped through the discharge service valve, indicated by the word "DISCHARGE" stamped on the cylinder head, and so into the system.

Removal

It will only be necessary to remove the compressor from the engine if any major repair work is carried out. Servicing to the cylinder head, valve and valve plate can be done with the unit in situ.

Remove the aluminium cap from the suction service valve, close the valve by rotating the square end of the valve stem anti-clockwise.

Remove the cap from the suction gauge port and connect a pressure gauge to the port union.

Open the valve and check the pressure recorded. If the gauge shows a pressure above zero, close the suction service valve and start the engine to operate the compressor until suction pressure is reduced to 2 lb. sq. in. (0·1406 kg./cm.F).

Stop the engine and close the discharge service valve.

Unscrew the large hexagon gland nuts and remove the service valves from the compressor unit.

Release the jockey pulley pivot bolt, swing the pulley bracket down and remove the drive belt.

Withdraw the two bolts securing the carrier bracket to the exhaust manifold and the cylinder block. Withdraw the carrier bracket lower pivot bolt. Disconnect the clutch cable and remove the unit. Detach the carrier bracket from the compressor.

Note: The left hand valve cover on the engine cannot be removed with the compressor in position. To give the necessary clearance to enable the cover to be removed proceed as follows:—
Release the jockey pulley pivot bolt, swing the pulley bracket down and remove the compressor

Page RY.s.4

AIR-CONDITIONING REFRIGERATION EQUIPMENT

drive belt.

Withdraw two bolts securing the carrier bracket to the exhaust manifold and the cylinder block. Release the lower bracket pivot bolt and swing the unit away from the engine.

DO NOT DISCONNECT THE COMPRESSOR HOSE CONNECTIONS FROM THE COMPRESSOR.

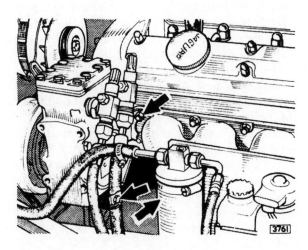

Fig. 3. *The compressor tie bracket and lower pivot mounting point.*

Refitting

Refitting is the reverse of the removal procedure.

Ensure that the pulleys are in line, check with a straight edge before finally tightening the securing bolts.

Re-charge the system as detailed on pages RY.S.00P RY.S.00.

When refitting the service valves, check that the mounting surfaces on the valves and the compressor are clean and that the "O" ring valve mounting gaskets are in good condition. If deformed, broken or split, replace with new gaskets.

Align the valves correctly to the cylinder head and tighten to a torque of 65–70 ft. lb. (8·983–9·674 kg./m.).

Special care should be taken to ensure that no dirt or foreign matter enters the compressor during installation.

A new replacement compressor should not be left unsealed to the atmosphere longer than is absolutely necessary for actual preparation and installation. In no case should the compressor be open to the atmosphere for longer than five minutes.

Check the oil level as detailed on page RY.s.2.

When replacing parts and re-inserting bolts and setscrews, the specified torque requirements should not be exceeded (see table below).

Bolts should always be run in so that the bolt heads make contact and then be rightened evenly to the correct torque figures.

Location	Torque
Cylinder head	20–24 ft. lb. (2·764–3·316 kg./m.)
Crankshaft end	15–20 ft. lb. (2·073–2·764 kg./m.)
Seal plate	6–10 ft. lb. (0·829–1·382 kg./m.)
Mounting	14–17 ft. lb. (1·934–2·349 kg./m.)
Front bearing lock nuts	6 ft. lb. (0·829 kg./m.)
Connecting rods	7 ft. lb. (0·967 kg./m.)
Oil filler plug	18–20 ft. lb. (2·487–3·040 kg./m.)
Service valve locknut	65–70 ft. lb. (8·983–9·674 kg./m.)

Shaft Seal Assembly (Replacement)

Remove the compressor unit as detailed previously in this section.

Remove the clutch assembly as detailed on page RY.s.7.

Place the compressor on the work bench with shaft uppermost.

Wash or clean the seal plate and adjoining surfaces to remove all dirt.

Remove seal plate assembly after withdrawing six bolts. Gently pry plate loose, being careful not to scratch or nick the crankcase mating surfaces or edges.

Remove the carbon nose and spring assembly by prying behind the drive ring. When removing the seal assembly from the shaft, care must be taken that the crankshaft is not scratched. If the rubber seal around the shaft does not come out with the carbon nose and spring assembly, remove the seal with long nosed pliers pulling on the edge of the grommet.

Remove all dirt and foreign material from crankshaft mating surfaces to seal plate, exposed crankshaft and adjacent surfaces.

AIR-CONDITIONING REFRIGERATION EQUIPMENT

Remove the new carbon shaft seal washer from the bellows seal assembly. (Parts are contained in the shaft seal assembly kit). Coat the exposed surface of the crankshaft with clean refrigerant oil. Dip the new bellows of the seal assembly and shaft seal washer in refrigerant oil. Place the bellows seal assembly over the shaft with the end for holding the shaft seal washer facing the shaft end.

Push the bellows seal assembly, by hand, on the shaft to a position beyond the taper of the shaft.

Assemble the shaft seal washer in the bellows seal assembly, checking that the assembly and the shaft are free from dirt. Assemble the seal washer so that the raised rim is away from the bellows seal and that the notches in the washer line up with the nibs in the bellows assembly. Cover exposed surfaces of the shaft seal washer with clean refrigerant oil.

Install new rectangular section "O" ring in the crankshaft mating surfaces for seal plate.

Place the new front seal plate over the shaft and line up the holes. Push the seal plate evenly against the crankcase and retain in this position while inserting and tightening two diametrically opposed bolts. Insert the remaining four bolts and tighten evenly to the correct torque figure.

Rotate the shaft by hand 15-20 revolutions to seat the seal.

Cylinder Head and Valve Plate Assembly

For the best results, the compressor should be removed from the engine as detailed previously in this section. Clean off all road dirt and dismantle on a clean bench.

Remove ten bolts from the cylinder head assembly and detach the valve plate and head assembly, by lightly tapping upwards with a soft-nosed mallet. Note the location of the valve plate in relation to the cylinder head for reference when refitting.

Remove the valve plate from the cylinder head by holding the plate and tapping sideways against the head with a soft mallet.

Remove all particles of gasket from the surface of the cylinder head and the cylinder, taking care not to scratch the mating surfaces or edges.

Fit the new valve plate gasket, contained in Valve Plate Kit. Keep the gasket dry.

Place the new valve plate assembly over the gasket so that the letter "S" stamped on the plate is visible, and on the same side as the word "Suction" cast on the front of the crankcase.

Locate the new cylinder head gasket on the valve plate as noted on removal. Fit the cylinder head. Check that the word "Suction" on the head is on the same side as the word "Suction" on the crankcase. Insert the bolts through the cylinder head, valve plate and gaskets. Tighten all bolts evenly in a diagonal sequence to the correct torque figure. Refit the magnetic clutch.

Refit the compressor unit to the engine and recharge the system

Leak test all joints on completion.

After a period of two hours from time of assembly re-torque the cylinder head bolts.

SERVICE REPLACEMENT UNITS

Replacement compressors are not available on an exchange basis and must be replaced with new units obtainable from JAGUAR DISTRIBUTORS or DEALERS.

THE COMPRESSOR SERVICE VALVES

DESCRIPTION

Two service valves, Suction and Discharge, are located on the crankcase and secured by locknuts.

Each of these valves has three ports or openings, one to which the refrigerant line is connected, one opening to the compressor and one known as the "service port", for connecting to the pressure gauge by means of a flexible hose.

The valve setting is accomplished by rotating the stem, normally covered by an aluminium cap, either clockwise or counter-clockwise.

With the discharge valve in the back-seated position —fully counter-clockwise—the refrigerant flow is

Page RY.s.6

AIR-CONDITIONING REFRIGERATION EQUIPMENT

from the compressor to the refrigerant line. In the suction valve this is in the opposite direction.

With the valve in the forward seated position (clockwise), the valve is blocking the seat to the refrigerant line and the flow is from the compressor through the service port.

The two valves are identical with the exception that the discharge valve has a $\frac{1}{2}$" refrigerant fitting connection and the suction valve has a $\frac{5}{8}$" fitting connection.

If the valves are positioned anywhere between the fully back seated and fully forward seated positions, all three ports would be opened and, therefore, there would be a passage to the compressor and also the service ports.

This would be the position in which the system could be operating and, at the same time, recording pressure reading.

No service is possible on the valves and, if faulty, should be replaced.

Removal

Remove all refrigerant from the system by "Pumping Down".

Disconnect all pipe lines, unscrew the locknuts and remove the valves.

Refitting

Ensure that the cylinder and valve mating faces are clean.

Apply a light film of refrigerant oil to both faces, fit the valve and tighten the locknut to the correct torque as detailed on page RY.s.5.

Position the valves at an angle of 45° to clear the bonnet when the pipe lines have been connected.

THE MAGNETIC CLUTCH

GENERAL DESCRIPTION

The magnetic clutch consists of two major parts—the holding coil and the rotor-pulley assembly. When an electrical circuit is completed through the holding or field coil to ground, the clutch coil is electrically energised, creating a magnetic field. This couples the plates against the tension of the spring which is connected to the pulley, forming an integral part with respect to motion. Inasmuch as the belt is driving the pulley and all parts are coupled, the compressor is driven at its tapered shaft through the key.

Upon breaking the coil electrical circuit, the magnetic field is broken and the pulley revolves freely around the compressor shaft and bearing. Shims are provided at the hub of the plate for adjusting the plate clearance. Too small a clearance can cause a scraping of the plates. Too great a clearance will cause short-circuiting of the magnetic clutch which will weaken the field.

The approximate current consumption of the magnetic clutch is three amps at 12 volts.

Very little maintenance is required on this assembly. Any clutch slipping should be traced to either incorrect clearance or low voltage to the clutch. If these are checked and found correct and the clutch is still inoperative, it should be replaced.

Removal

Disconnect the cable at the snap connector. Remove the compressor setscrew and flat washer and withdraw the clutch unit. If the rotor and pulley assembly will not release from the compressor shaft, insert a $\frac{5}{8}$" U.N.C. setscrew and tighten. The rotor assembly should then release.

Refitting

Refitting is the reverse of the removal procedure.

THE EXPANSION VALVE

GENERAL DESCRIPTION

The expansion valve is an automatic thermal valve located in the evaporator assembly case (see Fig. 1).

The valve is the dividing point between the high and low pressure sides of the system and automatically meters the high pressure, high temperature, liquid refrigerant through a small orifice controlled by a ball valve, into the low pressure, cold temperature, side of the evaporator coil.

AIR-CONDITIONING REFRIGERATION EQUIPMENT

The low pressure is created by the pull of the suction side of the compressor.

The expansion valve used in the Jaguar air conditioning equipment consists of a thermal bulb and capillary tube charged with vapour refrigerant; a diaphragm power element; balancing spring; external equalising pressure tube; valve seat actuating pin; ball seat; inlet port and screen and outlet port.

Liquid refrigerant, under high pressure, enters through the inlet port screen and tends to open the ball valve against the return spring tension.

The thermal bulb, which is clipped to the outlet line of the evaporator, is sensitive to the temperature of the vapour leaving the coil.

The last portion of the evaporator coil is absorbing additional heat from the air passing over, due to the action of the fans. The vapour of the coil becomes super-heated and the temperature rises.

The thermal bulb receives this temperature rise and the pressure of the vapour within the bulb correspondingly increases.

This pressure increase operating against the diaphragm and actuating needle, opens the valve against the spring pressure and allows a flow of liquid to the coil.

As the liquid in the coil progresses towards the outlet, the coil super-heat is reduced, the pressure on the diaphragm and the actuating needle is relieved and the spring again forces the ball seat to close the flow of liquid. In this way, the valve senses the demands of the system and meters the correct amount of liquid refrigerant and also prevents a liquid slug from entering the compressor.

In order to balance the liquid pressure on the ball seat plus the diaphragm and spring forces, an equal force is necessary, operating on the underside of the diaphragm. The external equaliser is used for this purpose.

This taps pressure from the outlet of the evaporator to the chamber behind the diaphragm of the valve.

By utilising the external equaliser an account is made of the small amount of pressure drop that occurs in the coil and the resultant drop in temperature. This relays the actual condition of the coil outlet to the expansion valve power element, more accurately provides metering of the fluid and more positively prevents liquid slugs from leaving the evaporator coil.

If the capillary tube should lose some of its vapour charge the power element will weaken so that the valve will close too frequently and, therefore, starve the coil, and the efficiency of the refrigeration system would be impaired. A low suction pressure of the compressor would result.

If the expansion valve screen were to become clogged with some foreign substance, flow of the liquid would be reduced and the coil would starve, the efficiency of the system would be impaired and a low suction pressure would result.

The effects of moisture or water in a refrigerant system will be covered more thoroughly under section "Receiver/Dryer" Assembly, however, one of its effects can be pointed out here. One of the characteristics of Refrigerant 12 is that it can carry in suspension minute droplets of water. These droplets remain as liquid water in the high pressure side because of the high pressure side's temperature. It now reasons that the discharge pressure of the compressor is maintained clear to the ball seat of the expansion valve. This point is the dividing point of the refrigeration system. Everything from it to the suction inlet of the compressor is maintained at the low pressure figure of the suction of the compressor. The small orifice at the ball valve seat is minute compared to the lines going to and passing from it, thereby producing this dividing point. At this point, of course, the suction pressure drops abruptly and the temperature of the liquid refrigerant correspondly drops. The small droplets of water that have been maintained as water in liquid form on the high side, are suddenly subjected to the extreme cold temperature of the low side. They become icy or an icy sludge and can either completely block the orifice at the ball seat of the valve or form an erratic sludge at that point that will go away and come back at frequent intervals. A solid plug of ice will render the system inoperative, and the suction pressure reading at the compressor will drop into a deep vacuum as the compressor will cause removal of everything from the ice block to the compressor. The sludge will reduce the effectiveness of the system and the suction pressure reading will be erratic.

An excellent way to determine whether an expansion valve difficulty is ice or something else, this, of course, after having observed a suction reading that is in vacuum, would be to heat the body of the valve in the vicinity of the ball seat, taking care not to overheat the diaphragm and observe the suction reading. If ice is there, the heating will immediately melt it, liquid will charge into the evaporator, the suction pressure will increase sharply.

Page RY.s.8

AIR-CONDITIONING REFRIGERATION EQUIPMENT

If the loss of the thermal bulb charge is suspect, it can be subjected to unusual heat by wrapping it with a person's hand. This unusual heat will cause the expansion valve to flood the coil if the thermal bulb charge is normal and an abnormally high suction reading will result. If the suction pressure does not rise, it can be assumed that the power element is weak.

The air conditioning valve is pre-set by the manufacturer of the valve for the correct opening and operating super heat and, therefore, spring tension field adjustments are not to be made. If the valve meters too much liquid, which would constantly maintain an excessive suction presure reading, this would be a result of a defective spring tension or adjustment and the valve should be replaced.

Erratic conditions sometimes occur at the expansion valve when the system is first started. These conditions result in erratic suction pressure readings that are unusually high. While a system lies dormant, through the night for example, the vapour in the evaporator coil can condense and when the refrigeration system is begun, a slug of liquid can make its way towards the compressor. At least the condition of the entering substance to the compressor is wet or dense. This, of course, results in unusually high suction pressure readings. As soon as the slug of liquid passes the thermal bulb point on the coil outlet, the expansion valve, as described, will close. These erratic conditions which are observed as pressure readings will cease whenever this stabilises and a constant production of liquid is produced to the expansion valve and they are quite normal.

Removal

Remove the Radio/Heater control panel as detailed in Section N.

The air-conditioning thermostat will remain in position after detaching the control knob.

Remove the gearbox console cover as detailed in Section N.

Remove all refrigerant from the system by pumping down.

Remove the clip securing the thermal bulb to the outlet pipe.

Disconnect the pipe unions and withdraw the valve.

Refitting

Refitting is the reverse of the removal procedure.
Re-charge the system on completion.

THE RECEIVER DRIER-SIGHTGLASS ASSEMBLY

GENERAL DESCRIPTION

The functions of the receiver drier assembly and sight-glass are to filter the liquid refrigerant, to absorb any water that may be prevalent in the system and to provide a storage tank of liquid refrigerant in which a pick-up tube submerged in the liquid will ensure that a solid column is available to supply the expansion valve.

A sight-glass is provided in the outlet line of the assembly to enable a visual check of the fluid flow to be made.

If the sight-glass is clear, this indicates that a solid column of pure colourless refrigerant is passing through; conversely if bubbles or foam are visible this indicates that the pick-up tube is not submerged in liquid refrigerant and is receiving a mixture of liquid and vapour and the vapour is, therefore, causing bubbles in the liquid, giving the appearance of foam.

This indicates that the liquid charge in the receiver is insufficient, and the system needs more refrigerant.

The cotton bobbin preparation used as a filter in the system, filters out all impurities and foreign matter. It is possible for refrigerant, under high temperature, to re-act with refrigerant oil in the compressor and form a precipitation which, if not effectively filtered out, will eventually clog the expansion screen.

The silica-gel molecular sieve desiccant absorbs and prevents passage of any moisture which may be present in the refrigerant.

Refrigerant "12" is a hydrocarbon containing the chlorine and fluorine halogens which is formed with methane.

The hydrogen found in water, can under certain temperature conditions, hydrolize with the chlorine and fluorine to form hydrochloric and hydrofluorine acids.

The hydrochloric acid will attack copper, of which all condenser and evaporator coil tubes are made, and carry the copper to steel portions of the system such as rod and valve plates.

Page RY.s.9

AIR-CONDITIONING REFRIGERATION EQUIPMENT

If the resultant copper plating becomes too thick it may seriously affect the efficiency of the compressor unit.

The hydrofluoric acid formed is an etching acid and can seriously attach and pit the finely polished surface of the compressor valve plate.

It is, therefore, of the utmost importance that the water is not allowed in the system and that small amounts be absorbed in a good drier.

Fig. 4. *The receiver/drier mounting.*

A. Thermostat switch (radiator fan).
B. Radiator strut.
C. Condenser top mounting.
D. Condenser bottom mounting.

SERVICING

If the drying agent in the receiver/drier unit becomes completely absorbed with water, the unit must be removed and returned for reclaiming.

The receiver/drier unit can only be completely serviced with the use of special equipment.

If the system is allowed to remain open for a long period of time, or for a shorter period in very humid conditions, the drier unit must always be changed before putting the care back in service.

DO NOT REMOVE the protective sealing caps from a new unit until it has been fitted and is ready for coupling to the pipe unions.

Removal

"Pump Down" the system.

Disconnect the pipe lines and blank off the unions and pipe lines.

Release the clip and withdraw the receiver unit.

Refitting

Refitting is the reverse of the removal procedure.
Recharge the system.

THE CONDENSER

GENERAL DESCRIPTION

The condenser is a single unit mounted in front of the radiator matrix.

The function of the condenser is to cause removal of the vapour super heat and to effect a change of state from vapour to liquid by passing the high latent heat off to the surrounding heat mediums.

Refrigerant "12" vapour at 120 lb. per sq. in. (8·43 kg./cm.^{2}) occupies approximately 23 cubic feet (·65 m.^{3}) per 1 lb. (0·454 kg.) of weight. The liquid refrigerant "12" occupies approximatley 1·3 cubic feet (·036 m.^{3}) per 1 lb. (0·454 kg.) of weight or the liquid takes approximately 1/18th as much space. Consequently, the quicker the condenser manufactures liquid the sooner there will be more space in which the compressor can unload its charge of vapour, and the compressor head pressure will be lower for a given heat load condition.

It is essential that the condenser is cooled efficiently by the passage of the car through the air.

Any obstruction such as dirt, mud, or any foreign matter, will prevent the lowering of the refrigerant temperature resulting in increased head pressure.

Normally the condenser will be of a lower temperature than the car radiator. Any lowering of the efficiency will increase the temperature of the condenser to a point where it may be higher than the car radiator and allow the engine to overheat.

No routine maintenance is necessary.

Removal

Remove the bonnet as detailed in Section N. "Body and Exhaust".

Remove all refrigerant from the system by "Pumping Down". Withdraw all securing screws and detach

Page RY.s.10

AIR-CONDITIONING REFRIGERATION EQUIPMENT

the condenser unit after disconnecting the hoses.

Refitting
Refitting is the reverse of the removal procedure. Recharge the system on completion.

Testing for Leaks
When removed from the car, the condenser unit can be leak tested if required as follows:—

Seal off the outlet pipe union with a suitable cap nut and sealing disc.

Connect a refrigerant container, one of the small capacity canisters is preferable, by a suitable length of flexible hose to the condenser inlet union.

Open the container valve and allow a quantity of refrigerant "12" to enter the condenser.

To test, pass the leak detector hose around all the condenser tubes, paying particular attention to the "U" bends at the ends of the tubes.

If any leaks are detected, it is advisable to replace the faulty unit as effective repairs are difficult to carry out without special equipment.

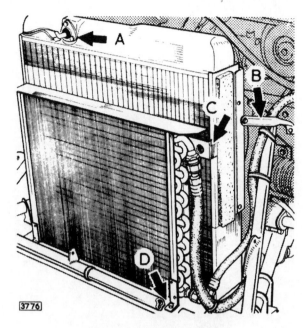

Fig. 5. *The condenser mounting.*

A Fan thermostat switch.
B Radiator mounting strap.
C Condenser upper monting.
D Condenser lower monting.

THE EVAPORATOR

GENERAL DESCRIPTION

The evaporator, of fin and tube construction contained in a case which also houses the blower fan, is mounted below the instrument panel.

A metered supply of low pressure, cold refrigerant is drawn through the evaporator coils by the suction side of the compressor.

Heat laden air from the car interior is pulled over the coil by the centrifugal fan, and the temperature difference between the hot air and cold refrigerant causes a heat transfer from the warm air to the cold liquid.

As the liquid is absorbing the heat from the air the refrigerant is caused to vaporise or "boil".

Refrigerant "12" boils at $-20°F$ ($-29°C$) at atmospheric pressure.

The vapour has more coil to pass through before its exit, and as the warm air is still imparting heat to the refrigerant vapour and as the vapour is in a saturated state, any further heat absorbed by it creates a superheat and its temperature begins to rise.

Condensation of the moisture in the air occurs simultaneously with the reduction of the air temperature. This water condensate is drained out of the evaporator assembly and discharged through drain pipes.

Frequently the condensate will drain from the evaporator case very soon after the car comes to rest and the blower is switched off and this will create a condensate puddle underneath the car.

This is a natural condition and no investigation as to cause is necessary.

No routine maintenance is required, and any repairs can only be carried out if the evaporator is removed from the car.

Page RY.s.11

AIR-CONDITIONING REFRIGERATION EQUIPMENT

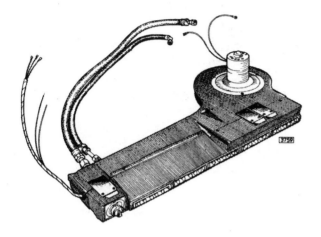

Fig. 6. *The evaporator unit.*

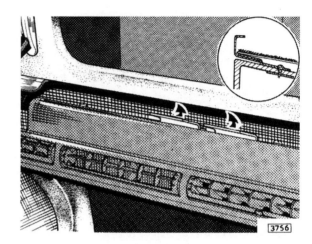

Fig. 7. *The evaporator facia mounting.*

Removal

Slide the seats back to the full extent of the slides.

Remove all refrigerant from the system by "Pumping Down"

Disconnect all pipe unions and blank off the ends to prevent the ingress of dirt or moisture.

Withdraw the drive screws securing the side fixing bracket to the body panel.

Carefully ease the unit forward until the top fixing clips are clear of the bottom edge of the facia panel and lower the unit to the floor of the car.

Extract the drive screws and remove the cover from the front of the case.

Lift the thermostat capillary tube away from the evaporator coils and withdraw the evaporator assembly away from the case.

Note: Care must be taken when the assembly has been removed, that the drain pipes and unions underneath are not damaged by contact with the floor or work-bench. Support the assembly on blocks to keep the pipes clear.

Refitting

Refitting is the reverse of the removal procedure.
Recharge the system on completion.

Testing for Leaks

When removed from the car the evaporator can be leak tested, if required, as follows:—

Seal off the outlet union with a suitable cap nut and sealing disc.

Connect a refrigerant container, one of the small capacity canisters obtainable is preferred, by a suitable length of flexible hose to the evaporator inlet union.

Open the container valve and allow a quantity of refrigerant "12" to enter the evaporator coil.

To test, pass the leak detector hose around all tubes, paying particular attention to the "U" bends at the ends of the tubes.

If any leaks are detected, it is advisable to replace the faulty unit, as effective repairs are difficult to carry out without special equipment.

AIR-CONDITIONING REFRIGERATION EQUIPMENT

THE ELECTRICAL SYSTEM

GENERAL DESCRIPTION

The electrical system consists of a combined rotary "ON/OFF" fan blower variable speed switch and thermostat control, a fan blower motor, one external resistor, an in-line fuse and a magnetic clutch unit with the necessary wiring.

A second relay is included in the radiator fan circuit when air-conditioning equipment is fitted, the purpose being to over-ride the "OTTER" thermostat switch located in the header tank when the system is switched on and so ensuring that the radiator cooling fan motors are operating.

This relay is additional to the one included in the radiator fan circuit fitted to all cars.

The rotary switch (outer control ring) controlling the blower motor speeds, is wired in series with the resistor unit to give "OFF", "L" (Low) "M" (Medium) and "H" (High) positions. See circuit diagram Fig. 5.

In the "L", "M", or "H" positions, the circuit is completed from the ignition switch via the fuse and relay, through the thermostatic switch (central control knob) to the compressor drive clutch.

Progressive rotation of the thermostat control in the direction of the arrow will result in the switch contacts remaining closed until the capillary tube from the switch, inserted into and in between the fins of the evaporator coils, senses a temperature that is below the manual setting.

When this temperature is reached, the contacts will open, and the circuit to the magnetic clutch will be broken.

The clutch will disengage and cease to drive the compressor.

As the temperature in the evaporator rises, the thermostat will again close and the clutch re-engage.

The manual setting is progressive in the direction of the arrow towards "COOLER".

Maximum clockwise rotation will give the coldest coil temperature without opening the clutch circuit.

Minimum rotation will open the clutch circuit most frequently as its sensing temperature is the highest, and will keep the evaporator coil at a warmer temperature.

The normal setting would be approximately three quarters of full rotation.

Icing of the evaporator coil, which restricts and can ultimately block the air flow into the car, occurs more during days of moderate temperature and relative high humidity.

Position the thermostat to a HIGHER TEMPERATURE setting to correct this condition.

HIGHER TEMPERATURE thermostat settings may have to be employed in moderate temperatures, cloudy days or night time driving.

These conditions are also conducive to icing as the relative humidity increases when the sun is obscured.

The extreme left thermostat setting, that is with the white indicator mark vertical, is **OFF** and the clutch circuit will always remain open.

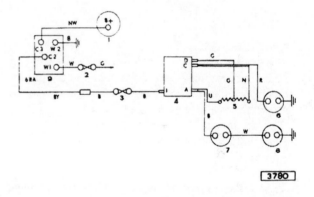

Fig. 8. *The evaporator circuit diagram.*
1. Alternator.
2. Fuse No. 6.
3. In-line fuse.
4. Control switch.
5. Resister.
6. Motor.
7. Thermostat switch.
8. Clutch.
9. Relay.

Page RY.s.13

AIR-CONDITIONING REFRIGERATION EQUIPMENT

THE EVAPORATOR FAN MOTOR
Checking
Switch on the fan motor. Check at the low, medium and high positions to ensure that the resistor is working correctly.

Removal
Remove two cap nuts and detach the radio/ashtray console.

Lower the evaporator unit as detailed on page RY.s.12.

Disconnect the feed and earth cables, withdraw the four retaining screws and remove the fan and motor assembly.

Check the fan for damage and renew if necessary.

The fan is a balanced assembly, and care must be taken to ensure that any balance pieces fitted are not displaced

Refitting
Refitting is the reverse of the removal procedure.

Renew the cork gasket if damaged, and ensure that the earth connection is clear and tight when re-assembling.

THE THERMOSTAT
If the thermostat is not automatically switching "OFF" the compressor drive clutch unit at the pre-set temperature, check that the capillary tube is positioned between the fins of the evaporator coil.

The thermostat and capillary tube assembly is a sealed unit and must be changed if faulty.

Removal
Lower the evaporator unit as detailed on page RY.s.12.

Withdraw the retaining screws and detach the top panel from the case.

Carefully remove the capillary tube from the evaporator coils.

With a small screwdriver used as a lever, remove the two control knobs.

Withdraw two screws and detach the unit from the mounting panel.

Remove two screws and detach the thermostat unit from the fan switch.

Refitting
Refitting is the reverse of the removal procedure.

"OTTER" THERMOSTAT (OVER-RIDING) RELAY
Checking
The relay is the top one of the two mounted on the carrier bracket attached to the sub-frame cross-member between the radiator fans.

Disconnect and remove the relay. Note the cable connections for reference when refitting.

Connect a 12 volt supply to terminals W1 and W2.

Connect a test lamp wired in series with a 12 volt battery to terminals C1 and C2.

If the relay contacts are closing the bulb will become illuminated.

Replace if faulty.

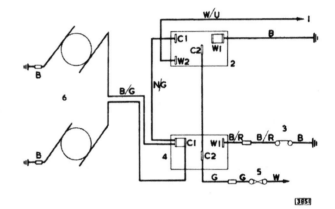

Fig. 9. *The radiator cooling fan circuit diagram with over-riding relay incorporated.*

1. Control switch feed.
2. Over-riding relay.
3. "Otter" thermostat switch.
4. Fan relay.
5. Fuse No. 6.
6. Radiator fan motors.

Page RY.s.14

© **Copyright Jaguar Cars Limited 1971 and
Brooklands Books Limited 1990, 1999 and 2014**

This book is published by Brooklands Books Limited and based upon text and illustrations protected
by copyright and first published in 1971 by Jaguar Cars Limited and may not be reproduced
transmitted or copied by any means without the prior written permission of
Jaguar Cars Limited, and Brooklands Books Limited.

Brooklands Books Ltd., PO Box 904,
Amersham, Bucks, HP6 9JA, UK
www.brooklandsbooks.com

ISBN: 9781855200203   Ref: J28WH   4W4/2999

OFFICIAL TECHNICAL BOOKS

Brooklands Technical Books has been formed to supply owners, restorers and professional repairers with official factory literature.

Workshop Manuals

Title	Code	ISBN
Jaguar Service Manual 1946-1948		9781855207844
Jaguar XK 120 140 150 150S & Mk 7, 8 & 9		9781870642279
Jaguar Mk 2 (2.4 3.4 3.8 240 340)	E121/7	9781870642958
Jaguar Mk 10 (3.8 & 4.2) & 420G	E136/2	9781855200814
Jaguar 'S' Type 3.4 & 3.8	E133/3	9781870642095
Jaguar E-Type 3.8 & 4.2 Series 1 & 2 E123/8, E123 B/3 & E156/1		9781855200203
Jaguar E-Type V12 Series 3	E165/3	9781855200012
Jaguar 420	E143/2	9781855201712
Jaguar XJ6 2.8 & 4.2 Series 1		9781855200562
Jaguar XJ6 3.4 & 4.2 Series	E188/4	9781855200302
Jaguar XJ12 Series 1		9781783180417
Jaguar XJ12 Series 2 / DD6 Series 2	E190/4	9781855201408
Jaguar XJ6 & XJ12 Series 3	AKM9006	9781855204010
Jaguar XJ6 OWM (XJ40) 1986-94		9781855207851
Jaguar XJS V12 5.3 & 6.0 Litre	AKM3455	9781855202627
Jaguar XJS 6 Cylinder 3.6 & 4.0 Litre	AKM9063	9781855204638

Owners Workshop Manuals

Title	ISBN
Jaguar E-Type V12 1971-1974	9781783181162
Jaguar XJ, Sovereign 1968-1982	9781783811179
Jaguar XJ6 Workshop Manual 1986-1994	9781855207851
Jaguar XJ12, XJ5.3 Double Six 1972-1979	9781783181186

Parts Catalogues

Title	Code	ISBN
Jaguar Mk 2 3.4	J20	9781855201569
Jaguar Mk 2 (3.4, 3.8 & 340)	J34	9781855209084
Jaguar Series 3 12 Cyl. Saloons		9781783180592
Jaguar E-Type 3.8	J30	9781869826314
Jaguar E-Type 4.2 Series 1	J37	9781870642118
Jaguar E-Type Series 2	J37 & J38	9781855201705
Jaguar E-Type V12 Ser. 3 Open 2 Seater	RTC9014	9781869826840
Jaguar XJ6 Series 1		9781855200043
Jaguar XJ6 & Daimler Sovereign Ser. 2	RTC9883CA	9781855200579
Jaguar XJ6 & Daimler Sovereign Ser. 3	RTC9885CF	9781855202771
Jaguar XJ12 Series 2 / DD6 Series 2		9781783180585
Jaguar 2.9 & 3.6 Litre Saloons 1986-89	RTC9893CB	9781855202993
Jaguar XJ-S 3.6 & 5.3 Jan 1987 on	RTC9900CA	9781855204003

Owners Handbooks

Title	Code	ISBN
Jaguar XK120		9781855200432
Jaguar XK140	E101/2	9781855200401
Jaguar XK150	E111/2	9781855200395
Jaguar Mk 2 (3.4)	E116/10	9781855201682
Jaguar Mk 2 (3.8)	E115/10	9781869826765
Jaguar E-Type (Tuning & prep. for competition)		9781855207905
Jaguar E-Type 3.8 Series 1	E122/7	9781870642927
Jaguar E-Type 4.2 2+2 Series 1	E131/6	9781869826383
Jaguar E-Type 4.2 Series	E154/5	9781869826499
Jaguar E-Type V12 Series 3	E160/2	9781855200029
Jaguar E-Type V12 Series 3 (US)	A181/2	9781855200036
Jaguar XJ (3.4 & 4.2) Series 2	E200/8	9781855201200
Jaguar XJ6C Series 2	E184/1	9781855207875
Jaguar XJ12 Series 3	AKM4181	9781855207868

Carburetters

Title	ISBN
SU Carburetters Tuning Tips & Techniques	9781855202559
Solex Carburetters Tuning Tips & Techniques	9781855209770
Weber Carburettors Tuning Tips and Techniques	9781855207592

Jaguar - Road Test Books

Title	ISBN
Jaguar and SS Gold Portfolio 1931-1951	9781855200630
Jaguar XK120 XK140 XK150 Gold Port. 1948-60	9781870642415
Jaguar Mk 7, 8, 9, 10 & 420G	9781855208674
Jaguar Mk 1 & Mk 2 1955-1969	9781855208599
Jaguar E-Type	9781855208360
Jaguar XJ6 1968-79 (Series 1 & 2)	9781855202641
Jaguar XJ12 XJ5.3 V12 Gold Portfolio 1972-1990	9781855200838
Jaguar XJS Gold Portfolio 1975-1988	9781855202719
Jaguar XJ-S V12 1988-1996	9781855204249
Jaguar XK8 & XKR 1996-2005	9781855207578
Road & Track on Jaguar 1950-1960	9780946489695
Road & Track on Jaguar 1968-1974	9780946489374
Road & Track On Jaguar XJ-S-XK8-XK	9781855206298

Available from Jaguar specialists, Amazon and all good motoring bookshops

Brooklands Books Ltd., PO Box 904, Amersham,
Bucks, HP6 9JA, UK

www.brooklandsbooks.com

Printed in Great Britain
by Amazon